LOGIC COLLOQUIUM '01

LECTURE NOTES IN LOGIC

A Publication of
THE ASSOCIATION FOR SYMBOLIC LOGIC

LECTURE NOTES IN LOGIC 20

LOGIC COLLOQUIUM '01

Proceedings of the Annual European Summer Meeting of the Association for Symbolic Logic, held in Vienna, Austria
August 6–11, 2001

Edited by

Matthias Baaz
Institute for Discrete Mathematics and Geometry
Vienna University of Technology

Sy-David Friedman
Kurt Gödel Research Center for Mathematical Logic
University of Vienna

Jan Krajíček
Mathematical Institute
Academy of Sciences of the Czech Republic

CRC Press
Taylor & Francis Group
Boca Raton London New York

CRC Press is an imprint of the
Taylor & Francis Group, an informa business

Addresses of the Editors of Lecture Notes in Logic and a Statement of Editorial Policy may be found at the back of this book.

Sales and Customer Service:
A K Peters, Ltd.
888 Worcester Street, Suite 230
Wellesley, Massachusetts 02482, USA

Association for Symbolic Logic:
C. Ward Henson, Publisher
Mathematics Department
University of Illinois
1409 West Green Street
Urbana, Illinois 61801, USA

Library of Congress Cataloging-in-Publication Data

Logic Colloquium (2001 : Vienna, Austria)
Logic Colloquium '01 : proceedings of the Annual European Summer Meeting of the Association for Symbolic Logic, held in Vienna, Austria, August 6–11, 2001 / edited by Matthias Baaz, Sy-David Friedman, Jan Krajíček.
p. cm. – (Lecture notes in logic ; 20)
Includes bibliographical references.
ISBN 1-56881-247-7 (acid-free paper) – ISBN 1-56881-248-5 (pbk. : acid-free paper)
1. Logic, Symbolic and mathematical–Congresses. I. Baaz, Matthias. II. Friedman, Sy D., 1953- III. Krajíček, Jan. IV. Title. V. Series.

QA9.A1L64 2001
511.3–dc22 2004060184

CRC Press
6000 Broken Sound Parkway, NW
Suite 300, Boca Raton, FL 33487
270 Madison Avenue
New York, NY 10016
2 Park Square, Milton Park
Abingdon, Oxon OX14 4RN, UK

PREFACE

The 2001 European Summer Meeting of the Association for Symbolic Logic, LOGIC COLLOQUIUM 2001 was held at the Vienna University of Technology in Vienna, Austria August 6–11, 2001. It was organized by the International Kurt Gödel Society on the occasion of the 70th anniversary of Gödel's famous publication *Über formal unentscheidbare Sätze der Principia Mathematica und verwandter Systeme I.*

The program of the LOGIC COLLOQUIUM 2001 included 17 invited one hour plenary addresses, 2 tutorials and 17 invited half-hour lecturers in parallel special sessions on Set Theory, Model Theory, Proof Theory and Proof Search, Computability Theory, Philosophy and History of Logic, Computer Science and Complexity Theory. In addition, there were 126 contributed talks plus 36 papers presented by title. The number of registered participants was 249 coming from around 25 countries.

The tutorial short courses were given by Alexander Leitsch, *Resolution theorem proving*, and Itay Neeman, *Determinacy of long games.*

Invited plenary addresses were given by Toshiyasu Arai, *Wellfoundedness proofs in reflecting universes*; Steve Awodey, *Modal and intensional types in categorical logic*; James Cummings, *Squares, scales and reflection*; Rod Downey, *Calibrating randomness*; Harvey Friedman, *Recent incompleteness*; Warren Goldfarb, *Gödel and Carnap: influence and opposition*; Olga Kharlampovich, *First order properties of free groups*; Julia Knight, *Quantifiers and complexity*; Michiel van Lambalgen, *Moschovakis's notion of meaning as applied to linguistics*; Chris Miller, *Tameness phenomena in expansions of the field of real numbers*; Ran Raz, *Resolution Lower bounds for the weak pigeonhole principle*; Saharon Shelah, $(\aleph_n, \aleph_0)$ *may fail* $\aleph_0$*–compactness*; Jouko Väänänen, *Models, trees and infinitary back-and-forth*; Frank Wagner, *Simplicity theory*; Stan Wainer, *Provable recursiveness and complexity.*

Special plenary addresses commemorating the anniversary of Gödel's incompleteness theorem were given by Georg Kreisel, *A consumer's (re)view around higher infinites*, and Gaisi Takeuti, *Incompleteness theorem and its frontier.*

The Programm Committee consisted of M. Arslanov, M. Baaz, W. Buchholz, S. Friedman, D. Isaacson, C. Jockusch, J. Krajíček (Chair), L. Newelski (resigned April 2000), J. Paris, J.-P. Ressayre, L. van den Dries, B. Velickovic, A. Visser.

The editors invited all pleanry, section and tutorial speakers to contribute to the proceedings. The present volume contains 19 papers. All papers were refereed. We appreciated very much the help of the referees (whose names we do not present for obvious reasons).

We would like to thank the following institutions for supporting the meeting: the Association for Symbolic Logic, the National Science Foundation, the Austrian Federal Ministry for Education, Science and Culture, the City Council of Vienna, the Vienna University of Technology, the University of Vienna, the Insitute for Algebra and Computational Mathematics (TU Wien) and the Institute for Computer Languages (TU Wien).

Finally, we are grateful to Norbert Preining for taking care of the main organizational tasks of LOGIC COLLOQUIUM 2001.

The Editors
Matthias Baaz, *Vienna*
Sy Friedman, *Vienna*
Jan Krajíček, *Prague*

TABLE OF CONTENTS

Preface v

TUTORIALS

Alexander Leitsch
Resolution theorem proving: a logical point of view 3

Itay Neeman
An introduction to proofs of determinacy of long games 43

ARTICLES

Ulrich Berger and Paulo Oliva
Modified bar recursion and classical dependent choice 89

Andrea Cantini
Choice and uniformity in weak applicative theories 108

James Cummings
Compactness and incompactness phenomena in set theory 139

Harvey M. Friedman
Selection for Borel relations 151

D. M. Gabbay and N. Olivetti
Interpolation in goal-directed proof systems 1 170

Julia F. Knight
Sequences of degrees associated with models of arithmetic 217

Pascal Koiran
The limit theory of generic polynomials 242

Michiel van Lambalgen and Fritz Hamm
Moschovakis's notion of meaning as applied to linguistics 255

Chris Miller
Tameness in expansions of the real field 281

Rahim N. Moosa
The model theory of compact complex spaces 317

Karl–Georg Niebergall
"Natural" representations and extensions of Gödel's second theorem 350

Jan Reimann and Frank Stephan
Effective Hausdorff dimension 369

Ralf Schindler
Mutual stationarity in the core model 386

Saharon Shelah
The pair $(\aleph_n, \aleph_0)$ may fail $\aleph_0$-compactness 402

Gaisi Takeuti
Incompleteness theorem and its frontier 434

Frank Wagner
Groups in Simple Theories 440

S. S. Wainer
Provable recursiveness and complexity 468

TUTORIALS

RESOLUTION THEOREM PROVING: A LOGICAL POINT OF VIEW

ALEXANDER LEITSCH

§1. **Introduction.** Logical calculi were invented to model mathematical thinking and to formalize mathematical arguments. The calculi of Boole [8] and of Frege [15] can be considered as the first *mathematical* models of logical inference. Their work paved the way for the discipline of metamathematics, where mathematical reasoning itself is the object of mathematical investigation. The early calculi, the so-called Hilbert type- and Gentzen-type calculi [25], [17] developed in the 20th century served the main purpose to *analyze* and to *reconstruct* mathematical proofs and to investigate *provability*. A practical use of these calculi, i.e., using them for solving actual problems (e.g., for proving theorems in "real" mathematics), was not intended and even did not make sense.

But the idea of a logical calculus as a *problem solver* is in fact much older than the origin of propositional and predicate logic in the 19th century. Indeed this idea can be traced back to G. W. Leibniz with his brave vision of a *calculus ratiocinator* [29], a calculus which would allow solution of arbitrary problems by purely mechanical computation, once they have been represented in a special formalism. Today we know that, even for restricted languages, this dream of a complete mechanization is not realizable—not even in principle (we just refer to the famous results of Gödel [20] and Turing [39]). That does not imply that we have to reject the idea altogether. Still it makes sense to search for a *lean version* of the calculus ratiocinator. Concerning the logical language, the ideal candidate is first-order logic; it is axiomatizable (and thus semidecidable), well-understood and sufficiently expressive to represent relevant mathematical structures. By Church's result [10] we know that there is no decision procedure for the validity problem of first-order logic; thus there is no procedure which is 1. capable of verifying the validity of all valid formulas and 2. terminating on all formulas. So, even in first-order logic, we have to be content with the *verification* of problems. The only thing we can hope for is a calculus which offers a basis for efficient *proof search*.

It is not surprising that the invention of the computer lead to a revival of Leibniz's dream. Indeed one of the first enterprises in the field we are

Logic Colloquium '01
Edited by M. Baaz, S. Friedman, and J. Krajíček
Lecture Notes in Logic, 20

calling Artificial Intelligence today was the problem of *automated theorem proving*. The very first systems were the geometry prover of Gelernter [16] and Gilmore's first-order prover [18]. Gilmore's approach was based on Herbrands's theorem and used a "projection" of predicate logic to propositional logic. By Herbrand's theorem, a universal formula F is unsatisfiable iff there exists a finite conjunction F' consisting of instances of the matrix which is unsatisfiable too. This gives a simple reduction of predicate logic to propositional logic which looks quite natural from the point of view of logical complexity. But *computationally* the method proved to be highly ineffective. In fact there are two main sources of inefficiency: 1. the size of F' and 2. the inference on F' itself. While the efficiency of 2. was considerably increased by the method of Davis and Putnam [11], the search for F' and also the large size of F' remained serious obstacles even in proving most simple first-order theorems. Obviously there was a need for a natural technique employed by humans in the finding of proofs, which did not receive much attention in mathematical logic so far. This technique is unification and the invention of the unification principle by J. A. Robinson [37] in the early sixties brought the decisive breakthrough and marks the very beginning of automated deduction. Indeed the consequent use of the unification principle lead to a substantially new type of calculus. Robinson's resolution principle, a combination of most general unification and atomic cut on conjunctive normal forms, opens the way for a large variety of computational calculi and still forms the logical basis of the most efficient theorem proving programs.

But even on the basis of the unification principle theorem proving programs turned out too weak to automatically handle substantial or even unsolved problems of real mathematics. One of the few exceptions is the solution of Robbin's problem by W. McCune [35]. This problem is purely equational and its solution requires special techniques of equational reasoning which will not be presented in this paper. On the other hand, the resolution principle had a strong impact on computer science and became the decisive tool in the development of logic programming. Other interesting applications of resolution can be found in mathematical logic itself. Two of them, the decision problem of first-order logic and cut-elimination, will be presented in this paper. Our aim is to illustrate that the resolution calculus is more than just a principle for a (partial) automatization of logical inference. In some sense resolution is the assembler language of deduction, which—by its simplicity—makes it possible to recognize new features of deductions and to develop a new *understanding* of formal proofs and of inference in general.

§2. The Logical Basis of Resolution. In contrast to most logic calculi for first-order logic resolution does not work on the full syntax of predicate logic but on *normal forms*. These normal forms, in fact clausal normal forms, have the advantage to admit simpler inference systems and thus a more efficient way of proof search. Roughly speaking the clausal form is a *logic free* form, where the whole logical structure of the original formula is coded on the level of atomic formulas. A proof of a first-order formula A via resolution is based on the refutation of a normal form of $\neg A$, the general principle being proof by contradiction on normal forms. So we distinguish two phases in a proof of a sentence A:

- Transform $\neg A$ to a clause form $\mathcal{C}$.
- Apply resolution to $\mathcal{C}$.

2.1. Normal Forms. In constructing normal forms we have to take care that the following conditions are fulfilled: 1. soundness, 2. efficiency and 3. preservation of structure. Point 1 is the most important one, though we will see that the concept of soundness need not be the usual one. Without taking care of point 2 the enterprise of computational inference does not really makes sense. Note that a sound transformation to normal forms would be to replace every unsatisfiable formula by falsum and every satisfiable one by itself; then proof by contradiction becomes trivial as falsum would be the only normal form for unsatisfiable formulas. The pathological aspect in such a transformation clearly is that it involves the whole complexity of theorem proving itself, i.e., to detect the unsatisfiability of a formula. Thus we have to take care that the *computation* of normal forms is simple, at best within polynomial time. Moreover it should be decidable within polynomial time whether a given formula is in normal form or not. Point 3 in the least trivial because is not completely clear which structure we mean and why it has to be preserved. We will address this point later and particularly in Section 3. Before we give a formal definition of a transformation to clause form we illustrate the main steps in an example:

EXAMPLE 2.1. Let A be the formula

$$\big[(\forall x)(\exists y)P(x,y) \wedge (\exists u)(\forall v)\big(P(u,v) \rightarrow Q(v)\big)\big] \rightarrow (\exists z)Q(z).$$

The first step consists in transforming $\neg A$ to a formula F_1 which does not contain $\rightarrow$ and where negation only occurs in front of atoms. To this aim we apply the transformations

$$(\mathcal{A} \rightarrow \mathcal{B}) \Rightarrow (\neg\mathcal{A} \vee \mathcal{B}),$$
$$\neg(\mathcal{A} \rightarrow \mathcal{B}) \Rightarrow (\mathcal{A} \wedge \neg\mathcal{B}),$$
$$\neg(\exists v)\mathcal{A} \Rightarrow (\forall v)\neg\mathcal{A}$$

and obtain the formula

$$F_1\colon\ \big[(\forall x)(\exists y)P(x,y) \wedge (\exists u)(\forall v)\big(\neg P(u,v) \vee Q(v)\big)\big] \wedge (\forall z)\neg Q(z).$$

Clearly all the transformations preserve logical equivalence and therefore F_1 is logically equivalent to $\neg A$.

In the next step we eliminate the existential quantifiers by a technique which is generally called "skolemization". Roughly speaking, we delete an existential quantifier and replace its variable by a functional term containing all variables of universal quantifiers which are "above" the existential one. By this technique we replace

$$(\forall x)(\exists y)P(x,y) \text{ by } (\forall x)P(x,f(x))$$

and

$$(\exists u)(\forall v)(\neg P(u,v) \vee Q(v)) \text{ by } (\forall v)(\neg P(a,v) \vee Q(v));$$

so we obtain

$$F_2\colon (\forall x)P(x,f(x)) \wedge (\forall v)(\neg P(a,v) \vee Q(v)) \wedge (\forall z)\neg Q(z).$$

Note that F_2 is not logically equivalent to F_1, the formulas are merely equivalent with respect to satisfiability (in this case they are both unsatisfiable)!

In F_2 the quantifiers do not carry much "information" anymore; in particular they can be shifted in front and arbitrarily commuted without destroying the logical equivalence to F_2. Thus we simply drop the quantifiers and obtain

$$F_3\colon P(x,f(x)) \wedge (\neg P(a,v) \vee Q(v)) \wedge \neg Q(z).$$

F_3 is quantifier free and any universal closure of F_3 is logically equivalent to F_2. In fact F_3 is already in conjunctive normal form, i.e., it consists of a conjunction of disjunctions where the disjunctions are composed of literals (i.e., of negated and unnegated atoms). We can easily represent this formula without any logical connective using a representation by sequents: we replace any disjunction containing the positive literals $A_1, \ldots, A_n$ and the negative literals $B_1, \ldots, B_m$ by the expression

$$B_1, \ldots, B_m \vdash A_1, \ldots, A_n.$$

These logic free expressions are called *clauses*. In particular F_3 yields the clauses

$$\begin{aligned} C_1 &= \ \vdash P(x,f(x)), \\ C_2 &= P(a,v) \vdash Q(v), \\ C_3 &= Q(z) \vdash . \end{aligned}$$

F_3 as a whole can eventually be represented by the set of clauses $C\colon \{C_1, C_2, C_3\}$. C is called a *clause form* of the original formula $\neg A$. ♯

The example above shows the typical three stages of a *standard* normal form transformation:

- Transformation to negation normal form,
- transformation to Skolem form,
- transformation to conjunctive normal form.

Below we give formal definitions of the various (intermediary) normal forms obtained in the transformation to clause form:

DEFINITION 2.1. Let A a closed first-order formula. A is in negation normal form (NNF) if it fulfils the following conditions:

1. the propositional connectives of A are in $\{\wedge, \vee, \neg\}$,
2. $\neg$ occurs only in front of atoms. ♮

Informally expressed, a formula is in NNF if only atoms are of negative polarity. An obvious way to transform a formula into an equivalent formula in NNF consists in 1. replacing $\rightarrow$ by $\vee$, 2. apply the rules of de Morgan, and 3. shift negations over quantifiers and dualize them, 4. eliminating multiple negations.

After transformation to NNF existential quantifiers are "really" existential in the semantic sense. For technical reasons we assume that the formulas are *rectified*, i.e., every variable is quantified at most once in the formula. In the next step the existential quantifiers of the formula are eliminated and a purely universal formula is produced. If (Qx) is a quantifier in A ((Qx) is uniquely determined as A is rectified) we write $A_{-(Qx)}$ for the formula A after omission of the occurrence of (Qx).

DEFINITION 2.2. Let A be a formula in negation normal form. We define an operator sk in the following way:

If A does not contain existential quantifiers then $sk(A) = A$. Now assume that A contains existential quantifiers and that $(\exists x)$ is the first one. We distinguish two cases:

- If $(\exists x)$ is not in the scope of universal quantifiers then

$$sk(A) = sk(A_{-(\exists x)}\{x \leftarrow a\})$$

  where a is a constant symbol not occurring in A.
- If $(\exists x)$ is in the scope of the universal quantifiers $(\forall y_1), \ldots, (\forall y_m)$ then

$$sk(A) = sk\left(A_{-(\exists x)}\{x \leftarrow f(y_1, \ldots, y_m)\}\right)$$

  where f is an m-ary function symbol not occurring in A.

$sk(A)$ is called the *Skolem form* of A. ♮

REMARK. Note that the recursive definition of sk is well-founded: if there are existential quantifiers in the formulas then the argument of sk on the right hand side has always one existential quantifier less. For illustration let A be $(\forall y)(\exists z)(\exists x)P(y, z, x)$. Then

$$\begin{aligned} sk(A) &= sk((\forall y)(\exists x)P(y, f(y), x) \\ &= sk((\forall y)P(y, f(y), g(y)) \\ &= (\forall y)P(y, f(y), g(y)). \end{aligned}$$

♮

As we already pointed out in Example 2.1 *sk* does usually not preserve logical equivalence. Nevertheless the transformation is *weakly sound* i.e., it preserves sat-equivalence:

THEOREM 2.1. *Let A be a formula in negation normal form. Then $sk(A) \sim_{sat} A$, i.e., $sk(A)$ is satisfiable iff A is satisfiable.*

PROOF. We only give the idea of the proof; a full formal proof can be found in [30].

By definition of *sk* it is easy to see that $sk(A) \rightarrow A$ is valid. Thus the satisfiability of $sk(A)$ implies that of A. For the other direction assume that A is satisfiable and that $\mathcal{M}$ is a model of A. Then a model $\mathcal{M}'$ for $sk(A)$ is obtained by extending $\mathcal{M}$ by an appropriate interpretation of the new function symbol. ⊣

In particular A is refutable iff $sk(A)$ is refutable. Thus, in some sense, we are not proving $\neg A$ itself, but rather $\neg sk(A)$. But such problem reductions are quite natural and standard in the practice of mathematical proofs.

Once we have constructed a Skolem form it remains to transform the formula to conjunctive normal form. Again there exists a straightforward method using the rules of distributivity. Although this method is quite simple it can lead to an exponential blow up of the formula size and, much worse, to an even nonelementary increase of proof length. We will come back to this problem in Section 3 where we present an alternative way of constructing clause forms.

DEFINITION 2.3. Let $A_1, \dots, A_n, B_1, \dots, B_m$ be atom formulas. Then the expression $C\colon A_1, \dots, A_n \vdash B_1, \dots, B_m$ is called a *clause*. C represents the formula $F\colon \neg A_1 \vee \cdots \vee \neg A_n \vee B_1 \vee \cdots \vee B_m$. The empty clause $\vdash$ represents falsum. Assume that the clauses C_i represent the formulas F_i for $i = 1, \dots, k$; then $\{C_1, \dots, C_n\}$ represents the universal closure of $F_1 \wedge \cdots \wedge F_k$. ♯

To sum up, in proving A by contradiction we reduce $\neg A$ to a set of clauses $\mathcal{C}$ representing the universal closure of a conjunctive normal form obtained via the transformation steps described above. These sets $\mathcal{C}$ are the very raw material for resolution.

2.2. The Resolution Principle. In some sense the resolution principle is the simplest first-order calculus which is possible at all. It is based on one single rule which combines substitution and atomic cut. Its most typical feature is a *binary* substitution rule computing a minimal (i.e., most general) substitution which makes two atoms equal.

EXAMPLE 2.2. Let A be the formula

$$[(\forall x)(\exists y)P(x,y) \wedge (\exists u)(\forall v)(P(u,v) \rightarrow Q(v))] \rightarrow (\exists z)Q(z).$$

of Example 2.1. We have shown that the clausal form of $\neg A$ is

$$\mathcal{C}\colon \{\vdash P(x,f(x)),\ P(a,v) \vdash Q(v),\ Q(z) \vdash\}.$$

We refute C by using two rules 1. substitution and 2. the atomic cut rule

$$\frac{\Gamma_1 \vdash \Delta_1, A \quad A, \Gamma_2 \vdash \Delta_2}{\Gamma_1, \Gamma_2 \vdash \Delta_1, \Delta_2}\ cut$$

But the substitution rule is tied to the cut and serves as preparatory step. The following derivation ϕ illustrates this property:

$$\dfrac{\dfrac{\dfrac{\vdash P(x, f(x))}{\vdash P(a, f(a))}\ S \quad \dfrac{P(a, v) \vdash Q(v)}{P(a, f(a)) \vdash Q(f(a))}\ S}{\vdash Q(f(a))}\ cut \quad \dfrac{Q(z) \vdash}{Q(f(a)) \vdash}\ S}{\vdash}\ cut$$

ϕ is indeed a refutation of the formula represented by C. The substitution rule is sound by the validity of the substitution axioms and by the fact that a clause represents the universal closure of a disjunction. The cut rule is sound because it "represents" the rule

$$\frac{B \vee A \quad \neg A \vee C}{B \vee C}$$

♯

In general there are infinitely many substitutions unifying two atoms, but only the *most general ones* are actually needed. The observation of this effect and the invention of the *unification principle* by J. A. Robinson (see [37]) is a landmark in the history of computational logic and automated deduction.

We may compare substitutions and expressions with regard to their "generality", i.e., whether they can be obtained from other substitutions (expressions) by instantiation.

DEFINITION 2.4 (generality). Let E_1 and E_2 be expressions. We say that E_1 is *more* (or equally) *general* than E_2 (notation $E_1 \leq_s E_2$) if there exists a substitution σ such that $E_1\sigma = E_2$. Let σ, τ be two substitutions. We define $\sigma \leq_s \tau$ (σ is more general than τ) if there exists a substitution ϑ such that $\sigma\vartheta = \tau$. ♯

We can go back now to the problem of unifying atoms, literals, and terms (i.e., expressions) by substitutions.

EXAMPLE 2.3. Consider the atoms

$$A_1 = P(x, f(y), f(y)) \text{ and}$$
$$A_2 = P(x', y', f(x')).$$

Let t be an arbitrary term different from x and x'. Then the substitution $\sigma_t\colon \{x \leftarrow t, y \leftarrow t, x' \leftarrow t, y' \leftarrow f(t)\}$ fulfils

$$\begin{aligned} P(x, f(y), f(y))\sigma_t &= P(x', y', f(x'))\sigma_t \\ &= P(t, f(t), f(t)). \end{aligned}$$

The same property holds for the substitution

$$\sigma = \{y \leftarrow x, x' \leftarrow x, y' \leftarrow f(x)\}.$$

Clearly $\sigma \leq_s \sigma_t$ by $\sigma\{x \leftarrow t\} = \sigma_t$. ♯

The example above motivates the following definition.

Definition 2.5 (unifier). Let E be a nonempty set of expressions. A substitution σ is called a *unifier* of E if for all $e, e' \in E$ $e\sigma = e'\sigma$. σ is called a *most general unifier* (m.g.u.) of E if for every unifier θ of E we have $\sigma \leq_s \theta$.

Remark. A unifier of a two-element set $\{e, e'\}$ is frequently called a unifier of e and e'. If the set of expressions E consists of one element only then every substitution is a unifier of E, and the identical substitution ε is the m.g.u. of E. ♯

Still two questions arise:

1. Is there always an m.g.u. of a unifiable set of expressions?
2. Are m.g.u.s effectively computable?

Fortunately the answer to both questions is *yes* as we will point out below. Note that computing a unifier of a set of atoms $\{A_1, \ldots, A_n\}$, i.e., a substitution θ with $A_1\theta = \cdots = A_n\theta$, can easily be reduced to the problem of unifying two atoms and even of unifying two terms. E.g., any unifier θ of $\{P(x), P(y), P(f(z))\}$ is also a unifier of the two terms $g(x, y, f(z))$ and $g(x, x, x)$ and vice versa.

We call any algorithm computing an m.g.u. in case of unifiability and deciding unifiability a *unification algorithm*. As the performance of theorem provers strongly depends on the efficiency of unification algorithms, many sophisticated algorithms for computing m.g.u.s have been developed so far. Besides the original algorithm in Robinson's paper [37] we just mention the algorithm of Martelli and Montanari [33]. In this algorithm, unification is considered as the problem of finding a *most general solution* for a system of term equations. We also follow this line and present a simple rule based approach like in [2].

Example 2.4. Let

$$A_1\colon P(g(x), f(x, z)) \text{ and}$$
$$A_2\colon P(g(g(u)), v).$$

The problem of unifying A_1 and A_2 can be reduced to solving the system of equations

$$\mathcal{E}_1\colon \{g(x) \doteq g(g(u)),\ v \doteq f(x, z)\},$$

i.e., we are searching for a substitution θ s.t.

$$g(x)\theta = g(g(u))\theta,\ v\theta = f(x, z)\theta.$$

(called a solution) s.t. for all other solutions η we have $\theta \leq_s \eta$.

Clearly the equations cannot be "read" as a substitution directly. But we observe that θ is a solution of $\mathcal{E}_1$ iff θ is a solution of $\mathcal{E}_2$ for

$$\mathcal{E}_2 = \{x \doteq g(u),\ v \doteq f(x, z)\}.$$

$\mathcal{E}_1$ and $\mathcal{E}_2$ are *equivalent* because substitutions are homomorphisms on terms. Thus we are allowed to *decompose* the equation $g(x) \doteq g(g(u))$ into $x \doteq g(u)$.

The system $\mathcal{E}_2$ can be "interpreted" as the substitution θ: $\{x \leftarrow g(u), v \leftarrow f(x, z)\}$; but θ is not a unifier. Indeed $A_1\theta \neq A_2\theta$. So we apply another transformation and interpret the equation $x \doteq g(u)$ as a substitution on the system $\mathcal{E}_2$. This gives us the equivalent system

$$\mathcal{E}_3\colon \{x \doteq g(u),\ v \doteq f(g(u), z)\}.$$

In $\mathcal{E}_3$ every equation is of the form $x \doteq t$, where x is a variable and t is a term, and the variables on the lefthandside of the equations occur only once in the whole system. What we have obtained is a system in *solved form*. Indeed, if we read the equations as substitution

$$\sigma\colon \{x \leftarrow g(u),\ v \leftarrow f(g(u), z)\}$$

then σ is indeed a most general solution of $\mathcal{E}_3$ and thus of $\mathcal{E}_1$. Clearly σ is also an m.g.u. of A_1 and A_2. ♯

Thus finding a most general unifier means to find a most general solution of a system of term equations. And solving such a system of equations means to transform it into an equivalent solved form. Clearly we must also define criteria for the unsolvability of such systems and to ensure that our transformations on the systems terminate.

DEFINITION 2.6. A system of term equations is a finite set of the form

$$\mathcal{E}\colon \{s_1 \doteq t_1, \ldots, s_n \doteq t_n\}$$

where s_i, t_i are terms for $i = 1, \ldots, n$. A substitution θ is called a *solution* of $\mathcal{E}$ if for all $i = 1, \ldots, n$: $s_i\theta = t_i\theta$. A solution σ is called a *most general solution* of $\mathcal{E}$ if for all solutions θ of $\mathcal{E}$ $\sigma \leq_s \theta$. Two systems $\mathcal{E}$ and $\mathcal{E}'$ are called *equivalent* if they have the same set of solutions.

We say that $\mathcal{E}$ is in *solved form* if

- $\{s_1, \ldots, s_n\} \subseteq V$ and
- every s_i occurs only (and thus exactly) once in $\mathcal{E}$ (for $i = 1, \ldots, n$).

By definition a solved form represents a substitution. It is easy to see that for a solved form

$$\mathcal{E}\colon \{x_1 \doteq t_1, \ldots, x_n \doteq t_n\}$$

the substitution σ: $\{x_1 \leftarrow t_1, \ldots, x_n \leftarrow t_n\}$ is a most general solution of $\mathcal{E}$. σ is also called the *substitution defined by* $\mathcal{E}$.

Equality is reflexive; thus any equation of the form $s \doteq s$ is redundant. By the symmetry of equality $s \doteq t$ is equivalent to $t \doteq s$. As we are interested in interpreting some equations as substitutions, equations of the form $t \doteq x$, for a term t which is not a variable, are transformed to $x \doteq t$. The observation above leads to the following two transformation rules:

(trivial): If the system $\mathcal{E}$ contains an equation $s \doteq s$ then replace $\mathcal{E}$ by $\mathcal{E} \setminus \{s \doteq s\}$.

(orient): If $\mathcal{E}$ contains an equation of the form $t \doteq v$, where v is a variable and t is not, then replace $\mathcal{E}$ by $(\mathcal{E} \setminus \{t \doteq v\}) \cup \{v \doteq t\}$.

There are systems which are unsolvable and thus cannot be transformed into solved form. As we have to avoid nonterminating procedures we have to detect some typical cases of unsolvability. If we recognize the unsolvability of a system S we reduce it to $\bot$ (where $\bot$ can be interpreted as a fixed unsolvable system).

DEFINITION 2.7 (failure rules). Let $\mathcal{E}$ be a system of term equations and $s \doteq t$ be an equation in $\mathcal{E}$.

(symbol clash): If s and t are functional terms having different leading symbols then $\mathcal{E}$ is replaced by $\bot$.

(occurs check): If s is a variable s.t. $s \neq t$ and s occurs in t then replace $\mathcal{E}$ by $\bot$. ♯

EXAMPLE 2.5. Let

$$\mathcal{E}\colon \{x \doteq y, x \doteq g(y)\}.$$

$\mathcal{E}$ is unsolvable: there exists no substitution θ s.t. $x\theta = y\theta$ and $x\theta = g(y)\theta$; for otherwise $y\theta = g(y)\theta = g(y\theta)$, which is impossible. However, neither (symbol clash) nor (occurs check) is actually applicable. But if we apply $x \doteq y$ as substitution $\{x \leftarrow y\}$ to the other equation we obtain the equivalent system

$$\mathcal{E}'\colon \{x \doteq y, y \doteq g(y)\},$$

which can be reduced to $\bot$ via (occurs check). ♯

The example above shows that we need additional rules for deciding unifiability and for computing most general unifiers. These rules, decomposition and replacement, are in fact the most important ones.

(decomposition): Let $s \doteq t$ be an equation in $\mathcal{E}$ s.t. there exists a function symbol $f \in \mathrm{FS}_n$ and terms $s_1, \ldots, s_n, t_1, \ldots, t_n$ with

$$s = f(s_1, \ldots, s_n),$$
$$t = f(t_1, \ldots, t_n).$$

Then replace $\mathcal{E}$ by the system

$$\mathcal{E}'\colon (\mathcal{E} \setminus \{s \doteq t\}) \cup \{s_1 \doteq t_1, \ldots, s_n \doteq t_n\}.$$

(replacement): Let $s \doteq t$ be an equation in $\mathcal{E}$ s.t.

(1) s is a variable and

(2) s does not occur in t and

(3) s occurs in $\mathcal{E} \setminus \{s \doteq t\}$.

Then replace $\mathcal{E}$ by the system

$$\mathcal{E}'\colon\ (\mathcal{E} \setminus \{s \doteq t\})\{s \leftarrow t\} \cup \{s \doteq t\}.$$

DEFINITION 2.8. Let $\mathfrak{R}$ be the set of rules {(trivial), (orient), (symbol clash), (occurs check), (decomposition), (replacement)}. If $\mathcal{E}$ is transformed to $\mathcal{E}'$ via a rule in $\mathcal{R}$ we write $\mathcal{E} \succ_{\mathfrak{R}} \mathcal{E}'$. For the reflexive and transitive closure of the relation $\succ_{\mathfrak{R}}$ we write $\succ_{\mathfrak{R}}^*$. A system $\mathcal{E}$ is called *irreducible* if none of the rules in $\mathfrak{R}$ is applicable to $\mathcal{E}$. $\perp$ is irreducible by definition. ♮

According to the definitions above, unification means nothing else than reduction of a system to an irreducible form (where an arbitrary order of applications of rules in $\mathfrak{R}$ is admitted).

EXAMPLE 2.6. We take the system $\mathcal{E}$ from Example 2.4 where

$$\mathcal{E} = \{g(x) \doteq g(g(u)),\ v \doteq f(x,z)\}.$$

Then $\mathcal{E} \succ_{\mathfrak{R}} \mathcal{E}'$ (via (decomposition)) where

$$\mathcal{E}' = \{x \doteq g(u),\ v \doteq f(x,z)\}.$$

Now (replacement) is applicable and $\mathcal{E}' \succ_{\mathfrak{R}} \mathcal{E}''$, where

$$\mathcal{E}'' = \{x \doteq g(u),\ v \doteq f(g(u),z)\}.$$

$\mathcal{E}''$ is irreducible and in solved form. ♮

REMARK. Note that every solved form is irreducible. On the other hand every irreducible form is either a solved form or $\perp$. ♮

THEOREM 2.2 (unification theorem). *$\mathfrak{R}$ is a unification system, i.e.,*

- *$\mathfrak{R}$ always terminates.*
- *If the system $\mathcal{E}$ is solvable then $\mathcal{E} \succ_{\mathfrak{R}}^* \mathcal{E}'$ s.t. $\mathcal{E}'$ is in solved form and the substitution defined by $\mathcal{E}'$ is a most general solution of $\mathcal{E}$.*

PROOF (SKETCH). For a full proof see [2]. The most involved part is termination.

If we know that the system is terminating our task is easier: all rules in $\mathfrak{R}$ preserve the equivalence of systems. So let $\mathcal{E}'$ be an irreducible system obtained from $\mathcal{E}$. If $\mathcal{E}' = \perp$ then $\mathcal{E}'$ and $\mathcal{E}$ itself are unsolvable; else $\mathcal{E}'$ must be in solved form (for otherwise it would not be irreducible). Thus the substitution defined by $\mathcal{E}'$ is a most general solution of $\mathcal{E}'$ and of $\mathcal{E}$ itself. ⊣

We have seen in Example 2.2 how an unsatisfiable set of clauses can be refuted via unification and cut. The example below shows that unification cannot be restricted to atoms of different clauses but must also be applied within clauses.

EXAMPLE 2.7. Let $\mathcal{C} = \{C_1, C_2\}$ for

$$C_1 = \vdash P(x), P(y),$$
$$C_2 = P(u), P(v) \vdash .$$

Clearly $\mathcal{C}$ is unsatisfiable, but we will see that without an additional technique $\vdash$ cannot be derived.

Even if we permute $P(x)$ and $P(y)$ within C_1 and $P(u), P(v)$ within C_2 and try all unifications between atoms in C_1 and C_2 we obtain a variant of the clause C_3: $P(x) \vdash P(y)$. But, modulo variable renaming, C_1 with C_3 gives C_1 and C_2 with C_3 gives C_2. We see that cutting out single atoms does not suffice to derive contradiction. What we need is unification within a clause followed by a contraction rule.

Let us apply the substitution λ: $\{x \leftarrow y\}$ to C_1; then we obtain the clause

$$\vdash P(y), P(y).$$

But $\vdash P(y), P(y)$ represents $(\forall y)(P(y) \vee P(y))$ which is logically equivalent to $(\forall y)P(y)$. Thus we may apply contraction within $\vdash P(y), P(y)$ and obtain C_1': $\vdash P(y)$. Similarly we obtain by μ: $\{u \leftarrow v\}$ the clause C_2': $P(v) \vdash$ from C_2. Now by unifying $P(y), P(v)$ via the m.g.u. σ: $\{y \leftarrow v\}$ we eventually obtain C_1'': $\vdash P(v)$ and C_2'': $P(v) \vdash$ which resolve to $\vdash$ via atomic cut. C_1' and C_2' are called factors of C_1 and of C_2 respectively. The steps are illustrated in the derivation below (S stands for substitution as preparation for the cut and F for factoring):

$$\dfrac{\dfrac{\dfrac{\vdash P(x), P(y)}{\vdash P(y)}\,F}{\vdash P(v)}\,S \qquad \dfrac{P(u), P(v) \vdash}{P(v) \vdash}\,F}{\vdash}\,cut$$

♮

DEFINITION 2.9 (factor). Let

$$C : \Gamma \vdash \Delta_1, A_1, \ldots, \Delta_n, A_n, \Delta_{n+1}$$

$(C : \Delta_1, A_1, \ldots, \Delta_n, A_n, \Delta_{n+1} \vdash \Gamma)$ be a clause where $n \geq 1$ and the Δ_i are (possibly empty) sequences of atoms. Moreover let σ be a most general unifier of $\{A_1, \ldots, A_n\}$. Then

$$C' : \Gamma\sigma \vdash \Delta_1\sigma, \ldots, \Delta_n\sigma, A_n\sigma$$

$(C' : A_n\sigma, \Delta_1\sigma, \ldots, \Delta_n\sigma \vdash \Gamma\sigma)$ is called a *factor* of C. ♮

REMARK. If the clause does not contain variables then a factor is obtained just by contraction of identical atoms. If in Definition 2.9 $n = 1$ then σ is the identical substitution and the atom A_1 is put to the extreme right, respectively left, of the clause. Thus we are using factoring also to position the atoms at the appropriate places for the following cut. ♮

DEFINITION 2.10 (binary resolvent). Let $C_1 : \Gamma_1 \vdash \Delta_1, A$ and $C_2 : B, \Gamma_2 \vdash \Delta_2$ be two clauses which are variable disjoint and A, B be unifiable with m.g.u. σ. Then the clause

$$\Gamma_1\sigma, \Gamma_2\sigma \vdash \Delta_1\sigma, \Delta_2\sigma$$

is called a *binary resolvent* of C_1 and C_2. ♯

After these preparations we are ready for the definition of the general resolution rule.

DEFINITION 2.11 (resolution). Let C_1, C_2 be two clauses and C_1', C_2' be variable disjoint variants of factors of C_1, C_2 and C be a binary resolvent of C_1' and C_2'. Then C is called a (general) *resolvent* of C_1 and C_2. ♯

By Definition 2.11 it is possible to define infinitely many resolvents of two clauses, as there are infinitely many possible variable renamings. We can avoid this effect by a standard renaming of the resolvent by variables $\{x_1, \ldots, x_n\}$. Using such a standard renaming there are indeed only *finitely many* resolvents of two clauses.

EXAMPLE 2.8. Let

$$C_1 = R(x, y) \vdash P(x), P(y),$$
$$C_2 = Q(x), P(f(x)) \vdash S(x, x).$$

Then, under a standard renaming, the following clauses are resolvents of C_1 and C_2:

$$C_3 = R(x_1, f(x_2)), Q(x_2) \vdash P(x_1), S(x_2, x_2),$$
$$C_4 = R(f(x_1), x_2), Q(x_1) \vdash P(x_2), S(x_1, x_1),$$
$$C_5 = R(f(x_1), f(x_1)), Q(x_1) \vdash S(x_1, x_1).$$ ♯

The following deduction principle on clause logic is characterized by using only one single inference rule, namely resolution.

DEFINITION 2.12 (resolution deduction).

A resolution deduction (R-deduction) of a clause C from a set of clauses $\mathcal{C}$ is a sequence of clauses $C_1, \ldots, C_n$ with the following properties:

1. $C_n = C$.
2. For every i with $i \in \{1, \ldots, n\}$ we have: either C_i is a clause in $\mathcal{C}$ or C_i is a resolvent of two clauses C_j, C_k with $j, k < i$.

An R-deduction of $\vdash$ from $\mathcal{C}$ is called an R-refutation of $\mathcal{C}$. ♯

EXAMPLE 2.9. Let

$$\mathcal{C} = \{ \vdash P(x, f(x));\ P(a, v) \vdash Q(v);\ Q(z) \vdash \}.$$

Then

$$\phi\colon\ \vdash P(x, f(x));\ P(a, v) \vdash Q(v);\ \vdash Q(f(a));\ Q(z) \vdash;\ \vdash$$

is an R-refutation of $\mathcal{C}$. In tree format ϕ is of the form

$$\dfrac{\dfrac{\vdash P(x, f(x)) \quad P(a, v) \vdash Q(v)}{\vdash Q(f(a))}\,R \qquad Q(z) \vdash}{\vdash}\,R$$

Modulo the clause form transformation ϕ can be considered as a proof by contradiction of the formula

$$[(\forall x)(\exists y)P(x, y) \wedge (\exists u)(\forall v)(P(u, v) \rightarrow Q(v))] \rightarrow (\exists z)Q(z). \qquad \sharp$$

Resolution is a complete inference principle on clause logic and, together with normal form transformations, a refutationally complete inference system for first-order logic.

THEOREM 2.3. *Resolution is* complete, *i.e., if $\mathcal{C}$ is an unsatisfiable set of clauses then there exists an R-refutation of $\mathcal{C}$.*

PROOF. We merely give the main line of the proof; a detailed proof can be found in [30].

By Herbrand's theorem a set of clauses $\mathcal{C}$ is unsatisfiable iff there exists a finite set $\mathcal{C}'$ of ground (i.e., variable-free) clauses, obtained by instantiation of clauses in $\mathcal{C}$, which is unsatisfiable. Thus the unsatisfiability of the first-order problem $\mathcal{C}$ is reduced to that of the propositional problem $\mathcal{C}'$.

Then it is proved that resolution is complete on sets of ground clauses. As a consequence there exists an R-refutation ϕ' of $\mathcal{C}'$.

Finally the refutation ϕ' is *lifted* to a refutation ϕ of $\mathcal{C}$. Lifting is a technique replacing a resolvent C' of instances C_1', C_2' of clauses C_1, C_2 by a resolvent C of C_1, C_2 s.t. C' is an instance of C. ⊣

The lifting principle is the most significant feature of first-order resolution. Instead of producing infinitely many resolvents of instances of two clauses it is enough to use the general resolvents which are based on most general unification only. We give a simple example illustrating this principle.

EXAMPLE 2.10. Let

$$\mathcal{C} = \{ \vdash P(x);\ P(u) \vdash Q(u);\ Q(f(z)) \vdash \}.$$

Let t be an arbitrary ground term over the signature $\{f, a\}$ and

$$\mathcal{C}' = \{ \vdash P(f(t));\ P(f(t)) \vdash Q(f(t));\ Q(f(t)) \vdash \}.$$

Then

$$\phi_t\colon\ \vdash P(f(t));\ P(f(t)) \vdash Q(f(t));\ \vdash Q(f(t));\ Q(f(t)) \vdash;\ \vdash$$

is an R-refutation of $\mathcal{C}'$. But

$$\phi_t = \phi\{x \leftarrow f(t), u \leftarrow f(t), z \leftarrow t\}$$

for

$$\phi = \vdash P(x);\ P(u) \vdash Q(u);\ \vdash Q(u);\ Q(f(z)) \vdash;\ \vdash .$$

But ϕ is an R-refutation of $\mathcal{C}$. $\sharp$

§3. Resolution and LK.

3.1. The calculus LK. Traditional logic calculi work on the full syntax of logic and not on normal forms. This has the advantage that the logical structure of the theorems is preserved and the semantical meaning is well-presented in the different steps of a derivation. The resolution calculus, in contrast, works on simple normal forms which favor efficient inference, but on the other hand its derivations look ugly and "unnatural". Moreover we even risk to lose structural information which is useful in defining good and/or short proofs. We will show in this section that there is a remedy for this defect of resolution, namely structural clause form transformation. This technique allows us to define an "interface" to standard logic calculi and to *store* the structure of full first-order logic without sacrificing efficiency. As "traditional" logic calculus we choose Gentzen's **LK** because it is the most elegant and "semantic" calculus for classical first-order logic. The elements **LK** is working with are sequents:

DEFINITION 3.1 (sequent). A sequent is an expression of the form $\Gamma \vdash \Delta$ where Γ and Δ are finite multisets of PL-formulas (i.e., two sequents $\Gamma_1 \vdash \Delta_1$ and $\Gamma_2 \vdash \Delta_2$ are considered equal if the multisets represented by Γ_1 and by Γ_2 are equal and those represented by Δ_1, Δ_2 are also equal). ♯

Note that clauses are sequents containing only atomic formulas.

DEFINITION 3.2 (the calculus **LK**). The initial sequents are $A \vdash A$ for first-order formulas A. In the rules of **LK** we always mark the auxiliary formulas (i.e., the formulas in the premise(s) used for the inference) and the principal (i.e., the inferred) formula using different marking symbols. Thus, in our definition, $\wedge$-introduction to the right takes the form

$$\frac{\Gamma_1 \vdash A^+, \Delta \qquad \Gamma_2 \vdash \Delta_2, B^+}{\Gamma_1, \Gamma_2 \vdash \Delta_1, A \wedge B^*, \Delta_2} \wedge : r$$

We usually avoid markings by putting the auxiliary formulas at the leftmost position in the antecedent of sequents and in the rightmost position in the consequent of sequents. The principal formula mostly is identifiable by the context. Thus the rule above will be written as

$$\frac{\Gamma_1 \vdash \Delta_1, A \qquad \Gamma_2 \vdash \Delta_2, B}{\Gamma_1, \Gamma_2 \vdash \Delta_1, \Delta_2, A \wedge B} \wedge : r$$

The version of **LK** we are using here is that in [4] and slightly deviates from Gentzen's original version [17]. The differences however are without importance to the results presented in this section. Readers who are interested in a detailed definition of **LK** are referred to [17] or to [38]. ♯

The main result of Gentzen's famous paper [17] was the cut-elimination theorem. It shows that, in arbitrary **LK**-proofs, the cut rule

$$\frac{\Gamma_1 \vdash \Delta_1, A \quad A, \Gamma_2 \vdash \Delta_2}{\Gamma_1, \Gamma_2 \vdash \Delta_1, \Delta_2}\ cut$$

can be eliminated; the result is a proof with the *subformula property*, i.e., the whole proof is made of the syntactic material of the end sequent. By this property Gentzen's cut-free **LK** can be used as a basis for proof search and automated deduction when combined with the unification principle; the corresponding calculus is the *tableaux-calculus* [22]. For illustration we give a simple cut-free **LK**-proof of the sequent

$$P(a), (\forall x)(P(x) \to P(f(x))) \vdash P(f^2(a))$$

which (semantically) stands for the formula

$$[P(a) \wedge (\forall x)(P(x) \to P(f(x)))] \to P(f^2(a)).$$

$$\dfrac{\dfrac{\dfrac{\dfrac{P(a) \vdash P(a) \quad \dfrac{P(f(a)) \vdash P(f(a)) \quad P(f^2(a)) \vdash P(f^2(a))}{P(f(a)), P(f(a)) \to P(f^2(a)) \vdash P(f^2(a))}\ \to: l}{P(f(a)) \to P(f^2(a)), P(a) \to P(f(a)), P(a) \vdash P(f^2(a))}\ \to: l}{(\forall x)(P(x) \to P(f(x))), P(a) \to P(f(a)), P(a) \vdash P(f^2(a))}\ \forall: l}{(\forall x)(P(x) \to P(f(x))), (\forall x)(P(x) \to P(f(x))), P(a) \vdash P(f^2(a))}\ \forall: l}{(\forall x)(P(x) \to P(f(x))), P(a) \vdash P(f^2(a))}\ c: l$$

3.2. Structural Clause Form. The key to the *simulation* of cut-free **LK** by resolution is the *structural normal form* transformation. In contrast to the standard normal form transformation presented in Subsection 2.1 the structural one is based on introductions of new predicate symbols. Let F be the formula to be transformed into clause form. The idea is the following one: For any subformula A in F create a new predicate symbol P_A and "define" P_A by A:

Suppose, for illustration, that $A = A_1 \circ A_2$ for $\circ \in \{\wedge, \vee, \to\}$ and that X is the set of variables free in A. Furthermore let Y be the set of variables free in A_1 and Z the set of variables free in A_2. Then clearly $X = Y \cup Z$. Define vectors of variables $\bar{x}, \bar{y}, \bar{z}$ from the sets X, Y, Z. Then the *defining formula* for P_A is:

$$(\forall \bar{x})\left[P_A(\bar{x}) \leftrightarrow (P_{A_1}(\bar{y}) \circ P_{A_2}(\bar{z}))\right].$$

Similarly the defining formula for $A = \neg B$ is $(\forall \bar{x})(P_A(\bar{x}) \leftrightarrow \neg P_B(\bar{x}))$ and for $A = (Qy)B$ it is

$$(\forall \bar{x})(P_A(\bar{x}) \leftrightarrow (Qy)P_B(\bar{x}, y)).$$

Note that the defining formulas just represent a mathematical tool which is widely used in mathematics, namely *extension by definition*. The extension formulas directly yield the structural clause form.

DEFINITION 3.3. Let F be a closed formula and $\mathcal{F}$ be the set of all defining formulas obtained from F. Then the *structural clause form* of F is the set $\gamma_{\text{struc}}(F)$ defined by

$$\gamma_{\text{struc}}(F) = \bigcup \{\gamma_0(B) \mid B \in \mathcal{F}\} \cup \{\vdash A_F\}$$

where γ_0 is the operator of the standard clause form transformation and in all defining formulas $\gamma_0(G \leftrightarrow H)$ is defined by $\gamma_0((G \to H) \wedge (H \to G))$. ♯

Like γ_0 also γ_{struc} is satisfiability preserving and thus weakly correct (a formal proof can be found in [12]). For practical reasons one can avoid to construct defining formulas for atoms. These would have the form $(\forall \bar{x})(P_Q(\bar{x}) \leftrightarrow Q(\bar{x}))$. It is easy to see that this results only in a renaming of the atoms themselves making the corresponding clauses redundant. The structural normal form subjected to this improvement is denoted by $\gamma'_{\text{struc}}(F)$.

EXAMPLE 3.1. Let $F = (\forall x)P(x) \vee (P(a) \wedge Q(b))$.

The defining formulas (omitting the atomic level) are the following ones:

$$E_1 = A_F \leftrightarrow (A_1 \vee A_2),$$
$$E_2 = A_1 \leftrightarrow (\forall x)P(x),$$
$$E_3 = A_2 \leftrightarrow (P(a) \wedge Q(b)).$$

We now compute the standard clause forms $C_i : \gamma_0(E_i)$ and obtain

$$C_1 = \{A_F \vdash A_1, A_2;\ A_1 \vdash A_F;\ A_2 \vdash A_F\},$$
$$C_2 = \{A_1 \vdash P(x);\ P(c) \vdash A_1\},$$
$$C_3 = \{A_2 \vdash P(a);\ A_2 \vdash Q(b);\ P(a), Q(b) \vdash A_2\}.$$

Now

$$\gamma'_{\text{struc}}(F) = \{\vdash A_F\} \cup C_1 \cup C_2 \cup C_3.$$

Note that in computing C_2 we have to apply skolemization and thus obtain the new constant symbol c. ♯

In Example 3.1 the structural clause form consists of nine clauses, while $\gamma_0(F) = \{\vdash P(x), P(a);\ \vdash P(x), Q(b)\}$ and thus consists of two clauses only. It might appear that structural transformation is much more expensive, but this is deceptive: the worst-case complexity of γ_{struc} is quadratic (in propositional logic even linear), but that of γ_0 is exponential (even for propositional problems). Moreover γ_{struc} preserves the structure of the formula and may lead to much shorter proofs. In fact it is proven in [3] that there exists a sequence of clause sets C_n fulfilling the following conditions:

- $\gamma_{\mathrm{struc}}(C_n)$ has resolution refutations of length $\leq 2^{2^{dn}}$,
- all resolution refutations of $\gamma_0(C_n)$ are longer than $s(n-1)$, where s is a nonelementary function defined by $s(0) = 1$, $s(n+1) = 2^{s(n)}$.

Note that $s(n)$ grows faster than any fixed iteration of the exponential function. The result above is obtained by using the complexity of cut-elimination (which is nonelementary) and the simulation of cut-free **LK** (and of **LK** with analytic cut) shown in the next subsection.

In [13] Egly and Rath proved that structural clause form transformation is not only superior in theory but also in practice. They gave a thorough experimental comparison with the standard transformation: the version of the prover using structural transformation did not only behave better in average but could handle problems unsolvable with the standard version. This shows that *preservation of logical structure* in normal form computations pays out also in practice.

3.3. Simulation of LK by Resolution. With γ_{struc} we have an operator which does not only map a formula into a set of clauses but also encodes all of its subformulas by new labels represented by new predicate symbols. This makes it possible to simulate inferences in **LK** by resolution deductions. For simplicity we only present the propositional case, for the full first-order simulation see [12] or [3].

In the first step we map sequents into clauses; all we have to do is to use atomic labels for formulas like in structural clause transformation. So a sequent of the form

$$F_1, \dots, F_m \vdash G_1, \dots, G_n$$

is represented by the clause

$$A_{F_1}, \dots, A_{F_m} \vdash A_{G_1}, \dots, A_{G_n}.$$

Now assume that we have a cut-free proof ψ of the sequent $F \vdash$ (which proves that F is unsatisfiable). Then we construct $\gamma_{\mathrm{struc}}(F)$. The simulation of logical introduction rules then is performed via resolution using the clauses encoding these introductions. We show the procedure for $\wedge\colon r$ and $\neg\colon l$ which suffices to illustrate the nature of this transformation.

Let π_1, π_2 **LK**-proofs and π be a subderivation of ψ of the form:

$$\dfrac{\begin{matrix}(\pi_1) & & (\pi_2)\\ S_1\colon \Gamma_1 \vdash \Delta_1, G & & S_2\colon \Gamma_2 \vdash \Delta_2, H\end{matrix}}{S_3\colon \Gamma_1, \Gamma_2 \vdash \Delta_1, \Delta_2, G \wedge H}\ \wedge\colon r$$

Now let $C_1\colon \Lambda_1 \vdash \Theta_1, A_G$ and $C_2\colon \Lambda_2 \vdash \Theta_2, A_H$ be the clauses corresponding to S_1, S_2. Assume inductively that resolution derivations λ_1 of C_1 and λ_2 of C_2 already exist. We construct a resolution derivation of C, the clause representing S_3. By definition of $\gamma_{\mathrm{struc}}(F)$ and by the fact that the formula

$G \wedge H$ is a subformula of F (note that we have the subformula property!) the set $\gamma_{\mathrm{struc}}(F)$ contains the clauses

$$A_G, A_H \vdash A_{G\wedge H};\ A_{G\wedge H} \vdash A_G;\ A_{G\wedge H} \vdash A_H.$$

Using these clauses we can construct the following resolution derivation λ:

$$\dfrac{\begin{matrix}(\lambda_2)\\ \Lambda_2 \vdash \Theta_2, A_H\end{matrix} \quad \dfrac{\begin{matrix}(\lambda_1)\\ \Lambda_1 \vdash \Theta_1, A_G\end{matrix} \quad A_G, A_H \vdash A_{G\wedge H}}{A_H, \Lambda_1 \vdash \Theta_1, A_{G\wedge H}}\, R}{\Lambda_2, \Lambda_1 \vdash \Theta_1, \Theta_2, A_{G\wedge H}}\, R$$

which is a resolution derivation of a clause representing S_3. So λ simulates π.

Now we illustrate the case of the negation introduction to the left. Let π be a subproof of ψ of the form

$$\dfrac{\begin{matrix}(\pi')\\ S' : \Gamma \vdash \Delta, G\end{matrix}}{S : \neg G, \Gamma \vdash \Delta}\, \neg : l$$

and λ' be a resolution proof of the clause $\Lambda \vdash \Theta, A_G$ representing S'. As $\neg G$ is a subformula of F the set $\gamma_{\mathrm{struc}}(F)$ contains the clauses $\vdash A_G, A_{\neg G}$ and $A_G, A_{\neg G} \vdash$. But then the following resolution derivation λ

$$\dfrac{\begin{matrix}(\lambda)\\ \Lambda \vdash \Theta, A_G\end{matrix} \quad A_G, A_{\neg G} \vdash}{A_{\neg G}, \Lambda \vdash \Theta}\, R$$

yields a representation of S.

Similar transformations can be constructed for the other connectives and inferences. Finally we obtain a resolution derivation ρ simulating ψ. ρ is a resolution derivation of the clause $A_F \vdash$. The final step uses the clause $\vdash A_F$ which is in $\gamma_{\mathrm{struc}}(F)$ by definition and eventually yields the resolution refutation

$$\dfrac{\begin{matrix}(\rho)\\ A_F \vdash\end{matrix} \quad \vdash A_F}{\vdash}\, R$$

which is a refutation of $\gamma_{\mathrm{struc}}(F) \cup \mathcal{D}$ where $\mathcal{D}$ is a set of tautological clauses of the form $A_G \vdash A_G$ representing the initial sequents of ψ. Note that every resolution refutation can be transformed into another (even shorter) one which is tautology-free (see e.g., [30]); therefore we can get rid of the set $\mathcal{D}$ and eventually obtain a refutation ρ' of $\gamma_{\mathrm{struc}}(F)$ alone. Note that the whole simulation is *polynomial*: the length of the resolution proof (counting the number of occurrences of formulas in a proof) is linear in the length of the **LK**-proof; the symbolic length can be quadratic due to the introduction of Skolem terms.

With γ_{struc} we can not only simulate the cut-free **LK** but also **LK** with *analytic* cut (a cut is called analytic if the cut formula occurs as subformula in the end sequent). Indeed if G is the cut formula in an inference of an **LK**-proof ψ of $F \vdash$ and G is subformula of F we have an atom $A_G(\bar{x})$ and the defining formula for G; thus the cut with G can be simulated by resolution with $A_G(\bar{t})$. If we add arbitrary formulas G and construct $\gamma_{\text{struc}}(G)$ we can even polynomially simulate full **LK**. This stronger form of structural transformation is called the *extension principle*; it was introduced (for first-order logic) by E. Eder [12]. Thus resolution + extension is capable of simulating virtually every logic calculus in an easy way: once we have simulated **LK** we can use Gentzen's transformations in [17] to simulate natural deduction and Hilbert type calculi.

§4. **Resolution Refinements.** As resolution is a calculus for proof search, a main direction of improvement was (since the very beginning of resolution theorem proving) the reduction of possible resolution deductions under preservation of completeness. Generally we call any restriction of resolution deduction a *resolution refinement.* Restrictions of the resolution principle did not only improve the efficiency of theorem provers but also lead to the development of logic programming and decision procedures; the latter application area will be treated in Section 5. In many applications of resolution we are not interested in the deductions themselves but rather in the deductive closure (i.e., in the set of derivable clauses). This aspect of deduction is best described by *resolution operators.*

DEFINITION 4.1 (resolution operator). Let $\mathcal{C}$ be a set of clauses and $\mathcal{R}es(\mathcal{C})$ denote the set of all resolvents definable from $\mathcal{C}$ and subjected to a normalization under a standard renaming of variables. Then we define the operator R, the *operator of unrestricted resolution* and its deductive closure R^* by

$$R(\mathcal{C}) = \mathcal{C} \cup \mathcal{R}es(\mathcal{C}), \qquad R^0(\mathcal{C}) = \mathcal{C},$$
$$R^{i+1}(\mathcal{C}) = R(R^i(\mathcal{C})), \qquad R^*(\mathcal{C}) = \bigcup_{i \in \mathbb{N}} R^i(\mathcal{C}).$$

♯

Note that, by the completeness of the resolution principle, we have $\vdash \in R^*(\mathcal{C})$ for unsatisfiable sets of clauses $\mathcal{C}$. By the undecidability of clause logic $R^*(\mathcal{C})$ must be infinite for some satisfiable sets of clauses $\mathcal{C}$.

Refinement operators can be defined in a similar way; we only have to replace $\mathcal{R}es$ by another operator.

DEFINITION 4.2 (refinement operator). A *resolution refinement operator* R_x is a mapping from sets of clauses to sets of clauses defined in the following way: There exists a (one-step) operator ϱ_x s.t.

- $R_x(\mathcal{C}) = \mathcal{C} \cup \varrho_x(\mathcal{C})$,
- ϱ_x is recursive,
- $\varrho_x(\mathcal{C})$ is a finite subset of $R^*(\mathcal{C})$.

Again the deductive closure is defined as

$$R_x^0(\mathcal{C}) = \mathcal{C},\ R_x^{i+1}(\mathcal{C}) = R_x(R_x^i(\mathcal{C})),$$
$$R_x^*(\mathcal{C}) = \bigcup_{i\in\mathbb{N}} R_x^i(\mathcal{C}).$$

♯

The requirement $\varrho_x(\mathcal{C}) \subseteq R^*(\mathcal{C})$, instead of $\rho_x(\mathcal{C}) \subseteq \mathcal{R}es(\mathcal{C})$, serves the purpose to encompass methods of macro-inference (several resolution steps may be considered as primitive inference step). Such a principle is hyperresolution which will be described in this section. Not all refinements can be formulated within the framework of operators; indeed linear refinements are defined via the restriction of the resolution tree, which cannot be formalized by operators (for details see [30]).

4.1. Ordering Refinements. The use of ordering refinements is based on the idea to keep resolvents "small". This does not only lead to smaller clauses during deduction but also to an improved termination behavior of operators. The most common ordering principle of this type is *atom ordering*; in this restriction no resolvents are admitted which contain atoms more complex than the resolved one.

DEFINITION 4.3. An A-ordering (atom ordering) $\prec$ is a binary relation on atoms with the following properties:

A.1 $\prec$ is irreflexive,

A.2 $\prec$ is transitive,

A.3 For all atoms A, B and for all substitutions θ:

$$A \prec B \text{ implies } A\theta \prec B\theta.$$

♯

The property A.3 guarantees the lifting property for A-orderings which is vital to completeness.

EXAMPLE 4.1. Let A and B be two arbitrary atoms. We define the *depth ordering* $\prec_d$ by

$A \prec_d B$ iff

d1. $\tau(A) < \tau(B)$,

d2. $\text{var}(A) \subseteq \text{var}(B)$ and $\tau_{\max}(x, A) < \tau_{\max}(x, B)$ for $x \in \text{var}(A)$.

where τ is the term depth and $\tau_{\max}(x, A)$ the maximal depth of the variable x in A.

It is easy to see that the depth ordering $\prec_d$ fulfils A.1,A.2,A.3 and thus is indeed an A-ordering. Note that by condition d1. alone this would not be the case:

Assume that $P(x) \prec_d P(f(a))$. Then by A.3 (using $\theta = \{x \leftarrow f(a)\}$) we have $P(x)\theta \prec_d P(f(a))\theta$, i.e., $P(f(a)) \prec_d P(f(a))$. But the last relation contradicts A.1. Generally $A \prec B$ is impossible for unifiable atoms A and B.

Some further examples for the $\prec_d$-relation:

$P(x,y) \prec_d P(f(x), f(y))$, $P(x,x) \prec_d Q(g(x,y))$,
$Q(x) \not\prec_d Q(f(a))$: d2. is violated,
$Q(f(a)) \not\prec_d Q(x)$: d1. is violated. ♯

Another A-ordering which will turn out useful in the next section is comparing rather the size than the depth of terms.

EXAMPLE 4.2. We define an ordering by first comparing functional terms: two functional terms are called *similar* if

$$s = g(r_1, \ldots, r_n) \text{ and } t = f(w_1, \ldots, w_n) \text{ and } \{r_1, \ldots, r_n\} = \{w_1, \ldots, w_n\}.$$

Similar terms may have different top symbols but their arguments are equal under permutations.

A functional term s *dominates* a functional term t if t has less arguments than s or more formally

- $s = f(r_1, \ldots, r_n)$, $t = g(w_1, \ldots, w_m)$,
- $n > m$,
- $\{w_1, \ldots, w_m\} \subseteq \{r_1, \ldots, r_n\}$.

The concepts above can be used in defining the term ordering $\prec_1$, where $s \prec_1 t$ iff

- s properly occurs in t or
- t dominates s or
- t contains a proper subterm which dominates s or is similar to s.

Some examples for $\prec_1$:

$f(x,y) \prec_1 h(x,y,z)$: the second term dominates the first one.
$f(x,y) \not\prec_1 g(x,y)$: the terms are similar.
$g(y,x) \prec_1 g(f(x,y),x)$: the second term contains a proper subterm which is similar to the first term.

The ordering $\prec_1$ can easily be extended to an atom ordering $\prec_2$ defined by:

$A \prec_2 B$ iff there exists an argument t of B such that for all arguments s of A we have $s \prec_1 t$.

Some examples for $\prec_2$:

$P(x,y) \prec_2 Q(f(x,y))$: by $x \prec_1 f(x,y)$ and $y \prec_1 f(x,y)$.
$P(x,z) \not\prec_2 Q(f(x,y))$: by $z \not\prec_1 f(x,y)$. ♯

We define the A-ordering refinement via $\varrho_\prec(\mathcal{C})$, which describes the set of all $\prec$-resolvents in $\mathcal{C}$. The *resolved atom* of a resolution is the atom cut out by resolution (after application of the m.g.u.s).

DEFINITION 4.4 (ordered resolution). Let $\mathcal{C}$ be a set of clauses and $\prec$ be an A-ordering. We define $C \in \varrho_\prec(\mathcal{C})$ iff $C \in \mathcal{R}es(\mathcal{C})$ and for no atom B in C: $A \prec B$, where A is the resolved atom of the corresponding resolution.

REMARK. Definition 4.4 restricts resolvents in a way that only maximal atoms in a clause may be resolved. But note that the ordering has to be considered after application of the m.g.u. (this ordering principle is called *aposteriori ordering*). ♯

EXAMPLE 4.3. Let $\prec_d$ be the ordering from Example 4.1 and

$$\mathcal{C} = \{ \vdash P(a);\ P(x) \vdash R(f(x));\ R(y) \vdash R(f(y));\ R(f^2(a)) \vdash \}.$$

Then

$$R_{\prec_d}(\mathcal{C}) = \mathcal{C} \cup \{R(f(a)) \vdash,\ P(f(a)) \vdash \},$$
$$R^2_{\prec_d}(\mathcal{C}) = R_{\prec_d}(\mathcal{C}) \cup \{R(a) \vdash,\ P(a) \vdash \},$$
$$R^3_{\prec_d}(\mathcal{C}) = R^*_{\prec_d}(\mathcal{C}) = R^2_{\prec_d}(\mathcal{C}) \cup \{ \vdash \}.$$

Note that there are more clauses in $R^*(\mathcal{C})$. E.g., $\vdash R(f(a)) \in \mathcal{R}es(\mathcal{C})$, but here $P(a)$ is the resolved atom and $P(a) \prec_d R(f(a))$, thus $\vdash R(f(a)) \notin \varrho_{\prec_d}(\mathcal{C})$. Similarly $P(x_1) \vdash R(f^2(x_1)) \in \mathcal{R}es(\mathcal{C}) - \varrho_{\prec_d}(\mathcal{C})$, as the resolved atom is $R(f(x_1))$ and $R(f(x_1)) \prec_d R(f^2(x_1))$. ♯

Kowalski and Hayes have shown in [28] that $R_\prec$ is complete for any A-ordering $\prec$, i.e., $\vdash \in R^*_\prec(\mathcal{C})$ whenever $\mathcal{C}$ is unsatisfiable. We will show in Section 5 that completeness and good termination properties make ordering refinements the ideal tool for resolution decision procedures.

4.2. Hyperresolution. While ordered resolution uses the complexity of atoms to restrict resolution, hyperresolution concentrates on the sign of clauses. We call a clause *positive* if it is of the form $\vdash A_1, \ldots, A_n$. Roughly speaking hyperresolution is the deduction principle where only positive clauses (and $\vdash$) are derivable; this is only possible if one-step resolution is replaced by many-step inferences. The following example motivates this principle of macro-inference:

EXAMPLE 4.4. Let $\mathcal{C}$ be the set of clauses $\{C_1, C_2, C_3, C_4\}$ for

$$\{C_1 = \vdash P(a,b), C_2 = \vdash P(b,a),$$
$$C_3 = P(x,y), P(y,z) \vdash P(x,z), C_4 = P(a,a) \vdash \}.$$

Then in the following resolution refutation one of the resolving clauses is always positive:

$$\dfrac{\dfrac{\vdash P(a,b) \qquad \dfrac{\vdash P(b,a) \quad P(x,y), P(y,z) \vdash P(x,z)}{P(x,b) \vdash P(x,a)}\,R}{\vdash P(a,a)}\,R \qquad P(a,a) \vdash}{\vdash}\,R$$

The clause $P(x,b) \vdash P(x,a)$ can be interpreted an intermediary result leading to the clause $\vdash P(a,a)$. So we say that $\vdash P(a,a)$ is a *hyperresolvent* of the clash sequence $(C_3; C_1, C_2)$. In this sense the only "macro-resolvents" are $\vdash P(a,a)$ and $\vdash$. Note that C_3 may not be resolved with C_4 as none of them is positive.

DEFINITION 4.5. Let C be a nonpositive clause and $D_1, \dots, D_n$ be positive clauses; then S: $(C; D_1, \dots, D_n)$ is called a clash sequence. Let $C_0 = C$ and $C_{i+1} \in \mathcal{R}es(\{C_i, D_{i+1}\})$ for $i = 1, \dots, n-1$. If C_n is defined and positive then it is called a *hyperresolvent* of S. We define the set of all hyperresolvents from a set of clauses $\mathcal{C}$ as $\varrho_H(\mathcal{C})$. The corresponding resolution operator R_H is called the operator of hyperresolution. ♯

REMARK. The operator R_H plays an important role in logic programming. In particular R_H coincides with the operator T_P (P being a logic program) in Horn logic; T_P defines the least fixed point of a logic program in the declarative semantics [31]. ♯

Hyperresolution was shown complete by J. A. Robinson in [36]. The construction of hyperresolvents can be subjected to various additional restrictions, like restriction of factoring to positive clauses and strict ordering of nonpositive clauses (see [30]).

If we take $\mathcal{C}$ as in Example 4.4 then $R_H^*(\mathcal{C}) = \mathcal{C} \cup \{\vdash P(a,a), \vdash\}$ and all produced clauses contain at most one atom. Indeed, on *Horn logic*, R_H produces only positive unit clauses and (possibly) $\vdash$ (a clause is Horn if it is of the form $A_1, \dots, A_n \vdash$ or $A_1, \dots, A_n \vdash B$ for $n \geq 0$). On satisfiable sets of Horn clauses R_H can be interpreted as a *model builder*: If $R_H(\mathcal{C}) = \mathcal{C}$ and $\vdash \notin \mathcal{C}$ then the positive clauses in $R_H(\mathcal{C})$ represent a (minimal) Herbrand model of $\mathcal{C}$ (for a proof see e.g., [30]).

EXAMPLE 4.5. Let

$$\mathcal{C} = \{\vdash P(a);\ P(x) \vdash P(f(x));\ P(b) \vdash\}.$$

Then $\mathcal{C}$ is satisfiable and

$$R_H^*(\mathcal{C}) = \mathcal{C} \cup \{\vdash P(f^n(a)) \mid n \geq 1\}.$$

Note that $R_H^*(\mathcal{C})$ is a fixed point of the operator R_H and $\vdash \notin R_H^*(\mathcal{C})$.

The set of positive clauses in $R_H^*(\mathcal{C})$ is just $\mathcal{A}$: $\{\vdash P(f^n(a)) \mid n \geq 0\}$. The corresponding atoms define a Herbrand model Γ where a ground atom A is true in Γ iff $\vdash A \in \mathcal{A}$. Note that $P(f^n(b))$ is false in Γ for all $n \geq 0$.

Unfortunately $R_H^*(\mathcal{C})$ is infinite and thus this model is not produced in finitely many steps. If we apply $R_{\prec_d}$ to $\mathcal{C}$ then it is easily verified that $R_{\prec_d}^*(\mathcal{C}) = \mathcal{C}$; this gives us the answer that $\mathcal{C}$ is satisfiable, but we do not have any information about models.

§5. Resolution and the Decision Problem.

5.1. The Decision Problem. The decision problem (or the "Entscheidungsproblem") of first-order logic can be traced back into the early years of the 20th century. Around 1920 Hilbert formulated the problem to find an algorithm which decides the validity of formulas in first-order predicate logic (see e.g., [24]). He called this decision problem the *fundamental problem of mathematical logic*.

Between 1920 and 1930 a positive solution of the decision problem seemed to be merely a question of mathematical invention. Indeed some progress was achieved soon as decidable subclasses of predicate logic were found. The decision algorithms provided for these classes were clearly effective in any intuitive sense of the word (note that before publication of Turing's landmark paper, no formal concept of algorithm was available). One of the first results (achieved even before the general problem was formulated by Hilbert) was the decidability of the monadic class [32] (i.e., the class of first order formulas containing only unary predicate—and no function symbols). In the same paper Löwenheim proved that dyadic logic (i.e., the subclass where all predicate symbols are binary) is a *reduction class*, i.e., a class of first-order formulas effectively encoding full predicate logic. In the time between world war I and world war II prominent logicians attacked this problem. The initial strategy (probably) was to enlarge the decidable classes and to "shrink" the reduction classes till they eventually meet at some point (the outcome would have been the decidability of first-order logic). But in 1936 A. Church succeeded to prove the undecidability of first-order logic and thus the *unsolvability* of the (general) decision problem [10]. An immediate consequence of Church's result was the undecidability of all reduction classes. Despite this negative result, the interest in the decision problem was kept alive, the focus shifting to the exploration of the borderline between decidable and undecidable classes.

In more recent research on the decision problem the satisfiability problem instead of the validity problem is investigated. Basically this is just a matter of taste as A is valid iff $\neg A$ is unsatisfiable.

Above we mentioned the monadic and the dyadic classes which are characterized by the arity of predicate symbols. Another type of syntax restriction concerns the quantifier prefix of prenex formulas. Some of the prenex classes shown decidable (i.e., the satisfiability problem of these classes was proved decidable) before publication of Church's result are: $\forall\exists^*$ (the *Ackermann class* [1]), $\forall\forall\exists^*$ (the *Gödel class* [21]), and $\exists^*\forall^*$ (the *Bernays–Schönfinkel class* [7]).

Slight changes in the prefixes above lead to undecidable classes, e.g., $\forall\exists\forall$ and $\forall\forall\forall\exists$ define prenex classes with undecidable satisfiability problems. For a thorough treatment of the decision problem as a whole see [9].

The methods applied to prove decidability of classes are at least as interesting as the classes themselves. In particular, the decidability of the Bernays–Schönfinkel class can be proved via the finite–model–property of this class (i.e., there exists a finite model iff there exists a model at all). A class enjoying this property is called *finitely controllable*. Most of the original proofs of decidability for the classes mentioned above were based on the finite–model–property. In fact the set of all first-order formulas having finite models is recursively enumerable. Thus in performing search for a refutation and for a finite model in parallel, we obtain a decision procedure. Clearly these model-theoretic methods were designed to *prove* decidability rather than to give efficient decision algorithms. In fact, the algorithms extracted from this method are based on exhaustive search and hardly are candidates for an even modest *calculus ratiocinator*. It turned out that the satisfiability problem of decidable classes can be handled by *proof theoretic* means on a larger scale. The most appropriate candidate for such a proof theoretic approach is *resolution*.

5.2. Resolution Decision procedures. Suppose that we start a theorem prover (i.e., a complete resolution refinement R_x) on a set of clauses C which may be satisfiable or unsatisfiable. Obviously there are three possibilities:

1. R_x terminates on C and refutes C.
   Because R_x is correct and $\vdash \in R_x(C)$ we know that C is unsatisfiable.
2. R_x terminates on C without deriving $\vdash$.
   By the completeness of R_x C must be satisfiable.
3. R_x does not terminate on C:
   In this case $R_x^*(C)$ (the set of all clauses derivable by R_x from C) is infinite and $\vdash \notin R_x^*(C)$ (we assume that the production of new clauses is stopped as soon as $\vdash$ is derived). Like in case (2) C is satisfiable, but we cannot detect this property just by computing $R_x^*(C)$.

As clause logic—being a reduction class of first-order logic—is undecidable we know that for every complete refinement operator R_x there must exist a (finite) set of clauses C s.t. $R_x^*(C)$ is *infinite*. That means it is, in principle, impossible to avoid nontermination on all sets of clauses. Let us investigate this point in somewhat more detail:

Let F be a sentence of (first-order) predicate logic. Using a normal form transformation we can transform F into a sat–equivalent set of clauses C. By the arguments above R_x must be nonterminating on some finite, satisfiable sets of clauses; therefore case (3) mentioned above cannot be avoided in general. But avoiding case (3) for specific *subclasses* of clause logic is precisely the principle of *resolution decision procedures!* It leads to the following method for proving decidability of (the satisfiability problem of) a first-order class Γ:

a: Transform the formulas in Γ into their sat-equivalent clause forms (resulting in a clausal class Γ' corresponding to Γ).

b: Find a complete resolution refinement which terminates on Γ'.

This principle is quite general and can be applied with other calculi than resolution and other normal forms than clause form. In 1968 S. Y. Maslov proved the decidability of the so called K–class (a decision class properly containing the Gödel class) using a similar approach; it is based on the so called inverse method, which is a resolution–type method formulated within the framework of a sequent calculus [34].

In the same spirit as Maslov, but on the basis of the resolution calculus, Joyner showed in his thesis [26] that resolution theorem provers can be used as decision procedures for some classical prenex classes (e.g., the Ackermann— and the Gödel class). His idea to find complete resolution refinements R_x which terminate on clause classes corresponding to prenex classes, lead to the general methodology of resolution decision procedures developed in [14].

Even unrestricted resolution can be useful in proving the decidability of first-order classes. An example is Herbrand's class **HC** [23], where **HC** is the class of all first-order formulas of the form

$$(Q_1x_1)\dots(Q_nx_n)(L_1 \wedge \dots \wedge L_m)$$

for function-free literals $L_1, \dots, L_m$.

THEOREM 5.1. *The satisfiability problem of* **HC** *is decidable.*

PROOF. Skolemization of formulas in **HC** directly yields a clausal class **HC'** consisting of finite sets of unit clauses. Now let

$$\mathcal{C} = \{\vdash A_1, \dots, \vdash A_m, B_1 \vdash, \dots, B_n \vdash\}$$

for some atoms $A_1, \dots, A_m, B_1, \dots, B_n$; then, clearly, $R^*(\mathcal{C}) = \mathcal{C}$ or $R^*(\mathcal{C}) = \mathcal{C} \cup \{\vdash\}$. Thus for all $\mathcal{C} \in$ **HC'** $R^*(\mathcal{C})$ is finite and, consequently, unrestricted resolution decides **HC'** and thus **HC**. ⊣

The proof of the theorem above shows the power of the unification principle which renders an originally complicated problem trivial.

However unrestricted resolution fails on very simple (satisfiable) sets of clauses:

EXAMPLE 5.1. Let $F = P(a) \wedge (\forall x)(P(x) \leftrightarrow \neg P(f(x)))$. Then F is satisfiable and (via the standard transformation) yields the set of clauses

$$\mathcal{C} = \{\vdash P(a);\ \vdash P(x), P(f(x));\ P(x), P(f(x)) \vdash\}.$$

As $\mathcal{C}$ is satisfiable $\vdash\ \notin R^*(\mathcal{C})$. Moreover, $\vdash P(x), P(f^{2n+1}(x)) \in R^*(\mathcal{C})$ for all $n \in \mathbb{N}$, thus $R^*(\mathcal{C})$ is infinite and resolution does not terminate on $\mathcal{C}$.

Applying the A-ordering refinement $R_{\prec_d}$ (see Example 4.1) to the set of clauses $\mathcal{C}$ in Example 5.1 we just obtain $R^*_{\prec_d}(\mathcal{C}) = \mathcal{C} \cup \{P(x) \vdash P(x)\}$. So $\vdash\ \notin R^*_{\prec_d}(\mathcal{C})$ and, by the completeness of $R_{\prec_d}$, we conclude that $\mathcal{C}$ is satisfiable. ♯

$R_{\prec_d}$ does not only work in Example 5.1 but gives a decision procedure of some well-known first-order classes, in particular of the Ackermann class:

Let us consider a formula of the form

$$F\colon (\exists x_1)\dots(\exists x_m)(\forall y)(\exists z_1)\dots(\exists z_k)M(x_1, \dots, x_m, z_1, \dots, z_k, y)$$

where $k, m \geq 0$ and M is a function—and constant free matrix. By skolemizing F we obtain a formula

$$F': (\forall y) M(c_1, \ldots, c_m, f_1(y), \ldots, f_k(y), y),$$

where $c_1, \ldots, c_m$ are (different) constant symbols and $f_1, \ldots, f_k$ are (different) one–place function symbols. In transforming the matrix of F' into conjunctive normal form (via the standard transformation) we obtain a set of clauses $\mathcal{C}$ fulfilling the following properties:

(1) All clauses contain at most one variable,
(2) all function symbols occurring in $\mathcal{C}$ are unary,
(3) the term depth of all clauses C in $\mathcal{C}$ is ≤ 1.

In particular all sets of clauses obtained from the Ackermann class belong to the more general *one–variable class* introduced in the following definition:

DEFINITION 5.1. The class **VARI** (also called the one–variable class) is the set of all finite sets of clauses $\mathcal{C}$ fulfilling the following condition: All $C \in \mathcal{C}$ contain at most one variable. ♯

We have seen that the clause forms of the formulas of the Ackermann class belong to **VARI**; on the other hand there exist sets of clauses in **VARI** which cannot be obtained by transforming Ackermann formulas into clause form. Clearly a decision procedure for **VARI** yields a decision procedure for Ackermann's class.

THEOREM 5.2. *The class* **VARI** *can be decided by the A–ordering* $<_d$ *or more exactly*: $R^*_{\prec_d}(\mathcal{C})$ *is finite for all* $\mathcal{C} \in$ **VARI**.

PROOF IN [30]. In fact a more general result is actually proven as **VARI** is not invariant under $R_{\prec_d}$. So **VARI** is extended to a class **K** which is invariant under $R_{\prec_d}$ (even under unrestricted resolution R). Then it is shown that $R_{\prec_d}$ terminates on **K**. ⊣

REMARK. The termination of $R_{\prec_d}$ on **VARI** only yields its decidability because $R_{\prec_d}$ is complete! But the completeness of A-ordering refinements is a general result in automated deduction. Also the proof that $\prec_d$ is indeed an A-ordering is quite simple. Thus the the major complexity of the proof lies in showing termination. ♯

It is not just **VARI** and the Ackermann class which can be decided by ordered resolution. In fact there is a broad range of traditional and new classes which can be decided by some ordering refinement (see [14]). Another example is the monadic class which can be decided via the A-ordering $\prec_2$ defined in Example 4.2; we only have to apply $\mathcal{R}_{\prec_2}$ to the clausal representation **MON'** of the monadic class:

DEFINITION 5.2. Let **MON** be the monadic class. Then **MON'** is the class of all sets of clauses $\mathcal{C}$ obtained via the standard clause form transformation from **MON**. ♯

THEOREM 5.3. *$R_{\prec_2}$ decides* **MON′**, *i.e., for all $C \in$* **MON′** *$R^*_{\prec_2}(C)$ is finite.*

PROOF IN [27] AND [30]. ⊣

EXAMPLE 5.2. Let F be the monadic formula

$$(\exists x_1)(\exists x_2)(\forall y_1)(\forall y_2)(\exists x_3)\left(P(x_1) \wedge \left(\neg P(y_1) \vee P(x_3)\right) \wedge \neg P(x_2)\right).$$

Then the corresponding set of clauses is

$$\mathcal{C}\colon \left\{ \vdash P(a);\ P(y_1) \vdash P(f(y_1, y_2));\ P(b) \vdash \right\}.$$

It is easy to see that $R^*_{\prec_2}(\mathcal{C}) = \mathcal{C}$, which gives a trivial proof of the satisfiability of F. On the other hand unrestricted resolution does not terminate on $\mathcal{C}$. ♯

Note that *termination* of resolution decision procedures is the key for proving decidability of classes. But the termination of resolution refinements is much less dependent on the signature of problems than the applicability of model theoretic methods. Therefore resolution decision theory leads to several syntactic extensions of the traditional first-order decision classes [14]. Concerning the decision problem itself resolution refinements thus offer an elegant tool to prove decidability of first-order classes; therefore they can be considered as a *general theoretical methodology* in mathematical logic. Moreover they are useful for solving *concrete satisfiability problems* and are most essential to the efficiency of theorem proving programs.

§6. Cut-elimination by Resolution. Cut-elimination is one of the most important techniques of proof transformation. Roughly speaking, eliminating cuts from a proof generates a new proof without lemmas, which essentially consists of the syntactic material of the proven theorem; i.e., we obtain proofs fulfilling the *subformula property*. Traditionally cut-elimination served the purpose to show consistency of calculi and thus played a central role in metamathematics. In this traditional context the aim is to define just a *constructive method* for eliminating cuts, its actual use as an algorithm is of minor importance. But in more recent time J. Y. Girard demonstrated that cut-elimination on real mathematical proofs may produce valuable mathematical information [19]. In particular he showed, how a proof of van der Waerden's theorem using concepts of topology can be transformed into an elementary combinatorial proof by means of cut-elimination. Thus it makes sense to investigate cut-elimination for *single proofs*, in order to obtain additional mathematical information on a theorem. But then an *algorithmic* approach to cut-elimination becomes more interesting.

The standard method of cut-elimination is that of Gentzen defined in his famous "Hauptsatz" [17]. The method is essentially a nondeterministic algorithm extracted from his (constructive) proof. Its characteristic feature is a *stepwise reduction* of cut complexity. In this reduction the cut formulas are decomposed w.r.t. their outermost logical operator (leading to a decrease of the logical complexity). Moreover, the cut formulas to be eliminated must be rendered principal formulas of inferences by adequate proof transformations (leading to a reduction of the rank). Despite its elegance, Gentzen's method is algorithmically very costly (of course we cannot blame Gentzen, as his aim was not to define an algorithm!). The reason is that the method is largely independent of the derivations and of the *inner* structure of the cut formulas.

The availability of resolution theorem proving and the fact that resolution is in some sense a "subcalculus" of **LK** (see Section 3) makes resolution a natural candidate in the investigation of cut-elimination. In this section we will informally present an algorithmic method of *cut-elimination by resolution*, an exact and exhaustive treatment can be found in [5]. The resolution method substantially differs from Gentzen's one. In the first step a set of clauses is generated from the derivations of the cut formulas. These sets of clauses are always unsatisfiable and thus have a resolution refutation. The construction of the resolution refutation is the second step of the procedure. Note that this step represents a direct application of automated theorem proving. The resolution refutation obtained from the theorem prover then serves as a *skeleton* of an **LK**-proof with only atomic cuts; this **LK**-proof is obtained by filling the skeleton with parts of the original **LK**-proof (actually with proof projections). The last step consists of the elimination of atomic cuts.

Although cut-elimination gave the original motivation to the development of the resolution method, the approach is far more general: indeed, the elimination of cuts appears as a special case of redundancy-elimination in **LK**-proofs. E.g., it suffices that the left cut formula logically implies the right one; they need not be syntactically equal. In fact the resolution method is largely a *semantic* one. Furthermore the method can be generalized to a method of occurrence elimination in **LK**-proofs which sheds more light on the role of redundancy in proofs.

In the first step we reduce cut-elimination to formula-elimination: that means we transform a proof φ with cuts into a cut-free proof ψ of an extended end-sequent; this transformation (unlike "real" cut-elimination) is harmless in the sense that the time complexity is linear in the size of φ.

DEFINITION 6.1. We define a mapping T_{cut} which transforms an **LK**-proof ψ of a sequent $S\colon \Gamma \vdash \Delta$ with cut formulas $A_1, \dots A_n$ into an **LK**-proof ψ^* of

$$\forall(A_1 \rightarrow A_1) \wedge \cdots \wedge \forall(A_n \rightarrow A_n), \Gamma \vdash \Delta$$

in the following way: Take an uppermost cut and its derivation χ:

$$\frac{\begin{array}{cc}(\chi_1) & (\chi_2)\\ \Pi_1 \vdash \Lambda_1, A & A, \Pi_2 \vdash \Lambda_2\end{array}}{\Pi_1, \Pi_2 \vdash \Lambda_1, \Lambda_2}\ cut$$

occurring in ψ and replace it by χ'

$$\frac{\begin{array}{cc}(\chi_1) & (\chi_2)\\ \Pi_1 \vdash \Lambda_1, A & A, \Pi_2 \vdash \Lambda_2\end{array}}{A \to A, \Pi_1, \Pi_2 \vdash \Lambda_1, \Lambda_2} \to: l$$

Afterwards apply $\forall: l$-inferences to the end-sequent of χ' on the free variables in $A \to A$ resulting in a proof χ'' of $\forall(A \to A), \Pi_1, \Pi_2 \vdash \Lambda_1, \Lambda_2$. Iterate the procedure on the next uppermost cuts till all cuts are eliminated and keep all other inferences unchanged. The result is a proof ψ' of the sequent S':

$$\forall(A_1 \to A_1), \ldots \forall(A_n \to A_n), \Gamma \vdash \Delta.$$

Finally ψ^* is obtained by contractions and $\wedge: l$.

We call the new sequent S': the *cut-extension* of S w.r.t. ψ. ♯

It is easy to see that $T_{cut}(\psi)$ is indeed a cut-free proof of the cut-extension of S w.r.t. ψ. The only nontrivial point is the preservation of the eigenvariable conditions.

After transformation of the proof ψ of S to $T_{cut}(\psi)$ of the cut-extension S' the problem of cut-elimination in ψ can be reduced to the construction of a cut-free proof of S from $T_{cut}(\psi)$. The new problem then consists in the elimination of the formula $B\colon \forall(A_1 \to A_1) \wedge \cdots \wedge \forall(A_n \to A_n)$ on the left-hand-side of the end-sequent. For technical reasons we assume that the end sequent of S is skolemized. Note that **LK**-proofs can be skolemized by a polynomial transformation defined in [4].

The first step in the formula-elimination procedure consists in the construction of a set of clauses. This set corresponds to a left occurrence of a (valid) formula in the end-sequent of an **LK**-proof. Roughly speaking we trace the derivation of the formula B (encoding the cut formulas of the original proof) back to the initial sequents. In the initial sequents we separate the parts which are ancestors of B and obtain a set of clauses $\mathcal{C}$ where each $C \in \mathcal{C}$ is of the form $\vdash$, $A \vdash$, $\vdash A$ or $A \vdash A$. Going down in the proof we look whether the corresponding inference works on ancestors of B or not. In the first case we have to subject the sets of clauses to union, in the second one to a product. The formal definition is given below:

Definition 6.2. Let ψ be a cut-free proof of S and α be an occurrence of a formula in S. We define the set of *characteristic clauses* $\mathrm{CL}(\psi, \alpha)$ inductively:

Let η be an occurrence of a sequent S' in ψ; by $anc(\eta, \alpha)$ we denote the subsequent S'' of S' which consists exactly of the formulas with occurrences being ancestors of the occurrence α in S. Let η be the occurrence of an initial sequent $A \vdash A$ in ψ and η_1 (η_2) be the left (right) occurrence of A in $A \vdash A$. If neither η_1 nor η_2 is an ancestor of α then $C_\eta = \{\vdash\}$; If both η_1 and η_2 are ancestors of α then $C_\eta = \emptyset$. Otherwise (exactly one of η_1, η_2 is ancestor of α)

$\mathcal{C}_\eta = \{anc(\eta, \alpha)\}$, i.e., $\mathcal{C}_\eta = \{A \vdash\}$ if η_1 is ancestor of α and $\mathcal{C}_\eta = \{\vdash A\}$ if η_2 is ancestor of α.

Let us assume that the clause sets $\mathcal{C}_\lambda$ are already constructed for all sequent–occurrences λ in ψ with $depth(\lambda) \leq k$ (where the depth of an occurrence λ is the length of the path in the proof tree from the root to λ).

Now let λ be an occurrence with $depth(\lambda) = k + 1$. We distinguish the following cases:

a: λ is the consequent of μ, i.e., a unary rule applied to μ gives λ. Here we simply define $\mathcal{C}_\lambda = \mathcal{C}_\mu$.

b: λ is the consequent of μ_1 and μ_2, i.e., a binary rule X applied to μ_1 and μ_2 gives λ.

- **b1**: The auxiliary formulas of X are ancestors of α, i.e., the formulas occur in $anc(\mu_1, \alpha), anc(\mu_2, \alpha)$. Then $\mathcal{C}_\lambda = \mathcal{C}_{\mu_1} \cup \mathcal{C}_{\mu_2}$.
- **b2**: The auxiliary formulas of X are not ancestors of α. In this case we define $\mathcal{C}_\lambda = \mathcal{C}_{\mu_1} \otimes \mathcal{C}_{\mu_2}$ where

$$\{\bar{P}_1 \vdash \bar{Q}_1, \dots \bar{P}_m \vdash \bar{Q}_m\} \otimes \{\bar{R}_1 \vdash \bar{T}_1, \dots \bar{R}_n \vdash \bar{T}_n\} = \{\bar{P}_i, \bar{R}_j \vdash \bar{Q}_i, \bar{T}_j \mid i \leq m, j \leq n\}$$

Finally $\mathrm{CL}(\psi, \alpha)$ is set to $\mathcal{C}_\nu$ where ν is the occurrence of the end-sequent. Note that α is an occurrence in ν and its own ancestor. ♯

EXAMPLE 6.1. Let ψ be the proof (for u, v free variables, a a constant symbol)

$$\dfrac{\psi_1 \qquad \psi_2}{(\forall x)(P(x) \to Q(x)) \vdash (\exists y)(P(a) \to Q(y))}\ cut$$

where ψ_1 is the **LK**-proof:

$$\dfrac{\dfrac{\dfrac{\dfrac{\dfrac{P(u)^* \vdash P(u) \quad Q(u) \vdash Q(u)^*}{P(u)^*, P(u) \to Q(u) \vdash Q(u)^*}\ \to: l}{P(u) \to Q(u) \vdash (P(u) \to Q(u))^*}\ \to: r}{P(u) \to Q(u) \vdash (\exists y)(P(u) \to Q(y))^*}\ \exists: r}{(\forall x)(P(x) \to Q(x)) \vdash (\exists y)(P(u) \to Q(y))^*}\ \forall: l}{(\forall x)(P(x) \to Q(x)) \vdash (\forall x)(\exists y)(P(x) \to Q(y))^*}\ \forall: r$$

and ψ_2 is:

$$\dfrac{\dfrac{\dfrac{\dfrac{\dfrac{P(a) \vdash P(a)^* \quad Q(v)^* \vdash Q(v)}{P(a), (P(a) \to Q(v))^* \vdash Q(v)}\ \to: l}{(P(a) \to Q(v))^* \vdash P(a) \to Q(v)}\ \to: r}{(P(a) \to Q(v))^* \vdash (\exists y)(P(a) \to Q(y))}\ \exists: r}{(\exists y)(P(a) \to Q(y))^* \vdash (\exists y)(P(a) \to Q(y))}\ \exists: l}{(\forall x)(\exists y)(P(x) \to Q(y))^* \vdash (\exists y)(P(a) \to Q(y))}\ \forall: l$$

The ancestors of the cut formula in ψ_1 and ψ_2 are marked by $*$. From ψ we construct the cut-extension ψ', where A denotes the cut formula $(\forall x)(\exists y)(P(x) \to Q(y))$ of ψ:

$$\frac{\psi_1 \qquad \psi_2}{A \to A, (\forall x)(P(x) \to Q(x)) \vdash (\exists y)(P(a) \to Q(y))} \to : l$$

Let α be the occurrence of $A \to A$ in the end sequent S' of ψ'. We compute the characteristic clauses $\mathrm{CL}(\psi', \alpha)$:

From the $*$-marks in the proofs ψ_1 and ψ_2 (which indicate the ancestors of α) we first get the sets of clauses corresponding to the initial sequents:

$$C_1 = \{P(u) \vdash \},\ C_2 = \{ \vdash Q(u)\},\ C_3 = \{ \vdash P(a)\},\ C_4 = \{Q(v) \vdash \}.$$

The first inference in ψ_1 (it is $\to : l$) takes place on nonancestors of α—the auxiliary formulas of the inference are not marked by $*$. Consequently we apply $\otimes$ and obtain the set $C_{1,2} = \{P(u) \vdash Q(u)\}$. The following inferences in ψ_1 are all unary and so we obtain

$$\mathrm{CL}(\psi_1, \alpha_1) = \{P(u) \vdash Q(u)\}$$

for α_1 being the occurrence of the ancestor of α in the end-sequent of ψ_1.

The first inference in ψ_2 takes place on ancestors of α (the auxiliary formulas are $*$-ed) and we have to apply the $\cup$ on C_3, C_4. We obtain $C_{3,4} = \{\vdash P(a),\ Q(v) \vdash\}$. Like in ψ_1 all following inferences in ψ_2 are unary leaving the set of clauses unchanged. Let α_2 be the ancestor of α in the end-sequent of ψ_2. Then the corresponding set of clauses is

$$\mathrm{CL}(\psi_2, \alpha_2) = \{ \vdash P(a),\ Q(v) \vdash \}.$$

The last inference $\to : l$ in ψ' takes place on ancestors of α and we have to apply $\cup$ on $C_{1,2}$ and $C_{3,4}$. This eventually yields

$$\mathrm{CL}(\psi', \alpha) = \{P(u) \vdash Q(u),\ \vdash P(a),\ Q(v) \vdash \}. \qquad \sharp$$

It is easy to verify that the set of characteristic clauses $\mathrm{CL}(\psi', \alpha)$ constructed in the example above is unsatisfiable. This is not merely a coincidence, but a general principle expressed in the next proposition.

PROPOSITION 6.1. *Let ψ be a cut-free proof of the sequent S and α be a left-occurrence of a valid formula occurring in S. Then the set of clauses* $\mathrm{CL}(\psi, \alpha)$ *is unsatisfiable.*

PROOF IN [4]. Basically it is shown that $B \vdash$, for B occurring at α, (which is an unsatisfiable sequent) is derivable in **LK** from the initial axioms $\mathrm{CL}(\psi, \alpha)$. ⊣

REMARK. The proof of Proposition 6.1 might suggest that the set of clauses $\mathrm{CL}(\psi, \alpha)$ is just a clausal normal form of the formula $\neg B$ corresponding to the sequent $B \vdash$; but this is not the case! As a simple counterexample consider the following derivation ψ:

$$\dfrac{\dfrac{Q(b) \vdash Q(b)}{\vdash Q(b) \to Q(b)} \to: r}{P(a) \to P(a) \vdash Q(b) \to Q(b)}\ w: l$$

The only initial sequent of ψ is $Q(b) \vdash Q(b)$. Neither the left- nor the right occurrence of $Q(b)$ in this sequent is an ancestor of the occurrence α of $P(a) \to P(a)$ in the end sequent. Thus the set of clauses $\mathcal{C}$ corresponding to the node of the initial sequent is $\{\vdash\}$. As there are only unary rules in ψ we finally obtain $\mathrm{CL}(\psi, \alpha) = \mathcal{C} = \{\vdash\}$. On the other hand no traditional transformation to normal form (like standard- or structural transformation) transforms the formula $\neg(P(a) \to P(a))$ into $\{\vdash\}$. In particular the standard transformation gives the set of clauses $\mathcal{D}$: $\{\vdash P(a),\ P(a) \vdash\}$. The example above illustrates that the set of clauses $\mathrm{CL}(\psi, \alpha)$ strongly depends on the derivation ψ and not only on the form of the formula on position α! We will see in the following presentation of the method that the construction of $\mathrm{CL}(\psi, \alpha)$ from the proof ψ plays a central role in the so-called proof projection (which cannot be performed on the basis of ordinary clause forms). ♯

Now let $\mathrm{CL}(\psi, \alpha)$ be the (unsatisfiable) set of clauses extracted from the **LK**-proof of the extended sequent S. By the completeness of the resolution principle there exists a resolution refutation γ (in form of a tree) of the set of clauses $\mathrm{CL}(\psi, \alpha)$. γ can be transformed into a ground refutation of $\mathrm{CL}(\psi, \alpha)$:

PROPOSITION 6.2. *Let γ be a tree resolution refutation of a set of clauses $\mathcal{C}$. Then there exists a ground instance γ' of γ s.t. γ' is a tree resolution refutation of $\mathcal{C}'$ where $\mathcal{C}'$ is a set of ground instances from $\mathcal{C}$.*

PROOF. Let λ be the simultaneous most general unifier of all the resolutions in γ. Then $\gamma\lambda$ is also a resolution refutation where the resolution rule reduces to atomic cut and contractions (i.e., to a mix, see [17]). Let σ be an arbitrary ground substitution of the variables of $\gamma\lambda$; then γ': $\gamma\lambda\sigma$ is the desired resolution refutation. ⊣

REMARK. We call the refutation γ' defined above a ground refutation *corresponding to γ*. ♯

Now let γ' be a ground refutation corresponding to a resolution refutation γ of $\mathrm{CL}(\psi, \alpha)$. By our definition of resolution γ' can easily be transformed to an **LK**-proof of $\vdash$ from $\mathcal{C}'$ with atomic cuts. Indeed, only additional contractions are necessary to simulate factoring. The resulting **LK**-proof γ' will serve as a *skeleton* of an **LK**-proof ϕ of $\Gamma \vdash \Delta$ with atomic cuts. Recall that S may be a cut-extension $B, \Gamma \vdash \Delta$ of $\Gamma \vdash \Delta$.

Thus ϕ corresponds (modulo the transformation T_{cut}) to a reduction of a proof with cuts to a proof with atomic cuts. The construction of ϕ from γ' is based on so called *projections* replacing the proof ψ of the cut-extension S by proofs $\psi[C]$ of $\bar{P}, \Gamma \vdash \Delta, \bar{Q}$ for clauses $C: \bar{P} \vdash \bar{Q}$ in C', where C' is the set of ground instances refuted by γ'. The existence of such projections of ψ w.r.t. clauses in C', guaranteed by the lemma below, is the most important property of the cut-elimination method based on resolution.

LEMMA 6.1. *Let ψ be a cut-free proof of a sequent $S: B, \Gamma \vdash \Delta$ s.t. $\Gamma \vdash \Delta$ is skolemized, B is valid and α is the occurrence of B in S. Let $C: \bar{P} \vdash \bar{Q}$ be a clause in $\mathrm{CL}(\psi, \alpha)$. Then there exists a cut-free proof $\psi[C]$ of $\bar{P}, \Gamma \vdash \Delta, \bar{Q}$ with $l(\psi[C]) \leq l(\psi)$ (where l denotes the length of the proof, i.e., the number of nodes occurring in the derivation tree).*

PROOF. We only give a proof sketch; a full formal proof can be found in [5].

The proof goes by induction on the depth of inference nodes in ψ. In constructing $\psi[C]$ we skip all inferences in ψ leading to the extension formula B. In the other inferences with auxiliary formulas which are not ancestors of B we select two clauses from the corresponding set of clauses and construct the corresponding projections via the induction hypothesis. We concentrate on a binary inference rule X and the following proof χ:

$$\dfrac{\begin{array}{cc}(\rho_1) & (\rho_2) \\ (\mu_1)\, \Gamma_1 \vdash \Delta_1 & (\mu_2)\, \Gamma_2 \vdash \Delta_2\end{array}}{(\lambda)\, \Gamma_1', \Gamma_2' \vdash \Delta_1', \Delta_2'}\ X$$

where μ_1, μ_2, λ are the nodes in the proof tree ψ and $\bar{P} \vdash \bar{Q}$ is a clause in C_λ. We assume that the auxiliary formulas of X are not ancestors of α and that the subsequents of $\Gamma_i \vdash \Delta_i$ defined by formulas which are not ancestors of B are $\Pi_i \vdash \Lambda_i$ for $i = 1, 2$. Then, by Definition 6.2, we have $C_\lambda = C_{\mu_1} \otimes C_{\mu_2}$. Therefore there are clauses $\bar{P}_1 \vdash \bar{Q}_1 \in C_{\mu_1}$ and $\bar{P}_2 \vdash \bar{Q}_2 \in C_{\mu_2}$ s.t.

$$\bar{P} \vdash \bar{Q} = \bar{P}_1, \bar{P}_2 \vdash \bar{Q}_1, \bar{Q}_2.$$

By induction hypothesis we obtain proofs ρ_1' of $\bar{P}_1, \Pi_1 \vdash \Lambda_1, \bar{Q}_1$ and ρ_2' of $\bar{P}_2, \Pi_2 \vdash \Lambda_2, \bar{Q}_2$ with $l(\rho_1') \leq l(\rho_1)$ and $l(\rho_2') \leq l(\rho_2)$. Then the projection corresponding to the node λ is χ':

$$\dfrac{\begin{array}{cc}(\rho_1') & (\rho_2') \\ \bar{P}_1, \Pi_1 \vdash \Lambda_1, \bar{Q}_1 & \bar{P}_2, \Pi_2 \vdash \Lambda_2, \bar{Q}_2\end{array}}{\bar{P}_1, \bar{P}_2, \Pi_1', \Pi_2' \vdash \Lambda_1', \Lambda_2', \bar{Q}_1, \bar{Q}_2}\ X$$

Clearly $l(\chi') \leq l(\chi)$. ⊣

REMARK. For the projections $\psi[C]$ we need the set of clause $\mathrm{CL}(\psi, \alpha)$ as defined in Definition 6.2. Here it is important that $C: \mathrm{CL}(\psi, \alpha)$ is constructed *from the proof* ψ itself! Thus C is not an ordinary clause form of $\neg B$ constructed from the syntax of B, but a clause form belonging to the derivation of $B \vdash$ within ψ. ♯

Once we we have constructed the projections $\psi[C]$ we can "insert" them into the resolution refutation γ. The formal procedure is defined below:

DEFINITION 6.3. Let ψ be a cut-free proof of $S\colon B, \Gamma \vdash \Delta$ s.t. B is valid, $\Gamma \vdash \Delta$ closed and skolemized and α the occurrence of B in S. Let γ' a ground refutation corresponding to a resolution refutation γ of $\mathrm{CL}(\psi, \alpha)$ s.t. $\gamma' = \gamma\sigma$. We define an **LK**-proof $\gamma'[\psi]$ inductively:

Let N be a leaf node in γ labelled with a clause $C\sigma$ for $C \in \mathrm{CL}(\psi, \alpha)$ and let $C\sigma = \bar{P} \vdash \bar{Q}$. To N we assign the proof $\omega_N\colon \psi[C]\sigma$, where $\psi[C]$ is the projection of ψ to C as defined in Lemma 6.1 By definition ω_N is a cut-free proof of the sequent $\bar{P}, \Gamma\sigma \vdash \Delta\sigma, \bar{Q}$. By assumption S is closed and thus ω_N is a cut-free proof of $\bar{P}, \Gamma \vdash \Delta, \bar{Q}$.

Assume that N is a node in γ' labelled with C and with parent nodes N_1 labelled with C_1 and N_2 labelled with C_2. Then, by definition of a resolution derivation, C is a (ground) resolvent of C_1 and C_2. Therefore $C_1 = \bar{P} \vdash \bar{Q}, A^r$, $C_2 = A^s, \bar{R} \vdash \bar{T}$ and $C = \bar{P}, \bar{R} \vdash \bar{Q}, \bar{T}$ for multisets of atoms $\bar{P}, \bar{Q}, \bar{R}, \bar{T}$ and an atom A occurring r-times in C_1 and s-times in C_2

Let ω_{N_1} and ω_{N_2} be the **LK**-proofs corresponding to N_1 and N_2, respectively. Assume that ω_{N_1} is a proof of $\bar{P}, \Gamma^k \vdash \Delta^k, \bar{Q}, A^r$ and ω_{N_2} of $A^s, \bar{R}, \Gamma^l \vdash \Delta^l, \bar{T}$ for $k, l \in \mathbb{N}$. Then ω_N, the **LK**-proof corresponding to N, is defined as

$$\dfrac{\dfrac{\begin{matrix}(\omega_{N_1})\\ \bar{P}, \Gamma^k \vdash \Delta^k, \bar{Q}, A^r\end{matrix}}{\bar{P}, \Gamma^k \vdash \Delta^k, \bar{Q}, A}\ c: r^* \qquad \dfrac{\begin{matrix}(\omega_{N_2})\\ A^s, \bar{R}, \Gamma^l \vdash \Delta^l, \bar{T}\end{matrix}}{A, \bar{R}, \Gamma^l \vdash \Delta^l, \bar{T}}\ c: l^*}{\bar{P}, \bar{R}, \Gamma^{k+l} \vdash \Delta^{k+l}, \bar{Q}, \bar{T}}\ cut$$

Let N_r be the root node of γ'; then $\gamma'[\psi]$ is defined as ω_{N_r}. ♯

If ψ is a cut-free proof of $B, \Gamma \vdash \Delta$ then $\gamma'[\psi]$ in Definition 6.3 is a proof with atomic cuts of $\Gamma, \ldots, \Gamma \vdash \Delta, \ldots, \Delta$ (note that the clause belonging to the root node of γ' is $\vdash$). Only additional contractions are necessary for getting a proof $\hat{\gamma}[\psi]$ with atomic cuts of $\Gamma \vdash \Delta$ itself. It remains only to eliminate the atomic cuts; to this aim any cut-elimination procedure (e.g., this in [17]) does the job. The length of the proofs with atomic cuts is essentially defined by the length of γ'.

THEOREM 6.1. *Let ψ be a cut-free proof of a closed sequent $S\colon B, \Gamma \vdash \Delta$, where B is a valid formula occurring at α in S and $\Gamma \vdash \Delta$ is skolemized. Furthermore let γ' be a ground refutation which corresponds to a resolution refutation of* $\mathrm{CL}(\psi, \alpha)$ *and* $\|\gamma\| = \max\{\|C\| \mid C \text{ in } \gamma\}$. *Then there exists a proof $\hat{\gamma}[\psi]$ of $\Gamma \vdash \Delta$ with atomic cuts and $l(\hat{\gamma}[\psi]) \leq 2 \cdot l(\psi) l(\gamma)(2\|\gamma\| + 1)$.*

PROOF. See [5]. ⊣

To illustrate the whole procedure described above we continue with Example 6.1.

EXAMPLE 6.2. Let ψ' be the proof of the sequent

$$S\colon A \to A, (\forall x)\big(P(x) \to Q(x)\big) \vdash (\exists y)\big(P(a) \to Q(y)\big)$$

as defined in Example 6.1. We have shown that

$$\mathrm{CL}(\psi', \alpha) = \big\{P(u) \vdash Q(u),\ \vdash P(a),\ Q(v) \vdash\ \big\}$$

where α is the occurrence of $A \to A$ in S.

We first define the projections of ψ' w.r.t. clauses in $\mathrm{CL}(\psi', \alpha)$:

We start with $\psi'[C_1]$, the projection of ψ' to $C_1\colon P(u) \vdash Q(u)$:

The problem can be reduced to the construction of $\psi_1[C_1]$ because of

$$\mathrm{CL}(\psi_1, \alpha_1) = \big\{P(u) \vdash Q(u)\big\}.$$

By definition of ψ_1 and of the projection, $\psi_1[C_1]$ is a proof of

$$P(u), (\forall x)\big(P(x) \to Q(x)\big) \vdash Q(u).$$

The last inference in ψ' applies to ancestors of α and thus $\psi'[C_1]$ is defined as

$$\frac{\begin{matrix}(\psi_1[C_1])\\ P(u), (\forall x)(P(x) \to Q(x)) \vdash Q(u)\end{matrix}}{P(u), (\forall x)(P(x) \to Q(x)) \vdash (\exists y)(P(a) \to Q(y)), Q(u)}\ w\colon r$$

We proceed "inductively" and construct $\psi_1[C_1]$:

$$\frac{\dfrac{P(u) \vdash P(u) \quad Q(u) \vdash Q(u)}{P(u), P(u) \to Q(u) \vdash Q(u)}\ \to\colon l}{P(u), (\forall x)(P(x) \to Q(x)) \vdash Q(u)}\ \forall\colon l$$

Putting the parts together we eventually obtain $\psi'[C_1]$:

$$\frac{\dfrac{\dfrac{P(u) \vdash P(u) \quad Q(u) \vdash Q(u)}{P(u), P(u) \to Q(u) \vdash Q(u)}\ \to\colon l}{P(u), (\forall x)(P(x) \to Q(x)) \vdash Q(u)}\ \forall\colon l}{P(u), (\forall x)(P(x) \to Q(x)) \vdash (\exists y)(P(a) \to Q(y)), Q(u)}\ w\colon r$$

For $C_2 = \vdash P(a)$ we obtain the projection $\psi'[C_2]$:

$$\frac{\dfrac{\dfrac{\dfrac{P(a) \vdash P(a)}{P(a) \vdash P(a), Q(v)}\ w\colon r}{\vdash P(a) \to Q(v), P(a)}\ \to\colon r}{\vdash (\exists y)(P(a) \to Q(y)), P(a)}\ \exists\colon l}{(\forall x)(P(x) \to Q(x)) \vdash (\exists y)(P(a) \to Q(y)), P(a)}\ w\colon l$$

In the next step we take a resolution refutation γ of $\mathrm{CL}(\psi, \alpha)$, construct a ground instance $\gamma\sigma$ via a ground substitution σ and insert appropriate instances of $\psi[C]\sigma$ into $\gamma\sigma$. The result is a proof with (only) atomic cuts of a sequent S' in which the occurrence α is eliminated.

Recall that

$$\mathrm{CL}(\psi',\alpha) = \{C_1\colon P(u) \vdash Q(u),\ C_2\colon \vdash P(a),\ C_3\colon Q(u) \vdash \}.$$

First we define a resolution refutation δ of $\mathrm{CL}(\psi',\alpha)$:

$$\dfrac{\dfrac{\vdash P(a) \quad P(u) \vdash Q(u)}{\vdash Q(a)}\,R \qquad Q(v) \vdash}{\vdash}\,R$$

and a corresponding ground refutation γ:

$$\dfrac{\dfrac{\vdash P(a) \quad P(a) \vdash Q(a)}{\vdash Q(a)}\,R \qquad Q(a) \vdash}{\vdash}\,R$$

The ground substitution defining the ground refutation is

$$\sigma = \{u \leftarrow a, v \leftarrow a\}.$$

Let $\chi_1 = \psi'[C_1]\sigma$, $\chi_2 = \psi'[C_2]\sigma$ and $\chi_3 = \psi'[C_3]\sigma$. For a more compact representation let us write B for $(\forall x)(P(x) \to Q(x))$ and C for $(\exists y)(P(a) \to Q(y))$.

Then $\hat{\gamma}[\psi']$ is of the form

$$\dfrac{\dfrac{\dfrac{\overset{(\chi_2)}{B \vdash C, P(a)} \quad \overset{(\chi_1)}{P(a), B \vdash C, Q(a)}}{B, B \vdash C, C, Q(a)}\,cut \qquad \overset{(\chi_3)}{Q(a), B \vdash C}}{B, B, B \vdash C, C, C}\,cut}{B \vdash C}\,\text{contractions}$$

$\hat{\gamma}[\psi']$ can be considered as the result of a transformation eliminating the occurrence of $A \to A$ in S. ψ' was defined as $T_{cut}(\psi)$ where ψ is a proof of $B \vdash C$. Therefore $\hat{\gamma}[\psi']$ is a proof of the same end-sequent with only atomic cuts. ♯

To put things together we obtain a procedure for occurrence-elimination, which can be transformed into a cut-elimination procedure via T_{cut}. We call this procedure OCERES (*OCcurence-Elimination by RESolution*) and display the main steps below:

procedure OCERES(ψ):

input: A (skolemized) proof ψ, a left-occurrence α of a valid formula in the end-sequent S of ψ.

output: A cut-free proof χ of the end-sequent S without the formula occurring at α:

1. Compute $\mathrm{CL}(\psi,\alpha)$.
2. Compute a ground resolution refutation γ of $\mathrm{CL}(\psi,\alpha)$.

3. Compute ϕ: $\hat{\gamma}[\psi]$.
4. Eliminate the atomic cuts in ϕ.

Then the cut-elimination procedure itself is simply defined as

$$\mathrm{CERES}(\psi) = \mathrm{OCERES}\big(T_{cut}(\psi)\big).$$

REMARK. As the worst-case complexity of cut-elimination is nonelementary (i.e., its time complexity cannot be bounded by a fixed iteration of the exponential function) we cannot expect CERES to be simply a fast algorithm. However it is shown in [5] that, for some sequences of **LK**-proofs, CERES gives a nonelementary speed-up of Gentzen's procedure. This speed-up is based on a "redundancy" at the atomic level of the proofs which can be detected in the construction of the set of clauses $\mathrm{CL}(\psi, \alpha)$, but not by Gentzen's procedure. By the availability of efficient refinements and search strategies for resolution CERES also performs quite well in experiments (see [6]). ♯

REFERENCES

[1] W. ACKERMANN, *Über die Erfüllbarkeit gewisser Zählausdrücke*, ***Mathematische Annalen***, vol. 100 (1928), pp. 638–649.

[2] F. BAADER and W. SNYDER, *Unification theory*, ***Handbook of automated reasoning*** (A. Robinson and A. Voronkov, editors), vol. I, Elsevier Science, 2001, pp. 445–532.

[3] M. BAAZ, C. FERMÜLLER, and A. LEITSCH, *A non-elementary speed-up in proof length by structural clause form transformation*, ***Proceedings of the symposium on Logic in Computer Science***, 1994, pp. 213–219.

[4] M. BAAZ and A. LEITSCH, *Cut normal forms and proof complexity*, ***Annals of Pure and Applied Logic***, vol. 97 (1999), pp. 127–177.

[5] ———, *Cut-elimination and redundancy-elimination by resolution*, ***Journal of Symbolic Computation***, vol. 29 (2000), pp. 149–176.

[6] M. BAAZ, A. LEITSCH, and G. MOSER, *System description: CutRes 01, cut elimination by resolution*, ***Conference on automated deduction, CADE-16***, Lecture Notes in Artificial Intelligence, Springer, 1999, pp. 212–216.

[7] P. BERNAYS and M. SCHÖNFINKEL, *Zum Entscheidungsproblem der mathematischen Logik*, ***Mathematische Annalen***, vol. 99 (1928), pp. 342–372.

[8] G. BOOLE, ***An investigation on the laws of thought***, Dover Publications, 1958.

[9] E. BÖRGER, E. GRÄDEL, and Y. GUREVICH, ***The Classical Decision Problem***, Perspectives in Mathematical Logic, Springer, 1997.

[10] A. CHURCH, *A note on the Entscheidungsproblem*, ***The Journal of Symbolic Logic***, vol. 1 (1936), pp. 40–44.

[11] MARTIN DAVIS and HILARY PUTNAM, *A computing procedure for quantification theory*, ***Journal of the ACM***, vol. 7 (1960), no. 3, pp. 201–215.

[12] E. EDER, ***Relative complexities of first-order calculi***, Vieweg, 1992.

[13] U. EGLY and T. RATH, *On the practical value of different definitional translations to normal form*, ***Proceedings of CADE-13***, Lecture Notes in Artificial Intelligence, vol. 1104, Springer, 1996, pp. 403–417.

[14] C. FERMÜLLER, A. LEITSCH, T. TAMMET, and N. ZAMOV, ***Resolution methods for the decision problem***, Lecture Notes in Artificial Intelligence, vol. 679, Springer, 1993.

[15] G. Frege, *Begriffsschrift, eine der arithmetischen nachgebildete Formelsprache des reinen Denkens*, ***From Frege to Gödel: A sourcebook in mathematical logic 1879–1931*** (J. van Heijenoort, editor), Harvard University Press, 1967.

[16] H. Gelernter, *Realization of a geometry theorem proving machine*, ***Proceedings of the international conference on information processing***, June 1959, pp. 273–282.

[17] G. Gentzen, *Untersuchungen über das logische Schließen*, ***Mathematische Zeitschrift***, vol. 39 (1934), pp. 405–431.

[18] P. C. Gilmore, *A proof method for quantification theory: its justification and realization*, ***IBM Journal of Research and Development***, (1960), pp. 28–35.

[19] Y. Girard, ***Proof theory and logical complexity***, Studies in Proof Theory, Bibliopolis, 1987.

[20] K. Gödel, *Über formal unentscheidbare Sätze der Principia Mathematica und verwandter Systeme*, ***Monatshefte für Mathematik und Physik***, vol. 38 (1931), pp. 175–198.

[21] ———, *Ein Spezialfall des Entscheidungsproblems der theoretischen Logik*, ***Ergebnisse Mathematischer Kolloquien***, vol. 2 (1932), pp. 27–28.

[22] R. Hähnle, *Tableaux and related methods*, ***Handbook of automated reasoning*** (A. Robinson and A. Voronkov, editors), vol. I, Elsevier Science, 2001, pp. 101–178.

[23] J. Herbrand, *Sur le probleème fondamental de la logique mathématique*, ***Sprawozdania z posiedzen Towarzysta Naukowego Warszawskiego, Wydzial III***, vol. 24 (1931), pp. 12–56.

[24] D. Hilbert and W. Ackermann, ***Grundzüge der theoretischen Logik***, Springer, Berlin, 1928.

[25] D. Hilbert and P. Bernays, ***Grundlagen der Mathematik***, Springer, 1970.

[26] W. Joyner, *Automated theorem proving and the decision problem*, ***Ph.D. thesis***, Harvard University, 1973.

[27] W. H. Joyner, *Resolution strategies as decision procedures*, ***Journal of the ACM***, vol. 23 (1976), pp. 398–417.

[28] R. Kowalski and P. Hayes, *Semantic trees in automatic theorem proving*, ***Machine intelligence*** (B. Meltzer and D. Michie, editors), vol. 7, American Elsevier, 1969, pp. 87–101.

[29] G. W. Leibniz, *Calculus ratiocinator*, ***Sämtliche Schriften und Briefe*** (Preussische Akademie der Wissenschaften, editor), Reichel, Darmstadt, 1923.

[30] A. Leitsch, ***The resolution calculus***, Texts in Theoretical Computer Science, Springer, 1997.

[31] John W. Lloyd, ***Foundations of logic programming***, second ed., Springer, 1987.

[32] L. Löwenheim, *Über Möglichkeiten im Relativkalkül*, ***Mathematische Annalen***, vol. 68 (1915), pp. 169–207.

[33] A. Martelli and U. Montanari, *Unification in linear time and space: a structured presentation*, ***Technical Report B76-16***, Instituto di elaborazione della informazione, 1976.

[34] S. Y. Maslov, *The inverse method for establishing deducibility for logical calculi*, ***Proceedings of the Steklov Institute of Mathematics***, vol. 98 (1968), pp. 25–96.

[35] W. McCune, *Solution of the Robbins problem*, ***Journal of Automated Reasoning***, vol. 19 (1997), no. 3, pp. 263–276.

[36] J. A. Robinson, *Automatic deduction with hyperresolution*, ***International Journal of Computer Mathematics***, vol. 1 (1965), pp. 227–234.

[37] ———, *A machine-oriented logic based on the resolution principle*, ***Journal of the ACM***, vol. 12 (1965), pp. 23–41.

[38] G. Takeuti, ***Proof theory***, second ed., North-Holland, 1987.

[39] A. Turing, *On computable numbers with an application to the Entscheidungsproblem*, ***Proceedings of the London Mathematical Society, Series 2***, vol. 42 (1936/37), pp. 230–265.

TU-VIENNA, FAVORITENSTRASSE 9-11
1040 VIENNA, AUSTRIA
E-mail: leitsch@logic.at

AN INTRODUCTION TO PROOFS OF DETERMINACY OF LONG GAMES

ITAY NEEMAN

Abstract. We present the basic methods used in proofs of determinacy of long games, and apply these methods to games of continuously coded length.

From the dawn of time women and men have aspired upward. The development of determinacy proofs is no exception to this general rule. There has been a steady search for higher forms of determinacy, beginning with the results of Gale–Stewart [2] on closed length ω games and continuing to this day. Notable landmarks in this quest include proofs of Borel determinacy in Martin [5]; analytic determinacy in Martin [4]; projective determinacy in Martin–Steel [8]; and $\mathsf{AD}^{L(\mathbb{R})}$ in Woodin [17].[1] Those papers consider length ω games with payoff sets of increasing complexity. One could equivalently fix the complexity of the payoff and consider games of increasing length. Such "long games" form the topic of this paper.

Long games form a natural hierarchy, the hierarchy of increasing length. This hierarchy can be divided into four categories: games of length less than $\omega \cdot \omega$; games of fixed countable length; games of variable countable length; and games of length ω_1.

Games in the first category can be reduced to standard games of length ω, at the price of increasing payoff complexity. The extra complexity only involves finitely many real quantifiers. Thus the determinacy of games of length less than $\omega \cdot \omega$, with analytic payoff say, is the same as projective determinacy.

Games in the second category can be reduced to combinations of standard games of length ω, with increased payoff complexity, and some additional strength assumptions. The first instance of this is given in Blass [1]. The techniques presented there can be used to prove the determinacy of length $\omega \cdot \omega$ games on natural numbers, with analytic payoff say, from $\mathsf{AD}^{L(\mathbb{R})}$ + "$\mathbb{R}^\#$ exists." In another, choiceless reduction to standard games, Martin and

Partially supported by NSF grant DMS 00-94174.

[1] $\mathsf{AD}^{L(\mathbb{R})}$ is the statement that all standard length ω games with payoff in $L(\mathbb{R})$ are determined.

Logic Colloquium '01
Edited by M. Baaz, S. Friedman, and J. Krajíček
Lecture Notes in Logic, 20

Woodin independently showed that AD + "all sets of reals admit scales" implies that all games in the second category are determined.

It is in the third category that the methods presented here begin to yield new determinacy principles. (The one previously known determinacy proof for games in the third category is a theorem of Steel [16], which applies to games of the kind described in Remark 1.1.)

Neeman [15] concentrates on third category games. Our goal here is to provide an introduction to the methods of [15]. We illustrate these methods with one game of the first category, one of the second, and one of the third. The proofs, like the results, form a hierarchy.

The proofs in the first category are closely related to the main construction of Martin–Steel [8].

The proofs in the second category can be viewed as combinations of (1) a construction which reduces one side of a given game to an *iteration game*; and (2) an appeal to a winning strategy for the good player in the iteration game. (Iteration games are described in Section 1.2.) This is a general pattern that continues higher up. Determinacy is thus dependent upon *iterability*—the existence of winning strategies for the good player in iteration games. We say more on this at the end of Section 1.2.

The construction for part (1) above is a matter of breaking the construction of the first category into blocks, and reassembling the blocks spreading them over countably many stages. In some ways this is analogous to the way scale propagation under infinitely many real quantifiers relates to the basic propagation under one quantifier. Readers interested in a side tour may check Sections 6C and 6E of Moschovakis [10], Moschovakis [11], and Martin [6] for results on scale propagation.

Third category proofs use the techniques of the second category, but the reassembling of the blocks is not done at the outset. Instead the decisions on how to spread the blocks of the construction are taken during the game and depend on the players' moves. Similar methods apply to open games of length ω_1 (the low end of the fourth category). Beyond that determinacy is not known.

We try to make this progression of ideas evident through the organization of the paper. In Section 2 we present the basic tools. One of the two lemmas there, Lemma 2.8, draws heavily on the techniques of Martin–Steel [8]. In Section 3 we use the basic tools to prove the determinacy of standard length ω games with Σ^1_2 payoff. In Section 4 we prove the determinacy of games of fixed length $\omega \cdot \omega$, with Σ^1_2 payoff. The proof involves breaking and reassembling the previous construction of Section 3. In Section 6 we prove the determinacy of games of continuously coded length. (These are games of the third category; of variable countable length. We define these games in Section 1.1.) Again the proof involves breaking and reassembling a construction of the kind done in

Section 3. But now the break line is not fixed at the outset; it varies depending on the actual moves during the game.

Sections 3 and 4 are included for their role in the development of methods which lead to Section 6. The results stated in those two sections, Theorem 3.1 and Theorem 4.12 are not new. Both are due to Woodin by methods different from ours. Theorem 3.1 in slightly weakened form was first proved by Martin–Steel [8].

As chance would have it the methods of Section 6 are also useful for longer games of the third category, specifically games ending at ω_1 in L of the play. These in turn are useful for the determinacy proof for open games of length ω_1. But we shall not reach that far here. Our discussion ends with Theorem 6.15, which establishes determinacy for games of continuously coded length.

§1. Preliminaries. We take this section to define precisely the long games which we intend to prove determined and sketch the large cardinal notions needed for the proofs, mainly iteration trees and iteration games. Our sketch of the large cardinal notions is informal, maybe even superficial, but it suffices for our needs.

1.1. The games. Following standard abuse we let $\mathbb{R}$ denote Baire space, namely the space $\mathbb{N}^\omega$. Let $C \subset \mathbb{R}^{<\omega_1}$ be given. Let $\nu\colon \mathbb{R} \to \mathbb{N}$, a partial function, be given. $G_{\text{cont}-\nu}(C)$ is played according to Diagram 1.

I	....	$y_\alpha(0)$		$y_\alpha(2)$		
II			$y_\alpha(1)$		$y_\alpha(3)$	$\cdots$

Diagram 1. The game $G_{\text{cont}-\nu}(C)$.

In mega-round α, players I and II alternate playing natural numbers $y_\alpha(i)$, $i < \omega$, producing a real y_α. If $\nu(y_\alpha)$ is not defined, the game ends. I wins iff $\langle y_0, y_1, \ldots\ldots, y_\alpha\rangle \in C$. Otherwise we set $n_\alpha = \nu(y_\alpha)$. If there exists $\xi < \alpha$ so that $n_\alpha = n_\xi$, the game ends. Again I wins iff $\langle y_0, y_1, \ldots\ldots, y_\alpha\rangle \in C$. Otherwise the game continues.

The end length of a run of $G_{\text{cont}-\nu}(C)$ may vary depending on the moves played by the two players. But the length is always countable. Indeed, a map witnessing that the length is countable is produced continuously—one extra bit of information at each mega-round—during the play. The game is said to have **continuously coded length.**

Remark 1.1. Our definition here generalizes the definition of continuously coded games in Steel [16], where ν acted on $\langle y_\xi \mid \xi < \alpha\rangle$, and n_α was set to be $\nu(y_\xi \mid \xi < \alpha)$. (Why is our definition a generalization? One could easily force one of the players to code $\langle y_\xi \mid \xi < \alpha\rangle$ into her moves for y_α. Thus in our settings too ν can refer to $\langle y_\xi \mid \xi < \alpha\rangle$.) The generalization is proper, in the sense that there are games which fall within our definition, but outside the definition of Steel [16].

We make a few simple observations about the game.

CLAIM 1.2. *Suppose that $\langle y_\xi \mid \xi < \lambda\rangle$ is a position of limit length. Then $n_\xi \to \infty$ as $\xi \to \lambda$.*

PROOF. This is immediate. The n_ξ-s are distinct, and so they cannot be forever bounded. ⊣

Suppose $\langle y_\xi \mid \xi < \alpha\rangle$ is a position reached during the game. The map $\xi \mapsto n_\xi = \nu(y_\xi)$ embeds α into $\mathbb{N}$, and can be used to code the position by a real. We let $\ulcorner y_\xi \mid \xi < \alpha \urcorner$ denote this real code. The precise method of coding is not important, so long as it satisfies the following property:

PROPERTY 1.3. The real codes $\ulcorner y_\xi \mid \xi < \alpha \urcorner$ and $\ulcorner y_\xi \mid \xi < \alpha + 1 \urcorner$ agree to $n_\alpha = \nu(y_\alpha)$.

Any reasonable use of the map $\xi \mapsto n_\xi$ to code $\langle y_\xi \mid \xi < \alpha\rangle$ will have this property.

REMARK 1.4. Combining Claim 1.2 and Property 1.3 we see that the reals $x_\alpha = \ulcorner y_\xi \mid \xi < \alpha \urcorner$ converge to $x_\lambda = \ulcorner y_\xi \mid \xi < \lambda \urcorner$ as $\alpha \to \lambda$.

Remark 1.4 will be crucial later when we reach the determinacy proof in Section 6. Indeed continuity is important throughout this paper, starting already in the arguments of Sections 3.

Let us say that the payoff set C is **Γ in the codes**—where Γ is some pointclass, for example Σ^1_2—if there is a Γ set $A \subset \mathbb{R} \times \mathbb{R}$ so that

$$\langle y_\xi \mid \xi \le \alpha\rangle \in C \iff \langle x_\alpha, y_\alpha\rangle \in A$$

where $x_\alpha = \ulcorner y_\xi \mid \xi < \alpha \urcorner$.

Our goal is to give a proof of determinacy for the games $G_{\text{cont}-\nu}(C)$ when ν is continuous and C is Σ^1_2 in the codes. As in illustrative case we will first consider games of fixed length. We will handle games of two lengths: games of length ω, and then games of length $\omega \cdot \omega$. We remind the reader of the format of these games:

Let $C \subset \mathbb{R}^\omega = \mathbb{N}^{\omega\cdot\omega}$ be given. In $G_{\omega\cdot\omega}(C)$ players I and II play ω mega-rounds according to Diagram 2.

I	$y_0(0)$		......	$y_1(0)$		...
II		$y_0(1)$			$y_1(1)$	...

DIAGRAM 2. The game $G_{\omega\cdot\omega}(C)$.

In mega-round k the players alternate playing natural numbers $y_k(i)$, producing together a real y_k. Once ω mega-rounds are completed, I wins if $\langle y_k \mid k < \omega \rangle$ belongs to C. Otherwise II wins.

Let $C \subset \mathbb{R} = \mathbb{N}^\omega$ be given. In $G_\omega(C)$ the players play only one mega-round, alternating natural number moves $y(i)$ as in Diagram 3 to produce together the real y. I wins if $y \in C$. Otherwise II wins.

I	$y(0)$		$y(2)$		...
II		$y(1)$		$y(3)$	...

DIAGRAM 3. The game $G_\omega(C)$.

1.2. Iteration trees. We include here an informal description of iteration trees and the notions of iterability which we shall need. This description is far from complete, and even farther from precise. The reader who desires more thorough knowledge should consult Kanamori [3] and Martin–Steel [9].

An **extender** on κ is a directed system of measures on κ. For an exact definition see [3, §26]. We use dom(E) to denote κ. An extender E allows us to form an **ultrapower** of V, denoted Ult(V, E), and an elementary **ultrapower embedding** $\pi\colon V \to \text{Ult}(V, E)$. We refer the reader to [3, §26] or [8, §1] for the exact construction.

Let us say that two ZFC models Q^* and Q **agree** to κ if $\mathcal{P}(\kappa) \cap Q^* = \mathcal{P}(\kappa) \cap Q$. Suppose $Q \models$ "E is an extender on κ." Suppose Q^* and Q agree to κ. Then E measures all subsets of κ in Q^*, and can thus be used to form an **ultrapower** Ult(Q^*, E) of Q^*, and an elementary **ultrapower embedding** $\sigma\colon Q^* \to \text{Ult}(Q^*, E)$. Ult$(Q^*, E)$ needn't always be wellfounded, but if it is then we assume it is transitive.

An **iteration tree** $\mathcal{T}$ of length ω consists of:

- A tree order T on ω;
- A sequence of models $\langle M_k \mid k < \omega \rangle$; and
- Embeddings $j_{k,l}\colon M_k \to M_l$ for $k \; T \; l$.

An iteration tree **on** M is a tree with $M_0 = M$.

A sample iteration tree together with its tree order T is displayed in Diagram 5. A precise definition can be found in [8, §3]. Rather than reproduce this definition let us only explain how to *form* an iteration tree: Suppose $M_0, \ldots, M_l$ and the order $T \restriction l + 1$ are known. We wish to form M_{l+1} and extend the tree order to $T \restriction l + 2$. To do this, we pick some extender E_l in M_l, and pick some $k \leq l$ so that M_l and M_k agree to dom(E_l). (Note that taking $k = l$ gives this agreement for free.) Set $M_{l+1} = \text{Ult}(M_k, E_l)$, and extend T by letting k be the predecessor of $l + 1$. The result is presented in Diagram 4. An iteration tree of length ω is any object produced by ω repetitions of this process.

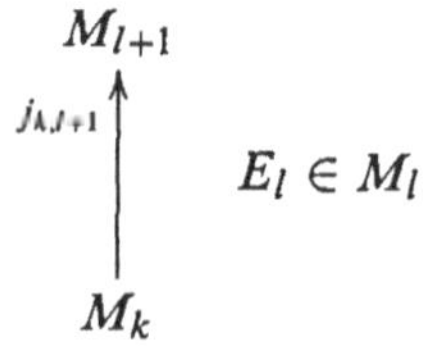

DIAGRAM 4. Forming M_{l+1}.

A cofinal **branch** through an iteration tree of length ω is an infinite set $b \subset \omega$ which is linearly ordered by T. The sample iteration tree of Diagram 5 has an **even** branch—the branch consisting of $\{0, 2, 4, 6, \dots\}$. Most of our iteration trees will have an even branch, and some complicated tree structure on the odd models. We use M_{even} to denote the direct limit of the models along the even branch. In general given a cofinal branch b we use M_b to denote the direct limit of the models along b.

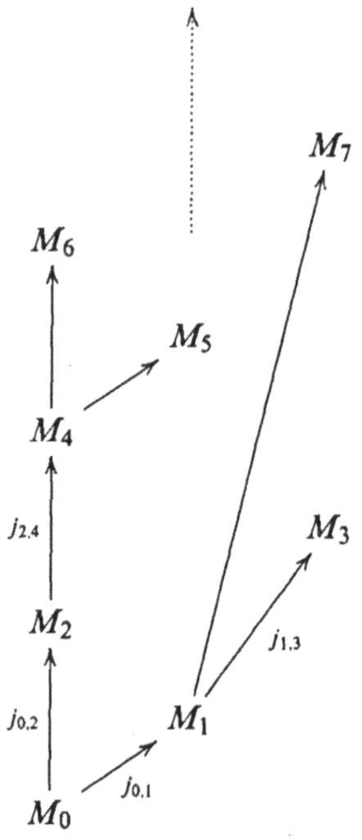

DIAGRAM 5. A sample iteration tree, with the tree order $0\ T\ 1, 0\ T\ 2, 1\ T\ 3, 1\ T\ 7, \dots$.

We shall need a couple of notions of iteration games. The notions we need are defined below. We call both of them "iteration games" though they correspond more closely to the standard notion of a "weak iteration game." Iteration games were first defined by Martin and Steel. The interested reader can find the general definition in [9].

Let M be a given model. In the first iteration game which we consider, players "good" and "bad" collaborate to produce a sequence of iteration trees as in Diagram 6.

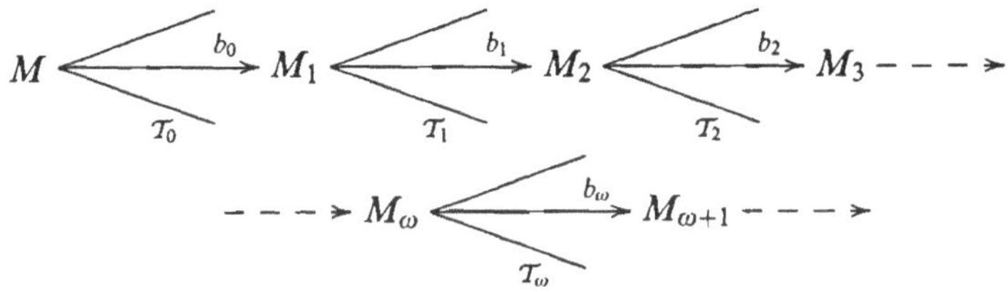

DIAGRAM 6. An iteration game.

In round ξ "bad" plays a length ω iteration tree $\mathcal{T}_\xi$ on M_ξ. "Good" plays a cofinal branch b_ξ through $\mathcal{T}_\xi$. We let $M_{\xi+1}$ be the direct limit model determined by b_ξ and proceed to the next round. For limit λ we let M_λ be the direct limit of the models M_ξ, $\xi < \lambda$. We start with the given model $M = M_0$. The game continues to ω_1.

If ever a model M_ξ, where $\xi < \omega_1$, is reached which is illfounded, "bad" wins. Otherwise "good" wins.

In the second iteration game which we consider, round ξ has the form presented in Diagram 7.

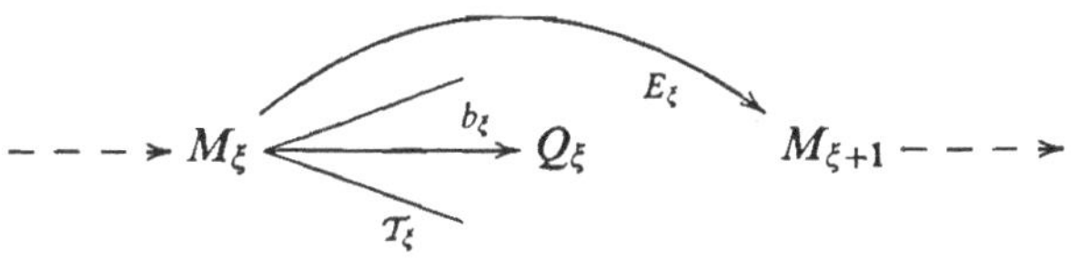

DIAGRAM 7. Round ξ of the second type iteration game.

"Bad" plays a length ω iteration tree $\mathcal{T}_\xi$ on M_ξ. "Good" plays a cofinal branch b_ξ, giving rise to the direct limit Q_ξ. Then "bad" plays an extender E_ξ in Q_ξ, with $\mathrm{dom}(E_\xi)$ within the level of agreement between M_ξ and Q_ξ. We set $M_{\xi+1} = \mathrm{Ult}(M_\xi, E_\xi)$ and continue to the next round.

As before the game continues to ω_1, taking direct limits at limit stages. If ever a model Q_ξ or M_ξ, where $\xi < \omega_1$, is reached which is illfounded, "bad" wins. Otherwise "good" wins.

M is **iterable** if the good player has a winning strategy for each of the iteration games described above and combinations thereof. We refer to such winning strategies as **iteration strategies**.

Typically in our constructions the iteration trees, but not the branches through them, will be produced by some mechanism which is part of the construction. To keep the construction going we will need a method of picking branches through the iteration trees we encounter. It will be important to maintain the wellfoundedness of all the models we construct. We will thus need an iteration strategy to carry our construction through.

The existence of winning strategies for the good player in *general* iteration games is one of the central problems facing large cardinalists. Our own definition of iterability is restricted to the weak iteration games described above. These weak games are easier for "good" than the general games, and we have the following theorem of Martin–Steel [9]:

THEOREM 1.5 (Martin–Steel). *Let* V_η *be some sufficiently closed rank initial segment of* V. *Then countable elementary substructures of* V_η *are iterable (in the weak sense described above).*

We note that once one tries to prove determinacy of games somewhat longer than the continuously coded, for example games ending at ω_1 in L of the play, the weak iteration games described above no longer suffice for the constructions. The kind of iterability needed for games ending at ω_1 in L of the play was proved in Neeman [13]. For longer games, for example open games of length ω_1, it seems that nothing short of *general* iterability could suffice for the determinacy proofs. This is one of several examples of the great importance of general iterability.

§2. Auxiliary moves. Fix throughout this section some ZFC model M which has a Woodin cardinal δ. Assume that in V there are M–generics for $\text{col}(\omega,\delta)$. Fix a name $\dot{A} \in M$ for a set of reals in $M^{\text{col}(\omega,\delta)}$.

Work with some $x = \langle x_n \mid n < \omega\rangle \in \mathbb{R}$. We work to define an auxiliary game, $\mathcal{A}[x]$, of ω moves taken from M. In this game I tries to witness that $x \in \dot{A}[h]$ for some generic h. II tries to witness the opposite. We shall use this method of "witnessing" later on in our determinacy proofs. What we present here is a gentle guide to the definition of $\mathcal{A}[x]$. We consider a more general version in Section 5.1. The actual definition (of the more general version) can be found in [15, Chapter 1].

The format of the auxiliary game $\mathcal{A}[x]$ is presented in Diagram 8. All moves belong to M, and each rule should be read relativized to M.

I	... $l_n, \mathcal{X}_n, p_n$		...
II		$\mathcal{F}_n, \mathcal{D}_n$	...

DIAGRAM 8. Round n of $\mathcal{A}[x]$.

In round n I plays:

- l_n, a number smaller than n, or $l_n =$ "new";
- $\mathcal{X}_n$, a set of names for reals of $M^{\text{col}(\omega,\delta)}$; and
- p_n, a condition in $\text{col}(\omega, \delta)$.

II plays:

- $\mathcal{F}_n$ a function from $\mathcal{X}_n$ into the ordinals; and
- $\mathcal{D}_n$, a function from $\mathcal{X}_n$ into $\{$dense sets in $\text{col}(\omega, \delta)\}$.

Set $l = l_n$. If $l_n =$ "new" we make no requirements on I. Otherwise we require:

1. p_n extends p_l;
2. $\mathcal{X}_n \subset \mathcal{X}_l$.

We further require that for every name $\dot{x} \in \mathcal{X}_n$:

3. p_n forces "$\dot{x} \in \dot{A}$";
4. p_n forces "$\dot{x}(0) = \check{x}_0$,"......,"$\dot{x}(l) = \check{x}_l$"; and
5. p_n belongs to $\mathcal{D}_l(\dot{x})$.

We make the following requirement on II when $l_n \neq$ "new":

6. For every name $\dot{x} \in \mathcal{X}_n$, $\mathcal{F}_n(\dot{x}) < \mathcal{F}_l(\dot{x})$.

This completes the rules for round n.

If there is an h which is $\text{col}(\omega, \delta)$-generic/$M$ and so that $x \in \dot{A}[h]$, then I can pick a name for x, play $\mathcal{X}_n$ containing this name, and play $p_n \in h$. Rule 6 ensures defeat for II. In other words, if there is an infinite run of $\mathcal{A}[x]$ where I played wisely enough, then there cannot be a name $\dot{x}$ and a generic h so that $x \in \dot{A}[h]$.

The game $\mathcal{A}[x]$ thus follows its stated intuitive goal—being a game in which II tries to witness that there is no generic h so that $x \in \dot{A}[h]$, while I tries to witness there is such h. This is consolidated below. In Section 2.1 we see that, if I plays wisely, then II's moves witness that $x \notin \dot{A}[h]$ for any generic h. Then in Section 2.2 we see that, if II plays wisely, then I's moves witness that x belongs to $j_b(\dot{A})[h]$, where $j_b(\dot{A})$ is some shifted image of $\dot{A}$, and h is generic for the collapse of the shifted δ.

REMARK 2.1. Rather than play the sets $\mathcal{X}_n$ directly, I plays their *type*. I plays $\kappa_n < \delta$, and a set u_n of formulae with parameters in $M \| \kappa_n \cup \{\kappa_n, \delta, \dot{A}\}$.[2] We take $\mathcal{X}_n$ to be the set of names which satisfy all these formulae. The fact that this still allows I enough control over her choice of $\mathcal{X}_n$ has to do with our

[2]By $M \| \kappa_n$ we mean $V^M_{\kappa_n}$.

assumption that δ is a Woodin cardinal. We refer the reader to [15, Chapter 1] for precise details. $\mathcal{F}_n$ and $\mathcal{D}_n$ are played similarly.

Observe that all moves in $\mathcal{A}[x]$ are therefore elements of $M\|\delta$.

Note that the association $x \mapsto \mathcal{A}[x]$ is continuous: the rules governing the first $n+1$ rounds of $\mathcal{A}[x]$ depend only on $x \restriction n$. We in fact defined an association $s \mapsto \mathcal{A}[s]$; for $s \in \omega^{<\omega}$ we have $\mathcal{A}[s]$, a game of $\mathrm{lh}(s)+1$ many rounds.

DEFINITION 2.2. $\mathcal{A}$ denotes the map $(s \mapsto \mathcal{A}[s])$.

Our definition of $\mathcal{A}[s]$ from s takes place entirely in M. It follows that the map $\mathcal{A}$ belongs to M. This is important; it allows us to shift $\mathcal{A}$ using elementary embeddings which act on M. Given an elementary $j: M \to M^*$ we have the map $j(\mathcal{A})$ defined on $s \in \omega^{<\omega}$. For a real x (in V) we can then define $j(\mathcal{A})[x]$ in the natural way: $j(\mathcal{A})[x] = \bigcup_{n<\omega} j(\mathcal{A})[x \restriction n]$. We shall use such shiftings later on, see for example Section 2.2.

2.1. Generic runs. Fix some g which is $\mathrm{col}(\omega,\delta)$–generic/$M$. We alternate between thinking of g as a generic enumeration of δ, and as a generic enumeration of $M\|\delta$. (δ and $M\|\delta$ have the same cardinality in M.)

Working in $M[g]$ define $\sigma_{\mathrm{gen}}[x]$, a strategy for I in $\mathcal{A}[x]$, as follows: $\sigma_{\mathrm{gen}}[x]$ plays in each round the *first* (with respect to the enumeration g) legal move. (Remember that moves in $\mathcal{A}[x]$ are elements of $M\|\delta$; see Remark 2.1.)

The association $x \mapsto \sigma_{\mathrm{gen}}[x]$ is continuous; we are in fact defining a map $s \mapsto \sigma_{\mathrm{gen}}[s]$ for $s \in \omega^{<\omega}$. This map belongs to $M[g]$.

DEFINITION 2.3. (Made with respect to a fixed g.) σ_{gen} denotes the map $(s \mapsto \sigma_{\mathrm{gen}}[s])$.

LEMMA 2.4. *Suppose that there exists an infinite run of $\mathcal{A}[x]$, played according to $\sigma_{\mathrm{gen}}[x]$. Then $x \notin \dot{A}[g]$. (This is only useful if $x \in M[g]$.)*

PROOF SKETCH. Suppose for contradiction that $x \in \dot{A}[g]$. In particular $x \in M[g]$. We have some name $\dot{x}$ so that $\dot{x}[g] = x$ and $g \Vdash$ "$\dot{x} \in \dot{A}$."

We have some infinite run of $\mathcal{A}[x]$, as displayed in Diagram 8. The run splits into **branches**: a branch is a sequence $\{n_k\}_{k<\omega}$ so that $l_{n_0} =$ "new" and $l_{n_k} = n_{k-1}$ for $k > 0$.

Note that $\dot{x}$ and conditions $p \in g$ satisfy rules 3–5 of $\mathcal{A}[x]$. The *genericity* of I's moves allows us to find a branch which realizes $\dot{x}$ and g. More precisely, a branch so that (a) $\dot{x} \in \mathcal{X}_{n_k}$ for all k; and (b) p_{n_k} belongs to g for all k. But then using rule 6 we get an infinite decreasing sequence of ordinals, a contradiction. ⊣

The key to the proof of Lemma 2.4 is the use of genericity in the last paragraph. We refer the reader to [15, Chapter 1] for a precise argument. The same proof can be used to show that in fact there is no generic h so that $x \in \dot{A}[h]$.

2.2. Pivots. We wish to phrase a lemma similar to Lemma 2.4, but now with a method of playing for II so that infinite runs put x *in* (something like) $\dot{\mathcal{A}}\lfloor h\rfloor$. We cannot directly come up with moves for II in $\mathcal{A}[x]$. Instead we phrase another game which is similar to $\mathcal{A}[x]$ but easier for II, and come up with a method of playing for II in the easier game. This easier game is denoted $\mathcal{A}_{\text{piv}}[x]$. Its format is presented in Diagrams 9 and 10.

I	...	$l_n, \mathcal{X}_n, p_n$		...
II			$E_{2n}, E_{2n+1}, \mathcal{F}_n, \mathcal{D}_n$	...

DIAGRAM 9. Round n of $\mathcal{A}_{\text{piv}}[x]$.

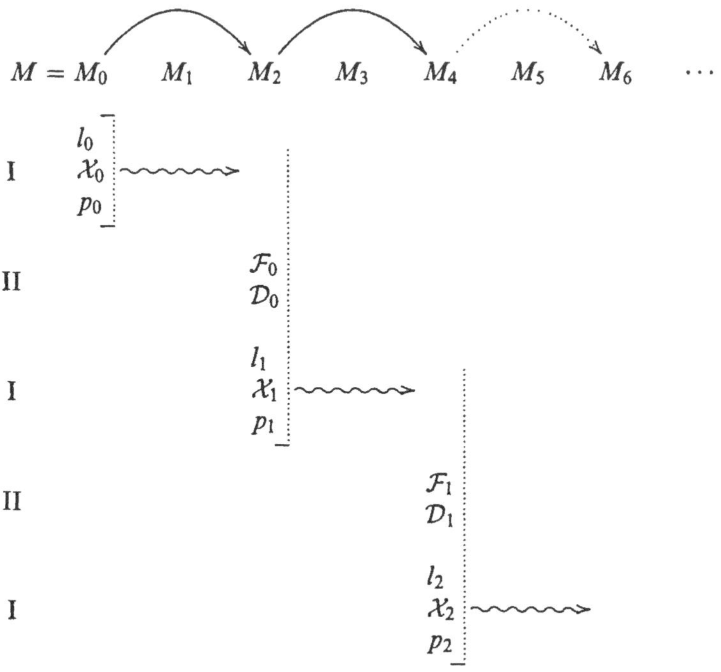

DIAGRAM 10. $\mathcal{A}_{\text{piv}}[x]$, the dynamic view.

At the start of round n we have a finite iteration tree $\mathcal{T}\restriction 2n+1$ on M ending with a model M_{2n}, an embedding $j_{0,2n}\colon M \to M_{2n}$, and a position P_n of n rounds in $j_{0,2n}(\mathcal{A})[x]$. During the round:

- I plays $l_n, \mathcal{X}_n, p_n$, a legal move in $j_{0,2n}(\mathcal{A})[x]$ following P_n.

We extend the tree order $T \restriction 2n+1$ by setting $(2l_n+1)\ T\ (2n+1)$ if $l_n \neq$ "new" and $(2n)\ T\ (2n+1)$ otherwise. We set further $(2n)\ T\ (2n+2)$. We have now $T \restriction 2n+3$.

- II plays extenders E_{2n}, E_{2n+1}, which combined with our definition of $T \restriction 2n+3$ give rise to models M_{2n+1} and M_{2n+2}. (It is II's responsibility to make sure the domains of the extenders are within the level of agreement of the relevant models.)

We have an embedding $j_{2n,2n+2}\colon M_{2n} \to M_{2n+2}$. Let Q_n be the position $j_{2n,2n+2}(P_n\frown, l_n, \mathcal{X}_n, p_n)$.[3] This "shifting" of $P_n\frown, l_n, \mathcal{X}_n, p_n$ from M_{2n} to M_{2n+2} is indicated in squiggly arrows in Diagram 10.

- II plays $\mathcal{F}_n, \mathcal{D}_n$, a legal move in $j_{0,2n+2}(\mathcal{A})[x]$ following Q_n.

This completes the round. We let $T \restriction 2n+3$ be the extended iteration tree (ending with M_{2n+2}), let $P_{n+1} = Q_n\frown, \mathcal{F}_n, \mathcal{D}_n$, and proceed to round $n+1$.

REMARK 2.5. We make one extra, technical demand on player II. We demand that all extenders used are taken from below δ, and have critical points larger than some pre-specified ordinal $\lambda < \delta$. For one example of how this is used (and which λ is specified) see Remark 4.6. For another example see Remark 4.9. Similar uses are made later, in Section 6.

We note as usual that the association $x \mapsto \mathcal{A}_{\text{piv}}[x]$ is continuous; for $s \in \omega^{<\omega}$ we get $\mathcal{A}_{\text{piv}}[s]$, a game of $\text{lh}(s)+1$ many round.

DEFINITION 2.6. We use $\mathcal{A}_{\text{piv}}$ to denote the map $(s \mapsto \mathcal{A}_{\text{piv}}[s])$.

As usual the map $\mathcal{A}_{\text{piv}} = (s \mapsto \mathcal{A}_{\text{piv}}[s])$ belongs to M.

DEFINITION 2.7. A **pivot** for x is a pair $\mathcal{T}, \vec{a}$ so that:

1. $\mathcal{T}$ is a length ω iteration tree on M, with an even branch.
2. $\vec{a}$ is a run of $j_{\text{even}}(\mathcal{A})[x]$.
3. For every cofinal odd branch b of $\mathcal{T}$, there exists some h so that:
   (a) h is $\text{col}(\omega, j_b(\delta))$–generic/$M_b$; and
   (b) $x \in j_b(\dot{A})[h]$.

Any run of $\mathcal{A}_{\text{piv}}[x]$ produces $\mathcal{T}, \vec{a}$ which satisfy conditions 1 and 2. To be a pivot the run must further satisfy the crucial condition 3. Intuitively condition 3 states that x belongs to interpretations of "shifts" of the name $\dot{A}$. Our goal here is to phrase a lemma which complements Lemma 2.4 and the notions of the previous subsection. We can now say precisely what this means: we need a strategy which plays for II in $\mathcal{A}_{\text{piv}}[x]$ and always secures condition 3.

[3]We write $P_n\frown, l_n, \mathcal{X}_n, p_n$ to indicate that P_n is a *sequence* while $l_n, \mathcal{X}_n$, and p_n are singleton objects. Formally we should write $P_n{}^\frown\langle l_n, \mathcal{X}_n, p_n\rangle$, but we prefer to avoid the additional brackets.

Fix some map $\varrho\colon \omega \to M\|\delta+1$. Applying techniques of the kind used in Neeman [14]—which in turn builds on Martin–Steel [8]—it is possible to construct a strategy $\sigma_{\text{piv}}[\varrho, x]$ which plays for II in $\mathcal{A}_{\text{piv}}[x]$, and so that:

LEMMA 2.8. *Suppose ϱ is* onto $M\|\delta+1$. *Then all runs according to $\sigma_{\text{piv}}[\varrho, x]$ are pivots.*

As usual the map $x \mapsto \sigma_{\text{piv}}[\varrho, x]$ is continuous in x. But we cannot expect this map to belong to M, since ϱ need not belong to M. This is why we include the extra variable ϱ. The map $\varrho, x \mapsto \sigma_{\text{piv}}[\varrho, x]$ is continuous, not only in x, but also in ϱ. For $s \in \omega^n$ and $\vartheta\colon n \to M\|\delta+1$ we get a strategy $\sigma_{\text{piv}}[\vartheta, s]$ which plays for II in $\mathcal{A}_{\text{piv}}[s]$. We have

$$\sigma_{\text{piv}}[\varrho, x] = \bigcup_{n<\omega} \sigma_{\text{piv}}[\varrho\restriction n, x\restriction n].$$

DEFINITION 2.9. σ_{piv} denotes the map $(\vartheta, s \mapsto \sigma_{\text{piv}}[\vartheta, s])$.

The construction of $\sigma_{\text{piv}}[\vartheta, s]$, indeed of the map $\vartheta, s \mapsto \sigma_{\text{piv}}[\vartheta, s]$, is phrased entirely in M. The map σ_{piv}, taken as a function in *two* variables, therefore belongs to M. This is important—it will allow us to shift this map using elementary embeddings which act on M. For an example of this see Section 4, particularly Remark 4.5.

For details on the construction of σ_{piv} and the proof of Lemma 2.8 we refer the reader to [15, Chapter 1]. Let us here only say that the construction draws heavily on the techniques of Martin–Steel [8], and that the assumption (earlier in this section) that δ is a Woodin cardinal is crucial.

§3. A first application, Σ^1_2 determinacy.

§3. A first application, Σ^1_2 determinacy. As a first example we use the methods of Section 2 to prove Σ^1_2 determinacy. The result we obtain, Theorem 3.1, was previously proved by Woodin using different methods. It strengthens a result of Martin–Steel [8]. For more information on determinacy within the projective hierarchy we refer the reader to Neeman [12] and [14].

THEOREM 3.1. *Suppose there is an iterable class model M with a Woodin cardinal δ. Suppose further that $M\|\delta+1$ is countable in* V. *Then Σ^1_2 determinacy holds.*

PROOF. Fix $A \subset \mathbb{R}$ a Σ^1_2 set, say the set of reals which satisfy a given Σ^1_2 statement ϕ. We wish to show that the standard game $G_\omega(A)$ is determined.

Fix M and δ which satisfy the hypothesis of Theorem 3.1. Let $\dot{A} \in M$ name the set of reals of $M^{\text{col}(\omega,\delta)}$ which satisfy ϕ in $M^{\text{col}(\omega,\delta)}$. We have the corresponding maps $\mathcal{A}$, σ_{gen}, $\mathcal{A}_{\text{piv}}$, and σ_{piv} of Section 2.

Working inside M we define a game G^*, played according to Diagram 11.

I	x_0	$l_0, \mathcal{X}_0, p_0$			$l_1, \mathcal{X}_1, p_1$		x_2 ...
II			$\mathcal{F}_0, \mathcal{D}_0$	x_1		$\mathcal{F}_1, \mathcal{D}_1$	

DIAGRAM 11. The game G^*.

I and II alternate playing natural numbers x_n, producing together $x = \langle x_n \mid n < \omega \rangle \in \mathbb{R}$. In addition they play auxiliary moves subject to the rules of $\mathcal{A}[x]$. If a player cannot follow these rules, she loses. Infinite runs of G^* are won by II.

REMARK 3.2. Our definition of G^* implicitly uses the continuity of the map $x \mapsto \mathcal{A}[x]$; in round n of G^* we only know $x \restriction n$, but this is enough to figure the rules for round n of $\mathcal{A}[x]$. Similarly, the fact that G^* exists *inside* M follows from the fact that $\mathcal{A} = (s \mapsto \mathcal{A}[s])$ belongs to M.

We will show that if I wins G^* in M then I wins $G(A)$ in V. Later on we will phrase a mirror image game H^*, and show that if II wins H^* in M then II wins $G(A)$ in V. Then we will use the determinacy of G^* and H^* in M—note G^* is an open game and H^* will be a closed game—to argue that one of these cases must hold.

CASE 1, if I wins G^* in M. Fix $\sigma^* \in M$ a winning strategy for I (the open player) in G^*. We wish to show that I wins $G_\omega(A)$ in V. Let us play $G_\omega(A)$ against an imaginary opponent. We describe how to play, and win.

In V fix a surjection $\varrho\colon \omega \to M \| \delta + 1$. Our description takes the form of a construction in V. We construct a run $x \in \mathbb{R}$ of $G_\omega(A)$. At the same time we construct $\mathcal{T}$, $\vec{a}$, a run of $\mathcal{A}_{\text{piv}}[x]$. The participants in our construction are:

- The imaginary opponent: playing x_n for odd n.
- The strategy $\sigma_{\text{piv}}[\varrho, x]$: playing for II in $\mathcal{A}_{\text{piv}}[x]$.
- The strategy σ^* and its shifts along the even branch of $\mathcal{T}$: playing x_n for even n and playing for I in $\mathcal{A}_{\text{piv}}[x]$ (i.e., playing for I in shifts of $\mathcal{A}[x]$).

The time line of the construction is presented in Diagram 12. At the start of round n we have $x \restriction n$, $\mathcal{T} \restriction 2n+1$ ending with the model M_{2n}, and a position P_n of n rounds in $j_{0,2n}(\mathcal{A})[x \restriction n]$. If n is odd, our opponent opens the round playing x_n. If n is even $j_{0,2n}(\sigma^*)$ plays x_n. Then $j_{0,2n}(\sigma^*)$ plays an auxiliary move $l_n, \mathcal{X}_n, p_n$, according to the rules of $j_{0,2n}(\mathcal{A})[x \restriction n]$ following the position P_n. At this point we apply $\sigma_{\text{piv}}[\varrho, x \restriction n]$ which creates the models M_{2n+1}, M_{2n+2}. Let $Q_n = j_{2n,2n+2}(P_n —, l_n, \mathcal{X}_n, p_n)$. $\sigma_{\text{piv}}[\varrho, x \restriction n]$ further plays $\mathcal{F}_n, \mathcal{D}_n$, a legal move for II in $j_{0,2n+2}(\mathcal{A})[x \restriction n]$ following Q_n. We let $P_{n+1} = Q_n —, \mathcal{F}_n, Q_n$. This completes round n.

Once the construction is completed we let

$$a_n = j_{2n,\text{even}}(l_n, \mathcal{X}_n, p_n) —, j_{2n+2,\text{even}}(\mathcal{F}_n, \mathcal{D}_n).$$

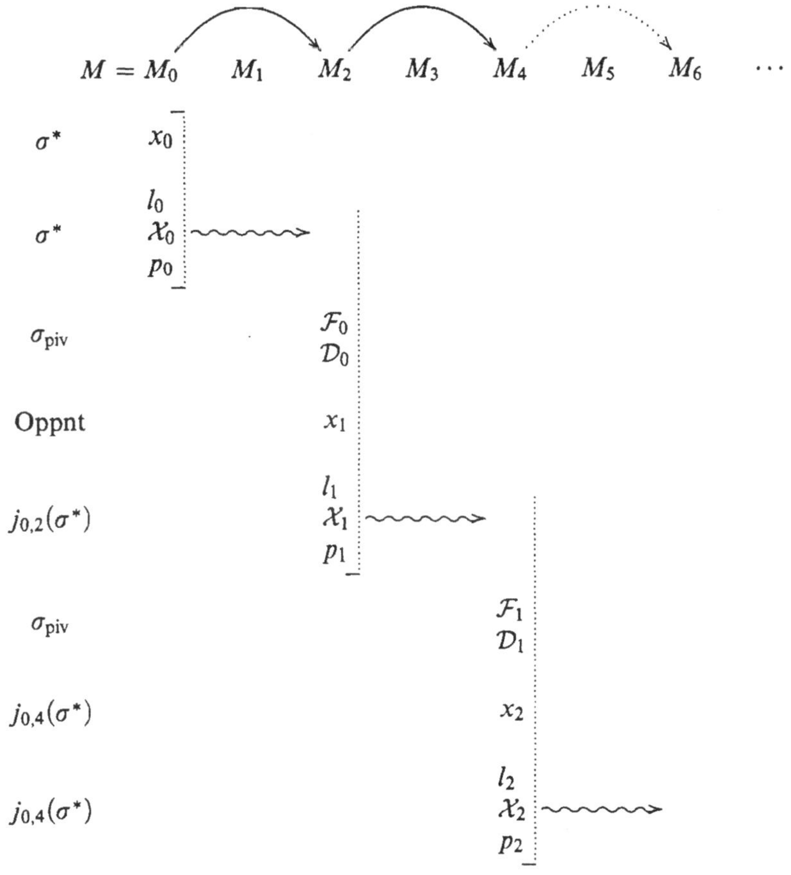

DIAGRAM 12. The construction in case 1.

We let $\vec{a} = \langle a_n \mid n < \omega \rangle$. Our construction is such that x and $\vec{a}$ form an infinite play of $j_{\text{even}}(G^*)$, which is played according to $j_{\text{even}}(\sigma^*)$. This play is created in V, since our opponent lives in V. If M_{even} were wellfounded the existence of such a play could be reflected into M_{even}. It could then be pulled back via j_{even} to yield the existence in M of an infinite play of G^* which is according to σ^*. But σ^* is a winning strategy for I, the open player in G^*; so there are no infinite plays according to σ^*. We conclude that M_{even} is illfounded.

Since M is iterable there must exist some cofinal wellfounded cofinal branch b through $\mathcal{T}$. b must be an odd branch. Our use of $\sigma_{\text{piv}}[\varrho, x]$ during the construction guarantees that $\mathcal{T}, \vec{a}$ is a pivot. Applying condition 3 of

Definition 2.7 we conclude that there exists some h which is $\mathrm{col}(\omega, j_b(\delta))$–generic/$M_b$ and so that:

(∗) $x \in j_b(\dot{A})[h]$.

This means that in $M_b[h]$ x satisfies the Σ^1_2 statement ϕ. By Shoenfield absoluteness x must also satisfy ϕ in V. (We are using here the wellfoundedness of M_b.) So $x \in A$ as required. This completes case 1. ⊣

REMARK 3.3. Note the importance of continuity throughout our construction. In round n we are able to use $\sigma_{\mathrm{piv}}[\varrho, x]$ despite only having knowledge of $x \restriction n+1$.

REMARK 3.4. Note further the importance of having G^* and σ^* *inside* M. During the construction we shifted G^* and σ^* along the even branch of $\mathcal{T}$, using $j_{0,2n}(\sigma^*)$ in round n.

Let $\dot{B} \in M$ name the set of reals which do **not** satisfy ϕ in $M^{\mathrm{col}(\omega,\delta)}$. Define $x \mapsto \mathcal{B}[x]$ and $x \mapsto \mathcal{B}_{\mathrm{piv}}[x]$ as in Section 2, but changing $\dot{A}$ to $\dot{B}$ and **interchanging** I **and** II. We have strategies $\tau_{\mathrm{gen}}[x]$ and $\tau_{\mathrm{piv}}[\varrho, x]$ as before, but with the roles of I and II switched. (In particular, τ_{gen} is a strategy for II and τ_{piv} is a strategy for I.) These strategies satisfy Lemmas 2.4 and 2.8, but with $\dot{A}$ (in Lemma 2.4 and in condition 3b of Definition 2.7) changed to $\dot{B}$.

Working inside M we define a game H^*, played according to Diagram 13.

I	x_0		$\mathcal{F}_0, \mathcal{D}_0$			$\mathcal{F}_1, \mathcal{D}_1$	x_2	...
II		$l_0, \mathcal{X}_0, p_0$		x_1	$l_1, \mathcal{X}_1, p_1$			

DIAGRAM 13. The game H^*.

As before I and II alternate playing natural numbers x_n, producing together $x = \langle x_n \mid n < \omega \rangle \in \mathbb{R}$. This time they play auxiliary moves subject to the rules of $\mathcal{B}[x]$. If a player cannot follow these rules, she loses. This time infinite runs of the game are won by I.

CASE 2, if II wins H^* in M. Then an argument similar to that of case 1 shows that (in V) II has a strategy to get into $B = \mathbb{R} - A$. In other words, II wins $G_\omega(A)$ in V. ⊣

So far we showed:

- (In case 1.) If I wins G^* in M, then I wins $G_\omega(A)$ in V.
- (In case 2.) If II wins H^* in M, then II wins $G_\omega(A)$ in V.

It is now enough to check that one of these cases must occur. Suppose not, i.e., assume that in M II wins G^* and I wins H^*. Fix strategies σ^* and τ^* in M witnessing this. We intend to derive a contradiction.

We work in $M[g]$ to construct a real $x = \langle x_n \mid n < \omega \rangle$, an infinite play $\vec{a} = \langle a_{n-\mathrm{I}}, a_{n-\mathrm{II}} \mid n < \omega \rangle$ of $\mathcal{A}[x]$ ($a_{n-\mathrm{I}}$ denotes I's auxiliary move in round n; $a_{n-\mathrm{II}}$ denotes II's auxiliary move in round n), and an infinite play $\vec{b} = \langle b_{n-\mathrm{II}}, b_{n-\mathrm{I}} \mid n < \omega \rangle$ of $\mathcal{B}[x]$. We construct as follows:

- σ^* (playing for II in G^*) produces x_n for odd n, and $a_{n-\mathrm{II}}$ for all n.
- $\sigma_{\mathrm{gen}}[x]$ produces $a_{n-\mathrm{I}}$ for all n.
- τ^* (playing for I in H^*) produces x_n for even n and $b_{n-\mathrm{I}}$ for all n.
- $\tau_{\mathrm{gen}}[x]$ produces $b_{n-\mathrm{II}}$ for all n.

As usual continuity is important. Our use of $\sigma_{\mathrm{gen}}[x]$ and $\tau_{\mathrm{gen}}[x]$ in round n can be carried through since it only requires knowledge of $x \restriction n$. We note that the maps σ_{gen} and τ_{gen} exist in $M[g]$. Since σ^* and τ^* exist in M the entire construction can be carried inside $M[g]$.

Our use of σ_{gen} guarantees that $x \notin \dot{A}[g]$ (see Lemma 2.4). Since x belongs to $M[g]$ this means that x fails to satisfy ϕ in $M[g]$. Similarly our use of τ_{gen} guarantees that $x \notin \dot{B}[g]$, and this means that x fails to **not** satisfy ϕ in $M[g]$. But this is a contradiction. ⊣

§4. **Games of length $\omega \cdot \omega$.** Fix $C \subset \mathbb{R}^\omega$ a Σ^1_2 set, say the set of all sequences $\langle y_n \mid n < \omega \rangle \in \mathbb{R}^\omega$ which satisfy a given Σ^1_2 statement ϕ. Fix M and an increasing sequence $\langle \delta_1, \delta_2, \ldots, \delta_\omega \rangle$ in M so that:

- M is a class model;
- M is iterable;
- Each δ_ξ, $1 \leq \xi \leq \omega$, is a Woodin cardinal in M; and
- $M \| \delta_\omega + 1$ is countable in V.

The existence of such an M is our large cardinal assumption. We work under this assumption to prove that $G_{\omega \cdot \omega}(C)$ is determined.

We work to define auxiliary games in M, analogous to the games G^* and H^* of Section 3. These games will be open and closed respectively, and hence determined. If in M I wins the analogue of G^*, we will show that in V I wins $G_{\omega \cdot \omega}(C)$. This is an analogue to case 1 in Section 3. If in M II wins the analogue of H^*, then by a parallel argument II wins $G_{\omega \cdot \omega}(C)$ in V. This is an analogue to case 2 in Section 3. Determinacy will follow once we verify, in Section 4.4, that one of these cases must occur. This is an analogue to the final argument in Section 3.

4.1. Names. Let δ_∞ denote δ_ω. Let $\dot{A}_\infty \in M$ name the set of sequences $\langle y_n \mid n < \omega \rangle \in \mathbb{R}^\omega$ in $M^{\mathrm{col}(\omega, \delta_\infty)}$ which satisfy ϕ in $M^{\mathrm{col}(\omega, \delta_\infty)}$. For each $\langle y_n \mid n < \omega \rangle \in \mathbb{R}^\omega$ we have the associated auxiliary game $\mathcal{A}_\infty[y_n \mid n < \omega]$ of Section 2 corresponding to the name $\dot{A}_\infty$ and the Woodin cardinal δ_∞. (There is a slight abuse of notation here; formally we should think of $\langle y_n \mid n < \omega \rangle$ as coded by some real x.) We remind the reader that moves in $\mathcal{A}_\infty[y_n \mid n < \omega]$

are arranged so that I tries to witness $\langle y_n \mid n < \omega\rangle \in \dot{A}_\infty[h]$ for some h, while II tries to witness the opposite.

The association $\langle y_n \mid n < \omega\rangle \mapsto \mathcal{A}_\infty[y_n \mid n < \omega]$ is continuous, given by the map $\mathcal{A}_\infty$. This map belongs to M. We will talk about $\mathcal{A}_\infty[y_0, \dots, y_{k-1}]$, which we take to be a game of $k+1$ rounds. (Only a finite part of the reals $y_0, \dots, y_{k-1}$ is needed to determine the rules of this game.) We use $a^\infty_{0-\mathrm{I}}$, $a^\infty_{0-\mathrm{II}}$, $a^\infty_{1-\mathrm{I}}$, etc. to refer to moves in the games $\mathcal{A}_\infty[y_n \mid n < \omega]$, and use a^∞_n to denote $\langle a^\infty_{n-\mathrm{I}}, a^\infty_{n-\mathrm{II}}\rangle$. We use $\vec{a}_\infty = \langle a^\infty_n \mid n < \omega\rangle$ to refer to infinite runs of $\mathcal{A}_\infty[y_n \mid n < \omega]$.

Definition 4.1. A **k-sequence** is a sequence

$$S = \langle y_0, \dots, y_{k-1}, a^\infty_0, \dots, a^\infty_{k-1}, \gamma\rangle$$

so that:

1. Each y_i is a real;
2. $a^\infty_0, \dots, a^\infty_{k-1}$ is a position in the auxiliary game $\mathcal{A}_\infty[y_0, \dots, y_{k-1}]$; and
3. γ is an ordinal.

Definition 4.2. A **valid extension** for a k-sequence S as in Definition 4.1 is a triple $y_k, a^\infty_k, \gamma^*$ so that:

1. y_k is a real;
2. $a^\infty_k = \langle a^\infty_{k-\mathrm{I}}, a^\infty_{k-\mathrm{II}}\rangle$ where $a^\infty_{k-\mathrm{I}}$ and $a^\infty_{k-\mathrm{II}}$ are legal moves for I and II respectively in the game $\mathcal{A}_\infty[y_0, \dots, y_{k-1}]$ following $a^\infty_0, \dots, a^\infty_{k-1}$; and
3. γ^* is an ordinal smaller than γ.

We use $S\frown, y_k, a^\infty_k, \gamma^*$ to denote the $k+1$-sequence

$$\langle y_0, \dots, y_{k-1}, y_k, a^\infty_0, \dots, a^\infty_{k-1}, a^\infty_k, \gamma^*\rangle.$$

We remind the reader that $\mathcal{A}_\infty[y_0, \dots, y_{k-1}]$ is a game of $k+1$ rounds, so condition 2 of Definition 4.2 makes sense.

For expository simplicity fix for each $k < \omega$ some g_k which is $\mathrm{col}(\omega, \delta_k)$-generic/$M$. Below we define sets in $M[g_k]$ where strictly speaking we should be defining names in $M^{\mathrm{col}(\omega,\delta_k)}$. For a tuple $a^\infty_0, \dots, a^\infty_{k-1}$ and an ordinal γ we work to describe $A_k[a^\infty_0, \dots, a^\infty_{k-1}, \gamma]$, a subset of $\mathbb{R}^k$ in $M[g_k]$. We shall then let $\dot{A}_k[a^\infty_0, \dots, a^\infty_{k-1}, \gamma]$ be the canonical name for this set.

We use the notation

$$\langle y_0, \dots, y_{k-1}, a^\infty_0, \dots, a^\infty_{k-1}, \gamma\rangle \in A_k$$

to mean that $\langle y_0, \dots, y_{k-1}\rangle$ belongs to $A_k[a^\infty_0, \dots, a^\infty_k, \gamma]$. We similarly think of $\dot{A}_k$ as a (class) name for the collection of tuples S so that $S \in A_k$. Thus we say

$$S = \langle y_0, \dots, y_{k-1}, a^\infty_0, \dots, a^\infty_{k-1}, \gamma\rangle \in \dot{A}_k[h]$$

to mean that $\langle y_0, \dots, y_{k-1}\rangle$ belongs to $\dot{A}_k[a^\infty_0, \dots, a^\infty_{k-1}, \gamma][h]$.

Let $\mathcal{A}_k[y_0, \ldots, y_{k-1}, a_0^\infty, \ldots, a_{k-1}^\infty, \gamma]$ be the auxiliary games corresponding to the name $\dot{A}_k[a_0^\infty, \ldots, a_{k-1}^\infty, \gamma]$ and the Woodin cardinal δ_k. We use $\mathcal{A}_k[S]$ to refer to these games, and use $a_{0-\mathrm{I}}^k$, $a_{0-\mathrm{II}}^k$ etc. to denote moves in the games. These moves are arranged so that I tries to witness that S belongs to $\dot{A}_k[h]$ for some generic h, while II tries to witness the opposite.

Given $S = \langle y_0, \ldots, y_{k-1}, a_0^\infty, \ldots, a_{k-1}^\infty, \gamma\rangle$ a k-sequence, we define a game $G_k^*(S)$ played inside M according to Diagram 14.

I	$\gamma^*, a_{k-\mathrm{I}}^\infty$		$y_k(0)$	$a_{0-\mathrm{I}}^{k+1}$			$a_{1-\mathrm{I}}^{k+1}$		$y_k(2) \ldots$
II		$a_{k-\mathrm{II}}^\infty$			$a_{0-\mathrm{II}}^{k+1}$	$y_k(1)$		$a_{1-\mathrm{II}}^{k+1}$	

DIAGRAM 14. The game $G_k^*(S)$.

I and II play

- γ^*,
- $a_k^\infty = \langle a_{k-\mathrm{I}}^\infty, a_{k-\mathrm{II}}^\infty\rangle$, and
- $y_k = \langle y_k(n) \mid n < \omega\rangle$

which form a valid extension of S. To be more precise we require:

1. γ^* is smaller than γ, in line with condition 3 of Definition 4.2.
2. $a_{k-\mathrm{I}}^\infty$ and $a_{k-\mathrm{II}}^\infty$ are legal moves for I and II respectively in $\mathcal{A}_\infty[y_0, \ldots, y_{k-1}]$ following $a_0^\infty, \ldots, a_{k-1}^\infty$. This is in line with condition 2 of Definition 4.2. Note that knowledge of y_k is not needed here.
3. $y_k(n)$ are natural numbers.

In addition I and II play auxiliary moves in the game $\mathcal{A}_{k+1}[S\frown, y_k, a_k^\infty, \gamma^*]$. If a player cannot follow these rules she loses. Infinite runs of the game are won by II.

DEFINITION 4.3. For $S \in M[g_k]$ set $S \in A_k$ iff S is a k-sequence and in $M[g_k]$ I has a winning strategy in $G_k^*(S)$.

Definition 4.3 at last specifies the sets $A_k[a_0^\infty, \ldots, a_{k-1}^\infty, \gamma]$, and by extension the names $\dot{A}_k[a_0^\infty, \ldots, a_{k-1}^\infty, \gamma]$.

REMARK 4.4. Our definition of the sets $A_k[a_0^\infty, \ldots, a_{k-1}^\infty, \gamma]$ is by induction on γ, not on k. To figure out whether $\langle y_0, \ldots, y_{k-1}\rangle \in M[g_k]$ belongs to the set $A_k[a_0^\infty, \ldots, a_{k-1}^\infty, \gamma]$ we need knowledge of the game $G_k^*(S)$ where $S = \langle y_0, \ldots, y_{k-1}, a_0^\infty, \ldots, a_{k-1}^\infty, \gamma\rangle$. For this we require knowledge of the auxiliary games $\mathcal{A}_{k+1}[S\frown, y_k, a_k^\infty, \gamma^*]$, but only for γ^* which are **smaller** than γ because of rule 1. Thus to determine $\dot{A}_k[a_0^\infty, \ldots, a_{k-1}^\infty, \gamma]$ we need knowledge of the names $\dot{A}_{k+1}[a_0^\infty, \ldots, a_{k-1}^\infty, a_k^\infty, \gamma^*]$, but only for $\gamma^* < \gamma$. We have this knowledge by induction.

Some words of motivation are due on Definition 4.3. Suppose that $S \in M[g_k]$ is a k-sequence and belongs to A_k. So I wins $G^*_k(S)$. Let us for a moment ignore the first round of $G^*_k(S)$. The remaining rounds essentially follow the rules of G^* of Section 3. (See Diagram 11 and the rules below it.) Our experience from Section 3 tells us that if I has a winning strategy for these rounds, then in V I has a strategy to enter some shift of $\dot{A}_{k+1}$. In other words, if S belongs to $A_k = \dot{A}_k[g_k]$ we expect to be able to produce y_k (working against an imaginary opponent who plays the odd half of y_k) so that $S\frown\langle y_k, a^\infty_k, \gamma^*\rangle$ belongs to $j_b(\dot{A}_{k+1})[h]$ for some iteration map j_b and some generic h.

This is a process of *perpetuation*. Membership in A_k allows us to aim for membership in a shift of $\dot{A}_{k+1}$.

And what about the first round of $G^*_k(S)$? This round too is related to the game G^* of Section 3, this time with the name $\dot{A}_\infty$. It is just one round out of this game, and our experience from Section 3 tells us that a winning strategy for I will allow us to aim into a shift of the name $\dot{A}_\infty$.

In short, membership in A_k allows us to (a) advance one round in witnessing that our sequence of reals belongs to a shift of $\dot{A}_\infty$; and (b) produce the next real, y_k, so that the resulting sequence belongs to a shift of $\dot{A}_{k+1}$. Once we entered a shift of $\dot{A}_{k+1}$ we can repeat the process, advancing an extra round towards $\dot{A}_\infty$ and entering a shift of $\dot{A}_{k+2}$, etc. At the end we make the full sequence of advances needed to witness membership in $\dot{A}_\infty$. This means that our sequence of reals (produced with the collaboration of some imaginary opponent playing for II) satisfies the Σ^1_2 statement ϕ. So we win the long game $G_{\omega\cdot\omega}(C)$, playing for I.

This argument is made more precise in Section 4.2. Then in Section 4.3 we phrase the mirror image argument and show under reversed circumstances that II wins the long game. The relationship between Sections 4.2 and 4.3 is analogous to the relationship between cases 1 and 2 in Section 3. Finally in Section 4.4 we show that either the circumstances of Section 4.2 (where I ends up winning) or the circumstances of Section 4.3 (where II ends up winning) must hold. This establishes the determinacy of $G_{\omega\cdot\omega}(C)$.

4.2. I wins. Suppose that there exists some γ so that in M I wins $G^*_0(\gamma)$. (Note that γ by itself is a 0-sequence.) We will show that I wins the original long game $G_{\omega\cdot\omega}(C)$ in V.

Fix $\sigma^*_0 \in M$, a winning strategy for I (the open player) in $G^*_0(\gamma)$. Fix an imaginary opponent, playing for II in the long game $G_{\omega\cdot\omega}(C)$.

Recall that we have strategies $\sigma_{\text{piv}-\infty}[y_n \mid n < \omega]$ corresponding to the name $\dot{A}_\infty$ (see Section 2.2).[4] Similarly we have strategies

$$\sigma_{\text{piv}-k}[y_0, \ldots, y_{k-1}, a^\infty_0, \ldots, a^\infty_{k-1}, \gamma]$$

[4]We are suppressing here and below the extra parameter ϱ. We shall comment on this in the discussion which follows Claim 4.10.

(which we denote $\sigma_{\text{piv}-k}[S]$) corresponding to the names $\dot{A}_k[a_0^\infty, \dots, a_{k-1}^\infty, \gamma]$. These strategies are given by maps $\sigma_{\text{piv}-\infty}$ and $\sigma_{\text{piv}-k}$, continuous in the relevant reals and in the suppressed variable ϱ. The maps belong to M.

We will use σ_0^*, the strategies $\sigma_{\text{piv}-1}, \sigma_{\text{piv}-2}, \dots$, the strategy $\sigma_{\text{piv}-\infty}$, and an iteration strategy for M, to play against the imaginary opponent and win.

Let us begin playing $G_{\omega\cdot\omega}(C)$. We divide the game into ω mega-rounds. In mega-round k we construct (among other things) the real y_k. At the start of mega-round k we will have:

(A) Reals $y_0, \dots, y_{k-1}$;
(B) An iterate M_k of M (the result of k iteration trees stacked one after the other) with iteration embedding $j_k : M \to M_k$;
(C) (If $k > 0$.) h_k which is $\text{col}(\omega, j_k(\delta_k))$-generic/$M_k$;
(D) A position of k rounds in the game $j_k(\mathcal{A}_{\text{piv}-\infty})[y_0, \dots, y_{k-1}]$, played according to $j_k(\sigma_{\text{piv}-\infty})[y_0, \dots, y_{k-1}]$; and
(E) An ordinal γ_k.

REMARK 4.5. With respect to (D) it is important to remember that the maps $\mathcal{A}_{\text{piv}-\infty}$ and $\sigma_{\text{piv}-\infty}$ belong to M. These maps can therefore be shifted via the embedding $j_k : M \to M_k$.

The position indicated in (D) includes an iteration tree $\mathcal{U}_k$ on M_k of length $2k + 1$. We use $W_0^k, \dots, W_{2k}^k$ to denote the models of this tree, and $\pi_{*,*}^k$ to denote the embeddings. The position indicated in (D) further includes a position P_k^∞ of k rounds in the shift of $\mathcal{A}_\infty[y_0, \dots, y_{k-1}]$ to W_{2k}^k, namely in $(\pi_{0,2k}^k \circ j_k)(\mathcal{A}_\infty)[y_0, \dots, y_{k-1}]$.

Note that $\langle y_0, \dots, y_{k-1}, P_k^\infty, \gamma_k \rangle$ is a k-sequence in the sense of W_{2k}^k—this is simply a restatement of the fact that P_k^∞ is a position of k rounds in $(\pi_{0,2k}^k \circ j_k)(\mathcal{A}_\infty)[y_0, \dots, y_{k-1}]$. We use S_k to denote this k-sequence. We shall make sure that

(i) S_k belongs to $W_{2k}^k[h_k]$ (to $M = W_0^0$ if $k = 0$); and
(ii) In $W_{2k}^k[h_k]$ (in $M = W_0^0$ if $k = 0$) I wins $(\pi_{0,2k}^k \circ j_k)(G_k^*)(S_k)$.

In condition (i) we are saying that the reals $y_0, \dots, y_{k-1}$ belong to $W_{2k}^k[h_k]$. The rest of S_k is just a finite list of objects from W_{2k}^k.

REMARK 4.6. The game $\mathcal{A}_{\text{piv}-\infty}$ is played in the vicinity of the Woodin cardinal δ_∞, and all critical points used in the game are larger than some pre-specified ordinal $\lambda < \delta_\infty$. (See Remark 2.5.) The ordinal we specify is $\lambda = \sup\{\delta_k \mid k < \omega\}$. We know then that the models W_{2k}^k and $M_k = W_0^k$ agree beyond $j_k(\delta_k)$, so that our reference to $W_{2k}^k[h_k]$ in conditions (i) and (ii) makes sense. Moreover, the iteration tree $\mathcal{U}_k$ can then be regarded not just as a tree on M_k, but also as a tree on $M_k[h_k]$.

We begin with $M_0 = M$, and $\gamma_0 = \gamma$. Condition (ii) holds because of our case assumption, that I wins $G_0^*(\gamma)$ in M. Let us handle mega-round k.

Our models at the start of mega-round k are presented in Diagram 15. Our situation at the end of the mega-round is presented in Diagram 16.

$$M_k = W_0^k \overset{\pi_{0,2k}^k}{\dashrightarrow} W_{2k}^k$$

DIAGRAM 15. At the start of mega-round k.

Using condition (ii) we have $\sigma_k^* \in W_{2k}^k[h_k]$, a winning strategy for I (the open player) in $(\pi_{0,2k}^k \circ j_k)(G_k^*)[S_k]$.

(a) To open mega-round k this strategy plays y_{k+1}^*.

The game $(\pi_{0,2k}^k \circ j_k)(G_k^*)[S_k]$ now proceeds with one round—round k—from the shifted $\mathcal{A}_\infty$, followed by all ω rounds from the shifted $\mathcal{A}_{k+1}$ together with natural number moves to produce y_k. Folding in constructions of the kind done in Section 3 we create:

(b) The models W_{2k+1}^k and W_{2k+2}^k extending $\mathcal{U}_k$, and the embedding $\pi_{2k,2k+2}^k$;
(c) A position of $k+1$ rounds in the shift of $\mathcal{A}_\infty$ to W_{2k+2}^k, extending $\pi_{2k,2k+2}^k(P_k^\infty)$;
(d) The real y_k;
(e) A length ω iteration tree $\mathcal{T}_k$ on W_{2k+2}^k, a cofinal odd branch b_k through it, the direct limit model W_{b_k}, and the direct limit embedding j_{b_k}; and
(f) h_{k+1} which is generic over W_{b_k} for the collapse of $(j_{b_k} \circ \pi_{0,2k+2}^k \circ j_k)(\delta_{k+1})$.

We let P_{k+1}^∞ denote the position of (c), shifted to W_{b_k} via j_{b_k}. We let y_{k+1} denote y_{k+1}^* of (a) shifted to W_{b_k} via $j_{b_k} \circ \pi_{2k,2k+2}^k$. As in Section 3 we get

(∗) $\langle y_0, \ldots, y_k \rangle \in \dot{A}_{k+1}^s[P_{k+1}^\infty, y_{k+1}][h_{k+1}]$,

where $\dot{A}_{k+1}^s$ denotes $\dot{A}_{k+1}$ shifted to W_{b_k}.

REMARK 4.7. Our construction in (b) and (c) simply extends the position of (D) to $k+1$ rounds. This is nothing more than an adaptation of the construction in round k of case 1 in Section 3, using the shifts of δ_∞ and $\mathcal{A}_\infty$.

REMARK 4.8. Our construction in (d)–(f) is an adaptation of the entire argument of case 1 in Section 3, using the shifts of δ_{k+1} and $\mathcal{A}_{k+1}$. We make the following notes:

The real y_k is constructed as a collaborative process involving our imaginary opponent and shifts of the strategy $\pi_{2k,2k+2}^k(\sigma_k^*)$ along the even branch of $\mathcal{T}_k$. As a reminder of this we refer the reader to Diagram 12.

Remember that we have not a single name $\dot{A}_{k+1}$, but a whole class of them. Our construction in (d)–(f) uses the strategy $\sigma_{\text{piv}-k+1}$ which corresponds to $\dot{A}_{k+1}[X]$ (shifted to W_{2k+2}^k) where "X" consists of the position created in (b)

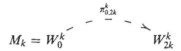

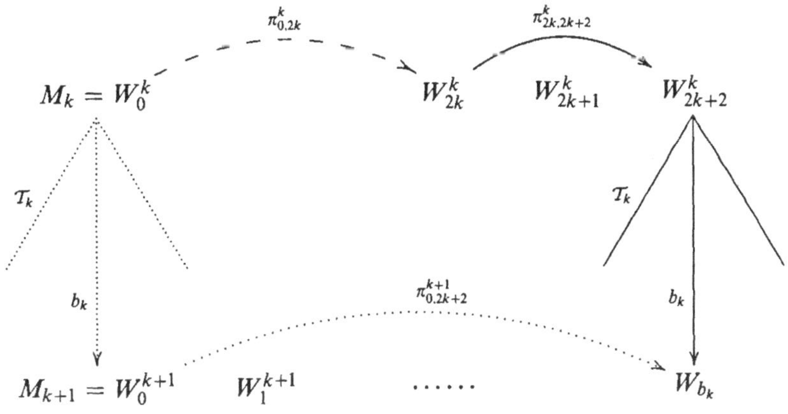

DIAGRAM 16. Mega-round k.

and the shift to W^k_{2k+2} of the ordinal of (a). It is this use of $\sigma_{\text{piv}-k+1}$ which gives h_{k+1} for (f) and secures (*), see Section 2.2.

The branch b_k is the work of an iteration strategy for M which we pick at the outset. (Recall that our initial assumptions on M included iterability.) In particular the wellfoundedness of W_{b_k} is guaranteed.

REMARK 4.9. The starting model for (d)–(f) is $W^k_{2k+2}[h_k]$. The generic extension here is *important*. Remember that as part of the construction we must shift the strategy $\pi^k_{2k,2k+2}(\sigma^*_k)$ along the even branch of the iteration tree $\mathcal{T}_k$. (See Diagram 12 and Remark 3.4.) W^k_{2k+2} does not contain this strategy; $\mathcal{T}_k$ must therefore act on $W^k_{2k+2}[h_k]$.

This is where we use the fact that δ_{k+1} is greater than δ_k, so that h_k is a "small generic" compared to the shift of δ_{k+1}. With Remark 2.5 this allows us to make sure that all extenders used in $\mathcal{T}_k$—a tree created in the vicinity of δ_{k+1}—have critical points above the shift of δ_k. $\mathcal{T}_k$ then extends to act on $W^k_{2k+2}[h_k]$.

We have so far the embeddings indicated in solid lines in Diagram 16. The top horizontal line represents the tree $\mathcal{U}_k$ and its extension by two extra models. This tree has critical points above the shift of $\sup\{\delta_1, \delta_2, \dots\}$ (see Remark 4.6), and hence certainly above the shift of δ_{k+1}. The vertical tree on the right is our $\mathcal{T}_k$. It has critical points below its Woodin cardinal, the shift of δ_{k+1}.

Using these relations between the critical points, standard commutativity allows us to switch the order of $\mathcal{T}_k$ and the extended $\mathcal{U}_k$. We can first apply $\mathcal{T}_k$—which we may regard as a tree on M_k—and then apply the image of the extended $\mathcal{U}_k$. This new order is represented in dotted lines in Diagram 16.

We let $\mathcal{U}_{k+1}$ be the image of the extended $\mathcal{U}_k$ (this image is presented in dots on the lower line of Diagram 16). The final model of this tree, W^{k+1}_{2k+2}, is precisely equal to W_{b_k}. We are now in a position to start mega-round $k+1$. Conditions (i) and (ii) hold because of $(*)$ above.

Two points about our construction in mega-round k should be recorded for future reference. We have:

(†) $\mathcal{U}_{k+1}$ extends $j_{b_k}(\mathcal{U}_k)$; and
(‡) $\gamma_{k+1} < (j_{b_k} \circ \pi^k_{2k,2k+2})(\gamma_k)$.

(‡) follows from our use of I's strategy σ^*_k, because of rule 1 in the game G^*_k (see also condition 3 of Definition 4.2). This is now our second use of this rule. The first one was in Remark 4.4.

Once the construction is over we are left with a sequence of reals $\langle y_n \mid n < \omega\rangle$, and a sequence of iteration trees $\mathcal{T}_k$ presented in Diagram 17 giving rise to a direct limit M_∞. Our use of an iteration strategy to pick the branches b_k during the construction guarantees the wellfoundedness of M_∞.

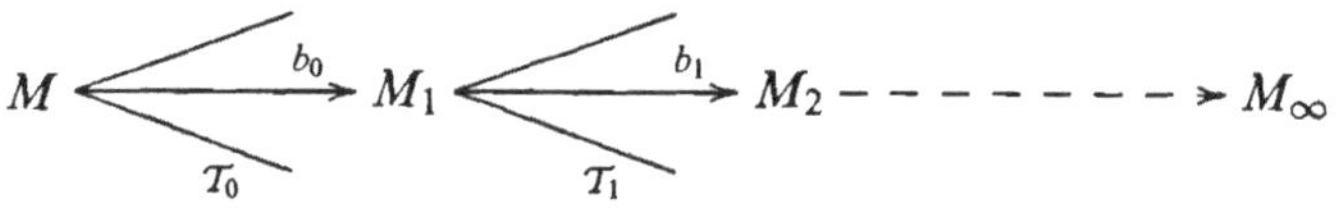

DIAGRAM 17. At the end.

We have further for each k the finite tree $\mathcal{U}_k$ on M_k. Let $\mathcal{U}_\infty$ on M_∞ be the natural limit of these trees, specifically the union of the trees $j_{k,\infty}(\mathcal{U}_k)$. This makes sense because of (†). $\mathcal{U}_\infty$ has an even branch, consisting of the models W^∞_{2k}. Let W^∞_{even} denote the direct limit along this branch. W^∞_{even} is in fact equal to the direct limit of the models W^k_{2k} under the embeddings $j_{b_k} \circ \pi^k_{2k,2k+2} \colon W^k_{2k} \to W^{k+1}_{2(k+1)}$. (We use here the same kind of commutativity that allowed us to "switch order" from the solid and broken lines to the dotted lines in Diagram 16.) Condition (‡) tells us that this last direct limit is illfounded.

So we have $\mathcal{U}_\infty$, a length ω iteration tree on M_∞, with an illfounded even branch. The iteration strategy for M, faced with $\mathcal{U}_\infty$, is forced to produces a cofinal odd branch c. Let W^∞_c be the direct limit, and let $\pi_c \colon M_\infty \to W^\infty_c$ be the direct limit embedding. Note W^∞_c, played by an iteration strategy, is wellfounded.

Now $\mathcal{U}_\infty$ is part of a play according to $j_{0,\infty}(\sigma_{\text{piv}-\infty})[y_n \mid n < \omega]$—this was part of our construction, see (D) above. Our use of $j_{0,\infty}(\sigma_{\text{piv}-\infty})[y_n \mid n < \omega]$ guarantees that there exists some h_∞ so that:

1. h_∞ is $\text{col}(\omega, (\pi_c \circ j_{0,\infty})(\delta_\infty))$–generic/$W^\infty_c$; and
2. $\langle y_n \mid n < \omega\rangle \in (\pi_c \circ j_{0,\infty})(\dot{A}_\infty)[h_\infty]$.

From condition 2 we see that $\langle y_n \mid n < \omega \rangle$ satisfies the Σ^1_2 statement ϕ, inside $W_c^\infty[h_\infty]$. By absoluteness ϕ is satisfied in V. This means that $\langle y_n \mid n < \omega \rangle$ belongs to the payoff set C, and is won by I, as required. This completes the argument. We proved:

CLAIM 4.10. *Suppose that there exists* γ *so that* I *wins* $G_0^*(\gamma)$ *in* M. *Then* I *wins* $G_{\omega \cdot \omega}(C)$ *in* V. ⊣

Before closing let us comment on our suppression throughout of the parameter ϱ. We worked during the construction with the map $\sigma_{\text{piv}-\infty}$, which exists in M and could thus be shifted via embeddings acting on M. This map takes two parameters: x, which was interpreted by the sequence $\langle y_n \mid n < \omega \rangle$ in our construction; and ϱ, which was suppressed.

To be precise we should add the following to the list (A)–(D) of objects constructed:

(F) A function $\vartheta_k : k \to M_k \| j_k(\delta_\infty) + 1$.

We should also insert the parameter ϑ_k in (D), replacing the occurrence of "$j_k(\sigma_{\text{piv}-\infty})[y_0, \dots, y_{k-1}]$" with "$j_k(\sigma_{\text{piv}-\infty})[\vartheta_k, y_0, \dots, y_{k-1}]$."

The functions ϑ_k should be constructed so that ϑ_{k+1} extends $j_k(\vartheta_k)$. This allows us at the end to set

$$\varrho = \bigcup_{k<\omega} j_{k,\infty}(\vartheta_k).$$

ϱ is then a function from ω into $M_\infty \| j_{0,\infty}(\delta_\infty) + 1$, and $\mathcal{U}_\infty$ is part of a play according to $j_{0,\infty}(\sigma_{\text{piv}-\infty})[\varrho, y_n \mid n < \omega]$.

Most importantly, we should (using standard book-keeping) construct the functions ϑ_k so that ϱ ends up being *onto*. This is necessary for our application of Lemma 2.8. It was Lemma 2.8 that gave us conditions 1 and 2 above.

4.3. II wins. Here we mirror the development of Section 4.2, just as case 2 of Section 3 mirrored case 1. Let $\dot{B} \in M$ name the set of sequences $\langle y_n \mid n < \omega \rangle \in \mathbb{R}^\omega$ in $M^{\text{col}(\omega,\delta_\infty)}$ which do **not** satisfy ϕ. Let $\mathcal{B}_\infty[y_n \mid n < \omega]$ be the associated auxiliary games, but with the roles of **I and II interchanged.**

We use $\dot{B}_\infty$ and $\mathcal{B}_\infty$ as our starting points here, instead of $\dot{A}_\infty$ and $\mathcal{A}_\infty$. Define names $\dot{B}_k$ and games $H_k^*(T)$ to parallel the names $\dot{A}_k$ and games $G_k^*(S)$ of Section 4.1, only switching the roles of I and II, and using $\mathcal{B}_\infty$ (which corresponds to the negation of ϕ) instead of $\mathcal{A}_\infty$. We very briefly outline these definitions.

The game $H_k^*(T)$ is played according to Diagram 18.

I		$b^\infty_{k-\text{I}}$	$y_k(0)$		$b^{k+1}_{0-\text{I}}$			$b^{k+1}_{1-\text{I}}$	$y_k(2)\dots$
II	$\gamma^*, b^\infty_{k-\text{II}}$			$b^{k+1}_{0-\text{II}}$		$y_k(1)$	$b^{k+1}_{1-\text{II}}$		

DIAGRAM 18. The game $H_k^*(T)$, which mirrors $G_k^*(S)$ of Diagram 14.

y_k, $b_k^\infty = \langle b_{k-\mathrm{II}}^\infty, b_{k-\mathrm{I}}^\infty \rangle$, and γ^* must form a valid extension of T. (See Definition 4.2, with $\mathcal{A}_\infty$ changed to $\mathcal{B}_\infty$.) $b_{n-\mathrm{II}}^{k+1}$ and $b_{n-\mathrm{I}}^{k+1}$ are auxiliary moves in $\mathcal{B}_{k+1}[T\frown, y_k, b_k^\infty, \gamma^*]$. Infinite runs of the game are won by I.

For $T \in M[g_k]$ set $T \in B_k$ iff T is a k-sequence (with $\mathcal{A}_\infty$ changed to $\mathcal{B}_\infty$) and II wins $H_k^*(T)$ in $M[g_k]$. This definition mirrors Definition 4.3. It determines the names $\dot{B}_k$ and by extension the auxiliary games $\mathcal{B}_k$. The definition is by induction on γ, see Remark 4.4.

An argument which mirrors that of Section 4.2 gives:

CLAIM 4.11. *Suppose that there exists γ so that* II *wins $H_0^*(\gamma)$ in M. Then* II *wins $G_{\omega\cdot\omega}(C)$ in* V. ⊣

4.4. Otherwise. To prove that $G_{\omega\cdot\omega}$ is determined it is now enough to verify that the hypotheses of Claims 4.10 and 4.11 cannot both fail.

Suppose for contradiction that they do, i.e., assume that for every γ II wins $G_0^*(\gamma)$ in M and I wins $H_0^*(\gamma)$ in M. We intend to derive a contradiction. Our argument here is similar to the final argument in Section 3, where we constructed a real x which neither satisfied, nor failed to satisfy, the statement ϕ. Here we shall construct a sequence $\langle y_n \mid n < \omega \rangle \in \mathbb{R}^\omega$ which neither satisfies nor fails to satisfy ϕ. The reader may wish to compare our construction here with the final construction in Section 3.

Fix $g_\infty \in$ V which is $\mathrm{col}(\omega, \delta_\infty)$–generic/$M$. Replacing the generics g_k if needed, we may assume that each g_k belongs to $M[g_{k+1}]$, and that the sequence $\langle g_k \mid k < \omega \rangle$ belongs to $M[g_\infty]$.

Pick ordinals $\gamma_{\min} < \gamma_{\max}$, substantially larger than δ_∞, so that

$$M \| (\gamma_{\min} + \omega) \models \varphi[\vec{c}, \gamma_{\min}] \iff M \| (\gamma_{\max} + \omega) \models \varphi[\vec{c}, \gamma_{\max}]$$

for any formula φ and any parameter $\vec{c} \in (M \| \delta_\infty + \omega)^{<\omega}$. These ordinals will serve as *indiscernibles.*

We work in $M[g_\infty]$ to construct $\langle y_n \mid n < \omega \rangle \in \mathbb{R}^\omega$; an infinite play $\vec{a}_\infty = \langle a_{n-\mathrm{I}}^\infty, a_{n-\mathrm{II}}^\infty \mid n < \omega \rangle$ of $\mathcal{A}_\infty[y_n \mid n < \omega]$; and an infinite play $\vec{b}_\infty = \langle b_{n-\mathrm{II}}^\infty, b_{n-\mathrm{I}}^\infty \mid n < \omega \rangle$ of $\mathcal{B}[y_n \mid n < \omega]$. We use the following notation:

$$S_k = \langle y_0, \ldots, y_{k-1}, a_0^\infty, \ldots, a_{k-1}^\infty, \gamma_{\min} \rangle$$
$$S_k' = \langle y_0, \ldots, y_{k-1}, a_0^\infty, \ldots, a_{k-1}^\infty, \gamma_{\max} \rangle.$$

(Note the switch from $\gamma_{\min}$ in S_k to $\gamma_{\max}$ in S_k'.) We use T_k and T_k' similarly, with $\vec{b}_\infty$ instead of $\vec{a}_\infty$.

We intend to maintain the following conditions:

1. (For $k \geq 1$.) $y_0, \ldots, y_{k-1}$ belong to $M[g_k]$;
2. In $M[g_k]$ (in M if $k = 0$) II wins $G_k^*(S_k)$; and
3. In $M[g_k]$ (in M if $k = 0$) I wins $H_k^*(T_k)$.

We construct in mega-rounds. At the start of mega-round k we will have conditions 1–3 for k. Note that for $k = 0$ conditions 2 and 3 hold because of our initial case assumption in this subsection.

Let us begin mega-round k. Using the indisernibility of $\gamma_{\min}$ and $\gamma_{\max}$ conditions 2 and 3 tell us that II wins $G_k^*(S_k')$ (note the switch to S_k') and I wins $H_k^*(T_k')$. Fix strategies σ_k^* and τ_k^* in $M[g_k]$ (in M if $k = 0$) witnessing this. We play the games $G_k^*(S_k')$ and $H_k^*(T_k')$. Both games start with an ordinal move, γ^*. In both games we play $\gamma^* = \gamma_{\min}$. Note that this is a legal move since $\gamma_{\min} < \gamma_{\max}$. We continue the games as follows:

- $\sigma_{\text{gen}-\infty}[y_0, \dots, y_{k-1}]$ plays $a_{k-\text{I}}^{\infty}$ in $G_k^*(S_k')$, and σ_k^* plays $a_{k-\text{II}}^{\infty}$.
- Similarly, $\tau_{\text{gen}-\infty}[y_0, \dots, y_{k-1}]$ plays $b_{k-\text{II}}^{\infty}$ in $H_k^*(T_k')$, and τ_k^* plays $b_{k-\text{I}}^{\infty}$.

This completes the first round. We pass to the remaining ω rounds which involve auxiliary moves from $\mathcal{A}_{k+1}$ and $\mathcal{B}_{k+1}$.

- σ_k^*, playing for II in $G_k^*(S_k')$, produces $y_k(n)$ for odd n, and $a_{n-\text{II}}^{k+1}$ for all n.
- $\sigma_{\text{gen}-k+1}[S_k'\frown, y_k, a_k^{\infty}, \gamma_{\min}]$ produces $a_{n-\text{I}}^{k+1}$ for all n.
- τ_k^*, playing for I in $H_k^*(T_k')$, produces $y_k(n)$ for even n and $b_{n-\text{I}}^{k+1}$ for all n.
- $\tau_{\text{gen}-k+1}[T_k'\frown, y_k, b_k^{\infty}, \gamma_{\min}]$ produces $b_{n-\text{II}}^{k+1}$ for all n.

$\sigma_{\text{gen}-\infty}$, $\sigma_{\text{gen}-k+1}$, $\tau_{\text{gen}-\infty}$, and $\tau_{\text{gen}-k+1}$ are the generic strategies defined in Section 2.1. As usual continuity is important; for example in the last item we are using $\tau_{\text{gen}-k+1}[T_k'\frown, y_k, b_k^{\infty}, \gamma_{\min}]$ at a stage where we only know $y_k \restriction n+1$.

The reader should consult Diagrams 14 and 18 to verify that the above strategies between them cover all moves in the games $G_k^*(S_k')$ and $H_k^*(T_k')$. (Well, except for the first move $\gamma^* = \gamma_{\min}$ which we decided on ourselves.) The conditions above therefore complete the construction in mega-round k. The reader may consult the final stages of Section 3 for a simpler example of a similar argument.

In mega-round k we used σ_k^* and τ_k^*, which exist in $M[g_k]$ (in M if $k = 0$); and the maps $\sigma_{\text{gen}-k+1}$ and $\tau_{\text{gen}-k+1}$, which exist in $M[g_{k+1}]$. The real y_k produced in mega-round k therefore belongs to $M[g_{k+1}]$.

Our use of the generic strategy $\sigma_{\text{gen}-k+1}[S_k'\frown, y_k, a_k^{\infty}, \gamma_{\min}]$ guarantees that $S_{k+1} = S_k'\frown, y_k, a_k^{\infty}, \gamma_{\min}$ does not belong to $\dot{A}_{k+1}[g_{k+1}]$. Since y_k and hence S_{k+1} belong to $M[g_{k+1}]$ we conclude (see Definition 4.3) that I does not win $G_{k+1}^*(S_{k+1})$. Now $G_{k+1}^*(S_{k+1})$ is an open game, hence determined. Thus II must win $G_{k+1}^*(S_{k+1})$. This secures condition 2 for $k+1$.

Similarly our use of $\tau_{\text{gen}-k+1}[T_k'\frown, y_k, b_k^{\infty}, \gamma_{\min}]$ guarantees that $T_{k+1} = T_k'\frown, y_k, b_k^{\infty}, \gamma_{\min}$ does not belong to $\dot{B}_{k+1}[g_{k+1}]$, and this secures condition 3 for $k+1$. We are now in a position to start mega-round $k+1$.

Once completed the construction leaves us with $\langle y_n \mid n < \omega \rangle \in \mathbb{R}^{\omega}$ and infinite plays $\vec{a}_{\infty}$ of $\mathcal{A}_{\infty}[y_n \mid n < \omega]$ and $\vec{b}_{\infty}$ of $\mathcal{B}_{\infty}[y_n \mid n < \omega]$. Note that everything we did took place in $M[g_\infty]$. (Here we are using the fact that $\langle g_k \mid k < \omega\rangle \in M[g_\infty]$.) These sequences therefore belong to $M[g_\infty]$.

Our use of $\sigma_{\text{gen}-\infty}[y_n \mid n < \omega]$ during the construction ensures that $\langle y_n \mid n < \omega\rangle$ does not belong to $\dot{A}_\infty[g_\infty]$. Since $\langle y_n \mid n < \omega\rangle$ belongs to $M[g_\infty]$ we conclude that $\langle y_n \mid n < \omega\rangle$ fails to satisfy ϕ, our original Σ^1_2 statement, inside $M[g_\infty]$. Similarly our use of $\tau_{\text{gen}-\infty}[y_n \mid n < \omega]$ ensures that $\langle y_n \mid n < \omega\rangle$ fails to **not** satisfy ϕ. This is a contradiction.

4.5. Summary. Claim 4.10, Claim 4.11, and the construction of Section 4.4 together give the following theorem:

THEOREM 4.12. *Suppose that there exist M and an increasing sequence $\langle \delta_1, \delta_2, \ldots, \delta_\omega\rangle$ in M so that:*

- *M is a class model;*
- *M is iterable;*
- *Each δ_ξ, $1 \le \xi \le \omega$, is a Woodin cardinal of M; and*
- *$M \| \delta_\omega + 1$ is countable in* V.

Then all games $G_{\omega\cdot\omega}(C)$ where C is Σ^1_2 are determined. ⊣

§5. **Pivots revisited.** In this section we return to our definition of auxiliary moves, and make some adjustments. These adjustments will be needed later on, in Section 6. We begin in Section 5.1 with a minor modification to the games $\mathcal{A}[x]$. We describe the modification and its effect on the notions of generic runs and pivots. Then in Section 5.2 we handle the more serious adjustment. We describe a game $\mathcal{A}_{\text{mix}}$, a variant of $\mathcal{A}_{\text{piv}}$, and use this game to define the notion of a *mixed* pivot. Mixed pivots will be used in the proof of determinacy of continuously coded games.

5.1. Modified auxiliary moves. Work as in Section 2 with a model M which has a Woodin cardinal δ. Fix $\dot{A} \in M$, a name for a subset of $(M\|\delta)^\omega \times \omega^\omega$ in $M^{\text{col}(\omega,\delta)}$. Note already here the change from Section 2, where we had a name for a set of reals, i.e., a subset of ω^ω.

Work with $x \in \mathbb{R}$. We define an auxiliary game $\mathcal{A}[x]$ displayed in Diagram 19. We use a_n to denote $\langle l_n, u_n, p_n, w_n\rangle$, the sequence of moves in round n, and let $\vec{a} = \langle a_n \mid n < \omega\rangle$. Moves in $\mathcal{A}[x]$ are elements of $M\|\delta$, so that $\vec{a}$ belongs to $(M\|\delta)^\omega$. A run $\vec{a}$ of $\mathcal{A}[x]$ is arranged so that I tries to witness that $\langle \vec{a}, x\rangle \in \dot{A}[h]$ for some generic h, while II tries to witness the opposite. Note the change from Section 2, where we dealt with "x" rather than $\langle \vec{a}, x\rangle$.

I	l_0, u_0, p_0		l_1, u_1, p_1		$\ldots$
II		w_0		w_1	$\ldots$

DIAGRAM 19. Outline of $\mathcal{A}[x]$.

Moves in $\mathcal{A}[x]$ are elements of M, and each rule should be read relativized to M. In round n I plays:

- $l = l_n$, a number smaller than n, or $l_n =$ "new";
- a *type* u_n which codes $\mathcal{X}_n$, a set of **pairs** of $M^{\mathrm{col}(\omega,\delta)}$–names; and
- p_n, a condition in $\mathrm{col}(\omega, \delta)$.

II plays a *type* w_n which codes $\mathcal{F}_n, \mathcal{D}_n$ where:

- $\mathcal{F}_n$ is a function from $\mathcal{X}_n$ into the ordinals; and
- $\mathcal{D}_n$ is a function from $\mathcal{X}_n$ into {dense sets in $\mathrm{col}(\omega, \delta)$}.

We remind the reader of Remark 2.1. Already in Section 2 the moves $\mathcal{X}_n$ and $\mathcal{F}_n, \mathcal{D}_n$ were coded by types. This part is not new. We didn't say much about the coding in Section 2, referring the reader to [15, Chapter 1] instead. We adopt the same attitude here. Let us only note that the types u_n and w_n are essentially elements of $M \| \delta$. This is important. It means that $a_n = \langle l_n, u_n, p_n, w_n \rangle$ is an element of $M \| \delta$, so that $\vec{a}$ is an element of $(M \| \delta)^\omega$.

If $l_n =$ "new" we make no requirements on I. Otherwise we demand that p_n extends p_l, that $\mathcal{X}_n \subset \mathcal{X}_l$, and that for every pair $\langle \dot{a}, \dot{x} \rangle \in \mathcal{X}_n$:

1. p_n forces "$\langle \dot{a}, \dot{x} \rangle \in \dot{A}$";
2. p_n forces "$\dot{a}(0) = \check{a}_0$,"....,"$\dot{a}(l) = \check{a}_l$";
3. p_n forces "$\dot{x}(0) = \check{x}_0$,"....,"$\dot{x}(l) = \check{x}_l$"; and
4. p_n belongs to $\mathcal{D}_l(\dot{a}, \dot{x})$.

We make the following demand on II when $l_n \neq$ "new":

5. $\mathcal{F}_n(\dot{a}, \dot{x}) < \mathcal{F}_l(\dot{a}, \dot{x})$ for every pair $\langle \dot{a}, \dot{x} \rangle \in \mathcal{X}_n$.

REMARK 5.1. Note the addition of condition 2, stating that $\dot{a}$ must name the actual run of $\mathcal{A}[x]$, $\vec{a}$. This is the condition which distinguishes our game here from the game in Section 2. Other than this the rules are essentially the same.

Condition 2 makes sense; $\vec{a}$ is an element of $(M \| \delta)^\omega$ and may potentially be named by $\dot{a}$. Observe that condition 2 in round n only involves $a_0, \ldots, a_l$, which are already known. It poses no greater hardship to the players than condition 3. The arguments (not) presented in Section 2 thus go through essentially unmodified. The curious reader can find these arguments in [15, Chapter 1]. Let us briefly go over the results of these arguments.

Fix some g which is $\mathrm{col}(\omega, \delta)$–generic/$M$. As in Section 2.1 we let $\sigma_{\mathrm{gen}}[x]$ be the strategy which plays in each round the first, with respect to g, legal move. The map $x \mapsto \sigma_{\mathrm{gen}}[x]$ is continuous, given by some $\sigma_{\mathrm{gen}} = (s \mapsto \sigma_{\mathrm{gen}}[s])$ which belongs to $M[g]$. We have:

LEMMA 5.2. *Suppose that $\vec{a}$ is an infinite run of $\mathcal{A}[x]$ played according to $\sigma_{\mathrm{gen}}[x]$. Then $\langle \vec{a}, x \rangle \notin \dot{A}[g]$. (This is only useful if both $\vec{a}$ and x belong to $M[g]$.)*

This should be compared with Lemma 2.4. Where now we have $\langle \vec{a}, x\rangle \notin \dot{A}[g]$, Lemma 2.4 had $x \notin \dot{A}[g]$.

Definition 2.7 can be adapted to our new game by changing condition 3 to:

3. For every cofinal odd branch b of $\mathcal{T}$ there exists some h so that:
   (a) h is $\text{col}(\omega, j_b(\delta))$–generic/$M_b$; and
   (b) $\langle \vec{a}, x\rangle \in j_b(\dot{A})[h]$.

(Note the change from x to $\langle \vec{a}, x\rangle$ in condition 3b.)

As in Section 2 there are strategies $\sigma_{\text{piv}}[\varrho, x]$ which are guaranteed to produce pivots. But when proving determinacy of continuously coded games this is not enough. We shall need stronger strategies than those given by σ_{piv}, capable of handling what we call *mixing*.

5.2. Mixed pivots. Instead of working with a single name $\dot{A}$ as before, we work here with a collection of names. Fix some ordinal ν. Fix a map $\dot{A} = (\gamma \mapsto \dot{A}[\gamma])$ assigning to each ordinal $\gamma < \nu$ a name $\dot{A}[\gamma]$ for a subset of $(M\| \delta)^\omega \times \omega^\omega$ in $M^{\text{col}(\omega,\delta)}$. We assume that the map $\dot{A}$ belongs to M.

We shall henceforth suppress mention of ν. When we say "for each γ" we mean for each $\gamma < \nu$. We generally think of ν as some very large ordinal. Indeed, if it weren't for our desire to work with sets rather than classes we would take $\nu = \text{ON}$.

For each γ we have the map $x \mapsto \mathcal{A}[\gamma, x]$ of Section 5.1, associated to the name $\dot{A}[\gamma]$. We regard it now as a map $\gamma, x \mapsto \mathcal{A}[\gamma, x]$. This map, which belongs to M, is continuous in x.

Working with reference to the map $\dot{A}$, we define for each $x \in \mathbb{R}$ the game $\mathcal{A}_{\text{mix}}[x]$ played according to Diagram 20. As usual the association is continuous, given by a map $\mathcal{A}_{\text{mix}} = (s \mapsto \mathcal{A}_{\text{mix}}[s])$ which belongs to M.

I		$f(n), \mathcal{T}\restriction f(n)+1, \gamma_n$	l_n, p_n, u_n		$\cdots$
II	$\cdots\cdots$			$E_{f(n)}, E_{f(n)+1}, w_n$	$\cdots$

DIAGRAM 20. Round n of the game $\mathcal{A}_{\text{mix}}[x]$.

At the start of round n we have a number $e(n)$, an iteration tree $\mathcal{T}\restriction e(n)+1$ ending with the model $M_{e(n)}$, and a position $P_n = \langle a_0, \ldots, a_{n-1}\rangle$ in $M_{e(n)}$. For $n = 0$ we set $e(0) = 0$, $M_0 = M$, $P_0 = \emptyset$.

The time line of round n is presented in Diagrams 20 and 21. At the start of round n player I:

- Plays some $f(n) \geq e(n)$;
- Extends $\mathcal{T}\restriction e(n)+1$ to $\mathcal{T}\restriction f(n)+1$;
- Plays an ordinal γ_n so that P_n is a legal position in $j_{0,f(n)}(\mathcal{A})[\gamma_n, x]$.

The rest of the round follows the usual rules of $\mathcal{A}_{\text{piv}}$ (see Section 2.2): I plays a move in $j_{0,f(n)}(\mathcal{A})[\gamma_n, x]$ following the position P_n; II shifts this move to the

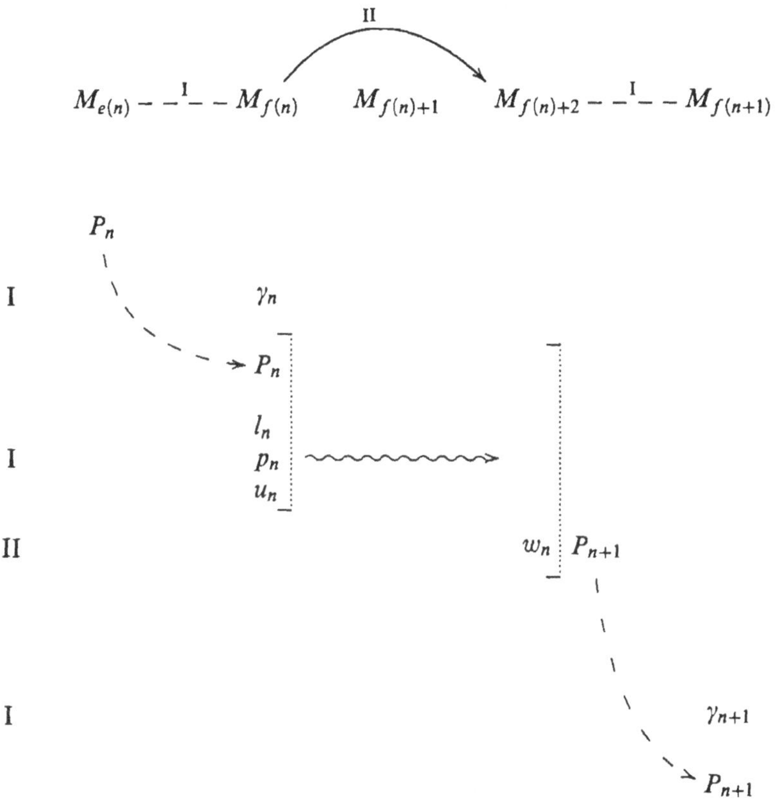

DIAGRAM 21. Round n of $\mathcal{A}_{\text{mix}}[x]$ and the beginning of round $n+1$.

model $M_{f(n)+2}$—this is illustrated by the squiggly arrow in Diagram 21—and replies there. We let $a_n = j_{f(n),f(n)+2}(l_n, u_n, p_n)$—, w_n. Note the shifting of l_n, u_n, p_n from $M_{f(n)}$ to $M_{f(n)+2}$. We let $P_{n+1} = P_n$—, $a_n = \langle a_0, \ldots, a_{n-1}, a_n \rangle$, let $e(n+1) = f(n)+2$, and proceed to the next round.

REMARK 5.3. Suppose I fixes some $\gamma_0 \in M$ and always plays $f(n) = e(n)$ (so that no extension of $\mathcal{T} \restriction e(n)+1$ is needed) and $\gamma_n = j_{0,f(n)}(\gamma_0)$. Then the game degenerates into $\mathcal{A}_{\text{piv}}$ associated to the real x and the name $\dot{A}[\gamma_0]$.

$\mathcal{A}_{\text{mix}}$ is thus a variant of $\mathcal{A}_{\text{piv}}$ which gives some extra control to player I: I *may* play $f(n) > e(n)$, inserting her own interval of models into the tree $\mathcal{T}$, and I *may* pick a new ordinal γ_n to work with.

In line with Remark 5.3 we make the following definition:

Definition 5.4. Round n is said to **contain mixing** if $f(n) > e(n)$; or (when $n > 0$) $f(n) = e(n)$ but $\gamma_n \neq i_{f(n-1),e(n)}(\gamma_{n-1})$.

There are some technical restrictions on the moves by the two players, not explained above. (For example the critical point of $j_{f(n),f(n)+2}$ must be large enough that $a_0, \ldots, a_{n-1}$ are not moved. This is why we take $P_{n+1} = P_n$—$, a_n$, and not $P_{n+1} = j_{f(n),f(n)+2}(P_n)$—$, a_n$.) The reader may find the exact rules in [15, Chapter 1]. Here we only comment on how these rules affect the branch structure of $\mathcal{T}$.

Suppose $\vec{a} = \langle a_n \mid n < \omega \rangle$ and $\mathcal{T}$ are given by a run of $\mathcal{A}_{\text{mix}}[x]$. A cofinal branch of $\mathcal{T}$ is **even** if it contains arbitrarily high nodes from $\{f(n) \mid n < \omega\}$.[5] Otherwise the branch is **odd**. Note that a mixed $\mathcal{T}$ may have many cofinal even branches; this has to do with *not* requiring $e(n)\ T\ f(n)$ in the rules of the game. How about the odd branches? The predecessors of nodes $\{f(n)+1, f(n)+2 \mid n < \omega\}$ are determined by the moves $\{l_n \mid n < \omega\}$ in the manner of Section 2.2 (see the rules following Diagram 9). There are extra rules now on player I limiting the way she may choose predecessors for nodes in $\bigcup_{n<\omega}(e(n), f(n)]$. The main point of these rules is to make sure that the following condition holds:

(o) Suppose b is a cofinal odd branch of $\mathcal{T}$. Then there is a sequence $\langle n_k \mid k < \omega \rangle$ so that:
  - $l_{n_0} =$ "new";
  - $l_{n_k} = n_{k-1}$ for $k > 0$; and
  - $\langle f(n_k) + 1 \mid k < \omega \rangle$ is a tail-end of b.

Thus the tree structure on the odd models is essentially the same structure we had in Section 2.2.

We use $n(b)$ to denote the n_0 given by condition (o). We use $f(b)$ to denote $f(n_0)$. Note that $f(b)$ is the largest node in b which belongs to $\{f(n) \mid n < \omega\}$. We think of $f(b)$ as the **even root** of the odd branch b (though it needn't be an even number, see footnote 5).

Definition 5.5. A **mixed pivot** for x is a run of $\mathcal{A}_{\text{mix}}[x]$ (given by $\vec{a}$, $\vec{f}$, $\vec{\gamma}$, and $\mathcal{T}$ say) with the property that for every cofinal odd branch b of $\mathcal{T}$ there exists some h so that:

1. h is $\text{col}(\omega, j_b(\delta))$–generic/$M_b$; and
2. $\langle \vec{a}, x \rangle \in j_b(\dot{A})[\gamma_b][h]$, where $\gamma_b = j_{f(b),b}(\gamma_{n(b)})$.

The reader should compare the conditions of Definition 5.5 to condition 3 listed immediately following Lemma 5.2. The difference is that here we work not with a single name but with a collection of names. So we have to say

[5] If there is no mixing in $\mathcal{T}$ this gives precisely the branch $0, 2, 4, \ldots$, and hence the terminology. The terminology may be slightly confusing since in general the numbers $f(n)$ needn't actually be even. One can avoid the confusion by adding the requirement "$f(n)$ is even" on I, and canceling the extra hardship by letting player I "pad" in her iteration trees.

which γ to use in condition 2 of Definition 5.5. The γ we take is the one which corresponds to I's move at the even root of b.

Recall that the main point in Section 2.2 was the existence of strategies $\sigma_{\text{piv}}[\varrho, x]$ which produced pivots. Similar strategies exist in our current situation. For each real x and each map $\varrho\colon \omega \to M\|\delta + 1$ there is a strategy $\sigma_{\text{mix}}[\varrho, x]$, playing for II in $\mathcal{A}_{\text{mix}}[x]$. The association is continuous, given by a map $\sigma_{\text{mix}} = (\vartheta, s \mapsto \sigma_{\text{mix}}[\vartheta, s])$. This map belongs to M. Most importantly we have:

LEMMA 5.6. *Suppose $\varrho\colon \omega \to M\|(\delta+1)$ is onto. Then all runs according to $\sigma_{\text{mix}}[\varrho, x]$ are mixed pivots for x.*

In the future we shall use Lemma 5.6 as before we had used Lemma 2.8. Note that Lemma 2.8 is really a special case of our current Lemma 5.6. This follows from Remark 5.3. We refer the reader to [15, Chapter 1] for more details on the construction of σ_{mix}. The construction involves only minor modifications to the construction of σ_{piv}.

§6. **Games of continuously coded length.** Fix a continuous function $\nu\colon \mathbb{R} \to \mathbb{N}$. Fix $C \subset \mathbb{R}^{<\omega_1}$ which is Σ^1_2 in the codes (see Section 1.1). We work to prove, or at least sketch a proof of, the determinacy of $G_{\text{cont}-\nu}(C)$. Our proof will build on the constructions presented in Sections 3 and 4, and will use the notions of Section 5.

Let us say that an extender E **overlaps** δ if dom(E) is smaller than δ, and the ultrapower embedding by E sends dom(E) above δ.

Fix $M, \delta < \delta_\infty$ in M, and an extender $E \in M$ which overlaps δ, so that:

1. M is a class model;
2. M is iterable;
3. δ and δ_∞ are Woodin cardinals of M;
4. $M\|\delta_\infty + 1$ is countable in V; and
5. E is strong enough that $M\|\delta + 1 \subset \text{Ult}(M, E)$.

The existence of such a model is our large cardinal assumption.

Let N denote Ult(M, E), and let $\pi\colon M \to N$ denote the ultrapower embedding. Let δ' denote $\pi(\delta)$. For expository simplicity fix g which is col(ω,δ)–generic/M, and g_∞ which is col(ω,δ_∞)–generic/M.

CLAIM 6.1. *g is also* col(ω,δ)*–generic over N =* Ult(M, E)*. If x is a* real *which belongs to $M[g]$, then x belongs also to $N[g]$.*

PROOF. The proof is immediate. We only note that condition 5 is crucial for the second part. ⊣

Remember that in Section 4 we needed an increasing sequence of Woodin cardinals. The reason was explained in Remark 4.9. Roughly speaking we wanted $y_0, \dots, y_{k-1}$ to belong to a small generic extension relative to the Woodin cardinal used in mega-round k. Here we have the single Woodin cardinal δ, but Claim 6.1 tells us that we can use E to manufacture a "next" Woodin cardinal δ' *and* have the current real x belong to a small generic extension relative to δ'.

6.1. Names. Recall that C, the payoff set, is assumed to be Σ^1_2 in the codes. Fix a Σ^1_2 set $A \subset \mathbb{R} \times \mathbb{R}$ so that $\langle y_\xi \mid \xi \leq \alpha \rangle \in C$ iff $\langle \ulcorner y_\xi \mid \xi < \alpha \urcorner, y_\alpha \rangle \in A$. Fix a Σ^1_2 statement ϕ so that $\langle x, y \rangle \in A$ iff $\langle x, y \rangle$ satisfies ϕ. Recall that $v \colon \mathbb{R} \to \mathbb{N}$ is assumed to be continuous. Fix a function $\bar{v} \colon \omega^{<\omega} \to \mathbb{N}$ so that $v(y) = n$ iff $\exists i\ \bar{v}(y \restriction i) = n$. Without loss of generality $\bar{v}$, which is essentially a real number, belongs to M. (We can always absorb $\bar{v}$ into a generic extension of an iterate of M of size much less than δ.)

Let $\dot{A}_\infty$ be the canonical name for the set of pairs $\langle x, y \rangle \in \mathbb{R}^2$ in $M^{\mathrm{col}(\omega,\delta_\infty)}$ which satisfy the Σ^1_2 statement ϕ. We have the associated auxiliary games $\mathcal{A}_\infty[x, y]$, of the kind presented in Section 2, where I tries to witness $\langle x, y \rangle \in \dot{A}[h_\infty]$ for some generic h_∞ and II tries to witness the opposite.

For each ordinal γ we define a name $\dot{A}[\gamma]$ for a subset of $(M \| \delta)^\omega \times \omega^\omega$ in $M^{\mathrm{col}(\omega,\delta)}$. Following notation similar to that of Section 4.1 we write $\langle \vec{a}, x, \gamma \rangle \in \dot{A}[h]$ to mean that $\langle \vec{a}, x \rangle \in \dot{A}[\gamma][h]$. We use $\mathcal{A}[\gamma, x]$ to denote the auxiliary game of Section 5, associated to the name $\dot{A}[\gamma]$ and the real x. A run $\vec{a}$ of this game is an attempt by I to witness that $\langle \vec{a}, x \rangle \in \dot{A}[\gamma][h]$—in other words that $\langle \vec{a}, x, \gamma \rangle \in \dot{A}[h]$—for some h, and an attempt by II to witness the opposite.

Definition 6.2. A **code** is any real x which has the form $\ulcorner y_\xi \mid \xi < \alpha \urcorner$ for some α and some position $\langle y_\xi \mid \xi < \alpha \rangle$ in $G_{\mathrm{cont}-v}$.

Following the ideas of Section 4.1 we work in $M[g]$ to define open games, denoted here $G^*(\vec{a}, x, \gamma)$. We then set:

Definition 6.3. For $\langle \vec{a}, x \rangle \in (M \| \delta)^\omega \times \omega^\omega$ in $M[g]$ put $\langle \vec{a}, x \rangle \in A[\gamma]$ iff x is a code and I wins $G^*(\vec{a}, x, \gamma)$ in $M[g]$. Let $\dot{A}[\gamma]$ be the canonical name for $A[\gamma]$.

As in Section 4 the definition is by induction on γ. The game $G^*(\vec{a}, x, \gamma)$, which we define shortly, will make reference to the names $\dot{A}[\gamma^*]$—indeed to the map $\gamma^* \mapsto \dot{A}[\gamma^*]$—but only for $\gamma^* < \gamma$.

Fix an ordinal γ, a code $x = \ulcorner y_\xi \mid \xi < \alpha \urcorner$, and a sequence $\vec{a} \in (M \| \delta)^\omega$. Suppose that $\vec{a}$ and x belong to $M[g]$. The game $G(\vec{a}, x, \gamma)$ is played in two parts, parts (F) and (M) described below. (F) stands for "finishing" and (M) stands for "main." In part (M) we use $\mathcal{A}$ to denote the map $\gamma^*, x^* \mapsto \mathcal{A}[\gamma^*, x^*]$, which we assume known for $\gamma^* < \gamma$. The map is continuous in x^* and belongs to M. Recall that $N = \mathrm{Ult}(M, E)$ and $\pi \colon M \to N$ is the ultrapower embedding. We use $\mathcal{A}'$ to denote $\pi(\mathcal{A})$. Similarly we use δ' to denote $\pi(\delta)$ and γ' to denote $\pi(\gamma)$.

(F) I and II collaborate as usual playing a real $y_\alpha = \langle y_\alpha(i) \mid i < \omega \rangle$. In addition they play auxiliary moves subject to the rules of $\mathcal{A}_\infty[x, y_\alpha]$.

The players stay in part (F) until, if ever, $i < \omega$ is reached so that $\bar{v}(y_\alpha \restriction i)$ is defined. If and when this happens we set $n_\alpha = \bar{v}(y_\alpha \restriction i)$. If there exists $\xi < \alpha$ so that $n_\alpha = v(y_\xi)$, the players simply continue with part (F). Otherwise they set $a' = \pi(\vec{a} \restriction n_\alpha)$ and pass to part (M):

(M) 1. I plays γ^* so that $\gamma^* < \gamma'$ and a' is a legal position in $\mathcal{A}'[\gamma^*, x]$.

2. The players collaborate to form the real y_α, continuing from the point they left in part (F).

We set $x^* = \ulcorner \langle y_\xi \mid \xi < \alpha \rangle {—}, y_\alpha \urcorner$. x^* is obtained continuously as y_α is played out. Regardless of the end value of y_α we know by Property 1.3 that x and x^* agree to n_α. Using the continuity of $\mathcal{A}'$ and rule (M1) we see that a' is a legal position in $\mathcal{A}'[\gamma^*, x^*]$, again regardless of the end value of y_α.

3. While forming y_α the players play auxiliary moves subject to the rules of $\mathcal{A}'[\gamma^*, x^*]$, starting from the position a'.

If a player cannot follow these rules she loses. Infinite runs are won by II.

Remark 6.4. As a whole, the game $G^*(\vec{a}, x, \gamma)$ is defined inside $M[g]$. But part (M) can be defined in the smaller model $N[g]$. (The parameters needed to phrase part (M) are x, which is used in defining the continuous $y_\alpha \mapsto x^*$, $\mathcal{A}'$, γ', and a'. The last three parameters belong to N, and x belongs to $N[g]$ by Claim 6.1.)

This completes the inductive definition of $G^*(\vec{a}, x, \gamma)$, and with it the inductive definition of the names $\dot{A}[\gamma]$. We make the following notes on motivation:

$G^*(\vec{a}, x, \gamma)$ consists of two separate parts. So long as it seems that α is the last round of the long game $G_{\text{cont}-v}(C)$—so long as $v(y_\alpha)$ is not defined or defined and equal to a previous n_ξ—the players follow the "finishing" part. What they do in this part is aim for the Σ^1_2 payoff set. I tries to witness that $\langle x, y_\alpha \rangle$ satisfies the Σ^1_2 statement ϕ, while II tries to witness the opposite.

Once (if ever) it becomes clear that α is not the last round in the game, the players pass to the "main" part. What they do is pass to the ultrapower $N = \text{Ult}(M, E)$ where they have the next Woodin cardinal $\pi(\delta)$. They play auxiliary moves in the vicinity of $\pi(\delta)$. We use $\vec{a}^*$ to denote these auxiliary moves. Rule (M3) is such that $\vec{a}^*$ must extend $a' = \pi(\vec{a} \restriction n_\alpha)$.

Note that I's goal in $\vec{a}^*$ is to witness that $\langle \vec{a}^*, x^*, \gamma^* \rangle \in \pi(\dot{A})[h^*]$ for some h^* which is generic over N for the collapse of $\pi(\delta)$. II's goal is to witness the opposite. We draw the reader's attention to the similarity with Section 4. Here too we have a process of *perpetuation*. Membership in $\dot{A}[\gamma][h]$ allows I to aim for membership in a shift of $\pi(\dot{A})[\gamma^*]$. But here we have an additional ingredient. The witness $\vec{a}^*$ agrees with the shift of the witness $\vec{a}$ up to n_α. Using Claim 1.2 this will allow us to argue that the witnesses converge at limit stages.

6.2. I wins. Suppose that there exists some γ so that in M I wins the open game $G(\emptyset, \ulcorner\emptyset\urcorner, \gamma)$. We claim that in this case I wins the long game $G_{\text{cont}-\nu}(C)$ in V.

Fix an imaginary opponent playing for II in $G_{\text{cont}-\nu}(C)$. Working against the imaginary opponent we construct:

(A) Reals $y_\xi \in \mathbb{R}$. We set $x_\alpha = \ulcorner y_\xi \mid \xi < \alpha \urcorner$;
(B) Iterates M_α of M, with embeddings $\tau_{0,\alpha}\colon M \to M_\alpha$;
(C) Mixed pivots $\vec{a}_\alpha, \vec{f}_\alpha, \vec{\gamma}_\alpha, \mathcal{T}_\alpha$ for x_α over the model M_α, played according to $\tau_{0,\alpha}(\sigma_{\text{mix}})[x_\alpha]$; and
(D) Sequences $\vec{\eta}_\alpha = \{\eta_i^\alpha\}_{i<\omega}$ witnessing that $\mathcal{T}_\alpha$ is continuously illfounded on its "even nodes," namely on nodes in $\{f_\alpha(n) \mid n < \omega\}$.

We use $\mathfrak{P}_\alpha$ to denote the mixed pivot of (C). In (D) we mean that $j^\alpha_{k,l}(\eta^\alpha_k) > \eta^\alpha_l$ whenever k, l both belong to $\{f_\alpha(n) \mid n < \omega\}$ and $k\ T_\alpha\ l$. ($j^\alpha_{*,*}$ here are the iteration embeddings forming part of the tree $\mathcal{T}_\alpha$.) The existence of a sequence $\vec{\eta}_\alpha$ of this kind implies that all the cofinal even branches of $\mathcal{T}_\alpha$ lead to illfounded direct limits, forcing the iteration strategy to pick an odd branch.

The construction of the objects (A)–(D) is similar to the previous constructions in Section 4.2 and in case 1 of Section 3. We shall not present it in great detail. Instead we concentrate on the two points which are new. We explain how to carry the construction through limits, and how and why *mixed* pivots appear in the construction.

Let us first consider the matter of limit stages. Fix a limit ordinal λ, and suppose that all objects up to λ were constructed. This includes the models M_ξ and reals y_ξ for $\xi < \lambda$. Let M_λ be the direct limit of the models M_ξ, $\xi < \lambda$. Let $x_\lambda = \ulcorner y_\xi \mid \xi < \lambda \urcorner$. Our construction below λ will satisfy the following *agreement* condition, which traces to the inclusion of a' in rule (M3) above.

(i) $\mathfrak{P}_{\alpha+1}$ agrees with the shifted image of $\mathfrak{P}_\alpha$ up to n_α, and similarly for the sequence $\vec{\eta}_{\alpha+1}$. To be more precise:

$$\vec{a}_{\alpha+1} \restriction n_\alpha = \tau_{\alpha,\alpha+1}(\vec{a}_\alpha \restriction n_\alpha);$$
$$\vec{f}_{\alpha+1} \restriction n_\alpha = \vec{f}_\alpha \restriction n_\alpha;$$
$$\vec{\gamma}_{\alpha+1} \restriction n_\alpha = \tau_{\alpha,\alpha+1}(\vec{\gamma}_\alpha \restriction n_\alpha);$$
$$\mathcal{T}_{\alpha+1} \restriction e_\alpha(n_\alpha) + 1 = \tau_{\alpha,\alpha+1}(\mathcal{T}_\alpha \restriction e_\alpha(n_\alpha) + 1); \text{ and}$$
$$\vec{\eta}_{\alpha+1} \restriction e_\alpha(n_\alpha) + 1 = \tau_{\alpha,\alpha+1}(\vec{\eta}_\alpha \restriction e_\alpha(n_\alpha) + 1).$$

($e_\alpha(n_\alpha)$ is 0 if $n_\alpha = 0$, and $f(n_\alpha - 1) + 2$ otherwise. See Section 5.2.)

It is this agreement condition that carries us through the limit. By Claim 1.2 $n_\alpha \to \infty$ as $\alpha \to \lambda$. This, (i), and our pending definition at limit stages imply that the mixed pivots $\tau_{\alpha,\lambda}(\mathfrak{P}_\alpha)$ *converge* as $\alpha \to \lambda$. We let $\mathfrak{P}_\lambda$ be their limit. By Remark 1.4 the reals x_α converge to x_λ as $\alpha \to \lambda$. Each $\tau_{\alpha,\lambda}(\mathfrak{P}_\alpha)$ is a play according to $\tau_{0,\lambda}(\sigma_{\text{mix}})[x_\alpha]$, because of (C). The plays $\tau_{\alpha,\lambda}(\mathfrak{P}_\alpha)$ must thus converge to a play according to $\tau_{0,\lambda}(\sigma_{\text{mix}})[x_\lambda]$. In other words $\mathfrak{P}_\lambda$ is a play according to $\tau_{0,\lambda}(\sigma_{\text{mix}})[x_\lambda]$. In particular $\mathfrak{P}_\lambda$ is a mixed pivot for x_λ over M_λ.

A similar limit construction allows us to define $\vec{\eta}_\lambda$, and argue that (D) is satisfied. This completes the construction at the limit stage λ.

In sum, several factors combine to carry us through limit stages. One is the convergence given by Remark 1.4. Another is the continuity of all the different maps we defined in Sections 2 and 5. A third is the agreement between $\mathfrak{P}_{\alpha+1}$ and $\tau_{\alpha,\alpha+1}(\mathfrak{P}_\alpha)$.

REMARK 6.5. The reader should compare the formation of $\mathfrak{P}_\lambda$ to the formation of $\mathcal{U}_\infty$ in Section 4.2. $\mathcal{U}_\infty$ was formed in parts spread over previous stages of the construction. Each stage contributed an extra round to the formation. $\mathfrak{P}_\lambda$ too is formed in parts spread over previous stages of the construction. But now the exact contribution of each stage is not set in advance. It depends on the behavior of the n_α-s, which in turn depends on the players. This extra flexibility in setting the break lines in the formation of limit pivots is the key to handling games of *variable* length.

Let us now consider the successor stage. We have the model M_α; the mixed pivot $\mathfrak{P}_\alpha$ of (C); and the ordinal sequence $\vec{\eta}_\alpha$ of (D). Our goal is to construct $M_{\alpha+1}$; the real y_α, which gives rise to the code $x_{\alpha+1}$; the mixed pivot $\mathfrak{P}_{\alpha+1}$; and the sequence $\vec{\eta}_{\alpha+1}$.

To start we use the iteration strategy to pick a cofinal branch b_α through $\mathcal{T}_\alpha$. We let Q_α denote the direct limit along b_α. The sequence $\vec{\eta}_\alpha$ of (D) forces the iteration strategy to pick an **odd** branch. We have the models presented in Diagram 22.

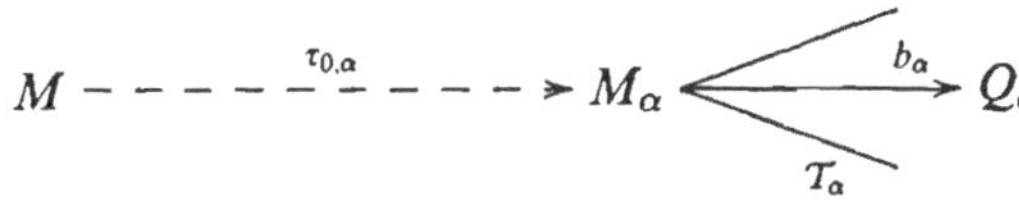

DIAGRAM 22. At the start of round α.

For simplicity assume that $\mathfrak{P}_\alpha$ does not contain any mixing (see Remark 5.3 and Definition 5.4). So $f_\alpha(n) = 2n$ and there is some single y^α so that $y_n^\alpha = j_{0,2n}^\alpha(y^\alpha)$ for all n. Since b_α is an odd branch we know that there is some h_α so that:

1. h_α is $\text{col}(\omega, (j_{b_\alpha} \circ \tau_{0,\alpha})(\delta))$–generic/$Q_\alpha$; and
2. $\langle \vec{a}_\alpha, x_\alpha\rangle \in (j_{b_\alpha} \circ \tau_{0,\alpha})(\dot{A})[j_{b_\alpha}(y^\alpha)][h_\alpha]$.

Using condition 2 and Definition 6.3 we get

3. In $Q_\alpha[h_\alpha]$, I wins the game $(j_{b_\alpha} \circ \tau_{0,\alpha})(G^*)(\vec{a}_\alpha, x_\alpha, j_{b_\alpha}(y^\alpha))$.

Fix $\sigma_\alpha^* \in Q_\alpha[h_\alpha]$ witnessing condition 3. Let us use G_α^* to denote the game $(j_{b_\alpha} \circ \tau_{0,\alpha})(G^*)(\vec{a}_\alpha, x_\alpha, j_{b_\alpha}(y^\alpha))$ of condition 3.

We divide now into two cases. Suppose first that in playing G_α^* we stay within part (F)—the "finishing" part. In this case we are essentially playing the game G^* of Section 3. Our construction in this case is similar to the construction in case 1 of Section 3. We use σ_α^* together with the appropriate image of $\sigma_{\text{piv}-\infty}$ to play against the imaginary opponent. The construction produces the real y_α, and makes sure that $\langle x_\alpha, y_\alpha \rangle$ satisfies the Σ_2^1 statement ϕ. The fact that we stayed within part (F) tells us that α is the last round in our run of $G_{\text{cont}-\nu}(C)$. The fact that $\langle x_\alpha, y_\alpha \rangle$ satisfies ϕ tells us that $\langle y_\xi \mid \xi \leq \alpha \rangle \in C$. $\langle y_\xi \mid \xi \leq \alpha \rangle$ is thus won by I, and our task for this subsection has been achieved.

So suppose that while playing G_α^* we enter part (M)—the "main" part. Let P denote our position when entering part (M). P determines $y_\alpha \restriction i$ for some i, and $y_\alpha \restriction i$ suffices to determine n_α. Let E_α denote $(j_{b_\alpha} \circ \tau_{0,\alpha})(E)$, where E is our original extender fixed at the beginning of Section 6. Let $N_\alpha = \text{Ult}(Q_\alpha, E_\alpha)$ and let π_α be the ultrapower embedding. This is presented in the lower line of Diagram 23. Let $a'_\alpha = \pi_\alpha(\vec{a}_\alpha \restriction n_\alpha)$. Note how we follow the definitions listed just before the rules of part (M). Let G_α^{**} be the game obtained from G_α^* by starting from the position P. G_α^{**} is played according to the rules of part (M).

REMARK 6.6. G_α^{**} exists in $N_\alpha[h_\alpha]$. This follows from Remark 6.4, which in turn traces back to the strength of E assumed in condition 5 at the start of Section 6.

Note that G_α^{**} is an open game. Note further that G_α^{**}, being a tail-end of G_α^* played from a position according to σ_α^*, is won by I. Using Remark 6.6 we may fix a winning strategy σ_α^{**} for player I so that:

(♯) σ_α^{**} belongs to $N_\alpha[h_\alpha]$.

We wish to use σ_α^{**} in much the same way we had used similar strategies in the past, combining it with moves given by some σ_{piv} or in our case σ_{mix}. Our problem is this: The starting position in G_α^{**} already includes auxiliary moves, the moves in a'_α. But we do not know that these auxiliary moves correspond to any starting position in the formation of a pivot. One particular aspect of our problem is the following: Auxiliary moves which correspond to pivots have some odd models around them. a'_α belongs to N_α and we have no odd models around N_α. N_α was not created as part of an iteration tree, it is simply the ultrapower of Q_α by E_α.

To solve this problem we try to look at N_α from a different perspective. Let $M_{\alpha+1}$ be the ultrapower by E_α of the model M_α rather than Q_α. Let $\tau_{\alpha,\alpha+1}\colon M_\alpha \to M_{\alpha+1} = \text{Ult}(M_\alpha, E_\alpha)$ be the ultrapower map. This is presented in the upper left part of Diagram 23.

REMARK 6.7. To form this ultrapower we need some agreement between M_α and Q_α. Now T_α is part of a pivot corresponding to $\delta_\alpha = \tau_{0,\alpha}(\delta)$. We can arrange that the critical points in T_α are larger than any pre-specified λ

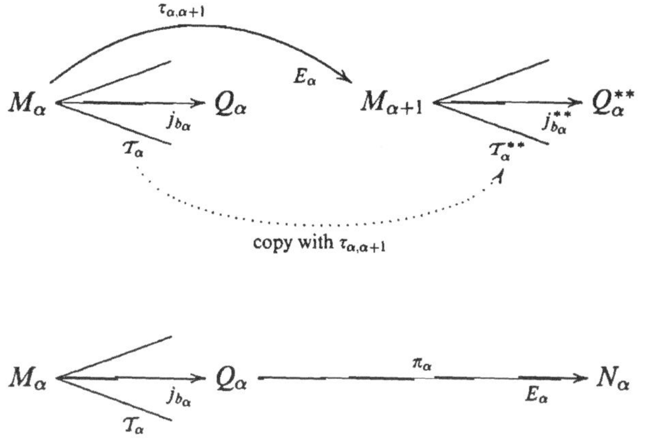

DIAGRAM 23. E_α applied to Q_α (lower line); and E_α applied to M_α (upper line) followed by copying.

below δ_α (see Remark 2.5). We take $\lambda = \tau_{0,\alpha}(\text{dom}(E))$. This ensures that all critical points in $\mathcal{T}_\alpha$ are above $\text{dom}(E_\alpha)$, and so M_α and Q_α are in sufficient agreement that $E_\alpha \in Q_\alpha$ can be applied to M_α.

Use $\tau_{\alpha,\alpha+1}$ to copy $\mathcal{T}_\alpha$, a tree on M_α, to a tree on $M_{\alpha+1}$. Let $\mathcal{T}_\alpha^{**}$ denote the copied tree. Let $\vec{a}_\alpha^{**}$ denote the result of copying $\vec{a}_\alpha$, which is formed in models of $\mathcal{T}_\alpha$, to the models of $\mathcal{T}_\alpha^{**}$. While we are at it, let $\mathfrak{P}_\alpha^{**}$ be the result of copying the entire pivot $\mathfrak{P}_\alpha$ via $\tau_{\alpha,\alpha+1}$. Let Q_α^{**} be the direct limit of the models of $\mathcal{T}_\alpha^{**}$ along b_α, and let $j_{b_\alpha}^{**}$ be the direct limit embedding. These copies of Q_α and j_{b_α} are presented in the upper right part of Diagram 23.

FACT 6.8. Q_α^{**} equals N_α. Moreover $\vec{a}_\alpha^{**}$ equals $\pi_\alpha(\vec{a}_\alpha)$ and $\pi_\alpha \circ j_{b_\alpha} = j_{b_\alpha}^{**} \circ \tau_{\alpha,\alpha+1}$.

REMARK 6.9. Fact 6.8 assumes some closure conditions on the extenders used in $\mathcal{T}_\alpha$. One can build these closure conditions into the construction of σ_{mix}. Alternatively one can use a weaker version of Fact 6.8 which holds in general. We refer the reader to [15, Chapter 3] for details.

Fact 6.8 is the answer to our problem. It tells us that $\pi_\alpha(\vec{a}_\alpha)$ does correspond to a pivot, the pivot $\mathfrak{P}_\alpha^{**}$. It follows that $a'_\alpha = \pi_\alpha(\vec{a}_\alpha \restriction n_\alpha)$ corresponds to $\mathfrak{P}_\alpha^{**} \restriction n_\alpha$.

Let $\mathfrak{P}_\alpha^{***}$ denote $\mathfrak{P}_\alpha^{**} \restriction n_\alpha$. This includes $\mathcal{T}_\alpha^{***} = \mathcal{T}_\alpha^{**} \restriction 2n_\alpha + 1$ and $a'_\alpha = \vec{a}_\alpha^{**} \restriction n_\alpha$. $\mathfrak{P}_\alpha^{***}$ represents a position of n_α rounds in $\tau_{0,\alpha+1}(\mathcal{A}_{\text{mix}})[x_\alpha]$, played according to $\tau_{0,\alpha+1}(\sigma_{\text{mix}})[x_\alpha]$. Since x_α and $x_{\alpha+1}$ agree to n_α (regardless of the end value of y_α) $\mathfrak{P}_\alpha^{***}$ is also a position in $\tau_{0,\alpha+1}(\mathcal{A}_{\text{mix}})[x_{\alpha+1}]$, played according to $\tau_{0,\alpha+1}(\sigma_{\text{mix}})[x_{\alpha+1}]$.

$\mathfrak{P}^{***}_\alpha$ will be our starting position when using $\tau_{0,\alpha+1}(\sigma_{\text{mix}})[x_{\alpha+1}]$ for the construction of $\mathfrak{P}_{\alpha+1}$.

REMARK 6.10. Note that starting the construction of $\mathfrak{P}_{\alpha+1}$ from $\mathfrak{P}^{***}_\alpha$—a restriction of $\tau_{\alpha,\alpha+1}(\mathfrak{P}_\alpha)$—has the pleasant side effect of securing the agreement condition, condition (i) above, which was used at limit stages.

Fix some $k < \omega$ which belongs to the odd branch b_α, is larger than $2n_\alpha$, and is large enough that $\vec{a}^{**}_\alpha \restriction n_\alpha$ has a pre-image in Q^{**}_k. (We use $Q^{**}_{...}$ to denote the models of $\mathcal{T}^{**}_\alpha$.) Let $\vec{a}^{***}_\alpha$ be this pre-image. Pick k large enough that G^{**}_α and σ^{**}_α, which belong to $N_\alpha[h_\alpha] = Q^{**}_\alpha[h_\alpha]$, have pre-images in $Q^{**}_k[h_\alpha]$. Let G^{***}_α and σ^{***}_α be these pre-images. Note our use here of Remark 6.6 and the condition (♯) following it.

G^{***}_α is an open game played according to rules (M1)–(M3), from the starting position $\vec{a}^{***}_\alpha$. σ^{***}_α is a winning strategy for I in this game.

From this point onward we continue along the lines of past constructions. We combine σ^{***}_α, $\tau_{0,\alpha+1}(\sigma_{\text{mix}})[x_{\alpha+1}]$, and the imaginary opponent to create y_α and $\mathfrak{P}_{\alpha+1}$. There is one difference though. We don't start the construction from zero. We start it from $\mathfrak{P}^{***}_\alpha$ which already contains n_α rounds according to $\tau_{0,\alpha+1}(\sigma_{\text{mix}})[x_{\alpha+1}]$.

The construction starts at round n_α. Let us go over this round. σ^{***}_α, in accordance with rule (M1), plays an ordinal γ^*. We have by that rule:

(†) $\gamma^* < (j^{**}_{0,k} \circ \tau_{\alpha,\alpha+1})(\gamma^\alpha)$; and
(‡) $\vec{a}^{***}_\alpha$ is a position in the auxiliary game $(j^{**}_{0,k} \circ \tau_{0,\alpha+1})(\mathcal{A})[\gamma^*, x_{\alpha+1}]$.

We now play round n_α of the mixed game $\tau_{0,\alpha+1}(\mathcal{A}_{\text{mix}})[x_{\alpha+1}]$ (see Diagram 21), continuing from the position given by $\mathfrak{P}^{***}_\alpha$. We play for I, and we intend to *mix*.

To begin, we play $f_{\alpha+1}(n_\alpha) = k$ and $\mathcal{T}_{\alpha+1} \restriction k+1 = \mathcal{T}^{**}_\alpha \restriction k+1$. Note that here already we have mixing, since k is larger than $2n_\alpha$. Next we play $\gamma^{\alpha+1}_{n_\alpha} = \gamma^*$. This is a legal move because of (‡).

The rest of the construction follows precisely the lines of case 1 in Section 3, except that the starting point is the model Q^{**}_k. σ^{***}_α and its shifts provide moves for I, $\tau_{0,\alpha+1}(\sigma_{\text{mix}})$ provides auxiliary moves for II, and the imaginary opponent provides natural number moves for II. These characters combined produce $\mathfrak{P}_{\alpha+1}$. We omit further details, and only point out that in shifting σ^{***}_α we use the fact that it belongs to $Q^{**}_k[h_\alpha]$, and the fact that $Q^{**}_k[h_\alpha]$ is a small extension relative to $\tau_{0,\alpha+1}(\text{dom}(E))$—hence relative to the critical points used in $\mathcal{T}_{\alpha+1}$, see Remark 6.7. The first fact traces back to condition (♯) above, which in turn traces back to condition 5 at the start of this section. The second fact traces back to our initial assumption that E overlaps δ.

REMARK 6.11. At the start of the construction we made the simplifying assumption that $\mathfrak{P}_\alpha$ does not contain any mixing. Still, we ended with $\mathfrak{P}_{\alpha+1}$

which does contain mixing. Mixed pivots are therefore an essential part of the construction.

We point out that $\mathfrak{P}_{\alpha+1}$ contains mixing in round n_α, but does not contain mixing in any round above n_α. (Rounds below n_α depend on $\mathfrak{P}_\alpha \restriction n_\alpha$, which in general may contain mixing.) This is a general pattern at successor stages.

The case of a limit ordinal λ is different. $\mathfrak{P}_\lambda$ is the limit of the mixed pivots $\tau_{\alpha,\lambda}(\mathfrak{P}_\alpha)$, and can contain mixing in cofinally many rounds.

Finally, note that every time a mixing is initiated, some "smaller ordinal" is produced by (†) above. (Without the simplifying assumption that $\mathfrak{P}_\alpha$ does not contain mixing, the statement of (†) becomes more involved. $(j^{**}_{0,k} \circ \tau_{\alpha,\alpha+1})(\gamma^\alpha)$ is replaced by the pre-image to Q^{**}_k of the ordinal corresponding to the even root of b_α, see condition 2 of Definition 5.5.) These ordinals are used to create the sequences $\vec{\eta}_\alpha$ of (D), ensuring the agreement in the last item of (i) so that (D) holds at limits.

REMARK 6.12. The base case of $\alpha = 0$ is similar to the successor case; our initial assumption, that I wins $G^*(\emptyset, \ulcorner\emptyset\urcorner, \gamma)$, is similar to condition 3 above and the construction starts from there. We leave this case to the reader.

6.3. Closing arguments. So far we defined the games $G^*(\vec{a}, x, \gamma)$, open games played in $M[g]$. We showed that if there exists γ so that I wins $G^*(\emptyset, \ulcorner\emptyset\urcorner, \gamma)$ then I wins $G_{\text{cont}-\nu}(C)$ in V. This work is analogous to the developments of Sections 4.1 and 4.2. To complete the proof of determinacy we must:

1. Define the mirror image games $H^*(\vec{b}, x, \gamma)$;
2. Show that if there exists γ so that II wins $H^*(\emptyset, \ulcorner\emptyset\urcorner, \gamma)$ then II wins $G_{\text{cont}-\nu}(C)$ in V; and
3. Derive a contradiction from the assumption that for all ordinals γ, II wins $G^*(\emptyset, \ulcorner\emptyset\urcorner, \gamma)$ and I wins $H^*(\emptyset, \ulcorner\emptyset\urcorner, \gamma)$.

The first two tasks are routine. Task 3 is an analogue of our work in Section 4.4. Working with σ_{gen} and τ_{gen} we construct a run of $G_{\text{cont}-\nu}$ in $M[g_\infty]$ which fails to satisfy ϕ, and fails to satisfy $\neg\phi$. The argument is an adaptation of the one in Section 4.4, but the adaptation is not entirely straightforward; some additional work is necessary. The precise details can be found in [15, Chapter 3]. Once task 3 is completed we get:

THEOREM 6.13. *Suppose that there exist M, $\delta < \delta_\infty \in M$, and $E \in M$ overlapping δ, which satisfy conditions 1–5 listed at the beginning of Section 6. Then all games $G_{\text{cont}-\nu}(C)$ where ν is continuous and C is Σ^1_2 in the codes are determined.* ⊣

6.4. Summary. We end with several observations about the proof of Theorem 6.13. Two of these observations show how the theorem can be improved somewhat.

In some sense our construction is a method for converting an iteration strategy into a winning strategy for I in $G_{\text{cont}-\nu}(C)$. (The mirror image construction of task 2 converts an iteration strategy into a winning strategy for II.) Note that the iteration trees in Section 6.2 are of the kind presented in Diagram 7, the "second" kind. The iteration strategy we use during the construction must therefore apply to the second kind iteration game. In contrast, Section 4 only used games of the first kind.

Next we note that the large cardinal assumption in Theorem 6.13 can be weakened without forcing great change to the proof. Suppose there exist M and $\delta < \delta_\infty$ which satisfy conditions 1–4 listed at the beginning Section 6 and satisfy the following weakened version of condition 5:

w5. For every $X \in M \| \delta + 1$ there exists an extender E in M overlapping δ and strong enough that $X \in \text{Ult}(M, E)$.

(In the original condition 5 one extender E worked for all $X \in M \| \delta + 1$.)

The proof of Theorem 6.13 can be repeated, almost verbatim, under this weaker assumption. Our main use of condition 5 was in Remark 6.6 and the condition (♯) which followed it. This use traced back to condition 5 through Claim 6.1; we needed to know that the real $x_\alpha \in Q_\alpha[h_\alpha]$ belonged also to $\text{Ult}(Q_\alpha, E_\alpha)[h_\alpha]$. Given a real $x_\alpha \in Q_\alpha[h_\alpha]$, the weak condition 5 can also be used to find an extender E_α so that $x_\alpha \in \text{Ult}(Q_\alpha, E_\alpha)[h_\alpha]$. So we can adjust the construction to only use the weak condition 5. (Note that with the weak condition we cannot expect a single extender to handle all reals. Thus we cannot at the outset fix $E \in M$ and always let $E_\alpha = \tau_{0,\alpha}(E)$. Instead we must let the extenders vary.)

Theorem 6.13 applies to games $G_{\text{cont}-\nu}(C)$ where ν is continuous, i.e., $\mathbf{\Sigma}^0_1$ measurable. Our final note is that the theorem can be strengthened to apply to ν which are $\mathbf{\Sigma}^0_2$ measurable.

Fix $\nu\colon \mathbb{R} \to \mathbb{N}$ which is $\mathbf{\Sigma}^0_2$ measurable. For each $n \in \mathbb{N}$ the pre-image $\nu^{-1}\{n\}$ is $\mathbf{\Sigma}^0_2$. Let C include all the closed sets which participate in the unions defining the sets $\nu^{-1}\{n\}$, $n < \omega$. Without loss of generality the real parameter which defines ν belongs to M. Working in $M[g]$ and using the unraveling techniques of Martin [7], find a covering (R, π, φ) of $\omega^{<\omega}$ which unravels each of the sets in C. Moves in the game on R are subsets of ω^ω in $M[g]$. For each $n \in \mathbb{N}$, the pre-image $(\pi^{-1} \circ \nu^{-1})\{n\}$ is *open*. Revise the rule of part (F) in Section 6.1 so that instead of forming $y_\alpha = \langle y_\alpha(i) \mid i < \omega\rangle$ by directly playing on $\omega^{<\omega}$, the players play on R. Part (F) continues until, if ever, the players enter one of the sets $(\pi^{-1} \circ \nu^{-1})\{n\}$, $n < \omega$. Note that the revision makes sense because these sets are open. If the players enter $(\pi^{-1} \circ \nu^{-1})\{n\}$ we set $n_\alpha = n$ and, if n_α is new, pass to part (M). The rules of part (M) are as before, except for rule (M2). What does it mean now to form y_α "continuing from the point" left in part (F)? The moves in part (F) give us some position in R to continue from. A position in R includes some initial segment $y_\alpha \restriction i$

of y_α, and some **commitment** T; T is a subtree of $\omega^{<\omega}$ and both players are committed to staying inside T. Revise rule (M2) to say that I and II play on $\omega^{<\omega}$ continuing from $y_\alpha \restriction i$ and must stay inside the tree T.

These revisions to parts (F) and (M) redefine the games $G^*(\vec{a}, x, \gamma)$. Using the techniques of Martin [7] one can adapt the construction of Section 6.2 to the new games, and complete the determinacy proof.

REMARK 6.14. In adapting the construction of Section 6.2 we must take care to preserve Remark 6.6 and the subsequent condition ($\sharp$). Tracing back we must preserve Remark 6.4. Let us check that Remark 6.4 applies to the revised part (M). The revised part (M) is defined from the parameters listed in Remark 6.4, plus the additional parameter T. T, a commitment in the covering R, is a subset of $\omega^{<\omega}$ in $M[g]$. The strength given by condition 5 is enough to make sure that it belongs to $N[g]$, as required.

Note that we have here a limitation on the size of moves permitted in R. This in turn limits the complexity of functions v which we can handle. To handle functions in pointclasses above Σ_2^0 we would need stronger agreement between $M[g]$ and $N[g]$ than the one given by condition 5.

Combining the observations above we get the following strengthening of Theorem 6.13:

THEOREM 6.15. *Suppose that there exist M and $\delta < \delta_\infty \in M$ which satisfy the following conditions*:

1. M *is a class model*;
2. M *is iterable*;
3. δ *and δ_∞ are Woodin cardinals of M*;
4. $M \| \delta_\infty + 1$ *is countable in* V; *and*
w5. *For every $X \in M \| \delta + 1$ there exists an extender E in M overlapping δ and strong enough that $X \in \mathrm{Ult}(M, E)$.*

Then all games $G_{\mathrm{cont}-v}(C)$ where v is Σ_2^0 measurable and C is Σ_2^1 in the codes are determined. ⊣

REFERENCES

[1] ANDREAS BLASS, *Equivalence of two strong forms of determinacy*, **Proceedings of the American Mathematical Society**, vol. 52 (1975), pp. 373–376.

[2] DAVID GALE and FRANK M. STEWART, *Infinite games with perfect information*, **Contributions to the theory of games**, vol. 2, Annals of Mathematics Studies, no. 28, Princeton University Press, 1953, pp. 245–266.

[3] AKIHIRO KANAMORI, **The higher infinite**, Perspectives in Mathematical Logic, Springer-Verlag, Berlin, 1994.

[4] DONALD A. MARTIN, *Measurable cardinals and analytic games*, **Fundamenta Mathematicae**, vol. 66 (1970), pp. 287–291.

[5] ———, *Borel determinacy*, **Annals of Mathematics**, vol. 102 (1975), pp. 363–371.

[6] ———, *The real game quantifier propagates scales*, ***Cabal seminar 79–81***, Lecture Notes in Mathematics, vol. 1019, Springer-Verlag, 1983, pp. 157–171.

[7] ———, *A purely inductive proof of Borel determinacy*, ***Proceedings of Symposia in Pure Mathematics***, vol. 42 (1985), pp. 303–308.

[8] Donald A. Martin and John Steel, *A proof of projective determinacy*, ***Journal of the American Mathematical Society***, vol. 2 (1989), no. 1, pp. 71–125.

[9] ———, *Iteration trees*, ***Journal of the American Mathematical Society***, vol. 7 (1994), no. 1, pp. 1–73.

[10] Yiannis Moschovakis, ***Descriptive set theory***, Studies in Logic and the Foundations of Mathematics, vol. 100, North Holland Publishing Company, Amsterdam-New York, 1980.

[11] ———, *Scales on coinductive sets*, ***Cabal seminar 79–81***, Lecture Notes in Mathematics, vol. 1019, Springer-Verlag, 1983, pp. 77–85.

[12] Itay Neeman, *Optimal proofs of determinacy*, ***The Bulletin of Symbolic Logic***, vol. 1 (1995), no. 3, pp. 327–339.

[13] ———, *Inner models in the region of a Woodin limit of Woodin cardinals*, ***Annals of Pure and Applied Logic***, vol. 116 (2002), pp. 67–155.

[14] ———, *Optimal proofs of determinacy II*, ***Journal of Mathematical Logic***, vol. 2 (2002), no. 2, pp. 227–260.

[15] ———, ***Long games***, in preparation.

[16] John Steel, *Long games*, ***Cabal seminar 81–85***, Lecture Notes in Mathematics, vol. 1333, Springer-Verlag, 1988, pp. 56–97.

[17] W. Hugh Woodin, unpublished work.

DEPARTMENT OF MATHEMATICS
UNIVERSITY OF CALIFORNIA LOS ANGELES
LOS ANGELES, CA 90095-1555, USA
E-mail: ineeman@math.ucla.edu

ARTICLES

MODIFIED BAR RECURSION AND CLASSICAL DEPENDENT CHOICE

ULRICH BERGER AND PAULO OLIVA

Abstract. We introduce a variant of Spector's bar recursion in finite types (which we call "modified bar recursion") to give a realizability interpretation of the classical axiom of dependent choice allowing for the extraction of witnesses from proofs of $\forall\exists$-formulas in classical analysis. As another application, we show that the fan functional can be defined by modified bar recursion together with a version of bar recursion due to Kohlenbach. We also show that the type structure $\mathcal{M}$ of strongly majorizable functionals is a model for modified bar recursion.

§1. Introduction. In [22], Spector extended Gödel's Dialectica Interpretation of Peano Arithmetic [10] to classical analysis using bar recursion in finite types. Although considered questionable from an intuitionistic point of view ([1], 6.6), there has been considerable interest in bar recursion, and several variants of this definition scheme and their interrelations have been studied by, e.g., Schwichtenberg [19], Bezem [8] and Kohlenbach [14]. In this paper we add another variant of bar recursion and use it to give a realizability interpretation of the negatively translated axiom of dependent choice that can be used to extract witnesses from proofs of $\forall\exists$-formulas in full classical analysis. Our interpretation is inspired by a paper by Berardi, Bezem and Coquand [2] who use a similar kind of recursion in order to interpret dependent choice. The main difference to our paper is that in [2] a rather ad-hoc infinitary term calculus and a non-standard notion of realizability are used whereas we work with a straightforward combination of negative translation, A-translation, modified realizability, and Plotkin's adequacy result for the partial continuous functional semantics of PCF [18].

As a second application of bar recursion, we show that the definition of the fan functional within PCF given in [3] and [17] can be derived from Kohlenbach's and our variant of bar recursion. Furthermore, we prove that our version of bar recursion exists in the model of majorizable functions. The relation between modified bar recursion and Spector's original definition is established in [5].

Funded by the British EPSRC, Swansea, United Kingdom

Funded by the Danish National Research Foundation Aarhus C, Denmark

Logic Colloquium '01
Edited by M. Baaz, S. Friedman, and J. Krajíček
Lecture Notes in Logic, 20

§2. Bar recursion in finite types. We work in a suitable extension of Heyting Arithmetic in finite types, HA^ω, with equality in all types. For convenience, we enrich the type system by the formation of finite sequences. So, our *Types* are $\mathbb{N}$, function types $\rho \to \sigma$, product types $\rho \times \sigma$, and finite sequences ρ^*. We set $\rho^\omega :\equiv \mathbb{N} \to \rho$. The *level* of a type is defined by $\mathsf{level}(\mathbb{N}) = 0$, $\mathsf{level}(\rho \times \sigma) = \max(\mathsf{level}(\rho), \mathsf{level}(\sigma))$, $\mathsf{level}(\rho^*) = \mathsf{level}(\rho)$, $\mathsf{level}(\rho \to \sigma) = \max(\mathsf{level}(\rho) + 1, \mathsf{level}(\sigma))$. By o we will denote an arbitrary but fixed type of level 0, and by ρ, τ, σ arbitrary types. The terms of our version of HA^ω are a suitable extension of the terms of Gödel's system T [10] in lambda calculus notation. We use the variables $i, j, k, l, m, n\colon \mathbb{N}$ and $s, t\colon \rho^*$; $\alpha, \beta\colon \rho^\omega$, where ρ is an arbitrary type. Other letters will be used for different types in different contexts. By $\stackrel{\tau}{=}$ we denote equality of type τ for which we assume the usual equality axioms. However, equality between functions is *not* assumed to be extensional. We also do *not* assume decidability for $\stackrel{\tau}{=}$, when $\mathsf{level}(\tau) > 0$ (if $\mathsf{level}(\tau) = 0$ one can, of course, *prove* decidability). Type information will be frequently omitted when it is irrelevant or inferable from the context. We let k^ρ denote the canonical lifting of a number $k \in \mathbb{N}$ to type ρ, e.g., $k^{\rho\to\sigma} :\equiv \lambda x^\rho.k^\sigma$. By an $\exists$*-formula* respectively $\forall\exists$*-formula* we mean a formula of the form $\exists y^\tau B$ respectively $\forall z^\sigma \exists y^\tau B$, where B is provably equivalent to an atomic formula. We will also use the following notations:

$$
\begin{aligned}
\langle x_0, \dots, x_{n-1}\rangle &:\equiv \text{the finite sequence with elements } x_0, \dots, x_{n-1},\\
|s| &:\equiv \text{the length of } s \text{, i.e., } |\langle x_0, \dots, x_{n-1}\rangle| = n,\\
s_k &:\equiv \text{the } k\text{-th element of } s \text{ for } k < |s|,\\
&\qquad \text{i.e., } \langle x_0, \dots, x_{n-1}\rangle_k = x_k,\\
s * t &:\equiv \text{the concatenation of } s \text{ and } t,\\
s * x &:\equiv s * \langle x\rangle,\\
s * \alpha &:\equiv \text{appending } \alpha \text{ to } s \text{, i.e.,}\\
&\qquad s * \alpha :\equiv \lambda k.\left[\text{if } k<|s| \text{ then } s_k \text{ else } \alpha(k-|s|)\right],\\
s \mathbin{@} \alpha &:\equiv \text{overwriting } \alpha \text{ with } s \text{, i.e.,}\\
&\qquad s \mathbin{@} \alpha :\equiv \lambda k.\left[\text{if } k < |s| \text{ then } s_k \text{ else } \alpha(k)\right],\\
\overline{\alpha}k &:\equiv \langle \alpha(0), \dots, \alpha(k-1)\rangle,\\
\beta \in \overline{\alpha}k &:\equiv \overline{\beta}k \stackrel{\rho^*}{=} \overline{\alpha}k.
\end{aligned}
$$

DEFINITION 1. *Spector's definition of bar recursion* [22] *reads in our notation as follows:*

$$
(1)\quad \Phi(Y, G, H, s) \stackrel{\tau}{=} \begin{cases} G(s) & \text{if } Y(s \mathbin{@} 0^{\rho^\omega}) < |s|,\\ H(s, \lambda x^\rho.\Phi(Y, G, H, s * x)) & \text{otherwise.}\end{cases}
$$

In his thesis [14] *Kohlenbach introduced the following kind of bar recursion which differs from Spector's only in the stopping condition*:

(2)

$$\Phi(Y, G, H, s) \stackrel{\tau}{=} \begin{cases} G(s) & \text{if } Y(s \,@\, 0^{\rho^\omega}) \stackrel{o}{=} Y(s \,@\, 1^{\rho^\omega}), \\ H(s, \lambda x^\rho.\Phi(Y, G, H, s * x)) & \text{otherwise.} \end{cases}$$

Finally, we define Modified bar recursion at type ρ:

$$\Phi(Y, H, s) \stackrel{o}{=} Y(s \,@\, H(s, \lambda x^\rho.\Phi(Y, H, s * x))). \tag{3}$$

Note that each of the equations above defines a family of functionals $\Phi_{\rho,\tau}$ (Φ_ρ *in the case of modified bar recursion) as* ρ *and* τ *range over arbitrary finite types. We shall often omit the parameters* Y, G *and* H *when defining a functional* Φ *using the equations above. We say a model* $\mathcal{S}$ *satisfies one of the respective variants of bar recursion if in* $\mathcal{S}$ *a functional exists satisfying the corresponding equation* (1), (2), *or* (3) *for all possible values of* Y, G, H *and* s.

Recursive definitions similar to (3) occur in [2], and, in a slightly different form, in [3] and [17] in connection with the fan functional (cf. Section 4).

REMARK. Note that replacing in equation (3) the operation @ by $*$ would be an inessential change. However it *is* essential that the type of $\Phi(s)$ is of level 0. If, for example, the type of $\Phi(s)$ were $\mathbb{N} \to \mathbb{N}$ we could set $Y(\alpha)(m) :\stackrel{\mathbb{N}}{\equiv} \alpha(m) + 1$ and $H(s, F)(k) :\stackrel{\mathbb{N}}{\equiv} F(0)(|s| + 1)$, and obtain the equation

$$\Phi(s)(m) \stackrel{\mathbb{N}}{=} (s \,@\, \lambda k.\Phi(s * 0)(|s| + 1))(m) + 1$$

implying

$$\Phi(\langle\,\rangle)(0) \stackrel{\mathbb{N}}{=} \Phi(\langle 0\rangle)(1) + 1 \stackrel{\mathbb{N}}{=} \Phi(\langle 0, 0\rangle)(2) + 2 \stackrel{\mathbb{N}}{=} \cdots$$

which is unsatisfiable in $\mathbb{N}$.

The structures of primary interest to interpret bar recursion are the model $\mathcal{C}$ of *total continuous functionals* of Kleene [13] and Kreisel [15], the model $\widehat{\mathcal{C}}$ of *partial continuous functionals* of Scott [20] and Ershov [9] (see also [17]), and the model $\mathcal{M}$ of (strongly) *majorizable functionals* introduced by Howard [11] and Bezem [7].

THEOREM 1. *The models* $\mathcal{C}$ *and* $\widehat{\mathcal{C}}$ *satisfy all three variants of bar recursion.*

PROOF. In the model $\widehat{\mathcal{C}}$ all three forms of bar recursion can simply be defined as the least fixed points of suitable continuous functionals. For $\mathcal{C}$ we use Ershov's result in [9] according to which the model $\mathcal{C}$ can be identified with the total elements of $\widehat{\mathcal{C}}$. Therefore it suffices to show that all three versions of bar recursion are total in $\widehat{\mathcal{C}}$. For Spector's version this has been shown by Ershov [9], and for the other versions similar argument apply. For example, in order to see that $\Phi(s)$ defined recursively by equation (3) is total for given total Y, H and s one uses bar induction on the bar

$$P(s) :\Leftrightarrow Y(s @ \bot_\rho) \text{ is total}$$

where $\bot_\rho$ denotes the undefined element of type ρ. $P(s)$ is a bar because Y is continuous. ⊣

THEOREM 2. *$\mathcal{M}$ satisfies Spector's bar recursion* (1), *but not Kohlenbach's* (2).

PROOF. See [7] and [14]. ⊣

In Section 5 we will show that $\mathcal{M}$ satisfies modified bar recursion (3).

§3. Using bar recursion to realize classical dependent choice. The aim of this section is to show how modified bar recursion can be used to extract witnesses from proofs of ∀∃-formulas in classical arithmetic plus the axiom (scheme) of dependent choice [12]

DC $\forall n, x^\rho\, \exists y^\rho\, A(n, x, y) \to \forall x \exists f\, (f(0) = x \wedge \forall n\, A(n, f(n), f(n+1)))$.

Actually we will need only the following *weak modified bar recursion* which is the special case of equation (3) where H is constant:

$$\Phi(Y, H, s) \overset{o}{=} Y(s @ \lambda k.H(s, \lambda x.\Phi(Y, H, s * x))). \tag{4}$$

Note that in (4) the returning type of H is ρ, i.e., the argument of Y consists of s followed by an infinite sequence with constant value of type ρ.

Before dealing with dependent choice we discuss our extraction method in general and then give a realizer for the (simpler) classical axiom of countable choice.

3.1. Witnesses from classical proofs. The method we use to extract witnesses from classical proofs is a combination of Gödel's negative translation (translation P^o in [16] page 42, see also [23]), the Dragalin/Friedman/Leivant trick, also called A-translation [25], and Kreisel's (formalized) modified realizability [24]. The method works in general for proofs in PA^ω, the classical variant of HA^ω. In order to extend it to PA^ω plus extra axioms Γ (e.g., $\Gamma \equiv$ **DC**) one has to find realizers for Γ^N, the negative translation of Γ [1], where $\bot$ is replaced by an ∃-formula (regarding negation, $\neg C$, is defined by $C \to \bot$). However, it is more direct and technically simpler to follow [6] and combine the Dragalin/Friedman/Leivant trick and modified realizability: instead of replacing $\bot$ by a ∃-formula we slightly change the definition of modified realizability by regarding y **mr** $\bot$ as an (uninterpreted) atomic formula. More formally we define

$$y^\tau\, \mathbf{mr}_\tau \bot :\equiv P_\bot(y),$$

[1] The negative translation double-negates atomic formulas, replaces $\exists x$ by $\neg\forall x\neg$ and $A \vee B$ by $\neg(\neg A \wedge \neg B)$.

where $P_\perp$ is a new unary predicate symbol and τ is the type of the witness to be extracted. Therefore, we have a modified realizability for each type τ, according to the type of the existential quantifier in the $\forall\exists$-formula we are realizing. The other clauses of modified realizability are as usual, e.g.,

$$f\ \mathbf{mr}_\tau(A \to B) :\equiv \forall x\,(x\ \mathbf{mr}_\tau\ A \to fx\ \mathbf{mr}_\tau\ B).$$

In the following proposition Δ is an axiom system possibly containing $P_\perp$ and further constants, which has the following closure property: If $D \in \Delta$ and B is a quantifier free formula with decidable predicates, then also the universal closure of $D[\lambda y^\tau.B/P_\perp]$ is in Δ, where $D[\lambda y^\tau.B/P_\perp]$ is obtained from D by replacing any occurrence of a formula $P_\perp(L)$ in D by $B[L/y]$.

PROPOSITION 1. *Assume there is a vector Φ of closed terms such that*

$$\mathsf{HA}^\omega + \Delta \vdash \Phi\, \mathbf{mr}_\tau\, \Gamma^N.$$

Then from any proof

$$\mathsf{PA}^\omega + \Gamma \vdash \forall z^\sigma\, \exists y^\tau\, B(z, y),$$

where $\forall z^\sigma\, \exists y^\tau\, B(z, y)$ is a $\forall\exists$-formula in the language of HA^ω, one can extract a closed term $M^{\sigma\to\tau}$ such that

$$\mathsf{HA}^\omega + \Delta \vdash \forall z\, B(z, Mz).$$

PROOF. The proof is folklore. The main steps are as follows. Assuming w.l.o.g. that $B(z, y)$ is atomic, we obtain from the hypothesis $\mathsf{PA}^\omega + \Gamma \vdash \forall z^\sigma\, \exists y^\tau\, B(z, y)$ via negative translation

$$\mathsf{HA}^\omega + \Gamma^N \vdash_m \forall y\,(B(z, y) \to \perp) \to \perp,$$

where $\vdash_m$ denotes derivability in minimal logic, i.e., ex-falso-quodlibet is not used. Now, soundness of modified realizability (which holds for our abstract version of modified realizability and minimal logic [6]), together with the assumption on Φ allows us to extract from this proof a closed term M such that

$$\mathsf{HA}^\omega + \Delta \vdash Mz\ \mathbf{mr}_\tau(\forall y\,(B(z, y) \to \perp) \to \perp)$$

i.e.,

$$\mathsf{HA}^\omega + \Delta \vdash \forall f^{\tau\to\tau}\,(\forall y\,(B(z, y) \to P_\perp(fy)) \to P_\perp(Mzf)).$$

Replacing $P_\perp$ by $\lambda y.B(z, y)$ respectively, and instantiating f by the identity function it follows

$$\mathsf{HA}^\omega + \Delta \vdash \forall z\, B(z, Mz(\lambda y.y)). \qquad \dashv$$

We will apply this proposition with $\tau :\equiv o$ (writing **mr** instead of $\mathbf{mr}_o$), $\Gamma :\equiv$ **DC**, or $\Gamma :\equiv$ **AC** (countable choice, see below), and an axiom system Δ consisting of the defining equation (3) for modified bar recursion, where the defined functionals Φ are new constants, together with the axiom of continuity and the scheme of relativized quantifier free bar induction which are defined as follows:

Continuity $\qquad \forall F^{\rho^\omega\to o}, \alpha\, \exists n\, \forall \beta\, (\overline{\alpha}n = \overline{\beta}n \to F(\alpha) = F(\beta)).$

We call any n such that $\forall\beta\,(\overline{\alpha}n = \overline{\beta}n \rightarrow F(\alpha) = F(\beta))$ a point of continuity of F at α.

Relativized quantifier free bar induction

$$\forall\alpha \in S\,\exists n\,P(\overline{\alpha}n) \wedge \forall s \in S\,\big(\forall x\,[S(s*x) \rightarrow P(s*x)] \rightarrow P(s)\big) \wedge S(\langle\rangle) \rightarrow P(\langle\rangle).$$

Here $S(s)$ is an arbitrary, and $P(s)$ a quantifier free predicate in the language of $\mathsf{HA}^\omega[P_\bot]$, and $\alpha \in S$ and $s \in S$ are shorthands for $\forall n\,S(\overline{\alpha}n)$ and $S(s)$ respectively. Clearly the condition on Δ in Proposition 1 is satisfied.

In order to make sure that realizers can indeed be used to compute witnesses one needs to know that, 1. the axioms of $\mathsf{HA}^\omega + \Delta$ hold in a suitable model—here we can choose the model C of continuous functionals—and, 2. every closed term of type level 0 (e.g., of type $\mathbb{N}$) can be reduced to a numeral in an effective and provably correct way. In [2] this is solved by building the notion of reducibility to normal form into the definition of realizability. In our case we solve this problem by applying Plotkin's adequacy result [18] as follows: each term in the language of HA^ω plus the bar recursive constants can be naturally viewed as a term in the language PCF [18], by defining the bar recursors by means of the general fixed point combinator. In this way our term calculus also inherits PCF's call-by-name reduction, i.e., if M is bar recursive and M reduces to M' then M' is bar recursive. Furthermore reduction is provably correct in our system, i.e., if M reduces to M' then $M = M'$ is provable. Now let M be a closed term of type $\mathbb{N}$. By Theorem 1, M has a total value, which is a natural number n, in the model of partial continuous functionals. Hence, by Plotkin's adequacy theorem M reduces to the numeral denoting n.

3.2. Realizing $\mathbf{AC}^N$. We now construct a realizer of the negatively translated axiom of countable choice

$$\mathbf{AC} \qquad \forall n^{\mathbb{N}}\,\exists y^\rho\,A(n,y) \rightarrow \exists f\,\forall n\,A(n,f(n)).$$

The realizer for $\mathbf{AC}^N$ is similar to the one for $\mathbf{DC}^N$, but technically simpler, so that the essential idea underlying the construction is more visible. Moreover we only need the following special case of relativized quantifier free bar induction:

Relativized quantifier free pointwise bar induction

$$\forall\alpha \in S\,\exists n\,P(\overline{\alpha}n) \wedge \forall s \in S\,\big(\forall x\,\big[S(x,|s|) \rightarrow P(s*x)\big] \rightarrow P(s)\big) \rightarrow P(\langle\,\rangle),$$

where $S(x,n)$ is arbitrary, $P(s)$ is quantifier free, and $\alpha \in S$, $s \in S$ are shorthands for $\forall n\,S(\alpha(n),n)$ and $\forall i < |s|\,S(s_i,i)$, respectively. The principles of relativized quantifier free bar induction respectively pointwise bar induction are similar to Luckhardt's general bar induction over species for quantifier free formulas, $(\mathrm{aBI})^\rho_{\mathrm{D}}$, respectively higher bar induction over species, $(\mathrm{hBI})^\rho_{\mathrm{D}}$ ([16], page 144).

The negative translation of $\mathbf{AC}$ is $\mathbf{AC}^N$

$$\mathbf{AC}^N \quad \forall n\,(\forall y\,(A(n,y))^N \rightarrow \bot) \rightarrow \bot) \rightarrow \forall f\,(\forall n\,A(n,f(n)))^N \rightarrow \bot) \rightarrow \bot.$$

Following Spector [22] we reduce $\mathbf{AC}^N$ to the double negation shift

DNS $\quad \forall n\,((B(n) \to \bot) \to \bot) \to (\forall n\, B(n) \to \bot) \to \bot$

observing that $\mathbf{AC} + \mathbf{DNS} \vdash_m \mathbf{AC}^N$, where **DNS** is used with the formula $B(n) :\equiv \exists y\, A(n,y)^N$ [2]. Therefore it suffices to show that this instance of **DNS** is realizable. The following lemma, whose proof is trivial, is necessary to see that the weak form (4) of modified bar recursion suffices to realize **AC** and **DC**.

LEMMA 1. *Let B be a formula such that all of its atomic subformulas occur in negated form. Then there is a closed term H such that $\forall \vec{z}\, H\, \mathbf{mr}(\bot \to B)$ is provable (in minimal logic), where $\vec{z}$ are the free variables of B (it is important here that H is closed, i.p. does not depend on $\vec{z}$).*

Note that the formula $B(n) :\equiv \exists y\, A(n,y)^N$ to which we apply **DNS** is of the form specified in Lemma 1.

THEOREM 3. *The double negation shift **DNS** for a formula $B(n)$ is realizable using the weak form (4) of modified bar recursion provided $B(n)$ is of the form specified in Lemma 1.*

PROOF. In order to realize the formula

$$\forall n((B(n) \to \bot) \to \bot) \to (\forall n B(n) \to \bot) \to \bot$$

we assume we are given realizers

$$Y^{\rho^\omega \to o}\, \mathbf{mr}(\forall n B(n) \to \bot)$$
$$G^{\mathbb{N} \to (\rho \to o) \to o}\, \mathbf{mr}\, \forall n((B(n) \to \bot) \to \bot)$$

and try to build a realizer for $\bot$. Using weak modified bar recursion (4) we define

$$\Psi(s) = Y(s\, @\, \lambda n.H(G(|s|, \lambda x^\rho.\Psi(s * x))))$$

where $H^{o \to \rho}$ is a closed term such that $\forall n\, H\, \mathbf{mr}(\bot \to B(n))$ is provable, according to Lemma 1. We set

$$S(x,n) :\equiv x\, \mathbf{mr}\, B(n),$$
$$P(s) :\equiv \Psi(s)\, \mathbf{mr}\, \bot,$$

and, by quantifier free pointwise bar induction relativized to S, we show $P(\langle\,\rangle)$, i.e., $\Psi(\langle\,\rangle)\, \mathbf{mr}\, \bot$.

(i) $\forall \alpha \in S\, \exists n\, P(\overline{\alpha}n)$. Let $\alpha \in S$, i.e., $\alpha\, \mathbf{mr}\, \forall n B(n)$. Let n be the point of continuity of Y at α, according to the continuity axiom. By assumption on Y, we get $\forall \beta\, (Y(\overline{\alpha}n\, @\, \beta)\, \mathbf{mr}\, \bot)$, which implies $\Psi(\overline{\alpha}n)\, \mathbf{mr}\, \bot$.

(ii) $\forall s \in S(\forall x\, [S(x,|s|) \to P(s*x)] \to P(s))$. Let $s \in S$ be fixed. Suppose $\forall x\, [S(x,|s|) \to P(s*x)]$, i.e., $\forall x\, [x\, \mathbf{mr}\, B(|s|) \to \Psi(s*x)\, \mathbf{mr}\, \bot]$, in other words

$$\lambda x^\rho.\Psi(s*x)\, \mathbf{mr}(B(|s|) \to \bot).$$

---

[2]The reduction is obvious because $\mathbf{AC}^N$ is equivalent in minimal logic to $\forall n\, \neg\neg\exists y\, A(n,y)^N \to \neg\neg\exists f\, \forall n\, A(n, f(n))^N$.

Using the assumption on G we obtain

$$G(|s|, \lambda x^\rho.\Psi(s * x)) \,\mathbf{mr}\, \bot,$$

and from that, setting $w :\stackrel{\rho}{\equiv} H(G(|s|, \lambda x^\rho.\Psi(s * x)))$, we obtain $w \,\mathbf{mr}\, B(n)$, for all n. Because $s \in S$ it follows that $s \mathbin{@} \lambda n.w \,\mathbf{mr}\, \forall n\, B(n)$ and therefore

$$Y(s \mathbin{@} \lambda n.w) \,\mathbf{mr}\, \bot.$$

Since $\Psi(s) = Y(s \mathbin{@} \lambda n.w)$ we have $P(s)$. ⊣

As explained above Theorem 3 yields

COROLLARY 1. *The negative translation of the countable axiom of choice, AC^N is realizable using the weak form* (4) *of modified bar recursion.*

3.3. Realizing DC^N. With a similar but technically more involved construction we now prove

THEOREM 4. *The negative translation of the axiom of dependent choice,* $\mathbf{DC}^N$, *is realizable using the weak form* (4) *of modified bar recursion.*

PROOF. Let σ be the type of realizers of $A(n, x, y)^N$. Given x_0^ρ and realizers

$$G^{\mathbb{N}\to\rho\to(\rho\to\sigma\to o)\to o} \,\mathbf{mr}\, \forall n, x\, (\forall y\, (A(n, x, y)^N \to \bot) \to \bot),$$
$$Y^{\rho^\omega\to\sigma^\omega\to o} \,\mathbf{mr}\, \forall f\, (f(0) = x_0 \wedge \forall n\, A(n, f(n), f(n+1))^N \to \bot),$$

we have to construct a realizer of $\bot$. In the rest of this proof the variables β and t have the types $(\rho \times \sigma)^\omega$ and $(\rho \times \sigma)^*$ respectively. First we perform a trivial transformation on Y defining

$$\tilde{Y}^{(\rho\times\sigma)^\omega\to o}(\beta) :\equiv Y(x_0 * (\pi_0 \circ \beta), \pi_1 \circ \beta),$$

where π_0, π_1 are the left and right projection and $\circ$ is composition of functions. Using weak bar recursion (4) we now define

$$\Psi(t) = \tilde{Y}(t \mathbin{@} \lambda n.\pi(0^\rho, H(G(|t|, (x_0 * (\pi_0 \circ t))_{|t|}, \lambda y^\rho \lambda z^\sigma.\Psi(t * \pi(y, z)))))),$$

where $\forall n, x, y\, H \,\mathbf{mr}(\bot \to A(n, x, y)^N)$ according to Lemma 1, $\pi(.,.)$ is pairing, and $\pi_0 \circ t :\equiv \langle \pi_0(t_0), \dots, \pi_0(t_{|t|-1})\rangle$ (hence $(\pi_0 \circ t)_i = \pi_0(t_i)$ for $i < |t|$). We define predicates

$$S(t) :\equiv \forall i < |t|\ (\pi_1(t_i) \,\mathbf{mr}\, A(i, (\langle x_0\rangle * (\pi_0 \circ t))_i, (\pi_0 \circ t)_i)^N)$$
$$P(t) :\equiv \Psi(t) \,\mathbf{mr}\, \bot.$$

We show $P(\langle\,\rangle)$ by quantifier free bar induction relativized to S. Obviously $S(\langle\,\rangle)$ holds.

(i) $\forall \beta \in S\, \exists n\, P(\overline{\beta}n)$. Let $\beta \in S$. Set $f^{\rho^\omega} :\equiv \langle x_0\rangle * (\pi_0 \circ \beta)$ and $\gamma^{\sigma^\omega} :\equiv \pi_1 \circ \beta$. Then $f(0) = x_0$ and $\forall n\, \gamma(n) \,\mathbf{mr}\, A(n, f(n), f(n+1))^N$. Therefore $Y(f, \gamma) \,\mathbf{mr}\, \bot$. Let n be a point of continuity of $\tilde{Y}$ at β. Then

$$\Psi(\overline{\beta}n) = \tilde{Y}(\beta) = Y(f, \gamma)$$

and therefore $\Psi(\overline{\beta}n) \,\mathbf{mr}\, \bot$, i.e., $P(\overline{\beta}n)$.

(ii) $\forall t \in S\,(\forall q^{\rho\times\sigma}[S(t*q) \to P(t*q)] \to P(t))$. Let $t \in S$ where, say, $t = \langle \pi(x_1, z_0), \dots, \pi(x_n, z_{n-1})\rangle$. Assume further $\forall q\,[S(t*q) \to P(t*q)]$, i.e.,

$$\forall x_{n+1}, z_n\,\big[\forall i \le n\; z_i\,\mathbf{mr}\,A(i, x_i, x_{i+1})^N \to \\ \Psi(\langle \pi(x_1, z_0), \dots, \pi(x_{n+1}, z_n)\rangle)\,\mathbf{mr}\,\bot\big].$$

Because $t \in S$ it follows that

$$\forall x_{n+1}, z_n\,\big[z_n\,\mathbf{mr}\,A(n, x_n, x_{n+1})^N \to \Psi(\langle \pi(x_1, z_0), \dots, \pi(x_{n+1}, z_n)\rangle)\,\mathbf{mr}\,\bot\big]$$

i.e.,

$$\lambda y \lambda z.\Psi(t * \pi(y, z))\,\mathbf{mr}\,\forall y\,(A(n, x_n, y)^N \to \bot).$$

By the assumption on G it follows $G(n, x_n, \lambda y\lambda z.\Psi(t * \pi(y, z)))\,\mathbf{mr}\,\bot$. Hence, for $w \overset{\sigma}{:\equiv} H(G(n, x_n, \lambda y \lambda z.\Psi(t * \pi(y, z))))$, we have $\forall n, x, x'$ $(w\,\mathbf{mr}\,A(n, x, x')^N)$. Now we set $f^{\rho^\omega} :\equiv \langle x_0, x_1, \dots, x_n\rangle @ \lambda k.0^\rho$ and $\gamma^{\sigma^\omega} :\equiv \langle z_0, \dots, z_{n-1}\rangle @ \lambda k.w$. Then $\forall n\, \gamma(n)\,\mathbf{mr}\,A(n, f(n), f(n+1))^N$ and therefore $Y(f, \gamma)\,\mathbf{mr}\,\bot$. But, because $x_n = (x_0 * (\pi_0 \circ t))_{|t|}$ we have

$$\Psi(t) = \tilde{Y}(t @ \pi(0^\rho, \lambda k.w)) = Y(f, \gamma).$$

Hence $\Psi(t)\,\mathbf{mr}\,\bot$, i.e., $P(t)$. ⊣

§4. Bar recursion and the fan functional. A functional $\mathrm{FAN}^{(\mathbb{N}^\omega \to o)\to\mathbb{N}}$ is called *fan functional* if it computes a modulus of uniform continuity for every continuous functional $Y^{\mathbb{N}^\omega \to o}$ restricted to infinite 0, 1-sequences, i.e., if FAN satisfies

$$\forall Y\,\forall \alpha, \beta \le \lambda x.1\,\big(\overline{\alpha}(\mathrm{FAN}(Y)) = \overline{\beta}(\mathrm{FAN}(Y)) \to Y\alpha \overset{o}{=} Y\beta\big).$$

A recursive algorithm for FAN(Y) that was given in [3] and [17] uses two procedures,

$$(5)\quad \Phi(s^{\mathbb{N}^*}, v^o) \overset{\mathbb{N}^\omega}{=} \\ s @ \big[\text{if } Y(\Phi(s*0, v)) \ne v \text{ then } \Phi(s*0, v) \text{ else } \Phi(s*1, v)\big]$$

$$(6)\quad \Psi(Y, s) \overset{\mathbb{N}}{=} \\ \begin{cases} 0 & \text{if } Y(\alpha) = Y(s @ \lambda k.0), \\ & \text{where } \alpha = \Phi(s,\ Y(s @ \lambda k.0)), \\ 1 + \max\{\Psi(Y, s*0), \Psi(Y, s*1)\} & \text{otherwise.} \end{cases}$$

The first functional, $\Phi(s, v)$, returns an infinite path α having s as a prefix, such that $Y(s @ \alpha) \ne v$, if such a path exists, and returns s extended by $\lambda x.1$, otherwise, i.e., if Y is constant v on all paths extending s. The second functional, $\Psi(Y, s)$, returns the least point of uniform continuity for Y on all extension of s. Therefore, a fan functional can be defined as FAN(Y) :≡

$\Psi(Y, \langle\,\rangle)$. A more formal proof that $\lambda Y.\Psi(Y, \langle\,\rangle)$ is indeed a fan functional can be found in [3] and [17] [3].

THEOREM 5. *The functional* FAN *can be defined using bar recursions* (3) *and* (2) *together.*

Before we give the proof of the theorem we prove two lemmas.

LEMMA 2. *Modified bar recursion* (3) *is equivalent to*

$$\Phi(s^{\rho^*}) \stackrel{o}{=} Y(s @ H(s, \lambda t^{\rho^*} \lambda x^{\rho}.\Phi(s * t * x))) \tag{7}$$

and also to

$$\Phi(s^{\rho^*}) \stackrel{\rho^\omega}{=} s @ H(s, \lambda t^{\rho^*} \lambda x^{\rho}.Y^{\rho^\omega \to o}(\Phi(s * t * x))). \tag{8}$$

PROOF. Obviously equation (7) subsumes modified bar recursion. It is also easy to see that equations (7) and (8) are equivalent: Given Φ satisfying (7) we define $\Phi'(s) :\equiv s @ H(s, \lambda t \lambda x.\Phi(s * t * x))$ which satisfies (8), provably by relativized bar induction. Conversely, if Φ' satisfies (8) then Φ defined by $\Phi(s) :\equiv Y(\Phi'(s))$ satisfies (7). Furthermore it is clear that we can replace the operation @ in each of the equations (3), (7) and (8) by $*$, i.e., we prefix with s instead of overwriting (see the definitions at the beginning of Section 2). Hence it suffices to show that we can define a functional Φ satisfying

$$\Phi(s^{\rho^*}) \stackrel{o}{=} Y(s * H(s, \lambda t^{\rho^*} \lambda x^{\rho}.\Phi(s * t * x))) \tag{9}$$

by modified bar recursion. To this end we will use equation (3) (where @ is replaced by $*$) at type ρ^*. We define freeze: $\rho^* \to \rho^{**}$ and melt: $\rho^{**} \to \rho^*$ by $\mathsf{freeze}(\langle x_0, \ldots, x_{n-1}\rangle) = \langle\langle x_0\rangle, \ldots, \langle x_{n-1}\rangle\rangle$, $\mathsf{melt}(\langle s_0, \ldots, s_{n-1}\rangle) = s_0 * \cdots * s_{n-1}$, so that $\mathsf{melt}(\mathsf{freeze}(s)) = s$. Given $Y^{\rho^\omega \to o}$ and $H^{\rho^* \to (\rho^* \times \rho \to o) \to \rho^\omega}$ we define using modified bar recursion (3)

$$\Psi(q) = Y(\mathsf{melt}(q) * H(\mathsf{melt}(q), \lambda t \lambda x.\Psi(q * (t * x)))).$$

By relativized bar induction one easily proves

$$\forall q, q' \, (\mathsf{melt}(q) = \mathsf{melt}(q') \to \Psi(q) = \Psi(q')),$$

which implies, again by relativized bar induction, that Φ, defined by $\Phi(s) :\equiv \Psi(\mathsf{freeze}(s))$, satisfies (9). ⊣

LEMMA 3. *Kohlenbach's bar recursion* (2) *is equivalent to*

$$\Phi(s) \stackrel{\tau}{=} \begin{cases} G(s) & \text{if } Y(s @ 0^{\rho^\omega}) \stackrel{o}{=} Y(s @ J(s)), \\ H(s, \lambda x^{\rho}.\Phi(s * x)) & \text{otherwise}, \end{cases} \tag{10}$$

where the new parameter J is of type $\rho^* \to \rho^\omega$ and, as usual, $\Phi(s)$ is shorthand for the more accurate $\Phi(Y, G, H, J, s)$.

---

[3]The authors were informed that Robin Gandy knew a recursive definition of the fan functional in $\widehat{\mathcal{C}}$ already around 1973.

PROOF. Our proof is based on the proof of Theorem 3.66 in [14]. The fact that (2) can be defined from (10) is trivial. To define (10) from (2) one uses the following trick. For s^{ρ^*}, $s + (\dot{-})k$ denotes pointwise addition (cut-off subtraction) of appropriate type, and $\kappa(n) :\equiv n$, $\kappa(f^{\rho\to\sigma}) :\equiv \kappa(f(0^\rho))$, $\kappa(z^{\rho\times\sigma}) :\equiv \kappa(\pi_0(z))$, so $\kappa(x^\rho + 2) > 1$ and $\kappa(n^\rho) = n$. Define

$$\eta(\beta^{\rho^\omega})(n) :\equiv \begin{cases} \beta(n)\dot{-}2 & \text{if } \kappa(\beta(n)) > 1, \\ J(\phi(\overline{\beta}n))(n) & \text{if } \kappa(\beta(n)) = 1, \\ 0 & \text{if } \kappa(\beta(n)) = 0, \end{cases}$$

where $\phi(s) :\equiv \langle s_0, \dots, s_{k-1}\rangle$ with $k < |s|$ minimal such that $\kappa(s_k) = 1$ (if $s = \langle\,\rangle$ then k is zero). Clearly

$$\eta((s+2) @ 0^{\rho^\omega}) = s @ 0^{\rho^\omega},$$
$$\eta((s+2) @ 1^{\rho^\omega}) = s @ J(s).$$

Now we can define using Kohlenbach's bar recursion (2)

$$\tilde{\Phi}(s) \overset{\tau}{=} \begin{cases} G(s\dot{-}2) & \text{if } Y(\eta(s @ 0^{\rho^\omega})) = Y(\eta(s @ 1^\rho)), \\ H(s\dot{-}2, \lambda x^\rho.\tilde{\Phi}(s * (x+2))) & \text{otherwise.} \end{cases}$$

Then clearly $\Phi(s) :\equiv \tilde{\Phi}(s+2)$ satisfies (10). ⊣

PROOF OF THEOREM 5. We show that procedures Φ and Ψ satisfying the equations (5) and (6) respectively can be defined using equations (3) and (2).

For defining the functional $\Phi(s, v)$ we use equation (8) of Lemma 2.

$$\Phi(s, v) \overset{0^\omega}{=} s @ H(s, v, \lambda t \lambda x.Y(\Phi(s * t * x, v)))$$

where H is defined by course of value primitive recursion as

$$H(s, v, F)(n) \overset{0}{=} \begin{cases} s_n & \text{if } n < |s|, \\ 0 & \text{if } n \geq |s| \wedge F(c, 0) \neq v, \\ 1 & \text{if } n \geq |s| \wedge F(c, 0) = v, \end{cases}$$

with $c :\equiv \langle H(s, v, F)(|s|), \dots, H(s, v, F)(n-1)\rangle$. Clearly Φ satisfies equation (5) at all $n < |s|$. For $n \geq |s|$ we first observe that

$$\Phi(s, v)(n) \overset{0}{=} \begin{cases} 0 & \text{if } Y(\Phi(s * c_{s,n} * 0, v)) \neq v, \\ 1 & \text{if } Y(\Phi(s * c_{s,n} * 0, v)) = v, \end{cases}$$

where $c_{s,n} :\equiv \langle \Phi(s, v)(|s|), \dots, \Phi(s, v)(n-1)\rangle$. Now if $Y(\Phi(s * 0, v)) \neq v$ then $\Phi(s, v)(|s|) = 0$ and therefore $s * c_{s,n} = s * 0 * c_{s*0,n}$. Hence $\Phi(s, v)(n) = \Phi(s * 0, v)(n)$ as required by (5). The case $Y(\Phi(s * 0, v)) = v$ is similar.

One immediately sees that a functional Ψ satisfying (6) can be defined from an instance of equation (10) using the functional Φ above. ⊣

§5. Modified bar recursion and the model $\mathcal{M}$. The model $\mathcal{M}$ ($= \bigcup \mathcal{M}_\rho$) of strongly majorizable functionals (introduced in [7] as a variation of Howard's majorizable functionals [11]) and the strongly majorizability relation s-maj$_\rho \subseteq \mathcal{M}_\rho \times \mathcal{M}_\rho$ are defined simultaneously by induction on types as follows [4]

$$n \text{ s-maj}_{\mathbb{N}}\, m :\equiv n, m \in \mathbb{N} \wedge n \geq m, \qquad \mathcal{M}_{\mathbb{N}} :\equiv \mathbb{N},$$

$$F^* \text{ s-maj}_{\rho\to\tau} F :\equiv F^*,\ F \in \mathcal{M}_\rho \to \mathcal{M}_\tau \wedge$$
$$\forall G^*,\ G \in \mathcal{M}_\rho \left[G^* \text{ s-maj}_\rho\, G \to F^*G^* \text{ s-maj}_\tau\, F^*G, FG\right],$$
$$\mathcal{M}_{\rho\to\tau} :\equiv \left\{F \in \mathcal{M}_\rho \to \mathcal{M}_\tau \colon \exists F^* \in \mathcal{M}_\rho \to \mathcal{M}_\tau\ F^* \text{ s-maj}_{\rho\to\tau} F\right\}.$$

In the following we abbreviate s-maj$_\rho$ by maj$_\rho$ and by "majorizable" we always mean "strongly majorizable". We often omit the type in the relation maj$_\rho$. We shall sometimes write "$F\colon \rho \to \sigma$" for "$F \in \mathcal{M}_{\rho\to\sigma}$" (as opposed to "$F\colon \mathcal{M}_\rho \to \mathcal{M}_\sigma$" which just means that F is a set-theoretic function from $\mathcal{M}_\rho$ to $\mathcal{M}_\sigma$, i.e., $F \in \mathcal{M}_\rho \to \mathcal{M}_\sigma$).

In [14] it is shown that the scheme of bar recursion (2) is provably not primitive recursively definable from (1), since (1) yields a well defined functional in the model of (strongly) majorizable functionals $\mathcal{M}$ (cf. [7]) and (2) does not. Equation (1), however, can be primitive recursively defined from (2) (cf. [14]). In [5] it is shown that a functional

$$\Phi\colon \mathcal{M}_{\rho^\omega\to\mathbb{N}} \times \mathcal{M}_{\rho^*\times(\rho\to\mathbb{N})\to\rho^\omega} \times \mathcal{M}_{\rho^*} \to \mathcal{M}_{\mathbb{N}},$$

exists satisfying equation (3). We now show that any such Φ indeed lives in $\mathcal{M}$, i.e., we show that there is a functional Φ^* majorizing Φ. Recall that for continuous functionals Y of type $\rho^\omega \to \mathbb{N}$ it is the case that from some initial segment of α the value of $Y(\alpha)$ is determined. For the majorizable functionals this does not hold, but a "weak continuity" property does hold. It says that a bound on the value of $Y(\alpha)$ can be determined from an initial segment of α. We prove this result in Lemma 5. This turned out to be an important tool for proving the main theorem of this section. For the rest of this section all variables (unless stated otherwise) are assumed to range over the type structure $\mathcal{M}$. We first recall from [7] the following lemma:

LEMMA 4 ([7], 1.4, 1.5). *For* $F_0, \dots, F_n\colon \rho$ *we define* $\max^\rho\langle F_0, \dots, F_n\rangle\colon \rho$, *also written* $\max_{i\leq n}{}^\rho F_i\colon \rho$, *as*

$$\max_{i\leq n}{}^{\mathbb{N}} m_i :\equiv \max\{m_0, \dots, m_n\},$$
$$\max_{i\leq n}{}^{\tau\to\rho} F_i :\equiv \lambda x^\tau. \max_{i\leq n}{}^\rho F_i(x),$$

and for α^{ρ^ω}, *define* $\alpha^+(n) :\equiv \max_{i\leq n}{}^\rho \alpha(i)$. *Then,*

$$\forall n(\alpha(n) \operatorname{maj} \beta(n)) \to \alpha^+ \operatorname{maj} \beta^+, \beta.$$

We also use pointwise addition in all types ρ, denoted $x +_\rho y$.

[4]For simplicity, we only consider the base type $\mathbb{N}$ and functional types. Later we extend the definition of majorizability for types ρ^*.

LEMMA 5 (Weak continuity for $\mathcal{M}$). $\forall Y^{\rho^{\omega}\to\mathbb{N}}, \alpha\, \exists n^{\mathbb{N}}\, \forall \beta \in \overline{\alpha}n\, (Y(\beta) \leq n)$.

PROOF. Let Y and α be fixed, α^* maj α and Y^* maj Y. From the assumption

$$(*) \qquad \forall n\, \exists \beta \in \overline{\alpha}n (Y(\beta) > n)$$

we derive a contradiction. For any n, let β_n be the functional whose existence we are assuming in $(*)$. Let

$$\beta_n^*(i) :\equiv \begin{cases} 0^\rho & i < n, \\ \beta_n(i)^* & i \geq n, \end{cases}$$

where $\beta_n(i)^*$ denotes some majorant of $\beta_n(i)$. Having defined the functional β_n^* we note two of its properties,

(i) $\forall i < n(\beta_n^*(i) = 0^\rho)$,
(ii) $(\alpha^* +_{\rho^\omega} \beta_n^*)^+$ maj β_n (by Lemma 4).

Consider the functional $\hat{\alpha}$ defined as $\hat{\alpha}(n) :\equiv \alpha^*(n) +_\rho \sum_{i\in\mathbb{N}} \beta_i^*(n)$. Since at each point n only finitely many β_i^* are non-zero, α^* is well defined. Let $Y^*(\hat{\alpha}^+) = l$. Note that $\hat{\alpha}^+$ maj β_i, for all $i \in \mathbb{N}$, and from $(*)$ we should have $l < Y(\beta_l) \leq l$, a contradiction. ⊣

We extend, for convenience, the definition of majorizability to finite sequences, i.e., for sequences $s^*, s \in \mathcal{M}_\rho^*$ we define

$$s^* \operatorname{maj}_{\rho^*} s :\equiv |s^*| \geq |s| \wedge \forall i \leq j < |s^*| \big(s_j^* \operatorname{maj} s_i^* \wedge (i < |s| \to s_j^* \operatorname{maj} s_i)\big).$$

It is clear that for any sequence $s \in \mathcal{M}_\rho^*$ we can find an $s^* \in \mathcal{M}_\rho^*$ such that s^* maj s. Therefore, we define $\mathcal{M}_{\rho^*}$ as $\mathcal{M}_\rho^*$. Majorizability for functionals involving the type ρ^* is extended accordingly, e.g., for $F^*, F \in \mathcal{M}_{\rho^*} \to \mathcal{M}_\mathbb{N}$

$$F^* \operatorname{maj}_{\rho^*\to\mathbb{N}} F :\equiv \forall s^*, s \in \mathcal{M}_{\rho^*}\, (s^* \operatorname{maj}_{\rho^*} s \to F^*(s^*) \geq F^*(s), F(s)).$$

LEMMA 6. *Let s^* and s s.t. $|s^*| = |s|$ be fixed. If s^* maj s then*

$$\forall \beta \in s \exists \beta^* \in s^*\, (\beta^* \operatorname{maj} \beta).$$

PROOF. Let s^*, s and $\beta \in s$ be fixed. Moreover, assume $|s^*| = |s| = n$ and s^* maj s. We define β^* recursively as

$$\beta^*(i) :\equiv \begin{cases} s_i^* & \text{if } i < n, \\ \max^\rho(\overline{\beta^*}(i) * \beta(i)^*) & \text{otherwise,} \end{cases}$$

where $\beta(i)^*$ is some majorant of $\beta(i)$. First note that, for all i, $\beta^*(i)$ maj $\beta(i)$. We show that β^* maj β. Let $k \geq i$.

If $k < n$ then $\beta^*(k) = s_k^* \operatorname{maj} s_i^* \operatorname{maj} s_i = \beta(i)$.
If $k \geq n$ then $\beta^*(k) = \max^\rho\{\max_{j<k}{}^\rho \beta^*(j), \beta(k)^*\} \operatorname{maj} \beta^*(i) \operatorname{maj} \beta(i)$. ⊣

In the following we shall make use of two functionals Ω and Γ defined below. The functional Ω was first introduced in [14], 3.40.

LEMMA 7 ([14], 3.41). *Define functionals* $\min^\rho$ *(from non-empty sets* $X \subseteq \mathcal{M}_\rho$ *to elements of* $\mathcal{M}_\rho$*) and* $\Omega : \mathcal{M}_\rho \to \mathcal{M}_\rho$ *as*

$$\min{}^{\mathbb{N}} X :\equiv \min X, \textit{ for } \emptyset \neq X \subseteq \mathbb{N},$$
$$\min{}^{\rho\to\tau} X :\equiv \lambda y^\rho . \min{}^\tau\{Fy : F \in X\}, \textit{ for } \emptyset \neq X \subseteq \mathcal{M}_{\rho\to\tau},$$
$$\Omega(F) :\equiv \min{}^\rho\{F^* : F^* \operatorname{maj} F\}.$$

Then,

(i) *For all* $F, \Omega(F) \operatorname{maj} F$,
(ii) $\Omega \operatorname{maj} \Omega$. *(Therefore,* $\Omega \in \mathcal{M}$.*)*

LEMMA 8. *Define* $\Gamma\colon \mathcal{M}_{\rho^\omega\to\mathbb{N}} \to (\mathcal{M}_{\rho^\omega} \to \mathcal{M}_{\mathbb{N}})$

$$\Gamma(Y)(\alpha) :\equiv \min n\, [\forall \beta \in \overline{\alpha}n(\Omega(Y)(\beta) \leq n)].$$

Then,

(i) $\Gamma(Y) \operatorname{maj} Y$ *(therefore* $\Gamma(Y) \in \mathcal{M}_{\rho^\omega\to\mathbb{N}}$*),*
(ii) $\Gamma(Y)$ *is continuous and* $\Gamma(Y)(\alpha)$ *is a point of continuity for* $\Gamma(Y)$ *at* α,
(iii) $\Gamma \operatorname{maj} \Gamma$ *(therefore,* $\Gamma \in \mathcal{M}$*).*

PROOF. First of all, we note that, by Lemma 5, the functional Γ is well defined. By Lemma 7 (i), $\Omega(Y) \operatorname{maj} Y$.

(i) Let $\alpha^* \operatorname{maj} \alpha$. We have to show $\Gamma(Y)(\alpha^*) \geq \Gamma(Y)(\alpha), Y(\alpha)$. By the definition of $\Gamma(Y)$, and Lemma 7 (i), we have $\Gamma(Y)(\alpha^*) \geq \Omega(Y)(\alpha^*) \geq Y(\alpha)$. It is only left to show that $\Gamma(Y)(\alpha^*) \geq \Gamma(Y)(\alpha)$. Suppose that $n = \Gamma(Y)(\alpha^*) < \Gamma(Y)(\alpha) = m$. Note that there exists a $\beta \in \overline{\alpha}(m-1)$ such that $\Omega(Y)(\beta) \geq m$ (otherwise we get a contradiction to the minimality in the definition of $\Gamma(Y)$). But since $m > n$, by Lemma 6, there exists a $\beta^* \in \overline{\alpha^*}n$ such that $\beta^* \operatorname{maj} \beta$. Therefore, $\Omega(Y)(\beta^*) \leq n < m \leq \Omega(Y)(\beta)$. But by Lemma 7 (i) also $\Omega(Y)(\beta^*) \geq \Omega(Y)(\beta)$, a contradiction.

(ii) Let α be fixed and take $n = \Gamma(Y)(\alpha)$. Suppose there exists a $\beta \in \overline{\alpha}n$ such that $\Gamma(Y)(\beta) \neq n$. If $\Gamma(Y)(\beta) < n$ we get, since $\alpha \in \overline{\beta}n$, that $\Gamma(Y)(\alpha) < n$, a contradiction. Suppose $\Gamma(Y)(\beta) > n$. Since $\beta \in \overline{\alpha}n$ we have, $\forall \gamma \in \overline{\beta}n(\Omega(Y)(\gamma) \leq n)$, also a contradiction.

(iii) Assume $Y^* \operatorname{maj} Y$ and $\alpha^* \operatorname{maj} \alpha$. We show $\Gamma(Y^*)(\alpha^*) \geq \Gamma(Y)(\alpha)$. By the self majorizability of $\Gamma(Y)$ we have $\Gamma(Y)(\alpha^*) \geq \Gamma(Y)(\alpha)$. We now show $\Gamma(Y^*)(\alpha^*) \geq \Gamma(Y)(\alpha^*)$. Let $n = \Gamma(Y^*)(\alpha^*)$ and suppose $m = \Gamma(Y)(\alpha^*) > n$. By the definition of $\Gamma(Y)$, there exists a $\beta \in \overline{\alpha^*}(m-1)$ s.t. $\Omega(Y)(\beta) \geq m$. But, since $m > n$, by Lemma 6, there exists a $\beta^* \in \overline{\alpha^*}n$ s.t. $\beta^* \operatorname{maj} \beta$, and by Lemma 7 (ii), $\Omega(Y^*)(\beta^*) \geq m > n$, a contradiction. ⊣

LEMMA 9. *Let* $Y^* \operatorname{maj} Y$ *of type* $\rho^\omega \to \mathbb{N}$ *and* α *of type* ρ^ω *be fixed. Set* $n = \Gamma(Y^*)(\alpha)$. *If* $\overline{\alpha}n \operatorname{maj} s$ *and* $|s| = n$ *then for all sequences* β *we have*

$$\Gamma(Y^*)(s @ \beta), \Gamma(Y)(s @ \beta), Y(s @ \beta) \leq n.$$

PROOF. We prove just that $\Gamma(Y^*)(s @ \beta) \leq n$. The other two cases follow similarly. Suppose there exists a β such that $n < \Gamma(Y^*)(s @ \beta)$. Since

$\overline{\alpha}n$ maj s, by Lemma 6, there exists a β^* such that $\overline{\alpha}n * \beta^*$ maj s @ β. Therefore, by Lemma 8 (iii), we must have $n < \Gamma(Y^*)(\overline{\alpha}n * \beta^*)$. And by the fact that n is a point of continuity for $\Gamma(Y^*)$ on α we get $\Gamma(Y^*)(\overline{\alpha}n * \beta^*) = n$, a contradiction. ⊣

We extend the $(\cdot)^+$ operator of Lemma 4 to functionals $F : \mathcal{M}_{\rho^*} \to \mathcal{M}_{\mathbb{N}}$ by

$$F^+ :\equiv \lambda s. \max_{s' \preceq s} F(s'),$$

where $s' \preceq s :\equiv |s'| \leq |s| \wedge \forall i < |s'| \, (s'_i = s_i)$.

LEMMA 10. *Let F and G be of type $\mathcal{M}_{\rho^*} \to \mathcal{M}_{\mathbb{N}}$. If*

$$\forall s^*, s \, [s^* \text{ maj } s \wedge |s^*| = |s| \to F(s^*) \geq F(s), G(s)]$$

then F^+ maj G^+, G.

PROOF. Let s^* maj s be fixed. For all prefixes t^* (of s^*) and t (of s) of the same length, by the assumption of the lemma, we have $F(t^*) \geq F(t), G(t)$. Therefore,

$$\max_{s' \preceq s^*} F(s') \geq \max_{s' \preceq s} F(s'), \max_{s' \preceq s} G(s').$$

Therefore, F^+ maj G^+, G. ⊣

THEOREM 6. *If Φ is a functional of type*

$$\mathcal{M}_{\rho^\omega \to \mathbb{N}} \times \mathcal{M}_{\rho^* \times (\rho \to \mathbb{N}) \to \rho^\omega} \times \mathcal{M}_{\rho^*} \to \mathcal{M}_{\mathbb{N}},$$

which for any given $Y, H, s \in \mathcal{M}$ (of appropriate types) satisfies equation (3), *then $\Phi \in \mathcal{M}$.*

PROOF. Our proof is based on the proof of the main result of [7]. The idea is that, if Φ satisfies equation (3) then the functional

$$\Phi^* :\equiv \lambda Y, H. \big[\lambda s. \Phi(\hat{Y}, \hat{H}, s)\big]^+ \text{ maj } \Phi,$$

where

$$\hat{Y}(\alpha) :\equiv \Gamma(Y)(\alpha^+) \text{ and}$$
$$\hat{H}(s, F) :\equiv H(s, \lambda x. F(\{x\}_s)),$$

and $\{x\}_s$ abbreviates $\max^\rho(s * x)$. Let Y^* maj Y and H^* maj H be fixed. For the rest of the proof s^* maj s is a shorthand for s^* maj $s \wedge |s^*| = |s|$, i.e., majorizability is only considered for sequences of equal length. The fact that Φ^* maj Φ follows from,

$$\big[\lambda s. \Phi(\hat{Y}^*, \hat{H}^*, s)\big]^+ \text{ maj } \big[\lambda s. \Phi(\hat{Y}, \hat{H}, s)\big]^+, \lambda s. \Phi(Y, H, s),$$

which follows, by Lemma 10, from $\forall s^* \, P(s^*)$ where

$$P(s^*) :\equiv \forall s \, \big[s^* \text{ maj } s \to \Phi(\hat{Y}^*, \hat{H}^*, s^*) \geq \Phi(\hat{Y}^*, \hat{H}^*, s), \Phi(\hat{Y}, \hat{H}, s), \Phi(Y, H, s)\big].$$

We prove $\forall s^* \, P(s^*)$ by bar induction:

(i) $\forall\alpha\exists n\, P(\overline{\alpha}n)$. Let α be fixed and $n :\equiv \hat{Y}^*(\alpha) = \Gamma(Y^*)(\alpha^+)$. If $\overline{\alpha}n$ does not majorize any sequence s we are done. Let s be such that $\overline{\alpha}n$ maj s. Note that $\overline{\alpha^+}n = \overline{(\overline{\alpha}n @ \beta)^+}n$, for all β. Therefore, by Lemma 8 (ii) and our assumption that Φ satisfies (3) we get $\Phi(\hat{Y}^*, \hat{H}^*, \overline{\alpha}n) = n$. Since $\overline{\alpha^+}n$ maj $\overline{(s @ \beta)^+}n$ (for all β), by Lemma 9, we have $n \geq \Phi(\hat{Y}^*, \hat{H}^*, s), \Phi(\hat{Y}, \hat{H}, s), \Phi(Y, H, s)$.

(ii) $\forall s^*(\forall x\, P(s^* * x) \to P(s^*))$. Let s^* be fixed. Assume that $\forall x\, P(s^* * x)$, i.e.,

$$\forall x, s\, \big[s^* * x \text{ maj } s \to \Phi(\hat{Y}^*, \hat{H}^*, s^* * x) \geq \Phi(\hat{Y}^*, \hat{H}^*, s), \Phi(\hat{Y}, \hat{H}, s), \Phi(Y, H, s)\big].$$

We derive $P(s^*)$. Note that if s^* does not majorize any sequence we are again done. Assume s is such that s^* maj s. If x^* maj x then (by $\forall x\, P(s^* * x)$),

$$\underbrace{\Phi(\hat{Y}^*, \hat{H}^*, s^* * \{x^*\}_{s^*})}_{\equiv:\, \Phi_1(\{x^*\}_{s^*})} \geq \underbrace{\Phi(\hat{Y}^*, \hat{H}^*, s * \{x\}_s)}_{\equiv:\, \Phi_2(\{x\}_s)}, \underbrace{\Phi(\hat{Y}, \hat{H}, s * \{x\}_s)}_{\equiv:\, \Phi_3(\{x\}_s)}, \underbrace{\Phi(Y, H, s * x)}_{\equiv:\, \Phi_4(x)}.$$

and also $\Phi_1(\{x^*\}_{s^*}) \geq \Phi_1(\{x\}_{s^*})$, which implies

$$\lambda x.\Phi_1(\{x\}_{s^*}) \text{ maj } \lambda x.\Phi_2(\{x\}_s), \lambda x.\Phi_3(\{x\}_s), \lambda x.\Phi_4(x),$$

and by the definition of majorizability

$$\underbrace{H^*(s^*, \lambda x.\Phi_1(\{x\}_{s^*}))}_{\hat{H}^*(s^*, \lambda x.\Phi_1(x))} \text{ maj } \underbrace{H^*(s, \lambda x.\Phi_2(\{x\}_s))}_{\hat{H}^*(s, \lambda x.\Phi_2(x))}, \underbrace{H(s, \lambda x.\Phi_3(\{x\}_s))}_{\hat{H}(s, \lambda x.\Phi_3(x))}, H(s, \lambda x.\Phi_4(x)),$$

which implies

$$\big(s^* @ \hat{H}^*(s^*, \lambda x.\Phi_1(x))\big)^+ \text{ maj } \big(s @ \hat{H}^*(s, \lambda x.\Phi_2(x))\big)^+, \big(s @ \hat{H}(s, \lambda x.\Phi_3(x))\big)^+, s @ H(s, \lambda x.\Phi_4(x)).$$

And finally, by Lemma 8 (i) and (iii),

$$\begin{aligned}(\Phi(\hat{Y}^*, \hat{H}^*, s^*) =)\ \hat{Y}^*(s^* @ \hat{H}^*(s^*, \lambda x.\Phi_1(x))) \geq\ & \hat{Y}^*(s @ \hat{H}^*(s, \lambda x.\Phi_2(x)))\ (= \Phi(\hat{Y}^*, \hat{H}^*, s)),\\ & \hat{Y}(s @ \hat{H}(s, \lambda x.\Phi_3(x)))\ (= \Phi(\hat{Y}, \hat{H}, s)),\\ & Y(s @ H(s, \lambda x.\Phi_4(x)))\ (= \Phi(Y, H, s)).\end{aligned}$$

In [5] we show that there exists a functional

$$\Phi : \mathcal{M}_{\rho^{\omega}\to\mathbb{N}} \times \mathcal{M}_{\rho^{*}\times(\rho\to\mathbb{N})\to\rho^{\omega}} \times \mathcal{M}_{\rho^{*}} \to \mathcal{M}_{\mathbb{N}}$$

which, for parameters Y, H, s in $\mathcal{M}$, satisfies equation (3). Therefore, by the theorem above, we obtain that $\mathcal{M}$ satisfies modified bar recursion.

§6. Conclusion. In this paper, we discussed modified bar recursion a variant of Spector's bar recursion that seems to be of some significance in proof theory and the theory and higher type recursion theory. Our main result was an abstract modified realizability interpretation (where realizability for falsity is uninterpreted) of the axioms of countable and dependent choice that can be used to extract programs from non-constructive proofs using these axioms. A similar result can be found in [2], however we claim that our solution is more accessible, since it builds on the well-known model of continuous functionals and the notion of modified realizability instead of an ad-hoc model and realizability as in [2]. It can be noted here that the weak form of modified bar recursion (4) used for the realization of dependent choice can be implemented quite efficiently by equipping the functional with an internal memory that records the value of $H(s, \lambda x.\Phi(s * x))$ and thus avoids its repeated computation. Such an optimization does not seem to be possible for the solution given in [2]. In order to make the realizability interpretation of dependent choice useful for program synthesis, it seems necessary to combine it with optimizations of the A-translation as development e.g., in [6] and [4]. To find out whether this is possible, will be a subject of further research.

Another important result was a definition of the fan functional using modified bar recursion and a version of bar recursion due to Kohlenbach, improving [3] and [17] where a PCF definition of the fan functional was given. In [21] this definition of the fan functional has been applied to give a purely functional algorithm for exact integration of real functions.

The paper concluded with some new results on the model $\mathcal{M}$ of strongly majorizable functionals, in particular, the fact that modified bar recursion exists in $\mathcal{M}$. In [5], further results on the relation between modified bar recursion and other bar recursive definitions can be found. One important result of [5] is that modified bar recursion defines Spector bar recursion primitive recursively and that the converse does not hold.

Acknowledgements. We would like to thank Ulrich Kohlenbach for pointing out some mistakes in an early formulation of Section 5, and for suggesting corrections.

REFERENCES

[1] J. Avigad and S. Feferman, *Gödel's functional ("Dialectica") interpretation*, ***Handbook of proof theory*** (S. R. Buss, editor), Studies in Logic and the Foundations of Mathematics, vol. 137, North-Holland, 1998, pp. 337–405.

[2] S. Berardi, M. Bezem, and T. Coquand, *On the computational content of the axiom of choice*, ***The Journal of Symbolic Logic***, vol. 63 (1998), no. 2, pp. 600–622.

[3] U. Berger, *Totale Objekte und Mengen in der Bereichstheorie*, ***Ph.D. thesis***, Mathematisches Institut der Universität München, 1990.

[4] U. Berger, W. Buchholz, and H. Schwichtenberg, *Refined program extraction from classical proofs*, ***Annals of Pure and Applied Logic***, vol. 114 (2002), pp. 3–25.

[5] U. Berger and P. Oliva, *Modified bar recursion*, ***BRICS Report Series RS-02-14***, BRICS, 2002, 23 pages, http://www.brics.dk/RS/02/14/BRICS-RS-02-14.ps.gz.

[6] U. Berger and H. Schwichtenberg, *Program extraction from classical proofs*, ***Logic and Computational Complexity workshop (LCC'94)*** (D. Leivant, editor), Lecture Notes in Computer Science, vol. 960, Springer, 1995, pp. 77–97.

[7] M. Bezem, *Strongly majorizable functionals of finite type: A model for bar recursion containing discontinuous functionals*, ***The Journal of Symbolic Logic***, vol. 50 (1985), pp. 652–660.

[8] ———, *Equivalence of bar recursors in the theory of functionals of finite type*, ***Archive for Mathematical Logic***, vol. 27 (1988), pp. 149–160.

[9] Y. L. Ershov, *Model C of partial continuous functionals*, ***Logic colloquium 1976*** (R. Gandy and M. Hyland, editors), North-Holland, 1977, pp. 455–467.

[10] K. Gödel, *Über eine bisher noch nicht benützte Erweiterung des finiten Standpunktes*, ***Dialectica***, vol. 12 (1958), pp. 280–287.

[11] W. A. Howard, *Hereditarily majorizable functionals of finite type*, ***Metamathematical investigation of intuitionistic Arithmetic and Analysis*** (A. S. Troelstra, editor), Lecture Notes in Mathematics, vol. 344, Springer, 1973, pp. 454–461.

[12] W. A. Howard and G. Kreisel, *Transfinite induction and bar induction of types zero and one, and the role of continuity in intuitionistic analysis*, ***The Journal of Symbolic Logic***, vol. 31 (1966), no. 3, pp. 325–358.

[13] S. C. Kleene, *Countable functionals*, ***Constructivity in mathematics*** (A. Heyting, editor), North–Holland, 1959, pp. 81–100.

[14] U. Kohlenbach, *Theorie der majorisierbaren und stetigen Funktionale und ihre Anwendung bei der Extraktion von Schranken aus inkonstruktiven Beweisen: Effektive Eindeutigkeitsmodule bei besten Approximationen aus ineffektiven Eindeutigkeitsbeweisen*, ***Ph.D. thesis***, Frankfurt, 1990.

[15] G. Kreisel, *Interpretation of analysis by means of constructive functionals of finite types*, ***Constructivity in mathematics*** (A. Heyting, editor), North–Holland, 1959, pp. 101–128.

[16] H. Luckhardt, ***Extensional Gödel functional interpretation—a consistency proof of classical analysis***, Lecture Notes in Mathematics, vol. 306, Springer, 1973.

[17] D. Normann, *The continuous functionals*, ***Handbook of computability theory*** (E. R. Griffor, editor), North-Holland, 1999, pp. 251–275.

[18] G. D. Plotkin, *LCF considered as a programming language*, ***Theoretical Computer Science***, vol. 5 (1977), pp. 223–255.

[19] H. Schwichtenberg, *On bar recursion of types 0 and 1*, ***The Journal of Symbolic Logic***, vol. 44 (1979), pp. 325–329.

[20] D. S. Scott, *Outline of a mathematical theory of computation*, ***4th annual Princeton conference on Information Sciences and Systems***, 1970, pp. 169–176.

[21] A. Simpson, *Lazy functional algorithms for exact real functionals*, ***Mathematical foundations of computer science*** (L. Brim, J. Gruska, and J. Zlatuska, editors), Lecture Notes in Computer Science, vol. 1450, Springer, 1998, pp. 456–464.

[22] C. Spector, *Provably recursive functionals of analysis: A consistency proof of analysis by an extension of principles in current intuitionistic mathmatics*, ***Recursive function theory: Proceedings of symposia in pure mathematics*** (F. D. E. Dekker, editor), vol. 5, American Mathematical Society, Providence, Rhode Island, 1962, pp. 1–27.

[23] A. S. Troelstra, ***Metamathematical investigation of intuitionistic Arithmetic and Analysis***, Lecture Notes in Mathematics, vol. 344, Springer, 1973.

[24] ———, *Realizability*, ***Handbook of proof theory*** (S. R. Buss, editor), vol. 137, North-Holland, 1998, pp. 408–473.

[25] A. S. Troelstra and D. van Dalen, ***Constructivism in mathematics. An introduction***, Studies in Logic and the Foundations of Mathematics, vol. 121, North-Holland, 1988.

DEPARTMENT OF COMPUTER SCIENCE
UNIVERSITY OF WALES SWANSEA
SINGLETON PARK
SWANSEA, SA2 8PP, UNITED KINGDOM
E-mail: u.berger@swansea.ac.uk
URL: http://www-compsci.swan.ac.uk/~csulrich/

BRICS, DEPARTMENT OF COMPUTER SCIENCE
UNIVERSITY OF AARHUS
AARHUS C, 8000, DENMARK
E-mail: pbo@dcs.qmul.ac.uk
URL: http://www.dcs.qmul.ac.uk/~pbo/
Current address: Department of Computer Science, Queen Mary, University of London, Mile End Road, London E1 4NS, United Kingdom

CHOICE AND UNIFORMITY IN WEAK APPLICATIVE THEORIES

ANDREA CANTINI

Abstract. We are concerned with first order theories of operations, based on combinatory logic and extended with the type W of binary words. The theories include forms of "positive" and "bounded" induction on W and naturally characterize primitive recursive and polytime functions (respectively).

We prove that the *recursive content* of the theories under investigation (i.e., the associated class of provably total functions on W) is invariant under addition of

1. an axiom of choice for operations and a uniformity principle, restricted to positive conditions;
2. a (form of) self-referential truth, providing a fixed point theorem for predicates.

As to the proof methods, we apply a kind of internal forcing semantics, non-standard variants of realizability and cut-elimination.

§1. Introduction. In this paper, we deal with theories of abstract computable operations, underlying the so-called explicit mathematics, introduced by Feferman in the midseventies as a logical frame to formalize Bishop's style constructive mathematics ([17], [18]). Following a common usage, these theories are termed *applicative*, since they primarily axiomatize structures, which are closed under a *general* binary operation of application (so that all objects represent abstract programs and self-application is allowed).

The most important feature of applicative systems is that they include forms of combinatory logic or untyped lambda calculus, and hence they are far more general than bounded arithmetical systems in the sense of Buss [8]. In particular, applicative theories have the strongest expressive power, as they comprise a Turing complete (functional) language, and they can justify suitably controlled recursion principles, without having to add them as primitive.

Although applicative systems are definitionally very strong, it has recently turned out that typical results from bounded arithmetic and the so called intrinsic (or implicit) approach to computational complexity,[1] can be lifted to these systems (see [28], [29], [11], [12]).

Our starting point is given by two natural applicative theories PR and PT, which are considered in [29]. There it is shown that the recursive content of PR

Research supported by Università di Firenze.

[1] Cf. the papers of Bellantoni, Leivant and others [5], [21], [22], [6].

Logic Colloquium '01
Edited by M. Baaz, S. Friedman, and J. Krajíček
Lecture Notes in Logic, 20

coincides with the class of primitive recursive functions, while PT characterizes the class of polytime operations.

We strengthen Strahm's results in two directions: (i) we include principles of choice and uniformity in weak applicative systems; (ii) we study intuitionistic applicative theories and the relations with their classical counterparts.

As to the first direction, we consider an axiom Pos-AC_W, stating that every positive[2] binary relation on binary words and total on W is uniformized by an *operation* of type $W \to W$. Pos-AC_W implies among others that our systems contain a weak version of second order arithmetic (recursive comprehension becomes interpretable in our theories). We also study a new principle Pos-UP of positive uniformity, which is characteristic of the present framework.[3] It states that if we have a positive relation R such that $(\forall x)(\exists y \in W)R(x, y)$, then $(\exists y \in W)(\forall x)R(x, y)$. As it will be seen below, Pos-UP holds in the term model of our theory. It conveys the (topological) idea that W is a *discrete subspace* within the full space of operations.[4]

Concerning the second direction of research, intuitionistic applicative systems having at least the strength of Peano arithmetic, are well known from the extant literature (besides Feferman's source work, cf. [31], [4]), but not much is known about *intuitionistic fragments of applicative theories*. In particular, if induction (on numbers or on binary notations) is restricted to special classes of formulas, e.g., positive or bounded formulas, there are additional genuine difficulties, which suggest a closer investigation of techniques for comparing classical with constructive systems. We will show that an elegant constructivization method, due to Coquand, Hofmann [14] and Avigad [1], can be lifted from fragments of arithmetical theories to applicative frameworks. Furthermore, we are led to investigate extensions to weak intuitionistic systems of *non-standard realizability techniques*, used so far in [21], [29], [12], but only for interpreting *positive* statements of *classical* theories.

Let us briefly survey the content of the paper. §§1–2 introduce an extension PRTC of Strahm's PR by means of positive choice and a truth predicate T, closed under $\wedge$, $\vee$, $\forall$, $\exists$ and satisfying Tarski's biconditional for atomic formulas. Roughly, the theory describes an extensional total combinatory algebra with an embedded structure, the algebra W of binary words. As to W, induction on binary words is admitted for positive conditions only.

---

[2] 'Positive' means 'definable by an implication free formula'.

[3] Of course, we have been guided by analogy with the uniformity principle for numbers in the context of the intuitionistic theory of species, cf. [31], p. 234–237.

[4] These ideas could be made precise by assigning to the term model of our applicative theories the so-called Visser topology ([3], ch. 17), under which the space becomes (hyper)connected, every definable operation is continuous, and W turns out to be a discrete subspace of the whole space. It would then immediately follow that definable operations sending arbitrary objects into elements of W are constant. Now Pos-UP obviously implies this fact and can also be regarded as a logical generalization of it.

§3 is devoted to the intended semantics: the ground structure is the so-called open term model (see [3]). It is shown that, although closed under universal quantification, T has a recursively enumerable interpretation, which is an important source for the realizability interpretation in later sections. This is also the key for observing that positive uniformity and an internal version of the classical law CD ("of constant domains") $(\forall x)(A \vee B) \rightarrow A \vee (\forall x)B$ (with $x \notin FV(A)$) hold in the term model.

§4 spells out a few consequences of the system; among them, it interprets the fragment RCA_0 of recursive analysis (in the sense of [27]; so it subsumes standard Σ_1-induction for arithmetical formulas) and it justifies a "second recursion theorem" for positive predicates.

§5 provides an interpretation of PRTC with positive uniformity into a "quasi-intuitionistic theory". The step requires a preliminary application of the double negation translation, on the assumption of a *stability principle* TS, having the form $\neg\neg Ta \rightarrow Ta$. In order to eliminate TS, we devise a generalized "Friedman-Dragalin" translation, which can be nicely presented as a forcing interpretation $f \Vdash A$. For fragments of arithmetic, this was discovered by Coquand and Hofmann, and Avigad ([14], and [1]). In our case, adapting the method is not straightforward; in particular, conditions must encode *arbitrary positive formulas* and we must be able to quantify over them. It is exactly at this point that we exploit the truth predicate T. Unfortunately, this step does not provide the conclusive constructive interpretation: it still depends on (positive instances) of the classical law CD in the case A is a universal formula.

Getting rid of CD, choice and uniformity is the main task of §6. We apply a somewhat non-standard formalized realizability interpretation $\rho\,\mathbf{r}\,A$, where, in particular, ρ counts as a realizer of a universal statement $(\forall x)A$ exactly when ρ realizes each instance $A(t)$, and hence it *does not depend* on t, as in the standard case. The $\forall$-clause is essential for eliminating the law of constant domains and the uniformity axiom. As a byproduct, the classical theory PRTC is reduced to an intuitionistic version of Strahm's PR, augmented with positive truth.

§7 yields the final upper bound on the recursive content of the theory; this can be obtained by combination of cut elimination with positive realizability. As a result, we have the expected invariance: provably total operations define only primitive recursive operations.

In §§8–9 we show that also the recursive content of the system PT is not altered by adding choice, uniformity and truth axioms. PT is characterized by notation induction for (generalized) Σ_1^b-formulas, i.e., those of the form: $(\exists x \leq t)B$, where B is positive combinatory (it does not contain W nor T). Formally, the principle mimicks a corresponding schema of bounded arithmetic (cf. [8]). But it should be stressed that the applicative principle genuinely extends its arithmetical counterpart: Σ_1^b-applicative formulas possibly include *universal unbounded quantifiers* on the ground universe and are *not decidable* in general. Consequently, the classification theorems obtained are *stronger* than the corresponding results for bounded arithmetical systems.

Computing the recursive content of PTTC is not entirely a routine repetition of the treatment for PRTC. There are two noticeable differences. First of all, we rely upon an external non-formalized realizability interpretation, which has a distinctive *infinitary* character. In addition, the forcing interpretation of Σ_1^b-notation induction does not translate into an instance of the same schema; but it requires an apparently stronger principle for conditions of the form $A(x) \vee C$, A being Σ_1^b, C positive (and $x \notin FV(C)$). Yet it is possible to carry out a modified realizability of the intuitionistic system PT with choice, uniformity, the law of constant domain CD and extended Σ_1^b-induction by means of the Cook-Urquhart feasible functionals. Since feasible functionals of type $W^k \to W$ are polytime, we have the expected upper bound for the system PTTC.

§2. Abstract operations and primitive recursion.

2.1. Syntactical preliminaries. The language $\mathcal{L}_W$ comprises: (i) countably many *individual variables* $x_1, x_2, x_3, \ldots$; (ii) logical constants $\to$, $\wedge$, $\vee$, $\exists$, $\forall$; (iii) *predicate symbols* W (binary words), T(abstract truth), $=$ (equality); (iv) *individual constants* K, S (basic combinators); $PAIR$ (ordered pair operation), $LEFT$ (left projection), $RIGHT$ (right projection); D (definition by cases on W), ϵ (empty sequence), S_0, S_1 (successors), P_W (predecessor), $c_\subseteq$ (testing the initial subsequence relation), $*$ (word concatenation) and $\times$ (word multiplication); (v) the *binary function symbol* Ap (application operation).

$\mathcal{L}_W$ will also comprise *dotted constants* $\dot\forall$, $\dot\exists$, $\dot\wedge$, $\dot\vee$, naming the *positive logical constants*, and $\doteq$ naming the equality predicate.

Terms are inductively defined from variables and constants by means of application Ap. We use x, y, z, u, v, w, f, g as syntactical variables; t, t', s, s', r, r' range over terms. We write (ts) instead of $Ap(t, s)$; outer brackets are usually omitted, while missing ones are restored by associating to the left. We adopt abbreviations for special terms: $(t, s) := PAIRts$ (the ordered pair composed by t and s); $(t)_0 := LEFTt$ (the left projection of t) and $(t)_1 := RIGHTt$ (the right projection of t). We let $S_0(\epsilon) := 0$, $S_1(\epsilon) := 1$ and in general $t^- := PRt$, $t0 := S_0t$, $t1 := S_1t$.

BIN is the least set X of terms, which comprises ϵ and is closed under successors S_0, S_1: if $t \in X$, then $(S_0t) \in X$ and $(S_1t) \in X$. In general, we can identify BIN with the set of binary words. Formally, if α is any binary word, $\overline{\alpha}$ will designate the corresponding term ('binary numeral') of BIN.

As usual [3], we can introduce lambda abstraction: for each term t and variable x, there is a term $\lambda x.t$ such that $(\lambda x.t)s := t[x := s]$ holds (provably in the theory to be defined below) and $FV(\lambda x.t) = FV(t - \{x\})$; $FV(t)$ is the set of variables occurring free in t.

Formulas are inductively generated by means of logical operators and quantifiers from atomic formulas (atoms in short) having the form $t = s$, Wt and Tt. We usually write $t \in W$ instead of Wt and $t \subseteq s$ ($\equiv$ "t is a subword of s") instead of $c_{\subseteq}ts = 0$.

Negation is introduced by letting $\neg A := A \rightarrow \bot$, where $\bot := K = S$.

A formula A is *positive* iff A is inductively generated from atoms of the form $t = s$, $t \in W$, Tt by means of $\wedge$, $\vee$, $\forall$, $\exists$ (so $\rightarrow$ does not occur in A).

If A is positive, we inductively define an operation $A \mapsto [A]$ such that $[A]$ is a term whose free variables are exactly the free variables of A:

$[t = s] = (\dot{=}ts)$;
$[Tt] = t$;
$[s \in W] = \dot{W}s$;
$[A \wedge B] = \dot{\wedge}[A][B]$;
$[A \vee B] = \dot{\vee}[A][B]$;
$[\forall x A] = \dot{\forall}(\lambda x[A])$;
$[\exists x A] = \dot{\exists}(\lambda x[A])$.

$\lambda x.[A]$ can be regarded as the term designating *the propositional function defined by the formula A*. This technical feature is important since it allows *direct* self-reference (to be contrasted with indirect self-reference, based on standard gödelnumberings and appropriate substitution functions).

The given language can interpret naive set theory, as follows:

$$x \in a := T(ax);$$
$$\{x \mid A\} := \lambda x.[A].$$

If $\vec{x} := x_0, \ldots, x_k$, $\vec{x} \in W^{k+1} := x_0 \in W \wedge \cdots \wedge x_k \in W$.
We also let:

$$(\forall x \in W)A := (\forall x)(x \in W \rightarrow A);$$
$$(\exists x \in W)A := (\exists x)(x \in W \wedge A);$$
$$f \in \mathcal{P}(W^{k+1}) := (\forall \vec{x} \in W^{k+1})(f\vec{x} = 0 \vee f\vec{x} = 1).$$

Assume that $A_1, \ldots A_{k+1}$ are arbitrary formulas, where the free variables of each A_i occur exactly in the list $\vec{x}_i$:

$$f : A_1 \rightarrow W := (\forall \vec{x}_1)(A_1(\vec{x}_1) \rightarrow f\vec{x}_1 \in W);$$

$$f : A_1 \times \cdots \times A_{k+1} \rightarrow W := (\forall \vec{x}_1)(A_1(\vec{x}_1) \rightarrow f(\vec{x}_1) : A_2 \times \cdots \times A_{k+1} \rightarrow W).$$

Of course, if $A(x) \equiv (x \in W)$, we have the standard notations for operations of binary words (e.g., $f : W^k \rightarrow W$).

2.2. The theory PRTC. PRTC is the first-order theory, comprising classical first-order logic and finitely many axioms listed below, which characterize (i) the structure W of binary words as embedded in a total combinatory algebra; (ii) a natural notion of positive self-referential truth. In addition, PRTC is equipped with a positive induction axiom and a principle of positive choice.

Our framework PRTC can be regarded as a natural extension of Strahm's system PR (see [29]).

Combinatory logic with extensionality

C.1 $Kxy = x$;
C.2 $Sxyz = xz(yz)$;
Ext $\forall x(fx = gx) \rightarrow f = g$;

Pairing and projections

[Pair] $\forall x \forall y((x, y)_0 = x \wedge (x, y)_1 = y)$;

Definition by cases on W

D.1 $x \in W \wedge y \in W \rightarrow x = y \rightarrow D_W abxy = a$;
D.2 $x \in W \wedge y \in W \rightarrow x = y \vee D_W abxy = b$;

Closure, binary successors and predecessor

B.1 $\epsilon \in W \wedge \forall x(x \in W \rightarrow S_0 x \in W \wedge S_1 x \in W)$;
B.2 $S_0 x \neq S_1 y \wedge S_0 x \neq \epsilon \wedge S_1 x \neq \epsilon$;
P.1 $P_W \colon W \rightarrow W$;
P.2 $P_W \epsilon = \epsilon$;
P.3 $x \in W \rightarrow P_W(S_0 x) = x \wedge P_W(S_1 x) = x$;
P.4 $x \in W \rightarrow x = \epsilon \vee S_0(P_W x) = x \vee S_1(P_W x) = x$;

Word Concatenation

S.1 $* \colon W^2 \longrightarrow W$;
S.2 $x \in W \rightarrow x * \epsilon = x$;
S.3 $x \in W \wedge y \in W \rightarrow x * (S_0 y) = S_0(x * y) \wedge x * (S_1 y) = S_1(x * y)$;

Word multiplication[5]

M.1 $\times \colon W^2 \longrightarrow W$;
M.2 $x \in W \rightarrow x \times \epsilon = \epsilon$;
M.3 $x \in W \wedge y \in W \rightarrow (x \times S_0 y) = (x \times y) * x \wedge (x \times S_1 y) = (x \times y) * x$;

Initial subword relation

I.1 $x \in W \wedge y \in W \rightarrow c_{\subseteq} xy = 0 \vee c_{\subseteq} xy = 1$;
I.2 $x \in W \rightarrow (x \subseteq \epsilon \leftrightarrow x = \epsilon)$;
I.3 $x \in W \wedge y \in W \rightarrow y = \epsilon \vee c_{\subseteq} xy = 1 \vee x \subseteq P_W y \vee x = y)$;
I.4 $x \in W \wedge y \in W \rightarrow y = \epsilon \vee (c_{\subseteq} x(P_W y) = 1 \wedge D01xy = 1) \vee x \subseteq y)$;
I.5 $x \in W \wedge y \in W \wedge z \in W \wedge (x \subseteq y) \wedge (y \subseteq z) \rightarrow x \subseteq z)$;

---

[5]The axioms concerning $*$ and $\times$ are redundant under the axiom of positive induction below, but we include them from the beginning, as they are needed later for the weaker system based on Σ_1^b-induction.

Positive induction Pos-I_W

$$(\epsilon \in a) \wedge (\forall x \in W)(x \in a \rightarrow x0 \in a \wedge x1 \in a) \rightarrow (\forall x \in W)(x \in a)$$

T-AC_W (Positive choice):

$$(\forall x \in W)(\exists y \in W)T(axy) \rightarrow (\exists f)(\forall x \in W)(fx \in W \wedge T(ax(fx)))$$

TA_W (Truth axioms):

$$\begin{array}{lll} (\mathrm{T.1}) & TA \leftrightarrow A & (A \equiv (x = y),\ x \in W) \\ (\mathrm{T.2}) & T(x \dot{\wedge} y) \leftrightarrow Tx \wedge Ty & \\ (\mathrm{T.3}) & T(x \dot{\vee} y) \leftrightarrow Tx \vee Ty & \\ (\mathrm{T.4}) & T(\dot{\forall} f) \leftrightarrow \forall x T(fx) & \\ (\mathrm{T.5}) & T\dot{\exists} f \leftrightarrow \exists x T(fx) & \end{array}$$

LOG (Independence) Let $LOG_1 = \{\dot{W}, \dot{\forall}, \dot{\exists}\}$ and $LOG_2 = \{\dot{\wedge}, \dot{\vee}, \dot{=}\}$. Then:

$$\begin{array}{ll} \neg L_1 = L_2 & (L_1, L_2 \in LOG_1,\ \neg L_1 \equiv L_2); \\ \neg L_1 x = L_2 yz & (L_1 \in LOG_1,\ L_2 \in LOG_2); \\ L_1 x = L_2 y \rightarrow L_1 = L_2 \wedge x = y & (L_1, L_2 \in LOG_1); \\ G_1 xy = G_2 x' y' \rightarrow G_1 = G_2 \wedge x = x' \wedge y = y' & (G_1, G_2 \in LOG_2) \end{array}$$

In the following, it will be important to consider a few subtheories of PRTC:

(i) PRTC_i is PRTC, where classical logic is replaced by *intuitionistic* logic;
(ii) PRT_i is PRTC_i without the axiom of choice;
(iii) PR_i is the subsystem of PRT_i, which omits the truth axioms.

Remark 1. That the provably total functions of PRT_i are the primitive recursive functions, already follows from [10]. There are two differences with the present context, however. The first one is inessential in presence of positive induction (in [10] the ground type is the set of natural numbers). The second one is more important; T can also be applied to *negated equations*, i.e., it is required that $T[x \neq y] \leftrightarrow x \neq y$.

The proof of [10] relies upon a combination of asymmetric interpretation, cut elimination and formalized semantics within $I\Sigma_1$ and the fact that this system proves induction also for boolean combinations of Σ_1-formulas.

As we shall see, the truth axioms ensure that we can define certain predicates by fixed point constructions, in harmony with what happens at the level of operations. In particular our system can be regarded as a fixed point theory, in the sense of Feferman [19], of strength PRA.

§3. **The term model.** We specify as intended semantics of the applicative (i.e., T-free) part of the system the so-called *open term model* $\mathcal{M}(\lambda\eta^+)$ of the extended untyped lambda calculus $\lambda\eta^+$ with extensionality

$\lambda\eta^+$ is the equational theory which extends standard $\lambda\eta$-conversion by means of the obvious natural equations for the additional constants[6] (predecessor, projections, concatenation, etc.); $\lambda\eta^+$ is known to be consistent by verifying the Church-Rosser theorem for an extended version of $\lambda\eta$-reduction (cf. [3], [11]).

The interpretation of the T-free language is fixed by the following clauses:

1. the universe is the set of all terms and application is simply syntactical application;
2. K and S are interpreted as the lambda terms $\lambda y \lambda x.y$ and $\lambda x \lambda y \lambda z.xz(yz)$ (respectively), while the remaining constants are "interpreted onto themselves";
3. $=$ is interpreted as the congruence relation $\approx$ induced by equational provability in $\lambda\eta^+$ (or, if you like, by the extended notion of 'reduction to a common reduct');
4. the predicate W is interpreted with the set $\mathcal{W}$ of all terms t, such that $\lambda\eta^+ \vdash t = \overline{\alpha}$, for some α ($\overline{\alpha}$ being the term canonically designating the binary word α).

In order to assign a denotation to T, let $\mathcal{T}(x, P)$ be the formula describing the recursive clauses for partial self-referential truth, i.e.,

$$\begin{aligned}
\exists u \exists v \big[&(x = [u = v] \wedge u = v) \vee \\
&\vee (x = [u \in W] \wedge u \in W) \vee \\
&\vee (x = u \mathbin{\dot\wedge} v \wedge (Pu \wedge Pv)) \vee \\
&\vee (x = u \mathbin{\dot\vee} v \wedge (Pu \vee Pv)) \vee \\
&\vee (x = \dot\forall u \wedge \forall y P(uy)) \vee \\
&\vee (x = \dot\exists u \wedge \exists y P(uy))\big]
\end{aligned}$$

Clearly $\mathcal{T}(x, P)$ is a positive operator form in the applicative language expanded with P (P being a unary fresh predicate) and hence it defines a monotone operator Γ from the open term model into itself; if X is a subset of (the universe of) $\mathcal{M}(\lambda\eta^+)$:[7]

$$\Gamma(X) = \{a \in \mathcal{M}(\lambda\eta^+) \mid \langle \mathcal{M}(\lambda\eta^+), X\rangle \models \mathcal{T}(a, X)\}.$$

It follows that the set *FIX* of all Γ-fixed points (those subsets X of the open term models satisfying $\Gamma(X) = X$) is non-empty. In particular, $T_{min} \in FIX$,

[6] Distinct from the combinators K, S.

[7] Below we are sloppy in our notations; we use the same symbols for elements of the underlying combinatory algebra and the corresponding constants.

where T_{min} is the least fixed point of *FIX*, and, by transfinite recursion:

$$T^{\delta} = \Gamma\Big(\bigcup_{\xi<\delta} T^{\xi}\Big);$$

$$T_{min} = \bigcup_{\delta \in ON} T^{\delta}.$$

ON is the class of ordinals, but we now show that *ON* can be collapsed to ω.

First of all, by adapting well-known theorems of pure lambda calculus, we have:

LEMMA 1. *$\mathcal{W}$ and the relation $\approx$ on $\mathcal{M}(\lambda\eta^{+})$ are recursively enumerable, but not recursive.*

Below we fix a polytime pairing function J with polytime projections π_0, π_1, under which W is closed.[8] For the sake of simplicity, we do not distinguish between the given functions J, π_0, π_1 and the closed terms representing them in the applicative language $\mathcal{L}_W$.

DEFINITION 1. We inductively define an abstract finitary derivability relation $d \vdash^m t$, where $d \in \mathcal{W}$, $m \in \omega$, t is an arbitrary term, by means of a set of introduction rules; m plays the role of a length measure:

- *ID*-rule:

$$\frac{t = [a = b] \quad a = b}{\epsilon \vdash^m t}$$

- W–rule:

$$\frac{t = [r \in W] \quad r = \overline{\alpha}}{\alpha \vdash^m t}$$

- $\vee$-rule: if $k < m$,

$$\frac{t = a \dot{\vee} b \quad d \vdash^k a \qquad \text{(or} \quad d \vdash^k b)}{J(0,d) \vdash^m t \qquad \text{(or} \quad J(1,d) \vdash^m t)}$$

- $\wedge$-rules: if $i < m$, $k < m$,

$$\frac{t = a \dot{\wedge} b \quad d \vdash^k a \quad e \vdash^i b}{J(d,e) \vdash^m t}$$

- $\forall$-rule: if $k < m$, $x \notin FV(at)$,

$$\frac{t = \dot{\forall} a \quad d \vdash^k ax}{d \vdash^m t}$$

- $\exists$-rule: if $k < m$,

$$\frac{t = \dot{\exists} a \quad d \vdash^k ar, \text{ for some } r}{d \vdash^m t}$$

[8] So J should not be confused with the pairing operation *PAIR*.

It is understood that the equalities occurring in the premises of the previous rules are assumed to be true in the open term model.

The independence properties of the dotted logical constants clearly ensure:

LEMMA 2 (Inversion). *The rules inductively generating $d \vdash t$ are invertible.*[9]

LEMMA 3 (Substitution). *If $d \vdash^m t[x]$, then $d \vdash^m t[x := s]$, for arbitrary s.*

PROOF. By induction on m such that $d \vdash^m t[x]$. We only consider two cases. If we have applied the ID-rule, the claim follows from the corresponding substitution property for $\lambda\eta^+$.[10] If we have applied the $\forall$-rule, we have for some $k < m$ and $e \in \mathcal{W}$:

$$(1) \qquad e \vdash^k a[x]u \quad (u \notin FV(a[x]t[x]));$$

$$(2) \qquad \mathcal{M}(\lambda\eta^+) \models t[x] = (\dot\forall a[x]).$$

Pick any term s and choose $z \notin FV(a[x]t[x]su)$. Since $k < m$ and u is not free in $a[x]$, we obtain by induction hypothesis and substituting u with z in $a[x]u$:

$$e \vdash^k a[x]z \quad (z \notin FV(a[x]))$$

Again by induction hypothesis:

$$e \vdash^k (a[x]z)[x := s] \equiv (a[x := s]z).$$

By the substitution lemma for $\lambda\eta^+$ applied to (2), we have:

$$t[x := s] = \dot\forall(a[x := s]).$$

Since $z \notin FV(a[x := s]t[x := s])$, we conclude:

$$e \vdash^m t[x := s]. \qquad \dashv$$

LEMMA 4. (i) *The relation $d \vdash t$ is recursively enumerable.*
(ii) $T_{min} = T_{re} = \{t \mid d \vdash^m t$, *for some* $d, m\}$.
(iii) *T_{min} is recursively enumerable, but not recursive.*

PROOF. (i) is immediate by inspection, while (iii) follows with (i) and (ii). As to (ii), we must check, for all m, d, t:

$$d \vdash^m t \Rightarrow t \in T_{min}.$$

This easily follows by induction on m with lemma 3 and the fact that T_{min} is Γ-closed. Similarly, using $T_{min} \subseteq \Gamma(T_{min})$, one verifies, for all m, t:

$$t \in T^m \Rightarrow t \in T_{re} \qquad \dashv$$

[9] For instance, if $d \vdash (t_0 \dot\wedge t_0)$, then $d = J(d_0, d_1)$ and $d_i \vdash t_i$, for $i = 0$ and $i = 1$; etc. ...
[10] Extending [3], 2.1.17.

By application of inversion, we easily obtain:[11]

LEMMA 5.

$$(*) \qquad d \vdash^m a \dot{\vee} bx \text{ and } x \notin FV(ab) \Rightarrow d \vdash^{m+1} a \dot{\vee} \dot{\forall} b;$$

$$(**) \qquad d \vdash^m \forall x.\exists y \in W.axy \Rightarrow d \vdash^m \exists y \in W.\forall x.axy.$$

The lemma implies that a truth-theoretic version T-CD of the (classically valid) logical law of $\forall\vee$-distributivity[12]

$$\text{(T-CD)} \qquad (\forall x)T(a \dot{\vee} bx) \to T(a \dot{\vee} \dot{\forall} b)$$

is constructively acceptable in our interpretation of self-referential truth as T_{re}. This is will be exploited at a later stage. Furthermore, $(**)$ implies that a principle of positive uniformity on W

$$\text{(T-UP)} \qquad (\forall x)(\exists y \in W)Taxy \to (\exists y \in W)(\forall x)Taxy$$

is validated by the open term model.

DEFINITION 2. $\mathcal{S}$ is the system PRTC+T-UP+T-CD. $\mathcal{S}_i$ is the intuitionistic version of $\mathcal{S}$.

THEOREM 1. *The structure $\langle \mathcal{M}(\lambda\eta^+), T_{min} \rangle$ is a model of $\mathcal{S}$(even with the notation induction schema on W for arbitrary formulas).*

PROOF. By construction and the previous lemmata, it remains to check the positive axiom of choice for W. Assume the term model satisfies

$$(\forall x \in W)(\exists y \in W)T(txy)$$

By lemma 4, the relation R on $\mathcal{W}$ such that

$$R(\alpha, \beta) \Leftrightarrow \langle \mathcal{M}(\lambda\eta^+), T_{min} \rangle \models Tt\overline{\alpha}\overline{\beta}$$

is recursively enumerable. Hence, by assumption on R and the selection theorem for r.e. sets, there exists a recursive choice function $F : \mathcal{W} \to \mathcal{W}$ such that

$$R(\alpha, F(\alpha)).$$

By λ-definability of recursive functions, there exists a term f such that

$$(\forall x \in W)(fx \in W \wedge T(tx(fx))) \qquad \dashv$$

---

[11] Below, for enhancing readability, we sometimes omit dots from logical constants used within terms and we write $Qx.t$ instead of the proper $\dot{Q}(\lambda x.t)$ (Q quantifier). For instance, $\forall x.\exists y \in W.axy$ will shorten

$$\dot{\forall}(\lambda x.\dot{\exists}(\lambda y.([y \in W] \dot{\wedge} axy))).$$

[12] Also known as law of constant domains, being complete for Kripke models with constant domains.

§4. Elementary consequences. That our system captures at least the primitive recursive operations, is easily implied by a straight interpretability argument. First of all, recall that we can use the standard fixed point theorem of combinatory logic to solve recursive equations:

LEMMA 6. *There exists a closed term Y such that (provably in pure combinatory logic):*

$$f(Yf) = Yf.$$

Hence by positive notation induction on W:

LEMMA 7 (Predicative primitive recursion). PRT_i *proves that there exists a closed term R_W such that, if $A(x)$ is an arbitrary formula, then:*

$$(f: A \to W) \wedge (g: A \times W \times W \to W) \to (R_W fg: A \times W \to W);$$

morevover, if we assume $A(x)$ and $y \in W$, and we let $hxy = R_W fgxy$,

$$hxy = \begin{cases} fx & \textit{if } y = \epsilon, \\ gxy(hxy^-) & \textit{provided } y \neq \epsilon. \end{cases}$$

DEFINITION 3. If $F: \mathcal{W}^k \to \mathcal{W}$, we say that F is *provably total* in a given theory $\mathcal{I}$ (in the language $\mathcal{L}_W$), if there exists a closed term t_F of $\mathcal{L}_W$ such that, forall $\alpha_1, \dots \alpha_k \in \mathcal{W}$:

$$\mathcal{I} \vdash t_F: W^k \to W;$$

$$\mathcal{M}(\lambda\eta^+) \models t_F\overline{\alpha_1}, \dots \overline{\alpha_k} = \overline{F(\alpha_1, \dots \alpha_k)}$$

Hence by lemma 7:

COROLLARY 1. *If a function $F: \mathcal{W}^k \to \mathcal{W}$ is primitive recursive, then F is provably total in* PRT_i.

The system PRT_i proves a corresponding second recursion theorem for (positive) predicates. According to the definitions of 2.1, membership and class abstraction (at least for positive formulas) make sense in our language.

PROPOSITION 1. (i) *If $A(x)$ is positive, then* PRT_i *proves:*

(T) $$T[A] \leftrightarrow A;$$

(Pos-CA) $$(\forall u)(u \in \{x \mid A\} \leftrightarrow A[x := u]).$$

(ii) *If $A(x, y)$ is positive , then there exists a term I_A, such that* PRT_i *proves:*

(Fix) $$(\forall u)(u \in I_A \leftrightarrow A(u, I_A)).$$

PROOF. (i) The truth axioms imply the schema (for A positive):

$$T[A] \leftrightarrow A.$$

Then positive comprehension follows by definition of class abstraction and β-conversion.

(ii): choose $I_A = Y(\lambda y.\{x \mid A(x, y)\})$. Then apply (i) and lemma 6. ⊣

Let

$$\Delta(f,g) \leftrightarrow (\forall u \in W)((\exists y \in W)(fuy = 0) \leftrightarrow \neg(\exists z \in W)(guz = 0)).$$

By Pos-AC$_W$ and classical logic, we easily obtain:

PROPOSITION 2 (Recursive Comprehension). PRTC *proves*

$$\Delta(f,g) \wedge f \in \mathcal{P}(W) \wedge g \in \mathcal{P}(W) \rightarrow$$
$$\rightarrow (\exists h \in \mathcal{P}(W))(\forall x \in W)(hx = 0 \leftrightarrow (\exists y \in W)(fxy = 0))$$

Proposition 2 establishes a link between PRTC and the well-known fragment RCA$_0$ of second order arithmetic, based on recursive comprehension and Σ_1-induction (for details, see Simpson's monograph [27]). Indeed, it is possible to prove with lemma 7 and proposition 2:

PROPOSITION 3. RCA$_0$ *is interpretable in* PRTC.

§5. **Interpreting the theory in a constructive one.** The result of this section is an interpretation of the theory $\mathcal{S}$ (definition 2, §3) in its intuitionistic counterpart $\mathcal{S}_i$. This step is carried out in two stages: we first embed $\mathcal{S}$ in the intuitionistic system $\mathcal{S}_i$ extended with a stability axiom TS. We then get rid of TS via a forcing interpretation, which essentially hinges upon the principle T-CD of the previous section. By a suitable version of abstract realizability, we finally get rid of T-CD itself, choice and T-UP.

5.1. Negative translation.

DEFINITION 4. The Gödel-Gentzen double-negation translation $A \mapsto A^N$ is inductively defined by stipulating:

1. $A^N = \neg\neg A$, if A is atomic;
2. the translation commutes with $\forall$, $\wedge$, $\rightarrow$;
3. $(\exists x B)^N = \neg\forall x \neg(B)^N$;
4. $(A \vee B)^N = \neg(\neg A^N \wedge \neg B^N)$.

We recall that PRTC$_i$ stands for the theory PRTC with *intuitionistic logic*.

We cannot simply interpret PRTC into its intuitionistic version PRTC$_i$ via the double negation translation, since the image of positive induction under the map $A \mapsto A^N$ is not an instance of positive induction. Thus we proceed with a preliminary step and introduce the axiom TS of *truth stability*:[13]

$$\neg\neg Tx \rightarrow Tx$$

LEMMA 8. *Positive formulas are stable and invariant under the negative translation, i.e.,* PRT$_i$+TS *proves*:

$$(*) \qquad \neg\neg A \leftrightarrow A,$$

$$(**) \qquad A^N \leftrightarrow A.$$

[13] It is a form of generalized Markov principle in disguise.

PROOF. $(*)$ is a consequence of TS and the Tarski schema (T) for positive formulas (cf. proposition 1). As to $(**)$, it follows by induction with intuitionistic logic and $(*)$. ⊣

If S is the system introduced in definition 2, we have:

LEMMA 9. *If* $S \vdash A$, *then* $S_i + \mathsf{TS} \vdash A^N$. *Moreover,* S *and* $S_i + \mathsf{TS}$ *have the same provably total functions.*

PROOF. The first claim is well known for the logical part. That the negative translations of the applicative axioms together with Pos-I_W, T-AC_W, TA_W, T-UP and T-CD are provable, is immediate by the previous lemma. The second claim follows from the first part with $(*)$. ⊣

5.2. Eliminating truth stability: the forcing interpretation.

DEFINITION 5. (i) Let g, h stand for arbitrary variables; we let

$$g \leq_T h \Leftrightarrow (Th \rightarrow Tg)$$

(ii) We inductively assign to every formula A a formula $f \Vdash A$, where f does not occur in A:

$$\begin{aligned} f \Vdash A &\quad \Leftrightarrow Tf \vee A, \text{ if A is atomic;} \\ f \Vdash A \rightarrow B &\quad \Leftrightarrow (\forall g \leq_T f)(g \Vdash A \rightarrow g \Vdash B) \\ f \Vdash A \circ B &\quad \Leftrightarrow (f \Vdash A \circ f \Vdash B) \quad (\circ = \wedge, \vee) \\ f \Vdash QxA &\quad \Leftrightarrow Qx(f \Vdash A) \quad (Q = \forall, \exists) \end{aligned}$$

Of course, by definition of negation, PRT_i proves:

$$f \Vdash \neg A \leftrightarrow (\forall g \leq_T f)(g \Vdash A \rightarrow Tg) \tag{3}$$

LEMMA 10 ($\leq_T$-properties, provable in PRT_i). (i) $\leq_T$ *is a partial ordering, which is closed under inf (greatest lower bound); in particular, for every* f, g, *there exists* h *such that* (1) $h \leq_T f$ *and* $h \leq_T g$; (2) *if* $l \leq_T f$ *and* $l \leq_T g$, *then* $l \leq_T h$.

(ii) *Downward persistence*:

$$(f \Vdash A) \wedge (g \leq_T f) \rightarrow (g \Vdash A).$$

The verification of (i) is immediate by standard (positive) logic, choosing $h = f \dot{\vee} g$ and applying the T-axiom for disjunction.

LEMMA 11 ("Law of constant domains" for positive formulas). *If* A, B *are positive and* $x \notin FV(A)$, PRT_i+T-CD *proves*:

$$(\forall x)(A \vee B(x)) \rightarrow (A \vee (\forall x)B) \tag{Pos-CD}$$

PROOF. Apply the positive Tarski schema (T), (T-CD), and (T.3). ⊣

LEMMA 12 (Provable in PRT_i+T-CD). *If A is positive,*

(i) $$f \Vdash A \leftrightarrow (Tf \vee A);$$

(ii) $$A \leftrightarrow (\forall f)(f \Vdash A);$$

(iii) $$f \Vdash \neg A \leftrightarrow (A \rightarrow Tf);$$

(iv) $$f \Vdash \neg\neg A \rightarrow f \Vdash A$$

If A is arbitrary,

(v) $$Tf \rightarrow f \Vdash A.$$

PROOF. Ad (i): inductive argument, with lemma 11.

Ad (ii): apply (i) from left to right. Conversely, assume $(\forall f)(f \Vdash A)$ and choose $f := [A]$. Then $T[A] \vee A$. Since the first disjunct implies A, we are done.

Ad (iii): apply (i).

Ad (iv): assume $f \Vdash \neg\neg A$, where A is positive. By the property (3) for negation, choosing $g \leq_T f$ such that

$$Tg \leftrightarrow T[A] \vee Tf$$

we successively obtain, with (iii) and the T-schema for positive truth:

$$(g \Vdash \neg A) \rightarrow Tg;$$
$$(A \rightarrow Tg) \rightarrow Tg;$$
$$A \rightarrow Tg;$$
$$T[A] \vee Tf \text{ (by choice of } g);$$
$$f \Vdash A.$$

Ad (iv): by induction on A. ⊣

THEOREM 2. (i) *If* $\mathcal{S}_i + \mathsf{TS} \vdash A$, *then*

$$\mathcal{S}_i \vdash (\forall f)(f \Vdash A).$$

(ii) *Moreover, if* $\mathcal{S}_i + \mathsf{TS}$ *proves* $(\forall x \in W)(\exists y \in W)T(axy)$, *the same formula is provable in* $\mathcal{S}_i$. *Hence* $\mathcal{S}$ *and* $\mathcal{S}_i$ *have the same provably total functions.*

PROOF. We first check (ii). By (i), $\mathcal{S}_i$ proves, for arbitrary f:

$$f \Vdash (\forall x \in W)(\exists y \in W)T(axy).$$

By the truth axioms, there exists h satisfying

$$Th \leftrightarrow (\exists y \in W)T(axy).$$

Choosing $f = h$, we easily obtain the conclusion.

(i): the proof is by induction on the length of the proof of A in $\mathcal{S}_i + \mathsf{TS}$. For definitess, we may assume that the logical axioms and rules are given in

Hilbert style and are those listed in [31], p. 69. Truth stability is validated by lemma 12 (iv), while logical rules and axioms are easily checked (in particular, $\bot \to A$ is forced with the help of lemma 12, (v)).

The interpretation of truth axioms, T-UP, T-CD, the applicative axioms (including those on combinators, pairing and projections, binary successors, predecessor, extensionality, etc.) and Pos-I_W essentially follows with the help of lemma 12 (i).

Let us verify T-AC_W. So assume

$$f \Vdash (\forall x \in W)(\exists y \in W)T(axy). \tag{4}$$

By (4) and lemma 12, for every x and $h \leq_T f$,

$$x \in W \vee Th \to ((\exists y \in W)T(axy)) \vee Th,$$

which implies by T-axioms, intuitionistic logic, since W is non-empty, for all $h \leq_T f$:

$$(\forall x \in W)(\exists y)((y \in W \wedge T(axy)) \vee Th),$$
$$(\forall x \in W)(\exists y)(y \in W \wedge (T(axy) \vee Th)).$$

By T-AC_W, choosing $h := f$, there exists g such that

$$(\forall x \in W)(gx \in W \wedge (T(ax(gx)) \vee Tf)), \tag{5}$$
$$(\forall x \in W)((gx \in W \wedge T(ax(gx))) \vee Tf). \tag{6}$$

Pick an arbitrary $h \leq_T f$ and assume $h \Vdash x \in W$. If $x \in W$, by (6),

$$(gx \in W \wedge T(ax(gx))) \vee Tf.$$

If the first disjunct holds, then $h \Vdash gx \in W \wedge T(ax(gx))$. If Tf holds, we have Th (as $h \leq_T f$), whence trivially

$$h \Vdash gx \in W \wedge T(ax(gx)).$$

Therefore we have proved that f forces the conclusion of the axiom of choice. ⊣

§6. Abstract internal realizability: eliminating AC_W, UP and CD. Henceforth we work in the system PRT_i, the intuitionistic subsystem of PRTC without the positive axiom of choice. We will define a realizability interpretation of $\mathcal{S}_i$ into PRT_i. In order to realize truth atoms and axioms, we need a formal version of the relation $d \vdash t$ (definition 1). Indeed, by the fixed point property of proposition 1, we have, with the help of the polytime pairing function J of section 3:

LEMMA 13. *There exists a formula* $d \vdash' a$ (*of the form* $Tt(d,a)$, *for some term* $t(x,y)$), *such that provably in* PRT_i:

$$d \vdash' a \leftrightarrow d \in W \wedge$$
$$(\exists u)(\exists v)\big[(a = [u = v] \wedge u = v \wedge d = \epsilon)$$
$$\vee\, (a = [u \in W] \wedge u = d)$$
$$\vee\, (a = [u \dot{\vee} v] \wedge ((d = J(0,d_1) \wedge d_1 \vdash' u)$$
$$\vee\, (d = J(1,d_1) \wedge d_1 \vdash' v)))$$
$$\vee\, (a = (u \dot{\wedge} v) \wedge d = J(d_0,d_1) \wedge (d_0 \vdash' u) \wedge (d_1 \vdash' v))$$
$$\vee\, (a = \dot{\forall} u \wedge (\forall x)(d \vdash' ax))$$
$$\vee\, (a = (\dot{\exists} u) \wedge (\exists x)(d \vdash' ax))\big].$$

The independence properties of the dotted logical operators immediately imply in PRT_i:

LEMMA 14 ($\vdash'$-Inversion).

$$\begin{aligned}
d \vdash' [t = s] &\leftrightarrow t = s \wedge d = \epsilon;\\
d \vdash' [t \in W] &\leftrightarrow d \in W \wedge d = t;\\
d \vdash' u \dot{\vee} v &\leftrightarrow (d = J(0,d_1) \wedge d_1 \vdash' u) \vee (d = J(1,d_1) \wedge d_1 \vdash' v);\\
d \vdash' u \dot{\wedge} v &\leftrightarrow d = J(d_0,d_1) \wedge d_0 \vdash' u \wedge d_1 \vdash' v;\\
d \vdash' \dot{\forall} u &\leftrightarrow (\forall x)(d \vdash' ux);\\
d \vdash' \dot{\exists} u &\leftrightarrow (\exists x)(d \vdash' ux)
\end{aligned}$$

DEFINITION 6. $x\,\mathbf{ur}\,A$,[14] where $x \notin FV(A)$ is defined by induction on the complexity of A:

$$\begin{aligned}
x\,\mathbf{ur}\,A &\Leftrightarrow x \vdash' [A] \qquad (A \text{ atomic});\\
x\,\mathbf{ur}\,A \wedge B &\Leftrightarrow x_0\,\mathbf{ur}\,A \wedge x_1\,\mathbf{ur}\,A \wedge x = (x_0, x_1);\\
x\,\mathbf{ur}\,A \vee B &\Leftrightarrow (x = (0,x_1) \wedge x_1\,\mathbf{ur}\,A) \vee (x = (1,x_1) \wedge x_1\,\mathbf{ur}\,B);\\
x\,\mathbf{ur}\,A \rightarrow B &\Leftrightarrow (\forall y)(\mathbf{ur}\,A \rightarrow xy\,\mathbf{ur}\,B);\\
x\,\mathbf{ur}\,(\exists z)A &\Leftrightarrow (\exists z)(x\,\mathbf{ur}\,A(z));\\
x\,\mathbf{ur}\,(\forall z)A &\Leftrightarrow (\forall z)(x\,\mathbf{ur}\,A(z)).
\end{aligned}$$

AC_W is the schema: if A is an arbitrary formula,

$$(\forall x \in W)(\exists y \in W)A(x,y) \rightarrow (\exists f)(\forall x)(x \in W \rightarrow fx \in W \wedge A(x, fx)).$$

UP is the schema: if A is an arbitrary formula,

$$(\forall x)(\exists y \in W)A(x,y) \rightarrow (\exists y \in W)(\forall x)A(x,y).$$

---

[14]**ur** should be reminiscent of the uniform infinitary character of this realizability.

CD is the classically valid schema: if A, B are arbitrary, $x \notin FV(A)$,

$$(\forall x)(A \vee B(x)) \rightarrow A \vee (\forall x)B(x).$$

LEMMA 15 (Realizing W-choice and UP). *If A is an instance of* AC_W *or* UP, *then* $(\lambda x.x)$ **ur** A, *provably in* PRT_i.

PROOF. We argue informally in PRT_i. We temporarily let ρ, φ range over realizers. In order to prove the choice schema, assume

$$\rho\ \mathbf{ur}\ (\forall x \in W)(\exists y \in W)A(x, y).$$

By the implication clause and the atomic clause for $(x \in W)$, we obtain, for arbitrary x:

$$x \in W \rightarrow \rho x\ \mathbf{ur}\ (\exists y \in W)A(x, y).$$

By definition of **ur**-realizability for $\exists$, $\wedge$, if $(x \in W)$, there exists some s such that:

$$(\rho x)_0\ \mathbf{ur}\ (s \in W) \wedge (\rho x)_1\ \mathbf{ur}\ A(x, s).$$

Then $s = (\rho x)_0$; hence, if we let $f = \lambda x.(\rho x)_0$, we obtain by substitution and β-conversion, for all $x \in W$:

$$(\rho x)_0\ \mathbf{ur}\ (fx \in W) \wedge (\rho x)_1\ \mathbf{ur}\ A(x, fx);$$
$$\rho\ \mathbf{ur}\ (x \in W \rightarrow fx \in W \wedge A(x, fx)).$$

By successive applications of realizability for $\forall$ and $\exists$, we finally obtain:

$$\rho\ \mathbf{ur}\ (\forall x)(x \in W \rightarrow fx \in W \wedge A(x, fx));$$
$$\rho\ \mathbf{ur}\ (\exists f)(\forall x)(x \in W \rightarrow fx \in W \wedge A(x, fx)).$$

As to the uniformity principle in its general form, assume that ρ satisfies

$$\rho\ \mathbf{ur}\ (\forall x)(\exists y \in W)A(x, y).$$

Apply realizability for $\forall$, $\exists$, and $\wedge$; then for every x, there exists some s, such that:

$$(\rho)_0\ \mathbf{ur}\ (s \in W) \wedge (\rho)_1\ \mathbf{ur}\ A(x, s).$$

Hence $s = (\rho)_0$, independently on x; by substitution, we obtain, *uniformly* in x:

$$(\rho)_1\ \mathbf{ur}\ A(x, (\rho)_0).$$

It immediately follows that ρ also realizes the consequent of UP. ⊣

Remark 2. Trivially, UP is inconsistent with the law of excluded middle; indeed $(x = 0 \vee x \neq 0)$ implies

$$(\forall x)(\exists y \in W)(x \neq y).$$

However, as we already know, the positive version Pos-UP of UP is true in the open term model. Pos UP is a consequence of a reflection schema W-Ref

$$(\forall x)(\exists y \in W)A(x,y) \to (\exists z \in W)(\forall x)(\exists y \leq z)(y \in W \wedge A(x,y))$$

where $\leq$ is the length ordering on binary words and A is positive in W, i.e., A is inductively generated from atoms of the form $t = s$, $\neg t = s$, $t \in W$ by means of quantifiers, $\wedge$, $\vee$. The analogue of W-Ref for natural numbers is studied in [10]. A principle related to Pos-UP is also considered by Minari [25] in the context of a suitable theory of types and names.

THEOREM 3 (Internal Realizability). *Assume* $\mathsf{PRT}_i + \mathsf{CD} + \mathsf{UP} + \mathsf{AC}_W \vdash A$. *Then there exists a closed term* φ_A *such that* $\mathsf{PRT}_i \vdash (\forall \vec{x})(\varphi_A\, \mathbf{ur}\, A(\vec{x}))$, *($\vec{x}$ being a list including all free variables of A).*

PROOF. The realizability of the logical axioms and rules is standard.[15] In particular, due to the trivial behaviour of realizability on quantifiers, the quantifier axioms and rules are trivially realized by $\lambda x.x$.

AC_W, UP: a fortiori from the previous lemma.

Truth axioms: this step essentially hinges upon the properties of $d \vdash' a$. Again we can choose the identity operation as realizer. We give an example;

$$\begin{aligned} \rho\, \mathbf{ur}\, T(\dot{\exists} a) &\leftrightarrow \rho \vdash' \dot{\exists} a \\ &\leftrightarrow (\exists x)(\rho \vdash' (ax)) \\ &\leftrightarrow (\exists x)(\rho\, \mathbf{ur}\, T(ax)) \\ &\leftrightarrow \rho\, \mathbf{ur}\, (\exists x) T(ax) \end{aligned}$$

CD: if ρ realizes the antecedent of CD, $(\rho)_0 = 0$ or $(\rho)_0 = 1$, independently on x (by the uniform realizability condition for $\forall$). This readily implies that ρ is also a realizer of the conclusion of CD.

Ad Pos-I_W: assume that ρ realizes the premiss of positive induction, and hence that there are ρ_0, ρ_1 realizing $Ta\epsilon$ and $(\forall x \in W)(Tax^- \to Tax)$. Define

$$F\rho x = \begin{cases} \rho_0 & \text{if } x = \epsilon, \\ \rho_1 x(F\rho x^-) & \text{otherwise.} \end{cases}$$

But $\rho \vdash' a$ has the form $Tt(a,x)$ and $\rho \vdash' a$ implies $\rho \in W$. Then Pos-I_W yields that $F\rho x\, \mathbf{ur}\, Tax$, for every $x \in W$ and hence F is the required realizer. ⊣

By induction on A, it is immediate to see that

---

[15]The propositional axioms are dealt with in much the same way as in [31]. E.g., in order to realize the disjunction axiom,

$$(A \to C) \to (B \to C) \to (A \vee B) \to C,$$

one exploits definition by cases on W and relies upon the fact that if $\rho\, \mathbf{ur}\, A \vee B$, then $\rho_0 \in W$.

LEMMA 16. *If A is T-free and positive,* $\mathsf{PRT}_i \vdash (x \mathbf{ur} A) \rightarrow A$.

The preceding theorem and the lemma imply

COROLLARY 2. (i) *If A is T-free positive and $(\forall \vec{x} \in W)A$ is a theorem of $\mathcal{S}$, then* $\mathsf{PRT}_i \vdash (\forall \vec{x} \in W)A$.

(ii) PRT_i+CD+UP+AC_W, $\mathcal{S}$ *and* PRT_i *have the same provably total functions.*

§7. **The recursive content of PRT_i.** We combine cut elimination and external positive realizability (in the style already used by [29], [12] and inspired by Leivant [21]). We first embed PRT_i into a sequent calculus, such that every derivation of a positive sequent can effectively be transformed into a quasi-normal derivation, i.e., it only contains cuts on positive formulas. This preliminary step has the effect of reducing the "higher type" realizers[16] implicit in the internal interpretation of the previous section to objects of ground type. We are then in the position of applying realizability to positive sequents and we get the conclusive estimate.

7.1. Proof-theoretic preliminaries. First of all, we consider an intuitionistic Gentzen-style version of PRT_i (for which we adopt the same name).

The Gentzen-style calculus PRT_i derives sequents of the form $\Gamma \Rightarrow \Delta$, where Γ, Δ are (possibly empty) multisets of formulas. It includes: (i) the standard logical inferences for intuitionistic connectives, quantifiers, cut, contraction (see the system $G2i$ of [30]); (ii) sequent style reformulations of the axioms for the basic constants (combinators, definition by cases on W, pairing and projections, binary successors and predecessor, word concatenation and multiplication, initial subword test); (iii) sequent style reformulations of the T-axioms; (iv) the Pos-I_W rule:

$$\frac{\Gamma \Rightarrow Tc\epsilon \quad \Gamma,\ a \in W,\ Tca \Rightarrow Tc(ai) \ \ (\text{where } i = 0, 1)}{\Gamma, t \in W \Rightarrow Tct};$$

Proviso: $a \notin FV(\Gamma \cup Tc(\epsilon))$.

It is not so important to specify the sequent style form of each PRT_i-axiom; the crucial property is that they can be presented in the form of sequents where the main formulas are positive. For instance, the axiom of extensionality and the axiom of truth elimination on disjunction, correspond to the sequents:

$$\Gamma,\ \forall x(tx = sx) \Rightarrow t = s, \quad \text{where } x \notin FV(t = s);$$

$$\Gamma,\ T(a \dot{\vee} b) \Rightarrow Ta \vee Tb.$$

[16]We can devise a direct realizability interpretation where realizers are functionals of higher type, in the predicative fragment of Gödel's $\widehat{T}$ (cf. [2]). Following this functional route, the formal interpretation of section 6 and cut elimination become superfluous, but we must rely on Parsons's theorem 5.1.1 in [2].

If we assign to each formula A a rank $rk(A)$, measuring the complexity of A over its positive subformulas (so that $rk(A) = 0$ if A is positive), we can prove by fairly standard methods:

THEOREM 4 (Partial cut elimination). *If* $\Gamma \Rightarrow A$ *is derivable in the sequent calculus* PRT_i, *then* $\Gamma \Rightarrow A$ *is already derivable in the same calculus with a derivation* $\mathcal{D}$ *in which all cut formulas are positive.*

If we call a sequent $\Gamma \Rightarrow A$ *positive* if all formulas occurring in $\Gamma \Rightarrow A$ are positive, we then have:

COROLLARY 3. *If* $\Gamma \Rightarrow A$ *is a positive sequent derivable in* PRT_i, *then there exists a derivation of* $\Gamma \Rightarrow A$ *in* PRT_i, *which contains only positive formulas.*

7.2. External positive realizability. In the following, when we claim that a sentence A is *true* (unqualifiedly), we understand that A holds in the open term model $\langle \mathcal{M}(\lambda\eta^+), T_{min}\rangle$. Let us remind that J is a fixed polytime pairing function with polytime projections π_0, π_1.

DEFINITION 7. We inductively define the relation ρ **pr** A,[17] where A is a *positive formula* and $\rho \in \mathcal{W}$:

$$\rho \,\mathbf{pr}\, A \Leftrightarrow \rho \vdash [A] \qquad (A \text{ is atomic});$$
$$\rho \,\mathbf{pr}\, A \wedge B \Leftrightarrow \rho = J(\rho_0, \rho_1) \text{ and } \rho_0 \,\mathbf{pr}\, A \text{ and } \rho_1 \,\mathbf{pr}\, B;$$
$$\rho \,\mathbf{pr}\, A \vee B \Leftrightarrow \rho = J(\rho_0, \rho_1) \text{ and } ((\rho_0 = 0 \text{ and } \rho_1 \,\mathbf{pr}\, A) \text{ or } (\rho_0 = 1 \text{ and } \rho_1 \,\mathbf{pr}\, B));$$
$$\rho \,\mathbf{pr}\, (\exists x)A \Leftrightarrow \rho \,\mathbf{pr}\, A(t), \qquad \text{for some term } t;$$
$$\rho \,\mathbf{pr}\, (\forall x)A \Leftrightarrow \rho \,\mathbf{pr}\, A[x := y], \quad y \notin FV(A).$$

From the substitution lemma for $d \vdash a$, we easily derive a corresponding property for realizability:

LEMMA 17 (Substitution). *If* A *is positive,*

$$\rho \,\mathbf{pr}\, A[a := t] \text{ and } t = s \Rightarrow \rho \,\mathbf{pr}\, A[a := s];$$
$$\rho \,\mathbf{pr}\, A(a) \Rightarrow \rho \,\mathbf{pr}\, A[a := t].$$

DEFINITION 8. (i) Let $\Gamma := \{A_1, \ldots, A_n\}$, where every A_i with $1 \leq i \leq n$ is positive. We define:

$$\vec{\rho} \,\mathbf{pr}\, \Gamma \Leftrightarrow \rho_1 \,\mathbf{pr}\, A_1, \ldots, \rho_n \,\mathbf{pr}\, A_n$$

where $\vec{\rho} := \rho_1, \ldots, \rho_n$.

(ii) $\Gamma_{\vec{a}} \Rightarrow C_{\vec{a}}$ means that the free variables of $\Gamma \Rightarrow C$ occur in the list $\vec{a}$.

By *'realizer'* we simply understand a binary word ρ; $\rho, \sigma, \rho', \sigma', \varphi$ etc. range over realizers, while $\vec{\rho}, \vec{\sigma}, \vec{\rho}', \vec{\sigma}', \vec{\varphi}$ etc., designate vectors of realizers.

---

[17] **pr** should be reminiscent of the restriction to positive formulas.

THEOREM 5. *If $\mathcal{D}$ is a derivation in* PRT_i *of a positive sequent* $\Gamma_{\vec{a}} \Rightarrow A_{\vec{a}}$*, then there exists a primitive recursive function* $\varphi_{\mathcal{D}}$*, whose range is* $\mathcal{W}^{q+1}$*, ($q+1$ being the number of formulas in Γ) such that, for all* $\vec{p}$ *and* $\vec{r}$:

$$\vec{p}\, \mathbf{pr}\, \Gamma_{\vec{r}} \Rightarrow \varphi_{\mathcal{D}}(\vec{p})\, \mathbf{pr}\, A_{\vec{r}}.$$

For the proof, cut elimination ensures that our sequent has a derivation which includes only positive formulas; so we can proceed by induction to realize the axioms and logical rules. Constructions are similar to those of the previous internal realizability (see also [29]); the only difference is that truth axioms are taken care by the relation of $d \vdash a$ and by its closure properties; Pos-I_W obviously requires primitive recursion. By the lower bound result and the previous section we therefore conclude:

COROLLARY 4. *The provably total functions of* $\mathsf{PRT}_i + \mathsf{CD} + \mathsf{UP} + \mathsf{AC}_W$ *and hence of* $\mathcal{S}$ *are exactly the primitive recursive functions.*

§8. Polytime induction, positive truth and choice.

We introduce a 'bounded' applicative analog PTTC of PRTC, which will play the role of the usual arithmetical systems based on NP-notation induction ([8], [20]) and will characterize polytime.

First of all, define

$$t \leq s \Leftrightarrow c_{\subseteq}(1 \times t)(1 \times s) = 0.$$

If t, s represent binary words, $t \leq s$ stands for the relation: "the length of t is at most the length of s". We also use standard abbreviations for bounded quantifiers; if x is not free in t,

$$(\forall x \leq t)A \Leftrightarrow (\forall x \in W)(x \leq t \rightarrow A);$$
$$(\exists x \leq t)A \Leftrightarrow (\exists x \in W)(x \leq t \wedge A).$$

A formula A is *W-free* if W does not occur in A; A is *combinatory* if it is W-free and does not contain T.

A is a Σ_1^b–formula, if it has the form: $(\exists x \leq t)B$, where B is positive combinatory.

DEFINITION 9. The system PTTC comprises all axioms of PRTC, except that:
positive notation induction Pos-I_W is replaced by the schema Σ_1^b-I_W:

$$(f : W \rightarrow W) \rightarrow$$
$$\rightarrow A(\epsilon) \wedge (\forall x \in W)(A(x) \rightarrow A(x0) \wedge A(x1)) \rightarrow (\forall x \in W)A(x),$$

where $A(x) \equiv (\exists y \leq fx)B(f, x, y)$ and B is positive combinatory.

As before, we shall also consider intuitionistic subtheories of PTTC:

(1) PTTC_i is PTTC based on intuitionistic logic;
(2) PTT_i is PTTC_i without choice;

(3) PT_i is PTT_i without truth axioms.

We also let $\mathcal{T} :=$ PTTC+T-UP+T-CD;[18] $\mathcal{T}_i$ is the corresponding intuitionistic theory.

Clearly, we have by induction on A:

LEMMA 18. *If A is positive, $T[A] \leftrightarrow A$ is provable in* PTT_i.

The intuitionistic fragment PT_i of PTTC proves that all polytime functions are total (in the sense that they assume binary words as values, if their inputs are binary words). Indeed, remind that a function F of binary words is defined by *bounded recursion on notation* iff there exist functions G, H, L such that:

$$F(x,y) = \begin{cases} G(x) & \text{if } y = \epsilon, \\ H(x,y,F(x,y^-))\lceil L(x,y) & \text{provided } y \neq \epsilon. \end{cases}$$

Here $a\lceil b$ is the truncation operator, satisfying (i) $a\lceil b = a$, if $a \leq b$; (ii) $a\lceil b = b$, else. $\lceil$ is definable in our language using definition by cases, subword test and word multiplication.

The collection of polytime functions, FPTIME, is the smallest set of functions (on $\mathcal{W}$) which is closed under composition, bounded primitive recursion on notation and comprises the constant function ϵ, binary successors, projections, word concatenation and word multiplication.

PROPOSITION 4. *Every polytime function is provably total in* PT_i.

For the proof, see [29]: induction for Σ_1^b-formulas yields a general form of recursion subsuming the schema above.

8.1. Negative interpretation and TS-elimination. We only sketch the main steps, pointing out the main differences with the primitive recursive system of the previous part.

In analogy with lemma 9, we have for the theory of definition 9:

LEMMA 19. (i) *$\mathcal{T} \vdash A$ implies $\mathcal{T}_i + \mathsf{TS} \vdash A^N$.*

(ii) *$\mathcal{T}$ and $\mathcal{T}_i + \mathsf{TS}$ have the same provably total functions.*

TS-elimination via forcing is slightly more delicate. Although the definition of forcing for $\mathcal{T}_i + \mathsf{TS}$, the formula

$$f \Vdash A \quad (f \notin FV(A)),$$

is defined as in subsection 5.2, we need a strengthened version of Σ_1^b-bounded induction,[19] which nevertheless does not alter the computational strength of the theories involved.

[18] T-CD is obviously derivable in PTT, but the redundant formulation is convenient for later statements.

[19] A strengthened bounded induction is also used by Coquand and Hofmann in the case of the arithmetical systems, where credit and reference to previous work of Buss is given.

DEFINITION 10. (i) $\Sigma_1^b\text{-I}_W^e$ is the axiom:

$$(\forall x \in W)(fx \in W \vee C) \to$$
$$[(A(\epsilon) \vee C) \wedge (\forall x \in W)(A(x) \to (A(x0) \vee C) \wedge (A(x1) \vee C))] \to$$
$$\to (\forall x \in W)(A(x) \vee C),$$

where $A(x) \equiv (\exists y \leq fx)B(f, x, y)$, B positive combinatory, C is positive *toutcourt* and $x \notin FV(C)$.

(ii) *The extended system* $\mathcal{F}_i^e$ is PTT_i together with the extended axioms $\Sigma_1^b\text{-I}_W^e$, and the usual $\mathsf{T\text{-}AC}_W$, T-UP and T-CD.

The main tools for interpreting $\mathcal{F}_i + \mathsf{TS}$ in $\mathcal{F}_i^e$ are condensed in the

LEMMA 20 (PTT_i+T-CD). *If A is positive,*

$$f \Vdash A \leftrightarrow (Tf) \vee A;$$
$$A \leftrightarrow (\forall f)(f \Vdash A);$$
$$f \Vdash \neg\neg Ta \to f \Vdash Ta$$

THEOREM 6. (i) *If $\mathcal{F}_i + \mathsf{TS} \vdash A$, then $\mathcal{F}_i^e \vdash (\forall f)(f \Vdash A)$.*

(ii) *Moreover, if $\mathcal{F}_i + \mathsf{TS}$ proves $(\forall x \in W)(\exists y \in W)T(axy)$, the same formula is provable in $\mathcal{F}_i^e$.*

(iii) *Hence $\mathcal{F}$ and $\mathcal{F}_i^e$ have the same provably total functions.*

The proof is similar to the one of theorem 2 (given the modifications required in interpreting induction).

§9. External realizability with feasible functionals. For the final step, we quickly recall the definition of the Cook-Urquhart feasible functionals (for details, cf. [13] or the survey in [2]).

First of all, we only consider functionals in $FT_{\to}$, the finite type hierarchy generated above the set of binary words by means of the function space constructor. So the collection of types is inductively represented by the standard symbols, generated from o by closing under application of $\to$. We keep using 'type' also for 'type symbol'.

To each type we inductively assign its *level*:

$$Lev(o) = 0; \quad Lev(\sigma \to \tau) = max\{Lev(\sigma) + 1, Lev(\tau)\}.$$

By PV^ω we understand the least class of functionals of finite type, which is closed under explicit definition, bounded recursion on notation and comprises a finite list of initial functionals (typically, binary predecessor and successors, definition by cases on the ground type, concatenation and word multiplication, subword test). Closure under bounded recursion on notation corresponds to the existence of a recursor R satisfying the equations:

$$R(a,g,f,\epsilon) = a;$$
$$R(a,g,f,x) = g(x^-, R(a,g,f,x^-))\lceil f(x)$$

where $x \neq \epsilon$, a, x have type o, while g, f have types $(o \to o \to o)$, $o \to o$ (in the given order).

The level of a functional is simply the level of its type. Then we can define:

$$1 - \mathsf{sec}(\mathsf{PV}^\omega) = \{f \mid f \ \mathsf{PV}^\omega\text{-functional of level } 1\}$$

Below we hinge upon the main result of [13](cf. also theorem 5.2.1 in [2]):

THEOREM 7. $1 - \mathsf{sec}(\mathsf{PV}^\omega) = \mathsf{FPTIME}$

While the inclusion from right to left is immediate since the schemata generating FPTIME are special cases of those generating the feasible functionals, the converse is non-trivial and makes use of normalization for the associated typed lambda calculus (see [13], theorems 6.16, 6.17).

Henceforth, when we claim that a sentence A is true (unqualifiedly), we understand that A holds in the open term model $\langle \mathcal{M}(\lambda\eta^t), T_{min}\rangle$.

$\rho, \sigma, \varphi, \Theta, \Phi, \Psi, \Omega$ will range over feasible functionals; the vector notation, e.g., $\vec{\rho}$, designates a (finite) sequence thereof. The uppercase symbols will be generally occur in functional position.

By a *realizer* we henceforth understand a sequence of feasible functionals. $\vec{\Phi}(\vec{\rho})$ means: $\Phi_1(\vec{\rho}), \dots \Phi_n(\vec{\rho})$, where $\vec{\Phi}$ is the sequence $\Phi_1, \dots \Phi_n$.[20] As usual, the length ordering on binary words will be extended pointwise to arbitrary sequences (of the same length). If $\vec{\rho} \equiv \rho_1, \dots, \rho_n$, $\vec{\sigma} \equiv \sigma_1, \dots, \sigma_n$, $\vec{\rho} \leq \vec{\sigma}$ means $\rho_j \leq \sigma_j$, for each j with $1 \leq j \leq n$.

We now extend the realizability interpretation of 7.2 to the new context; of course, since we deal with arbitrary formulas, we need an *infinitary* clause for universally quantified formulas. If A is an arbitrary formula, we inductively define:

$$\begin{array}{rl}
\rho \,\mathbf{mr}\, Tt & \Leftrightarrow \rho \vdash t; \\
\rho \,\mathbf{mr}\, t = s & \Leftrightarrow \rho = \epsilon \wedge t = s \text{ is true;} \\
\rho \,\mathbf{mr}\, t \in W & \Leftrightarrow t = \rho \text{ is true;} \\
\vec{\rho}, \vec{\sigma} \,\mathbf{mr}\, A \wedge B & \Leftrightarrow \vec{\rho} \,\mathbf{mr}\, A \text{ and } \vec{\sigma} \,\mathbf{mr}\, B; \\
i, \vec{\rho}, \vec{\sigma} \,\mathbf{mr}\, A \vee B & \Leftrightarrow (i = 0 \text{ and } \vec{\rho} \,\mathbf{mr}\, A) \text{ or} \\
 & \quad\; (i = 1 \text{ and } \vec{\sigma} \,\mathbf{mr}\, B); \\
\vec{\rho} \,\mathbf{mr}\, A \to B & \Leftrightarrow (\vec{\sigma} \,\mathbf{mr}\, A \text{ implies } \vec{\rho}\vec{\sigma} \,\mathbf{mr}\, B), \text{ for arbitrary } \vec{\sigma}; \\
\vec{\rho} \,\mathbf{mr}\, (\exists x)A & \Leftrightarrow \vec{\rho} \,\mathbf{mr}\, A(t), \text{ for some term } t; \\
\vec{\rho} \,\mathbf{mr}\, (\forall x)A & \Leftrightarrow \vec{\rho} \,\mathbf{mr}\, A[x := t], \text{ for all terms } t.
\end{array}$$

[20]Of course, we implicitly assume that all objects involved have the appropriate types.

LEMMA 21 (Reduction). (i) *Let* $\vec{\rho} \equiv \rho_1, \dots \rho_n$. *If* B *is either positive or an implication of positive formulas, and* $\vec{\rho}\,\mathbf{mr}\,B$, *then each* ρ_j *is either a binary word or a polytime function* $(1 \leq j \leq n)$.
(ii) *If* B *is positive combinatory and* $\vec{\rho}\,\mathbf{mr}\,B$, *then there exists a finite sequence* $\vec{\rho}_B$ *such that* $\vec{\rho} \leq \vec{\rho}_B$.

PROOF. As to (i), observe that the functionals involved must have level 1; then apply theorem 7. (ii) is immediate once we note that, if B is positive combinatory, all atomic formulas have the same trivial realizer ϵ. Hence $\vec{\rho}_B$ is a finite sequence of ϵ's, 0's and 1's, which only depends on the logical complexity of B. ⊣

We say that $F: W^k \to W$ is a polynomial iff F is definable (in the standard algebra of binary words) by a term $t(x_1, \dots, x_k)$, built up from variables by means of word concatenation, word multiplication, binary successors and the empty word. The following property (see [20]) is useful below:

LEMMA 22 (Bounding). *Let* $F: W^k \to W$ *be polytime. There exists a monotone polynomial* Bd_F *majorizing* F, *i.e.*,

$$\rho_1 \leq \sigma_1 \wedge \dots \wedge \rho_n \leq \sigma_n \Rightarrow Bd_F(\rho_1, \dots, \rho_n) \leq Bd_F(\sigma_1, \dots, \sigma_n)$$
$$F(\rho_1, \dots, \rho_n) \leq Bd_F(\rho_1, \dots, \rho_n).$$

LEMMA 23 (Substitution).

$$\vec{\rho}\,\mathbf{mr}\,A[a := t] \quad \text{and } t = s \text{ is true} \Rightarrow \vec{\rho}\,\mathbf{mr}\,A[a := s].$$

THEOREM 8 (Soundness). *Let* $\mathcal{F}_i^e + \mathrm{CD} + \mathrm{UP} + \mathrm{AC}_W \vdash A(\vec{x})$, *and assume that every free variable of* A *occurs in the list* $\vec{x}$. *Then there exists a realizer* Φ_A *such that, for all* $\vec{t}$,

$$\Phi_A\,\mathbf{mr}\,A[\vec{x} := \vec{t}].$$

PROOF. Although the definition of realizability is semantical, the proof is in many respects similar to that of the theorem 3. This holds in particular of the verifications of UP, AC and of CD. Note however that in verifying AC_W, we apply the fact that each feasible functional of type $o \to o$ is definable in the open term model by a closed term (this is clear by theorem 7 and combinatory definability of recursive functions).

In general, all operations involving realizers have to be admissible for feasible functionals, and this may require non-trivial additional work (for instance, in the verification of disjunction axioms; details can be obtained from [13], 8.6).

We only check that the extended induction schema $\Sigma_1^b\text{-}\mathrm{I}_W^e$ is realizable in the class of feasible functionals. Below, we do not explicitly mention parameters, since the construction is uniform in any given choice of $\vec{t}$.

Let $A(x) := (\exists y \leq fx)B(f, x, y)$, where B is positive combinatory. Let C be a positive formula with $x \notin FV(C)$. Assume that there are realizers $\vec{\Pi}$,

$\vec{\Lambda}$, $\vec{\Theta}_i$ (where $i = 0, 1$):

(Bound) $\qquad \vec{\Lambda}\,\mathbf{mr}\,(\forall x \in W)(fx \in W \vee C)$;

(Hyp) $\qquad \vec{\Pi}\,\mathbf{mr}\,A(\epsilon) \vee C$;

(Hypi) $\qquad \vec{\Theta}^i\,\mathbf{mr}\,(\forall x \in W)(A(x) \to A(xi) \vee C)$.

Observe that, due to the form of formulas involved, we have, for arbitrary binary words ρ, $\vec{\sigma}$[21] such that $\vec{\sigma}\,\mathbf{mr}\,A(\overline{\rho})$:

$$\vec{\Lambda}(\rho) \equiv \Lambda_0(\rho), \Lambda_1(\rho), \vec{\Lambda}_2(\rho)\,\mathbf{mr}\,(f\overline{\rho} \in W \vee C);$$

$$\vec{\Pi} \equiv \Pi_0, \vec{\Pi}_1, \vec{\Pi}_2\,\mathbf{mr}\,A(\epsilon) \vee C;$$

$$\vec{\Theta}^i(\rho, \vec{\sigma}) \equiv \Theta^i_0(\rho, \vec{\sigma}), \vec{\Theta}^i_1(\rho, \vec{\sigma}), \vec{\Theta}^i_2(\rho, \vec{\sigma})\,\mathbf{mr}\,A(\overline{\rho i}) \vee C.$$

We want a realizer $\vec{\Phi}$ such that $\vec{\Phi}(\rho)\,\mathbf{mr}\,A(\overline{\rho}) \vee C$, where ρ is an arbitrary binary word and $\vec{\Phi}$ has the form $\Phi_0, \vec{\Phi}_1, \vec{\Phi}_2$. This is done by simultaneous bounded recursion on notation (which is legal in PV^ω; see [13], 6.12) and repeated applications of definition by cases on $\mathcal{W}$:

Case 0: $\vec{\Phi}(\epsilon) = \vec{\Pi}$;

Case 1.0: $\Phi_0(\rho) = 1$. Choose:

$$\Phi_0(\rho i) = 1$$
$$\vec{\Phi}_1(\rho i) = \vec{\epsilon}$$
$$\vec{\Phi}_2(\rho i) = \vec{\Phi}_2(\rho)$$

Case 1.1: $\Phi_0(\rho) = 0$ and $\Lambda_0(\rho) = 1$. Choose:

$$\Phi_0(\rho i) = 1$$
$$\vec{\Phi}_1(\rho i) = \vec{\epsilon}$$
$$\vec{\Phi}_2(\rho i) = \vec{\Lambda}_2(\rho)$$

Case 1.2: $\Lambda_0(\rho) = 0 = \Phi_0(\rho)$. Choose:

$$\vec{\Phi}(\rho i) = \vec{\Theta}^i(\rho, \vec{\Phi}_1(\rho))$$

This ends the definition of the realizer.[22] It is straightforward to check by induction on $\rho \in \mathcal{W}$ that indeed $\vec{\Phi}(\rho)$ realizes $A(\overline{\rho}) \vee C$.

[21] The length of $\vec{\sigma}$ depends on the form of $A(x)$.

[22] Here is a possible alternative definition:

Case 0: $\vec{\Phi}(\epsilon) = \vec{\Pi}$;

Case 1.0: $\Lambda_0(\rho) = 1$. Choose:

$$\Phi_0(\rho i) = 1$$
$$\vec{\Phi}_1(\rho i) = \vec{\epsilon}$$
$$\vec{\Phi}_2(\rho i) = \vec{\Lambda}_2(\rho)$$

In order to check that $\vec{\Phi}$ is a sequence of feasible functionals, we prove that there exist bounding feasible functionals $\vec{\Omega}(\rho) \equiv \Omega_0(\rho), \vec{\Omega}_1(\rho), \vec{\Omega}_2(\rho)$, such that, for all $\rho \in \mathcal{W}$,

$$\vec{\Phi}(\rho) \leq \vec{\Omega}(\rho).$$

First of all, if we let $\Omega_0(\rho) = 0 * 1$, we immediately have that $\Phi_0(\rho) \leq \Omega_0(\rho)$.

As to the bounding functions for $\vec{\Phi}_1$ and $\vec{\Phi}_2$ at the successor stage, let us consider case 1.2, where $\Lambda_0(\rho) = 0 = \Phi_0(\rho)$. By definition of realizability, we then have, for some term t:

$$\Phi_1^0(\rho)\,\mathbf{mr}\,(t \in W)$$
$$\Phi_1^1(\rho)\,\mathbf{mr}\,(t \leq f\overline{p})$$
$$\vec{\Phi}_1^2(\rho)\,\mathbf{mr}\,B(f, \overline{p}, t)$$
$$\Lambda_1(\rho)\,\mathbf{mr}\,f\overline{p} \in W$$

where $\vec{\Phi}_1(\rho) \equiv \Phi_1^0(\rho), \Phi_1^1(\rho), \vec{\Phi}_1^2(\rho)$. Since $\Phi_0(\rho) = 0$ and B is positive combinatory, there exists a sequence $\vec{\rho}_B$ by lemma 21 (ii), such that

$$\Phi_1^1(\rho) = \epsilon,$$
$$\Phi_1^0(\rho) \leq \Lambda_1(\rho),$$
$$\vec{\Phi}_1^2(\rho) \leq \vec{\rho}_B.$$

We are now in the position of providing bounds for $\vec{\Phi}_k(\rho i)$ (where $k = 1, 2$), which do not depend on $\vec{\Phi}_k(\rho)$. Since A, C are positive,[23] we can apply lemmata 21 (i) and 22 to $\vec{\Theta}_k^i$, for $k = 1, 2$. Hence there exist monotone polynomials $\vec{\Psi}_k^i$, where $k = 1, 2$, such that

$$\vec{\Theta}_k^i(\rho, \vec{\sigma}) \leq \vec{\Psi}_k^i(\rho, \vec{\sigma}).$$

Choosing $\vec{J}_k^i(\rho) = \vec{\Psi}_k^i(\rho, \Lambda_1(\rho), \epsilon, \vec{\rho}_B)$,

$$\vec{\Phi}_k(\rho i) = \vec{\Theta}_k^i(\rho, \vec{\Phi}_1(\rho)) \leq \vec{J}_k^i(\rho).$$

Case 1.1: $\Lambda_0(\rho) = 0$ and $\Phi_0(\rho) = 1$. Choose:

$$\Phi_0(\rho i) = 1$$
$$\vec{\Phi}_1(\rho i) = \vec{\epsilon}$$
$$\vec{\Phi}_2(\rho i) = \vec{\Phi}_2(\rho)$$

Case 1.2: $\Lambda_0(\rho) = 0 = \Phi_0(\rho)$. Choose:

$$\vec{\Phi}(\rho i) = \vec{\Theta}^i(\rho, \vec{\Phi}_1(\rho))$$

[23]This is the only point where we use the restriction on the formula C. In the arithmetical case, there is no need of similar restriction (see [14]). We do not know whether it is possible to avoid it via a direct majorization argument for functionals.

We now exploit the fact that FPTIME (and hence PV^{ω}) is closed under the operation

$$T(\rho) = \sum_{\xi \subseteq \rho} M(\xi).$$

Define

$$\vec{\Omega}_1(\rho) = \vec{\Pi}_1 * \sum_{\xi \subseteq \rho^-} (\vec{\mathcal{J}}^0_1(\xi) * \vec{\mathcal{J}}^1_1(\xi)),$$

$$\vec{\Omega}_2(\rho) = \vec{\Pi}_2 * \sum_{\xi \subseteq \rho^-} (\vec{\mathcal{J}}^0_2(\xi) * \vec{\mathcal{J}}^1_2(\xi) * \vec{\Lambda}_2(\xi)).$$

A straightforward induction on ρ yields $\vec{\Phi}(\rho) \leq \vec{\Omega}(\rho)$.[24] This completes the verification that $\vec{\Phi}$ is an admissible realizer for the induction schema Σ^b_1-I^e_W. ⊣

Corollary 5. *The provably total functions of $\mathcal{F}^e_i$, extended with* CD, UP *and* AC_W *(and hence of $\mathcal{F}$) are exactly the polytime functions.*

§**10. Final comments and problems.** In this paper we have shown that adding suitable choice axioms and uniformity principles does not alter the recursive content of applicative theories naturally associated to primitive recursive functions and to polytime functions.

However, there is a *significant* difference between the results of §6 and those ending the previous section. In the first case, we have a formal elimination of choice and uniformity, which produces a conservativity theorem. This can even be sharpened; by cut elimination and asymmetric interpretation, we can show:

Theorem 9. *If A is T-free positive,*

$$\mathsf{PRT} \vdash (\forall \vec{x} \in W)A \Rightarrow \mathsf{PR} \vdash (\forall \vec{x} \in W)A.$$

It follows that $\mathcal{S}$ is a conservative extension of PR with respect to formulas of the form $(\forall \vec{x} \in W)A$, where A is T-free positive.

This should be contrasted with the result of §§8–9 and corollary 5, which only establish invariance with respect to the computational content of the theory.

Problem 1. Is $\mathcal{S}$ conservative over PR for all formulas in the common language?

Problem 2. Is $\mathcal{F}$ conservative over PT for all formulas in the common language, or at least for formulas of the form $(\forall \vec{x} \in W)A$, where A is T-free positive?

---

[24]The given bound is certainly quite large, but a sharp upper bound is not needed here.

Another question concerns extensions of the main theorems to systems with different induction principles. For instance, Strahm proves that adding lexicographic induction on binary words for Σ_1^b-formulas to PT provides a theory characterizing functions computable in polynomial space, while replacing Σ_1^b-notation induction with the corresponding 'lexicographic' schema and omitting $\times$ leads to the class of functions computable in linear space.

PROBLEM 3. Do Strahm's characterization results involving lexicographic induction still hold in presence of choice and uniformity?

It is not difficult to observe that the method of the final section can be lifted to different theories, as soon as we have higher type systems extending FPSPACE, FLINSPACE and playing the same role as PV^{ω}. [29] contains hints in this direction and we conjecture that the problem has a positive answer.

Latterly, it also remains to investigate to what extent the results of this paper hold for systems based on forms of ramified or safe induction in the sense of [5], [22]. If we restrict our consideration to polytime functions, we believe that a positive answer can be obtained by relying upon some higher type variant of the Bellantoni-Cook class (see [5], [6]).

REFERENCES

[1] J. AVIGAD, *Interpreting classical theories in constructive ones*, ***The Journal of Symbolic Logic***, vol. 65 (2000), pp. 1785–1812.

[2] J. AVIGAD and S. FEFERMAN, *Gödel's functional (Dialectica) interpretation*, In Buss [9], pp. 337–405.

[3] H. BARENDREGT, ***The Lambda Calculus. Its Syntax and Semantics***, Studies in Logic and the Foundations of Mathematics, Elsevier, Amsterdam, 1984.

[4] M. J. BEESON, ***Foundations of Constructive Mathematics***, Springer, Berlin, 1985.

[5] S. BELLANTONI, *Predicative Recursion and Computational Complexity*, ***Ph.D. thesis***, University of Toronto, 1992.

[6] S. BELLANTONI, K. H. NIGGL, and H. SCHWICHTENBERG, *Ramification, modality and linearity in higher types*, ***Annals of Pure and Applied Logic***, vol. 104 (2000), pp. 17–30.

[7] M. Boffa, D. van Dalen, and K. McAloon (editors), ***Logic Colloquium '78***, Studies in Logic and the Foundations of Mathematics, North Holland, Amsterdam, 1979.

[8] S. J. BUSS, ***Bounded Arithmetic***, Bibliopolis, Napoli, 1986.

[9] S. J. Buss (editor), ***Handbook of Proof Theory***, Studies in Logic and the Foundations of Mathematics, Elsevier, Amsterdam, 1998.

[10] A. CANTINI, *Proof theoretical aspects of self-referential truth*, In Dalla Chiara et al. [16], pp. 7–27.

[11] ———, *Feasible operations and applicative theories based on $\lambda\eta$*, ***Mathematical Logic Quarterly***, vol. 46 (2000), pp. 291–312.

[12] ———, *Polytime, combinatory logic and positive safe induction*, ***Archive for Mathematical Logic***, vol. 41 (2002), no. 2, pp. 169–189.

[13] S. A. COOK and URQUHART A., *Functional interpretations of feasibly constructive arithmetic*, ***Annals of Pure and Applied Logic***, vol. 63 (1993), pp. 103–200.

[14] T. Coquand and M. Hofmann, *A new method for establishing conservativity of classical systems over their intuitionistic version*, ***Mathematical Structures in Computer Science***, vol. 9 (1999), pp. 323–333.

[15] J. N. Crossley (editor), ***Algebra and logic***, Lecture Notes in Mathematics 450, Springer, Berlin, 1975.

[16] M. L. Dalla Chiara, H. Doets, D. Mundici, and J. Van Benthem (editors), ***Logic and Scientific Methods. LMPS '95***, Kluwer, Dordrecht, 1996.

[17] S. Feferman, *A language and axioms for explicit mathematics*, In Crossley [15], pp. 87–139.

[18] ———, *Constructive theories of functions and classes*, In Boffa et al. [7], pp. 159–225.

[19] ———, *Iterated inductive fixed point theories: application to Hancock's conjecture*, In Metakides [24], pp. 171–195.

[20] F. Ferreira, *Polynomial time computable arithmetic*, In Sieg [26], pp. 147–156.

[21] D. Leivant, *A foundational delineation of polytime*, ***Information and Computation***, vol. 110 (1994), pp. 391–420.

[22] ———, *Intrinsic theories and computational complexity*, In *Logic and Computational Complexity* [23], pp. 177–194.

[23] D. Leivant (editor), ***Logic and Computational Complexity***, Lecture Notes in Computer Science 960, Springer, Berlin-Heidelberg, 1995.

[24] G. Metakides (editor), ***Patras Logic Colloquium***, Studies in Logic and the Foundations of Mathematics, North Holland, Amsterdam, 1982.

[25] P. Minari, *Theories with types and names with positive stratified comprehension*, ***Studia Logica***, vol. 62 (1999), pp. 215–242.

[26] W. Sieg (editor), ***Logic and Computation***, Contemporary Mathematics, American Mathematical Society, Providence R. I., 1990.

[27] S. G. Simpson, ***Subsystems of Second Order Arithmetic***, Perspectives in Mathematical Logic, Springer, Berlin-Heidelberg, 1998.

[28] T. Strahm, *Polynomial time operations in explicit mathematics*, ***The Journal of Symbolic Logic***, vol. 62 (1997), pp. 575–594.

[29] ———, *Theories with self-application and computational complexity*, ***Information and Computation***, vol. 185 (2003), pp. 263–297.

[30] A. S. Troelstra and H. Schwichtenberg, ***Basic Proof Theory***, Cambridge University Press, 1997.

[31] A. S. Troelstra and D. van Dalen, ***Constructivism in Mathematics***, Studies in Logic and the Foundations of Mathematics, Elsevier-North Holland, Amsterdam, 1988.

DIPARTIMENTO DI FILOSOFIA
UNIVERSITÀ DEGLI STUDI DI FIRENZE
FIRENZE, ITALY
E-mail: cantini@philos.unifi.it

COMPACTNESS AND INCOMPACTNESS PHENOMENA IN SET THEORY

JAMES CUMMINGS

Abstract. We prove two results with a common theme: the tension between *compactness* and *incompactness* phenomena in combinatorial set theory. Theorem 1 uses PCF theory to prove a sort of "compactness" for a version of Džamonja and Shelah's strong non-reflection principle. Theorem 2 investigates Jensen's *subcompact* cardinals and their relationship with stationary set reflection and the failure of the square principle.

§1. Introduction. A persistent theme in combinatorial set theory is the tension between *compactness* and *incompactness*, or to put it another way between *reflection* and *non-reflection*. These are not very precise concepts, so we illustrate with two lists of examples; list one contains ideas and results which (in the view of the author) are instances of compactness, and list two contains some more or less complementary instances of incompactness.

- Compactness:
  1. The compactness theorem for first-order logic
  2. Large cardinals and generic embeddings
  3. Stationary reflection principles
  4. The tree property
  5. Silver's theorem that GCH can't first fail at $\aleph_{\omega_1}$
  6. Shelah's singular compactness theorem
- Incompactness:
  1. The failure of compactness for many infinitary logics
  2. The Axiom of Constructibility $V = L$ and its consequences (e.g., square principles)
  3. Non-reflecting stationary sets
  4. Aronszajn trees
  5. Magidor's theorem that GCH can fail first at $\aleph_\omega$
  6. Constructions by Shelah and others for "almost free non-free" objects (e.g., groups)

It is our thesis that many interesting problems arise from considering the extent of some form of compactness, or the tension between some compactness

Logic Colloquium '01
Edited by M. Baaz, S. Friedman, and J. Krajíček
Lecture Notes in Logic, 20

principle (for example the existence of large cardinals) and a different incompactness principle (for example the existence of square sequences). This way of thinking is one of the main motivations in our joint work on singular cardinal combinatorics with Foreman and Magidor [7], [8] to which this paper is in some sense a sequel.

To describe a favourite problem about the extent of compactness, we recall that a κ-*Aronszajn tree* is a tree of height κ, with every level of size less than κ and with no cofinal branches. The cardinal κ is said to have the *tree property* if there are no κ-Aronszajn trees.

It is provable in ZFC that $\aleph_0$ has the tree property and $\aleph_1$ does not, but by work of Mitchell [19] "$\aleph_2$ has the tree property" is independent of ZFC and has the strength of a weakly compact cardinal. Each of the statements "$\aleph_n$ has the tree property for all n with $2 \leq n < \omega$" [6] and "$\aleph_{\omega+1}$ has the tree property" [18] is known to be consistent and of very high consistency strength, but the consistency of their conjunction is open.

Of course there are ideas and results which do not fit nicely into this picture. For example the principle $\diamondsuit_\kappa$ seems to be a reflection principle, saying as it does that any subset of κ is anticipated by the diamond sequence at many points below κ; indeed $\diamondsuit_\kappa$ follows from the assumption that κ is a sufficiently large cardinal. On the other hand $\diamondsuit_\kappa$ can be used to construct non-compact objects.

Theorem 1 is an application of PCF theory, which shows that a version of a combinatorial principle called *Strong Non-Reflection* (SNR) does not fail first at $\aleph_{\omega+1}$. As the name suggests SNR is a principle which would not be out of place in our "Incompactness" list, so we have a sort of compactness for an incompactness property.

Theorem 2 is about the tension between large cardinal axioms and Jensen's square principle. We start by observing that Jensen's quasicompactness principle is sufficiently strong to derive the failure of $\square_\lambda$ for λ singular. Theorem 2 shows that a similar proof idea from the weaker subcompactness principle is doomed to failure (which is not to say that another proof may not succeed). An exact determination of the consistency strength of the failure of $\square_\lambda$ for λ singular will presumably have to await progress in the inner model program.

§2. Strong non-reflection. The following "Strong Non-Reflection" principle was introduced by Džamonja and Shelah [11]: here κ and λ are regular and $\omega < \kappa < \lambda$.

DEFINITION 1. SNR(λ, κ) holds *iff there is a function* $f : \lambda \to \kappa$ *such that for all* $\alpha \in \lambda \cap \operatorname{cof}(\kappa)$ *there is* $C \subseteq \alpha$ *club with* $f \restriction C$ *strictly increasing.*

It follows from SNR(λ, κ) that for every stationary $S \subseteq \lambda$ there is a stationary $T \subseteq S$ such that T reflects to no point of cofinality κ (where T *reflects at*

α if $T \cap \alpha$ is stationary in α). To see this we just apply Fodor's theorem to find T on which f is constant and argue that this set can not reflect.

REMARK 1. *Džamonja and Shelah observed that for any κ and λ,* SNR(λ, κ) *follows from Jensen's global square principle* [15]. *In particular* SNR(λ, κ) *holds in L.*

Džamonja, Shelah and the author [5], [9], [12] studied the SNR principles and used them to prove independence results about reflection. Džamonja and Shelah made the following definition, which can be seen as a measure of the extent of a certain kind of incompactness in the universe of set theory.

DEFINITION 2. *$u(\kappa)$ is the least regular $\lambda > \kappa$ such that* SNR(λ, κ) *fails, assuming that such a cardinal exists. If no such cardinal exists then by convention $u(\kappa) = \infty$, where $\lambda < \infty$ for all cardinals λ.*

Džamonja and Shelah showed by a hard forcing argument that $u(\kappa)$ can be the successor of a singular cardinal (other kinds of regular cardinal are considerably easier to achieve). We state and prove a result which sheds some light on the difficulties. Theorem 1 is closely related to a theorem from our joint work with Foreman and Magidor [8] which gives a similar kind of compactness for the square principle.

It is convenient to work with a variation on SNR: here $\mu < \kappa < \lambda$ and they are all regular cardinals.

DEFINITION 3. SNR(λ, κ, μ) holds *iff there exists $f : \lambda \cap \mathrm{cof}(\mu) \rightarrow \kappa$ such that for all $\alpha \in \lambda \cap \mathrm{cof}(\kappa)$ there is $C \subseteq \alpha$ club with $f \restriction C \cap \mathrm{cof}(\mu)$ strictly increasing.*

The principle SNR(λ, κ, μ) implies that every stationary subset of $\lambda \cap \mathrm{cof}(\mu)$ contains a stationary set reflecting at no point in $\lambda \cap \mathrm{cof}(\kappa)$. Principles of this type can be used [9] to separate the phenomena of stationary reflection at different cofinalities.

THEOREM 1 (CH). *If the principle* SNR$(\aleph_n, \aleph_2, \aleph_1)$ *holds for all finite $n > 2$, then the principle* SNR$(\aleph_{\omega+1}, \aleph_2, \aleph_1)$ *holds.*

Before proving Theorem 1 we need a few PCF-theoretic preliminaries. We note that the hypothesis of CH in Theorem 1 may not be necessary, and that this is related to an important open question in PCF theory which we discuss below. We refer the reader to the survey papers [4], [1] and Shelah's book [20] for background on PCF theory.

For an infinite set $A \subseteq \omega$, we let $\prod_{n \in A} \aleph_n$ be the set of functions f such that $\mathrm{dom}(f) = A$ and $f(n) \in \aleph_n$ for all $n \in A$. There are various relations on $\prod_{n \in A} \aleph_n$ which will concern us. Given $f, g \in \prod_{n \in A} \aleph_n$ we say that *f is dominated by g ($f < g$)* if and only if $f(n) < g(n)$ for all $n \in A$. We say *f is eventually dominated by g ($f <^* g$)* if and only if $f(n) < g(n)$ for all sufficiently large $n \in A$, and also define $f \leq^* g$ if and only if $f(n) \leq g(n)$ for all large $n \in A$. Similarly *f is eventually equal to g ($f =^* g$)* if and only if $f(n) = g(n)$ for all sufficiently large $n \in A$. Finally $f <^\infty g$ if and only if $f(n) < g(n)$ for unboundedly many $n \in A$.

We will use the following theorem by Shelah; apart from our appeal to this basic result, this section of the paper is essentially self-contained.

FACT 1. *There is an infinite $A \subseteq \omega$ and a sequence $\langle f_\alpha : \alpha < \aleph_{\omega+1} \rangle$ increasing and cofinal in $\prod_{n \in A} \aleph_n$ with the eventual domination ordering.*

For the rest of this section we fix A and $\langle f_\alpha : \alpha < \aleph_{\omega+1} \rangle$ as in Fact 1.[1]

If β is a limit ordinal less than $\aleph_{\omega+1}$, then we say that a function $g \in \prod_{n \in A} \aleph_n$ is an *exact upper bound (eub)* for $\langle f_\alpha : \alpha < \beta \rangle$ if and only if

1. For all $\alpha < \beta$, $f_\alpha <^* g$.
2. For all $h \in \prod_{n \in A} \aleph_n$, if $h <^* g$ then there is $\alpha < \beta$ such that $h <^* f_\alpha$.

This is equivalent to demanding that $\langle f_\alpha : \alpha < \beta \rangle$ is increasing and cofinal modulo finite in $\prod_{n \in A} g(n)$. It is not difficult to see that an exact upper bound, if it exists, is unique modulo finite.

We will need a result by Shelah guaranteeing many points where an eub exists. The result is an easy corollary[2] of Shelah's quite technical "trichotomy theorem", but in the interests of making this paper self-contained we sketch a more direct proof (we are really just working through the trichotomy construction and cutting some corners with the help of CH).

FACT 2. *Let* CH *hold. If* $\mathrm{cf}(\beta) = \aleph_2$ *then* $\langle f_\alpha : \alpha < \beta \rangle$ *has an eub H such that* $\mathrm{cf}(H(n)) = \aleph_2$ *for all but finitely many n.*

PROOF. We define a sequence of functions $H_\zeta \in \prod_{n \in A} \aleph_n$ which are attempts to build a *least upper bound (lub)*, that is to say a function H such that

1. $f_\alpha <^* H$ for all $\alpha < \beta$.
2. There is no $\overline{H} \leq^* H$ such that $\overline{H} <^\infty H$ and $f_\alpha <^* \overline{H}$ for all $\alpha < \beta$.

Our construction will be such that if $\zeta < \eta$ then $H_\eta \leq^* H_\zeta$ and $H_\eta <^\infty H_\zeta$. The construction will proceed for at most $\aleph_2$ many steps.

$H_0 = f_\beta$. If H_ζ is an lub then we stop the construction, otherwise we choose $H_{\zeta+1} \leq^* H_\zeta$ such that $H_{\zeta+1} <^\infty H_\zeta$ and $f_\alpha <^* H_{\zeta+1}$ for all $\alpha < \beta$. At limit $\lambda < \aleph_2$ we work as follows: let $X_n = \{H_\zeta(n) : \zeta < \lambda\}$ and for every $\gamma < \beta$ let $G_\gamma(n) = \min(X_n \setminus f_\gamma(n))$, or zero if $X_n \subseteq f_\gamma(n)$.

It is routine to check that $f_\alpha <^* G_\gamma$ for all $\alpha < \beta$, that $G_\gamma \leq^* H_\zeta$ for all $\zeta < \lambda$ and that $\gamma < \delta \Longrightarrow G_\gamma \leq^* G_\delta$. By CH there are only $\aleph_1^{\aleph_0} = \aleph_1$ possibilities for the $=^*$-equivalence class of G_γ, and since $\mathrm{cf}(\beta) = \aleph_2$ that equivalence class must stabilise: we choose H_λ so that $H_\lambda =^* G_\gamma$ for all large $\gamma < \beta$.

We claim that the constuction of the H_ζ must halt before $\aleph_2$ steps. Otherwise we may define a function F from $[\aleph_2]^2$ to ω, by $F(\zeta, \eta) = n$ for n minimal

[1] It is interesting to note that there is a canonical maximal choice for A which is well-defined modulo finite, though we will not use this.

[2] For the experts: we just observe that points of cofinality greater than the continuum can not fall into the Bad or Ugly cases of the trichotomy.

with $H_\eta(n) < H_\zeta(n)$. The Erdos-Rado theorem then gives a decreasing $\aleph_1$-sequence of ordinals, which is impossible.

We have constructed an lub H. We claim it is an eub. To see this let $g <^* H$ and suppose for a contradiction that $Y_\gamma = \{n : g(n) > f_\gamma(n)\}$ is infinite for all γ. Clearly $\gamma < \delta$ implies that Y_δ is contained in Y_γ mod finite, and so by CH again there is a fixed Z such that Y_γ is equal to Z mod finite for all large $\gamma < \beta$. Define $\overline{H}$ by $\overline{H}(n) = H(n)$ for $n \notin Z$, $\overline{H}(n) = g(n)$ for $n \in Z$; clearly $\overline{H} \leq^* H$, $\overline{H} <^\infty H$ and $f_\alpha <^* \overline{H}$ for all α, contradiction!

To finish we show that $\mathrm{cf}(H(n)) = \aleph_2$ for all but finitely many n. Suppose first that $\mathrm{cf}(H(n)) < \aleph_2$ for every n in some infinite subset B of A, and fix $A_n \subseteq H(n)$ cofinal with $\mathrm{ot}(A_n) = \mathrm{cf}(H(n))$ for every $n \in B$. Now by CH $\prod_{n\in B} A_n$ has cardinality $\aleph_1$, so we may find f_α such that for every $f \in \prod_{n\in B} A_n$ we have $f <^* f_\alpha \restriction B$. Since $f_\alpha <^* H$ we may find $g \in \prod_{n\in B} A_n$ such that $f_\alpha \restriction B <^* g$, which is a contradiction.

Now suppose that $\mathrm{cf}(H(n)) > \aleph_2$ for every n in some infinite subset B of A. Let $\langle \beta_i : i < \aleph_2 \rangle$ be increasing and cofinal in β and define $f \in \prod_{n\in B} H(n)$, by setting $f(n) = \sup_i f_{\beta_i}(n)$ for all $n \in B$. Then since H is an eub and the β_i are cofinal we may find i such that $f <^* f_{\beta_i} \restriction B$, which is a contradiction. ⊣

Fact 2 is our only use of CH. As we discuss further at the end of this section, it is unclear whether the assumption of CH can be removed from Fact 2 or indeed from Theorem 1. Next we characterise those points where an eub of uniform uncountable cofinality exists. Again this result is due to Shelah.

FACT 3. *The following are equivalent for β of uncountable cofinality.*

1. *There exists g an eub for $\langle f_\alpha : \alpha < \beta \rangle$ with* $\mathrm{cf}(g(n)) = \mathrm{cf}(\beta)$ *for all but finitely many n.*
2. *There exists g an eub for $\langle f_\alpha : \alpha < \beta \rangle$ and an uncountable regular λ with* $\mathrm{cf}(g(n)) = \lambda$ *for all but finitely many n.*
3. *There exists a sequence of functions $\langle h_\eta : \eta < \mathrm{cf}\,\beta \rangle$ in ${}^A ON$ which is pointwise increasing and is* cofinally interleaved *with $\langle f_\alpha : \alpha < \beta \rangle$ in the eventual domination ordering (which is to say that each function in each of the sequences is eventually dominated by some function from the other sequence).*
4. *For every $Y \subseteq \beta$ which is unbounded in β there is $Z \subseteq Y$ unbounded in β and $n < \omega$, such that* $\mathrm{ot}(Z) = \mathrm{cf}(\beta)$ *and $\langle f_\alpha(m) : \alpha \in Z \rangle$ is strictly increasing for $m \in A$ with $m > n$.*

PROOF. It is immediate that 1 implies 2. Given 2, define a pointwise increasing sequence of functions $\langle h_\eta : \eta < \lambda \rangle$ such that $\langle h_\eta(n) : \eta < \lambda \rangle$ is increasing and cofinal in $g(n)$ for all but finitely many n. Since $h_\eta <^* g$, $h_\eta <^* f_\alpha$ for some $\alpha < \beta$: conversely if $\alpha < \beta$ then $f_\alpha <^* g$, and since λ is uncountable and the h_η are pointwise increasing we may choose η such that $f_\alpha <^* h_\eta$. It follows that $\lambda = \mathrm{cf}(\beta)$ so 2 implies 3.

Given 3, let $Y \subseteq \beta$ be unbounded. We may clearly choose $\eta_j < \mathrm{cf}(\beta)$ and $\zeta_j \in Y$ such that $g_{\eta_j} <^* f_{\zeta_j} <^* g_{\eta_{j+1}}$ for $j < \mathrm{cf}(\beta)$, and then choose $n_j \in A$ such that $g_{\eta_j}(m) < f_{\zeta_j}(m) < g_{\eta_{j+1}}(m)$ for $m > n_j$. Since $\mathrm{cf}(\beta)$ is uncountable we may find $T \subseteq \mathrm{cf}(\beta)$ unbounded and n such that $n_j = n$ for all $j \in T$, and then let $Z = \{\zeta_j : j \in T\}$. If $j_1 < j_2$ are in T and $m > n$ then $f_{\zeta_{j_1}}(m) < g_{\eta_{j_1+1}}(m) \leq g_{\eta_{j_2}}(m) < f_{\zeta_{j_2}}(m)$, so 3 implies 4.

Given 4 we let $Y = \beta$, choose a suitable Z and n and define H by $H(m) = \sup_{\alpha \in Z} f_\alpha(m)$. Clearly $\mathrm{cf}(H(m)) = \mathrm{cf}(\beta)$ for all but finitely many m, and $f_\alpha <^* H$ for all H. Now if $f <^* H$, then since $\mathrm{cf}(\beta)$ is uncountable and $\langle f_\alpha(m) : \alpha \in Z \rangle$ is strictly increasing for $m \in A$ with $m > n$ we may choose $\alpha \in Z$ such that $f <^* f_\alpha$. It follows that H is an eub so 4 implies 1. ⊣

REMARK 2. *In the course of this proof, we saw that for any cofinally interleaved sequence $\langle h_\eta : \eta < \mathrm{cf}\,\beta \rangle$ as in 3, the pointwise supremum of $\langle h_\eta : \eta < \mathrm{cf}\,\beta \rangle$ is an exact upper bound for $\langle f_\alpha : \alpha < \beta \rangle$.*

Let G be the set of all those $\beta < \aleph_{\omega+1}$ of uncountable cofinality such that there is an eub g for $\langle f_\alpha : \alpha < \beta \rangle$ with $\mathrm{cf}(g(n)) = \mathrm{cf}(\beta)$ for all but finitely many n. We claim that if $\gamma \in G$ there is a club subset C of γ such that all points of C with uncountable cofinality are in G. To see this we use "1 implies 3" from Fact 3 to construct $\langle h_\eta : \eta < \mathrm{cf}(\gamma) \rangle$ which is cofinally interleaved with $\langle f_\alpha : \alpha < \gamma \rangle$, and then let C be the set of $\delta < \gamma$ such that $\langle f_\alpha : \alpha < \delta \rangle$ is cofinally interleaved with $\langle h_\eta : \eta < \bar{\eta} \rangle$ for some $\bar{\eta} < \eta$; C is clearly club, and by "3 implies 1" from Fact 3 every element of C with uncountable cofinality is in G.

The claim of the last paragraph and the fact that under CH every point of cofinality $\aleph_2$ is in G are the keys to the proof of Theorem 1. We will use the set G as a scaffolding on which to build a function witnessing $\mathrm{SNR}(\aleph_{\omega+1}, \aleph_2, \aleph_1)$.

Before starting the proof of Theorem 1, it will be convenient to make some cosmetic adjustments to A and the sequence $\langle f_\alpha : \alpha < \aleph_{\omega+1} \rangle$. It follows easily from the discussion above that we may assume that

1. The minimum element of A is at least 3.
2. The sequence $\langle f_\alpha : \alpha < \aleph_{\omega+1} \rangle$ is *continuous*, that is to say that whenever an exact upper bound for an initial segment $\langle f_\alpha : \alpha < \beta \rangle$ exists then f_β is an exact upper bound.
3. If $\beta \in G$ and $\mathrm{cf}(\beta) \leq \aleph_2$ then $\mathrm{cf}(f_\beta(n)) = \mathrm{cf}(\beta)$ for all $n \in A$.

PROOF OF THEOREM 1: We assume that CH holds. We fix $F_n : \aleph_n \cap \mathrm{cof}(\aleph_1) \to \aleph_2$ witnessing $\mathrm{SNR}(\aleph_n, \aleph_2, \aleph_1)$ for $2 < n < \omega$. We define F for $\gamma \in G \cap \mathrm{cof}(\aleph_1)$ by $F(\gamma) = \sup_n F_n(f_\gamma(n))$ and verify that F is a witness for $\mathrm{SNR}(\aleph_{\omega+1}, \aleph_2, \aleph_1)$.

We fix a point $\delta < \aleph_{\omega+1}$ of cofinality $\aleph_2$. Since CH holds Fact 2 tells us that $\delta \in G$, so that by our cosmetic work above f_δ is an exact upper bound for $\langle f_\alpha : \alpha < \delta \rangle$ and $\mathrm{cf}(f_\delta(n)) = \aleph_2$ for all $n \in A$. We fix for each n a club C_n in

$f_\delta(n)$ such that F_n is increasing on $C_n \cap \mathrm{cof}(\aleph_1)$, and C_n has order type $\aleph_2$. Then we define h_η for $\eta < \aleph_2$ by setting $h_\eta(n)$ to be the ηth point in C_n; by the argument for "2 implies 3" in Fact 3, $\langle h_\eta : \eta < \aleph_2 \rangle$ is pointwise increasing and is cofinally interleaved with $\langle f_\alpha : \alpha < \delta \rangle$.

We may now fix E club in δ with order type $\aleph_2$, such that for every point $\gamma \in E \cap \mathrm{cof}(\aleph_1)$ there is a (necessarily unique) $\zeta(\gamma) \in \aleph_2 \cap \mathrm{cof}(\aleph_1)$ such that the sequences $\langle f_\alpha : \alpha < \gamma \rangle$ and $\langle h_\eta : \eta < \zeta(\gamma) \rangle$ are cofinally interleaved. Thinning out E if necessary we may arrange that $\mathrm{ot}(\zeta(\gamma) \cap \mathrm{cof}(\aleph_1)) = \zeta(\gamma)$ for all $\gamma \in E \cap \mathrm{cof}(\aleph_1)$. Fixing for the moment some $\gamma \in E \cap \mathrm{cof}(\aleph_1)$, it follows from the discussion above that the pointwise supremum of $\langle h_\eta : \eta < \zeta(\gamma) \rangle$ is an exact upper bound for $\langle f_\alpha : \alpha < \gamma \rangle$, so by continuity and the fact that the sets C_n are closed we see that $f_\gamma =^* h_{\zeta(\gamma)}$.

Since F_n is strictly increasing on $C_n \cap \mathrm{cof}(\aleph_1)$, $F_n(h_\eta(n)) \geq \mathrm{ot}(\eta \cap \mathrm{cof}(\aleph_1))$ for all n and η, so in particular $F_n(f_\gamma(n)) = F_n(h_{\zeta(\gamma)}(n)) \geq \zeta(\gamma)$ for all large n. By the definition of F it follows that $F(\gamma) \geq \zeta(\gamma)$ for all $\gamma \in E \cap \mathrm{cof}(\aleph_1)$. Clearly $\zeta(\gamma)$ is strictly increasing with γ, and so we may thin out E to a club subset E^* of δ with the property that $\zeta(\gamma_1) > F(\gamma_0)$ for all $\gamma_1 \in E^* \cap \mathrm{cof}(\aleph_1)$ and $\gamma_0 \in E^* \cap \gamma_1 \cap \mathrm{cof}(\aleph_1)$.

If γ_0 and γ_1 are points of $E^* \cap \mathrm{cof}(\aleph_1)$ with $\gamma_0 < \gamma_1$ then $F(\gamma_0) < \zeta(\gamma_1) \leq F(\gamma_1)$. We have verified that F witnesses $\mathrm{SNR}(\aleph_{\omega+1}, \aleph_2, \aleph_1)$ and this concludes the proof. ⊣

The set G (the set of "good points") turns out to be an interesting invariant of the universe of set theory: for more about this point of view see our papers with Foreman and Magidor [7], [8]. To whet the reader's appetite we note that

- Modulo the club filter, G is independent of the choice of the scale $\langle f_\alpha : \alpha < \aleph_{\omega+1} \rangle$.
- Jensen's weak square principle $\square^*_{\aleph_\omega}$ implies that almost all points of uncountable cofinality are good.
- Starting from very large cardinals it is known [14], [8] that we may build models in which there are stationarily many ungood points of cofinality $\aleph_1$. In particular this statement follows from Martin's Maximum (which implies the negation of CH) and also from the strong Chang conjecture $(\aleph_{\omega+1}, \aleph_\omega) \twoheadrightarrow (\aleph_1, \aleph_0)$ (which is consistent with CH), so is consistent both with CH and its negation.
- It is open whether or not there may be stationarily many ungood points of cofinality $\aleph_2$. It is known that any point of cofinality greater than $2^{\aleph_0}$ is good.

§3. Quasicompact and subcompact cardinals. Jensen showed [15] that $\square_\lambda$ holds in L for all λ, and also [10] that if $0^\sharp$ does not exist then L computes correctly the successors of V-singular cardinals; it follows that if $0^\sharp$ does not exist then $\square_\lambda$ holds for all singular λ, and so combinatorial statements which are incompatible with $\square_\lambda$ (e.g., the non-existence of a λ^+-Aronszajn tree) must have a substantial consistency strength.

Workers in the inner model program have constructed L-like models (the so-called "$L[\vec{E}]$-models") which can contain substantial large cardinals. We refer the reader to the survey paper by Steel [21] for more details. The $L[\vec{E}]$-models are only known to exist up to a certain point in the large cardinal hierarchy (roughly a measurable limit of Woodin cardinals). It is anticipated they will be shown to exist at higher levels of the large cardinal hierarchy, going past the subcompact and quasicompact cardinals defined below; modulo the assumption of existence it is already possible to analyse the fine structure of these hypothetical models.

A natural problem is to prove that $\square_\lambda$ holds in $L[\vec{E}]$-models, but there are limits on what can be done in this direction. Jensen has identified a large cardinal property called *subcompactness* (qv) and has shown that if κ is subcompact then $\square_\kappa$ fails. Schimmerling and Zeman have closed the case in $L[\vec{E}]$-models, by showing that in such models $\square_\lambda$ holds exactly when λ is not subcompact.

We recall that H_λ is the set of those X such that the transitive closure of $\{X\}$ has cardinality less than λ. Informally it is often helpful to think of H_λ as the set of those X which can be coded by bounded subsets of λ.

DEFINITION 4 (Jensen). *Let κ be a cardinal.*

1. κ *is* quasicompact *iff for all $A \subseteq H_{\kappa^+}$ there are $\lambda > \kappa$, $\pi\colon H_{\kappa^+} \longrightarrow H_{\lambda^+}$ and $B \subseteq H_{\lambda^+}$ such that π is an elementary embedding from $(H_{\kappa^+}, \in, A)$ to $(H_{\lambda^+}, \in, B)$, $\pi(\kappa) = \lambda$ and the critical point of π is κ.*
2. κ *is* subcompact *iff for all $A \subseteq H_{\kappa^+}$ there are $\beta < \kappa$, $\pi\colon H_{\beta^+} \longrightarrow H_{\kappa^+}$ and $b \subseteq H_{\beta^+}$ such that π is an elementary embedding from $(H_{\beta^+}, \in, b)$ to $(H_{\kappa^+}, \in, A)$, $\pi(\beta) = \kappa$ and the critical point of π is β.*

REMARK 3. *The quasicompactness of κ is witnessed by the existence of a family of superstrong extenders with critical point κ, and similarly the subcompactness of κ is witnessed by the existence of a family of superstrong extenders with target κ.*

If κ is quasicompact then κ is measurable and subcompact, and the least subcompact cardinal is not measurable.

Theorem 2 is motivated by the problem of calibrating the large cardinal strength needed to make $\square_\lambda$ fail for λ singular. Before stating and proving Theorem 2 we note that a quasicompact cardinal will suffice for this.

FACT 4 (Foreman and Magidor [7]). *Let κ be measurable and let $S = \kappa^+ \cap \mathrm{cof}(< \kappa)$. Assume that every stationary subset of S reflects at a point of cofinality less then κ; then the same reflection property holds in any generic extension by Prikry forcing.*

FACT 5 (Jensen [16]). *If κ is quasicompact then stationary subsets of $\kappa^+ \cap \mathrm{cof}(< \kappa)$ reflect at some point of cofinality less than κ.*

SKETCH OF THE PROOF OF FACT 5: Suppose for a contradiction that $T \subseteq \kappa^+ \cap \mathrm{cof}(< \kappa)$ is stationary and non-reflecting, and let $\pi\colon (H_{\kappa^+}, \in, T) \to (H_{\lambda^+}, \in, U)$ be as in the definition of quasicompactness. U may not be stationary (stationarity is not first-order) but every initial segment is non-stationary, and so we can choose C disjoint from π"T. Pulling back C we get D club in κ^+ with π"$D \cap \mathrm{cof}(< \kappa) \subseteq C$, so $D \cap T = \emptyset$. Contradiction! ⊣

Since $\Box_\kappa$ implies that every stationary subset of κ^+ has a non-reflecting stationary subset, it follows that doing Prikry forcing at a quasicompact cardinal κ gives a model where κ is singular and $\Box_\kappa$ fails. The following result shows that the same scenario can not be made to work starting with a measurable subcompact cardinal.

We will use the idea of strategic closure [13] for a poset $\mathbb{P}$. Consider a game in which two players I and II collaborate to build a decreasing sequence of conditions p_i in $\mathbb{P}$, with player I playing p_i for i odd and player II playing p_i for i even; player II loses a run of the game when a position is reached in which she can not move. If κ is an uncountable regular cardinal and for every $\delta < \kappa$ II has a strategy which enables her to play for δ moves, then $\mathbb{P}$ adds no sequences of length less than κ.

THEOREM 2. *It is consistent (modulo the existence of a supercompact cardinal) that there exists κ which is measurable and subcompact, and every stationary subset of κ^+ contains a non-reflecting stationary subset.*

SKETCH OF THE PROOF OF THEOREM 2: We use arguments similar to those of Apter and Shelah's papers [2], [3]. Let κ be a *Laver indestructible* [17] supercompact cardinal, that is to say κ remains supercompact in any extension by κ-directed closed forcing. Let GCH hold at and above κ.

We define a poset $\mathbb{P}_0$ for adding a non-reflecting stationary set in κ^+. Conditions are functions f such that $\mathrm{dom}(f) < \kappa^+$, $\mathrm{rge}(f) \subseteq \{0, 1\}$ and for every $\delta \leq \mathrm{dom}(f)$ of uncountable cofinality there is a club subset C of δ such that $f \restriction C$ is constant with value zero. The ordering is extension.

It is easy to see that in the strategic closure game player II can keep going for κ^+ moves by extending I's play by a single zero at every successor step, and taking unions at limit steps. In the next paragraph we will check this in some detail. In particular $\mathbb{P}_0$ adds no κ-sequences.

We claim that $\mathbb{P}_0$ adds the characteristic function of a set which is stationary in every cofinality up to κ. To see this let $\delta \leq \kappa$ be regular and let $\dot{C}$ be a $\mathbb{P}_0$-name for a club subset of κ^+. Consider a run of the strategic closure game of length $\delta + 1$ where f_i is played at stage i. Player I plays so that for every even $i < \delta$, f_{i+1} forces that $\dot{C} \cap (\mathrm{dom}(f_i), \mathrm{dom}(f_{i+1})) \neq \emptyset$. For odd $i < \delta$ player II lets $f_{i+1} = f_i \cup \{(\mathrm{dom}(f_i), 0)\}$. For limit i she lets $\gamma_i = \sup_{j<i} \mathrm{dom}(f_j)$ and then sets $f_i = \bigcup_{j<i} f_j \cup \{(\gamma_i, 0)\}$ if $i < \delta$, $f_i = \bigcup_{j<i} f_j \cup \{(\gamma_i, 1)\}$ if

$i = \delta$. The key point is that for every limit i player II has arranged that f_i is zero on a club set in γ_i, thereby guaranteeing that f_i is a condition. It is routine to check that f_δ is a condition and that f_δ forces that γ_δ is in $\dot{C}$.

In fact the stationary subset added by $\mathbb{P}_0$ has stationary intersection with every stationary subset of κ^+ from the ground model. To see this fix $T \in V$ a stationary subset of κ^+ and $\dot{C}$ a name for a club. Build a run of length κ^+ of the strategic closure game, where I plays as in the last paragraph and II adjoins a single zero to the play so far at each of her turns. With the same notation as in the last paragraph the γ_i form a club subset of κ^+, and so we may find i limit with $\gamma_i \in T$. Then the function $f_i \cup \{(\gamma_i, 1)\}$ is a condition forcing that γ_i is in $\dot{C}$.

We now define in $V^{\mathbb{P}_0}$ a poset $\mathbb{Q}_0(S)$ to destroy the stationarity of the set S added by $\mathbb{P}_0$. Conditions are closed bounded subsets disjoint from S, ordered by end-extension. It is easy to check that $\mathbb{P}_0 * \mathbb{Q}_0$ has a dense κ^+-closed subset, consisting of those $(f, \check{c})$ such that $\max(c) + 1 = \mathrm{dom}(f)$ and $f \restriction c$ is constant with value zero. With more work it can be shown that $\mathbb{P}_0 * \mathbb{Q}_0$ is equivalent to the standard poset for adding a Cohen subset of κ^+.

Let $\mathbb{P}$ be the product of κ^{++} copies of $\mathbb{P}_0$ with supports of size κ, and let S_i be the stationary set added by the ith copy. Let $\mathbb{Q}$ be the product of $\mathbb{Q}_0(S_i)$ for $i < \kappa^{++}$, again with supports of size κ. As in the case of $\mathbb{P}_0$ and $\mathbb{Q}_0$, it can be checked that all the sets S_i are stationary in $V^{\mathbb{P}}$ and that $\mathbb{P} * \mathbb{Q}$ has a dense κ^+-closed subset (and is in fact equivalent to the poset for adding κ^{++} Cohen subsets of κ^+). The usual Δ-system argument shows that $\mathbb{P}$ is κ^{++}-c.c.

Let $V_1 = V[G]$ for some $\mathbb{P}$-generic G, and let $V_2 = V_1[H]$ for some $\mathbb{Q}$-generic H. V_1 is the model we want, V_2 is used in the proof that κ is subcompact in V_1. The following claims are immediate.

- The power set of κ is the same in each of V, V_1 and V_2. It follows that all these models compute H_{κ^+} in the same way, and also that κ is measurable in V_1.
- By the assumption of indestructibility, κ is supercompact in V_2.

We claim that in V_1 every stationary subset of κ^+ contains a non-reflecting stationary subset. Let $S \in V_1$ be such a stationary set. The forcing $\mathbb{P}$ is κ^{++}-c.c. and so S is determined by the first β coordinates in $\mathbb{P}$ for some $\beta < \kappa^{++}$; an argument like that given above for $\mathbb{P}_0$ now shows that if T is the stationary set added by copy β of $\mathbb{P}_0$ then $S \cap T$ is stationary in $V^{\mathbb{P}}$.

We need to check that κ is subcompact in V_1. Let A be a predicate on H_{κ^+} with $A \in V_1$. Let $j : V_2 \to M_2$ be an embedding in V_2 witnessing κ is κ^+-supercompact. Then in V_2 the map $j \restriction H_{\kappa^+}$ is elementary from $(H_{\kappa^+}, \in, A)$ to $(H^{M_2}_{j(\kappa)^+}, \in, j(A))$. What is more the map $j \restriction H_{\kappa^+}$ lies in M_2, and $H_{\kappa^+} = H^{M_2}_{\kappa^+}$.

By reflection there are in V_2 an ordinal $\beta < \kappa$, a predicate b on H_{β^+} and an elementary π from $(H_{\beta^+}, \in, b)$ to $(H_{\kappa^+}, \in, A)$. This map π lies in V_1, since it is a subset of H_{κ^+} with cardinality less than κ, and similarly $H_{\beta^+}^{V_2} = H_{\beta^+}^{V_1}$ and $b \in V_1$. It follows that κ is quasicompact in V_1. ⊣

We conclude with a list of problems:

1. Can the methods of Theorem 1 be used to "step up" some other combinatorial principles, for example the existence of a non-reflecting stationary set?
2. What is the largest $L[\vec{E}]$ model such that non-reflecting stationary sets exist (or are dense) in every successor cardinal?
3. Does a subcompact cardinal (or a measurable subcompact cardinal) suffice to force failure of $\square_\lambda$ for λ singular?
4. Do quasicompact cardinals suffice for any more of the notable applications of supercompactness?
5. In particular, is a quasicompact cardinal sufficient to produce a model where $\square_\lambda^*$ fails for λ singular?

I would like to thank Mirna Džamonja, Matt Foreman, Ronald Jensen, Menachem Magidor and Ernest Schimmerling for several helpful conversations. I would also like to thank the anonymous referee for their careful reading of the first version of this paper.

REMARK 4. *Matt Foreman has pointed out that if player II has a strategy which allows her to play for κ^+ moves in the strategic closure game on a poset $\mathbb{P}$, and $2^\kappa = \kappa^+$, then $\mathbb{P}$ preserves the subcompactness of κ. So the conclusion of Theorem 2 can be obtained more cheaply by applying the same forcing construction to a measurable subcompact κ with $2^\kappa = \kappa^+$.*

REFERENCES

[1] U. Abraham and M. Magidor, *PCF theory*, To appear in the Handbook of Set Theory.

[2] A. Apter and S. Shelah, *Menas' result is best possible*, ***Transactions of the American Mathematical Society***, vol. 349 (1997), no. 5, pp. 2007–2034.

[3] ———, *On the strong equality between supercompactness and strong compactness*, ***Transactions of the American Mathematical Society***, vol. 349 (1997), no. 1, pp. 103–128.

[4] M. Burke and M. Magidor, *Shelah's PCF theory and its applications*, ***Annals of Pure and Applied Logic***, vol. 50 (1990), pp. 207–254.

[5] J. Cummings, M. Džamonja, and S. Shelah, *A consistency result on weak reflection*, ***Fundamenta Mathematicae***, vol. 148 (1995), no. 1, pp. 91–100.

[6] J. Cummings and M. Foreman, *The tree property*, ***Advances in Mathematics***, vol. 133 (1998), no. 1, pp. 1–32.

[7] J. Cummings, M. Foreman, and M. Magidor, *Squares, scales and stationary reflection*, ***Journal of Mathematical Logic***, vol. 1 (2001), no. 1, pp. 35–99.

[8] ———, *Canonical structure in the universe of set theory*, ***Annals of Pure and Applied Logic***, (To appear).

[9] J. Cummings and S. Shelah, *Some independence results on reflection*, ***Journal of the London Mathematical Society***, vol. 59 (1999), no. 1, pp. 37–49.

[10] K. Devlin and R. Jensen, *Marginalia to a theorem of Silver*, ***Logic colloquium, Kiel, 1974*** (Berlin), Springer-Verlag, 1975, pp. 115–142.

[11] M. Džamonja and S. Shelah, *Saturated filters at successors of singular, weak reflection and yet another weak club principle*, ***Annals of Pure and Applied Logic***, vol. 79 (1996), no. 3, pp. 289–316.

[12] ———, *Weak reflection at the successor of a singular cardinal*, ***Journal of the London Mathematical Society***, vol. 67 (2003), pp. 1–15.

[13] M. Foreman, *Games played on Boolean algebras*, ***The Journal of Symbolic Logic***, vol. 48 (1983), pp. 714–723.

[14] M. Foreman and M. Magidor, *A very weak square principle*, ***The Journal of Symbolic Logic***, vol. 62 (1997), no. 1, pp. 175–196.

[15] R. Jensen, *The fine structure of the constructible hierarchy*, ***Annals of Mathematical Logic***, vol. 4 (1972), pp. 229–308.

[16] ———, Circulated notes, 1998.

[17] R. Laver, *Making the supercompactness of κ indestructible under κ-directed closed forcing*, ***Israel Journal of Mathematics***, vol. 29 (1978), pp. 385–388.

[18] M. Magidor and S. Shelah, *The tree property at successors of singular cardinals*, ***Archive for Mathematical Logic***, vol. 35 (1996), no. 5-6, pp. 385–404.

[19] W. Mitchell, *Aronszajn trees and the independence of the transfer property*, ***Annals of Mathematical Logic***, vol. 5 (1972), pp. 21–46.

[20] S. Shelah, ***Cardinal arithmetic***, Oxford University Press, Oxford, 1994.

[21] J. Steel, *An outline of inner model theory*, To appear in the Handbook of Set Theory.

DEPARTMENT OF MATHEMATICAL SCIENCES
WEAN HALL, ROOM 6113
CARNEGIE MELLON UNIVERSITY
PITTSBURGH, PA 15213-3890, USA
E-mail: jcumming@andrew.emu.edu

SELECTION FOR BOREL RELATIONS

HARVEY M. FRIEDMAN

Abstract. We present several selection theorems for Borel relations, involving only Borel sets and functions, all of which can be obtained as consequences of closely related theorems proved in [2], [3], [4], [5], involving coanalytic sets. The relevant proofs given there use substantial set theoretic methods, which were also shown to be necessary. We show that none of our Borel consequences can be proved without substantial set theoretic methods. The results are established for Baire space. We give equivalents of some of the main results for the reals.

§1. Introduction. Let S be a set of ordered pairs and A be a set. We say that f is a selection for S on A if and only if $dom(f) = A$ and for all $x \in A$, $(x, f(x)) \in S$.

Let N be the set of all nonnegative integers. $2^N = \{0,1\}^N$ is Cantor space, where $\{0,1\}$ is given the discrete topology. N^N is Baire space, where N is given the discrete topology.

We use $\Re$ for the reals with the usual topology. All results in sections 2–6 are formulated on the Baire space N^N. This is most convenient for the proofs. In section 7 we give equivalent formulations on of some of the main results.

Our involvement in showing that strong set theoretic methods are required to prove statements about Borel sets and Borel functions begain with [8] where we showed that Borel determinacy cannot be proved with only countably many iterations of the power set operation. D. A. Martin proved that uncountably many iterations suffices (see [13]). We gave the following descriptive set theoretic version of Borel determinacy in [9].

THEOREM 1. *Let $E \subseteq \Re^2$ be a Borel set symmetric about the line $y = x$. Then E has a Borel selection on $\Re$ or $\Re^2 \backslash E$ has a Borel selection on $\Re$.*

In [9] we prove this Theorem rather directly using Borel determinacy, and also show that the proof requires use of uncountably many iterations of the power set operation. In fact, we show within $ZFC \backslash P$ that this Theorem is equivalent to Borel determinacy.

The following result from [5] led to this research.

This research was partially supported by NSF Grant DMS-9970459.

Logic Colloquium '01
Edited by M. Baaz, S. Friedman, and J. Krajíček
Lecture Notes in Logic, 20

PROPOSITION 2. *Let $S \subseteq N^N \times N^N$ be coanalytic and $E \subseteq N^N$ be Borel. If there is a continuous selection for S on every compact subset of E, then there is a continuous selection for S on E.*

Here a function is continuous if and only if the inverse of every open subset of N^N is an open subset of the domain; i.e., the intersection of an open subset of N^N with the domain. This does not imply the existence of a continuous extension to the whole space N^N.

Proposition 2 is proved in [5] using set theoretic assumptions going beyond *ZFC*. In fact, [5] shows that Proposition 2 is provably equivalent to

COUNT. For all $f \in N^N$, $N^N \cap L[f]$ is countable over *ZFC*.

[5] also gives the comparatively simple proof of Proposition 2 using analytic determinacy, which works for all coanalytic E. Moreover, [5] shows that Proposition 2 for coanalytic E is equivalent to analytic determinacy.

The proof of Proposition 2 from *COUNT* in [5] is rather complicated, and we conjecture that the proof can be considerably simplified using additional methods from modern set theory.

In section 2, we present the simple [5] proof of Proposition 2 from analytic determinacy. It is obvious that if "coanalytic" is replaced by "Borel", then the proof goes through unmodified with Borel determinacy instead of analytic determinacy. Therefore we have the following theorem of *ZFC*.

THEOREM 3. *Let $S \subseteq N^N \times N^N$ be Borel and $E \subseteq N^N$ be Borel. If there is a continuous selection for S on every compact subset of E, then there is a continuous selection for S on E.*

In section 5, we show that Theorem 3 cannot be proved using only countably many iterations of the power set operation.

Moreover, we show that Theorem 3 with $E = N^N$ cannot be proved using only countably many iterations of the power set operation.

In fact, we go further and show that the following theorem cannot be proved using only countably many iterations of the power set operation.

THEOREM 4. *Let $S \subseteq N^N \times N^N$ be Borel. If there is a constant selection for S on every compact set, then there is a Borel selection for S on N^N.*

Here a function is Borel if and only if the inverse of every open subset of N^N is a Borel subset of the domain; i.e., an intersection of a Borel subset of N^N with the domain. For any countable ordinal α, we can place $\leq \alpha$ or $< \alpha$ in front of all three occurrences of "Borel". Here we can always extend a Borel function to a Borel function on the whole space (raising α by at most a finite amount).

We now discuss strengthenings of Theorem 4 involving only Borel selection.

PROPOSITION 5. *Let $S \subseteq N^N \times N^N$ be Borel. If there is a Borel selection for S on every compact set, then there is a Borel selection for S on N^N.*

This turns out to be independent of *ZFC*.

Moreover, we show in sections 4 and 6 that Proposition 5 is equivalent to the following principle *DOM* over *ZFC*:

$$DOM. \quad (\forall f \in N^N)(\exists g \in N^N)(\forall h \in N^N \cap L[f]) \quad (g \text{ eventually strictly dominates } h).$$

That Proposition 5 is provable in $ZFC + DOM$ follows from more general results in [3] and [5], using the fact that the set of all Borel selectors of a Borel relation can be coded by a Π^1_1 set of reals.

The "reason" Proposition 5 is independent of *ZFC* relates to the unbounded levels of the hypothesized Borel selections. Consider the following variants involving Borel selections of bounded ranks.

THEOREM 6. *Let $S \subseteq N^N \times N^N$ be Borel, and λ be a countable limit ordinal. If there is a $< \lambda$ Borel selection for S on every compact set, then there is a Borel selection for S on N^N.*

Theorem 6 follows from Theorem 8 (see below), which we prove in *ZFC* using Borel determinacy.

The two versions of Theorem 6 obviously imply Theorem 4, and hence cannot be proved with only countably many iterations of the power set operation.

We now bring the Borel set $E \subseteq N^N$ back into the discussion.

It follows from [3] and [5] that $ZFC + COUNT$ proves the following, again using the observation that the set of all Borel selectors of a Borel relation can be coded by a Π^1_1 set of reals.

PROPOSITION 7. *Let $S \subseteq N^N \times N^N$ be Borel and $E \subseteq N^N$ be Borel. If there is a Borel selection for S on every compact subset of E, then there is a Borel selection for S on E.*

Obviously Proposition 7 implies Proposition 5, and hence Proposition 7 implies *DOM* over *ZFC*. Proposition 7 has recently been proved in $ZFC + DOM$, in [6], section IV-2. Hence Propositions 5 and 7 are both provably equivalent to *DOM* over *ZFC* (and presumably over $ZFC \setminus P$).

Finally, we come to the analog of Proposition 6 with E.

THEOREM 8. *Let $S \subseteq N^N \times N^N$ be Borel, $E \subseteq N^N$ be Borel, and λ be a countable limit ordinal. If there is a $< \lambda$ Borel selection for S on every compact subset of E, then there is a $\leq \lambda$ Borel selection for S on E.*

In section 3, we give a proof of Proposition 8 in *ZFC* using Borel determinacy. This result is implicit in [5]. In fact, using the more precise arguments there, it follows that we can replace $< \lambda$ and $\leq \lambda$ by $< \alpha$ and $\leq \alpha$, where α is any countable ordinal.

In section 7 we give versions of some of the main results on the reals. The equivalences are rather straightforward.

THEOREM 9. *Let $S \subseteq \Re \times \Re$ be Borel and $E \subseteq \Re$ be Borel with empty interior. If there is a continuous selection for S on every compact subset of E, then there is a continuous selection for S on E.*

In section 7 we show that Theorem 9 is provably equivalent to Theorem 3 over ATR_0. Here ATR_0 is one of the basic systems of reverse mathematics (see [16], p. 37).

PROPOSITION 10. *Let $S \subseteq \Re \times \Re$ be Borel and $E \subseteq \Re$ be Borel. If there is a Borel selection for S on every compact subset of E, then there is a Borel selection for S on E.*

In section 7 we show that Proposition 10 is provably equivalent to Proposition 7 over ATR_0.

THEOREM 11. *Let $S \subseteq \Re \times \Re$ be Borel. If there is a constant selection for S on every compact set of irrationals, then there is a Borel selection for S on the irrationals.*

In section 7 we show that Theorem 11 is provably equivalent to Theorem 4 over ATR_0.

§2. **Proofs from analytic determinacy.** We present the proof in [5] of Proposition 2 from analytic determinacy:

Let $S \subseteq N^N \times N^N$ be coanalytic and $E \subseteq N^N$ be Borel. If there is a continuous selection for S on every compact subset of E, then there is a continuous selection for S on E.

As remarked in the Introduction, [5] also shows that Proposition 2 is provably equivalent to *COUNT* over *ZFC* by a much more difficult argument.

The proof below using analytic determinacy works for coanalytic E. In fact, in [4], Proposition 2 for coanalytic E is shown to be equivalent to analytic determinacy over *ZFC*.

PROOF. Let S, E be coanalytic. Assume there is a continuous selection for S on every compact subset of E.

For each $g \in 2^N$ with infinitely many 1's, let $g^* \in N^N$ be the successive positions of the 1's in g. Thus g^* is strictly increasing. Let $g^{**} \in N^N$ be given by $g^{**}(n) =$ the exponent of 2 in $g^*(n)$. It is convenient to extend this definition to any $g \in 2^N$, where we allow g^* and g^{**} to be defined on an initial segment of N.

Let G be the following game, where players I, II, successively play elements of N. Let I play $f \in N^N$ and II play $g \in N^N$.

II wins the game if and only if

1. $f \notin E$; or
2. $g \in 2^N \wedge g$ has infinitely many 1's $\wedge (f, g^{**}) \in S$.

Note that the set of wins for player II is the union of an analytic set and a coanalytic set. Hence by an argument of Donald Martin in [7], this game is determined just using analytic determinacy.

Suppose I wins this game with winning strategy T. Let $V \subseteq N^N$ be the set of all plays of I made using strategy T, where II plays elements of 2^N. Then obviously V is a compact subset of E.

Let J be a continuous selection for S on V.

We now show how player II can defeat player I even though player I uses winning strategy T. Player II can accomplish this by

1. first playing enough 0's until the value of $J(f)(0)$ has been determined;
2. then playing 0's followed by a 1 so that $g^{**}(0) = J(f)(0)$;
3. then playing enough 0's until the value of $J(f)(1)$ has been determined;
4. then playing 0's followed by a 1 so that $g^{**}(1) = J(f)(1)$;
5. continuing to play in this way.

This describes II's plays even if the values of the $J(f)(n)$ are not all determined during the game. This is because II would, according to (1)–(5), simply play a tail of 0's.

Note that II is always playing 0's and 1's. Hence player I, following strategy T, must play in V.

It is now clear that the values of the $J(f)(n)$ do get determined since I is playing in V.

In fact, if player I plays $f \in N^N$ then $f \in V$ and player II plays some $g \in 2^N$ such that g has infinitely many 1's and $g^{**} = J(f)$. This is a contradiction since player II wins this run of the game.

We have shown that I does not win the game, and hence II wins the game. A winning strategy for II defines a continuous selection for S on E as follows. Let $f \in E$. Have player I play f. Then II plays $g \in 2^N$, where there are infinitely many 1's in g and $(f, g^{**}) \in S$. The map that sends f to g^{**} is obviously continuous on E. ⊣

Proposition 7 is another result implicit in [5], that was proved in [5], even for coanalytic S, using only *COUNT*. Proposition 7 has recently been proved using only *DOM*, in [6], section IV-2.

Let $S \subseteq N^N \times N^N$ be Borel and $E \subseteq N^N$ be Borel. If there is a Borel selection for S on every compact subset of E, then there is a Borel selection for S on E.

In fact, [5] has a much more difficult proof of Proposition 7 even for coanalytic S, that uses only *COUNT*.

The proof below using analytic determinacy works for coanalytic S, and analytic E. In [4], Proposition 7 for coanalytic S, E is shown to be equivalent to analytic determinacy over *ZFC*.

Proof. Let S be coanalytic and E be analytic. Assume there is a Borel selection for S on every compact subset of E.

Consider the following game G. Players I, II successively play elements of N. Let I play $f \in N^N$ and II play $g \in N^N$. II wins if and only if

1. $f \notin E$; or
2. $g \in 2^N$ and codes a Borel function $H: N^N \to N^N$ with $(f, H(f)) \in S$.

The set of wins for II is a coanalytic set. The proof then follows that of Proposition 2 above, with less to verify. Again, assume I wins with winning strategy T, and let V be the set of plays of I using T, where II plays in 2^N. Then V is again a compact subset of E, and let $J: V \to N^N$ be a Borel selection for S on V. Let $g \in 2^N$ be a Borel code for J.

Have player II play g no matter what player I plays. Let player I play his winning strategy T. Then player I will play $f \in E$. But then player II will win this run of the game since $(f, J(f)) \in S$.

Again we conclude that player II wins the game. Using his winning strategy, player II must play a Borel code no matter what player I plays, provided player I plays an element of E. The set of plays of II against I's plays in E is analytic. Hence let λ be a countable limit ordinal such that player II always plays a rank $< \lambda$ Borel code against I's plays in E.

We now obtain the desired Borel selection for S on E using a winning strategy for player II in the game. For each $f \in N^N$, have player I play f, and let $F(f)$ be the response of player II using his winning strategy. According to the previous paragraph, if $f \in E$, then $F(f)$ is a $< \lambda$ rank Borel code.

Finally, let $F'(f)$ be the value of the Borel function coded by $F(f)$ at the argument f. Note that if $f \in E$ then $(f, F'(f)) \in S$. Hence F' is a Borel selection for S on E. Furthermore, F' is Borel. ⊣

§3. Proofs from Borel determinacy. Recall Theorem 3:

Let $S \subseteq N^N \times N^N$ be Borel and $E \subseteq N^N$ be Borel. If there is a continuous selection for S on every compact subset of E, then there is a continuous selection for S on E.

PROOF. The proof is identical to the proof of Proposition 2, except that Borel determinacy suffices. This is because the relevant game is Borel. ⊣

Recall Theorem 8:

Let $S \subseteq N^N \times N^N$ be Borel, $E \subseteq N^N$ be Borel, and λ be a countable limit ordinal. If there is a $< \lambda$ Borel selection for S on every compact subset of E, then there is a $\leq \lambda$ Borel selection for S on E.

PROOF. The proof follows the proof of Proposition 7, except that Borel determinacy suffices. This is because the relevant game is Borel. ⊣

§4. Proof from *ZFC + DOM*. For $f_1, \ldots, f_k \in N^N$ write $\omega_1(f_1, \ldots, f_k)$ for the first uncountable ordinal in the sense of $L[f_1, \ldots, f_k]$.

LEMMA 12. *Assume DOM and let $f \in N^N$. There exists $g, h \in N^N$ such that*

1. $\omega_1(f, g, h) = \omega_1(f, g)$;
2. *h eventually strictly dominates every $x \in N^N \cap L[f, g]$.*

PROOF. Assume DOM and let $f \in N^N$. If there exists $g \in N^N$ such that $\omega_1(g) = \omega_1$, then apply DOM to (f, g) to obtain the desired h.

Now assume that for all $g \in N^N$, $\omega_1(g) < \omega_1$. We set $g = f$ and force over $L_{\omega_1}[f]$.

We use a notion of forcing in $L_{\omega_1}[f]$ that has the countable chain condition in $L_{\omega_1}[f]$, where the generic object eventually strictly dominates all elements of N^N lying in $L_{\omega_1}[f]$. A convenient choice is Hechler forcing (see [10]), where the conditions are (s, G), where $G \in N^N \cap L_{\omega_1}[f]$, and s is a finite sequence indexed from 0, which is extended by G. We take $(s, G) \leq (s', G')$ if and only if s is an initial segment of s' and G' strictly dominates G after $dom(s)$.

Note that generic objects exist for this notion of forcing over $L_{\omega_1}[f]$ since the number of dense sets of conditions in $L_{\omega_1}[f]$ is countable. Let $h \in N^N$ be generic with respect to this notion of forcing. Then $L_{\omega_1}[f, h]$ preserves cardinals over $L_{\omega_1}[f]$, and h eventually strictly dominates every $x \in N^N \cap L[f]$. Since $L_{\omega_1}[f, h]$ contains all elements of N^N constructible in (f, h), we see that $\omega_1(f, h) = \omega_1(f)$. ⊣

Recall Proposition 5:

Let $S \subseteq N^N \times N^N$ be Borel. If there is a Borel selection for S on every compact set, then there is a Borel selection for S on N^N.

We now prove Proposition 5 from $ZFC + DOM$. This result follows from more general results in [3] and [5], using the fact that the set of all Borel selectors of a Borel relation can be coded by a Π^1_1 set of reals.

PROOF. Let $S \subseteq N^N \times N^N$ be Borel with Borel code $f \in N^N$. Assume there is a Borel selection for S on every compact set. By DOM and Lemma 12, let $g, h \in N^N, \omega_1(f, g, h) = \omega_1(f, g)$, and h eventually strictly dominate every $x \in N^N \cap L[f, g]$.

Let $V = \{x \in N^N\colon x$ is eventually strictly dominated by $h\}$. Then V is a countable union of compact subsets of N^N, and $N^N \cap L[f, g] \subseteq V$. Furthermore, the sequence of presentations of these compact subsets as trees lies in $L[f, g, h]$.

There is a Borel selection for S on V. This is a Σ^1_2 statement. Hence by Shoenfield absoluteness ([15]), there is a Borel selection for S on V whose code lies in $L[f, g, h]$. Let α be the rank of some Borel selection for S on V whose code lies in $L[f, g, h]$. Then $\alpha < \omega_1(f, g, h) = \omega_1(f, g)$.

We now claim that the following holds in $L[f, g]$: for every compact set B there is a Borel selection for S on B of rank α. Note that this property of B is Σ^1_2.

To see this, note that α is countable in $L[f, g]$, and for every compact set B coded in $L[f, g]$ there is in fact a Borel selection for S on B of rank α. (This is because $B \subseteq V$). The claim follows by absoluteness.

We now can apply Theorem 6 to $L[f, g]$ and obtain a Borel selection for S on N^N in the sense of $L[f, g]$. We then obtain an actual Borel selection for S on N^N by absoluteness. ⊣

We will find it useful to summarize what we have proved in the following way. Let $ZFC\backslash P$ be ZFC without the power set axiom. Note that Lemma 12 was proved in $ZFC\backslash P$.

THEOREM 13. *The implication, Theorem 6 $\Rightarrow$ Proposition 5, is provable in $ZFC\backslash P + DOM$. In particular, Proposition 5 is provable in $ZFC + DOM$.*

§5. Obtaining Borel determinacy. In this section, we focus on Theorem 4.

Let $S \subseteq N^N \times N^N$ be Borel. If there is a constant selection for S on every compact subset of N^N, then there is a Borel selection for S on N^N.

We show that Theorem 4 cannot be proved using only countably many iterations of the power set operation.

For this purpose, it is convenient to use $ZFC\backslash P$ as the base theory.

We make the following assumptions, sitting within $ZFC\backslash P$, until the proof of Lemma 19 is complete.

1. $V = L$;
2. α is a countable ordinal;
3. For all countable β, there is a set of rank $\leq \alpha$ lying in $L(\beta + 1)\backslash L(\beta)$.

Recall that the rank of a set is the strict sup of the ranks of its elements.

Our aim is to refute Theorem 4.

We let KP be the usual Kripke/Platek set theory (see [1], p. 11, Definition 2.5). Let KP' be KP together with the axiom of infinity.

Let W be the class of all (ω, R) such that

1. (ω, R) satisfies $KP' + V = L$;
2. there is an internal ordinal of (ω, R) whose set of predecessors is of order type α;
3. (ω, R) satisfies "for all β, there is a set of rank $\leq \alpha$ lying in $L(\beta + 1)\backslash L(\beta)$".

By an internal ordinal of (ω, R) satisfying KP', we mean an element of the domain ω that is satisfied to be an ordinal. Some internal ordinals may be nonstandard in the sense that its set of predecessors is not well ordered by R.

Obviously W is Borel. Note that W includes all the well founded (ω, R) satisfying $KP' + V = L$.

Let $(\omega, R) \in W$. For each internal ordinal b of (ω, R), we let b^* be the set of all sets of rank $\leq \alpha$ that are internal to the $L(b)$ of (ω, R). Finally, define $(\omega, R)'$ as the set of all b^* such that b is an internal ordinal of (ω, R). Then $(\omega, R)'$ consists of transitive sets of rank $\leq \alpha + 1$ which are linearly ordered under inclusion. By clause (3), there are no repetitions among the b^*. Hence the mapping that sends b to b^* is an order preserving isomorphism from the internal ordinals of (ω, R) under R, onto $(\omega, R)'$ under inclusion.

LEMMA 14. *The mapping that sends $(\omega, R) \in W$ to $(\omega, R)'$ is Borel.*

PROOF. This Lemma is stated precisely in terms of standard codes for countable sets of rank $\leq \alpha + 2$. Specifically, we mean that there is a Borel function which sends each $(\omega, R) \in W$ to some standard code for $(\omega, R)'$. This is left to the reader. ⊣

We say that (ω, R) is special if and only if it lies in W and every element of its domain is definable without parameters.

We say that $L(\beta)$ is special if and only if (ω, R) is special for any (ω, R) isomorphic to $(L(\beta), \in)$. We write $L(\beta)'$ for $(\omega, R)'$, where (ω, R) is isomorphic to $(L(\beta), \in)$.

LEMMA 15. *Let $L(\beta)$ satisfy KP', $\beta < \omega_1$, and $L(\beta + 1)\backslash L(\beta)$ meet N^N. Then $L(\beta)$ is special. There are arbitrarily large countable β such that $L(\beta)$ is special.*

PROOF. Let $L(\beta)$ be isomorphic to (ω, R). Then $(\omega, R) \in W$ by assumption (3). Also a well known Skolem hull and transitive collapse argument establishes that $(L(\beta), \in)$ has every element definable without parameters.

For the second claim, let $\gamma < \omega_1$ be so large that there is no countable $\delta > \gamma$ such that $L(\delta)$ satisfies KP' and has every element definable without parameters. The least such γ is Σ_2 definable over $(L(\omega_1), \in)$. Using $ZFC\backslash P$, let A be the least Σ_2 elementary substructure of $(L_{\omega_1}, \in)$. Then there exists $\delta < \omega_1$ such that $A = L(\delta)$. Also $\gamma < \delta$, $L(\delta)$ satisfies KP', and $(L(\delta), \in)$ has every element definable without parameters. This is a desired contradiction.

We have thus shown that there are arbitrarily large countable β such that $L(\beta)$ satisfies KP' and has every element definable without parameters. Note that if $L(\beta)$ satisfies KP' and has every element definable without parameters, and $\beta > \alpha$, then $L(\beta)$ is special. This is because of assumptions 1–3 at the beginning of this section. ⊣

Let $(\omega, R) \in W$ be special. The special code for (ω, R), written $(\omega, R)^*$, is constructed as follows. First let T be the set of all first order sentences that hold in (ω, R), coded as $T \subseteq N$ via Gödel numbers. We take $(\omega, R)^*$ to be the first element of N^N that is recursive in the double jump of T (in some fixed standard recursion theoretic indexing) such that

1. $(\omega, R)^*$ eventually strictly dominates every element of N^N that is recursive in T;
2. for each n, $(\omega, R)^*(n)$ is even if and only if $n \in T$.

A special code is an element of N^N that is the special code of some special (ω, R). Note that the "underlying" (ω, R) is unique up to isomorphism.

We also define $L(\beta)^*$ for special $L(\beta)$, in exactly the same way, using $(L(\beta), \in)$.

Note that the set of special codes is a Borel subset of N^N. In addition, observe that we can recover an underlying (ω, R) from the special code in a uniformly effective way.

We now define $S \subseteq N^N \times N^N$ as follows. Let $x, y \in N^N$. Then $(x, y) \in S$ if and only if either x is not a special code, or x, y are both special codes and the following holds:

(#) Let $x = (\omega, R)^*$ and $y = (\omega, Y)^*$, where (ω, R) and (ω, Y) are obtained by the uniformly effective way from x, y. Let U be the maximum common initial segment of $(\omega, R)'$ and $(\omega, Y)'$. Then U is the set of strict predecessors of some element of $(\omega, Y)'$.

Note that S is Borel.

Lemma 16. *There is no Borel selection for S on N^N.*

Proof. Let $F: N^N \to N^N$ be a Borel selection for S on N^N. Let u be a Borel code for F.

According to Lemma 15, let $\beta < \omega_1$ be such that $L(\beta)$ is special and $u \in L(\beta)$. Then u is recursive in $L(\beta)^*$, and so $F(L(\beta)^*)$ is hyperarithmetic in $L(\beta)^*$.

Let $x = L(\beta)^*$. Then $x, F(x)$ are special codes, and condition (#) holds for x, and $y = F(x)$. Let $x = (\omega, R)^*$ and $F(x) = (\omega, Y)^*$. Let U be the maximum common initial segment of $(\omega, R)'$ and $(\omega, Y)'$. Let U be the set of strict predecessors of some element v of $(\omega, Y)'$.

Suppose U is a proper initial segment of $(\omega, R)'$. Since U is well founded, v must lie in the well founded part of $(\omega, Y)'$. By this well foundedness, we have a unique isomorphism between the initial segment of the constructible hierarchy in (ω, R) corresponding to U and the initial segment of the constructible hierarchy in (ω, Y) up to v Hence at the very next stages, both produce the same sets of rank $\leq \alpha$. Hence U can be extended to a longer common initial segment of $(\omega, R)'$ and $(\omega, Y)'$, which is a contradiction.

Hence $U = (\omega, R)'$ and U has order type β. Therefore $(\omega, Y)'$ has an initial segment of order type $\beta + 1$. Hence the internal ordinals of $(\omega, Y)'$ have an initial segment of order type $\beta + 1$. So we have an internal copy of $L(\beta)$ in $(\omega, Y)'$.

Now $L(\beta)$ has every element definable without parameters. Hence $L(\beta)^*$ is internal to (ω, Y). Since (ω, Y) satisfies KP', we see that every element of N^N hyperarithmetic in $L(\beta)^*$ is recursive in $F(x)$. But $F(x) = F(L(\beta)^*)$ is hyperarithmetic in $x = L(\beta)^*$. This is a contradiction. ⊣

Lemma 17. *Let $V \subseteq N^N$ be compact. There is a countable ordinal bound to the lengths of the maximum well ordered initial segments of the internal ordinals of every $(\omega, R) \in W$ such that $(\omega, R)^*$ lies in V.*

Proof. Let $f \in N^N$ be such that every element of V is everywhere strictly dominated by f. Let $f \in L(\beta)$, $\beta < \omega_1$. Suppose $(\omega, R)^* \in V$ has maximum well ordered initial segment of order type $> \beta$. Then f is internal to (ω, R), and so f is recursive in the set of first order sentences that hold in (ω, R). Therefore f is eventually strictly dominated by $(\omega, R)^*$. Since $(\omega, R)^* \in V$, $(\omega, R)^*$ is everywhere strictly dominated by f. This is a contradiction. $\dashv$

LEMMA 18. *There is a constant Borel selection for S on every compact subset of N^N.*

PROOF. Let $V \subseteq N^N$ be compact. By Lemma 17, let $\beta < \omega_1$ be such that for every $(\omega, R)^* \in V$, the maximum well ordered initial segment of (ω, R) has type $< \beta$.

By Lemma 15, let $L(\gamma)$ be special, where $\beta < \gamma < \omega_1$. Let $(L(\gamma), \in)$ be isomorphic to (ω, Y). Let $(\omega, R)^* \in V$.

We claim that $((\omega, R)^*, (\omega, Y)^*) \in S$. To see this, let U be the maximum common initial segment of $(\omega, R)'$ and $(\omega, Y)'$. Then U is well ordered and must be of order type $< \beta$. Hence U is a proper initial segment of $(\omega, Y)'$ determined by a point (since (ω, Y) is well founded).

For $f \in V$ that is not a special code, $(f, (\omega, Y)^*) \in S$ by default. $\dashv$

LEMMA 19. *Theorem 4 is false.*

PROOF. By Lemmas 16 and 18. $\dashv$

We now release the assumptions (1)–(3) made at the beginning of this section. We have shown the following.

LEMMA 20. *$ZFC \backslash P + V = L +$ Theorem 4 proves*

$$(\forall \alpha < \omega_1)(\exists \beta < \omega_1)(L(\beta + 1) \backslash L(\beta) \textit{ contains no set of rank } \leq \alpha).$$

PROOF. By Lemma 19 and the assumptions (1)–(3) at the beginning of this section that we used to prove it. $\dashv$

LEMMA 21. *The following is provable in $ZFC \backslash P$. Let $\omega \leq \alpha < \beta$ and suppose that $L(\beta + 1) \backslash L(\beta)$ has an element of rank $\leq \alpha$. There exists a surjective function f from an element of $L(\beta)$ of rank $\leq \alpha$ onto $L(\beta)$, whose graph is first order definable over $(L(\beta), \in)$. There exists a surjective function f from an element of $L(\beta + 1)$ of rank $\leq \alpha$ onto $L(\beta + 1)$, whose graph is first order definable over $(L(\beta + 1), \in)$. $L(\beta + 2) \backslash L(\beta + 1)$ has an element of rank $\leq \alpha$.*

PROOF. Let α, β be as given. Let $rk(x) \leq \alpha \wedge x \in L(\beta + 1) \backslash L(\beta)$. Using the standard definable well ordering of $L(\beta)$ over $(L(\beta), \in)$, we can assume that x is definable with at most the parameter α, over $(L(\beta), \in)$. Fix $x = \{y \in L(\beta) \colon (L(\beta), \in)$ satisfies $\varphi(y, \alpha)\}$, where φ has at most the free variable y and the parameter α. Assume that φ is in prenex form with $p \geq 8$ quantifiers.

Now build the elementary submodel $(A, \in)$ of $(L(b), \in)$ generated by $TC(\{x\}) \cup \alpha + 1 \subseteq A$, with respect to all formulas with at most p quantifiers, where we always choose the first witness in a standard well ordering of $L(\beta)$ definable over $(L(\beta), \in)$.

Let $j: A \to B$ be the unique isomorphism from A onto a transitive set B. Then j is the identity on $TC(\{x\}) \cup \alpha + 1$. Furthermore, since $A, L(\beta)$ are sufficiently elementarily equivalent, we fix $\alpha \leq \gamma \leq \beta$ such that $B = L(\gamma)$.

By the construction of A, since $x = \{y \in A: (A, \in) \text{ satisfies } \varphi(y, \alpha)\}$, we have $x = \{y \in L(\gamma): (L(\gamma), \in) \text{ satisfies } \varphi(y, \alpha)\}$. Hence $x \in L(\gamma + 1)$. Since $x \notin L(\beta)$, we have $\beta \leq \gamma$, and so $\beta = \gamma$.

Note that every $y \in A$ is definable over $(L(\beta), \in)$ with parameters from $TC(\{x\}) \cup \alpha + 1$ in a certain way according to the construction of A. By the construction of A, we see that y is definable over $(A, \in)$ with parameters from $TC(\{x\}) \cup \alpha + 1$ in the same way. Since j is an isomorphism, we see that $j(y)$ is definable over $(L(\beta), \in)$ with parameters from $TC(\{x\}) \cup \alpha + 1$ in the same way. Hence $j(y) = y$. This establishes that j is the identity, and so $A = L(\beta)$.

We now verify the first claim. It is straightforward to code finite sequences of sets of rank $\leq \alpha$ as sets of rank $\leq \alpha$ in a suitable way, since $\alpha \geq \omega$. In particular, we need this coding to take place in $(L(\beta), \in)$, so that finite sequences of sets of rank $\leq \alpha$ that are elements of $L(\beta)$ are suitably coded by sets of rank $\leq \alpha$ that are also elements of $L(\beta)$.

We also use an appropriate coding of finite sequences from $L(\beta)$ by elements of $L(\beta)$. We can then redo the construction of $A = L(\beta)$ definably over $(L(\beta), \in)$, starting with $TC(\{x\}) \cup \alpha + 1$. This verifies the first claim.

For the second claim, let $f: E \to L(\beta)$, where f is according to the first claim. Every element of $L(\beta + 1)$ is defined over $(L(\beta), \in)$ with finitely many parameters from $L(\beta)$. Using f, these parameters can be replaced by elements of E. This gives the required $g: E' \to L(\beta + 1)$ verifying the second claim, by making use of partial truth definitions over $(L(\beta), \in)$.

For the final claim, note that $D = \{x \in E': x \notin g(x)\} \in L(\beta + 2)$ is of rank $\leq \alpha$. If $D \in L(\beta + 1)$ then let $D = g(y)$, $y \in E'$. Hence for all $x \in E'$, $x \notin g(x) \leftrightarrow x \in D \leftrightarrow x \in g(y)$. Setting $y = x$, we obtain a contradiction. ⊣

Lemma 22. *The following is provable in $ZFC \setminus P$. Suppose α, β are ordinals such that $\alpha \geq \omega$ and β is the least ordinal with the property that there is no element of $L(\beta + 1) \setminus L(\beta)$ of rank $\leq \alpha + 3$. Then $(L(\beta), \in)$ satisfies $ZFC \setminus P + V(\alpha)$ exists.*

Proof. Here "$V(\alpha)$ exists" means that there is a function from $\alpha + 1$ into sets which satisfies the usual inductive definition for the cumulative hierarchy. $V(\alpha)$ itself is taken to be the value of this function at α.

Let α, β be as given. By Lemma 21, β is a limit ordinal.

Suppose $\gamma < \alpha$ and $(L(\beta), \in)$ satisfies "$V(\gamma)$ exists". Let A be the $V(\gamma)$ of $L(\beta)$. Then $B = \{x \in L(\beta): x \subseteq A\}$ is an element of $L(\beta + 1)$ of rank $\leq \alpha + 3$, and hence $B \in L(\beta)$. Therefore $(L(\beta), \in)$ satisfies "$V(\gamma + 1)$ exists".

Suppose $\gamma \leq \alpha$ is a limit, and for all $\delta < \gamma$, $(L(\beta), \in)$ satisfies "$V(\delta)$ exists". For each $\delta < \gamma$, let $f_\delta \in L(\beta)$ be the relevant function with domain $\delta + 1$.

Let f be the union of the f_δ. The $f \subseteq L(\beta)$, and hence f is an element of $L(\beta+1)$ of rank $\leq \alpha+3$. Hence $f \in L(\beta)$, and so $L(\beta), \in)$ satisfies "$V(\gamma)$ exists".

We have thus shown by transfinite induction that $(L(\beta), \in)$ satisfies "$V(\alpha)$ exists".

We have only to verify replacement in $L(\beta), \in)$. Let $f : \gamma \to \beta$ be definable over $(L(\beta), \in)$, where $\omega \leq \gamma < \beta$. Assume the range of f is unbounded. Since $L(\gamma+1) \backslash L(\gamma)$ has an element of rank $\leq \alpha+3$, by Lemma 21, there is a surjective function g from an element of $L(\gamma)$ of rank $\leq \alpha+3$, onto $L(\gamma)$, where $g \in L(\beta)$. By the same reasoning, for each $\delta < \gamma$, let $h_\delta : x_\delta \to L(f(\delta))$ be surjective, x_δ of rank $\leq \alpha+3$, where h_δ is chosen to occur as early as possible in a standard well ordering of $L(\beta)$. We can combine g and the h_δ in order to obtain a surjective function $J : y \to L(\beta)$, $J \in L(\beta+1)$, and y is of rank $\leq \alpha+3$. Note that $\{b \in y : b \notin J(b)\} \in L(\beta+1)$ is a set of rank $\leq \alpha+3$ that does not lie in $L(\beta)$. This is a contradiction. $\dashv$

THEOREM 23. *The following are provably equivalent in* $ZFC \backslash P + V = L$.

1. Theorem 3;
2. Theorem 4;
3. Theorem 6;
4. Theorem 8;
5. Borel determinacy;
6. $(\forall \alpha < \omega_1)(\exists \beta < \omega_1)(L(\beta)$ satisfies $ZFC \backslash P + V(\alpha)$ exists).

PROOF. By Lemma 22, the conclusion of Lemma 20 is equivalent to 6. Hence 2 $\Rightarrow$ 6. The proofs given in section 3 of Theorems 3 and 8 are within $ZFC \backslash P$ + Borel determinacy, and Theorem 4 immediately follows from Theorem 3, and Theorem 6 immediately follows from Theorem 8. Each of Theorems 3, 6, 8 immediately imply Theorem 4. Also, it follows easily from [13] that 6 implies 5 over $ZFC \backslash P + V = L$. Hence $1 \Rightarrow 2 \Rightarrow 6 \Rightarrow 5 \Rightarrow 4 \Rightarrow 3 \Rightarrow 2 \Rightarrow 6 \Rightarrow 5 \Rightarrow 1$. $\dashv$

We remark that by [8], 5 $\Rightarrow$ 6 over $ZFC \backslash P$. In fact, by [8] and [13], Borel determinacy is provably equivalent to: $(\forall \alpha < \omega_1)(\forall x \subseteq \omega)(\exists \beta < \omega_1)(L(\beta, x)$ satisfies $ZFC \backslash P + V(\alpha)$ exists) over $ZFC \backslash P$.

THEOREM 24. *Let* $\varphi(x)$ *be a* Σ_1 *formula of set theory with only the free variable shown, where* $ZFC \backslash P$ *proves* $(\exists x)(\varphi(x) \wedge x$ *is an ordinal*). *Then Theorems* 3, 4, 6, 8 *cannot be proved in* $ZFC \backslash P + (\exists x)(\varphi(x) \wedge V(x)$ *exists*). *However, Theorems* 3, 4, 6, 8 *can all be proved in* $ZFC \backslash P + (\forall \alpha < \omega_1)(V(\alpha)$ *exists*).

PROOF. The last claim is obvious since these Theorems were proved from Borel determinacy without use of $V = L$, and [13] also proves Borel determinacy in $ZFC \backslash P + (\forall \alpha < \omega_1)(V(\alpha)$ exists), without the use of $V = L$.

We will prove a sharper form of the first claim with both instances of $ZFC \backslash P$ replaced by $ZFC \backslash P + V = L$ in the second sentence.

First note that by Σ_1 absoluteness, there exists a constructibly countable ordinal α such that $\varphi(\alpha)$ holds in L. Fix the least such α.

There exists $\lambda > \alpha$ such that $(L(\lambda), \in)$ satisfies $ZFC\backslash P + V(\alpha)$ exists, since we can take λ to be a sufficiently large successor cardinal.

Let λ be the least ordinal such that $(L(\lambda), \in)$ satisfies $ZFC\backslash P + V(\alpha)$ exists.

By hypotheses, $(L(\lambda), \in)$ satisfies that $\varphi(x)$ has an ordinal solution. Hence $(L(\lambda), \in)$ satisfies that $\varphi(x)$ has a countable ordinal solution. Any such internal solution must be an external solution, and hence must be at least as large as α. Therefore α is countable in $(L(\lambda), \in)$.

Suppose any of Theorems 3, 4, 6, 8 holds in $(L(\lambda), \in)$. By Theorem 23, 6 holds in $(L(\lambda), \in)$. By applying 6 to α, we see that $(L(\lambda), \in)$ satisfies "there exists β such that $(L(\beta), \in)$ satisfies $ZFC\backslash P + V(\alpha)$ exists". But this contradicts the choice of λ. ⊣

In light of Theorem 24, we say that Theorems 3, 4, 6, 8 cannot be proved using only countably many iterations of the power set operation, but can be proved using uncountably many iterations of the power set operation.

§6. **Obtaining *DOM*.** In this section we focus on Proposition 5:

Let $S \subseteq N^N \times N^N$ be Borel. If there is a Borel selection for S on every compact set, then there is a Borel selection for S on N^N.

In this section we show that Proposition 5 implies *DOM* over *ZFC*.

Until the proof of Lemma 29 is complete, we work in $ZFC\backslash P + \neg DOM$. Accordingly, fix $f \in N^N$ such that no $g \in N^N$ eventually strictly dominates every $h \in L[f] \cap N^N$.

Note that $L[f] \cap N^N$ must be uncountable. Hence $\omega_1(f) = \omega_1$.

We seek to refute Proposition 5. We modify the machinery introduced in section 5.

We use the constructible hierarchy relative to f. Let $L(\alpha, f)$ be the α-th stage of this hierarchy, where we start with $L(0, f) = TC(f)$. Let $L[f]$ be the class of constructible sets relative to f.

Let X be the class of all (ω, R, c) such that

1. (ω, R) is a model of KP' with standard integers;
2. $c \in \omega$ represents f in (ω, R);
3. (ω, R) satisfies $V = L[c]$.

Note that X is Borel.

Let $(\omega, R, c) \in X$. The well founded part of (ω, R, c) is the set of all n such that there is no infinite backwards R-chain starting with n.

We say that (ω, R, c) is well founded if and only if its well founded part is the entire domain.

We say that (ω, R, c) is unusual if and only if every element of its domain is definable in (ω, R, c) with only the parameter c.

We say that $L(\alpha, f)$ is unusual if and only if $L(\alpha, f)$ satisfies KP', and every element of $L(\alpha, f)$ is definable in $(L(\alpha, f), \in)$ with only the parameter f.

Note that the $(\omega, R, c) \in X$ that are well founded and unusual are exactly the $(\omega, R, c) \in X$ where (ω, R) is isomorphic to some unusual $(L(\alpha, f), \in)$.

Let $(\omega, R, c) \in X$ be unusual. The unusual code for (ω, R, c), written $(\omega, R, c)^*$, is constructed as follows. First let T be the set of all first order sentences that hold in (ω, R, c), coded as $T \subseteq N$ via Gödel numbers. Here c is treated as a constant symbol representing the function f.

We take $(\omega, R, c)^*$ to be the first element of N^N that is recursive in the double jump of (T, f), (in some fixed standard recursion theoretic indexing), such that

1. $(\omega, R, c)^*$ eventually strictly dominates every element of N^N that is recursive in T, f;
2. for each n, $(\omega, R, c)^*(n)$ is even if and only if $n \in T$.

An unusual code is an element of N^N that is the unusual code of some unusual element of X. Note that the "underlying" (ω, R, c) is unique up to isomorphism.

Also note that the set of unusual codes is a subset of N^N that is arithmetic in f. In addition, observe that we can recover an underlying (ω, R, c) from the unusual code in a uniformly effective way using f.

We now define $S' \subseteq N^N \times N^N$. Let $x, y \in N^N$. Then $(x, y) \in S'$ if and only if

1. x is not an unusual code; or
2. x is an unusual code and y is an infinite backward R chain. Here $x = (\omega, R, c)^*$, where (ω, R, c) is obtained by the uniformly effective way from x, f; or
3. x, y are unusual codes and x is internal to (ω, Y). Here $y = (\omega, Y, d)^*$, where Y is obtained by the uniformly effective way from y, f.

Note that S' is a subset of $N^N \times N^N$ that is arithmetic in f.

LEMMA 25. *Let $L(\alpha, f)$ satisfy KP', $\alpha < \omega_1$, and $L(\alpha + 1, f) \backslash L(\alpha, f)$ meet N^N. Then $L(\alpha, f)$ is unusual. There are arbitrarily large countable α such that $L(\alpha, f)$ is unusual.*

PROOF. Argue as in the proof of Lemma 15, but this time with the parameter f. Use the fact that $\omega_1(f) = \omega_1$. ⊣

LEMMA 26. *There is no Borel selection for S' on N^N.*

PROOF. First observe that the domain of S' is N^N. To see this, let $x = (\omega, R, c)^*$. If (ω, R, c) is not well founded, then $x \in dom(S')$ by clause 2. If (ω, R, c) is well founded, then $x \in L[f]$, and apply Lemma 25 and clause 3.

Now observe that being a Borel code for a Borel selection for S' on N^N is a Π^1_1 property with parameter f. By Shoenfield absoluteness, let $F : N^N \to N^N$

be a Borel selection for S' on N^N and let u be a Borel code for F, where $u \in L[f]$.

According to Lemma 25, let $\alpha < \omega_1$ be such that $L(\alpha, f)$ is unusual and $u \in L(\alpha, f)$. Let $x = (\omega, R, c)^*$, where (ω, R, c) is isomorphic to $(L(\alpha, f), \in, f)$. Let $y = F(x)$. Then y is an unusual code. Let $y = (\omega, Y, d)^*$.

Since $(x, y) \in S'$, we see that clause (3) applies in the definition of S' since (ω, R) is well founded. Hence x is internal to (ω, Y). Hence every set hyperarithmetic in x is recursive in y. But u is recursive in x and $y = F(x)$ is hyperarithmetic in (x, u). Hence y is hyperarithmetic in x. This is the desired contradiction. ⊣

LEMMA 27. *Let $V \subseteq N^N$ be compact. There is a countable ordinal bound to the lengths of the maximum well ordered initial segments of the internal ordinals of every $(\omega, R, c) \in V$ such that $(\omega, R, c)^*$ lies in V.*

PROOF. Let $g \in N^N$ be such that every element of V is everywhere strictly dominated by g. Suppose there are unusual $(\omega, R, c) \in X$ such that $(\omega, R, c)^* \in V$ with arbitrarily long countable well ordered initial segments of the internal ordinals. Then every $h \in L[f]$ will be internal to some unusual (ω, R, c) such that $(\omega, R, c)^* \in V$. Hence every $h \in L[f]$ will be eventually strictly dominated by g. This contradicts our assumption made at the beginning of this section, that no g eventually strictly dominates every element of $L[f] \cap N^N$. ⊣

LEMMA 28. *There is a Borel selection for S on every compact subset of N^N.*

PROOF. Let $V \subseteq N^N$ be compact. Let α be the strict bound given by Lemma 27. Recall that α is countable in $L[f]$. By Lemma 25, let $L(\beta, f)$ be unusual, $\beta > \alpha$, so that every unusual code of every unusual $L(\gamma, f)$, $\gamma < \alpha$, is internal to $L(\beta, f)$. Let $y = L(\beta, f)^*$.

For x that are not unusual codes, or $x \notin V$, set $F(x) = x$. For $x \in V$ that are unusual codes, let $x = (\omega, R, c)^*$, where (ω, R, c) is chosen by the uniformly effective procedure in x, f. Because of the bound, α, we can tell in a Borelian way whether (ω, R, c) is well founded. If (ω, R) is not well founded, then we can construct its well founded part in a Borelian way, and construct an infinite backwards R chain in a Borelian way. So $(x, y) \in S'$ by clause 2.

If (ω, R) is well founded, then its length is $< \alpha$, and we set $F(x) = y$. By the construction of y, $(x, y) \in S'$ by clause 3. ⊣

LEMMA 29. *Proposition 5 is false.*

PROOF. By Lemmas 26 and 28. ⊣

We now release our assumptions.

THEOREM 30. *The following are provably equivalent in ZFC.*

1. *Proposition* 5.
2. *Proposition* 7.
3. *DOM.*

In particular, none of 1–3 can be proved or refuted in ZFC, assuming ZFC is consistent.

PROOF. $2 \Rightarrow 1$ is obvious. $1 \Rightarrow 3$ is by Lemma 29 and the assumptions used to prove it. By Theorem 13, we have $3 \Rightarrow 1$. A proof of $3 \Rightarrow 2$ has recently been given in [6]. The relative consistency of DOM is by Hechler forcing as in the proof of Lemma 12. The relative consistency of $\neg DOM$ is clear since DOM demonstrably fails in L. ⊣

§7. On the reals. Recall Theorems 3 and 9:

Let $S \subseteq N^N \times N^N$ be Borel and $E \subseteq N^N$ be Borel. If there is a continuous selection for S on every compact subset of E, then there is a continuous selection for S on E.

Let $S \subseteq \Re \times \Re$ be Borel and $E \subseteq \Re$ be Borel with empty interior. If there is a continuous selection for S on every compact subset of E, then there is a continuous selection for S on E.

THEOREM 31. *Theorems 3 and 9 are provably equivalent over ATR_0.*

PROOF. Assume Theorem 3. Let S, E be as given in Theorem 9, and assume there is a continuous selection for S on every compact subset of E. Let A be a countable dense subset of such that $E \subseteq \Re \backslash A$. It is well known that $\Re \backslash A$ is homeomorphic to $\Re \backslash Q$, which is in turn homeomorphic to N^N.

Let $h: \Re \backslash A \rightarrow N^N$ be a homeomorphism. Then $hS \subseteq N^N \times N^N$ and $hE \subseteq N^N$ are Borel. We claim that there is a continuous selection for hS on every compact subset of hE. To see this, let $V \subseteq hE$ be compact. Write $hW = V$, $W \subseteq E$. Then W is compact. Hence there is a continuous selection for S on W. This conjugates to a continuous selection for hS on $hW = V$, establishing the claim.

By Theorem 3, there is a continuous selection for hS on hE. This conjugates to a continuous selection for S on E.

Assume Theorem 9. Let S, E be as given in Theorem 3, and assume there is a continuous selection for S on every compact subset of E. Let $h': N^N \rightarrow \Re \backslash Q$ be a homeomorphism. Then $h'S \subseteq \Re \backslash Q \times \Re \backslash Q$ and $h'E \subseteq \Re \backslash Q$ are Borel and hE has empty interior in $\Re$. We claim that there is a continuous selection for $h'S$ on every compact subset of $h'E$. To see this, let $V \subseteq h'E$ be compact. Write $h'W = V$, $W \subseteq E$. Then W is compact. Hence there is a continuous selection for S on W. This conjugates to a continuous selection for $h'S$ on $h'W = V$, establishing the claim.

By Theorem 9, there is a continuous selection for $h'S$ on $h'E$. This conjugates to a continuous selection for S on E. ⊣

Recall Propositions 7 and 10:

Let $S \subseteq N^N \times N^N$ be Borel and $E \subseteq N^N$ be Borel. If there is a Borel selection for S on every compact subset of E, then there is a Borel selection for S on E.

Let $S \subseteq \Re \times \Re$ be Borel and $E \subseteq \Re$ be Borel. If there is a Borel selection for S on every compact subset of E, then there is a Borel selection for S on E.

THEOREM 32. *Propositions 7 and 10 are provably equivalent over ATR_0.*

PROOF. Let Proposition 10′ be the same as Proposition 10, but with the added hypothesis that E has empty interior. The proof of the equivalence of Proposition 7 and 10′ is the same as the proof of the equivalence of Theorem 3 and 9, except that continuous replaced everywhere by Borel.

It remains to show that Proposition 10′ implies Proposition 10. Let S, E be as given, where there is a Borel selection for S on every compact subset of E. Then $E \backslash Q$ is Borel, and so by Proposition 10′, there is a Borel selection for S on $E \backslash Q$. Hence there is a Borel selection for S on E. ⊣

Recall Theorems 4 and 11:

Let $S \subseteq N^N \times N^N$ be Borel. If there is a constant selection for S on every compact set, then there is a Borel selection for S.

Let $S \subseteq \Re \times \Re$ be Borel. If there is a constant selection for S on every compact set of irrationals, then there is a Borel selection for S on the irrationals.

THEOREM 33. *Theorems 4 and 11 are provably equivalent over ATR_0.*

PROOF. This is clear using a homeomorphism from N^N onto the irrationals. ⊣

By Theorems 24, 31, and 33, Theorems 9 and 11 can be proved using uncountably many iterations of the power set operation but not with only countably many iterations of the power set operation.

By Theorems 30 and 32, Proposition 10 is equivalent to DOM over ZFC, and Proposition 10 is independent of ZFC.

REFERENCES

[1] JON BARWISE, ***Admissible sets and structures***, Perspectives in Mathematical Logic, Springer-Verlag, 1975.

[2] G. DEBS and J. SAINT RAYMOND, *Compact covering and game determinacy*, ***Topology and Its Applications***, vol. 68 (1996), pp. 153–185.

[3] ———, *Cofinal Σ_1^1 and Π_1^1 subsets of N^N*, ***Fundamenta Mathematicae***, vol. 159 (1999), pp. 161–193.

[4] ———, *Compact covering mappings and cofinal families of compact subsets of a Borel set*, ***Fundamenta Mathematicae***, vol. 167 (2001), pp. 213–249.

[5] ———, *Compact covering mappings between Borel sets and the size of constructible reals*, ***Transactions of the American Mathematical Society***, vol. 356 (2004), pp. 73–117.

[6] ———, *Borel liftings of Borel sets: some decidable and undecidable results*, `http://web.ccr.jussieu.fr/eqanalyse/Users/jsr`, to appear as an AMS Memoir.

[7] DERRICK DUBOSE, *The equivalence of determinacy and iterated sharps*, ***The Journal of Symbolic Logic***, vol. 55 (1990), pp. 502–525.

[8] H. M. FRIEDMAN, *Higher set theory and mathematical practice*, ***Annals of Mathematical Logic***, vol. 2 (1971), pp. 326–357.

[9] ———, *On the necessary use of abstract set theory*, ***Advances in Mathematics***, vol. 41 (1981), no. 3, pp. 209–280.

[10] S. S. HECHLER, *On the existence of certain cofinal subsets of* ${}^{\omega}\omega$, ***Proceedings of Symposia in Pure Mathematics***, vol. 13, Part II, 1974, pp. 155–173.

[11] T. JECH, ***Set theory***, Academic Press, 1978.

[12] A. S. KECHRIS, ***Classical descriptive set theory***, Graduate Texts in Mathematics, Springer-Verlag, 1995.

[13] D. A. MARTIN, *A purely inductive proof of Borel determinacy*, ***Proceedings of the AMS Symposium in Pure Mathematics***, Recursion Theory, vol. 42, 1985, pp. 303–308.

[14] R. L. SAMI, *Analytic determinacy and 0#: A forcing-free proof of Harrington's theorem*, ***Fundamenta Mathematicae***, vol. 160 (1999), pp. 153–159.

[15] J. R. SHOENFIELD, *The problem of predicativity*, ***Essays on the foundations of mathematics*** (Y. Bar-Hillel et al., editor), The Magnes Press, Jerusalem, 1961, pp. 132–142.

[16] S. G. SIMPSON, ***Subsystems of second order arithmetic***, Perspectives in Mathematical Logic, Springer, 1999.

DEPARTMENT OF MATHEMATICS
OHIO STATE UNIVERSITY
COLUMBUS, OHIO 43210, USA
E-mail: friedman@math.ohio-state.edu

INTERPOLATION IN GOAL-DIRECTED PROOF SYSTEMS 1

D. M. GABBAY AND N. OLIVETTI

Abstract. In this paper we study interpolation properties for a variety of implicational logics, covering intuitionisitic and classical logic, substructural logics, and modal logics of strict implication. We give a constructive proof of the interpolation property based on the goal-directed formulation of these logics. We conclude by discussing two generalization of the ordinary notion of interpolation, namely the notion of *chain interpolation* and the notion of *structural interpolation*.

§1. Introduction. This series of papers aims to study interpolation properties for implicational fragments of a variety of substructural, strict modal and intuitionistic and intermediate logics. The methodology is proof theoretical and makes use of a goal directed formulation of these fragments which follows the logic programming style of deduction (see [16]). The aim of this series is threefold:

- to demonstrate the power of the goal directed proof methods in obtaining metalevel properties of logics, such as complexity results, abduction, relevance, failure and, of course, interpolation.
- to obtain more refined as well as new kinds of interpolation theorems for our logics.
- to investigate new global methods for obtaining interpolation.

We insist that interpolants be found by metalevel induction on the goal directed proof algorithm. We are systematically going through the landscape of logics for which [16] gives a goal directed formulation and we show how to compute interpolants for them. In many cases the results are new and give a new fine tuning of interpolation. We also include the results for known cases, like intuitionistic or classical logics. The present paper deals with the $\{\to, \wedge\}$ propositional fragments. Subsequent papers will extend the results. This paper is the beginning of a research programme.

The rest of this section will provide background for our paper.

1.1. General background. Given two languages $\mathbb{L}_1$ and $\mathbb{L}_2$ (with $\mathbb{L}$ the common language) and a consequence relation $\vdash$ on $\mathbb{L}_1 \cup \mathbb{L}_2$, we can formulate the interpolation property as follows:

- If $A \vdash B$, with A a formula in $\mathbb{L}_1$ and B in $\mathbb{L}_2$, then there exists an H in $\mathbb{L}$ such that $A \vdash H \vdash B$.

Logic Colloquium '01
Edited by M. Baaz, S. Friedman, and J. Krajíček
Lecture Notes in Logic, 20

What we mean by common language $\mathbb{L}$ can vary.

(i) $\mathbb{L}$ may contain only constants, predicates and variables that are common to A and to B.
(ii) $\mathbb{L}$ may contain only connectives and quantifiers that are common to A and B.
(iii) Refinements about the form of occurrence of the common items in A and B and how they appear in H (positive, negative, free, etc.).

It was [3] who proposed and proved interpolation for classical predicate logic. Improvements along the lines of (iii) were put forward by [19]. See [27] for the most recent account.

The classical proof methods can be adapted to yield interpolation for the intuitionistic case.

Gabbay [7], [8], [11], [9] has used semantic methods to get interpolation for a variety of quantified modal logics and intuitionistic logic, among them **K**, **T**, **K4**, **S4**, **D**, **S4.1**, **S4.2** and more. His methods do not apply to logics of constant domains in the modal case, see [10] (but do apply to intuitionistic logic with constant domains). Indeed Kit Fine [4] has shown failure of interpolation for quantified **S5**.

Meanwhile L. Maksimova 1977–1979 [20], [22], [23], [21] studied interpolation and algebraic amalgamation of the corresponding algebras for propositional intermediate logics (also called superintutionistic logics) and companion modal logics. She has shown that there exist exactly seven consistent intermediate logics with interpolation. See the forthcoming book [14] for a full account of this work.

Fitting [5] used semantic tableaux to obtain interpolation for **K**, **T**, **S4**, **K4**, **D** and **D4**. His most recent paper [6] shows that the negative result of Kit Fine for **S5** can be remedied if one allows second order quantifiers. These ideas relate to the results of Gabbay and Ohlbach [15] on SCAN, where some interpolation results can be obtained by eliminating second order quantifiers. See also [25] for a survey.

Marx and Areces [24] as well as Gabbay [13] considered interpolating on connectives, for the case of combination of modalities or fibered logics.

Roorda [29] and Pentus [28] have proved various interpolation properties for several fragments of linear logic and Lambek logic.

The above brief summary gives the reader some background and references, providing a context for the present work.

There are, in fact, further developments which make a further study of interpolation more urgent.

In the past two decades many logics have become more prominent in view of their applications in computer science and Artificial Intelligence. New ways of presenting these logics have been put forward. Two aspects are especially relevant to interpolation:

1. The data-structures of theories Δ are more complex than sets, they can be lists, lists of lists, hypersequents, trees, etc.
2. A unifying proof theoretical method (such as labelled deductive systems or goal directed methodologies) have been developed for these logics, see [12], [16].

The above advances require us to revise the way we view interpolation and investigate more refined interpolation concepts.

Some ideas in this direction will be discussed in Section 5 of this paper.

We therefore offer a series of papers studying interpolation for the new logics, offering more general and more refined interpolation theorems. The present paper, the first in our series, will show how to get interpolation for strict implication and substructural logics formulated in the goal directed environment of [16].

Our strategy is to consider several case studies of the $\wedge, \rightarrow$ fragments of some important logics, before presenting, in a later paper, a general formulation of interpolation for Labelled Deductive Systems [12] and investigate how to enrich one's language to get interpolation in cases where interpolants are not available.

1.2. Specific background. This subsection will set the semantical and conceptual scheme for the logics of this paper. We begin with the semantics.

DEFINITION 1.1 (Semigroups). *Let* $\mathbf{m} = (S, \circ, \leq, e)$ *be an associative semigroup, with binary operation* $\circ$, *partial ordering* $\leq$ *and two sided unit* e, (*that is* $x \circ e = e \circ x = x$ *for* $x \in S$). *We also have*

- $x \leq x' \wedge y \leq y'$ *imply* $x \circ y \leq x' \circ y'$.

DEFINITION 1.2 (Various implications). *Semigroups* $\mathbf{m}$ *can be used as a basis for defining semantics for several kinds of implications. The general schema for defining the truth conditions has two clauses.*

(#1) *Some condition on the assignment* h.
(#2) $t \vDash A \rightarrow B$ *iff* $\forall s(\rho(t, s)$ *and* $s \vDash A$ *imply* $f(t, s) \vDash B)$.
where ρ *is a binary relation on* S *and* f *is a binary function on* S.

Different conditions (#1) *and* (#2) *define different logics.*

1. *Intuitionistic implication* Let h be an assignment to the atoms of the language (i.e., for each atom q, $h(q) \subseteq S$). Assume the following holds:
   (*1) $t \leq s$ and $t \in h(q)$ imply $s \in h(q)$.
   Ignore multiplication $\circ$ and let
   (*2) $t \vDash q$ if $t \in h(q)$, q atomic.
   (*3) $t \vDash A \rightarrow B$ iff $\forall s(t \leq s$ and $s \vDash A$ imply $s \vDash B)$.
   Then (*1), (*3) define intuitionistic $\rightarrow$. We have
   (*4) $\mathbf{m} \vDash A$ iff for all $t \in S$, $t \vDash A$.

2. *Right (and left) Lambek implications*
   We assume we have two implications $\rightarrow_r$ and $\rightarrow_l$ in the language. We ignore the condition (*1) and let
   (*5) $t \vDash A \rightarrow_r B$ iff $\forall s(s \vDash A$ implies $t \circ s \vDash B)$.
   (*6) $t \vDash A \rightarrow_l B$ iff $\forall s(s \vDash A$ implies $s \circ t \vDash B)$.
   (*7) $\mathbf{m} \vDash A$ iff $e \vDash A$.
3. *Linear Implication*
   Like Lambek implication assuming the semigroup is commutative.
4. *Relevant implication*
   Like linear implication assuming the semigroup satisfies $t \circ t = t$.
5. *Bunched implications* [26]
   Here we have both right $\rightarrow_r$ and intuitionistic implications.
6. *Strict* S4 *implication*
   Like intuitionistic implication without condition (*1).

REMARK 1.3. *As a rough guide to the proof theoretic meaning of these logics, consider the following list of Hilbert type axiom schemas and rules:*

1. $A \rightarrow A$
2. $(A \rightarrow B) \rightarrow ((B \rightarrow C) \rightarrow (A \rightarrow C))$

2a. $\dfrac{\vdash A \rightarrow B}{\vdash (B \rightarrow C) \rightarrow (A \rightarrow C)}$

3. $(A \rightarrow B) \rightarrow ((C \rightarrow A) \rightarrow (C \rightarrow B))$
4. $(A \rightarrow (B \rightarrow C)) \rightarrow ((B \rightarrow (A \rightarrow C))$

4a. $(A \rightarrow ((B_1 \rightarrow B_2) \rightarrow C) \rightarrow (((B_1 \rightarrow B_2) \rightarrow (A \rightarrow C))$

5. $(C \rightarrow D) \rightarrow (A \rightarrow (C \rightarrow D))$
6. $(A \rightarrow (A \rightarrow B)) \rightarrow (A \rightarrow B)$
7. $A \rightarrow (B \rightarrow A)$
8. $\dfrac{\vdash A, \vdash A \rightarrow B}{\vdash B}$
9. $((A \rightarrow B) \rightarrow A) \rightarrow A$

The right-hand Lambek implication is characterised by (1), (3), (2a) *and* (8).
Linear implication is characterised by (1), (2), (3), (4) *and* (8).
Relevance implication is characterised by (1), (2), (3), (4), (6) *and* (8).
Strict **S4** *implication is characterised by* (1), (2), (3), (4a), (5) *and* (8).
Intuitionistic implication is characterised by (1), (2), (3), (4), (6), (7) *and* (8).
Classical implication is characterised by adding (9) *to intuitionistic implication.*
See [12] *for details.*

Our aim is to investigate interpolation properties for some of the above logics. It will be convenient to formulate the semigroup semantics more carefully.

DEFINITION 1.4. *Let $\mathbb{N}$ be the set of all natural numbers. Let $\mathbb{N}^*$ be the set of all finite sequences of natural numbers including the empty sequence $\varnothing$. Let $*$ be concatenation. Define a relation R_0 on $\mathbb{N}^*$ by:*

- xR_0y *iff for some* $n \in \mathbb{N}$, $y = x * (n)$.
- *Let* $\bar{R}_0$ *be the transitive and reflexive closure of* R_0.

$(\mathbb{N}^*, R_0)$ *defines a big tree for us which we shall use to present semantical interpretations.*

1. *A modal logic model has the form* $(S, R, \varnothing)$, *where* $S \subseteq \mathbb{N}^*$ *and* R *satisfy the following:*
   - $x \in S \wedge \varnothing \bar{R}_0 y \wedge y \bar{R}_0 x \to y \in S$.
   - $R_0 \restriction S^2 \subseteq R \subseteq S^2$.

   So for modal **K** *we take* $R = R_0 \restriction S^2$. *For* **S4** *we take* $R = \bar{R}_0 \restriction S^2$, *for* **T** *we take* $R = R_0 \restriction S^2 \cup \text{Identity} \restriction S^2$, *etc., etc.*
2. *A Lambek calculus model is* $(\mathbb{N}^*, *)$.

It is possible to generate all the above logics uniformly in this particular $\mathbb{N}$ *model.*

3. *First define* $y \cdot x$ *as follows:*
   (a) *Let* $x \wedge y$ *be the largest sequence* z *such that there are* x' *and* y' *satisfying* $x = z * x'$ *and* $y = z * y'$.
   Let $x - y$ *be defined as* x' *and similarly* $y - x$ *is* y'.
   We have $x = (x \wedge y) * (x - y)$.
   (b) *Let* $y \cdot x$ *be defined as* $y * (x - y)$.
   Figure 1 *shows the meaning of* $\wedge$, $-$ *and* $\cdot$.

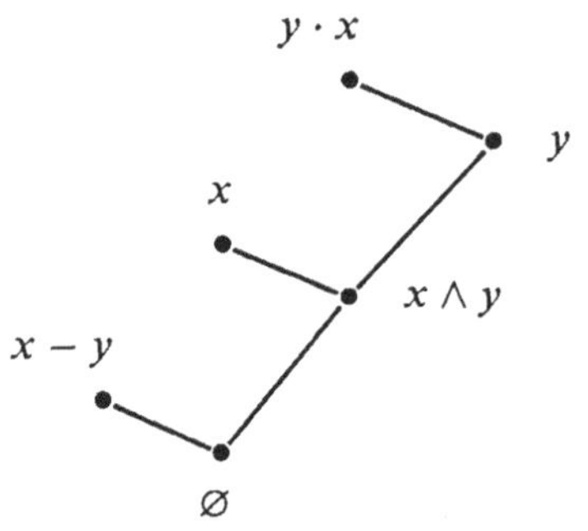

FIGURE 1

   (c) *Note that if* $\rho(x, y)$ *is defined as* $\exists z(x = y * z)$, *then we have that* $\rho(x, y) \to y \cdot x = x$.
4. *We can now define satisfaction for Lambek implication and* **S4** *implication uniformly by*
   - $y \vDash A \to B$ *iff* $\forall x(\rho(x, y)$ *and* $x \vDash A$ *imply* $y \cdot x \vDash B)$

where $\rho(x, y) = Truth$ for the Lambek case and $\rho(x, y) = \exists z(x = y * z)$, in the **S4** *case.*

We now explain the goal directed approach to these logics.

Each section below gives an exact definition of the goal directed algorithm for the logic discussed in the section, but it does help to briefly explain the basic idea to the reader at this point in time, before we embark on the full details.

EXAMPLE 1.5. *This example explains intuitively two notions of $\rightarrow$, the Lambek and concatenation $\rightarrow$ and the strict implication $\rightarrow$.*

Consider a PhD programme at a university for a student. Upon registration the following rule is activated: if the student submits a thesis (call this a) and then passes an oral exam (call this b) then he is awarded a PhD degree (call this q).

We write the rule as follows:

$$a \rightarrow (b \rightarrow q).$$

We have

$$a \rightarrow (b \rightarrow q), a, b \vdash q$$

but

$$a \rightarrow (b \rightarrow q), b, a \nvdash q$$

because we insist on the exam to be after the submission of the thesis.

Also there is no insistence on how much time it takes to write the thesis and how much time passes between the submission of the thesis and the exam.

If we want to be sensitive to time we need to interpret $x \rightarrow y$ differently, and take it to mean 'next year'. Thus $a \rightarrow (b \rightarrow q)$ would mean that next year if the thesis is submitted, then $b \rightarrow q$ holds, namely that if in the following year the exam is passed, then a PhD is awarded.

The first implication is the Lambek or concatenation logic implication, and the second one is **K** *strict implication (counting possible worlds as years).*

Let us now bring out the difference between the two implications. Suppose there is a rule $e \rightarrow a$ that says that a major practical implementation work (call this e) can count as a thesis submission, (write it as $e \rightarrow a$). Then we have

$$e \rightarrow a, e \vdash a$$

but

$$e, e \rightarrow a \nvdash a.$$

(The implementation must occur when the rule is valid.)

Thus we have

$$a \rightarrow (b \rightarrow q), e \rightarrow a, e, b \vdash q.$$

However, if $\rightarrow$ *means 'next year', the above will not hold because* $e \rightarrow a, e$ *gives* a *in two years and* $a \rightarrow (b \rightarrow q)$ *requires the* a *to be in one year.*

Let us now do this example again formally and in a goal directed way: consider the following list of data items and query:

$$a \rightarrow (b \rightarrow q), e \rightarrow a, e, d \rightarrow b, d \vdash^{?} q$$

Obviously the clause that can give us q *is* $a \rightarrow (b \rightarrow q)$*. The goal directed algorithm will tell us where we can expect to find a clause which can give* q *and where we can expect to find the rest of the data to prove* a *and to prove* b*.*

For the Lambek implication, the clause for q *must be the leftmost one, as it is indeed the case in our example, and the rest of the data must be divided into chunks, one to give* a *and one to give* b *in the correct order!*

Figure 2 shows the derivation, where $t_1, \ldots, t_5$ *name the clauses:*

t_1	t_2	t_3	t_4	t_5
•	•	•	•	•
$a \rightarrow (b \rightarrow q)$	$e \rightarrow a$	e	$d \rightarrow b$	d
	a		b	

FIGURE 2

The original problem reduces to Figure 3.

t_1	s_1	s_2
•	•	•
$a \rightarrow (b \rightarrow q)$	a	b

FIGURE 3

In the case of strict implication of modal **K**, $t_1, \ldots, t_5$ *are possible worlds and* $\vdash ?q$ *asks whether* q *must be true at* t_5*. We have* t_{i+1} *is a possible world for* t_i*. In this case the computation reduces to Figure 4. In other words,* t_2, t_3

t_1	t_2	t_3	t_4	t_5
•	•	•	•	•
$a \rightarrow (b \rightarrow q)$	$\top$	a	$\top$	b

FIGURE 4

do not collapse into one point s_1 *nor do* t_3, t_4 *collapse into one point* s_2*. They are possible worlds and they remain there. Thus* q *cannot be proved in the modal* *logic* **K**. *However, if the clause at* t_1 *were* $\top \to (a \to (\top \to (b \to q)))$ *to derive* q *at* t_5.

The different goal directed algorithms differ on how to run around the data looking for groups of clauses to provide antecedents for modus ponens.

A few words about notation. We are going to consider several implicational calculi. We use $\to$ for implication and build up the notion of wff in the usual way. $\top$, $\bot$ denote *top* and *bottom*. We use A, B, C to denote wffs and $\mathbb{A}$, $\mathbb{B}$, $\mathbb{C}$ to denote sets of wffs (thus allowing for implicit conjunctions in the language) and Δ, Γ, Θ, Φ etc. to denote databases (usually sequences $(\mathbb{A}_1, \dots, \mathbb{A}_n)$ of sets of wffs.

We consider the basic interpolation problem for a logic **L** as follows.

Given $\mathbb{A} \vdash_{\mathbf{L}} \mathbb{B}$, $\mathbb{A}$ in language 1 and $\mathbb{B}$ in language 2, is there an $\mathbb{H}$ in the common language such that $\mathbb{A} \vdash_{\mathbf{L}} \mathbb{H} \vdash_{\mathbf{L}} \mathbb{B}$.

In practice we shall need to prove a more general theorem. Let q be in language 1 and let $\Delta_{1,2}$ be a theory containing wffs in both language 1 and 2. Assume $\Delta_{1,2} \vdash q$. We are seeking ways to replace language 2 wffs in $\Delta_{1,2}$ by $\Delta_{1,2}$ provable wffs in the common language, to obtain a new theory Θ such that $\Delta_{1,2} \vdash \Theta \vdash q$.

We consider refinements like the complexity of the interpolant, restrictions on the sublanguages involved, number of elements in $\mathbb{A}$, $\mathbb{B}$ and $\mathbb{H}$, etc.

We shall write $\Delta \vdash_m A$ to mean A is provable from Δ in exactly (no more and no less than) m nested computational steps. We shall also write the obvious $\Delta \vdash_{\leq m} A$.

§2. Interpolation for linear and for intuitionistic implications. We begin, in this section, with the relatively more simple cases of linear and intuitionistic implications. Subsection 2.1 deals with linear implication while subsection 2.2 deals with intuitionistic implication.

In the case of linear implication we can deal with the $\to$ fragment alone (we do not need $\top$ as we do not have theorems like $p \to p \vdash q \to q$). In intuitionistic logic we need $\top$ and we may as well add $\bot$ and conjunctions $\wedge$. We will get interpolation for classical logic as a bonus.

2.1. Interpolation for linear implication.

DEFINITION 2.1. *Let* $\mathbb{L}$ *be a language with* $\to$ *only and atomic propositions* $\{q_1, q_2, \dots\}$.

1. *A wff has the form either* q, *for* q *atomic, or* $A_1 \to (A_2 \to \cdots \to (A_n \to q) \dots)$, *for* A_i *wffs and* q *atomic.*
2. *A database is a multiset of formulas.*
3. *Let* Δ *be a database and* $(A_1, \dots, A_n)$ *be a sequence of wffs. We let* $\Delta + (A_1, \dots, A_n)$ *be* $\Delta \cup \{A_1, \dots, A_n\}$, *where* $\cup$ *is multiset union and* $\{A_1, \dots, A_n\}$ *is the multiset based on* $(A_1, \dots, A_n)$.

DEFINITION 2.2. *Let $\Delta \vdash_{\leq m} A$ be defined in a goal directed way as follows. The index indicates maximal number of steps.*

1. $\Delta \vdash_{\leq 1} q$ *if* $\{q\} = \Delta$, *for q atomic.*[1]
2. $\Delta \vdash_{\leq m} (A_1 \to (\cdots \to (A_n \to q)\ldots)$ *iff* $\Delta + (A_1, \ldots, A_n) \vdash_{\leq m} q$
3. $\Delta \cup \{A_1 \to (\cdots \to (A_n \to q)\ldots)\} \vdash_{\leq(m+1)} q$ *iff* $\Delta = \Delta_1 + \cdots + \Delta_n$ *and for each* $1 \leq i \leq n$, $\Delta_i \vdash_{\leq m_i} A_i$, *for some* $m_i \leq m$, *and* Δ_i *are all pairwise disjoint and* $max(m_i) = m$.[2]
4. $\Delta \vdash A$ *iff for some n*, $\Delta \vdash_{\leq n} A$.
5. *We define* $\Delta \vdash_{\leq m} \{A_1, \ldots, A_n\}$ *iff for some* $\Delta_1, \ldots, \Delta_n$ *we have* $\Delta_i \vdash_{\leq m} A_i$, $1 \leq i \leq n$ *and* $\Delta_1 + \cdots + \Delta_n = \Delta$.
6. *In* (3) *above we say that* $A_1 \to (\cdots \to (A_n \to q)\ldots)$ *was* used *in this step of the computation.*
7. $\Delta \vdash_m B$ *iff* $\Delta \vdash_{\leq m} B$ *and for no* $m' < m$, *do we have* $\Delta \vdash_{\leq m'} B$.

THEOREM 2.3 (Completeness). *$\vdash$ above is linear logic consequence for $\to$.*

In particular $x \to (y \to z) \vdash y \to (x \to z)$ *and we can write* $\{A_1, \ldots, A_n\} \to q$ *for* $A_1 \to (A_2 \to \ldots (A_n \to q)\ldots)$.

PROOF. See Chapter 5 of [16]. ⊣

DEFINITION 2.4. *Let* $\mathbb{L}_i$, $i = 1, 2$ *be two languages for* $\to$, *based on atoms* $Q \cup Q_i$, $i = 1, 2$ *respectively, where* Q, Q_1, Q_2 *are pairwise disjoint.*

We say the logic based on $Q \cup Q_1 \cup Q_2$ *satisfies interpolation iff the following holds.*

If $\Delta^1 + \Delta^2 \vdash_{\leq m} A^1$ *where superscripts indicate language, i.e., where* Δ^1, A^1 *are in* $\mathbb{L}_1$ *and* Δ^2 *is in* $\mathbb{L}_2$, *then for some* Φ *in the common language (i.e., in the language* $\mathbb{Q}$ *we have* $\Delta^2 \vdash_{\leq m} \Phi$ *and* $\Delta^1 + \Phi \vdash_{\leq m} A^1$. *We say the interpolant is effectively computable iff* Φ *can be effectively obtained from* Δ^1, Δ^2 *and* A^1. *Note that we can actually build an interpolant up only from atoms appearing both in* Δ^2 *and* $\Delta^1 + A^1$)

THEOREM 2.5. *The implication fragment of linear logic has effectively computable interpolation.*[3]

PROOF. By induction on the number of steps in the proof.

1. CASE of $m = 1$.
   Since $\Delta^1 + \Delta^2 \vdash_1 A^1$ we must have $\Delta^1 + \Delta^2 = \{q\}$, where $q = A^1$ is atomic. If $\Delta^2 = \varnothing$ then the interpolant is $\varnothing$. Otherwise $\Delta^2 = \{q\}$ and

[1] For intuitionistic implication this condition is modified to $\{q\} \subseteq \Delta$ and databases are sets and not multisets. See next subsection.

[2] Note that for the case of $\to$ being intuitionistic implication, Δ_i are sets and need not be disjoint. In fact we can have $\Delta_i = \Delta \cup \{A_1 \to \cdots \to (A_n \to q)\ldots)\}$. See next subsection.

[3] We have to be pedantic when we talk about common language. In classical or intuitionistic logic, for example, we have $p \to p \vdash q \to q$, and yet, unless we have a constant $\top$ we cannot interpolate. We may need similar constants for the interpolation theorems of this and other sections.

$\Delta^1 = \varnothing$. Let the interpolant be $\{q\}$. It is in the common language, and $\Delta^2 \vdash_1 q$ and $q \vdash_1 q$.

2. CASE $m+1$.
We have that

$$\Delta^1 + \Delta^2 \vdash_{\leq(m+1)} (A_1^1 \to \dots (A_n^1 \to q^1)\dots).$$

This reduces to[4]

$$\Delta = (A_1^1, \dots, A_n^1) + \Delta^1 + \Delta^2 \vdash_{\leq(m+1)} q^1.$$

Hence for some

$$B = [B_1 \to (B_2 \to \dots \to (B_k \to q)\dots)] \in \Delta.$$

we have

$$(\Delta - B) = \Theta_1 + \dots + \Theta_k \text{ and for } 1 \leq i \leq k, \Theta_i \vdash_{\leq m} B_i, \text{ and } q = q^1.$$

We can assume for each i, $\Theta_i = \Theta_i^1 \cup \Theta_i^2$ with Θ_i^j in the language of Δ^j. Let $\Theta^j = \bigcup_i \Theta_i^j$ then $\{B\} \cup \Theta^1 \cup \Theta^2 = \Delta$.

We distinguish two subcases:

SUBCASE 1. $B \in (A_1^1, \dots, A_n^1) + \Delta^1$.

This case means that B is in the language of Δ^1.

Hence $B \in (A_1^1, \dots, A_n^1) + \Delta^1$ and $\{B\} \cup \Theta^1 = (A_1^1, \dots, A_n^1) + \Delta^1$ and $\Theta^2 = \Delta^2$.

Since $\Theta_i^1 + \Theta_i^2 \vdash_{\leq m} B_i$ by the induction hypothesis there exist $\Phi_1, \dots, \Phi_k$ in the common language such that $\Theta_i^2 \vdash_{\leq m} \Phi_i$ and $\Theta_i^1 + \Phi_i \vdash_{\leq m} B_i$ for $1 \leq i \leq k$.

Hence we have

$$\Delta^2 \vdash_{\leq m} \bigcup_{i=1}^{k} \Phi_i = \Phi.$$

We must now show that

$$(A_1, \dots, A_n) + \Delta^1 + \Phi \vdash_{\leq(m+1)} q^1.$$

We have that $\Theta_i^1 + \Phi_i \vdash_{\leq m} B_i$.

Hence

$$\bigcup_{i=1}^{k} \Theta_i^1 \cup \bigcup_{i=1}^{k} \Phi_i \vdash_{\leq m} \{B_1, \dots, B_k\},$$

namely $\Theta^1 \cup \Phi \vdash_{\leq m} \{B_1, \dots, B_k\}$, but $\Theta^1 = ((A_1^1, \dots, A_n^1) + \Delta^1) - B$.

[4]The data-structures in linear logic are multisets. Thus + is commutative. We added (A_j^i) from the left, but in linear logic it does not matter. The question of exactly where a wff is added or placed in the data-structure will be important when we address interpolation for the Lambek calculus.

Hence

$$((\{A_1^1, \ldots, A_n^1\} + \Delta^1) - \{B\}) + \Phi \vdash_{\leq m} \{B_1, \ldots, B_k\}$$

but $B = B_1 \to (\cdots \to (B_k \to q^1))$.

Hence by modus ponens of linear logic

$$\{A_1, \ldots, A_n\} + \Delta^1 + \Phi \vdash_{\leq (m+1)} q^1$$

as required.

Note that in this subcase we have $\Delta^2 \vdash_{\leq m} \Phi$ and not $\Delta^2 \vdash_{\leq (m+1)} \Phi$. We shall need this sharper result for subcase 2 coming next.

SUBCASE 2: $B \in \Delta^2$.

In this case, since $q = q^1$, q is in the common language, in particular in language 2. We have

$$(A_1^1, \ldots, A_n^1) + \Delta^1 + \Delta^2 \vdash_{\leq (m+1)} q.$$

We can view the above as a particular case of subcase 1, where we swap languages 1 and 2. We use the sharp result on m obtained for subcase 1. Hence there exists a Φ in the common language such that $(A_1^1, \ldots, A_n^1) + \Delta^1 \vdash_{\leq m} \Phi$ and $\Phi + \Delta^2 \vdash_{m+1} q$.

By the properties of linear implication we have

$$\Delta^2 \vdash_{\leq (m+1)} \Phi \to q.$$

$\Phi \to q$ is our interpolant. We need to show that $\{A_1^1, \ldots, A_n^1\} + \Delta^1 + \Phi \to q \vdash_{\leq (m+1)} q$.

But since $(A_1^1, \ldots, A_n^1) + \Delta^1 \vdash_{\leq m} \Phi$ we get the result. ⊣

COROLLARY 2.6. *If $A \vdash_{\leq m} B$, with A in language* 1 *and B in language* 2, *then for some Φ in the common language, we have $A \vdash_{\leq m} \Phi \vdash_{\leq m} B$.*

Observe that we have $A \vdash \{C_1, \ldots, C_n\}$ if and only if for some $i = 1, \ldots, n$ $A \vdash C_i$ and $\varnothing \vdash C_j$ for $j = 1, \ldots, i-1, i+1, \ldots n$. Thus, the corollary can be reinforced as follows:

COROLLARY 2.7. *If $A \vdash_{\leq m} B$, with A in language* 1 *and B in language* 2, *then for some* single *formula C in the common language, we have $A \vdash_{\leq m} C \vdash_{\leq m} B$.*

This result has been obtained by Roorda [29] using the standard Gentzen formulation of linear logic (see [29] Theorem 6.5.7 page 433). It is interesting to observe that Roorda's result is obtained by a more indirect argument, whose main part consists of turning any derivation (by permuting the application of sequent rules) into a 'good' derivation—from which the *single* interpolant formula can be extracted. In our case, the goal-directed derivations do not require such a transformation and from any derivation we can extract a single interpolant. We notice that Roorda's notion of interpolant is more refined than ours, in the sense it pays attention not only to the common variables, but also to the number and the polarity (positive or negative) of each occurrence of every variable. We think that our result can be strengthened by considering also these further paramethers. We shall investigate this point in future research. On the other hand Roorda's result does not give immeditaly a bound on the derivation depth of the interpolant as we do by considering the notion of $\vdash_{\leq m}$.

2.2. Interpolation for intuitionistic logic.

DEFINITION 2.8.

1. *A wff has the form of either q, for q atomic, or $\top$ or $\bot$, or is of the form $\Delta \to x$, where Δ is a finite set of wffs and x is atomic q or $\top$ or $\bot$.*
2. *A database Δ is a set of wffs.*
3. *Let $\Delta_1 + \Delta_2$ be $\Delta_1 \cup \Delta_2$ for databases Δ_1, Δ_2.*

DEFINITION 2.9. *Let $\Delta \vdash_{\leq m} A$ be defined in a goal directed way as follows:*

1. $\Delta \vdash_0 \top$.
2. $\Delta \vdash_1 q$ *iff* $q \in \Delta$, *for q atomic or $\bot$ or $\bot \in \Delta$.*
3. $\Delta \vdash_{\leq m} \Gamma \to q$ *iff* $\Delta + \Gamma \vdash_{\leq m} q$.
4. $\Delta \vdash_{\leq (m+1)} q$ *for q atomic or $q = \bot$ iff for some $\Gamma \to x \in \Delta$, where $x = q$ or $x = \bot$, and all $A \in \Gamma$ there exists $m_A \leq m$ such that $\Delta \vdash_{\leq m_A} A$ and $max\{m_A\} = m$. We say $\Gamma \to x$ was used in this step of the computation.*[5]
5. *We define $\Delta \vdash_{\leq m} \Gamma$ if for each $A \in \Gamma$ there exist m_A such that $\Delta \vdash_{\leq m_A} A$ and $max\{m_A\} = m$.*
6. $\Delta \vdash_m B$ *iff* $\Delta \vdash_{\leq m} B$ *and for no $m' < m$, do we have $\Delta \vdash_{\leq m'} B$.*
7. $\Delta \vdash B$ *iff for some n, $\Delta \vdash_{\leq n} B$.*

THEOREM 2.10 (Completeness). *$\vdash$ above is intuitionistic consequence for the fragment with $\to$, $\top$, $\bot$ (and $\wedge$ because we use sets as antecedents in wffs).*

PROOF. See [16]. $\dashv$

DEFINITION 2.11 (Interpolants). *Let indices indicate languages, as in Definition 2.4. Assume $\Delta^1 + \Delta^2 \vdash_{\leq m} A^1$. We say a set of wffs Φ of the common language is an interpolant for the above, if $\Delta^2 \vdash_{\leq m} \Phi$ and $\Delta^1 + \Phi \vdash_{\leq m} A^1$.*

EXAMPLE 2.12. *Let*

$$\Delta^1 = \{a_1, e, \{a_1, r\} \to q^1, \{a_1, s\} \to q^1\},$$
$$\Delta^2 = \{b_1, b_1 \to r, b_2, b_2 \to s, e\}.$$

We have $\Delta^1 + \Delta^2 \vdash_2 q^1$.

Possible interpolants are

$$\mathbb{H}_1 = \{r\}, \quad \mathbb{H}_2 = \{s\},$$
$$\mathbb{H}_3 = \{r, e\}, \quad \mathbb{H}_4 = \{s, e\},$$
$$\mathbb{H}_5 = \{r, s\}, \quad \mathbb{H}_6 = \{r, s, e\}.$$

[5]Note that q as a goal can unify with $\Gamma \to x$ in the database where $x = q$ or $x = \bot$. However, $\bot$ as a goal can unify only with $x = \bot$.

Of the above, $\mathcal{I} = \{\{r\}\{s\}\}$ *forms a* basis, *in the sense that any interpolant* Φ *s.t.* $\Delta^2 \vdash \Phi$ *and* $\Delta^1 + \Phi \vdash q^1$ *satisfies that for some* $\mathbb{H} \in \mathcal{I}$, *we have* $\Phi \vdash \mathbb{H}$.

THEOREM 2.13 (Intuitionistic Interpolation). *Let* $\Delta^1 + \Delta^2 \vdash_{\leq m} A^1$. *Then for some effectively computable* Φ *in the common language we have* $\Delta^2 \vdash_{\leq m} \Phi$ *and* $\Delta^1 + \Phi \vdash_{\leq m} A^1$.[6]

PROOF. The proof follows the same lines as the proof of Theorem 2.5 (for the case of linear logic) but is, in fact, simpler. We use induction on m.
Notice that for all m, we have that either $A^1 = \Delta' \to r$, with $\Delta' \neq \varnothing$, or $A^1 = q_1$ atomic, or $A^1 = \top$ or $A^1 = \bot$. Since A^1 is in language 1, in the first case we have:

$$\Delta^1 + \Delta^2 \vdash_{\leq m} \Delta' \to r \text{ iff } (\Delta^1 + \Delta') + \Delta^2 \vdash_{\leq m} r.$$

Thus this case reduces to the latter ones (A^1 atomic or $A^1 = \top$ or $A^1 = \bot$), and we give the inductive argument only for them.

CASE $m = 0$.
In this case $A^1 = q^1 = \top$ and we can let $\Phi = \{\top\}$.

CASE $m = 1$.
In this case $A^1 = q^1$ is either in Δ^1, in which case $\Phi = \{\top\}$ or $q^1 \in \Delta^2$, in which case $\Phi = \{q^1\}$, since q^1 is in the common language, or $\bot \in \Delta^1 + \Delta^2$ in which cases $\Phi = \{\top\}$.

CASE $m + 1$.
$A^1 = q^1$ unifies with some clause $\Gamma \to x \in \Delta^1 + \Delta^2$, where $x = q^1$ or $x = \bot$.[7]

SUBCASE 1: $\Gamma \to x$ *is in* Δ^1.
In this case we have $\Delta^1 + \Delta^2 \vdash_{\leq m} \Gamma$, and Γ is in language 1. For each $A \in \Gamma$, use the induction hypothesis to find an interpolant Φ_A such that $\Delta^2 \vdash_{\leq m} \Phi_A$ and $\Delta^1 + \Phi_A \vdash_{\leq m} A$.

Let $\Phi = \bigcup_{A \in \Gamma} \Phi_A$. Then Φ is our interpolant.

Notice the sharp result for this subcase, where we have $\Delta^2 \vdash_{\leq m} \Phi$ and $\Delta^1 + \Phi \vdash_{\leq (m+1)} q^1$.

SUBCASE 2: $\Gamma \to x$ *is in* Δ^2.

---

[6] This result is not new as a theorem and follows from proof theoretical demonstrations of interpolation for intuitionistic logic. The same applies for the case of classical logic. See [3], [8], [11], [19]. Of course, the goal directed proof in itself is new and shows the capability of the goal directed methodology.

[7] Note that for a consistent $\Delta^1 + \Delta^2$ for the cases $m = 0$ and $m = 1$, our interpolant Φ was *minimal*, i.e., any other interpolant Ψ had to satisfy $\Psi \vdash \Phi$. This is not necessarily true in general, as Example 2.12 shows. Indeed, for case $m + 1$, different choices of $\Gamma_i \to x_i$ (for which $x_i = q^1$ or $x_i = \bot$ and $\Delta^1 + \Delta^2 \vdash_{\leq m} \Gamma_i$) will yield possibly different interpolants. In fact, it may be that there are more clauses $\Theta \to y$ with $y = \bot$ or $y = q^1$ such that $\Delta^1 + \Delta^2 \vdash_{\leq n} \Theta$, for $n > m$, which give even more interpolants. The problem of identifying all the interpolants is yet to be addressed.

In this subcase, either $x = q^1$ is in the common language or $x = \bot$ and q^1 may or may not be in the common language. We have, at any rate, that Γ is in language 2.

We have $\Delta^1 + \Delta^2 \vdash_{\leq m} \Gamma$.

By the induction hypothesis we can interpolate language 1. Thus there exists a Φ in the common language such that $\Delta^1 \vdash_{\leq m} \Phi$ and $\Phi + \Delta^2 \vdash_{\leq m} \Gamma$.

We therefore also have $(\Gamma \to x) \vdash_{\leq(m+1)} (\Delta^2 + \Phi) \to x$.

Since $\Gamma \to x \in \Delta^2$, we get that $\Delta^2 \vdash_{\leq(m+1)} \Phi \to x$.

We now show that $\Delta^1 + \Phi \to x \vdash_{\leq(m+1)} q^1$.

This holds since q^1 can unify with x and $\Delta^1 \vdash_{\leq m} \Phi$. ⊣

REMARK 2.14. *The proof of the previous theorem shows that we can assume the interpolants are generated from common language subformulas of* Δ^1, Δ^2, q^1 *using conjunctions and* $\to$. *(By a subformula of* $\Gamma \to x$, *we mean* x, *any* $A \in \Gamma$ *and any* $\Theta \to x$ *for* $\Theta \subseteq \Gamma$).

Let us examine the complexity of the interpolants more closely by checking what interpolants are obtained from the proof.

We assume $\Delta^1 + \Delta^2 \vdash_{\leq m} q$. *Cases* $m = 0$ *and* $m = 1$ *give as interpolants* $\top, \bot$ *or* q^1, *i.e., subformulas of* $\Delta^1 + \Delta^2 + q^1$ *of nested implications complexity* $\leq m - 1$. *Assume for cases up to* m *we get interpolants with the following two properties*:

1. *They are generated from subformulas of* $\Delta^1 + \Delta^2 + q^1$ *of the common language.*
2. *The maximal number of nested implications in the interpolant is* $\leq m - 1$.

We now show (1) *and* (2) *for the interpolants obtained for case* $m + 1$. *We examine the inductive step of the proof.*

In SUBCASE 1: $\Gamma \to x$ is in Δ^1, *the interpolants are unions of interpolants for* $\Delta^1 + \Delta^2 \vdash_{\leq m} A, A \in \Gamma$. *By the induction hypothesis their nested complexity is* $\leq m - 1$ *and indeed they are generated by common language subformulas.*

In SUBCASE 2: $\Gamma \to x$ is in Δ^2, *the inductive interpolant is* Φ *for the case* $\Delta^1 + \Delta^2 \vdash_{\leq m} \Gamma$ *and the interpolant itself is* $\Phi \to x$. *Again, the complexity of* Φ *is* $\leq m - 1$ *and so the complexity of* $\Phi \to x$ *is* $\leq m$, *as required.*

2.3. Interpolation for classical logic. It is easy to get interpolation for classical logic if we have interpolation for intuitionistic logic. Consider the $\{\bot, \wedge, \to\}$ fragment of classical logic. We have $\mathbb{A} \vdash \mathbb{B}$ in classical logic iff $\mathbb{A} \vdash (\mathbb{B} \to \bot) \to \mathbb{B}$ in intuitionistic logic, where $\mathbb{D} \to \mathbb{E} = \{\mathbb{D} \to E \mid E \in \mathbb{E}\}$.

Let $\mathbb{H}$ be the interpolant in intuitionistic logic. Then $\mathbb{H}$ is an interpolant in classical logic.

The intuitionistic interpolation gives us more. If $\mathbb{A} \vdash_{\leq m} (\mathbb{B} \to \bot) \to \mathbb{B}$ then $\mathbb{A} \vdash_{\leq m} \mathbb{H}$ and $\mathbb{H} \vdash_{\leq m} (\mathbb{B} \to \bot) \to \mathbb{B}$, and furthermore all the members of $\mathbb{H}$ are subformulas of $\mathbb{A} \cup \mathbb{B}$.

§3. Interpolation for the Lambek Calculus. We now examine the interpolation theorem for the Lambek Calculus.

DEFINITION 3.1. 1. *The formulas of the Lambek calculus are built up from atoms, and from two implications $\rightarrow_r$ and $\rightarrow_l$ being right- and left-implications, respectively.*

2. *The data-structures (theories) are lists $(A_1, \ldots, A_n)$ and $+$ is concatenation of lists $(A_1, \ldots, A_n) + (B_1, \ldots, B_n) = (A_1, \ldots, A_n, B_1, \ldots, B_m)$*
3. *We have in the Lambek calculus the following:*

$$\begin{aligned} A + \Delta + B \vdash C \ &\text{iff } \Delta + B \vdash A \rightarrow_l C \\ &\text{iff } A + \Delta \vdash B \rightarrow_r C \\ &\text{iff } \Delta \vdash A \rightarrow_l (B \rightarrow_r C) \\ &\text{iff } \Delta \vdash B \rightarrow_r (A \rightarrow_l C) \end{aligned}$$

In fact we have

$$\Delta \vdash A_1 \rightarrow_{x_1} (\cdots \rightarrow_{x_{n-1}} (A_n \rightarrow_{x_n} B) \ldots)$$

iff

$$\sum_{x_i = l} A_i + \Delta + \sum_{x_i = r} A_i \vdash B.$$

We can thus write every wff in a normal form as $\varphi = (A_1, \ldots, A_n \mid B_1, \ldots, B_m) \rightarrow q$ where A_i, B_j are also *in a normal form and*

$$\Delta \vdash \varphi \textit{ iff } \sum A_i + \Delta + \sum B_j \vdash q.$$

DEFINITION 3.2 (Goal directed computation for normal form). *The definition is by induction*

1. $\Delta \vdash_1 q$ *iff* $\Delta = (q), q$ *atomic.*
2. $\Delta \vdash_{\leq n} (A_1, \ldots, A_k \mid B_1, \ldots, B_m) \rightarrow q$ *iff* $\sum A_i + \Delta + \sum B_j \vdash_{\leq n} q$.
3. $\Delta \vdash_{\leq (n+1)} q$ *iff for some* $C = (A_1, \ldots, A_k \mid B_1, \ldots, B_m) \rightarrow q$ *and some* $\Delta_1, \ldots, \Delta_k, \Gamma_1, \ldots, \Gamma_m$

$$\Delta = \sum \Delta_i + C + \sum \Gamma_j$$

and for $1 \leq i \leq k$, $1 \leq j \leq m$ *and* n_j, $n_j \leq n$ *and* n_i', $n_i' \leq n$, *and* $max(n_j, n_i') = n$, *we have* $\Delta_i \vdash_{\leq n_i'} A_i$ *and* $\Gamma_j \vdash_{\leq n_j} B_j$.

We say that C *was* used *in the computation at this step.*

4. $\Delta \vdash A$ *iff for some* n, $\Delta \vdash_{\leq n} A$.
5. $\Delta \vdash_{\leq m} (A_1, \ldots, A_n)$ *iff for some* $\Delta_1, \ldots, \Delta_n$, $\Delta_i \vdash_{\leq m_i} A_i$, $1 \leq i \leq n$, *and* $m_i \leq m$ *and* $max(m_i) = m$ *and* $\Delta = \Delta_1 + \cdots + \Delta_n$.[8]
6. $\Delta \vdash_m B$ *iff* $\Delta \vdash_{\leq m} B$ *and for no* $m' < m$, *do we have* $\Delta \vdash_{\leq m'} B$.

[8] In substructural logics there is a distinction whether Δ_i is allowed to be empty or not. For example $C = (p \rightarrow_l p) \rightarrow_r q \vdash q$ iff for some Δ, $C + \Delta = C$ and $\Delta \vdash p \rightarrow_l p$. We have $\Delta \vdash p \rightarrow_l p$ iff $p + \Delta \vdash p$. We need to allow $\Delta = \varnothing$ for this to work.

Soundness and completeness with respect to the standard sequent calculus formulation, or Hilbert style axiomatisation for Lambek logic can be proved using the methods of [16] or [12].

THEOREM 3.3 (Interpolation).

1. *Let superscripts denote languages. Let q^1 be an atom and assume*

$$\Delta = \Delta_1^1 + \Delta_0^2 + \Delta_2^1 \vdash_{\leq m} q^1$$

Then for some effectively computable Φ_0 in the common language we have $\Delta_0^2 \vdash_{\leq m} \Phi_0$ and $\Delta^1 = \Delta_1^1 + \Phi_0 + \Delta_2^1 \vdash_{\leq m} q^1$.[9]

PROOF. By induction on the goal directed computation.

CASE $n = 1$.
If $\Delta_0^2 = \varnothing$ then there is no need to interpolate. If $\Delta_0^2 \neq \varnothing$, then $\Delta_0^2 = (q^1)$ and $\Delta_1^1 = \Delta_2^1 = \varnothing$. Hence the statements of the theorem follow, (q^1) is the interpolant, and it is in the common language.

CASE $n + 1$.
Assume $\Delta = \Delta_1^1 + \Delta_0^2 + \Delta_2^1 \vdash_{\leq(n+1)} q^1$. Then for some x and y and C we have $C \in \Delta_x^y$, $y \in \{1, 2\}$, and $C = (C_1, \dots, C_k \mid C_{k+1}, \dots, C_m) \to q$ and for some $\Theta_1, \dots, \Theta_k, \Theta_{k+1}, \dots, \Theta_m$ we have $q = q^1$, $\Theta_j \vdash_{\leq n} C_j$, $1 \leq j \leq k + m$ and $\Delta = \sum_{i=1}^k \Theta_i + C + \sum_{i=k+1}^m \Theta_i$.

We wrote $C \in \Delta_x^y$ because there are two possibilities, $y = 1$ and C is in language 1 and $y = 2$ and C is in language 2.

Each theory Θ_j has the form $\Theta_j = \Theta_{j,1}^1 + \Theta_j^2 + \Theta_{j,2}^1$.

We are ready now to use the induction hypothesis.

SUBCASE $y = 1$.
In this case C is in language 1, the same language as q^1.

Assume $C_j = (C_{j,1} \dots, C_{j,k_j} \mid \dots C_{j,m_j}) \to q_j$.

Hence $\sum_{r=1}^{k_j} C_{j,r} + \Theta_j + \sum_{r=k_j+1}^{m_j} C_{j,r} \vdash_n q_j$.

$C_{j,r}$ and q_j are in the language 1.

We can interpolate on language 2 by the induction hypothesis.

For some Φ_j such that $\Theta_j^2 \vdash_{\leq n} \Phi_j$ we have $\Theta_j \vdash_n \Theta_j' = \Theta_{j,1}^1 + \Phi_j + \Theta_{j,2}^1 \vdash_{\leq n} C_j$. Hence $\Delta \vdash_{\leq(n+1)} \sum_{i=1}^k \Theta_i' + C + \sum_{i=k+1}^m \Theta_i' \vdash_{\leq(n+1)} q^1$.

The interpolants are the Φ_js, interpolating the Θ_j^2, which are all the language 2 theories.

SUBCASE $y = 2$.
In this case C is in language 2 and hence $q = q^1$ is in the common language.

Hence also $C_{j,r}$ and q_j are in language 2.

---

[9]This formulation is equivalent to the traditional one. If we have $\Delta_0^2 \vdash A^1$ and $A^1 = A_1^1 \to (\dots(A_k^1 \to q^1)\dots)$ where $\to$ is either $\to_r$ or $\to_l$, then by the deduction rule we get the formulation of the theorem.

We can therefore interpolate on language 1. We have to be careful because the interpolation in this case is very finely tuned.

Assume then that

$$\Delta = \Delta_1^1 + \Delta_1^2 + C + \Delta_2^2 + \Delta_2^1 \vdash_{\leq(n+1)} q$$

where q is in the common language and C is in language 2. We recall that we have

$$C = (C_1, \ldots, C_k \mid C_{k+1} \ldots C_m) \to q$$

and we have $\Theta_i \vdash_{\leq n} C_i$, and

$$\Delta = \sum_{i \leq k} \Theta_i + C + \sum_{i \geq k+1} \Theta_i.$$

The fine tuning comes in observing that $\sum_{i \leq k} \Theta_i = \Delta_1^1 + \Delta_1^2$ and $\sum_{i>k} \Theta_i = \Delta_2^2 + \Delta_2^1$.

Let $r \leq k$ and $s \geq k+1$ be the indices such that:

- $\Theta_i \subseteq \Delta_1^1$ for $i < r$
- $\Theta_i \subseteq \Delta_2^1$ for $i > s$
- $\Theta_i \subseteq \Delta_1^2$ for $r < i \leq k$
- $\Theta_i \subseteq \Delta_2^2$ for $k+1 \leq i < s$.

For the case of r and s we have

- $\Theta_r = \Theta_r^1 + \Theta_r^2$, $\Theta_r^1 \subseteq \Delta_1^1$ and $\Theta_r^2 \subseteq \Delta_1^2$
- $\Theta_s = \Theta_s^2 + \Theta_s^1$, $\Theta_s^2 \subseteq \Delta_2^2$ and $\Theta_s^1 \subseteq \Delta_2^1$.

Thus Θ_r, Θ_s are those theories that are of mixed languages.

Of course some of Θ_r^y, Θ_s^y may be empty, depending where the language division occurs. Notice the exact ordering of the language division. Θ_r starts with language 1 and ends in language 2 and Θ_s the other way around. To stress the languages involved we write Θ_i^1 for Θ_i for $i < r$, Θ_i^2 for $r < i < s$, and Θ_i^1 for $i > s$.

Thus by the induction hypothesis we have interpolants Φ_i such that $\Theta_i^1 \vdash_{\leq n} \Phi_i \vdash_{\leq n} C_i$ for $i < r$ or $i > s$ and interpolating on language 1 for the cases of index r and index s we get $\Theta_r^1 \vdash_{\leq n} \Phi_r$ and $\Phi_r + \Theta_r^2 \vdash_{\leq n} C_r$ and $\Theta_s^1 \vdash_{\leq n} \Phi_s$ and $\Theta_s^2 + \Phi_s \vdash_{\leq n} C_s$.

We want to use the theories Φ_i to interpolate on language 2, i.e., to find a Φ_0 such that $\Delta_0^2 = \Delta_1^2 + C + \Delta_2^2 \vdash_{\leq(n+1)} \Phi_0$ and $\Delta_1^1 + \Phi_0 + \Delta_2^1 \vdash_{\leq(n+1)} q$. We now proceed to achieve this goal and find Φ_0. It would help if we visualise what we are going to do and for this purpose, let us write Δ more explicitly:

$$\begin{aligned}\Delta = {} & \Theta_1^1 + \cdots + \Theta_{r-1}^1 + \Theta_r^1 + \Theta_r^2 + \Theta_{r+1}^2 + \cdots + \Theta_k^2 + C \\ = {} & (C_1, \ldots, C_{r-1}, C_r, C_{r+1}, \ldots, C_k \mid C_{k+1}, \ldots, C_{s-1}, C_s, C_{s+1} \ldots C_m) \to \\ & q + \Theta_{k+1}^2 + \cdots + \Theta_{s-1}^2 + \Theta_s^2 + \Theta_s^1 + \Theta_{s+1}^1 + \cdots + \Theta_m^1.\end{aligned}$$

We can write:

$$\Delta = \Theta_1^1 + \cdots + \Theta_{r-1}^1 + \Theta_r^1 + \Theta_r^2 + \Gamma + \Theta_s^2 + \Theta_s^1 + \Theta_{s+1}^1 + \cdots + \Theta_m^1$$

where

$$\Gamma = \Theta_{r+1}^2 + \cdots + \Theta_k^2 + C + \Theta_{k+1}^2 + \cdots + \Theta_{s-1}^2.$$

Note that $\Delta_0^2 = \Theta_r^2 + \Gamma + \Theta_s^2$.

We also have $\Gamma \vdash_{\leq(n+1)} C'$, where

$$C' = (C_1, \ldots, C_{r-1}, C_r \mid C_s, C_{s+1}, \ldots, C_m) \rightarrow q.$$

But we also have:

$$\Phi_r + \Theta_r^2 \vdash_{\leq n} C_r$$
$$\Theta_s^2 + \Phi_s \vdash_{\leq n} C_s$$
$$\Phi_i \vdash_{\leq n} C_i,$$

for $i < r$ and $i > s$.

Hence, we have that

$$C' \vdash_{\leq(n+1)} C''$$

where $C'' = (\Phi_1, \ldots, \Phi_{r-1}, \Phi_r, \Theta_r^2 \mid \Theta_s^2, \Phi_s, \ldots, \Phi_m) \rightarrow q$.

Thus we get that $\Gamma \vdash_{\leq(n+1)} C''$. Therefore $\Delta_0^2 = \Theta_r^2 + \Gamma + \Theta_s^2 \vdash_{\leq(n+1)} C'''$ where $C''' = (\Phi_1, \ldots, \Phi_r \mid \Phi_s, \ldots, \Phi_m) \rightarrow q$. C''' is in the common language.

We therefore found a $\Phi_0 = C'''$ such that $\Delta_0^2 \vdash_{\leq(n+1)} \Phi_0$. We need to show that

$$\Delta_1^1 + \Phi_0 + \Delta_2^1 \vdash_{\leq(n+1)} q.$$

This holds because

$$\Delta_1^1 = \Theta_1^1 + \cdots + \Theta_r^1$$
$$\Delta_2^1 = \Theta_s^1 + \cdots + \Theta_m^1$$

and for each $i \leq r$ or $i \geq s$, $\Theta_i^1 \vdash \Phi_i$.

This completes the proof for subcase $y = 2$ and thus the theorem is proved. ⊣

Let us now examine interpolation for the right hand arrow $\rightarrow$ alone of Lambek Calculus. This logic is called *concatenation logic.* We give the necessary definitions.

DEFINITION 3.4. *Let $\mathbb{L}$ be a language with $\rightarrow$ alone and atoms.*

1. *A wff is either an atom q or of the form $A_1 \rightarrow (A_2 \rightarrow \ldots (A_n \rightarrow q)\ldots)$, for q atomic and A_i, $i = 1, \ldots, n$ are formulas.*
2. *A database has the form $\Delta = (A_1, \ldots, A_n)$, a (possibly empty) sequence of formulas.*
3. *We can write $A_1 \rightarrow (A_2 \rightarrow \ldots (A_n \rightarrow q)\ldots)$ also as $(A_1, \ldots, A_n) \rightarrow q$.*

4. *Let* Δ_1, Δ_2 *be two databases. We let* $\Delta_1 + \Delta_2$ *be the database obtained by concatenating the two sequences.*

DEFINITION 3.5. *Let* $\Delta \vdash_{\leq m} A$ *be defined for concatenation logic in a goal directed way as follows:*

1. $\Delta \vdash_1 q$ *iff* $\Delta = (q)$, *for* q *atomic.*
2. $\Delta \vdash_{\leq m} (A_1, \ldots, A_n) \rightarrow q$ *iff* $\Delta + (A_1, \ldots, A_n) \vdash_{\leq m} q$
3. $((A_1, \ldots, A_n) \rightarrow q) + \Delta \vdash_{\leq (m+1)} q$ *iff* $\Delta = \Delta_1 + \cdots + \Delta_n$ *and for each* $i = 1, \ldots, n$ *we have* $\Delta_i \neq \varnothing$ *and* $\Delta_i \vdash_{\leq m_i} A_i$ *and* $max(m_i) = m$.
4. $\Delta \vdash A$ *iff for some* n, $\Delta \vdash_{\leq n} A$.
5. *We define* $\Delta \vdash_{\leq m} (A_1, \ldots, A_n)$ *iff for some* $\Delta_1 + \cdots + \Delta_n = \Delta$ *we have* $\Delta_i \neq \varnothing$ *and* $\Delta_i \vdash_{\leq m_i} A_i$ *for* $1 \leq i \leq n$ *and* $max(m_i) = m$.
6. *In* (3) *above we can say* $(A_1, \ldots, A_n) \rightarrow q$ *was used in the computation.*
7. $\Delta \vdash_m B$ *iff* $\Delta \vdash_{\leq m} B$ *and for no* $m' < m$, *do we have* $\Delta \vdash_{\leq m'} B$.

THEOREM 3.6 (Completeness). *The above consequence* $\vdash$ *characterises the right arrow fragment of the Lambek calculus.*

PROOF. See [16] or [12]. ⊣

THEOREM 3.7 (Interpolation). *Let superscripts denote languages. Let* A^1 *be a formula and assume* $\Delta = \Delta_1^1 + \Delta_0^2 + \Delta_2^1 \vdash_m A^1$.

Then for some effectively computable Φ_0 *in the common language we have* $\Delta_0^2 \vdash_{\leq m} \Phi_0$ *and* $\Delta' = \Delta_1^1 + \Phi_0 + \Delta_2^1 \vdash_{\leq m} A^1$.

PROOF. The proof is by induction on m.

Let us write the theories explicitly.

$$\Delta_1^1 = ((E_1, \ldots, E_r) \rightarrow p, A_2, \ldots, A_n)$$
$$\Delta_0^2 = (B_1, \ldots, B_k)$$
$$\Delta_2^1 = (A_1', \ldots, A_{n'}').$$

We have that $A^1 = \Delta_3^1 \rightarrow r$, with $\Delta_3^1 \neq \varnothing$, or $A^1 = q_1$ atomic. In the former case we have:

$$\Delta_1^1 + \Delta_0^2 + \Delta_2^1 \vdash_{\leq m} A^1 \text{ iff } \Delta_1^1 + \Delta_0^2 + (\Delta_2^1 + \Delta_3^1) \vdash_{\leq m} r.$$

no matter what m is. Since Δ_3^1 is in language 1, this case reduces to the case of atomic A^1, thus we give the inductive argument only for that case.

CASE $m = 1$.

If $\Delta_0^2 = \varnothing$, there is nothing to interpolate. Otherwise, in this case $\Delta_0^2 = (q^1)$ and Δ_1^1 and Δ_2^1 are empty. q^1 is then in the common language and it is the interpolant.

CASE $m + 1$.

Thus, let $A = q_1$ be an atom.

SUBCASE $\Delta_1^1 \neq \varnothing$.

In this subcase, we must have $q^1 = p$ and $r \leq n + k + n' - 1$ and for some theories $\Delta_1, \ldots, \Delta_t, \Theta_t, \Theta_{t+1}, \ldots, \Theta_s, \Delta_s, \Delta_{s+1}, \ldots, \Delta_r$, we have that

(i) $\Delta_i \vdash_{\leq m} E_i$, and $\Delta_i \neq \varnothing$, for $1 \leq i \leq t-1$.
(ii) $\Delta_t + \Theta_t \vdash_{\leq m} E_t$ and $\Delta_t + \Theta_t \neq \varnothing$.
(iii) $\Theta_i \vdash_{\leq m} E_i$ and $\Theta_i \neq \varnothing$ for $t+1 \leq i \leq s-1$.
(iv) $\Theta_s + \Delta_s \vdash_{\leq m} E_s$ and $\Theta_s + \Delta_s \neq \varnothing$.
(v) $\Delta_i \vdash_{\leq m} E_i$ and $\Delta_i \neq \varnothing$, for $s+1 \leq i \leq r$.

We also have

(vi) $(A_2, \ldots, A_n) = \Delta_1 + \cdots + \Delta_t$.
(vii) $(B_1, \ldots, B_k) = \Theta_t + \cdots + \Theta_s$.
(viii) $(A'_1, \ldots, A'_{n'}) = \Delta_s + \cdots + \Delta_r$.

By the induction hypothesis there exist Φ_i^0 for $t \leq i \leq s$ in the common language such that

(ix) $\Theta_i \vdash_{\leq m} \Phi_i^0$, for $t \leq i \leq s$

and such that

(x) $\Delta_t + \Phi_t^0 \vdash_{\leq m} E_t$.
(xi) $\Phi_i^0 \vdash_{\leq m} E_i$, for $t+1 \leq i \leq s-1$.
(xii) $\Phi_s^0 + \Delta_s \vdash_{\leq m} E_s$.

From (vii) and (ix) we get that $\Delta_0^2 \vdash_{\leq m} \Phi_t^0 + \cdots + \Phi_s^0$ and from (x) and (xii) we get that $\Delta_1^1 + \Phi_0 + \Delta_2^1 \vdash_{\leq(m+1)} q^1$ where $\Phi_0 = \Phi_t^0 + \cdots + \Phi_s^0$.

SUBCASE $\Delta_1^1 = \varnothing$.

In this case let $B_1 = (D_1, \ldots, D_r) \to q$.

We have that $q^1 = q$ is in the common language and $r \leq k - 1 + n'$ and for some theories $\Theta_1, \ldots, \Theta_t, \Delta_t, \Delta_{t+1}, \ldots, \Delta_r$ we have

1. $\Theta_i \vdash_{\leq m} D_i$ and $\Theta_i \neq \varnothing$, for $1 \leq i \leq t-1$.
2. $\Theta_t + \Delta_t \vdash_{\leq m} D_t$ and $\Theta_t + \Delta_t \neq \varnothing$.
3. $\Delta_i \vdash_{\leq m} D_i$ and $\Delta_i \neq \varnothing$, for $t+1 \leq i \leq r$.
4. $(B_2, \ldots, B_k) = \Theta_1 + \cdots + \Theta_t$
5. $(A'_1, \ldots, A'_{n'}) = \Delta_t + \cdots + \Delta_r$.

By the induction hypothesis, there exist Ψ_i, for $t+1 \leq i \leq r$, such that

6. $\Delta_i \vdash_{\leq m} \Psi_i \vdash_{\leq m} D_i$,

and there exists a Ψ_t such that

7. $\Delta_t \vdash_{\leq m} \Psi_t$

and

8. $\Theta_t + \Psi_t \vdash_{\leq m} D_t$.

We are now seeking to define the interpolant Φ_0, such that

- $(B_1, \ldots, B_k) \vdash_{\leq(m+1)} \Phi_0$

and

- $\Phi_0 \vdash_{\leq(m+1)} (A'_1, \ldots, A'_{n'}) \to q$.

From $\Psi_i \vdash_{\leq m} D_i, t+1 \leq i \leq r$ (see (6)), we get:

9. $(D_{t+1}, \ldots, D_r) \to q \vdash_{\leq(m+1)} (\Psi_{t+1} + \cdots + \Psi_r) \to q$

and from (8) and (9) we get

10. $(D_t, \ldots, D_r) \to q \vdash_{\leq(m+1)} (\Theta_t + \Psi_t + \cdots + \Psi_r) \to q$

We rewrite (10) as

11. $(D_t, \ldots, D_r) \to q + \Theta_t \vdash_{\leq(m+1)} (\Psi_t + \cdots + \Psi_r) \to q.$

We can also get from (1) that

12. $((D_1, \ldots, D_r) \to q) + \Theta_1 + \cdots + \Theta_{t-1} \vdash_{\leq(m+1)} (D_t, \ldots, D_r) \to q.$

Combining (11) and (12) we get

13. $((D_1, \ldots, D_r) \to q) + \Theta_1 + \cdots + \Theta_t \vdash_{\leq(m+1)} (\Psi_t + \cdots + \Psi_r) \to q.$

However, (6) and (7) tell us that $\Delta_i \vdash_{\leq m} \Psi_i$, $t \leq i \leq r$, and therefore we get

14. $(\Psi_t + \cdots + \Psi_r) \to q \vdash_{\leq(m+1)} (\Delta_t + \cdots + \Delta_r) \to q.$

We let $\Phi_0 = (\Psi_t + \cdots + \Psi_r) \to q$ and in view of (4), (5), (13) and (14) we get that Φ_0 is the interpolant.

This ends the proof of Theorem 3.7. ⊣

REMARK 3.8. *An examination of the proof reveals that if* $\mathbb{L}$ *is a fragment of the language satisfying the conditions below, then interpolation holds for the fragment.*

1. *atoms* q *are in the fragment.*
2. *If* $(\Delta_1 + \Delta_2 + \Delta_3) \to q$ *is in the fragment and* $\Delta_2 \vdash \Theta$ *then* $(\Theta \to q)$ *is in the fragment.*

Pentus [28] has proved various interpolation properties for several fragments of the Lambek calculus including also what we have called concatenation logic.[10] Of course our proof makes use of the goal-directed formulation, rather than the standard sequent axiomatization and hence it is entirely different. An exact comparison with his results will be done in future investigation.

§4. **Interpolation for strict implication.** We shall deal with **K**, **K4**, **T**, and **S4** strict implications. We start with the case of **K**.

First let us motivate and explain the kind of consequence relation we are defining for modal logics strict implication. Our data-structure are sequences of (sets of) wffs. Take such a sequence, say $\Delta = (A_1, \ldots, A_n)$. We understand $\Delta \vdash B$ to mean semantically that for every possible world model (S, R, a, h) and every sequence of worlds $(t_1, \ldots, t_n)$ such that $t_1 R t_2 \wedge t_2 R t_3 \wedge \cdots \wedge t_{n-1} R t_n$ and such that $t_i \vDash A_i$ for $i = 1, \ldots, n$, we also have that $t_n \vDash B$.

The reason we chose this kind of consequence is

(i) it gives us more structure in the data, which is good for application.

[10]Concatenation logic is the right hand $\to$ of the Lambek calculus. It is motivated in Section 1.2.

(ii) it compares with the list data-structures of other substructural logics.
(iii) it corresponds to the natural deduction proof theory: to show $x \to (y \to z)$, at t_1 we assume x (at world t_2) and show $y \to z$ at world t_2 with $t_1 R t_2$. To show $y \to z$ at t_2 we assume y at world t_3 (such that $t_2 R t_3$) and show z at t_3. Thus we have $x \to (y \to z), x, y \vdash z$ meaning exactly as we defined it. Compare with Example 1.5.

Completness of the goal directed algorithm for this consequence is given in [16].

DEFINITION 4.1.

1. *A data-structure is a sequence of sets of implicational formulas of the form* $(\mathbb{A}_1, \dots, \mathbb{A}_n)$ *(also written as* $\mathbb{A}_1 + \cdots + \mathbb{A}_n$*).*
2. *The notion of an implicational formula is defined by induction using the following clauses*
   - *an atom q is an atomic formula with head q*
   - *$\top$ is an atomic formula with head $\top$*
   - *if $(\mathbb{A}_1, \dots, \mathbb{A}_n)$ is a sequence of sets of wffs then $(\mathbb{A}_1, \dots, \mathbb{A}_n) \to q$ is a wff with head q, where q is an atom or $\top$.*
3. *Let (S, R, a, h) be a Kripke model with $a \in S$ the actual world and R the accessibility relation. Let $\vec{t} = (t_1, \dots, t_n)$ be such that $t_1 R t_2 \wedge t_2 R t_3 \wedge \cdots \wedge t_{n-1} R t_n$. We say that $(t_1, \dots, t_n) \vDash (\mathbb{A}_1, \dots, \mathbb{A}_n)$ iff $t_i \vDash A$ for all $A \in \mathbb{A}_i$ and for each $1 \le i \le n$.*
4. *We say $\Delta = (\mathbb{A}_1, \dots, \mathbb{A}_n) \vDash \Gamma = (\mathbb{B}_1, \dots, \mathbb{B}_n)$ for data-structures Δ and Γ iff in any model and any $(t_1, \dots, t_n)$ such that $\bigwedge_{i=1}^{n-1} t_i R t_{i+1}$ holds, we have that for all $1 \le m \le n$, $(\bigwedge_{i=1}^{m} t_i \vDash \mathbb{A}_i) \to (t_m \vDash \mathbb{B}_m)$.*

DEFINITION 4.2. (Computation for $\Delta \vdash B$ (intended to be the same as $\Delta \vDash B$)) *We define $\Delta \vdash_{\le m} B$ (reading: A follows from Δ in at most m unification steps) by the following clauses*

1. *$(\mathbb{A}_1, \dots, \mathbb{A}_n) \vdash_1 q$, q atomic, if $q \in \mathbb{A}_n$ or if $q = \top$. We can also say $(\mathbb{A}_1, \dots, \mathbb{A}_n) \vdash_0 \top$.*
2. *$(\mathbb{A}_1, \dots, \mathbb{A}_n) \vdash_{\le m} (\mathbb{B}_1, \dots, \mathbb{B}_k) \to q$ iff $(\mathbb{A}_1, \dots, \mathbb{A}_n, \mathbb{B}_1, \dots, \mathbb{B}_k) \vdash_{\le m} q$.*
3. *$(\mathbb{A}_1, \dots, \mathbb{A}_n) \vdash_{\le m} \mathbb{B}$ iff $(\mathbb{A}_1, \dots, \mathbb{A}_n) \vdash_{\le m} B$ for all $B \in \mathbb{B}$.*
4. *Let $\Delta = (\mathbb{A}_1, \dots, \mathbb{A}_n)$. Then $\Delta \vdash_{\le (m+1)} q$ iff for some $1 \le k \le n$ and $B_{k+1} = (\mathbb{B}_{k+1}, \dots, \mathbb{B}_n) \to q \in \mathbb{A}_k$ and for each $k+1 \le r \le n$ we have that $\Delta_r \vdash_{\le m_r} \mathbb{B}_r$ for some $m_r \le m$, where $\Delta_r = (\mathbb{A}_1, \dots, \mathbb{A}_r)$, and $max(m_r) = m$.*

   We say B_{k+1} is used *at this step.*
5. *$\Delta \vdash B$ iff for some m, $\Delta \vdash_{\le m} B$.*
6. *$\Delta \vdash_m B$ iff $\Delta \vdash_{\le m} B$ and for no $m' < m$, do we have $\Delta \vdash_{\le m'} B$.*

DEFINITION 4.3 ($\Delta \vdash B$ for the logics **K4**, **T** and **S4**).

1. *Let x, y be two numbers, $\mathbf{L}$ a logic. We define $x R_{\mathbf{L}} y$ as follows:*
   - *$x R_{\mathbf{K}} y$ iff $y = x + 1$,*

- $xR_{\mathbf{T}}y$ *iff* $y = x \vee y = x + 1$,
- $xR_{\mathbf{K4}}y$ *iff* $x < y$,
- $xR_{\mathbf{S4}}y$ *iff* $x \leq y$.

2. *In the previous definition* 4.2 *of* $\Delta \vdash B$ *for* $\Delta = (\mathbb{A}_1, \ldots, \mathbb{A}_n)$ *for the logic* **K**, *replace clause* 4 *by the following clause* (**4L**) *where* $\mathbf{L} = \mathbf{K}, \mathbf{T}, \mathbf{K4}, \mathbf{S4}$.

4L. $\Delta \vdash_{\leq(m+1)} q$ *iff for some* $1 \leq k \leq n$ *and* $B = (\mathbb{B}_{r_1}, \ldots, \mathbb{B}_{r_t}) \to q \in \mathbb{A}_k$ *we have that* $1 \leq r_i \leq n$, $i = 1, \ldots, t$ *and* $kR_{\mathbf{L}}r_1$ *and for each* $1 \leq i < t$, $r_iR_{\mathbf{L}}r_{i+1}$ *and* $r_t = n$, *and for each* r_i, $(\mathbb{A}_1, \ldots, A_{r_i}) \vdash_{\leq m_i} \mathbb{B}_{r_i}$, *for some* $m_i \leq m$, $max(m_i) = m$.

We now discuss interpolation for **K** in general. First consider the case of

$$(\mathbb{A}_1, \ldots, \mathbb{A}_n) \to q, \mathbb{B}_1, \ldots, \mathbb{B}_k \vdash_{\leq m} b.$$

Suppose we want to prove, using induction on m the existence of an interpolant $\mathbb{C}$ such that $(\mathbb{A}_1, \ldots, \mathbb{A}_n) \to q \vdash \mathbb{C}$ and $\mathbb{C} + \mathbb{B}_1 + \cdots + \mathbb{B}_k \vdash b$. The difficult case is when $b = q$, $k = n$ and b unifies with $(\mathbb{A}_1, \ldots, \mathbb{A}_n) \to q$. We get that for $i = 1, \ldots, n$ we must have

$$(\mathbb{A}_1, \ldots, \mathbb{A}_n) \to q, \mathbb{B}_1, \ldots, \mathbb{B}_i \vdash_{\leq(m-1)} \mathbb{A}_i.$$

We would like to get interpolants for this case but we need to assume an inductive hypothesis of interpolation for an arbitrary alternation of languages, i.e., the form

$$\mathbb{A}^1_1, \ldots, \mathbb{A}^1_{r_1}, \mathbb{B}^2_1, \ldots, \mathbb{B}^2_{r_2}, \ldots, \mathbb{A}^k_1, \ldots, \mathbb{A}^k_{r_k} \vdash q$$

and further, it is not clear at what points we want to interpolate, between $\mathbb{A}^1_{r_1}$ and $\mathbb{B}^2_1$ or ... or between $\mathbb{B}^{k-1}_{r_{k-1}}$ and $\mathbb{A}^k_1$.

An example would help.

EXAMPLE 4.4. *Consider*

$$(\{\top\}, \{\top\}) \to a, B_1, B_2 \vdash a$$

clearly we cannot interpolate at the junction before B_1. *However, the* B_1, B_2 *can be interpolated out.*

Obviously we need some formulation of interpolation which will work inductively and will include the general $A \vdash B$ case as a particular instance.

We now address the general case of interpolation. Assume our theory is a sequence of sets of wffs of the form

$$\mathbb{A}^1_1, \ldots, \mathbb{A}^1_{t_1}, \mathbb{B}^2_1, \ldots, \mathbb{B}^2_{t_2}, \ldots$$

where the notation $\mathbb{A}^x_y$ indicates the set is in language 1 ($\mathbb{A}$ language) and the notation $\mathbb{B}^x_y$ indicates the set is in language 2, ($\mathbb{B}$ language). We can use the letter $\mathbb{K}$ to indicate theories and write our sequence as

$$\Delta = \sum_{r=1}^{2a+1} \sum_{i=1}^{t_r} \mathbb{K}^r_i$$

where $\mathbb{K}_k^{2j} = \mathbb{B}_k^{2j}$, $1 \leq k \leq t_{2j}$ is in language 2 ($\mathbb{B}$ language) and $\mathbb{K}_k^{2j+1} = \mathbb{A}_k^{2j+1}$ $1 \leq k \leq t_{2j+1}$ is in language 1 ($\mathbb{A}$ language).

We assume that for all $2 \leq r \leq 2a+1$, $t_r \geq 1$. For the case $r = 1$, i.e., for $\mathbb{A}_1^1, \dots, \mathbb{A}_{t_1}^1$ we may allow for these $\mathbb{A}_j^1$ not to appear.

This representation is the most general. We get alternation of non-empty sequences of languages beginning (as an arbitrary naming choice) with language $\mathbb{A}$ and ending in either language $\mathbb{A}$ or language $\mathbb{B}$. To capture all options let us write the last two blocks explicitly and perform a case analysis. Thus our interpolation problem can be taken to be of the form

$$\Delta + \mathbb{B}_1 + \cdots + \mathbb{B}_n + \mathbb{A}_1 + \cdots + \mathbb{A}_k \vdash_{\leq m} q$$

with q atomic, where $\mathbb{A}_1, \dots, \mathbb{A}_k$ may not appear (case of last alternation being a $\mathbb{B}$ language) but we have $n \geq 1$. If $\mathbb{B}$ do not appear then $\mathbb{A}_1, \dots, \mathbb{A}_k$ are part of the block $\mathbb{A}_1^{2a+1}, \dots, \mathbb{A}_{t_{a+1}}^{2a+1}$.

We have to pay attention to q. q is assumed to be in the $\mathbb{A}$ language. It may, however, be in the common language. If it is in the common language and $\mathbb{A}_1, \dots, \mathbb{A}_k$ do not appear then we can view it as in the $\mathbb{B}$ language and swap the $\mathbb{A}$ and $\mathbb{B}$ languages in our minds. But then we must allow for the sequence $\mathbb{K}_y^x$ to start with the $\mathbb{B}$ language. This is why we allowed the first block $\mathbb{K}_1^1, \dots, \mathbb{K}_{t_1}^1$ to be empty, so that we can swap languages in our notation.

The interpolation proof will depend in its inductive step (induction on m) on what clause q unifies with. The important cases are

1. The clause is in $\mathbb{A}$ language.
2. The clause is in $\mathbb{B}$ language and is from $\mathbb{B}_n$.
3. The clause is in $\mathbb{B}$ language and is from earlier in the sequence.

Let us now state and prove the interpolation theorem. Since the goals may be clauses with non-empty body we give it in the more general formulation with an arbitrary goal formula $\mathbb{A}$ in language 1.

THEOREM 4.5 (Interpolation for **K**). *Let* $\Delta = \sum_{r=1}^{2a+1} \sum_{i=1}^{t_r} \mathbb{K}_i^r$, *be as discussed and* $\mathbb{B}_1, \dots, \mathbb{B}_n$, $\mathbb{A}_1, \dots, \mathbb{A}_k$, $n \geq 1$ *be as discussed and* $\mathbb{A}^*$ *a set of formulas in the* $\mathbb{A}$ *language and assume that* $\Delta + \sum_{i=1}^n \mathbb{B}_i + \sum_{i=1}^k \mathbb{A}_i \vdash_{\leq(m+1)} \mathbb{A}^*$.

Then there exist effectively computable sets of wffs $\mathbb{H}_1, \dots, \mathbb{H}_n$ *in the common language such that*

1. $\Delta + \mathbb{B}_1, \dots, \mathbb{B}_j \vdash_{\leq(m+1)} \mathbb{H}_j$ *for* $j = 1, \dots, n$ *and*
2. $\Delta + \sum_{i=1}^n \mathbb{H}_i + \sum_{i=1}^k \mathbb{A}_i \vdash_{\leq(m+1)} \mathbb{A}^*$.

PROOF. By induction on m.

CASE $m = 0$, $k \geq 1$, $\mathbb{A}^* = \{q\}$, *or* $\mathbb{A}^* = \{\top\}$.
In this case $q \in \mathbb{A}_k$. Let

$$\mathbb{H}_i = \{\top\},\ i = 1, \dots, n.$$

CASE $m = 0$, $\mathbb{A}_i$ *do not exist*, $\mathbb{A}^* = \{q\}$.
In this case $q \subset \mathbb{B}_n$. q is therefore in the common language. Let $\mathbb{H}_1, \ldots, \mathbb{H}_{n-1}$ be $\{\top\}$ and let $\mathbb{H}_n = \{q\}$.

CASE $m = 0$, $\mathbb{A}^* = \{\mathbb{A}_1^* \to q_1, \ldots, \mathbb{A}_r^* \to q_r\}$.
Then we have $\Delta + \sum_{i=1}^n \mathbb{B}_i + \sum_{i=1}^k \mathbb{A}_i \vdash_1 \mathbb{A}^*$ iff for $l = 1, \ldots, r$ $\Delta + \sum_{i=1}^n \mathbb{B}_i + \sum_{i=1}^k \mathbb{A}_i + \mathbb{A}_l^* \vdash_1 q_l$. Observe that every $\mathbb{A}_l^*$ is in $\mathbb{A}$-language. For each $l = 1, \ldots, r$ define the interpolants $\mathbb{H}_1^l, \ldots, \mathbb{H}_n^l$ as indicated above, then let for $i = 1, \ldots, n$, $\mathbb{H}_i = \bigcup_{l=1}^r \mathbb{H}_i^l$.

CASE m.
Consider first the case $\mathbb{A}^* = \{q\}$.
SUBCASE 1. q unifies with an element in the $\mathbb{A}$ language. For convenience we distinguish two subcases

(a) q unifies with an element $A_i \in \mathbb{A}_i$ for some $1 \leq i \leq k$. We let $A_i = (\mathbb{E}_{i+1}, \ldots, \mathbb{E}_k) \to q$.
(b) q unifies with an element in $\mathbb{A}_j^{2b+1}$, $b \leq a$, $1 \leq j \leq t_{2b+1}$. We let $A_j^{2b+1} \in \mathbb{A}_j^{2b+1}$ be of the form $(\mathbb{E}_{j+1}^{2b+1}, \ldots, \mathbb{E}_{t_{2a+1}}^{2a+1}, \mathbb{F}_1, \ldots, \mathbb{F}_n, \mathbb{E}_1, \ldots, \mathbb{E}_k) \to q$.

The two cases are not much different, the distinction being which side of $\mathbb{B}_1, \ldots, \mathbb{B}_n$ the unification occurs.

From the notation point of view, it is simpler to treat subcase (a) first.

We have in this case that

$$\text{(K1)} \qquad \Delta + \sum_{i=1}^{n} \mathbb{B}_i + \sum_{j=1}^{i} \mathbb{A}_j + \mathbb{A}_{i+1} + \cdots + \mathbb{A}_{i+r} \vdash_{\leq m} \mathbb{E}_{i+r}$$

for $r = 1, \ldots, k - i$.

By the induction hypothesis there exists interpolants $\mathbb{H}_1^r, \ldots, \mathbb{H}_n^r$, $r = 1, \ldots, k - i$ such that

$$\text{(K2)} \qquad \Delta + \sum_{i=1}^{j} \mathbb{B}_i \vdash_{\leq m} \mathbb{H}_j^r$$

for $1 \leq j \leq n$ and such that

$$\text{(K3)} \qquad \Delta + \sum_{i=1}^{n} \mathbb{H}_i^r + \sum_{j=1}^{i} \mathbb{A}_j + \mathbb{A}_{i+1} + \cdots + \mathbb{A}_{i+r} \vdash_{\leq m} \mathbb{E}_{i+r}$$

$r = 1, \ldots, k - i$.

Let

$$\text{(K4)} \qquad \mathbb{H}_j = \bigcup_r \mathbb{H}_j^r$$

Then we have:

$$\text{(K5)} \qquad \Delta + \sum_{i=1}^{n} \mathbb{H}_i + \sum_{j=1}^{i+r} \mathbb{A}_j \vdash_{\leq m} \mathbb{E}_{i+r}$$

for $r = 1, \ldots, k - i$.

Hence

$$\text{(K6)} \qquad \Delta + \sum_{i=1}^{n} \mathbb{H}_i + \sum_{i=1}^{k} \mathbb{A}_i \vdash_{\leq (m+1)} q$$

We also have

$$\text{(K7)} \qquad \Delta + \mathbb{B}_1 + \cdots + \mathbb{B}_j \vdash_{\leq m} \mathbb{H}_j$$

for $j = 1, \ldots, n$.

We now treat subcase (b), where q unifies with $A_j^{2b+1} \in \mathbb{A}_j^{2b+1}$.

We have that

$$\text{(K8)} \qquad \sum_{r=1}^{x-1} \sum_{i=1}^{t_r} \mathbb{K}_i^r + \sum_{i=1}^{y} \mathbb{K}_i^x \vdash_{\leq m} \mathbb{E}_y^x$$

for each $2b + 1 < x \leq 2a + 1$ and $1 \leq y \leq t_x$, and for $x = 2b + 1$ and $1 \leq s \leq t_{2b+1} - j$, we have that

$$\sum_{r=1}^{2b} \sum_{i=1}^{t_r} \mathbb{K}_i^r + \sum_{l=1}^{s} \mathbb{K}_{j+l}^{2b+1} \vdash_{\leq m} \mathbb{E}_{j+1}^{2b}$$

and we have

$$\text{(K9)} \qquad \Delta + \mathbb{B}_1 + \cdots + \mathbb{B}_i \vdash_{\leq m} \mathbb{F}_i$$

for $i = 1, \ldots, n$ and

$$\text{(K10)} \qquad \Delta + \mathbb{B}_1 + \cdots + \mathbb{B}_n + \mathbb{A}_1 + \cdots + \mathbb{A}_j \vdash_{\leq m} \mathbb{E}_j.$$

for $j = 1, \ldots, k$.

Using the induction hypothesis there exist, for $i = 1, \ldots, n$ and $1 \leq u \leq i$ and for $j = 1, \ldots, k$, $w = 1, \ldots, n$, sets in the common language $\mathbb{H}_u^i$ and $\mathbb{G}_w^j$ such that the following holds:

$$\text{(K11)} \qquad \begin{aligned} &\Delta + \mathbb{B}_1 + \cdots + \mathbb{B}_u \vdash_{\leq m} \mathbb{H}_u^i, \\ &\Delta + \mathbb{H}_1^i + \cdots + \mathbb{H}_i^i \vdash_{\leq m} \mathbb{F}_i, \\ &\Delta + \mathbb{B}_1 + \cdots + \mathbb{B}_w \vdash_{\leq m} \mathbb{G}_w^j, \\ &\Delta + \mathbb{G}_1^j + \cdots + \mathbb{G}_n^j + \mathbb{A}_1 + \cdots + \mathbb{A}_j \vdash_{\leq m} \mathbb{E}_j. \end{aligned}$$

Let $\mathbb{H}_u$, $u = 1, \ldots, n$ be

$$\text{(K12)} \qquad \mathbb{H}_u = \bigcup_{u \leq i} \mathbb{H}_u^i \cup \bigcup_{1 \leq j \leq k} \mathbb{G}_u^j.$$

The following holds, for $u = 1, \dots, n$

$$\text{(K13)}\qquad \begin{aligned} &\Delta + \mathbb{D}_1 + \cdots + \mathbb{D}_u \vdash_{\leq m} \mathbb{H}_u, \\ &\Delta + \mathbb{H}_1 + \cdots + \mathbb{H}_u \vdash_{\leq m} \mathbb{F}_u, \\ &\Delta + \sum_{u=1}^{n} \mathbb{H}_u + \sum_{i=1}^{j} \mathbb{A}_j \vdash_{\leq m} \mathbb{E}_j. \end{aligned}$$

Therefore we have

$$\text{(K14)}\qquad \Delta + \sum_{u=1}^{n} \mathbb{H}_u + \sum_{i=1}^{k} \mathbb{A}_k \vdash_{\leq (m+1)} q.$$

The latter follows since q can unify with A_j^{2b+1}.

SUBCASE 2.

q unifies with some elements in the $\mathbb{B}$ language. Hence q is in the common language. We distinguish three subcases:

(a) The element is $B_j \in \mathbb{B}_j$ for some j. We write the element as $B_j = (\mathbb{C}_{j+1}, \dots, \mathbb{C}_n, \mathbb{D}_1, \dots, \mathbb{D}_k) \to q$. This splits into two subcases, (a1) in which $j = n$ and (a2) in which $j < n$.

(b) The element is $B_j^{2b} \in \mathbb{B}_j^{2b}$, $b \leq a$ and $1 \leq j \leq t_{2b}$. We write the element as

$$B_j^{2b} = (\mathbb{Y}_{j+1}^{2b}, \dots, \mathbb{Y}_{t_{2a+1}}^{2a+1}, \mathbb{C}_1, \dots, \mathbb{C}_n, \mathbb{D}_1, \dots, \mathbb{D}_k) \to q.$$

The important distinction is whether q unifies with $B_n \in \mathbb{B}_n$ (subcase (a1)) or not. After treating subcase (a1) the proof method is slightly modified to treat subcases (a2) and (b).

Let us begin with subcase (a).

In this case we have for each $1 \leq r \leq n - j$ and each $1 \leq s \leq k$ (**K**15) and (**K**16) below:

$$\text{(K15)}\qquad \Delta + \sum_{i=1}^{j+r} \mathbb{B}_i \vdash_{\leq m} \mathbb{C}_{j+r},$$

$$\text{(K16)}\qquad \Delta + \sum_{i=1}^{n} \mathbb{B}_n + \sum_{i=1}^{s} \mathbb{A}_s \vdash_{\leq m} \mathbb{D}_s.$$

We distinguish two further subcases.

SUBCASE (a1). $j = n$

This subcase means $\mathbb{C}_{j+r}$ do not exist. It means $B_n \in \mathbb{B}_n$, $B_n = (\mathbb{D}_1, \dots, \mathbb{D}_k) \to q$ and we deal with (**K**16) only.

If $\mathbb{A}_1, \dots, \mathbb{A}_k$ also do not exist, then $B_n = q$ and we can use case 1 (by swapping languages). Otherwise we use the induction hypothesis to eliminate $\mathbb{A}_1, \dots, \mathbb{A}_k$, since $\mathbb{D}_1, \dots, \mathbb{D}_k$ are in the $\mathbb{B}$ language, and then use the result to perform further interpolations.

Therefore for some $\mathbb{H}_1^s, \ldots, \mathbb{H}_s^s$, $1 \leq s \leq k$ we have

$$\text{(K17)} \qquad \Delta + \sum_{i=1}^{n} \mathbb{B}_i + \sum_{i=1}^{j} \mathbb{A}_i \vdash_{\leq m} \mathbb{H}_j^s$$

for $1 \leq j \leq s$, and

$$\text{(K18)} \qquad \Delta + \sum_{i=1}^{n} \mathbb{B}_i + \sum_{i=1}^{s} \mathbb{H}_i^s \vdash_{\leq m} \mathbb{D}_s$$

for $s = 1, \ldots, k$.

Let $\mathbb{H}_i = \bigcup_{i \leq s} \mathbb{H}_i^s, i = 1, \ldots, k$. Then

$$\text{(K19)} \qquad \Delta + \sum_{i=1}^{n} \mathbb{B}_i + \sum_{i=1}^{j} \mathbb{A}_i \vdash_{\leq m} \mathbb{H}_j$$

$j = 1, \ldots, k$ and also

$$\text{(K20)} \qquad \Delta + \sum_{i=1}^{n} \mathbb{B}_i + \sum_{i=1}^{j} \mathbb{H}_i \vdash_{\leq m} \mathbb{D}_j.$$

Now since

$$\text{(K21)} \qquad B_n = (\mathbb{D}_1, \ldots, \mathbb{D}_k) \to q \in \mathbb{B}_n$$

we get in view of (**K**20) that

$$\text{(K22)} \qquad \Delta + \sum_{i=1}^{n} \mathbb{B}_i + \sum_{i=1}^{k} \mathbb{H}_i \vdash_{\leq (m+1)} q.$$

This means that

$$\text{(K22)} \qquad \Delta + \sum_{i=1}^{n} \mathbb{B}_i \vdash_{\leq (m+1)} (\mathbb{H}_1, \ldots, \mathbb{H}_k) \to q.$$

We want however to interpolate out the $\mathbb{B}_i$, not the $\mathbb{A}_i$. We can do that as follows:

Consider (**K**19). We can use the induction hypothesis and find interpolants for $\sum_{i=1}^{n} \mathbb{B}_i$, since $\sum_{i=1}^{k} \mathbb{A}_i$ is not empty. Let $\mathbb{H}'_1, \ldots, \mathbb{H}'_n$ be the interpolants. This means that

$$\text{(K23)} \qquad \Delta + \mathbb{B}_1 + \cdots + \mathbb{B}_i \vdash_{\leq m} \mathbb{H}'_i$$

$i = 1, \ldots, n$.

$$\text{(K24)} \qquad \Delta + \sum_{i=1}^{n} \mathbb{H}'_i + \sum_{i=1}^{j} \mathbb{A}_i \vdash_{\leq m} \mathbb{H}_j$$

$j = 1, \ldots, k$.

Note that we jumped a step. We first find $\mathbb{H}'_{i,j}$ for each $\mathbb{H}_j$, $1 \leq i \leq j$ and then let $\mathbb{H}'_j = \bigcup_{i \leq j} \mathbb{H}'_{i,j}$.

Now let $\mathbb{G}_j$, $j = 1, \ldots, n$ be defined as follows

$$\text{(K25)} \qquad \begin{aligned} \mathbb{G}_j &= \mathbb{H}'_j, \; j = 1, \ldots, n-1, \\ \mathbb{G}_n &= \mathbb{H}'_n \cup \{(\mathbb{H}_1, \ldots, \mathbb{H}_k) \to q\}. \end{aligned}$$

In view of (**K**23) and (**K**22) we get that

$$\text{(K26)} \qquad \Delta + \mathbb{B}_1 + \cdots + \mathbb{B}_j \vdash_{\leq (m+1)} \mathbb{G}_j$$

and also we get in view of (**K**24) and the fact that $((\mathbb{H}_1, \ldots, \mathbb{H}_k) \to q) \in \mathbb{G}_n$ that

$$\text{(K27)} \qquad \Delta + \sum_{i=1}^{n} \mathbb{G}_i + \sum_{i=1}^{k} \mathbb{A}_i \vdash_{\leq (m+1)} q.$$

This completes subcase (a1).

Strategy for subcase (a2)

In this case q unifies with $B_j \in \mathbb{B}_j$, $B_j = (\mathbb{C}_{j+1}, \ldots, \mathbb{C}_n, \mathbb{D}_1, \ldots, \mathbb{D}_k) \to q$.

The following holds.

$$\text{(K28)} \qquad \Delta + \sum_{i=1}^{j} \mathbb{B}_j + \mathbb{B}_{j+1} + \cdots + \mathbb{B}_{j+r} \vdash_{\leq m} \mathbb{C}_{j+r}$$

for $r = 1, \ldots, n - j$

$$\text{(K29)} \qquad \Delta + \sum_{i=1}^{n} \mathbb{B}_j + \mathbb{A}_1 + \cdots + \mathbb{A}_s \vdash_{\leq m} \mathbb{D}_s$$

for $s = 1, \ldots, k$.

To explain our strategy for this case and in fact for subcase (b) as well, observe the following. From (**K**28) we get that

$$\text{(K30)} \qquad \Delta + \mathbb{B}_1 + \cdots + \mathbb{B}_n \vdash_{\leq (m+1)} (\mathbb{D}_1, \ldots, \mathbb{D}_k) \to q.$$

We can therefore consider interpolation for the sequence

$$\Delta + \mathbb{B}_1 + \cdots + \mathbb{B}_{n-1} + \mathbb{B}_n \cup \{(\mathbb{D}_1, \ldots, \mathbb{D}_k) \to q\} + \mathbb{A}_1 + \cdots + \mathbb{A}_k \vdash_{\leq (m+1)} q$$

where q unifies with $(\mathbb{D}_1, \ldots, \mathbb{D}_k) \to q$.

This will put us back in subcase (a1).

We can get the interpolants $\mathbb{H}^*_1, \ldots, \mathbb{H}^*_n$ for this case and we will indeed have

$$\Delta + \mathbb{H}^*_1 + \cdots + \mathbb{H}^*_n + \mathbb{A}_1 + \cdots + \mathbb{A}_k \vdash_{\leq (m+1)} q$$

but we also have

$$\Delta + \mathbb{B}_1 + \cdots + \mathbb{B}_n \cup \{(\mathbb{D}_1, \ldots, \mathbb{D}_k) \to q\} \vdash_{\leq (m+1)} \mathbb{H}^*_n$$

but we need to have

$$\Delta + \mathbb{B}_1 + \cdots + \mathbb{B}_n \vdash_{\leq(m+1)} \mathbb{H}_n^*.$$

If we want to get our interpolation theorem for the original sequence we can indeed use the cut theorem in view of (**K**30). This course of action is feasible but not so good for complexity estimates on the interpolants. We choose a better course of action. We proceed with the proof as in subcase (a1) *pretending as if* we have $(\mathbb{D}_1, \ldots, \mathbb{D}_k) \to q$ available at $\mathbb{B}_n$. A crucial step in the proof of subcase (a1) is to show (**K**22), namely that

$$\Delta + \sum_{i=1}^{n} \mathbb{B}_i \vdash_{\leq(m+1)} (\mathbb{H}_1, \ldots, \mathbb{H}_k) \to q.$$

To show the latter we relied on the fact that (**K**20) and (**K**21) are available, because q can unify with the clause $(\mathbb{D}_1, \ldots, \mathbb{D}_k) \to q$.

In our case we an replace (**K**21) by (**K**31) below

(**K**31)
$$B_j = (\mathbb{C}_{j+1}, \ldots, \mathbb{C}_n, \mathbb{D}_1, \ldots, \mathbb{D}_k) \to q \text{ is in } \mathbb{B}_j,$$
and for $r = j+1, \ldots, n$,
$$\Delta + \sum_{i=1}^{r} \mathbb{B}_i \vdash_{\leq m} \mathbb{C}_r.$$

This shows (**K**22) for our case, since q can unify with B_j.

Having explained the idea, let us actually write the proof by going through the details step by step.

It is really the same proof as that for subcase (a1) and it will go through also for subcase (b).

Proof for subcases (a2) *and* (b)

Our situation is as follows. For some clause in the language $\mathbb{B}$ of the form

$$B_j = (\mathbb{C}_{j+1}, \ldots, \mathbb{C}_n, \mathbb{D}_1, \ldots, \mathbb{D}_k) \to q \in \mathbb{B}_j$$

in case of subcase (a2), or of the form

$$B_j^{2b} = (\mathbb{Y}_{j+1}^{2b}, \ldots, \mathbb{Y}_{t_{2a+1}}^{2a+1}, \mathbb{C}_1, \ldots, \mathbb{C}_n, \mathbb{D}_1, \ldots, \mathbb{D}_k) \to q \text{ in } \mathbb{K}_j^{2b}$$

in case of Subcase (b), we have the following holds:

(**K**32)
$$\sum_{r=1}^{2b-1} \sum_{i=1}^{t_r} \mathbb{K}_i^r + \mathbb{K}_1^{2b} + \cdots + \mathbb{K}_j^{2b} + \mathbb{K}_{j+1}^{2b} + \cdots + \mathbb{K}_{j+k}^{2b} \vdash_{\leq m} \mathbb{Y}_{j+k}^{2b}$$

$1 \leq k \leq t_{2b} - j$.

(**K**33)
$$\sum_{r=1}^{z} \sum_{i=1}^{t_z} \mathbb{K}_i^z + \mathbb{K}_1^{z+1} + \cdots + \mathbb{K}_k^{z+1} \vdash_{\leq m} \mathbb{Y}_k^{z+1}$$

for $1 \leq k \leq t_{z+1}$, $2b < z \leq 2a$.

(**K**34) $$\Delta + \mathbb{B}_1 + \cdots + \mathbb{B}_r \vdash_{\leq m} \mathbb{C}_r, 1 \leq r \leq j.$$

(**K**35) $$\Delta + \mathbb{B}_1 + \cdots + \mathbb{B}_r \vdash_{\leq m} \mathbb{C}_r,\ j < r \leq n.$$

(**K**36) $$\Delta + \sum_{i=1}^{n} \mathbb{B}_i + \mathbb{A}_1 + \cdots + \mathbb{A}_r \vdash_{\leq m} \mathbb{D}_r, \text{ for } 1 \leq r \leq k.$$

For case (b), (**K**32)–(**K**36) hold. For case (a2), (**K**35)–(**K**36) hold.

We now proceed as in Subcase (a1). Note that (**K**35) is (**K**15) and (**K**36) is (**K**16) of Subcase (a1).

We get sets $\mathbb{H}_1, \ldots, \mathbb{H}_k$ such that (**K**19) holds, namely

(**K**19) $$\Delta + \sum_{i=1}^{n} \mathbb{B}_i + \sum_{i=1}^{j} \mathbb{A}_i \vdash_{\leq m} \mathbb{H}_j.$$

We now show that

(**K**37) $$\Delta + \sum_{i=1}^{n} \mathbb{B}_i \vdash_{\leq (m+1)} (\mathbb{H}_1, \ldots, \mathbb{H}_k) \rightarrow q.$$

We need to show that

$$\Delta + \sum_{i=1}^{n} \mathbb{B}_i + \mathbb{H}_1 + \cdots + \mathbb{H}_k \vdash_{\leq (m+1)} q.$$

However we have $B_j \in \mathbb{B}_j$ in subcase (a2) or $B_j^{2b} \in \mathbb{K}_j^{2b}$ in subcase (b), with (**K**35)–(**K**36) and (**K**32)–(**K**36) holding respectively. So q can unify with this clause and (**K**32)–(**K**36) ensure that (**K**37) holds.

We now continue as in subcase (a1) and consider (**K**19). We interpolate and get $\mathbb{H}'_1, \ldots, \mathbb{H}'_n$ such that (**K**23) an (**K**24) hold.

We define $\mathbb{G}_j$ as before and in view of (**K**23) and (**K**37) we get (**K**26) and (**K**27).

This completes subcases (a2) and (b) and therefore the inductive case is complete for $\mathbb{A}^* = \{q\}$.

If $\mathbb{A}^* = \{\mathbb{A}_1^* \rightarrow q_1, \ldots, \mathbb{A}_r^* \rightarrow q_r\}$, we have $\Delta + \sum_{i=1}^{n} \mathbb{B}_i + \sum_{i=1}^{k} \mathbb{A}_i \vdash_{\leq (m+1)} \mathbb{A}^*$ iff for $l = 1, \ldots, r$, $\Delta + \sum_{i=1}^{n} \mathbb{B}_i + \sum_{i=1}^{k} \mathbb{A}_i + \mathbb{A}_l^* \vdash_{\leq (m+1)} q_l$. Observe that every $\mathbb{A}_l^*$ is in $\mathbb{A}$-language. For each $l = 1, \ldots, r$ we compute the interpolants $\mathbb{H}_1^l, \ldots, \mathbb{H}_n^l$ as indicated for the atomic case, then we let for $i = 1, \ldots, n$, $\mathbb{H}_i = \bigcup_{l=1}^{r} \mathbb{H}_i^l$.

The interpolation theorem is proved. ⊣

COROLLARY 4.6 (Interpolation for **K**).

1. *Let* $\mathbb{A} \vdash_{\leq m} \mathbb{B}$ *in* **K**, *where* $\mathbb{A}$ *is in language* 1 *and* $\mathbb{B}$ *is in language* 2. *Then for some* $\mathbb{H}$ *in the common language we have* $\mathbb{A} \vdash_{\leq m} \mathbb{H}$ *and* $\mathbb{H} \vdash_{\leq m} \mathbb{B}$.

2. *Let* $\sum_{i=1}^{n} \mathbb{K}_i^{x_i} \vdash_{\leq m} q$, *where* q *is atomic in language* 1 *and* $\mathbb{K}_i^{x_i}$ *is in language* $x_i \in \{1, 2\}$, *for* $i = 1, \ldots, n$. *Then for all* j *such that* $x_j = 2$ *there exist* $\mathbb{H}_j^2$ *in the common language such that*
   (a) $\sum_{i=1}^{n} [\mathbb{K}/\mathbb{H}]_i \vdash_{\leq m} q$
   where $[\mathbb{K}/\mathbb{H}]_i$ *denotes* $\mathbb{K}_i^{x_i}$ *if* $x_i = 1$ *and* $\mathbb{H}_i^{x_i}$ *if* $x_i = 2$.
   (b) *For each* $k \leq n$ *such that* $x_k = 2$ *we have*

$$\sum_{i=1}^{k} \mathbb{K}_i^{x_i} \vdash_{\leq m} \mathbb{H}_k^2.$$

PROOF.

1. Let $B = (\mathbb{D}_1, \ldots, \mathbb{D}_k) \to q$ be in $\mathbb{B}$. Then

$$\mathbb{A} + \mathbb{D}_1 + \cdots + \mathbb{D}_k \vdash_{\leq m} q.$$

   By the interpolation theorem for some $\mathbb{H}_B$ in the common language we have $\mathbb{A} \vdash_m \mathbb{H}_B$ and $\mathbb{H}_B + \mathbb{D}_1 + \cdots + \mathbb{D}_k \vdash_m q$.
   Let $\mathbb{H} = \bigcup_{B \in \mathbb{B}} \mathbb{H}_B$.
2. This follows from repeated applications of the main interpolation theorem. ⊣

THEOREM 4.7 (Interpolation for the n-fragment). *Consider the* n*-fragment of* **K** *defined as follows*:

1. *Atoms* q *or* $\top$ *are in the fragment.*
2. *If* $\mathbb{A}_1, \ldots, \mathbb{A}_r$, $r \leq n$ *are sets of wffs in the fragment and* x *is atomic or* $\top$ *then* $(\mathbb{A}_1, \ldots, \mathbb{A}_r) \to x$ *is in the fragment.*

Then Theorem 4.5 *holds in this fragment, provided the lengths of the blocks are no more than* n.

PROOF. We follow the same induction steps as the proof of Theorem 4.5 and further assume the interpolant is in the fragment. This additional assumption goes through easily in the induction argument for the case $m = 0$ and case m subcase 1. For subcase 2 observe that the key additional formula $(\mathbb{H}_1, \ldots, \mathbb{H}_k) \to q$, added to $\mathbb{G}_n$ in (**K**25), is also in the fragment, since k is the length of the last block. ⊣

THEOREM 4.8 (Interpolation for **K4**, **T** and **S4**). *The interpolation theorem as formulated in Theorem* 4.5 *holds for* **K4**, **T** *and* **S4** *defined in Definition* 4.3.

PROOF. We consider the proof of Theorem 4.5. Only small changes are required to accommodate the slightly different goal directed proof theory of **K4**, **T** and **S4**, as indicated in item (4L) of Definition 4.3. Let us follow the proof of the interpolation for **K** and indicate the small changes required. In the case of **K**, we numbered key steps by (**K**1) to (**K**37). If we make changes for the logic **L** (**L** is **K4**, **T** or **S4**), we indicate the new step as (**L**i), $1 \leq i \leq 37$.

The proof is by induction on m.

CASE $m = 0$.
Same as the case of **K**.

CASE m SUBCASE 1.
q unifies with an element of the $\mathbb{A}$ language. There are two subcases.

SUBCASE (a).
q unifies with $A_i \in \mathbb{A}_i$. In this subcase we write

$$A_i = (\mathbb{E}_1, \ldots, \mathbb{E}_\alpha) \to q.$$

Unlike the case of **K**, α need not be equal to $k - i$.

Therefore condition (**K**1) is replaced by (**L**1) (in accordance with item (4**L**) of Definition 4.3 as follows:

$$(\mathbf{L}1) \qquad \Delta + \sum_{i=1}^{n} \mathbb{B}_i + \sum_{j=1}^{i-1} \mathbb{A}_j + \sum_{j=i}^{\beta(r)} \mathbb{A}_j \vdash_{\leq m} \mathbb{E}_r$$

where $r = 1, \ldots, \alpha$ and $iR_{\mathbf{L}}\beta(1)R_{\mathbf{L}}\beta(2)\ldots R_{\mathbf{L}}\beta(\alpha) = k$.

Note that $R_{\mathbf{L}}$ is defined in Definition 4.3.

By the induction hypothesis there exist interpolants

$$\mathbb{H}_1^r, \ldots, \mathbb{H}_n^r, r = 1, \ldots, \alpha$$

such that (**k**2) holds and such that the following (**L**3) holds (replacing (**K**3)):

$$(\mathbf{L}3) \qquad \Delta + \sum_{i=1}^{n} \mathbb{H}_i^r + \sum_{j=1}^{n-1} \mathbb{A}_j + \sum_{j=n}^{\beta(r)} \mathbb{A}_j \vdash_{\leq m} \mathbb{E}_r$$

We define $\mathbb{H}_j = \bigcup_r \mathbb{H}_j^r$, as in the case of **K**, and get the following:

$$(\mathbf{L}5) \qquad \Delta + \sum_{i=1}^{n-1} \mathbb{H}_i + \sum_{i=n}^{\beta(r)} \mathbb{A}_i \vdash_{\leq m} \mathbb{E}_r$$

$r = 1, \ldots, \alpha$.

This gives us (**L**6) and (**L**7) as follows:

$$(\mathbf{L}6) \qquad \Delta + \sum_{i=1}^{n-1} \mathbb{H}_i + \sum_{i=n}^{k} \mathbb{A}_i \vdash_{\leq(m+1)} q$$

$$(\mathbf{L}7) \qquad \Delta + \mathbb{B}_1 + \cdots + \mathbb{B}_j \vdash_{\leq m} \mathbb{H}_j, j = 1, \ldots, n.$$

SUBCASE (b).
This subcase is again very similar. q unifies with some $A_j^{2b+1} \in \mathbb{A}_j^{2b+1}$. We can write the formula as

$$A = (\mathbb{E}_{\beta(1)}, \ldots, \mathbb{E}_{\beta(\alpha_1)}, \mathbb{E}_{\beta(\alpha_1+1)}, \ldots, \mathbb{E}_{\beta(\alpha_2)}, \mathbb{E}_{\beta(\alpha_2+1)}, \ldots, \mathbb{E}_{\beta(\alpha_3)}) \to q$$

We have that $\beta(1)$ is a theory $\mathbb{K}_y^x$ which is R_{L} related to $\beta(2) = \mathbb{K}_{y'}^{x'}$, and for each $1 \leq i \leq \alpha_3$ we have $\beta(i) R_{\mathrm{L}} \beta(i+1)$ and $\beta(\alpha_3) = \mathbb{A}_k$.[11]

The division α_1, α_2, α_3 give the divisions in the range of β. $\beta(1)$ is up to $\beta(\alpha_1)$, are theories from among $\mathbb{K}_1^1, \ldots, \mathbb{K}_{t_{2a+1}}^{2a+1}$, $\beta(\alpha_1+1), \ldots, \beta(\alpha_2)$ fall among $\mathbb{B}_1, \ldots, \mathbb{B}_n$ and $\beta(\alpha_2+1)$, $\beta(\alpha_3)$ fall among $\mathbb{A}_1, \ldots, \mathbb{A}_k$.

We have that (**K**8)–(**K**10) are replaced by the following single (**L**8–10):

$$(\mathbf{L}8\text{–}10) \qquad \mathbb{K}_1^1 + \cdots + \beta(i) \vdash_{\leq m} \mathbb{E}_i.$$

We can now continue and interpolate as we do in (**K**11)–(**K**14), paying attention to the $\alpha_1+1, \ldots, \alpha_2$ and $\alpha_2+1, \ldots, \alpha_3$ ranges.

SUBCASE 2.

Here cases (a1) and (a2) and (b) follow through steps similar to those used in the proofs for **K**. We just have to pay attention to ranges of indices. Let us do subcase (a1) in more detail.

In subcase (a1), B_n will have the form

$$B_n = (\mathbb{D}_{r_1}, \ldots, \mathbb{D}_{r_\alpha}) \to q,$$

with $\mathbb{B}_n R_{\mathrm{L}} \beta(r_1) \ldots R_{\mathrm{L}} \beta(r_\alpha) = \mathbb{A}_k$.

The following are the ($\mathbf{L}_i$) replacements of the respective ($\mathbf{K}_i$) steps:

$$(\mathbf{L}19) \qquad \Delta + \sum_{i=1}^{n} \mathbb{B}_i + \mathbb{A}_1 + \cdots + \mathbb{A}_j \vdash_{\leq m} \mathbb{H}_j,\ j = 1, \ldots, k,$$

$$(\mathbf{L}20) \qquad \Delta + \sum_{i=1}^{n} \mathbb{B}_i + \mathbb{H}_1 + \cdots + \mathbb{H}_j \vdash_{\leq m} \mathbb{D}_j,\ j = 1, \ldots, k.$$

Notice that (**L**19) and (**L**20) are the same as (**K**19) and (**K**20). This is because although e.g., $\beta(r_1)$ may be say $\mathbb{A}_5$, when we interpolate by the induction hypothesis, we get interpolants $\mathbb{H}_1^{r_1}, \ldots, \mathbb{H}_t^{r_1}$, i.e., for all the $\mathbb{A}_j$ preceding $\beta(r_1)$.

Thus (**L**22) will be the same as (**K**22) and similarly (**L**23)–(**L**27).

The proof of cases (a2) and (b) are modified in a similar way. ⊣

COROLLARY 4.9. *Corollary* 4.6 *and Theorem* 4.7 *also hold for* T, **K4** *and* **S4**.

PROOF. Follows from the proof of Theorem 4.8. ⊣

QUESTION 4.10. *Is it true that if A and B are single wffs such that $A \vdash_{\mathrm{L}} B$ then for some single wff H we have $A \vdash_{\mathrm{L}} H \vdash_{\mathrm{L}} B$? (We know an $\mathbb{H}$ exists!).*

We now sharpen our interpolation theorem for **K** (and similarly for **K4**, **T** and **S4**).

---

[11] Note that because of our special use of double indices and letters in $\mathbb{K}_y^x$, $\mathbb{B}_i$, $\mathbb{A}_j$, we are giving the theories as values of β instead of the indices and assume R_{L} is correctly understood.

We begin by explaining the weakness of the current version of our interpolation for **K**. Assume we have

$$\mathbb{A}_1 + \mathbb{A}_2 + \mathbb{B}_1 + \mathbb{B}_2 + \mathbb{A}_3 + \mathbb{A}_4 \vdash a$$

where a, $\mathbb{A}_1 - \mathbb{A}_4$ are in language 1 and $\mathbb{B}_1$ and $\mathbb{B}_2$ are in language 2. The current version of the interpolation theorem which we have for **K** (item 2 in Corollary 4.6) allows for interpolants $\mathbb{H}_1$, $\mathbb{H}_2$ in the common language such that

(a) $\mathbb{A}_1 + \mathbb{A}_2 + \mathbb{H}_1 + \mathbb{H}_2 + \mathbb{A}_3 + \mathbb{A}_4 \vdash a$

(b) $\mathbb{A}_1 + \mathbb{A}_2 + \mathbb{B}_1 \vdash \mathbb{H}_1$ and $\mathbb{A}_1 + \mathbb{A}_2 + \mathbb{B}_1 + \mathbb{B}_2 \vdash \mathbb{H}_2$.

The weakness in this formulation is (b) above. If we compare this version of interpolation with the version we have for the Lambek calculus, linear logic or concatenation logic we see that in these other logics we have

(b′) $\mathbb{B}_1 + \mathbb{B}_2 \vdash \mathbb{H}_1 + \mathbb{H}_2$.

In the version for **K** we need $\mathbb{A}_1$, $\mathbb{A}_2$ as well. So the interpolant follows from the *mixed* language $\mathbb{A}_1 + \mathbb{A}_2 + \mathbb{B}_1 + \mathbb{B}_2$ and not from the $\mathbb{B}_1$, $\mathbb{B}_2$ language alone!

The following example shows that we cannot hope for an improvement.

EXAMPLE 4.11. *Consider*

$$a_1 \to (\top \to (\top \to q)), a_1, \{b, q\} \to (\top \to (\top \to r)), b, r \to a, \top \vdash a$$

where a_1, a are in language 1, b is in language 2 and q, r, $\top$ in the common language, namely $\mathbb{A}_1 = \{a_1 \to (\top \to (\top \to q))\}$, $\mathbb{A}_2 = \{a_1\}$, $\mathbb{B}_1 = \{\{b, q\} \to (\top \to (\top \to r))\}$, $\mathbb{B}_2 = \{b\}$, $\mathbb{A}_3 = \{r \to a\}$, $\mathbb{A}_4 = \{\top\}$. The interpolants are $h_1 = \top$ and $h_2 = \top \to (\top \to r)$ but $\mathbb{B}_1 + \mathbb{B}_2$ on their own (without the help of $\mathbb{A}_1 + \mathbb{A}_2$ which give q) cannot prove the interpolants.

We do have, however, that for some common language formulas $g_1 = \top$, $g_2 = \top \to (\top \to q)$, the following holds:

$$\mathbb{A}_1 \vdash g_1$$
$$\mathbb{A}_1 + \mathbb{A}_2 \vdash g_2$$
$$g_1 + g_2 + \mathbb{B}_1 \vdash h_1$$
$$g_1 + g_2 + \mathbb{B}_1 + \mathbb{B}_2 \vdash h_2$$
$$\mathbb{A}_1 + \mathbb{A}_2 + h_1 + h_2 + \mathbb{A}_3 + \mathbb{A}_4 \vdash a$$

The above shows a chain of interpolants, first $g_1 g_2$ and then $h_1 h_2$, each proved from one *language alone.*

We call this kind of interpolation chain interpolation. *Here the chain is*

1. g_1, g_2
2. h_1, h_2.

We now formulate our chain interpolation theorem. We start with a lemma:

LEMMA 4.12. *Let* $\mathbb{K}_i^{x_i}$, $i = 1, \dots, n$, *be theories in language* x_i, $x_i \in \{1, 2\}$. *Assume that for each* r *s.t.* $x_r = 2$ *we are given* $\mathbb{H}_r$ *in the common language such that*

(1)
$$\sum_{i=1}^{r} \mathbb{K}_i^{x_i} \vdash_{\leq m} \mathbb{H}_r.$$

Then there exist $\mathbb{G}_s$, *in the common language, for all* s *such that* $x_s = 1$ *such that the following holds*

(2)
$$\sum_{i=1}^{s} \mathbb{K}_i^{x_i} \vdash_{\leq m} \mathbb{G}_s$$

(3)
$$\sum_{i=1}^{r} [\mathbb{K}/\mathbb{G}]_i \vdash_{\leq m} \mathbb{H}_r$$

where $[\mathbb{K}/\mathbb{G}]_i = \mathbb{K}_i^{x_i}$ *if* $x_i = 2$, $[\mathbb{K}/\mathbb{G}]_i = \mathbb{G}_i$ *if* $x_i = 1$.

PROOF. Consider $\mathbb{H}_r$ to be in language 2 (since it is in the common language). Then using (1) we get by the interpolation theorem of item 2 of Corollary 4.6 that for all s such that $s \leq r$ and $x_s = 1$ there exists a $\mathbb{G}_s^r$ such that

(1*) $\sum_{i=1}^{s} \mathbb{K}_i^{x_i} \vdash_{\leq m} \mathbb{G}_s^r$,
(2) $\sum_{i=1}^{r} [\mathbb{K}/\mathbb{G}]_i^r \vdash_{\leq m} \mathbb{H}_r$

where, as before

$$[\mathbb{K}/\mathbb{G}]_i^r = \begin{cases} \mathbb{K}_i^{x_i} & \text{if } x_i = 2, \\ \mathbb{G}_i^r & \text{if } x_i = 1. \end{cases}$$

Let $\mathbb{G}_s = \bigcup_{s \leq r} \mathbb{G}_s^r$.

Then we have for $\mathbb{G}_s$

(1**) $\sum_{i=1}^{s} \mathbb{K}_i^{x_i} \vdash_{\leq m} \mathbb{G}_s$,
(2*) $\sum_{i=1}^{r} [\mathbb{K}/\mathbb{G}]_i \vdash_{\leq m} \mathbb{H}_r$

where

$$[\mathbb{K}/\mathbb{G}]_i = \begin{cases} \mathbb{K}_i^{x_i} & \text{if } x_i = 2, \\ \mathbb{G}_i & \text{if } x_i = 1. \end{cases}$$

(1**) holds because (1*) holds for each $r \geq s$, and (2*) holds because (2) holds for $\mathbb{G}_i^r$ and $\mathbb{G}_i^r \subseteq \mathbb{G}_i$.

This proves Lemma 4.12. ⊣

THEOREM 4.13 (Chain interpolation for **K**). *Let* Δ *be the theory of the form*

$$\Delta = \mathbb{K}_1^2 + \cdots + \mathbb{K}_{s_1}^2 + \mathbb{K}_{s_1+1}^1 + \cdots + \mathbb{K}_{r_1}^1 + \dots$$
$$+ \mathbb{K}_{r_{t-1}+1}^2 + \cdots + \mathbb{K}_{s_t}^2 + \mathbb{K}_{s_t+1}^1 + \cdots + \mathbb{K}_{r_t}^1$$

where $\mathbb{K}_i^1$ is a theory of language 1 and $\mathbb{K}_i^2$ is a theory of language 2 and the indices $s_1 < r_1 < \ldots s_t < r_t$ indicate the blocks of language change.

We allow for the block $\mathbb{K}_1^2 + \cdots + \mathbb{K}_{s_1}^2$ not to exist (i.e., Δ starts with language 1) or the block $\mathbb{K}_{s_t+1}^1 + \cdots + \mathbb{K}_{r_t}^1$ not to exist (i.e., Δ ends with language 2) or both not to exist. Let q_1 be in language 1 and assume that

$$\Delta \vdash_{\leq n} q_1.$$

Then (the chain interpolation theorem says that) there exists a sequence $\mathbb{H}_i$ of theories in the common language, for $1 \leq i \leq s_t$ such that the following holds:

(s_1) $\mathbb{K}_1^2 + \cdots + \mathbb{K}_i^2 \vdash_{\leq n} \mathbb{H}_i$, *for* $1 \leq i \leq s_1$.

(r_1) $\mathbb{H}_1 + \cdots + \mathbb{H}_{s_1} + \mathbb{K}_{s_1+1}^1 + \cdots + \mathbb{K}_i \vdash_{\leq n} \mathbb{H}_i$, *for* $s_1 + 1 \leq i \leq r_1$.

(s_2) $\mathbb{K}_1^2 + \cdots + \mathbb{K}_{s_1}^2 + \mathbb{H}_{s_1+1} + \cdots + \mathbb{H}_{r_1} + \mathbb{K}_{r_1+1}^2 + \cdots + \mathbb{K}_i^2 \vdash_{\leq n} \mathbb{H}_i$, *for* $r_1 + 1 \leq i \leq s_2$.

(r_2) $\mathbb{H}_1 + \cdots + \mathbb{H}_{s_1} + \mathbb{K}_{s_1+1}^1 + \mathbb{K}_{r_1}^1 + \mathbb{H}_{r_1+1} + \ldots + \mathbb{H}_{s_2} + \mathbb{K}_{s_2+1}^1 + \cdots + \mathbb{K}_i^1 \vdash_{\leq n} \mathbb{H}_i$, *for* $s_2 + 1 \leq i \leq r_2$.

In general

(s_{m+1}) $\sum_{j=1}^{i} [\mathbb{K}/\mathbb{H}]_j^1 \vdash_{\leq n} \mathbb{H}_i$, *for* $r_m + 1 \leq i \leq s_{m+1}$ *and* $1 \leq m \leq t-1$.

(r_{m+1}) $\sum_{j=1}^{i} [\mathbb{K}/\mathbb{H}]_j^2 \vdash_{\leq n} \mathbb{H}_i$, *for* $s_{m+1} + 1 \leq i \leq r_{m+1}$ *and* $0 \leq m \leq t-1$

where

$$[\mathbb{K}/\mathbb{H}]_i^1 = \begin{cases} \mathbb{K}_i & \text{if } \mathbb{K}_i \text{ in language 2,} \\ \mathbb{H}_i & \text{if } \mathbb{K}_i \text{ in language 1,} \end{cases}$$

$$[\mathbb{K}/\mathbb{H}]_i^2 = \begin{cases} \mathbb{H}_i & \text{if } \mathbb{K}_i \text{ in language 2,} \\ \mathbb{K}_i & \text{if } \mathbb{K}_i \text{ in language 1.} \end{cases}$$

Note that $\mathbb{K}_i$ is in language 2 if $r_m + 1 \leq i \leq s_{m+1}$ and $\mathbb{K}_i$ is in language 1 if $s_m + 1 \leq i \leq r_m$ for some $0 \leq m \leq t$.

Furthermore, we get

$$\sum_{i=1}^{r_t} [\mathbb{K}/\mathbb{H}]_i^1 \vdash_{\leq n} q_1.$$

PROOF. The proof is by induction on the number of changes of language blocks in Δ. The junctions of language change in our notation are the s_m and r_m points, where $1 \leq s_1 \leq r_1 \leq \cdots \leq s_m \leq r_m \cdots \leq s_t \leq r_t$.

CASE ONE JUNCTION.
In this case we have

$$\mathbb{K}_1^2 + \cdots + \mathbb{K}_{s_1}^2 + \mathbb{K}_{s_1+1}^1 + \cdots + \mathbb{K}_{r_1}^1 \vdash_{\leq n} q_1.$$

The theorem for this case is the usual interpolation.

CASE TWO JUNCTIONS.[12]
In this case we have

$$\mathbb{K}_1^2 + \cdots + \mathbb{K}_{s_1}^2 + \mathbb{K}_{s_1+1}^1 + \cdots + \mathbb{K}_{r_1}^1 + \mathbb{K}_{r_1+1}^2 + \cdots + \mathbb{K}_{s_2}^2 \vdash_{\leq n} q_1.$$

By using interpolation we eliminate language 2, i.e., we find $\mathbb{H}_1, \ldots, \mathbb{H}_{s_1}$ and $\mathbb{H}_{r_1+1}, \ldots, \mathbb{H}_{s_2}$ such that

1. $\mathbb{K}_1^2 + \cdots + \mathbb{K}_i^2 \vdash_{\leq n} \mathbb{H}_i$ for $1 \leq i \leq s_1$.
2. $\mathbb{K}_1^2 + \cdots + \mathbb{K}_{s_1}^2 + \mathbb{K}_{s_1+1}^1 + \cdots + \mathbb{K}_{r_1}^1 + \mathbb{K}_{r_1+1}^2 + \cdots + \mathbb{K}_i^2 \vdash_{\leq n} \mathbb{H}_i$, for $r_1 + 1 \leq i \leq s_2$.
3. $\mathbb{H}_1 + \cdots + \mathbb{H}_{s_1} + \mathbb{K}_{s_1+1}^1 + \cdots + \mathbb{K}_{r_1}^1 + \mathbb{H}_{r_1} + \cdots + \mathbb{H}_{s_2} \vdash_{\leq n} q_1$.

Because of (1) and (2) using Lemma 4.12 we can find $\mathbb{G}_{s_1+1}, \ldots, \mathbb{G}_{r_1}$ such that

4. $\mathbb{K}_1^2 + \cdots + \mathbb{K}_{s_1}^2 + \mathbb{K}_{s_1+1}^1 + \cdots + \mathbb{K}_i^1 \vdash_{\leq n} \mathbb{G}_i$, for $s_1 + 1 \leq i \leq r_2$.
5. $\mathbb{K}_1^2 + \cdots + \mathbb{K}_{s_1}^2 + \mathbb{G}_{s_1+1} + \cdots + \mathbb{G}_{r_1} + \mathbb{K}_{r_1+1}^2 + \cdots + \mathbb{K}_i^2 \vdash_{\leq n} \mathbb{H}_i$, for $r_1 + 1 \leq i \leq s_2$.

Use the lemma again for the case (4) and get $\mathbb{F}_1, \ldots, \mathbb{F}_{s_1}$ such that

6. $\mathbb{K}_1^2 + \cdots + \mathbb{K}_i^2 \vdash_{\leq n} \mathbb{F}_i$, $1 \leq i \leq s_1$.
7. $\mathbb{F}_1 + \cdots + \mathbb{F}_{s_1} + \mathbb{K}_{s_1+1}^1 + \cdots + \mathbb{K}_i^1 \vdash_{\leq n} \mathbb{G}_i$ for $s_1 + 1 \leq i \leq r_2$.

Our chain interpolants $\mathbb{H}_i^*$, $1 \leq i \leq s_2$ are therefore as follows

$$\mathbb{H}_i^* = \mathbb{H}_i \cup \mathbb{F}_i, \text{ for } 1 \leq i \leq s_1,$$
$$\mathbb{H}_i^* = \mathbb{G}_i, \text{ for } s_1 + 1 \leq i \leq r_1,$$
$$\mathbb{H}_i^* = \mathbb{H}_i, \text{ for } r_1 + 1 \leq i \leq s_2.$$

The $\mathbb{H}^*$ are the chain interpolants for our case.

CASE ARBITRARY NUMBER OF JUNCTIONS.
Assume we have the situation at (0) below and that we have at our disposal the general interpolation theorem for **K**, (item 2 of Corollary 4.6) and chain interpolation for the inductive case of less junctions. We are seeking chain interpolants for the situation described in (0) below:

(0) $\mathbb{K}_1^2 + \cdots + \mathbb{K}_{s_1}^2 + \mathbb{K}_{s_1+1}^1 + \cdots + \mathbb{K}_{r_1}^1 + \cdots + \mathbb{K}_{r_{t-1}+1}^2 + \cdots + \mathbb{K}_{s_t}^2 + \mathbb{K}_{s_t+1}^1 + \cdots + \mathbb{K}_{r_t}^1 \vdash_{\leq n} q_1$.

We use interpolation, and get $\mathbb{H}_{r_{m-1}+1}, \ldots, \mathbb{H}_{s_m}$ for $1 \leq m \leq t$, (let $r_0 = 0$), such that (1) and (2) hold:

[12]This case can be done as a particular case of the induction step. However, to explain to the reader what is going on, we are doing it explicitly.

(1) $\mathbb{K}_1^2 + \cdots + \mathbb{K}_{r_{m-1}}^1 + \mathbb{K}_{r_{m-1}+1}^2 + \cdots + \mathbb{K}_j^2 \vdash_{\leq n} \mathbb{H}_j$
for all $r_{m-1} + 1 \leq j \leq s_m$, $1 \leq m \leq t$.

(2) $\sum_{1 \leq m \leq t}[\mathbb{K}_1^1 + \cdots + \mathbb{K}_{r_m}^1 + \mathbb{H}_{r_m+1} + \cdots + \mathbb{H}_{s_m} + \mathbb{K}_{s_m+1}^1 + \cdots + \mathbb{K}_{r_{m+1}}^1] \vdash_{\leq n} q_1$,

Consider (1). For each m and each j, $r_{m-1} + 1 \leq j \leq s_m$, we can view $\mathbb{H}_j$ as in language 2. This means we have an interpolation problem for less changes in languages at the junction $\mathbb{K}_{r_{m-1}}^1 + \mathbb{K}_{r_{m-1}+1}^2$.

Therefore for each $r_{m-1} + 1 \leq j \leq s_m$ and each $1 \leq m \leq t$ and for each $1 \leq i \leq r_{m-1}$ there exists $\mathbb{H}_i^j$ such that the statement of the theorem holds by induction.

Let

$$\mathbb{H}_i^* = \bigcup_{\substack{r_{m-1}+1 \leq j \leq s_m \\ 1 \leq m \leq t}} \mathbb{H}_i^j \cup \mathbb{H}_i.$$

Then the $\mathbb{H}_i^*$, $1 \leq i \leq s_t$ are our chain interpolants. ⊣

We now give further refinements of the interpolation theorem for **K**. The traditional formulation has the form:

Given $\mathbb{A} \vdash \mathbb{B}$, find an interpolant $\mathbb{H}$ such that $\mathbb{A} \vdash \mathbb{H} \vdash \mathbb{B}$ (see Corollary 4.6).

We would now like to consider a mixed case of the form:
given $\mathbb{A} \cup \mathbb{B}' \vdash \mathbb{B}$ find an interpolant $\mathbb{H}$ such that $\mathbb{A} \vdash \mathbb{H}$ and $\mathbb{H} \cup \mathbb{B}' \vdash \mathbb{B}$.

Note that $\mathbb{A} \cup \mathbb{B}'$ is not the same as $\mathbb{A} + \mathbb{B}'$. Also note that if we had classical implication (say $\supset$) in our language, we could have presented $\mathbb{A} \cup \mathbb{B}' \vdash \mathbb{B}$ as $\mathbb{A} \vdash \mathbb{B}' \supset \mathbb{B}$, where

$$\mathbb{B}' \supset \mathbb{B} = \{\mathbb{B}' \supset B \mid B \in \mathbb{B}\}.$$

The next theorem yields interpolation for this new case. We need some lemmas first.

LEMMA 4.14. *In modal* **K** *we have that* (1) *is equivalent to* (2)

1. $\mathbb{A} \cup \mathbb{B} \vdash \mathbb{C}$.
2. $\Box\mathbb{A} + \mathbb{B} \vdash \mathbb{C}$.

where $\Box\mathbb{A} = \{\top \to A \mid A \in \mathbb{A}\}$.

PROOF. Let $E \in \mathbb{C}$. We show that $\mathbb{A} \cup \mathbb{B} \nvdash E$ iff $\Box\mathbb{A} + \mathbb{B} \nvdash E$. The best way to show this is semantically.

CASE 1. Assume for some model and possible worlds t_1 and t_2 we have $t_1 R t_2$, $t_1 \models \Box\mathbb{A}$ and $t_2 \models \mathbb{B}$ and $t_2 \not\models E$. Then certainly $t_2 \models \mathbb{A}$ and hence $\mathbb{A} \cup \mathbb{B} \nvdash E$.

CASE 2. Assume for some model and some t, we have $t \models \mathbb{A} \cup \mathbb{B}$ and $t \not\models E$. Create a new model by adding a new point s and let sRt hold. Then, since s is connected only to t we have $s \models \Box\mathbb{A}$. This shows that $\Box\mathbb{A} + \mathbb{B} \nvdash E$.

This concludes our proof. ⊣

THEOREM 4.15. *Assume* $\mathbb{A} \cup \mathbb{B}' \vdash \mathbb{B}$. *Then for some* $\mathbb{H}$ *in the common language we have* $\mathbb{A} \cup \mathbb{B}' \vdash \mathbb{H}$ *and* $\mathbb{H} \cup \mathbb{B}' \vdash \mathbb{B}$.

PROOF. Consider $\Box\mathbb{B}' + \mathbb{A} \vdash \mathbb{B}$. By Theorem 4.5 there exists an $\mathbb{H}$ in the common language such that $\Box\mathbb{B}' + \mathbb{A} \vdash \mathbb{H}$ and $\Box\mathbb{B}' + \mathbb{H} \vdash \mathbb{B}$. Therefore again by our lemma we have $\mathbb{A} \cup \mathbb{B}' \vdash \mathbb{H}$ and $\mathbb{B}' \cup \mathbb{H} \vdash \mathbb{B}$. ⊣

§5. Concluding discussion, chain interpolation.

5.1. Structural Interpolation. Traditional available interpolation theorems, for classical, modal or intuitionistic logics have a very restricted form. We are given $\mathbb{A} \vdash \mathbb{B}$ where $\mathbb{A}$ is in language 1 and $\mathbb{B}$ is in language 2 and we are seeking an interpolant $\mathbb{H}$ in the common language such that

$$\mathbb{A} \vdash \mathbb{H} \vdash \mathbb{B}.$$

Non-classical logics allow for databases with a more complex structure than that of sets, and define consequence relations between such databases and single formulas. Consequently we have more scope for defining more structured versions of interpolation. We have seen one such version for modal **K**. The data-structure was a list, say

$$\Delta = (\mathbb{A}_1, \mathbb{A}_2, \mathbb{B}_1, \mathbb{B}_2, \mathbb{A}_3, \mathbb{A}_4) \vdash a$$

(as in Example 4.11) and the interpolation theorem seeks to replace the $\mathbb{B}$s with interpolants.

Let us formulate this example in a more general way. Consider a place-holder matrix $\mathcal{M}(x_1, x_2, x_3, y_1, y_2)$ of the form $(x_1, x_2, y_1, y_2, y_3)$ such that our theory Δ is obtained by substituting language 1 sets of wffs for x_i and language 2 sets of wffs for y_j. We have

$$\Delta = \mathcal{M}(x_i/\mathbb{A}_i, y_j/\mathbb{B}_j).$$

Assume $\Delta \vdash q_1$, q_1 in language 1. We want to interpolate the y_j substitutions of language 2 by a common language set of wffs $\mathbb{H}_j$. Written schematically for arbitrary logics, we ask the following:

Schematic structural interpolation problem.
Assume we are given a logic $\vdash$ and an acceptable structure matrix $\mathcal{M}(x_i, y_j)$ for the logic. That is to say that a consequence relation $\vdash$ is defined between structures $\Delta = \mathcal{M}(x_i/\mathbb{A}_i, y_j/\mathbb{B}_j)$ and wffs A. Further assume that the occurrence of each y_j in $\mathcal{M}$ is either clear positive or clear negative (i.e., $\mathcal{M}(-Y-) \vdash \mathcal{M}(-Y'-)$ if $Y \vdash Y'$ for positive (resp. $Y' \vdash Y$ for negative)). Let $\mathbb{A}_i$, $\mathbb{B}_j$ be sets of wffs in the respective languages and q an atom in the $\mathbb{A}$ language. Assume that $\Delta = \mathcal{M}(x_i/\mathbb{A}_i, y_j/\mathbb{B}_j) \vdash q$. Find theories $\mathbb{H}_j$ in the common language such that

1. $\mathcal{M}(x_i/\mathbb{A}_i, y_j/\mathbb{H}_j) \vdash q$.
2. Some connection between $\mathcal{M}(x_i/\mathbb{A}_i, y_j/\mathbb{B}_j)$ and $\mathcal{M}(x_i/\mathbb{A}_i, y_j/\mathbb{H}_j)$ is stipulated or a more strong connection between $\mathbb{B}_j$ and $\mathbb{H}_j$?

For example

(a) $\mathcal{M}(x_i/\mathbb{A}_i, y_j/\mathbb{D}_j) \vdash \mathcal{M}(x_i/\mathbb{A}_i, y_j/\mathbb{H}_j)$
or possibly

(b) $\mathbb{B}_j \vdash \mathbb{H}_j$ if y_j occurs positive in $\mathcal{M}$ and $\mathbb{H}_j \vdash \mathbb{B}_j$ if y_j occurs negatively in $\mathcal{M}$.

Going back to the example of **K**, the interpolation theorem of item 2 of Corollary 4.6 falls under option (2a). This is because we have $X_1 + \cdots + X_n \vdash_{\mathbf{K}} Y_1 + \cdots + Y_n$ iff for all $1 \leq i \leq n$, $X_1 + \cdots + X_i \vdash_{\mathbf{K}} Y_i$.

Option (2b) is false for **K**, as Example 4.11 shows, and instead we have the chain interpolation of Theorem 4.13.

It looks like that a good definition of interpolation condition (2) for the general case should be some form of chain interpolations.

Before we go into that, let us remark that this form of structural interpolation is a generalisation of the ordinary traditional interpolation.[13] Consider the case of modal **K**, Example 4.11. We have

$$\mathbb{A}_1, \mathbb{A}_2, \mathbb{B}_1, \mathbb{B}_2, \mathbb{A}_3, \mathbb{A}_4 \vdash_{\mathbf{K}} a.$$

This means, if we use $\Box$ and $\Diamond$ and $\neg$ and ignore for the moment that we are dealing with the strict implication fragment, that

$$\mathbb{A}_1 \wedge \Diamond(\mathbb{A}_2 \wedge \Diamond(\mathbb{B}_1 \wedge \Diamond(\mathbb{B}_2 \wedge \Diamond(\mathbb{A}_3 \wedge \Diamond(\mathbb{A}_4 \wedge \neg a))))) \vdash \bot.$$

Consider the matrix

$$\mathcal{M}(x_i, y_j) = x_1 \wedge \Diamond(x_2 \wedge \Diamond(y_1 \wedge \Diamond(y_2 \wedge \Diamond(x_3 \wedge \Diamond x_4)))).$$

Then we have

$$\mathcal{M}(\mathbb{A}_1, \mathbb{A}_2, \mathbb{A}_3, \mathbb{A}_4 \wedge \neg a, \mathbb{B}_1, \mathbb{B}_2) \vdash \bot$$

and we *can* interpolate on the $\mathbb{B}$s.

The traditional interpolation theorem for modal **K** is for $\mathbb{A} \vdash \mathbb{B}$. This version has been known for years [9], and probably does not give the above structural interpolation. We shall address this question in Part 2 of this series of papers. Meanwhile, consider it open.

5.2. Chain interpolation. Let us examine more closely the chain interpolation theorem for modal **K**. Imagine that we have, say:

- $\mathbb{A}_1 + \mathbb{B}_1 + \mathbb{A}_2 + \mathbb{B}_2 + \mathbb{A}_3 \vdash q^1$.

The chain interpolation theorem promises $\mathbb{H}_1, \mathbb{G}_1, \mathbb{H}_2, \mathbb{G}_2, \mathbb{H}_3$ such that

1. $\mathbb{A}_1 \vdash \mathbb{H}_1$,
2. $\mathbb{H}_1 + \mathbb{B}_1 \vdash \mathbb{G}_1$,
3. $\mathbb{A}_1 + \mathbb{G}_1 + \mathbb{A}_2 \vdash \mathbb{H}_2$,

[13] This is not a proven claim. We need to show a logic that has traditional interpolation for wffs $A \vdash B$ but not structural interpolation or at least that in **K** the structural interpolation does not follow from ordinary interpolation.

4. $\mathbb{H}_1 + \mathbb{B}_1 + \mathbb{H}_2 + \mathbb{B}_2 \vdash \mathbb{G}_2$

and finally

5. $\mathbb{A}_1 + \mathbb{G}_1 + \mathbb{A}_2 + \mathbb{G}_2 + \mathbb{A}_3 \vdash q^1$.

$\mathbb{G}_1$ and $\mathbb{G}_2$ are the interpolants and they are found through chain interpolation, each element in the chain is obtained by using *one* language only.

The above chain theorem is not in a general form. If we think of

$$\mathcal{M}(x_1, y_1, x_2, y_2, x_3) = x_1 + y_1 + x_2 + y_2 + x_3$$

as a data-structure, we are relying in our formulation of the chain interpolation theorem on the fact that in the case of **K**, $\mathcal{M}$ can be *approximated* by the partial increasing structures:

$$\begin{array}{c} x_1 \\ x_1 + y_1 \\ x_1 + y_1 + x_2 \\ x_1 + y_1 + x_2 + y_2 \\ x_1 + y_1 + x_2 + y_2 + x_3 \end{array}$$

This may seem a strong assumption on the logic involved. However, our guess is that we need such an assumption anyway in order to formulate good cut elimination theorems for any sequents type proof theory for the logics.

Furthermore it seems reasonable to expect to effectively know exactly how the databases of the logic are generated!

Let us see if we can formulate a general chain interpolation concept for an arbitrary logic, without relying on any properties of its data-structures. Let $\mathcal{M}(x_i, y_j)$ be a matrix data-structure and assume that every place z in the matrix is either positive or negative, namely we have either:

$$\frac{z \vdash z'}{\mathcal{M}(-z-) \vdash \mathcal{M}(-z'-)}$$

or

$$\frac{z \vdash z'}{\mathcal{M}(-z'-) \vdash \mathcal{M}(-z-)}$$

Assume $\mathcal{M}(x_i/\mathbb{A}_i, y_j/\mathbb{B}_j) \vdash q^1$, where q^1 is in the $\mathbb{A}$ language. We want to formulate a chain interpolation concept for language $\mathbb{B}$.

Ideally we would like interpolants as in option (2b), mentioned in Subsection 5.1. So let us try the approximate substitutions $y_j/\top$ for positive y and $y_j/\bot$ for negative y. Call this substitution $\mathbb{G}^0_j$. If $\mathcal{M}(x_i/\mathbb{A}_i, y_j/\mathbb{G}^0_j) \vdash q^1$, then we found our interpolants. Otherwise we seek $\mathbb{H}^0_i$, $\mathbb{G}^1_j$, such that the following chain step 1 holds:

(cs1) • $\mathcal{M}(x_i/\mathbb{A}_i, y_j/\mathbb{G}^0_j) \vdash \mathcal{M}(x_i/\mathbb{H}^0_i, y_j/\mathbb{G}^0_j)$,
• $\mathcal{M}(x_i/\mathbb{H}^0_i, y_j/\mathbb{B}_j) \vdash \mathcal{M}(x_i/\mathbb{H}^0_i, y_j/\mathbb{G}^1_j)$.

We now hope that

$$\mathcal{M}(x_i/\mathbb{A}_i, y_j/\mathbb{G}_j^1) \vdash q^1.$$

If yes, we found our chain interpolant. Otherwise we carry on:

The following is chain interpolation step $k+1$:

We find $\mathbb{H}_i^k$, $\mathbb{G}_j^{k+1}$ such that

(cs(k+1))
- $\mathcal{M}(x_i/\mathbb{A}_i, y_j/\mathbb{G}_j^k) \vdash \mathcal{M}(x_i/\mathbb{H}_i^k, y_j/\mathbb{G}_j^k)$
- $\mathcal{M}(x_i/\mathbb{H}_i^k, y_j/\mathbb{B}_j) \vdash \mathcal{M}(x_i/\mathbb{H}_i^k, y_j/\mathbb{G}_j^{k+1})$.

We hope that $\mathbb{G}_j^{k+1}$ are our interpolants, i.e., we have:

$$\mathcal{M}(x/\mathbb{A}_i, y_j/\mathbb{G}_j^{k+1}) \vdash q^1.$$

These interpolants are said to be obtained by a chain of length k.

A logic $\vdash$ has chain interpolation if there is always a k-chain interpolant for some k.

Let us now see whether the chain interpolation for **K** can be interpreted in this way.

We reconsider the case of $\mathbb{A}_1 + \mathbb{B}_1 + \mathbb{A}_2 + \mathbb{B}_2 + \mathbb{A}_3 \vdash q^1$.

Here $\mathcal{M} = x_1 + y_1 + x_2 + y_2 + x_3$.

All variables are positive. Let $\mathbb{G}_1^0 = \mathbb{G}_2^0 = \top$. Then

$$\begin{aligned}
&\mathbb{H}_1^0 = \mathbb{H}_1, \quad \mathbb{H}_2^0 = \mathbb{H}_3^0 = \top,\\
&\mathbb{H}_1^1 = \mathbb{G}_1, \quad \mathbb{G}_2^1 = \top,\\
&\mathbb{H}_1^1 = \mathbb{H}_1, \quad \mathbb{H}_2^1 = \mathbb{H}_2, \quad \mathbb{H}_3^1 = \top,\\
&\mathbb{G}_1^2 = \mathbb{G}_1, \quad \mathbb{G}_2^2 = \mathbb{G}_2.
\end{aligned}$$

these are our interpolants.

5.3. Beth definability. Interpolation can give a more refined version of Beth definability. We show two examples, intuitionistic $\to$ and strict $\to$. Let us first formulate a version for intuitionistic implications:

Assume we have:

(d) $$\varphi(q_1, \ldots, q_n, x) \wedge \varphi(q_1, \ldots, q_n, y) \vdash x \to y$$

for x, y atoms pairwise different, and different from q_i.

This condition means that $\varphi(q_1, \ldots, q_n, x)$ defines x implicitly, (i.e., $\varphi(q_1, \ldots, q_n, x) \wedge \varphi(q_1, \ldots, q_n, y)$ proves $x \leftrightarrow y$).

We can get, using interpolation, that x can be defined explicitly in terms of $q_1, \ldots, q_n$ in the same fragment as φ. Here is the proof.

Condition (d) implies

(*) $$\varphi(q_1, \ldots, q_n, x) \wedge x \vdash \varphi(q_1, \ldots, q_n, y) \to y$$

hence by interpolation there exists a $\alpha(q_1, \ldots, q_n)$ such that

$$\varphi(x) \wedge x \vdash \alpha \vdash \varphi(y) \to y$$

i.e., $\varphi(x) \vdash x \to \alpha$ and $\varphi(y) \vdash \alpha \to y$.

Since x, y are not in α and differnt from $q_1, \ldots, q_n$ we get that for arbitary z

(dd) $$\varphi(q_1, \ldots, q_n, z) \vdash z \leftrightarrow \alpha.$$

For the case of **K** strict implication we have to be more careful in our formulation of the definability theorem. Consider

(kd) $$\{\varphi(q_1, \ldots, q_n, x), \varphi(q_1, \ldots, q_n, y), x\} \vdash y.$$

(kd) says that at any possible world, $\varphi(q_1, \ldots, q_n, x)$ defines x uniquely.

We are seeking an explicit definition; we use Theorem 4.15.

For some $\mathbb{H}$ in the common language (i.e., using $q_1, \ldots, q_n$ only) we have

1. $\{\varphi(q_1, \ldots, q_n, x), x\} \vdash \mathbb{H}$

and

2. $\{\varphi(q_1, \ldots, q_n, y)\} \cup \mathbb{H} \vdash y$.

To see what this means, allow for a moment material implication $\supset$ and conjunctions and let $H = \bigwedge \mathbb{H}$. (1) and (2) become

1. $\varphi(q_1, \ldots, q_n, x) \vdash x \supset H$,
2. $\varphi(q_1, \ldots, q_n, y) \vdash H \supset y$.

This means that ther exists an explicit definition H being a conjunction of strict implicational sentences.

Obviously one needs to systematically go through our logics and see what fine tuning we can get for the definability theorem. This we shall do in our next paper.

5.4. Standard interpolation in classical logic. Let us now turn to classical logic.

Can classical logic give structural interpolation? Classical logic has interpolation and has disjunctive/conjunctive normal forms and so any formula embedded inside a structure can be brought to the surface? Let us see.

EXAMPLE 5.1. *Let a be in language 1, b in language 2 and q_1, q_2, q be in the common language. Consider the following database Δ*

$$(a \wedge b \to q), (a \to q_1) \to a, (b \to q_2) \to b$$

we have $\Delta \vdash q$.

We want to interpolate on b in classical logic.

We have

$$\vdash (a \wedge b \to q) \wedge ((a \to q_1) \to a) \wedge ((b \to q_2) \to b) \to q.$$

Since $(x \wedge y \to q) \equiv (x \to q) \vee (y \to q)$ we get that the above is equivalent to the two statements

$$(a \to q) \wedge ((a \to q_1) \to a) \vdash ((b \to q_2) \to b) \to q,$$
$$(b \to q) \wedge ((b \to q_2) \to b) \vdash ((a \to q_1) \to a) \to q.$$

Thus the structural interpolation problem split into two traditional problems. We got two interpolants in the general case, h_1 and h_2.

QUESTION 1. *How do we find (in the general case) a structural interpolant for eliminating b in Δ?*

In this particular case we have $h_1 = h_2 = q$ and q is also the structural interpolant for b in Δ.

QUESTION 2. *Can the interpolants h_1, h_2 give us some sort of a chain interpolation for Δ?*

5.5. Concluding remarks. In this paper we have shown how to get interpolation properties for several non-classical logics in the realm of substructural logics and modal logics by making use of a goal-directed formulation. Interpolants are found by metalevel induction on the goal directed proof algorithm, as to each goal-directed derivation corresponds one interpolant (non necessarily distinct). We have obtained these results for the implication or (implication/conjunction) fragment of each logic. Moreover from the goal-directed computation one can also extract a bound on the derivation depth of the interpolant.

The investigation has lead us to consider more general notions of interpolation, namely the notion of *chain interpolation* and the notion of *structural interpolation*, whose relation to the ordinary notion of interpolation in each specific case remains to be investigated. These notions arise from the consequence relation associated to proof configurations or structured sequents specific for each logic.

In subsequent research we shall investigate how to extend these results along several directions. On the one hand, we shall investigate how to extend these results to broader fragments. On the other hand we shall study how to extend them to other logical systems. Finally we shall consider how to refine these results by paying attention to the internal structure of interpolants, as the number and position of every variable occurrence.

Acknowledgements. We are grateful to the referee for his meticulously careful reading of the paper, his helpful suggestions, and constructive criticisms.

REFERENCES

[1] C. ARECES, P. BLACKBURN, and M. MARX, *Repairing the interpolation theorem in quantified modal logic*, 2003, pp. 287–299.

[2] J. BICARREGUI, D. M. GABBAY, T. DIMITRAKOS, and T. MAIBAUM, *Interpolation in practical formal development*, ***Logic Journal of the IGPL***, vol. 9 (2001), pp. 231–244.

[3] W. CRAIG, *Linear reasoning. A new form of the Herbrand–Gentzen theorem and three uses of the Herbrand–Gentzen theorem*, ***The Journal of Symbolic Logic***, vol. 22 (1957), pp. 250–268 and 269–285.

[4] K. FINE, *Failure of the interpolation lemma in quantified modal logic*, ***The Journal of Symbolic Logic***, vol. 44 (1979), pp. 201–206.

[5] M. Fitting, ***Proof methods for modal and intuitionistic logic***, D. Reidel, 1983.

[6] ———, *Interpolation theorem for first order* S5, ***The Journal of Symbolic Logic***, vol. 67 (2002), pp. 621–634.

[7] D. M. Gabbay, *Semantic proof of the Craig interpolation theorem for intuitionistic logic and extensions, part I*, ***Proceedings of the 1969 Logic Colloquium in Manchester***, North-Holland, 1969, pp. 391–401.

[8] ———, *Semantic proof of the Craig interpolation theorem for intuitionistic logic and extensions, part II*, ***Proceedings of the 1969 Logic Colloquium in Manchester***, North-Holland, 1969, pp. 403–410.

[9] ———, *Craig's interpolation theorem for modal logics*, ***Proceedings of Logic Conference*** (W. Hodges, editor), Springer, 1970, pp. 111–128.

[10] ———, ***Investigations in modal and tense logics with applications***, D. Reidel, 1976.

[11] ———, *Craig's theorem for intuitionistic logic, III*, ***The Journal of Symbolic Logic***, vol. 42 (1977), pp. 269–271.

[12] ———, ***Labelled deductive systems***, Oxford University Press, 1996.

[13] ———, ***Fibring logics***, Oxford University Press, 1998.

[14] D. M. Gabbay and L. Maksimova, ***Treatise on interpolation and definability***, Oxford University Press, Forthcoming.

[15] D. M. Gabbay and H. J. Ohlbach, *Quantifier elimination in second order predicate logic*, ***Proceedings of KR '92*** (B. Nebel, C. Rich, and W. Swartout, editors), 1992, pp. 425–435.

[16] D. M. Gabbay and N. Olivetti, ***Goal directed algorithmic proof***, Kluwer, 2000.

[17] D. M. Gabbay, S. Schlobach, and H. J. Ohlbach, *A note on interpolation by translation*, In preparation.

[18] E. G. K. Lopez-Escobar, *On the interpolation theorem for the logic of constant domains*, ***The Journal of Symbolic Logic***, vol. 46 (1981), pp. 87–88.

[19] R. Lyndon, *An interpolation theorem in the predicate calculus*, ***Pacific Journal of Mathematics***, vol. 9 (1959), pp. 155–164.

[20] L. Maksimova, *Craig's theorem in superintuitionistic logic and amalgamable varieties of pseudo Boolean algebras*, ***Algebra i Logika***, vol. 16 (1977), pp. 643–681.

[21] ———, *Interpolation properties of superintuitionistic logic*, ***Studia Logica***, vol. 38 (1979), pp. 419–428.

[22] ———, *Interpolation theorem in modal logic and amalgamated varieties of topoBoolean algebras*, ***Algebra i Logika***, vol. 18 (1979), pp. 556–586.

[23] ———, *Failure of interpolation property in modal companions of Dummett's logic*, ***Algebra i Logika***, vol. 21 (1982), pp. 690–694.

[24] M. Marx and C. Areces, *Failure of interpolation in combined modal logics*, ***Notre Dame Journal of Formal Logic***, vol. 39 (1998), pp. 253–273.

[25] A. Nonnengart, H. J. Ohlbach, and A. Szalas, *Elimination of predicate quantifiers*, ***Logic, language and reasoning, Essays in honour of D. M. Gabbay*** (H. J. Ohlbach and U. Reyle, editors), Kluwer, 1999, pp. 149–172.

[26] P. W. O'Hearn and D. Pym, *The logic of bunched implications*, ***The Bulletin of Symbolic Logic***, vol. 5 (1999), pp. 215–244.

[27] M. Otto, *An interpolation theorem*, ***The Bulletin of Symbolic Logic***, vol. 6 (2000), pp. 447–462.

[28] M. Pentus, *Lambek calculus and formal grammars*, ***Provability, complexity, grammar***, American Mathematical Society Translations–Series 2, 1999, pp. 57–86.

[29] D. Roorda, *Interpolation in fragments of classical linear logic*, ***The Journal of Symbolic Logic***, vol. 59 (1994), pp. 419–444.

GROUP OF LOGIC, LANGUAGE AND COMPUTATION
DEPARTMENT OF COMPUTER SCIENCE
KING'S COLLEGE LONDON
STRAND, LONDON WC2R 2LS, UK
E-mail: dg@dcs.kcl.ac.uk

DIPARTIMENTO DI INFORMATICA
UNIVERSITA' DEGLI STUDI DI TORINO
CORSO SVIZZERA, 185
I-10149 TORINO, ITALY
E-mail: olivetti@di.unito.it

SEQUENCES OF DEGREES ASSOCIATED WITH MODELS OF ARITHMETIC

JULIA F. KNIGHT

The main result of the present paper is a characterization of the sequences of Turing degrees of "n-diagrams" for non-standard models of PA. We consider only models $\mathcal{A}$ whose universe is a subset of ω, and we think of the elements as constants. The *complete diagram of* $\mathcal{A}$, denoted by $D^c(\mathcal{A})$, is the set of all sentences true in the expanded structure $(\mathcal{A}, (a)_{a\in\mathcal{A}})$. A Σ_n (or Π_n) *formula* is a formula in prenex normal form, with n blocks of like quantifiers, starting with $\exists$ (or $\forall$). A B_n *formula* is a Boolean combination of Σ_n formulas. The *n-diagram* of $\mathcal{A}$ is $D_n(\mathcal{A}) = D^c(\mathcal{A}) \cap \Sigma_n$. Note that $D_n(\mathcal{A}) \equiv_T D^c(\mathcal{A}) \cap B_n$.

It is easy to establish some necessary conditions on the sequence of degrees of n-diagrams of a non-standard model of PA. First, if $\mathcal{A}$ is a non-standard model of PA, then it follows from a result of Scott [10] that there is a completion of PA computable in $D_0(\mathcal{A})$. The set of degrees of completions of PA is closed upwards. Therefore, $D_0(\mathcal{A})$ is Turing equivalent to a completion of PA. Second, $D_n(\mathcal{A})$ is computable in $D_{n+1}(\mathcal{A})$, and $D_{n+1}(\mathcal{A})$ is c.e. in $D_n(\mathcal{A})$, uniformly in n. These two necessary conditions turn out to be sufficient. The following definitions simplify the statement of the main theorem.

DEFINITION. An *ω-table* is a sequence of sets $(C_n)_{n\in\omega}$ such that C_{n+1} is c.e. in and above C_n, uniformly in n. The ω-table is said to be *over* the set X if $C_0 = X$.

THEOREM 0.1 (Main Theorem). *For a sequence of Turing degrees $(\boldsymbol{d}_n)_{n\in\omega}$, the following are equivalent:*

1. *there is a non-standard model $\mathcal{A}$ of PA such that for all n, $deg(D_n(\mathcal{A})) = \boldsymbol{d}_n$,*
2. *there is an ω-table $(C_n)_{n\in\omega}$, over a completion of PA, such that for all n, $deg(C_n) = \boldsymbol{d}_n$.*

We have already indicated how $1 \Rightarrow 2$. The proof that $2 \Rightarrow 1$ involves an independence result from [7], together with an infinitely nested priority construction. The construction has some interesting features. While we may assign initial priorities according to some computable list of requirements, higher priority requirements at lower levels (associated with C_n for smaller n)

Logic Colloquium '01
Edited by M. Baaz, S. Friedman, and J. Krajíček
Lecture Notes in Logic, 20

must be allowed to crowd in front of high priority requirements at higher levels.

The result below is obviously related to the main theorem, but the proof is simpler—the priority construction is only finite-injury.

THEOREM 0.2 (Warm-up). *Suppose C_0 is a completion of PA, and C_1 is c.e. in and above C_0. Then there is a non-standard model $\mathcal{A}$ of PA such that $D_0(\mathcal{A}) \equiv_T C_0$ and $D_1(\mathcal{A}) \equiv_T D^c(\mathcal{A}) \equiv_T C_1$.*

In Section 1, we give some background on Scott sets. In Section 2, we describe the independence result from [7]. In Section 3, we prove Theorem 0.2. In Sections 4 and 5, we describe machinery for infinitely nested priority constructions with special features for coding and enumeration. In Section 6, we use this machinery to prove Theorem 0.1.

REMARK. Throughout, we identify objects such as finite sequences and sentences with their Gödel numbers.

§1. Scott sets. Scott [10] investigated the families of sets naturally coded in completions of *PA* and in non-standard models of *PA*. In this section, we recall some facts about these families. For a more comprehensive discussion of the algorithmic properties of completions of *PA* and models of *PA*, gathering together material from [10] and other sources, see [8], or Chapter 19 of [5].

DEFINITION. A *Scott set* is a non-empty set $\mathcal{S} \subseteq P(\omega)$ such that

1. if $A \in \mathcal{S}$ and $B \leq_T A$, then $B \in \mathcal{S}$,
2. if $A, B \in \mathcal{S}$, then $A \oplus B = \{2x\colon x \in A\} \cup \{2x+1\colon x \in B\} \in \mathcal{S}$,
3. if $\mathcal{T}$ is an infinite subtree of $2^{<\omega}$ in $\mathcal{S}$, then $\mathcal{T}$ has a path in $\mathcal{S}$; equivalently, if A is a consistent set of axioms in $\mathcal{S}$, then A has a completion in $\mathcal{S}$.

DEFINITION. If $\mathcal{A}$ is a non-standard model of *PA*, then the *Scott set of $\mathcal{A}$* is $SS(\mathcal{A}) = \{d_a\colon a \in \mathcal{A}\}$, where $d_a = \{n \in \omega\colon \mathcal{A} \models p_n \mid a\}$. Here $p_n \mid x$ is a formula saying that x is divisible by the nth prime.

Note that for each $a \in \mathcal{A}$, $d_a \leq_T D_0(\mathcal{A})$. To decide whether $\mathcal{A} \models p_n \mid a$ (for standard n), we enumerate the atomic diagram, searching for a sentence of the form $p_n b + S^r 0 = a$, for $r = 0, \dots, p_n - 1$.

THEOREM 1.1 (Scott). *If $\mathcal{A}$ is a non-standard model of PA, then $SS(\mathcal{A})$ is a Scott set.*

By Property 3, every Scott set contains a completion of *PA*, so we obtain the following statement from the introduction.

COROLLARY 1.2. *If $\mathcal{A}$ is a non-standard model of PA, then there is a completion T of PA such that $T \leq_T D_0(\mathcal{A})$.*

Scott [10] considered the families of sets coded in completions of *PA*, as well as those coded in models.

DEFINITION. Let T be a completion of PA. A set $X \subseteq \omega$ is *representable* with respect to T if there is a formula $\varphi(x)$ such that

$$T \vdash \varphi(S^k 0) \text{ for } k \in X,$$
$$T \vdash \neg\varphi(S^k 0) \text{ for } k \notin X.$$

The family of sets representable with respect to T is denoted by $Rep(T)$.

THEOREM 1.3 (Scott). *For a family $\mathcal{S} \subseteq P(\omega)$, the following are equivalent:*

1. *there is a completion T of PA such that $Rep(T) = \mathcal{S}$,*
2. *$\mathcal{S}$ is a countable Scott set.*

DEFINITION. Let $\mathcal{S} \subseteq P(\omega)$ be countable. An *enumeration* of $\mathcal{S}$ is a relation $R \subseteq \omega \times \omega$ such that $\mathcal{S} = \{R_n : n \in \omega\}$, where $R_n = \{x : (n, x) \in R\}$. If $R_n = X$, then n is called an *R-index* for X.

There are natural enumerations of the Scott sets $SS(\mathcal{A})$ and $Rep(T)$. If $\mathcal{A}$ is a non-standard model of PA with universe ω, then the *canonical enumeration* of $SS(\mathcal{A})$ is $R = \{(a, k) : k \in d_a\}$. If T is a completion of PA, then the *canonical enumeration* of $Rep(T)$ is $R = \{(n, k) : T \vdash \varphi_n(S^k)\}$.

DEFINITION. An *effective enumeration* of a Scott set $\mathcal{S}$ is an enumeration R, equipped with functions f, g, and h, which witness that $\mathcal{S}$ is a Scott set; i.e.,

1. if $R_i = A$ and $\varphi_e^A = \chi_B$, then $R_{f(i,e)} = B$,
2. if $R_i = A$ and $R_j = B$, then $R_{g(i,j)} = A \oplus B$,
3. if R_i is an infinite subtree of $2^{<\omega}$, then $R_{h(i)}$ is a path.

The effective enumeration is said to be *computable relative to X* if R, f, g, and h are all computable relative to X.

It is clear how to equip the canonical enumeration of $Rep(T)$ with functions to produce an effective enumeration of $Rep(T)$ that is computable in T. We may take the functions to be *computable*, not just computable in T.

PROPOSITION 1.4. *If T is a completion of PA, then there is an effective enumeration of a Scott set, computable in T.*

Marker [9] showed that any enumeration of a Scott set can be replaced by an effective one.

THEOREM 1.5 (Marker). *For any enumeration R of a Scott set $\mathcal{S}$, there is an effective enumeration of $\mathcal{S}$ computable in R. Moreover, we may take the functions witnessing effectiveness to be computable.*

§2. **Independence.** The Gödel-Rosser sentence says, "for any proof of me from axioms of PA, there is a smaller proof of my negation (from the same kinds of sentences)". We may take this to be a Π_1 sentence φ, and we may consistently add φ or $\neg\varphi$ to PA. The Gödel-Rosser sentence can be varied in more than one way. First, the axioms of PA may be replaced by a larger consistent set of axioms that is still computable. We can refer to a computable set of axioms using a Σ_1 or Π_1 representing formula. We can also refer to the indefinite set of "true" Σ_{n-1} and Π_{n-1} sentences, using the Σ_{n-1} formula that defines truth for Σ_{n-1} sentences. We arrive at the following—this was the starting point for Scott [10].

THEOREM 2.1. *There is a computable sequence of sentences $(\varphi_n)_{n\geq 1}$ such that for each n, φ_n is Π_n and for any set Γ of B_{n-1} sentences consistent with PA, we may consistently add either φ_n or $\neg\varphi_n$.*

Using Theorem 2.1, we can justify the following statement, from the introduction.

COROLLARY 2.2. *The set of degrees of completions of PA is closed upwards.*

PROOF. Let T be a completion of *PA* and suppose $T <_T X$. We determine another completion $T^* \equiv_T X$. Let $(\varphi_n)_{n\geq 1}$ be the sequence of sentences from Theorem 2.1. There is an effective enumeration R of a Scott set, where R is computable in T, and hence, computable in X. Using X, we determine a sequence of R-indices for the B_n fragments of T^*. We proceed by induction, letting $\varphi_n \in T^* \cap B_n \Leftrightarrow n \in X$.

Arana [1] has developed some further variants of the Gödel-Rosser sentence. The immediate goal was to produce completions of *PA* whose Σ_n part could be properly extended.

THEOREM 2.3 (Arana). *There exists a computable sequence of sentences $(\psi_n)_{1\leq n<\omega}$ such that ψ_n is Π_n and*

1. *for any set Γ of B_{n-1} sentences consistent with PA, we can consistently add ψ_n,*
2. *for any set Λ of Σ_n sentences consistent with PA, we may consistently add $\neg\psi_n$.*

IDEA OF PROOF. The sentence ψ_n says "for any p and u, if p is a proof of me from axioms of *PA*, true Σ_{n-1} and Π_{n-1} sentences, and a true Σ_n sentence with witness u, there exist $q < p$ and $v \leq u$ such that q is a proof of my negation from the same kinds of sentences, with witness v".

In [8], there are variants of Arana's sentences. These were used to produce Σ_n types that can be properly extended. Let $\overline{a}$ be a tuple of constants, and let c be a further constant. If $\Sigma(\overline{a})$ is a set of sentences in the language of arithmetic, possibly with added constants $\overline{a}$, let $Code_{\Sigma(\overline{a})}(c)$ be the set of sentences

$$p_n \mid c \ \text{ for } n \in \Sigma(\overline{a}),$$
$$p_n \nmid c \ \text{ for } n \notin \Sigma(\overline{a}).$$

Note that if $\Sigma(\overline{a})$ is consistent with *PA*, and c is a new constant, not in $\overline{a}$, then $\Sigma(\overline{a}) \cup Code_{\Sigma(\overline{a})}(c)$ is also consistent with *PA*.

THEOREM 2.4. *Let $\overline{a}$ be a tuple of constants, and let c be a further constant. For each $n \geq 1$, we can find a Π_n sentence $\theta_n(\overline{a}, c)$, in constants $\overline{a}, c$, such that for any set $\Sigma(\overline{a})$ of sentences in constants $\overline{a}$,*

1. *for any set $\Gamma(\overline{a}, c)$ of B_{n-1} sentences in constants $\overline{a}, c$, if*
$$PA \cup \Sigma(\overline{a}) \cup Code_{\Sigma(\overline{a})}(c) \cup \Gamma(\overline{a}, c)$$
is consistent, then we can consistently add $\theta_n(\overline{a}, c)$,
2. *for any set $\Lambda(\overline{a}, c)$ of Σ_n sentences in constants $\overline{a}, c$, if*
$$PA \cup \Sigma(\overline{a}) \cup Code_{\Sigma(\overline{a})}(c) \cup \Lambda(\overline{a}, c)$$
is consistent, then we can consistently add $\neg\theta_n(\overline{a}, c)$.

IDEA OF PROOF. The sentence $\theta_n(\overline{a}, c)$ says "for any p and u, if p is a proof of me from sentences coded in c, true sentences of the form $p_n \mid c$ or $p_n \nmid c$, and a further Σ_n sentence true of $\overline{a}, c$, with witness u, then there exist $q < p$ and $v \leq u$ such that q is a proof of my negation (from these same kinds of sentences), with witness v".

§3. The Warm-up Theorem. In this section, we prove Theorem 0.2. Recall the statement.

THEOREM. Let C_0 be a completion of PA, and suppose C_1 is c.e. in and above C_0. Then there is a non-standard model $\mathcal{A}$ of PA such that $D_0(\mathcal{A}) \equiv_T C_0$ and $D_1(\mathcal{A}) \equiv_T D^c(\mathcal{A}) \equiv_T C_1$.

PROOF. Let $\mathcal{S}$ be a Scott set with an effective enumeration R, f, g, h that is computable in C_0. Let T be a completion of PA such that $T \in \mathcal{S}$. Let A be an infinite computable set of constants, for the universe of $\mathcal{A}$. We produce a model of T using a modified Henkin construction. We assign R-indices for complete types to tuples of constants, and if we have made $\exists x\ \psi(x)$ true, we also choose a constant c to satisfy $\psi(x)$.

For convenience, we include in the language of arithmetic a predicate E, for the even numbers. Let $(\psi_n(x))_{n\in\omega}$ be a computable list of all formulas, with just x free, in the language of arithmetic with the constants from A. Let $(\alpha_n)_{n\in\omega}$ be a computable list of all atomic sentences. We may suppose that the constants in α_n and $\psi_n(x)$ are among the first n from A.

REQUIREMENTS. We have requirements of the following forms:

1. determine the complete type of a tuple $\overline{a}$ that includes the first n constants,
2. choose a constant to satisfy $\psi_n(x)$, if this is consistent with the type of the constants in $\psi_n(x)$,
3. code the presence or absence of n in C_1,
4. code the presence or absence of n in C_0.

We satisfy a requirement of type 3, coding $\chi_{C_1}(n)$, by choosing an appropriate Π_1 sentence θ_n, as in Theorem 2.4, and making θ_n true just in case $n \notin C_1$. We satisfy a requirement of type 4, coding $\chi_{C_0}(n)$, by making Ea_{2n} true just in case $n \in C_0$.

We first consider the construction from the top level, where we have the oracle C_1. We imagine an infinite sequence of steps, computable in C_1. At step 0, we assign to $\emptyset$ an R-index for the theory T. Suppose that in steps $0, \ldots, n$, we have determined an R-index for the complete type of a tuple $\overline{a}_n$, so as to satisfy the first n requirements of types 1, 2, and 3. Suppose also that we have satisfied the requirements of type 4 for all "even" constants a_{2k} in the tuple $\overline{a}_n$.

We carry out step $n+1$ as follows. First, we choose a new "odd" constant c_n ($c_n = a_{2k+1}$ for some k) to code the complete type of $\overline{a}_n$, and we add sentences saying that c_n codes the complete type of $\overline{a}_n$. Then we form the Π_1 sentence $\theta_n(\overline{a}_n, c_n)$ as in Theorem 2.4. We add $\theta_n(\overline{a}_n, c_n)$ if $n \notin C_1$, and $\neg\theta_n(\overline{a}_n, c_n)$ otherwise. We choose a constant to witness $\psi_n(x)$, if this is consistent with the type of $\overline{a}_n$, either using one of the constants already mentioned, or taking a new odd one. Finally, we extend the set of sentences to a complete type for a tuple $\overline{a}_{n+1}$ that includes the first $n+1$ constants and all others mentioned so far. There is such a type in S, and we can use the effective enumeration to find an R-index for it.

This completes our description of the construction at the top level. The important thing to note is that given C_0 and $C_1 \cap n$, we can carry out the first n steps, coding $C_1 \cap n$, and assigning a complete type to the first n constants. In addition, we can determine the sentence $\theta_n(\overline{a}_n, c_n)$ used to code $\chi_{C_1}(n)$.

Next, we consider the construction from the bottom level, where we have oracle C_0. At stage s, we have an approximation to the sequence of steps at the top level, based on a stage s approximation of C_1. For the requirements of type 4, when we first mention an even constant a_{2n}, we make sure that Ea_{2n} holds just in case $n \in C_0$. We enumerate the atomic diagram as we go along, including the sentences that involve E.

At stage 0, we let $\overline{a}_0 = \emptyset$. At stage 1, we attempt the first top level step. We choose an odd constant c_0 to code T, and we make $\theta_0(c_0)$ hold iff 0 has not appeared in C_1. We make Ea_0 hold just in case $0 \in C_0$. We extend, choosing a complete type for the constants we have used. If $\psi_0(x)$ is consistent with the chosen type, and we do not have a witness already, then we take as a witness a new odd constant, and extend the type to the full tuple of constants, call it $\overline{a}_1$. We enumerate into the atomic diagram $\pm\alpha_0$, as determined by the type. We also enumerate the formulas $\pm Ea$, for $a \in \overline{a}_1$.

Suppose that at stage s, we have tentatively carried out the first n top level steps, assigning an increasing sequence of types to tuples $\overline{a}_k$ for $k \leq n$, designating elements c_k and coding sentences $\theta_k(\overline{a}_k, c_k)$ (all Π_1). We have

$\theta_k(\overline{a}_k, c_k)$ in the type of $\overline{a}_{k+1}$ iff k has not appeared in C_1 by stage s. We have enumerated into the atomic diagram a finite set of atomic sentences and negations of atomic sentences, including $\pm\alpha_k$ for $k < s$, and $\pm Ea$, for all constants $a \in \overline{a}_s$.

At stage $s+1$, if there is no change in C_1 for $k < n$, then we determine a new c_n and $\theta_n(\overline{a}_n, c_n)$. We choose an appropriate type for a tuple $\overline{a}_{n+1}$ that includes $\overline{a}_n, c_n$, the first $n+1$ constants, and a witness for $\psi_n(x)$, if $\exists x\ \psi_n(x)$ is in the type. If some $k < n$ enters C_1 at stage $s+1$, then we change the type of $\overline{a}_k, c_k$ to include $\neg\theta_k(\overline{a}_k, c_k)$, preserving what we have enumerated into the diagram. We choose a new type for a tuple $\overline{a}_{k+1}$ that includes the old $\overline{a}_n$ and a witness for $\psi_k(x)$, if this is consistent. We drop our work on step m for all $m > k$. In either case, we complete stage $s+1$ by adding sentences to the atomic diagram, so that we have $\pm\alpha_s$, and $\pm Ea$, for each $a \in \overline{a}_{s+1}$, always maintaining consistency with the chosen type.

The bottom level sequence yields, in the limit, a sequence of the kind we want at the top level, coding C_1. At the bottom level, we enumerate the atomic diagram, never removing sentences from it, and we code C_0. The result is a model $\mathcal{A}$ of T such that $D_0(\mathcal{A}) \equiv_T C_0$. We can determine C_1 from $D_1(\mathcal{A})$ and C_0. Of course, $D_0(\mathcal{A}) \leq_T D_1(\mathcal{A})$, so we have $C_1 \leq_T D_1(\mathcal{A})$. Using C_1, we can determine the sequence of steps at the top level, assigning permanent types to tuples. Therefore, $D^c(\mathcal{A}) \leq_T C_1$. Of course, $D_1(\mathcal{A}) \leq_T D^c(\mathcal{A})$, so C_1, $D_1(\mathcal{A})$, and $D^c(\mathcal{A})$ are all Turing equivalent.

§4. Difficulties in proving the Main Theorem. The easier half of Theorem 0.1 is stated below. We have already indicated why it is true.

Proposition 4.1. *If $\mathcal{A}$ is a non-standard model of PA, then there is a completion T of PA, with an ω-table $(C_n)_{n\in\omega}$ over T, such that $D_n(\mathcal{A}) \equiv_T C_n$ for all n.*

To complete the proof of Theorem 0.1, we need the converse.

Theorem 4.2. *If $(C_n)_{n\in\omega}$ is an ω-table over a completion of PA, then there is a non-standard model $\mathcal{A}$ of PA such that for $n \in \omega$, $D_n(\mathcal{A}) \equiv_T C_n$, uniformly in n.*

The proof is similar in outline to that of Theorem 0.2, but it is more complicated because of the added levels. In the remainder of this section, we first describe the requirements in an informal way. We then define a tree and other objects, and we say, in terms of these objects, what is needed to complete the proof of Theorem 4.2. The proof will be completed in Section 7, after we have developed appropriate machinery.

By hypothesis, C_0 is a completion of *PA*. Then there is a Scott set $\mathcal{S}$ with an effective enumeration R, f, g, h that is computable in C_0. Let T be a completion of *PA* in $\mathcal{S}$. Our aim is to produce a non-standard model $\mathcal{A}$ of T such that $D_n(\mathcal{A}) \equiv_T C_n$ for all n. As in the proof of Theorem 0.2, let A be an infinite computable set of constants, for the universe of $\mathcal{A}$. Let $(\psi_n(x))_{n\in\omega}$ be a computable list of formulas, with just x free, in the language of arithmetic with the added constants. We may suppose that the constants in $\psi_n(x)$ are among the first n.

REQUIREMENTS. The requirements have the same forms as in Theorem 0.2.

1. Determine the complete type of a tuple $\overline{a}$ including the first n constants.
2. Choose a constant to satisfy $\psi_n(x)$, if this is consistent with the type of the first n constants.
3. For $n \geq 1$, and $x \in \omega$, code the presence or absence of x in C_n.
4. For $x \in \omega$, code the presence or absence of x in C_0.

The strategy for the requirements of type 4 is as in Theorem 0.2. We let Ea_{2n} hold just in case $n \in C_0$. The strategies for requirements of types 1 and 2 are also the same as before. We assign indices for complete types to tuples of constants. Everything except the requirements of type 3 could be satisfied just using the effective enumeration of the Scott set, which is computable in C_0. The strategy for a requirement of type 3, coding the presence or absence of k in C_n, is to choose an appropriate Π_n sentence $\theta_{(n,k)}$, as a "coding witness", and make $\theta_{(n,k)}$ hold just in case $k \notin C_n$. We must make sure that given C_{n-1} and $C_n \cap k$, we can effectively determine $\theta_{(n,k)}$.

There is a computable list of all requirements. At level ω, where we have oracle $\oplus C_n$, we want a sequence of steps such that at step k, at least the first k requirements of types 1, 2, and 3 have been carried out. At level n, where we have oracle C_{n-1}, we want a sequence of steps that code C_n, eventually correctly. For each k, once we have designated a coding witness $\theta_{(n,k)}$, we make $\theta_{(n,k)}$ true until/unless k appears in C_n, at which point we make it false. We hold to this plan unless some $k' < k$ appears in C_n.

Step k of level $n+1$ uses at most $C_n \cap k$. Assuming that we know the procedure used for computing the step, we can approximate it at level n, using our approximation to $C_n \cap k$. The approximation is based on the part of C_{n-1} previously coded, plus information about the next element—if we have previously coded $C_{n-1} \cap r$, then we use $C_{n-1} \cap (r+1)$. Finally, at level n, we enumerate the B_{n-1} diagram, with no changes allowed. For $n > 1$, we preserve whole B_{n-1} types.

At level 1, we code C_1, eventually correctly. In addition, we use the accumulated information passed down from above to satisfy higher level requirements. We assign complete types to tuples, we choose constants to satisfy $\psi_n(x)$, if consistent, and we determine coding witnesses, all tentatively. We code C_0 by making Ea_{2n} hold just in case $n \in C_0$. We preserve what we have enumerated into the atomic diagram.

The description we have just given is rather vague. Below, we describe in a precise way the object of the construction, in terms of a tree P and other objects.

NOTATION. Instead of saying that $f(x)$ is defined, or undefined, we may write $f(x)\downarrow$, or $f(x)\uparrow$.

Let D be the set of all finite sets $d \subseteq \omega \times \omega$ such that if $(n, x) \in d$ and $y < x$, then $(n, y) \in d$. The pairs (n, x) represent coding requirements, and the elements of D represent sets of coding requirements that may be attempted by some stage in the construction.

Next, we define a set L, consisting of triples $\ell = (\overline{a}, t, w)$. The first component, $\overline{a}$, is a tuple of constants from A, and the second component, t, is the index of a complete type, consistent with T, in variables corresponding to $\overline{a}$. We may write $R_t(\overline{a})$ for the result of substituting $\overline{a}$ for the variables in the type with index t. The third component, w, is a function with domain some $d \in D$. For each $(n, x) \in d$, $w(n, x)$ is a pair of form (θ, u), where u, called the "action", is 0 or 1, and θ, called the "coding witness", is a sentence involving constants from $\overline{a}$. For $n \geq 1$, if $(n, x) \in d$, then the coding witness is a Π_n sentence $\theta(\overline{b}, c)$, chosen as in Theorem 2.4, and $\theta(\overline{b}, c) \in R_t(\overline{a})$ iff the action is 0. For $n = 0$, we require that $(0, x) \in d$ iff $a_{2x} \in \overline{a}$. In $w(0, x)$, the coding witness is Ea_{2x}, and the action is 1 iff $Ea_{2x} \in R_t(\overline{a})$.

We associate with L three functions. First, the "coding function", denoted by w, is defined on L such that if $\ell = (\overline{a}, t, w)$, then $w(\ell) = w$. We write $w_n(\ell)$ for the restriction of $w(\ell)$ to pairs of form (n, x). Second, the "enumeration function", denoted by E, is defined on $\omega \times L$ such that for $n \geq 1$, $E(n, (\overline{a}, t, w))$ is the set of B_n sentences in $R_t(\overline{a})$ with Gödel number bounded by $length(\overline{a})$. For $n = 0$, $E(0, (\overline{a}, t, w))$ is a finite subset of $R_t(\overline{a})$ consisting of the open sentences with Gödel number bounded by $length(\overline{a})$, plus all sentences of form $\pm Ea$. We write $E_n(\ell)$ for $E(n, \ell)$. Third, the "witness-choosing function", b, is defined on ω such that $b(n)$ is Ea_{2n}.

Next, we define a family of binary relations on L. If $\ell = (\overline{a}, t, w)$ and $\ell' = (\overline{a}', t', w')$ are in L, then

1. $\ell \leq_0 \ell'$ iff $E_0(\ell) \subseteq E_0(\ell')$—this implies that $\overline{a} \subseteq \overline{a}'$ and $w_0(\ell) \subseteq w_0(\ell')$,
2. for $n \geq 1$, $\ell \leq_n \ell'$ iff the following hold:
   (a) $R_t(\overline{a}) \cap B_n \subseteq R_{t'}(\overline{a}')$,
   (b) for all $m \leq n$, $w_m(\ell) \subseteq w_m(\ell')$,
   (c) if $w(\ell)(m, x)\uparrow$ and $w(\ell')(m, x)\downarrow = (\theta(\overline{b}, c)), u)$, then $\overline{a} \subseteq \overline{b}$,
3. $\ell \subseteq \ell'$ iff for all n, $\ell \leq_n \ell'$.

We fix an element $\hat{\ell} = (\emptyset, t_0, \emptyset)$ of L, where t_0 is an R-index for T. We define a tree P, consisting of the finite non-empty sequences $\sigma = \ell_0\ell_1\ell_2\ldots$ such that $\ell_0 = \hat{\ell}$, and letting $\ell_n = (\overline{a}_n, t_n, w_n)$, we have the following:

1. $\ell_k \subseteq \ell_{k+1}$,
2. $\overline{a}_k$ includes the first k odd constants,
3. if i is first such that the constants of ψ_i are included in $\overline{a}_i$, $\psi_i(x)$ is consistent with $R_{t_{k+1}}(\overline{a}_k)$, and there is no $c \in \overline{a}_k$ such that $\psi_i(c) \in R_{t_k}(\overline{a}_k)$, then for some $c \in \overline{a}_{k+1}$, we have $\psi_i(c) \in R_{t_{k+1}}(\overline{a}_{k+1})$.

NOTATION. If $\pi = \hat{\ell}\ell_1\ell_2\ldots$ is a path through P, we write $E_n(\pi)$ for $\cup_k E_n(\ell_k)$, $w(\pi)$ for $\cup_k w(\ell_k)$, and $w_n(\pi)$ for $\cup_k w_n(\ell_k)$.

LEMMA 4.3. *Suppose* $\pi = \hat{\ell}\ell_1\ell_2\ldots$ *is a path through* P, *where* $\ell_n = (\overline{a}_n, t_n, w_n)$. *If each even constant is included in some* $\overline{a}_k$, *then* $\cup_k R_{t_k}(\overline{a}_k)$ *is the complete diagram of a model* $\mathcal{A}$ *of* T, *and for all* n, $E_n(\pi) = D^c(\mathcal{A}) \cap B_n$.

We can now say, in terms of P, E, and w, what is needed to prove the main result.

CLAIM. To prove Theorem 4.2, it is enough to produce a path π through P such that

1. $E_n(\pi)$ is c.e. relative to C_n, uniformly in n,
2. for all (n,x), $w(\pi)(n,x)\downarrow$, with action $\chi_{C_n}(x)$,
3. for all (n,x) with $n \geq 1$, we can determine the coding witness in $w(\pi)(n,x)$ using C_{n-1} and $C_n \cap x$. The coding witness for $w(\pi)(0,x)$ is $b(x)$.

PROOF OF CLAIM. Suppose that $\pi = \hat{\ell}\ell_1\ell_2\ldots$ is a path through P satisfying 1, 2, 3. Say $\ell_n = (\overline{a}_n, t_n, w_n)$. By Lemma 4.3, $\cup_n R_{t_n}(\overline{a}_n)$ is the complete diagram of a model $\mathcal{A}$ of T, and $E_n(\pi) = D^c(\mathcal{A}) \cap B_n$. We show that $D_n(\mathcal{A}) \equiv_T C_n$, uniformly in n. By 1, $E_n(\pi)$ is c.e. relative to C_n, so we have

$$D_n(\mathcal{A}) \leq_T D^c(\mathcal{A}) \cap B_n \leq_T C_n,$$

uniformly in n. We must show that $C_n \leq_T D_n(\mathcal{A})$, also uniformly. By 2, $w(\pi)(n,x)\downarrow$, with action $\chi_{C_n}(x)$, for all n and x.

If $n = 0$, then the coding witness for $(0,x)$ is Ea_{2x}, and the definitions of L and P guarantee that $\mathcal{A} \models Ea_{2x}$ iff the action is 1. Therefore, $C_0 \leq_T D_0(\mathcal{A})$. Now, suppose $n \geq 1$. If $\theta_{(n,x)}$ is the coding witness for (n,x), then the definitions of L and P guarantee that $\mathcal{A} \models \theta_{(n,x)}$ iff the action is 0. Assuming that $C_{n-1} \leq_T D_{n-1}(\mathcal{A})$, and also assuming that we have decoded $C_n \cap n$, then by 3, we can determine $\theta_{(n,x)}$. Then using $D_n(\mathcal{A})$, we can decide whether $x \in C_n$. This completes the proof of the claim.

We give four lemmas. The first three are clear.

LEMMA 4.4. *For all* n, $\leq_n$ *is transitive and reflexive.*

LEMMA 4.5. *For all* n, $\leq_{n+1}$ *implies* $\leq_n$.

LEMMA 4.6. *If* $\ell \leq_n \ell'$, *then* $E_n(\ell) \subseteq E_n(\ell')$ *and* $w_n(\ell) \subseteq w_n(\ell')$.

To state the fourth lemma, we need more definitions. First, we say what is a "picture".

DEFINITION. A *picture* is a triple $c = (\sigma\ell^0, \tau, a)$ such that $\sigma\ell^0 \in P$ and τ and a satisfy one of the following:

1. $\tau = \emptyset$, and a maps (n,x) to $\{0,1\}$, where for some n, x is first such that $w(\ell^0)(n,x)\uparrow$,

2. $\tau = n_0\ell^1 \ldots \ell^{k-1} n_{k-1}\ell^k$ and a maps $\{(n_0, x_0), \ldots, (n_{k-1}, x_{k-1}), (n_k, x_k)\}$ to $\{0, 1\}$, where
   (a) $n_k < n_{k-1} < \cdots < n_0$,
   (b) $\ell^0 \leq_{n_0} \ell^1 \leq_{n_1} \cdots \leq_{n_{k-2}} \ell^{k-1} \leq_{n_{k-1}} \ell^k$,
   (c) for $i \leq k$, x_i is first such that $w(\ell^i)(n_i, x_i) \uparrow$,
   (d) for $i < k$, if $w(\ell^{i+1})(n_i, x_i) \downarrow$, then the action is 0 while $a(n_i, x_i) = 1$.

Next, we say what it means to "complete" a picture.

DEFINITION. Let $c = (\sigma\ell^0, \tau, a)$ be a picture.

CASE 1: Suppose $\tau = \emptyset$ and a is defined on (n, x). Then ℓ^* *completes* c if

1. $\sigma\ell^0\ell^* \in P$,
2. for $m > n$, $w_m(\ell^*) = w_m(\ell^0)$,
3. for $m < n$, $w_m(\ell^0) \subseteq w_m(\ell^*)$,
4. $w(\ell^*)(n, x) \downarrow$, with action $a(n, x)$,
5. $w_n(\ell^*)(n, y) = w_n(\ell^0)(n, y)$, for $y < x$,
6. $w(\ell^*)(n, y) \uparrow$ for $y > x$.

CASE 2: Suppose $\tau = n_0\ell^1 \ldots \ell^{k-1} n_{k-1}\ell^k$, and for $i \leq k$, $a(n_i, x_i) = u_i$. Then ℓ^* *completes* c if

1. $\sigma\ell^0\ell^* \in P$,
2. for $0 < i \leq k$, $\ell^i \leq_{n_i} \ell^*$,
3. for $i \leq k$, $w(\ell^*)(n_i, x_i) \downarrow$, with action $a(n_i, x_i)$,
4. for $i \leq k$, $w_{n_i}(\ell^*)$ agrees with $w_{n_i}(\ell^i)$ on pairs (n_i, y) for $y < x_i$, and is undefined on pairs (n_i, y) for $y > x_i$.
5. for $m > n_0$, $w_m(\ell^*) = w_m(\ell^0)$,
   and for $n_{i+1} < m < n_i$, $w_m(\ell^*) = w_m(\ell^{i+1})$,
6. for $m < n_k$, $w_m(\ell^k) \subseteq w_m(\ell^*)$.

LEMMA 4.7. *Every picture has a completion. Moreover, there is an effective procedure for finding completions, using* C_0.

PROOF. Let $c = (\sigma\ell^0, \tau, a)$ be a picture. Suppose $\sigma\ell^0$ has length k.

CASE 1: Suppose $\tau = \emptyset$.

SUBCASE 1(a): Suppose a is defined on $(0, x)$, say the value is u.

Say $\ell^0 = (\overline{a}, t, w)$. To $R_t(\overline{a})$, we add $\psi_i(c)$, where i is first such that $\exists x\, \psi_i(x) \in R_t(\overline{a})$ but there is no $a \in \overline{a}$ such that $\psi_i(a) \in R_t(\overline{a})$, and c is the first new odd constant. We also add Ea_{2x}, if $u = 1$, or $\neg Ea_{2x}$, if $u = 0$. Let $\overline{a}' = \overline{a}, c, a_{2x}$, and take t' such that $R_{t'}(\overline{a'})$ completes our extension of $R_t(\overline{a})$. Let w' be the extension of w mapping $(0, x)$ to (Ea_{2x}, u). Then $\ell^* = (\overline{a}', t', w')$ completes c.

SUBCASE 1(b): Suppose a is defined on (n, x), for $n \geq 1$, say the value is u.

To $R_t(\overline{a})$, we add $\psi_i(c)$, where i is first such that $\exists x\ \psi_i(x) \in R_t(\overline{a})$ but there is no $a \subset \overline{a}$ such that $\psi_i(a) \in R_t(a)$, and c is the first new odd constant. Let $\overline{a}' = \overline{a}, c$, and form a completion $R_{t'}(\overline{a}')$. Let c' be a further new odd constant, and let $\theta(\overline{a}', c')$ be a Π_n sentence, as in Theorem 2.4. To $R_{t'}(\overline{a}')$, we add $Code_{R_{t'}(\overline{a}')}(c')$. We also add $\theta(\overline{a}', c')$, if $u = 0$, and $\neg\theta(\overline{a}, c)$, if $u = 1$. Let $\overline{a}'' = \overline{a}', c'$. Form a completion $R_{t''}(\overline{a}'')$. Let w' be the extension of w taking (n, x) to $(\theta(\overline{a}', c'), u)$. Then $\ell^* = (\overline{a}'', t'', w')$ completes c.

CASE 2: Suppose $\tau = n_0\ell^1 \ldots \ell^{k-1}n_{k-1}\ell^k$, and $a(n_i, x_i) = u_i$.

Let $\ell^i = (\overline{a}_i, t_i, w_i)$. We work our way up from the bottom, forming $m^i \supseteq \ell^i$ such that for $i < k$, $m^{i+1} \leq_{n_i - 1} m^i$, and for $m \leq n_i$, $w_m(m^i)$ has the values we want for $w_m(\ell^*)$. For m^k, there are two cases.

(a) Suppose $n_k = 0$.

To $R_{t_k}(\overline{a}_k)$, we add Ea_{2x_k}, if $u_k = 1$, or $\neg Ea_{2x_k}$, if $u_k = 0$. Let $\overline{a}'_k = \overline{a}_k, a_{2x_k}$, and let $R_{t'}(\overline{a}'_k)$ be a completion of the set of sentences. Let w'_k be the extension of w_k mapping $(0, x_k)$ to (Ea_{2x_k}, u_k). Then $m_k = (\overline{a}'_k, t'_k, w'_k)$.

(b) Suppose $n_k \geq 1$.

Let c be a new odd constant, and let $\theta(\overline{a}_k, c_k)$ be a Π_{n_k} sentence as in Theorem 2.4. Let $\overline{a}'_k = \overline{a}_k, c$. We extend $R_{t_k}(\overline{a}_k)$, adding $Code_{R_{t_k}(\overline{a}_k)}(c)$. We also add $\theta(\overline{a}, c)$, if $u_k = 0$, or $\neg\theta(\overline{a}'_k, c)$, if $u_k = 1$. Let $\overline{a}'_k = \overline{a}_k, c$, and let $R_{t'_k}(\overline{a}'_k)$ be a completion of the set of sentences. Let w'_k be the extension of w_k mapping (n_k, x_k) to $(\theta(\overline{a}'_k, c), u_k)$. Let $m^k = (\overline{a}'_k, t'_k, w'_k)$.

Having determined $m^k = (\overline{a}'_k, t'_k, w'_k)$, we turn to m^{k-1}. Again there are two cases.

(a) First, suppose $w(\ell^k)(n_{k-1}, x_{k-1}) \uparrow$.

Since $\ell^{k-1} \leq_{n_{k-1}} m^k$, we have $R_{t_{k-1}}(\overline{a}_{k-1}) \cap B_{n_{k-1}} \subseteq R_{t'_k}(\overline{a}'_k)$. It follows that $R_{t_{k-1}}(\overline{a}_{k-1}) \cup (R_{t'_k}(\overline{a}'_k) \cap \Sigma_{n_{k-1}})$ is consistent. Let $\Gamma(\overline{a}'_k)$ be a completion of our set of sentences. Take a new odd constant c', and let $\theta'(\overline{a}'_k, c')$ be a $\Pi_{n_{k-1}}$ sentence as in Theorem 4.2. We extend $\Gamma(\overline{a}'_k)$, adding $Code_{\Gamma(\overline{a}'_k)}(c')$. We also add $\theta'(\overline{a}'_k, c')$ if $u_{k-1} = 0$, and $\neg\theta'(\overline{a}'_k, c')$ otherwise. Let $\overline{a}'_{k-1} = \overline{a}'_k, c'$, and let $R_{t'_{k-1}}(\overline{a}'_{k-1})$ be a completion of the set of sentences. Let w'_{k-1} be the extension of w_{k-1} that agrees with w'_k on any new pairs (m, y) with $m \leq n_k$, and maps (n_{k-1}, x_{k-1}) to $(\theta_{k-1}(\overline{a}'_k, c_{k-1}), u_{k-1})$. Let $m^{k-1} = (\overline{a}'_{k-1}, t'_{k-1}, w'_{k-1})$.

(b) Now, suppose $w(\ell^{k-1}, x_{k-1}) \downarrow= (\theta, u)$.

We must have $u = 0$, while $u_{k-1} = 1$. Also, θ is a $\Pi_{n_{k-1}}$ sentence in constants $\overline{b}, c$, chosen as in Theorem 2.4. Let $\Gamma(\overline{b})$ be the restriction of $R_{t_k}(\overline{a}_k)$, or $R_{t'_k}(\overline{a}'_k)$, to sentences in the constants $\overline{b}$. Let $\Sigma(\overline{b}, c)$ be the restriction of $R_{t_k}(\overline{a}_k)$, or $R_{t'_k}(\overline{a}'_k)$, to $\Sigma_{n_{k-1}}$ sentences in the constants $\overline{b}, c$. Theorem 2.4 says that $\Gamma(\overline{b}) \cup \Sigma(\overline{b}, c) \cup \{\neg\theta\}$ is consistent. For any $\Sigma_{n_{k-1}}$ sentence $\varphi(\overline{b}, c, \overline{b}_1)$ in

$R_{t'_k}(\overline{a}'_k)$, $\exists \overline{u}\, \varphi(\overline{b}, c, \overline{u}) \in \Sigma(\overline{b}, c)$. Therefore, $\Gamma(\overline{b}) \cup (R_{t'_k}(\overline{a}'_k) \cap \Sigma_{n_{k-1}}) \cup \{\neg\theta\}$ is consistent. Let $\Lambda(\overline{a}'_k)$ be a completion.

Since $\ell^{k-1} \leq_{n_{k-1}} m^k$, $\overline{a}_{k-1} \subseteq \overline{b}$. Also, $R_{t_{k-1}}(\overline{a}_{k-1}) \cap B_{n_{k-1}} \subseteq \Gamma(\overline{b}) \subseteq \Lambda(\overline{a}'_k)$. Therefore, $R_{t_{k-1}}(\overline{a}_{k-1}) \cup (\Lambda(\overline{a}'_k) \cap \Sigma_{n_{k-1}})$ is a consistent set of sentences. Let $\overline{a}'_{k-1} = \overline{a}'_k$, and let $R_{t'_{k-1}}(\overline{a}'_{k-1})$ be a completion of the set of sentences. Let w'_{k-1} be the extension of w_{k-1} that agrees with w'_k on any new pairs (m, y) with $m \leq n_k$, and maps (n_{k-1}, x_{k-1}) to $(\theta, 1)$. Let $m^{k-1} = (\overline{a}'_{k-1}, t'_{k-1}, w'_{k-1})$.

We continue in this way until we arrive at $m^0 = (\overline{a}'_0, t'_0, w'_0)$. Now, $w(m^0)$ has exactly the values we want for $w(\ell^*)$. We may need to extend $\overline{a}'_0$ and R'_t so that $\sigma\ell^0\ell^* \in P$. All of our completions come from $\mathcal{S}$, and, using the effective enumeration, we can effectively determine the indices, using C_0. This completes the proof of Lemma 4.7.

We have reduced the proof of Theorem 4.2 to the proof that for a certain tree P, and other objects L, E, w, and $\leq_n$, there is a path π with properties listed in the claim. In the next section, we prove a metatheorem with conditions on abstract objects P, L, E, w, and $\leq_n$, guaranteeing the existence of a path that satisfies these properties. The specific objects L, P, etc., defined above provide the motivating example.

There is a further difficulty to overcome. In the metatheorem, L, P, and $\leq_n$ are supposed to be uniformly c.e. If we are relativizing to C_0, then L, P, $\leq_n$ are supposed to be uniformly c.e. relative to C_0. As defined above, L, P, and $\leq_n$ are Δ^0_2 relative to C_0, but not c.e. In Section 6, we shall give a general result that allows us to apply the metatheorem anyway.

§5. The metatheorem. Priority constructions, especially those with more than two levels, are complicated. For that reason, it seems worthwhile, when we have succeeded in thinking through a construction with new features, to record those features abstractly, so that if we encounter another construction with the same features, we can use what we have done. Ash [2] formulated in an abstract way the object of a nested priority construction in which the information for satisfying the requirements is Δ^0_α. He proved a metatheorem for "α-systems", with conditions guaranteeing the success of the construction.

There are now several further metatheorems for constructions with special features not covered by Ash's original result. In [4], there is a metatheorem for "witnessing α-systems", which handles certain constructions involving coding. In [3], there is a metatheorem for "ramified α-systems", for constructions in which we enumerate not only a c.e. set, but also a Σ^0_2 set, a Σ^0_3 set, etc. In the proof of Theorem 4.2, we need both features. We give a metatheorem for "coding and enumerating α-systems", where $\alpha \leq \omega$. In the present paper, we need the result for $\alpha = \omega$. Moreover, we apply the metatheorem in relativized form, so we determine the coding witnesses for C_0 in advance. We have in mind another application in [6], where α may be either finite or ω. In [6], we do not relativize, so the coding witnesses for C_0 need not be chosen in advance.

We begin by describing the basic setting. We fix a set L. An *enumeration function* on $\alpha \times L$ is a function E mapping elements of $\alpha \times L$ to finite subsets of ω. For $n < \alpha$, we write $E_n(\ell)$ for $E(n, \ell)$. Let D be the set of all finite sets $d \subseteq \{(n,x) \colon n < \alpha \; \& \; x \in \omega\}$ such that if $(n,x) \in d$ and $y < x$, then $(n,y) \in d$. A *coding function* on L is a partial computable function w such that for $\ell \in L$, $w(\ell)$ is a function from some $d \in D$ to $\omega \times \{0,1\}$. If $w(\ell)(n,x) = (b,u)$, then we call b the *coding witness*, and we call u the *action*. For $n < \alpha$, we write $w_n(\ell)$ for the restriction of $w(\ell)$ to pairs (n,x) with first component n. Let $\hat{\ell} \in L$, where $w(\hat{\ell}) = \emptyset$.

By a *tree*, we mean a set P of non-empty finite sequences, closed under non-empty initial segments. The tree is said to be *on* L if the terms in the finite sequences come from L. A *path* through P is a function π on ω such that for all $n \geq 1$, $\pi \mid n \in P$. We consider a tree P on L, in which all sequences start with $\hat{\ell}$. If $\pi = \hat{\ell}\ell_1\ell_2\ldots$ is a path and $n < \alpha$, let $E_n(\pi) = \cup_k E_n(\ell_k)$. Let $w(\pi) = \cup_k w(\ell_k)$, and for $n < \alpha$, let $w_n(\pi) = \cup_k w_n(\ell_k)$.

Object of the construction. Let $L, w, E, \hat{\ell}$, and P be as described above, where L and P are c.e., and w and E are partial computable. Let $(C_n)_{n<\alpha}$ be an α-table. We want to produce a path π through P such that

1. $E_n(\pi)$ is c.e. relative to C_n, uniformly in n, for $n < \alpha$,
2. $w(\pi)(n,x) \downarrow$, with action $\chi_{C_n}(x)$, for $n < \alpha$ and $x \in \omega$,
3. for $1 \leq n < \alpha$, we can determine the coding witness in $w(\pi)(n,x)$ using C_{n-1} and $C_n \cap x$, and we can determine the coding witness in $w(\pi)(0,x)$ using either nothing at all, or just $C_0 \cap x$.

The conditions guaranteeing success of the construction involve a further function b and binary relations $\leq_n$ on L, for $n < \alpha$. There are two versions of b. In the first version, the coding witnesses for the pairs of form $(0,x)$ (with first component 0) are all chosen in advance. For this version, we use a *strong witness-choosing function* b, where b maps ω to ω—$b(x)$ gives the coding witness for $(0,x)$. In the second version, the coding witnesses for the pairs of form $(0,x)$ are chosen during the construction, using a *weak witness-choosing function* b, where b maps L to ω—if x is first such that $w(\ell)(0,x) \uparrow$, then $b(\ell)$ is a coding witness for $(0,x)$.

The binary relations $\leq_n$ appear in various "preservation" conditions. We suppose that if $\ell \leq_n \ell'$, then for $m \leq n$, $w_m(\ell) \subseteq w_m(\ell')$ and $E_m(\ell) \subseteq E_m(\ell')$. We write $\ell \subseteq \ell'$ if $\ell \leq_n \ell'$ for all $n < \alpha$. If $\alpha = N$, then $\subseteq \; = \; \leq_{N-1}$. We suppose that for $\sigma = \ell_0\ell_1\ldots$ in P, $\ell_0 = \hat{\ell}$, and for all k, $\ell_k \subseteq \ell_{k+1}$. We shall make a further assumption involving the relations $\leq_n$—saying that "pictures" can be completed.

The pairs (n,x) represent requirements for coding the set C_n. For $n \geq 1$, the strategy for requirement (n,x)—viewed at level n, where we have oracle

C_{n-1}—is to designate a coding witness b, and take action 0 until/unless x appears in C_n, then switch to action 1, dropping any coding witnesses that have been designated for requirements (n, y), for $y > x$. Once a coding witness is designated for (n, x), it is not changed unless the action changes on (n, y), for some $y < x$.

If we are using a strong witness-choosing function b, then the coding witness for $(0, x)$ is $b(x)$. If, instead, we are using a weak witness-choosing function b, then at level 1, we use only $C_0 \cap x$ at step x, where we code $C_0 \cap x$. Having obtained $\ell \in L$ at step x, we take $b(\ell)$ as the coding witness for $(0, x)$. In either case, to satisfy the requirement, we take action $\chi_{C_0}(x)$.

The following definition indicates what the strategy for level n requirements says about successive steps.

DEFINITION. Suppose $\ell, \ell' \in L$.

CASE I: Let $1 \leq n < \alpha$.

Then ℓ' *follows* ℓ, with action u on (n, x), provided that

1. $w(\ell)(n, y) = w(\ell')(n, y)$, for all $y < x$,
2. $w(\ell')(n, x) = (b, u)$, where one of the following holds:
   (a) x is first such that $w(\ell)(n, x) \uparrow$—in this case, u may be 0 or 1,
   (b) $w(\ell)(n, x) = (b, 0)$ and $w(\ell')(n, x) = (b, 1)$—in this case, $u = 1$,
3. $w(\ell')(n, y) \uparrow$ for all $y > x$.

CASE II: Let $n = 0$.

First, suppose that we are using a strong witness-choosing function b. Then ℓ' *follows* ℓ, with action u on $(0, x)$ if we have 1, 2a, and 3 above and the coding witness is $b(x)$. Now, suppose that we are using a weak witness-choosing function b. Then ℓ' *follows* ℓ, with action u on $(0, x)$ if we have 1, 2 (a), and 3 above, and $w(\ell')(0, x)$ has coding witness $b(\ell)$.

The next definition represents information being passed down, with additions, through a sequence of levels. The witness-choosing function is not involved.

DEFINITION. A *picture* is a triple $c = (\sigma\ell^0, \tau, a)$, where $\sigma\ell^0 \in P$, and τ and a satisfy one of the following:

1. $\tau = \emptyset$, and a maps some (n, x) to $\{0, 1\}$, where $n < \alpha$, and x is first (for this n) such that $w(\ell^0)(n, x) \uparrow$,
2. τ has form $n_0\ell^1 \dots \ell^{k-1} n_{k-1} \ell^k$ and a maps $\{(n_0, x_0), \dots, (n_{k-1}, x_{k-1}), (n_k, x_k)\}$ to $\{0, 1\}$, where
   (a) $\alpha > n_0 > \dots > n_{k-1} > n_k$,
   (b) $\ell^0 \leq_{n_0} \dots \leq_{n_{k-1}} \ell^k$,
   (c) for $i \leq k$, x_i is first such that $w(\ell^i)(n_i, x_i) \uparrow$,
   (d) for $i < k$, if $w(\ell^{i+1})(n_i, x_i) \downarrow$, with action u, then $u = 0$ while $a(n_i, x_i) = 1$.

The next definition indicates how the information in a picture is used, once it has been passed down.

DEFINITION. Let $c = (\sigma\ell^0, \tau, a)$ be a picture.

CASE 1: Suppose $\tau = \emptyset$, and $a(n, x) = u$. Then ℓ *completes* c if

1. $\sigma\ell^0\ell \in P$,
2. for $m > n$, $w_m(\ell) = w_m(\ell^0)$,
3. ℓ follows ℓ^0 with action u on (n, x).

CASE 2: Suppose $\tau = \ell^0 n_0 \ldots \ell^{k-1} n_{k-1} \ell^k$, and for $i \leq k$, $a(n_i, x_i) = u_i$. Then ℓ *completes* c if

1. $\sigma\ell^0\ell \in P$,
2. for $0 < i \leq k$, $\ell^i \leq_{n_i} \ell$,
3. for $i \leq k$, ℓ follows ℓ^i, with action u_i on (n_i, x_i),
4. for $i < k$, ℓ follows ℓ^{i+1}, with action u_i on (n_i, x_i),
5. for $m > n_0$, $w_m(\ell) = w_m(\ell^0)$,
6. for $n_{i+1} < m < n_i$, $w_m(\ell) = w_m(\ell^i)$.

REMARK. The definition of *completion* depends on the type of witness-choosing function we are using, since it affects the definition of *follows*. In Case 1, if $n = 0$, we want ℓ to follow ℓ^0 with action u on $(0, x)$, and in Case 2, if $n_k = 0$, we want ℓ to follow ℓ^k with action u on $(0, x_k)$. If we are using a strong witness-choosing function, then the coding witness depends only on x. If we are using a weak witness-choosing function, then the coding witness depends on ℓ^k.

We are ready to define a variant of Ash's α-systems. There are two versions, depending on the type of witness-choosing function.

DEFINITION. Let $1 \leq \alpha \leq \omega$. A *coding and enumerating α-system* with a *strong*, or *weak*, *witness-choosing function* is a structure

$$(L, w, E, b, \hat{\ell}, P, (\leq_n)_{n<\alpha}),$$

where P is a tree on L, w is a coding function on L, E is an enumeration function on $\alpha \times L$, b is a strong, or weak, witness-choosing function, $(\leq_n)_{n<\alpha}$ is a family of binary relations on L, P is a tree on L, if ℓ and ℓ' are successive terms in some sequence $\sigma \in P$, then $\ell \subseteq \ell'$, where this means $\ell \leq_n \ell'$ for all $n < \alpha$, and the following four conditions hold:

1. for $n < \alpha$, $\leq_n$ is reflexive and transitive,
2. for $n + 1 < \alpha$, if $\ell \leq_{n+1} \ell'$, then $\ell \leq_n \ell'$,
3. for $n < \alpha$, if $\ell \leq_n \ell'$, then $E_n(\ell) \subseteq E_n(\ell')$ and $w_n(\ell) \subseteq w_n(\ell')$,
4. if $c = (\sigma, \tau, a)$ is a picture, then c has a completion ℓ^* (the type of witness-choosing function affects the definition just here).

REMARK. In the presence of Condition 2, Condition 3 says that if $\ell \leq_n \ell'$, then for $m \leq n$, $E_m(\ell) \subseteq E_m(\ell')$ and $w_m(\ell) \subseteq w_m(\ell')$.

Here is the metatheorem for coding and enumerating α-systems with a strong witness-choosing function.

THEOREM 5.1. *Suppose* $1 \leq \alpha \leq \omega$, *and let* $(L, w, E, b, \hat{\ell}, P, (\leq_n)_{n<\alpha})$ *be a coding and enumerating* α*-system with a strong witness-choosing function. Suppose* L *and* P *are c.e.,* w *and* E *are partial computable,* b *is computable, and the relations* $\leq_n$ *are c.e., uniformly in* n. *Let* $(C_n)_{n<\alpha}$ *be an* α*-table. Then* P *has a path* π *such that*

1. *for* $n < \alpha$, $E_n(\pi)$ *is c.e. relative to* C_n, *uniformly in* n,
2. *for* $(n, x) \in \alpha \times \omega$, $w(\pi)(n, x) \downarrow$, *with action* $\chi_{C_n}(x)$, *and*
3. *we can effectively determine the coding witness in* $w(\pi)(n, x)$ *using* $C_n \cap x$ *and* C_{n-1}, *for* $n \geq 1$; *for* $n = 0$, *the coding witness in* $w(\pi)(0, x)$ *is* $b(x)$.

Here is the metatheorem for coding and enumerating α-systems with a weak witness-choosing function.

THEOREM 5.2. *Suppose* $1 \leq \alpha \leq \omega$, *and let* $(L, w, E, b, \hat{\ell}, P, (\leq_n)_{n<\alpha})$ *be a coding and enumerating* α*-system with a weak witness-choosing function. Suppose* L *and* P *are c.e.,* w, E, *and* b *are partial computable, and the relations* $\leq_n$ *are c.e., uniformly in* n. *Let* $(C_n)_{n<\alpha}$ *be an* α*-table. Then* P *has a path* π *such that*

1. *for* $n < \alpha$, $E_n(\pi)$ *is c.e. relative to* C_n, *uniformly in* n,
2. *for* $(n, x) \in \alpha \times \omega$, $w(\pi)(n, x) \downarrow$, *with action* $\chi_{C_n}(x)$, *and*
3. *we can effectively determine the coding witness in* $w(\pi)(n, x)$ *using* C_{n-1} *and* $C_n \cap x$, *for* $1 \leq n < \alpha$, *and we can effectively determine the coding witness in* $w(\pi)(0, x)$ *using just* $C_0 \cap x$.

The proofs of the theorems are essentially the same. We shall point out the differences when we come to them. Let C be the set of all pictures. For $1 \leq n < \alpha$, let C^n be the set of pictures $c = (\sigma, \tau, a)$ such that the numbers in τ are all $\geq n$. Let C^α be the set of pictures of the form $c = (\sigma, \emptyset, a)$. We shall describe a family of "alternating trees", with corresponding "instruction functions"—these notions are defined below.

DEFINITION. An *alternating tree*, on sets L and U, is a tree P consisting of non-empty finite sequences $\ell_0 u_1 \ell_1 \ldots$, where $u_k \in U$ and $\ell_k \in L$.

DEFINITION. For an alternating tree P on L and U, an *instruction function* q takes each sequence $\sigma \in P$ of odd length (ending in L) to some $u \in U$ such that $\sigma u \in P$.

DEFINITION. Let P be an alternating tree on L and U, and let q be an instruction function for P. A *run* of (P, q) is a path through P in which the terms from U are chosen by q.

We shall define alternating trees P^β for $\beta \leq \alpha$, with corresponding instruction functions q_β. We begin at the top. The tree P^α, which is alternating on L and C^α, consists of the finite sequences $\hat{\ell} c_1 \ell_1 c_2 \ell_2 \ldots$ such that the following hold:

1. for each k, c_k has form $(\sigma_k, \emptyset, a_k)$, where $\sigma_k = \hat{\ell}\ell_1 \dots \ell_{k-1}$,
2. a_k maps the first pair (n, x) such that $w(\ell_{k-1})(n, x) \uparrow$ to $\{0, 1\}$,
3. ℓ_k completes c_k.

The instruction function q_α is defined as follows. For $\rho = \hat{\ell}c_1\ell_1 \dots c_{k-1}\ell_{k-1}$ in P^α, let (n, x) be the first pair in $\alpha \times \omega$ such that $w(\ell_{k-1})(n, x) \uparrow$. Then $q_\alpha(\rho) = (\sigma, \emptyset, a)$, where $\sigma = \hat{\ell}\ell_1 \dots \ell_{k-1}$, and a maps (n, x) to $\chi_{C_n}(x)$.

NOTATION. For a sequence σ, finite or infinite, with some terms in L, $L(\sigma)$ denotes the subsequence consisting of those terms. For a finite sequence σ, $\ell(\sigma)$ denotes the last term of $L(\sigma)$.

LEMMA 5.3. *For any run π^* of (P^α, q_α), then $\pi = L(\pi^*)$ is a path through P, and for all $(n, x) \in \alpha \times \omega$, $w(\pi)(n, x) \downarrow$.*

NOTATION. Earlier, we defined $w(\pi)$, $E_n(\pi)$ in the case where π is a path through P. We now extend the use of these notations. If π is a sequence with some terms in L, then $w(\pi) = \cup_{\ell \in L(\pi)} w(\ell)$, and $E_n(\pi) = \cup_{\ell \in L(\pi)} E_n(\ell)$.

We define P^n and q_n, for $1 \leq n < \alpha$, so that if π^n is a run of (P^n, q_n), then $L(\pi^n)(m, x) \downarrow$, for all $m < n$ and all x, and at least for the pairs of form $(n-1, x)$, the action matches $\chi_{C_{n-1}}(x)$. Since E is partial computable, $E(\pi^n)$ is c.e. relative to π^n.

The tree P^n, which is alternating on L and C^n, consists of the finite sequences $\rho = \hat{\ell}c_1\ell_1c_2\ell_2 \dots$ such that the following hold:

1. c_k has the form (σ, τ, a), where either $\tau = \emptyset$ and $\ell(\sigma) = \ell_{k-1}$, or else $\ell(\tau) = \ell_{k-1}$,
2. ℓ_k completes c_k,
3. $\ell_{k-1} \leq_{n-1} \ell_k$.

We shall define instruction functions q_n, computable in C_{n-1}, uniformly in n, such that a run of (P^n, q_n) that is computable in C_{n-1} yields, as a limit, a run of (P^{n+1}, q_{n+1}) that is computable in C_n. For $\rho \in P^n$, of odd length, in addition to $q_n(\rho)$, we determine a sequence ρ^+, where this represents an approximation of a run of (P^{n+1}, q_{n+1}), with the terms in L coming from ρ. The definition will proceed by induction. We describe, for pairs $\langle n, x\rangle$, how to compute ρ^+ and $q_n(\rho)$, for $\rho \in P^n$ of length $2x+1$, assuming that we have previously said how to compute these objects for the pairs $\langle n, y\rangle$ and $\langle n+1, y\rangle$ where $y < x$. (We use angular brackets for the pairs associated with this induction, since we are already using pairs with round brackets to represent requirements.)

Oracle restraint and approximations. Given an index for the restriction of q_{n+1} to sequences having the length of ρ^+, given also the correct approximation for the part of C_n used in computing $q_{n+1}(\rho^+)$, we obtain the correct value for $q_{n+1}(\rho^+)$. We put some restraints on the use of the oracles.

1. For $n \geq 2$, if $\rho \in P^n$, $\ell(\rho) = \ell$, and k is first such that $w(\ell)(n-1,k)\uparrow$, then $q_n(\rho)$ is computed using at most $\chi_{C_{n-1}} \mid (k+1)$. For $n = 1$, if we are using a weak witness-choosing function, then we have the same restriction—if $\ell(\rho) = \ell$ and k is first such that $w(\ell)(0,k)\uparrow$, then $q_1(\rho)$ is computed using at most $\chi_{C_0} \mid (k+1)$. If we are using a strong witness-choosing function, then there is no restriction on the use of C_0.

Let (m, k) be the first pair such that $w(\ell)(m,k)\uparrow$.

2. If $m < n$, then we use no approximation in computing $q_n(\rho)$. The definition is like that of q_α. One effect of this is that if $\ell(\rho^+) = \ell'$ and (m', k') is the first pair not in $dom(w(\ell'))$, then either $length(\rho^+) = length(\rho)$, so $\ell(\rho) = \ell'$, or else $m' \geq n$.
3. If $m \geq n$, then we use an approximation of C_n. For the given α-table $(C_n)_{n<\alpha}$, there is some e such that $C_r = W_e^{C_{r-1}}$ for $0 < r < \alpha$. Let s be the length of ρ, and let σ be the restriction of $\chi_{C_{n-1}}$ described in 1 above. We compute $W^\sigma_{e,s}$, assuming non-convergence unless σ provides all of the oracle information. The restriction to s of the characteristic function of $W^\sigma_{e,s}$ will be called the *ρ-approximation* for C_n.
4. If $m > n$, then we approximate q_{n+1}. For sequences $\rho' \in P^{n+1}$ of length $< s$, we have an index for computing $q_{n+1}(\rho')$, using C_n. We suppose that the procedure from the index will converge, with a "sensible" value, given any "reasonable" approximation τ of χ_{C_n}. Suppose $\ell(\rho') = \ell$, and let y be first such that $w(\ell)(n,y)\uparrow$. A *reasonable* approximation v has the feature that for all $y' < y$, $v(y')$ matches the action in $w(\ell)(n,y')$, and $v(y) = u$, where u is 0 or 1. A value c is *sensible* if $\rho' c \in P^{n+1}$, and if $c = (\sigma, \tau, a)$, then $a(n,y) = u$. For the ρ-approximation for $q_{n+1}(\rho')$, we apply the procedure with the given index to input ρ' using the ρ-approximation for C_n. We always get sensible value c, and if our ρ-approximation for C_n is correct on all $y' \leq y$, then $c = q_{n+1}(\rho')$.

Before giving precise inductive definitions of ρ^+ and $q_n(\rho)$, we state some conditions that will be maintained.

Conditions to be maintained.

1. For $\rho \in P^n$, $\rho^+ \in P^{n+1}$,
2. $L(\rho^+)$ is a subsequence of $L(\rho)$,
3. $\ell(\rho^+) \leq_n \ell(\rho)$.

Case-by-case definitions. We say how to compute ρ^+ and $q_n(\rho)$, for $\rho \in P^n$ of length $2x+1$. First suppose $x = 0$. Then $\rho = \hat{\ell}$. We let $\hat{\ell}^+ = \hat{\ell}$. The first pair not in $dom(w(\hat{\ell}))$ must have the form $(m, 0)$. If $m < n$, then we let $q_n(\hat{\ell}) = (\hat{\ell}, \emptyset, a)$, where a maps the single pair $(m, 0)$ to $\chi_{C_m}(0)$—this is the same as $q_\alpha(\hat{\ell})$. If $m \geq n$, then we let $q_n(\hat{\ell}) = (\hat{\ell}, \emptyset, a)$, where a maps $(n-1, 0)$ to $\chi_{C_{n-1}}(0)$.

Now, suppose $x > 0$. Let $\rho = \hat{\ell} c_1 \ell_1 \ldots c_x \ell_x$ be an element of P^n of length $2x + 1$. For $k \leq x$, let ρ_k be the restriction of ρ of length $2k + 1$. We suppose that for $k < x$, ρ_k^+ and $q_n(\rho_k)$ have been determined. We say that ρ *follows* q_n if for all $k < x$, $c_{k+1} = q_n(\rho_k)$. First, suppose ρ does not follow q_n. Say k is first such that $c_{k+1} \neq q_n(\rho_k)$. Then $\rho^+ = \rho_k^+$, and $q_n(\rho)$ is an arbitrary c such that $\rho c \in P^n$. ¿From now on, we suppose that ρ follows q_n. We describe ρ^+ first, and then $q_n(\rho)$. Our ρ^+ will be an initial segment of a *tentative* version, ρ^{+T}, which is defined below.

First, suppose that $length(\rho_{x-1}^+) = length(\rho_{x-1})$. It follows that $\ell(\rho_{x-1}^+) = \ell_{x-1}$. Let (r, k) be the first pair in $\alpha \times \omega$ such that $w(\ell_{x-1})(r, k) \uparrow$. If $r \geq n$, then $\rho^{+T} = \rho_{x-1}^+$, and if $r < n$, then $\rho^{+T} = \rho_{x-1}^+ c_x \ell_x$. Now, suppose that $length(\rho_{x-1}^+) < length(\rho_{x-1})$. Then $\rho^{+T} = \rho_{x-1}^+ d \ell_x$, where d is the ρ_{x-1}-approximation for $q_{n+1}(\rho_{x-1}^+)$. We assume that that ρ^{+T} "follows" the ρ_{x-1}-approximation of q_{n+1}.

Having defined ρ^{+T}, we let ρ^+ be the greatest initial segment that follows the ρ_x-approximation of q_{n+1}. Now that we have ρ^+, we are ready to define $q_n(\rho)$. Let $\ell(\rho^+) = \ell$. Let (m, z) be the first pair such that $w(\ell_x)(m, z) \uparrow$. Let y be first such that $w(\ell)(n, y) \uparrow$, and let y' be first such that $w(\ell_x)(n - 1, y') \uparrow$.

CASE 1: Let $m < n$.

In this case, we ignore ρ^+. If $c_x = (\sigma, \tau, a)$, then $q_n(\rho) = (\sigma \ell_x, \emptyset, a')$, where $a'(m, z) = \chi_{C_m}(z)$.

In the remaining cases, we suppose $m \geq n$. Let (m', z') be the first pair such that $w(\ell)(m', z') \uparrow$. We should have $m' \geq n$, or ρ^+ would be longer.

CASE 2: Let $m' = n$.

CASE 2 (a): Suppose there exists $x' < x$ such that $\rho_{x'}^+ = \rho^+$.

This means that our approximation of $\chi_{C_n}(y)$ has changed, so that $w(\ell_x)(n, y)$ has action 0, while y is in the ρ-approximation for C_n. In this case, we let $q_n(\rho) = (\sigma, n\ell_x, a)$, where σ is the first component of $q_n(\rho_{x'})$, $a(n, y) = 1$, and $a(n - 1, y') = \chi_{C_{n-1}}(y')$.

CASE 2 (b): Suppose there is no x' as in Case 2 (a).

Then $\ell = \ell_x$ and $w(\ell_x)(n, y) \uparrow$. In this case, we let $q_n(\rho) = (\sigma \ell_x, n\ell_x, a)$, where σ is the first component of c_n, $a(n, y)$ is the ρ-approximation for $\chi_{C_n}(y)$, and $a(n - 1, y') = \chi_{C_{n-1}}(y')$.

CASE 3: Let $m' > n$.

CASE 3 (a): Suppose there exists $x' < x$ such that $\rho_{x'}^+ = \rho^+$ and $length(\rho^+) < length(\rho_{x'})$.

Again, this means that our approximation of $\chi_{C_n}(y)$ has changed, so that $w(\ell_x)(n, y)$ has action 0, while y is in the ρ-approximation for C_n. Say the ρ-approximation for $q_{n+1}(\rho^+)$ is (σ, τ, a), where $a(n, y) = 1$. We let

$q_n(\rho) = (\sigma, \tau n\ell_x, a')$, where a' is the extension of a mapping $(n-1, y')$ to $\chi_{C_{n-1}}(y')$.

CASE 3 (b): Suppose there is no x' as in Case 3 (a), but $length(\rho^+) < length(\rho)$.

Then $w(\ell_x)(n, y) \uparrow$. Say the ρ-approximation for $q_{n+1}(\rho^+)$ is (σ, τ, a)—$a(n, y)$ is the ρ-approximation for $\chi_{C_n}(y)$. We let $q_n(\rho) = (\sigma, \tau n\ell_x, a')$, where a' is the extension of a mapping $(n-1, y')$ to $\chi_{C_{n-1}}(y')$.

CASE 3 (c): Suppose there is no x' as in Case 3 (a), and $length(\rho^+) = length(\rho)$.

We have no ρ-approximation for $q_{n+1}(\rho^+)$, so we do not want to act on (n, y). Say σ is the first component of c_x. We let $q_n(\rho) = (\sigma\ell_x, \emptyset, a)$, where $a(n-1, y') = \chi_{C_{n-1}}(y')$.

We have described the instruction functions q_n, for $1 \leq n < \alpha$.

The next lemma is clear from the definitions.

LEMMA 5.4. *If π^n is a run of (P^n, q_n), then for all x, $w(\pi^n)(n-1, x) \downarrow$, with action $\chi_{C_{n-1}}(x)$. If π^n is computable in C_{n-1}, then $E_{n-1}(\pi^n)$ is c.e. relative to C_{n-1}. Moreover, given an index for π^n, we can find an index for $E_{n-1}(\pi^n)$.*

The next three lemmas give the existence of a special family of runs of our trees and instruction functions. For $1 \leq \beta \leq \alpha$, π^β will be a run of (P^β, q_β). The whole family will be "coherent", by which we mean that if $1 \leq \beta < \gamma \leq \alpha$, then $L(\pi^\gamma)$ is a subsequence of $L(\pi^\beta)$. In addition, for $1 \leq n < \alpha$, π^n will be computable in C_{n-1}, uniformly in n. The lemma below yields π^1.

LEMMA 5.5. *There is a run $\pi^1 = \hat{\ell} c_1 \ell_1 c_2 \ell_2$ of (P^1, q_1), computable in C_0, such that for all x, $w(\pi^1)(0, x) \downarrow$, with action $\chi_{C_0}(x)$. If we are using a strong witness-choosing function b, then the coding witness for $(0, x)$ is $b(x)$. If we are using a weak witness-choosing function b, then the coding witness for $(0, x)$ is $b(\ell_x)$, where this is computed effectively using just $C_0 \cap x$.*

PROOF. First, suppose we are using a strong witness-choosing function b. We form a run $\pi^1 = \hat{\ell} c_1 \ell_1 c_2 \ell_2 \ldots$, getting c_x from q_1, which is computable in C_0, and using the fact that P^1 is c.e. relative to C_0 to find a suitable completion ℓ_x. The definitions guarantee that for each x, $w(\ell_{x+1})(0, k) \downarrow$, with coding witness $b(x)$, and with action $\chi_{C_0}(x)$.

Now, suppose we are using a weak witness-choosing function b. Again, we form a run $\pi^1 = \hat{\ell} c_1 \ell_1 c_2 \ell_2 \ldots$, getting c_x from q_1, and using the fact that P^1 is c.e. to find ℓ_x. The part of C_0 coded in ℓ_x is $C_0 \cap x$, and c_{x+1} is computed using $C_0 \cap (x+1)$. The coding witness for $(0, x)$ is $b(\ell_x)$, or $b(\hat{\ell})$, if $x = 0$, and ℓ_{x+1} is computed effectively, given c_{x+1}.

The next lemma lets us pass from a run π^n of (P^n, q_n) to a run π^{n+1} of (P^{n+1}, q_{n+1}).

Lemma 5.6. *Let $1 \leq n < \alpha$. If π^n is a run of (P^n, q_n), computable in C_{n-1}, then there is a run π^{n+1} of (P^{n+1}, q_{n+1}), computable in C_n, such that $L(\pi^{n+1})$ is a subsequence of $L(\pi^n)$. Moreover, given an index for π^n, relative to C_{n-1}, we can find an index for π^{n+1}, relative to C_n.*

Proof. Let $\pi^n = \hat{\ell} c_1^n \ell_1^n \ldots$. Let ρ_s be the initial segment of π^n of length $2s+1$. We define a sequence of numbers $(s(k))_{k \in \omega}$, computable in C_n, such that $\rho^+_{s(k)}$ has length $2k+1$, and for all $s > k$, $\rho^+_s \supseteq \rho^+_{s(k)}$. Let $s(0) = 0$. Suppose we have $s(k)$, where $\ell(\rho^+_{s(k)}) = \ell$. Let x be first such that $w(\ell)(n, x) \uparrow$. Then $s(k+1)$ is the first $s > s(k)$ such that the stage s approximation of $C_n(x)$ is correct, and for the first (m, y) such that $w(\ell_{s-1}) \uparrow$, we have $m \geq n$. Then π^{n+1} is the infinite sequence whose initial segment of length $2k+1$ is $\rho^+_{s(k)}$.

Lemma 5.7. *Let $1 \leq n < \alpha$. If π^n and π^{n+1} are as in Lemma 5.6, then $w(\pi^{n+1})$ is defined on all pairs $(n, x) \in \alpha \times \omega$, with action $\chi_{C_n}(x)$, and with coding witness that can be effectively determined using C_{n-1} and $C_n \cap x$.*

Proof. The fact that $w(\pi^{n+1})$ is defined on all pairs (n, x) is clear from the definition of q_{n+1}. To determine the coding witness, we look for the first term ℓ in $L(\pi^n)$ such that $w(\ell)(n, y)$ is defined for all $y \leq x$, with action $\chi_{C_n}(y)$ for all $y < x$. The coding witness in $w(\ell)(n, x)$ will be preserved, so that it is the one in $w(\pi^{n+1})(n, x)$.

We are aiming for a coherent family $(\pi^\beta)_{1 \leq \beta \leq \alpha}$ of runs. First, suppose that α is finite, say $\alpha = N$. Combining Lemmas 5.5, 5.6, and 5.7, we obtain a coherent family $(\pi^n)_{1 \leq n \leq N}$ such that π^n is a run of (P^n, q_n), computable in C_{n-1}. By Lemma 5.7, $w(\pi^{n+1})(n, x) \downarrow$ for all pairs $(n, x) \in N \times \omega$, with action $\chi_{C_n}(x)$, and with coding witness that we can determine, via a uniform effective procedure, using C_{n-1} and $C_n \cap x$—if we are using a strong witness-choosing function b, then the witness for $(0, x)$ is $b(x)$. Now, suppose $\alpha = \omega$. Lemmas 5.5, 5.6, and 5.7 yield a family $(\pi_n)_{1 \leq n < \omega}$ such that for all $n < \omega$, $L(\pi^{n+1})$ is a subsequence of $L(\pi^n)$. The next lemma lets us add a run π^ω of (P^ω, q_ω) to complete the coherent family.

Lemma 5.8. *Suppose $\alpha = \omega$, and let $(\pi^n)_{1 \leq n < \omega}$ be a family of sequences such that for all n, π^n is a run of (P^n, q_n) and $L(\pi^{n+1})$ is a subsequence of $L(\pi^n)$. Then for each k, the initial segment of π^n of length $2k+1$ is the same for all sufficiently large n. Moreover, the limit is a run π^ω of (P^ω, q_ω).*

Proof. Let σ^n_k be the initial segment of π^n of length $2k+1$. We show, by induction on k, that for all sufficiently large n, σ^n_k is an element of P^ω following q_ω. For $k = 0$, this is clear. Suppose it holds for k, and let σ_k be the limit of the sequences σ^n_k. Say $\ell(\sigma_k) = \ell$. If (m, x) is the first pair not in $dom(w(\ell))$, then for all $n > m$, $q_n(\sigma_k) = q_\omega(\sigma_k) = (L(\sigma_k), \emptyset, a)$, where a maps (m, x) to $\chi_{C_m}(x)$. Moreover, for all sufficiently large n, $q(\sigma_k)$ is followed, in π^n, by the same completion.

We are ready to complete the proof of Theorem 5.1. Let $\pi = L(\pi^\alpha)$, where π^α is the top sequence in our coherent family. We can show that π has all of the properties we need. By Lemmas 5.7 and 5.3, π is a path through P, and for all pairs $(n, x) \in \alpha \times \omega$, $w(\pi^{n+1})(n, x) \downarrow$, with action $\chi_{C_n}(x)$, and with coding witness that we can determine, by a uniform effective procedure, using C_{n-1} and $C_n \cap x$—if we are using a strong witness-choosing function b, then we get the coding witness for $(0, x)$ just from b. Since π is a subsequence of π^{n+1} and w_n is preserved in π^{n+1}, $w(\pi)(n, x) = w(\pi^{n+1})(n, x)$. By Lemma 5.5, $E_n(\pi^{n+1})$ is c.e. relative to C_n, uniformly in n. Since π is a subsequence of π^{n+1}, and $\leq_n$ is preserved in π^{n+1}, $E_n(\pi) = E_n(\pi^{n+1})$. This completes the proof of the theorem.

We relativize Theorem 5.1—the version using a strong witness-choosing function—to an arbitrary set X in the following way.

THEOREM 5.9 (Relativization). *Suppose* $1 \leq \alpha \leq \omega$. *Let*

$$(L, w, E, b, \hat{\ell}, P, (\leq_n)_{n \in \alpha})$$

be a coding and enumerating α-system, with a strong witness-choosing function b. Suppose L and P are c.e. relative to X, w and E are partial computable, b is computable, and the relations $\leq_n$ are c.e. relative to X, uniformly in n. Let $(C_n)_{n \in \alpha}$ be an α-table, where $X \leq_T C_0$. Then P has a path π such that $E_n(\pi)$ is c.e. relative to C_n, uniformly in n, for $(n, x) \in \alpha \times \omega$, $w(\pi)(n, x) \downarrow$, with action $\chi_{C_n}(x)$. If $n \geq 1$, then the coding witness in $w(\pi)(n, x)$ is effectively determined using C_{n-1} and $C_n \cap x$, and if $n = 0$, then the coding witness in $w(\pi)(0, x)$ is $b(x)$.

We shall apply Theorem 5.9, taking $X = C_0$, to prove Theorem 4.2. We have already described a coding and enumerating ω-system. In this system, L, P, and $\leq_n$ are not c.e. relative to C_0. However, we have an effective procedure for completing pictures, using C_0. In the next section, we show that this is enough.

§6. The c.e. core. In this section, we show how, given a coding and enumerating system in which the set L, the tree P, and the relations $\leq_n$ are not c.e., it is possible to extract a new system in which the corresponding objects are c.e. We use a strong witness-choosing function.

THEOREM 6.1. *Let $(L, w, E, b, \hat{\ell}, P, (\leq_n)_{n \in \alpha})$ be a coding and enumerating α-system, with a strong witness-choosing function b. Suppose w and E are the restrictions to L, $\alpha \times L$, of partial computable functions, and suppose b is computable. Finally, suppose that there is a partial computable function which, when applied to a picture, yields a completion. Then there is a coding and enumerating α-system $(L^*, w^*, E^*, b, \hat{\ell}, P^*, (\leq_n^*)_{n \in \alpha})$, with the same strong witness-choosing function b, where $L^* \subseteq L$, $w^* = w \mid L^*$, $E^* = E \mid (\alpha \times L^*)$, $P^* \subseteq P$, and $\leq_n^* \subseteq \leq_n$, and L^*, P^*, and the relations $\leq_n^*$ are uniformly c.e.*

PROOF. We generate L^*, P^*, $\leq_n^*$, and a subset C^* of the set C of pictures, according to the following rules:

1. $\hat{\ell} \in L^*$, and $\hat{\ell} \in P^*$.
2. Let $c = (\sigma\ell^0, \tau, a)$, where $\sigma\ell^0 \in P^*$ and τ and a satisfy (a) or (b) below:
   (a) $\tau = \emptyset$, a maps (n, x) to $\{0, 1\}$, and x is first such that $w(\ell^0)(n, x) \uparrow$,
   (b) $\tau = n_0\ell^1 n_1 \ldots n_{k-1}\ell^k$ and a maps $\{(n_0, x_0), \ldots, (n_k, x_k)\}$ to $\{0, 1\}$, where
      (i) $\ell^i \in L^*$,
      (ii) $n_k < \cdots < n_0 < \alpha$,
      (iii) $\ell^0 \leq_{n_0}^* \cdots \leq_{n_{k-1}}^* \ell^k$,
      (iv) for $i \leq k$, x_i is first such that $w(\ell^i)(n_i, x_i) \uparrow$, and if $i < k$ and $w(\ell^{i+1})(n_i, x_i) \downarrow$, then $w(\ell^{i+1}(n_i, x_i))$ has action 0, while $a(n_i, x_i) = 1$.

   Then $c \in C^*$.
3. Suppose $c \in C^*$, and let ℓ be the effectively determined completion of c. Then $\ell \in L^*, \sigma\ell \in P^*, \ell^0 \leq_n^* \ell$ for all $n < \alpha$, and for $0 < i \leq k, \ell^i \leq_{n_i}^* \ell$.
4. If $\ell \in L^*$, then $\ell \leq_n^* \ell$.
5. If $\ell, \ell', \ell'' \in L^*$ and $\ell \leq_n^* \ell' \leq_n^* \ell''$, then $\ell \leq_n^* \ell''$.
6. If $\ell, \ell' \in L^*$ and $\ell \leq_n^* \ell'$, where $m < n$, then $\ell \leq_m^* \ell'$.

Each element of L^*, each sequence in P^*, each triple in C^*, and each pair in $\leq_n^*$ has a derivation involving finitely many applications of the rules. It follows that these objects are uniformly c.e. We can show by induction on the derivations that $L^* \subseteq L$, $P^* \subseteq P$, $C^* \subseteq C$, and for all n, $\leq_n^* \subseteq \leq_n$, and $\leq_n^*$ is a binary relation on L^*. We should check the four conditions for a coding and enumerating α-system.

1. Rules 4 and 5 guarantee that $\leq_n^*$ is reflexive and transitive.
2. Rule 6 guarantees that $\leq_{n+1}^*$ implies $\leq_n^*$.
3. Suppose $\ell \leq_n^* \ell'$. Since $\leq_n^* \subseteq \leq_n$, and E^* is the restriction of E to $\alpha \times L^*$, we have $E_n^*(\ell) \subseteq E_n^*(\ell')$. Similarly, since w^* is the restriction of w to L^*, we have $w_n^*(\ell) \subseteq w_n^*(\ell')$.
4. Rule 2 guarantees that if $c = (\sigma, \tau, a)$ is a picture for the new system, then it is in C^*. Rule 3 guarantees that all pictures in C^* have completions.

We can relativize Theorem 6.1 to an arbitrary set X. The relativized statement requires a procedure for completing pictures that is computable in X. Then the derived coding and enumerating system has components that are c.e. relative to X, uniformly.

REMARK. The ideas in Theorem 6.1 can be used not just in the current setting, but also with Ash's original metatheorem, or other variants.

§7. Completing the proof of Theorem 4.2. Theorem 4.2, the non-trivial half of our main theorem, is re-stated below.

THEOREM. If $(C_n)_{n\in\omega}$ is an ω-table over a completion of PA, then there is a non-standard model $\mathcal{A}$ of PA such that $D_n(\mathcal{A}) \equiv_T C_n$, uniformly in n.

PROOF. In Section 4, we defined the components of a coding and enumerating ω-system $(L, w, E, b, \hat{\ell}, P, (\leq_n)_{n\in\omega})$, with a strong witness-choosing function b. The four conditions are given in Lemmas 4.4, 4.5, 4.6, and 4.7. The functions w and E are restrictions to L, $L \times \omega$, of partial computable functions, and the function b is computable. The components L, P, and $\leq_n$ are not c.e. relative to C_0. However, by Lemma 4.7, we have an effective procedure, using C_0, for completing pictures. Therefore, we can apply Theorem 6.1, relativized to C_0. We obtain a new coding and enumerating ω-system

$$(L^*, w^*, E^*, b, \hat{\ell}, P^*, (\leq_n^*)_{n\in\omega}),$$

with the same strong witness-choosing function b, where $L^* \subseteq L$, $P^* \subseteq P$, and $\leq_n^* \subseteq \leq_n$, but L^*, P^*, and $\leq_n^*$ are c.e. relative to C_0, uniformly. Any path through P^* is a path through P.

We apply Theorem 5.9 to the derived system, taking $X = C_0$. We obtain a path π such that for all $n < \alpha$, $E_n(\pi)$ is c.e. relative to C_n, uniformly in n. For all pairs $(n, x) \in \omega \times \omega$, $w(\pi)(n, x) \downarrow$, with action $\chi_{C_n}(x)$. The coding witness in $w(\pi)(n, x)$ is effectively determined using C_{n-1} and $C_n \cap x$, for $n \geq 1$, and for $n = 0$, the coding witness in $w(\pi)(0, x)$ is $b(x)$. In Section 4, we saw that this is enough to prove Theorem 4.2.

REFERENCES

[1] A. ARANA, *Solovay's Theorem cannot be simplified*, ***Annals of Pure and Applied Logic***, vol. 112 (2001), pp. 27–41.

[2] C. J. ASH, *Recursive labeling systems and stability of recursive structures in hyperarithmetical degrees*, ***Transactions of the American Mathematical Society***, vol. 298 (1986), pp. 497–514, Corrections, *ibid*, vol. 310 (1988), p. 851.

[3] C. J. ASH and J. F. KNIGHT, *Ramified systems*, ***Annals of Pure and Applied Logic***, vol. 70 (1994), pp. 205–221.

[4] ———, *Coding a family of sets*, ***Annals of Pure and Applied Logic***, vol. 94 (1998), pp. 127–142.

[5] ———, ***Computable structures and the hyperarithmetical hierarchy***, Elsevier, 2000.

[6] V. S. HARIZANOV, J. F. KNIGHT, and A. S. MOROZOV, *Sequences of n-diagrams*, ***The Journal of Symbolic Logic***, vol. 67 (2002), pp. 1227–1247.

[7] J. F. KNIGHT, *True approximations and models of arithmetic*, ***Models and computability*** (B. Cooper and J. Truss, editors), Cambridge University Press, 1999, pp. 255–278.

[8] ———, *Models of arithmetic: quantifiers and complexity*, ***Reverse mathematics*** (S. Simpson, editor), Lecture Notes in Logic, vol. 21, AK Peters, to appear.

[9] A. MACINTYRE and D. MARKER, *Degrees of recursively saturated models*, ***Transactions of the American Mathematicsl Society***, vol. 282 (1984), pp. 539–554.

[10] D. SCOTT, *Algebras of sets binumerable in complete extensions of arithmetic*, ***Recursive function theory*** (Dekker, editor), American Mathematical Society, 1962, pp. 117–122.

UNIVERSITY OF NOTRE DAME, DEPARTMENT OF MATHEMATICS
255 HURLEY HALL, NOTRE DAME, IN 46556-4618, USA
E-mail: Julia.F.Knight.1@nd.edu

THE LIMIT THEORY OF GENERIC POLYNOMIALS

PASCAL KOIRAN

Abstract. We show that the set T of first-order sentences satisfied by all generic polynomials of sufficiently high degree forms a complete theory. As a consequence, the set of complex polynomials of even degree is not first-order definable. Another consequence of the results in this paper is that T admits an analytic model.

§1. Introduction. A generic polynomial is a polynomial $f: \mathbb{C} \to \mathbb{C}$ of the form

$$f(x) = c_1 x + \cdots + c_d x^d$$

where the coefficients $c_1, \ldots, c_d$ are algebraically independent over $\mathbb{Q}$. One could work in the language of fields expanded with a unary function symbol f. Here we prefer to work in the language of "curved fields": the language of fields expanded with a binary predicate C (which stands for $y = f(x)$). This choice emphasizes the connection with generic curves [2], [5] and turns out to be more convenient for the explicit construction of a model by an amalgamation method à la Hrushovski (see [8] for a survey on this topic).

In this paper we show that the parity of the degree of complex polynomials is not definable. This means that there does not exist a first-order sentence F of the language of curved fields such that:

(i) F is true in $\mathbb{C}$ whenever C is interpreted by the graph of a polynomial of even degree.

(ii) F is false in $\mathbb{C}$ whenever C is interpreted by the graph of a polynomial of odd degree.

Note that the parity of the degree of *real* polynomials is definable. This follows from the fact that a real polynomial is of odd degree if and only if it has different limits at $\pm\infty$ (more on this in section 2).

Our undefinability result follows immediately from the following main result: a sentence of the language of curved fields is either true for all generic polynomials of sufficiently high degree or false for all generic polynomials of sufficiently high degree. The set T of sentences which are true for all

Logic Colloquium '01
Edited by M. Baaz, S. Friedman, and J. Krajíček
Lecture Notes in Logic, 20

generic polynomials of sufficiently high degree therefore forms a consistent and complete theory: the "limit theory of generic polynomials".

Starting from somewhat different motivations, Zilber has independently constructed a related theory [12]. He works in a richer language with countably many binary predicates, which are to be thought of as the graphs of the higher order derivatives of the original function.

In section 3 we give an axiomatization of a theory which is shown to be consistent and (more crucially) complete. It will turn out in section 4 that this theory really is the limit theory of generic polynomials.

By elementary equivalence the above results apply not only to the field of complex numbers but to any algebraically closed field of characteristic 0. They seem quite likely to hold also in positive characteristic (this is definitely true for the consistency and completeness results of section 3).

One could naively imagine that generic polynomials are easier to understand than generic curves. Even though the results and the general proof strategy are similar, the opposite is true: the axiomatization is slightly more complicated for generic polynomials, and it is significantly more difficult to show that the axioms are satisfied when the degree is sufficiently high. The proof given in section 4 relies on the intersection theorems of section 5 and of [6].

It has been shown in [4] that the complex field expanded by a Liouville function as defined by Wilkie [11] is an analytic model of the limit theory of generic polynomials.

§2. Case of the real field. We have just seen that the set of real polynomials of even degree is definable. Given $p \geq 3$, is it also possible to define the set of real polynomials whose degree is a multiple of p? It turns out that the answer to this question is negative.

PROPOSITION 1. *Let $p \geq 3$ be an integer. There exists no sentence F of the language of curved fields such that when the curve C is interpreted by the graph of a degree d polynomial, $(\mathbb{R}, C) \models F$ if and only if d is a multiple of p.*

PROOF. Assume by contradiction that such a sentence exists for some even integer $p \geq 4$. Write $p = 2q$. When we interpret C by the graph of the polynomial $y = x^{2n}$, $(\mathbb{R}, C) \models F$ if and only if n is a multiple of q. Let $G(n, x, y)$ be the following formula of language of the real exponential field:

$$(x = 0 \wedge y = 0) \vee \exists z\ (\exp(z) = x^2 \wedge y = \exp(nz)).$$

When n is an integer, $G(n, ., .)$ defines the graph of the polynomial $y = x^{2n}$. Let $H(n)$ be the formula of the real exponential field obtained from F by replacing each instance $C(x, y)$ of C by $G(n, x, y)$. This formula should be true whenever the integer n is a multiple of q, and false otherwise. This is impossible for $q \geq 2$ since the real exponential field is o-minimal [10].

Assume now that $p \geq 3$ is an odd integer, and that it is possible to define the set of real polynomials of degree multiple of p. Since the set of real polynomials of even degree is definable, one could also define the set of real polynomials of degree multiple of $2p$. We have just shown that this is impossible. ⊣

Several other undefinability results can be obtained by the same o-minimality argument [3], [9]. Unfortunately, for algebraically closed fields we do not know any theory which could play the same "universal" role. More ad-hoc constructions are therefore necessary. The limit theory of generic curves is a tool which makes it possible to obtain several undefinability results for algebraic curves [2]. The theory constructed and studied in the rest of this paper plays a similar role for definability problems involving polynomials over an algebraically closed field of characteristic 0.

§3. Construction of the limit theory. In this section we work in an algebraically closed field K of arbitrary characteristic. We denote by Q the prime field of K. Given n elements $x_1, \dots, x_n$ of a model, we denote the tuple $(x_1, \dots, x_n)$ by x. When there is a risk of confusion between tuples and elements we may also write $\overline{x}$ instead of x.

We consider the theory T defined by the following axioms.

1. The axioms of algebraically closed fields of a fixed characteristic.
2. The functional axiom

$$\forall x, y_1, y_2\ C(x, y_1) \wedge C(x, y_2) \rightarrow y_1 = y_2.$$

3. The function is everywhere defined: $\forall x \exists y C(x, y)$.
4. $C(0, 0)$.
5. The universal axioms. Let $\phi(x_1, y_1, \dots, x_n, y_n)$ be a conjunction of polynomial equations with coefficients in the prime field. If the subset of K^{2n} defined by ϕ is of dimension $< n$, we add the axiom

$$\forall x_1, y_1, \dots, x_n, y_n\ A(x_1, y_1, \dots, x_n, y_n) \rightarrow \neg\phi(\overline{x}, \overline{y}) \qquad (1)$$

where $A(x_1, y_1, \dots, x_n, y_n)$ stands for

$$\bigwedge_{i \neq j} (x_i, y_i) \neq (x_j, y_j) \wedge \bigwedge_{i=1}^{n} x_i \neq 0 \wedge \bigwedge_{i=1}^{n} C(x_i, y_i).$$

Instead of these axioms, one could also use an equivalent set of universal axioms in the style of [2].

6. The inductive axioms. Let $\phi(x_1, y_1, \dots, x_n, y_n, \overline{z})$ be a conjunction of polynomial equations with coefficients in the prime field. For any fixed value of the parameter $\overline{z}$, ϕ defines an algebraic subset $V_{\overline{z}}$ of K^{2n}. Let $\xi(\overline{z})$ be a formula of the language of fields which states that $V_{\overline{z}}$ is irreducible and is not contained in a subspace of the form $x_i = x_j$ for some $i \neq j$, or of the form $x_i = c$ for some element c in the model.

Let ε be a function which chooses one variable $u_i^\varepsilon \in \{x_i, y_i\}$ for every $i \in \{1, \ldots, n\}$. For each value of the parameter $\overline{z}$, the formula $\exists u_1^\varepsilon, \ldots, u_n^\varepsilon \phi(\overline{x}, \overline{y}, \overline{z})$ defines a constructible set $C_{\overline{z}}^\varepsilon \subseteq K^n$. As pointed out in [2], there is a formula $\psi_\varepsilon(\overline{z})$ of the language of fields which states that $C_{\overline{z}}^\varepsilon$ is dense in K^n. Let $\psi(\overline{z})$ be the disjunction of the 2^n formulas $\psi_\varepsilon(\overline{z})$. Let θ be the conjunction of ξ and ψ. We add the following axiom:

$$\forall \overline{z}\, \exists x_1, y_1, \ldots, x_n, y_n\ \theta(\overline{z}) \rightarrow \bigwedge_{i=1}^{n} C(x_i, y_i) \wedge \phi(\overline{x}, \overline{y}, \overline{z}). \tag{2}$$

3.1. Consistency. The consistency of T follows from the results of section 4. In this subsection we give a simpler and more direct proof. We shall construct a model K of T which is the union of an increasing sequence of "curved fields" K_i. Each K_i is an algebraically closed field of finite transcendence degree endowed with a "curve" C_i made up of finitely many points. We start from $K_0 = \overline{Q}$ with the curve $C_0 = \{(0,0)\}$. One goes from K_i to K_{i+1} by application of one of the two following steps:

(i) We choose some $a \in K_i$ such that $K_i \models \forall y\ \neg C(a, y)$. K_{i+1} is the algebraic closure of $K_i \cup \{(a, b)\}$ where b is transcendental over K_i, and $C_{i+1} = C_i \cup \{(a, b)\}$.

(ii) We choose a conjunction of polynomial equations $\phi(\overline{x}, \overline{y}, \overline{z})$ and a tuple $\overline{z}$ of parameters from K_i which satisfies formula θ in (2). K_{i+1} is the algebraic closure of $K_i \cup \{(a_1, b_1), \ldots, (a_n, b_n)\}$ where $(a_1, b_1, \ldots, a_n, b_n)$ is a generic point over K_i of the irreducible variety $V_{\overline{z}}$, and $C_{n+1} = C_n \cup \{(a_1, b_1), \ldots, (a_n, b_n)\}$. Note that the $a_1, \ldots, a_n$ all lie outside K_i and are distinct from each other due to condition $\xi(\overline{z})$ in the inductive axioms. Moreover, for any subset Γ of $\{(a_1, b_1), \ldots, (a_n, b_n)\}$ the transcendence degree of Γ over K_i is at least equal to the cardinality of Γ due to condition $\psi(\overline{z})$.

A straightforward induction on i shows that n points on C_i with nonzero first coordinates must have transcendence degree at least n over the prime field. Hence all the K_i (and K itself) satisfy the universal axioms. It is also true that the K_i, and therefore K, satisfy the functional axiom. This follows from the fact that in step (i) we only consider a point a whose image is not already defined, and that in step (ii) $a_1, \ldots, a_n$ are distinct elements of $K_{i+1} \setminus K_i$. Finally, it follows from a standard diagonal argument that we can alternate steps (i) and (ii) in order to obtain a function which is defined everywhere and satisfies the inductive axioms.

3.2. Completeness. We start with a strengthening of the inductive axioms.

LEMMA 1. *Let M be a model of T and $\phi(x_1, y_1, \ldots, x_n, y_n, \overline{z})$ a conjunction of polynomial equations where the tuple $\overline{z}$ of parameters satisfies the formula θ in the inductive axioms. Let S be a finite subset of M^2. There exist n distinct points in $C \setminus S$ whose $2n$ coordinates satisfy ϕ.*

PROOF. Let $S = \{(a_1, b_1), \dots, (a_s, b_s)\}$. We add a pair (x_0, y_0) of new variables and consider the formula ϕ' in $2n+2$ free variables (and the parameters $\bar{z}' = \bar{z}\hat{}(a_1, b_1, \dots, a_s, b_s)$) which is the conjunction of ϕ and of the formula

$$y_0 \prod_{0<i<j} (x_i - x_j) \times \prod_{i,j>0} (x_i - a_j) = 1.$$

Note that x_0 does not appear in ϕ'. We just have to check that $\bar{z}'$ satisfies the condition $\theta' = \xi' \wedge \psi'$ which is associated to ϕ' in the inductive axioms.

The variety V' defined by ϕ' is irreducible since the same is true of ϕ. Let $(\alpha_0, \beta_0, \dots, \alpha_n, \beta_n)$ be a point of V' which is generic over $k' = Q(\bar{z}')$. By the hypothesis on ϕ each α_i is transcendental over k', and these $n+1$ coordinates are distinct from each other. Moreover there exist n coordinates $u_1 \in \{\alpha_1, \beta_1\}, \dots, u_n \in \{\alpha_n, \beta_n\}$ such that $(\alpha_0, u_1, \dots, u_n)$ is of transcendence degree $n+1$ over k'. Hence we can indeed apply the inductive axiom associated to ϕ'. ⊣

Let M be a model of T and k an algebraically closed subfield of M of finite transcendence degree. We define $\delta(k) = \text{tr.deg}(k) + 1 - |C \cap k^2|$. The universal axioms imply that $\delta(k) \geq 0$ for all k. We say that k is self-sufficient if for all algebraically closed subfields $l \subseteq M$ of finite transcendence degree, $k \subseteq l$ implies $\delta(k) \leq \delta(l)$. Any algebraically closed subfield of M of finite transcendence degree has a self-sufficient extension (consider an extension of δ as small as possible).

An extension $k \subset l$ with k and l self-sufficient is said to be minimal if there exists no self-sufficient structure k' with $k \subset k' \subset l$. Any extension $k \subset l$ with k and l self-sufficient can be broken down into a tower of minimal extensions.

LEMMA 2. *Let M be an ω-saturated model of T and k a self-sufficient substructure of M. There exists $\alpha \in M \setminus k$ such that the algebraic closure l of $k \cup \{\alpha\}$ is self-sufficient and there is no point on the curve in $l^2 \setminus k^2$.*

PROOF. Let $(x_1, y_1), \dots, (x_n, y_n)$ be the (finitely many) points on the curve in k^2, and $(a_1, \dots, a_r)$ a transcendence basis of k. We define three families $\mathcal{F}, G, H$ of first-order formulas in one free variable c. The first family is made of all formulas of the form $P(a_1, \dots, a_r, c) \neq 0$ where P is a nonzero polynomial with integer coefficients. Obviously, any point α which satisfies this family lies outside k. The second family is made of all formulas of the form

$$\neg\exists x, y \ \ [\phi(c) \wedge P(x, a_1, \dots, a_r, c) = 0 \wedge Q(y, a_1, \dots, a_r, c) = 0$$
$$\wedge\, C(x, y) \wedge \bigwedge_{i=1}^{n} (x, y) \neq (x_i, y_i)]$$

where P and Q are nonzero polynomials with integer coefficients, and ϕ states that none of the two univariate polynomials $x \mapsto P(x, a_1, \dots, a_r, c)$ and $y \mapsto Q(y, a_1, \dots, a_r, c)$ are identically zero. If α satisfies this family, any point on the curve with both coordinates in the algebraic closure of $k \cup \{\alpha\}$ is in fact in k^2. Moreover, this family is satisfied by all $\alpha \in k$ since $(x_1, y_1), \ldots, (x_r, y_r)$ are the only points on the curve with both coordinates in this algebraically closed field.

The third family is made of all formulas of the form

$$\neg\exists \overline{u}, \overline{v} \bigwedge_{i=1}^{m} C(u_i, v_i) \wedge \psi(\overline{u}, \overline{v}, \overline{a}, c) \wedge \theta(c) \wedge \bigwedge_{i,j} (u_i, v_i) \neq (x_j, y_j)$$

where $\overline{u} = (u_1, \ldots, u_m)$, $\overline{v} = (v_1, \ldots, v_m)$, $\overline{a} = (a_1, \ldots, a_r)$, ψ is a conjunction of polynomial equations with integer coefficients, and θ states that the subset of all $(u_1, \ldots, u_m, v_1, \ldots, v_m) \in M^{2m}$ which satisfy $\psi(\overline{u}, \overline{v}, \overline{a}, c)$ has dimension at most $m - 1$. If α satisfies this family, the algebraic closure of $k \cup \{\alpha\}$ is self-sufficient. Moreover, all $\alpha \in k$ satisfy this family since k is self-sufficient.

By ω-saturation we just need to check that $\mathcal{F} \cup \mathcal{G} \cup \mathcal{H}$ is finitely satisfiable. This is clear since any finite subset of $\mathcal{F}$ is satisfiable by an element of k, and any element of k satisfies all of $\mathcal{G}$ and $\mathcal{H}$. ⊣

We also need the "independent marriage lemma" from [2].

LEMMA 3. *Let K be a field. Fix an arbitrary set $\{(x_1, y_1), \ldots, (x_n, y_n)\}$ of n points of K^2, and a subfield $k \subseteq K$. The two following properties are equivalent:*

(i) *There exist n coordinates $t_1 \in \{x_1, y_1\}, \ldots, t_n \in \{x_n, y_n\}$ which are algebraically independent over k.*

(ii) *For any integer $m \leq n$ and any subset $\{(x_{i_1}, y_{i_1}), \ldots, (x_{i_m}, y_{i_m})\}$ of our set of n points, the transcendence degree of $(x_{i_1}, y_{i_1}, \ldots, x_{i_m}, y_{i_m})$ over k is at least m.*

The completeness of T follows immediately from the next proposition.

PROPOSITION 2. *Let M and M' be ω-saturated models of T. The family of partial isomorphisms between self-sufficient substructures of M and M' has the back-and-forth property.*

PROOF. Let $\sigma : k \to k'$ be such an isomorphism. By symmetry it suffices to show that for every $\alpha \in M \setminus k$ there exists a partial isomorphism τ between self-sufficient substructures which extends σ and is defined in α. Two cases can be distinguished:

(i) There exists an extension l of $k \cup \{\alpha\}$ such that $\delta(l) = \delta(k)$.

(ii) There is no such extension.

In the first case l is self-sufficient since k is self-sufficient. Without loss of generality we may assume that the extension $k \subset l$ is minimal. This leaves only two possibilities:

(a) There exists a point (a, b) on the curve with $a \in k$ and $b \in l \setminus k$.

By minimality of l, (a, b) is unique and l is the algebraic closure of $k \cup \{b\}$. There is a unique $b' \in M'$ such that $M' \models C(\sigma(a), b')$. Let l' be the algebraic closure of $k' \cup \{b'\}$. Since b is transcendental over k and b' over k', there exists an isomorphism of fields $\tau: l \to l'$ which extends σ and maps b to b'. By self-sufficiency (a, b) is the only point on the curve of M in $l^2 \setminus k^2$, and $(\sigma(a), b')$ is the only point on the curve of M' in $l'^2 \setminus k'^2$. This isomorphism of fields is therefore an isomorphism of curved fields.

(b) The points $(a_1, b_1), \dots, (a_n, b_n)$ on the curve in $l^2 \setminus k^2$ all have their first coordinates outside k.

We can view $(a_1, b_1, \dots, a_n, b_n)$ as a generic point over k of an irreducible variety $V_{\bar{z}}$ defined by a conjunction $\phi(x_1, y_1, \dots, x_n, y_n, \bar{z})$ of polynomial equations. Here $\bar{z}$ is a tuple of parameters from k. This variety has dimension n since $\delta(l) = \delta(k)$. Our tuple of parameters $\bar{z}$ satisfies the formula ξ which is associated to ϕ in the inductive axioms since $a_1, \dots, a_n$ are pairwise distinct (by the functional axiom) and transcendental over k. Since k is self-sufficient, by Lemma 3 this tuple satisfies formula ψ as well. The tuple $\sigma(\bar{z})$ must therefore satisfy $\xi \wedge \psi$. By application of an inductive axiom there exist n points $(a'_1, b'_1), \dots, (a'_n, b'_n)$ on the curve of M' whose $2n$ coordinates satisfy $\phi(\bar{x}, \bar{y}, \sigma(\bar{z}))$. In fact, by Lemma 1 we can choose these points so that they are all distinct from the (finitely many) points of the curve in k'^2, and distinct from each other. This choice implies that $(a'_1, b'_1, \dots, a'_n, b'_n)$ is a generic point of the algebraic subset of M'^{2n} defined by ϕ since this variety has dimension n and the transcendence degree over k' of $(a'_1, b'_1, \dots, a'_n, b'_n)$ must be at least n by the self-sufficiency of k'. It follows that exists an isomorphism of fields $\tau: l \to l'$ which extends σ and maps $a_1, b_1, \dots, a_n, b_n$ to $a'_1, b'_1, \dots, a'_n, b'_n$. Here l' is the algebraic closure of $k' \cup \{a'_1, b'_1, \dots, a'_n, b'_n\}$. This isomorphism of fields is an isomorphism of curved fields since $(a'_1, b'_1), \dots, (a'_n, b'_n)$ are the only points on the curve in $l'^2 \setminus k'^2$ by self-sufficiency of k'. The analysis of case (i) is now completed.

In case (ii) we set $l = \overline{k(\alpha)}$. Since we are not in case (i), l is self-sufficient and there is no point on the curve in $l^2 \setminus k^2$. Hence we just need to find $\alpha' \in M' \setminus k'$ such that $l' = \overline{k'(\alpha')}$ is self sufficient, and there is no point on the curve in $l'^2 \setminus k'^2$. Any isomorphism of fields $\tau: l \to l'$ will then be good for our purposes. The existence of α' is given by Lemma 2 (applied to M'). ⊣

§4. Convergence to the limit theory. In this section we show that any axiom of T is satisfied by all generic polynomials of sufficiently high degree. This implies immediately that any sentence F in the language of curved fields is satisfied by all generic polynomials of sufficiently high degree if $T \vdash F$. Conversely, if F is not a sentence of T then $T \vdash \neg F$ by completeness of T, so that F is false for all generic polynomials of sufficiently high degree.

4.1. Satisfaction of the universal axioms. Given two sequences $x = (x_1, \dots, x_n)$ and $y = (y_1, \dots, y_n)$ of elements of K, we denote by $W(x, y)$ the affine subspace of all $(\alpha_1, \ldots, \alpha_d) \in K^d$ such that $\sum_{j=1}^{d} x_i^j \alpha_j = y_i$ for $i = 1, \ldots, n$. This set can be interpreted as the set of degree d polynomials without constant term which go through the n points $(x_1, y_1), \ldots, (x_n, y_n)$.

LEMMA 4. *If $d \geq n$ and $x_1, \ldots, x_n$ are distinct and nonzero then $W(x, y)$ has dimension $d - n$.*

PROOF. If the x_i's are pairwise distinct and nonzero the matrix of the system defining $W(x, y)$ has rank n by the well-known property of Vandermonde determinants. The solution space is thus of dimension $d - n$. ⊣

PROPOSITION 3. *If $d \geq n$ and a generic polynomial of degree d goes through n distinct points $(x_1, y_1), \ldots, (x_n, y_n)$ with $x_i \neq 0$ for all i, their $2n$ coordinates have transcendence degree at least n.*

PROOF. Since these $2n$ points are distinct and lie on the same polynomial curve, their first coordinates are also distinct. It then follows from Lemma 4 that the coefficients of the generic polynomial, which are of transcendence degree d over the prime field, have transcendence degree at most $d - n$ over $\{x_1, y_1, \ldots, x_n, y_n\}$. These $2n$ coordinates must therefore be of transcendence degree at least n over the prime field. ⊣

It follows that a generic polynomial of degree d satisfies axiom (1) as soon as $d \geq n$.

4.2. Satisfaction of the inductive axioms. From now on we assume that $K = \mathbb{C}$. Let $\varphi(r, n) = n(nr + n + r)$. We first state two intersection theorems.

THEOREM 1. *Let V be an algebraic subset of $\mathbb{C}^d$ of codimension r, defined over an algebraically closed subfield $k \subseteq \mathbb{C}$. Then $V \cap W(x, y) \neq \emptyset$ if x and y satisfy the following two conditions and if $d \geq \varphi(r, n)$:*

(i) *The x_i's are pairwise distinct and all lie outside k.*
(ii) *There exists $u_1 \in \{x_1, y_1\}, \ldots, u_n \in \{x_n, y_n\}$ such that $(u_1, \ldots, u_n)$ is of transcendence degree n over k.*

It is clear that the theorem fails if condition (i) is removed. We have shown in [6] that condition (ii) is also necessary. We do not known whether the bound $n(nr + n + r)$ in this theorem is optimal. A precise analysis [6] of the case where V is an affine subspace yields the lower bound $n(r + 1)$. The consideration of more general algebraic sets could perhaps yield better lower bounds.

The second main ingredient is the following result.

THEOREM 2. *Let V be an irreducible algebraic subset of $\mathbb{C}^d$ of codimension r, defined over an algebraically closed subfield $k \subseteq \mathbb{C}$.*

Assume that $d \geq n(2n + r + 1)$ and that condition (ii) *is satisfied as well as:*

(i′) *The x_i's are pairwise distinct and are all nonzero.*

If $V \cap W(x, y) \neq \emptyset$ then this intersection contains a generic point of V (i.e., a point of transcendence degree over k equal to the dimension of V).

The proof of these two theorems is postponed to the next section. Our immediate goal is to show that any inductive axiom is satisfied by all generic polynomials of sufficiently high degree.

Given $r \in \mathbb{N}$, consider any inductive axiom such that the tuple $\bar{z}$ of parameters in (2) is of length at most r. Let C be the graph of a generic polynomial of degree d. We claim that C satisfies the inductive axiom as soon as $d \geq \varphi'(r,n)$, where $\varphi'(r,n) = n(nr+n+r+1)$. Indeed, fix any $\bar{z}$ such that $\mathbb{C} \models \theta(\bar{z})$. The tuple $\alpha = (\alpha_1,\ldots,\alpha_d)$ of coefficients of C can be viewed as a generic point over $\mathbb{Q}(\bar{z})$ of some irreducible algebraic set V of $\mathbb{C}^d$ (defined over the algebraic closure of $\mathbb{Q}(\bar{z})$). Since $\bar{z}$ is of length at most r, V is of codimension at most r.

Let $V_{\bar{z}}$ be the subset of $\mathbb{C}^{2n}$ defined by $\phi(.,.,\bar{z})$. Let $(x_1,y_1,\ldots,x_n,y_n)$ be a point of $V_{\bar{z}}$ which is generic over $\mathbb{Q}(\bar{z})$. Conditions (i) and (ii) of Theorem 1 are satisfied by $x=(x_1,\ldots,x_n)$ and $y=(y_1,\ldots,y_n)$ since $\mathbb{C} \models \theta(\bar{z})$. The intersection $V \cap W(x,y)$ is therefore nonempty since $d \geq \varphi'(r,n) \geq \varphi(r,n)$. By Theorem 2 this intersection contains a generic point β of V since $\varphi'(r,n) \geq n(2n+r+1)$. We conclude that the graph of the generic polynomial of coefficient vector β contains n points whose $2n$ coordinates lie on $V_{\bar{z}}$ (and the same is true of C since α and β have same type over $\bar{z}$). We have thus proved that the inductive axiom is satisfied as soon as $d \geq \varphi'(r,n)$.

§5. **The intersection theorems.** In this section we give the proofs of Theorems 1 and 2. We start with Theorem 2 since it is needed for the proof of Theorem 1.

5.1. Proof of Theorem 2. It is almost identical to the proof of the corresponding result in [2]. In particular, the proof of Lemma 5 below differs from the proof of Lemma 7 from [2] only by minor details. Here the main difference with that previous paper does not lie in the proof of the theorem but in its hypotheses: we need to assume that $V \cap W(x,y)$ is nonempty whereas that condition was automatically satisfied in [2] due to the projective nature of the problem.

Let k' be the algebraic closure of $k(x,y)$. We denote the transcendence degree of k' over k by $n+m$. Note that $m \geq 0$ by condition (ii) and that $W(x,y)$ is of codimension n by condition (i′). Let us assume that $W(x,y)$ intersects V and let β be a point of $V \cap W(x,y)$ which is generic over k'. "Generic" again means that the transcendence degree of β over k' is equal to the dimension of $V \cap W(x,y)$, which is at least $d-(n+r)$.

Lemma 5. *If $d \geq n(2n+r+1)$, $x \,\hat{}\, y$ is of transcendence degree at most m over $k(\beta)$.*

Proof. Let n_1 be the cardinality of the set P_1 of the points (x_i,y_i) which are algebraic over $k(\beta)$. We denote by P_2 the set of the remaining $n_2 = n - n_1$ points. Note that x_i is transcendental over $k(\beta)$ for every point $(x_y,y_i) \in P_2$.

Indeed, if x_i was algebraic over $k(\beta)$ the same would be true of y_i since $\beta \in W(x, y)$.

Let θ be the transcendence degree of P_2 over $k(\beta)$. We need to show that $\theta \leq m$ if d is large enough. We assume that $n_2 \neq 0$ (otherwise $\theta = 0$ and we are done). Note that $\text{tr.deg}(k(P_1)/k) \geq n_1$ by condition (ii). Since $\text{tr.deg}(k(P_1, P_2)/k) = n_1 + n_2 + m$ by definition of m, this implies that $\text{tr.deg}(k(P_1, P_2)/k(P_1)) \leq n_2 + m$. Now we consider $l = \lfloor (d - n_1)/n_2 \rfloor$ independent copies $P_{2,1}, \ldots, P_{2,l}$ of P_2. More precisely, these new points should satisfy the two following conditions:

1. Each set $P_{2,i}$ has same type over $k(\beta)$ as P_2.
2. The transcendence degree over $k(\beta)$ of their union is equal to $l\theta$.

Let Q be the union of P_1 and of the set of new points. The first coordinates of the new points are all transcendental over $k(\beta)$ and distinct from each other. It follows that the first coordinates of the points of Q are all nonzero and distinct from each other. In particular, these points are all distinct so that $|Q| = n_1 + ln_2 \in [d - n_2 + 1, d]$. We can now apply Lemma 4 since the polynomial curve of coefficient vector β goes through all the points of Q. It follows that that β is of transcendence degree at most $n_2 - 1$ over $k(Q)$. We can now compute the transcendence degree of $\beta \,\hat{}\, Q$ over k in two different ways. On the one hand,

$$\begin{aligned}\text{tr.deg}(k(\beta \,\hat{}\, Q)/k) &= \text{tr.deg}(k(\beta)/k) + \text{tr.deg}(k(\beta \,\hat{}\, Q)/k(\beta)) \\ &\geq d - (n + r) + l\theta.\end{aligned}$$

On the other hand,

$$\begin{aligned}\text{tr.deg}(k(\beta \,\hat{}\, Q)/k) &= \text{tr.deg}(k(Q)/k) + \text{tr.deg}(k(\beta \,\hat{}\, Q)/k(Q)) \\ &< \text{tr.deg}(k(P_1)/k) + l(n_2 + m) + n_2.\end{aligned}$$

Comparing these two expressions, we obtain the inequality

$$d - (n + r) + l\theta < l(n_2 + m) + 2n_1 + n_2 \tag{3}$$

since $\text{tr.deg}(k(P_1)/k) \leq 2n_1$. Letting d (and consequently l) go to infinity in this inequality then shows that $\theta > m$ is impossible. The specific bound $d \geq n(2n + r + 1)$ can be derived from a straightforward calculation which we do not detail completely. The first step is to assume that $\theta \geq m + 1$. We obtain from (3) that $l < 2n + r$, which implies in turn $d < n_1 + n_2(2n + r)$. We finally conclude that the (probably rough) bound $d < n(2n + r + 1)$ holds if $\theta > m$. ⊣

In order to complete the proof of Theorem 2 we compute the transcendence degree of $k(\beta, x, y)$ over k in two different ways. On the one hand, it is upper bounded by $\text{tr.deg}(k(\beta)/k) + \text{tr.deg}(k(\beta, x, y)/k(\beta))$. By the above lemma this is at most $\text{tr.deg}(k(\beta)/k) + m$ if $d \geq n(2n + r + 1)$. On the other hand, this transcendence degree is equal to $\text{tr.deg}(k(x, y)/k) + \text{tr.deg}(k(x, y, \beta)/k(x, y)) =$

$n+m+\dim(V\cap W(x,y))$ which is at least $n+m+d-(n+r)=m+d-r$. A comparison of these two expressions shows that $\text{tr.deg}(k(\beta)/k)\geq d-r=\dim V$. Since $\beta\in V$ this inequality must be an equality, and β is indeed a generic point of V.

5.2. Pseudocylinders. Before moving on to the proof of Theorem 1 we need some geometric preliminaries. In this subsection we work with the Euclidean topology on $\mathbb{C}^p$. We denote by $Z(f_1,\dots,f_s)$ the zero set of s polynomials $f_1,\dots,f_s\in\mathbb{C}[X_1,\dots,X_p]$. We denote by $\mathcal{P}(d_1,\dots,d_s)$ the space of all tuples $(f_1,\dots,f_s)$ of polynomials such that $\deg(f_i)\leq d_i$ for $i=1,\dots,s$. This space can be identified to $\mathbb{C}^N$ for $N=\prod_{i=1}^{s}\binom{p+d_i}{p}$, and it is also endowed with the Euclidean topology.

Lemma 6 (continuity of roots). *Assume that $s\leq p$ and that a point α of $\mathbb{C}^p$ lies on an irreducible component of $Z(f_1,\dots,f_s)$ of codimension s. For any neighbourhood O of α, there exists a neighbourhood O' of $f_1,\dots,f_s$ in $\mathcal{P}(\deg(f_1),\dots,\deg(f_s))$ such that $Z(f_1',\dots,f_s')\cap O\neq\emptyset$ for any $(f_1',\dots,f_s')$ in O'.*

Proof. There exist $n-s$ affine functions $l_1,\dots,l_{n-s}$ such that $Z(f_1,\dots,f_s,l_1,\dots,l_{n-s})=\{\alpha\}$. We can now apply the "extended geometric version" of Bézout's Theorem from [1] since we have as many equations as unknowns. We conclude that there exists a neighbourhood O' of $(f_1,\dots,f_s)$ in $\mathcal{P}(\deg(f_1),\dots,\deg(f_s))$ and a neighbourhood O'' of $l_1,\dots,l_{n-s}$ in $\mathcal{P}(1,\dots,1)$ such that $Z(f_1',\dots,f_s',l_1',\dots,l_{n-s}')\cap O\neq\emptyset$ whenever $(f_1',\dots,f_s')\in O'$ and $(l_1',\dots,l_s')\in O''$. ⊣

Let V be an irreducible algebraic subset of $\mathbb{C}^p$ of codimension r. Recall that V is said to be a complete intersection if it can be defined by exactly r polynomial equations. This is not always possible, but it is always possible to find a variety V' defined by r equations such that V is an irreducible component of V' (see for instance Proposition 2.7 of [7]).

Corollary 1. *Let V be an irreducible algebraic subset of $\mathbb{C}^p$ of codimension r and W an affine subspace of codimension n where $r+n\leq p$. Let α be a point of V which does not lie on any other component of V', where V' is as above. If α lies on an irreducible component of $V\cap W$ of codimension $r+n$ then $V\cap W'\neq\emptyset$ for any affine subspace W' of codimension n which is sufficiently close to W.*

Proof. By choice of α, there is a neighbourhood O of this point such that $V'\cap O\subseteq V$. Let us apply Lemma 6 to the system of $s=r+n$ equations defining $V'\cap W$. We conclude that $V'\cap W'\cap O\neq\emptyset$ if W' is sufficiently close to W, but $V'\cap W'\cap O\subseteq V\cap W'$ by choice of O. ⊣

Proposition 4. *Let V be an algebraic subset of $\mathbb{C}^p$ of codimension r. For any $n\leq p-r$, let I_n be the set of points $\alpha\in\mathbb{C}^p$ such that there exists an affine subspace of codimension n containing α which does not intersect V. If I_n is* *dense in* $\mathbb{C}^p$, *there exists an affine subspace* W *of codimension* n *in which* V *is of codimension* $< r$ (*i.e.*, $V \cap W$ *has codimension at most* $n + r - 1$ *in* $\mathbb{C}^p$).

PROOF. Assume first that V is irreducible. Let α be a point of V which does not lie on any other component of V', where V' is as in Corollary 1. If I_n is dense we can find a sequence of points α_k which converges to α and a sequence W_k of affine subspaces of codimension n containing α_k such that $V \cap W_k = \emptyset$. By compactness of the set of directions of affine subspaces of codimension n, we may assume (extracting a subsequence if necessary) that W_k converges to some affine subspace W of codimension n. We claim that $V \cap W$ has codimension at most $n + r - 1$. Indeed, assume by contradiction that $V \cap W$ has codimension at least $n + r$. Since $V \cap W \neq \emptyset$ (α is in the intersection) the codimension should be exactly $n+r$. This is in contradiction with Corollary 1 (take $W' = W_k$ for a large enough k). This completes the proof of Proposition 4 for irreducible varieties. If V is not irreducible, we can apply the result to an irreducible component. ⊣

For instance, in the case $r = 1$ the conclusion of this proposition is that V contains an affine subspace of codimension n. As an example of this situation, take for V a cylinder in $\mathbb{C}^3$ (in this case $p = 3$, $r = 1$ and $n = 2$).

5.3. Proof of Theorem 1. This theorem was obtained in [6] in the special case where V is an affine subspace. More precisely, we proved the following result.

THEOREM 3. *Let* V *be an affine subspace of* $\mathbb{C}^d$ *of codimension* r, *defined over an algebraically closed subfield* $k \subseteq \mathbb{C}$.

If x *satisfies condition* (i) *then* $V \cap W(x, y) \neq \emptyset$ *as soon as* $d \geq n(r + 1)$, *and this intersection is of codimension* $r + n$ *in* $\mathbb{C}^d$.

The proof of Theorem 1 relies on Theorem 3 as well as on an inductive argument which is contained in the next lemma.

LEMMA 7. *Let* V *be an algebraic subset of* $\mathbb{C}^d$ *defined over an algebraically closed subfield* $k \subseteq \mathbb{C}$. *Assume that there exists an affine subspace* A *of* $\mathbb{C}^d$ *of codimension* s *in which* V *is of codimension* $r \geq 1$. *Assume also that* x *and* y *satisfy conditions* (i) *and* (ii) *and that* $V \cap W(x, y) = \emptyset$. *There exists an affine subspace* A' *of* $\mathbb{C}^d$ *of codimension* $s + n$ *in which* V *is of codimension* $r - 1$ *as soon as*:

$$d \geq \max(n(s + 1), n(2n + s + 1), n + r + s).$$

PROOF. We shall assume that A is defined over k (there must exist such an A since V is defined over k). If $d \geq n(s + 1)$ the intersection $A \cap W(x, y)$ is nonempty by Theorem 3, and it is of codimension n in A. If additionally $d \geq n(2n+s+1)$ this intersection contains a generic point of A by Theorem 2. Obviously, $A \cap W(x, y)$ does not intersect $A \cap V$ since $W(x, y)$ does not intersect V.

Since $n + r \leq \dim(A) = d - s$, we can now apply Proposition 4 with $p = d - s$, A in place of $\mathbb{C}^p$ and $V \cap A$ in place of A. We conclude that there exists an affine subspace $A' \subseteq A$ of codimension n in A such that $V \cap A'$ is of codimension at most $r - 1$ in A'. ⊣

Theorem 1 follows almost immediately from the above lemma. Indeed, assume by contradiction that $d \geq \varphi(r, n)$ but $V \cap W(x, y) = \emptyset$. We can apply the lemma iteratively for $s = 0, n, 2n, \ldots$ up to $s = (r-1)n$. This is legitimate due to the inequalities $\varphi(r, n) \geq n((r-1)n+1)$, $\varphi(r, n) \geq n(2n+(r-1)n+1)$ and $\varphi(r, n) \geq n+r+(r-1)n$. We conclude that V contains an affine subspace of codimension nr. Without loss of generality we can assume again that this subspace is defined over k. Since $\varphi(r, n) \geq n(nr+1)$ we can apply Theorem 3 one more time. This yields the contradiction $V \cap W(x, y) \neq \emptyset$.

Acknowledgments. Thanks are due to the anonymous referee for carefully reading the manuscript, and to Bruno Poizat for useful discussions on the axiomatization of the limit theory.

REFERENCES

[1] L. Blum, F. Cucker, M. Shub, and S. Smale, ***Complexity and real computation***, Springer-Verlag, 1998.

[2] O. Chapuis, E. Hrushovski, P. Koiran, and B. Poizat, *La limite des théories de courbes génériques*, ***The Journal of Symbolic Logic***, vol. 67 (2002), pp. 24–34.

[3] P. Koiran, *On defining irreducibility*, ***Comptes Rendus de l'Académie des Sciences***, vol. 330 (2000), pp. 529–532.

[4] ———, *The theory of Liouville functions*, ***The Journal of Symbolic Logic***, vol. 68 (2003), no. 2, pp. 353–365.

[5] P. Koiran and N. Portier, *Back-and-forth systems for generic curves and a decision algorithm for the limit theory*, ***Annals of Pure and Applied Logic***, vol. 111 (2001), pp. 257–275.

[6] P. Koiran, N. Portier, and G. Villard, *A rank theorem for Vandermonde matrices*, ***Linear Algebra and Its Applications***, vol. 378 (2004), pp. 99–107.

[7] D. Perrin, ***Géométrie algébrique: une introduction***, Interéditions / CNRS Editions, 1995.

[8] B. Poizat, *Amalgames de Hrushovski: une tentative de classification*, ***Tits buildings and the model theory of groups*** (***Würzburg, 2000***), London Mathematical Society Lecture Note Series, 291, Cambridge University Press, Cambridge, 2002.

[9] A. J. Wilkie, *On defining* C^∞, ***The Journal of Symbolic Logic***, vol. 59 (1994), no. 1, p. 344.

[10] ———, *O-minimality*, ***Documenta Mathematica***, vol. I (1998), pp. 633–636, `http://www.mathematik.uni-bielefeld.de/documenta`, Extra Volume ICM 1998.

[11] ———, *Liouville functions*, ***Logic colloquium 2000*** (R. Cori et al., editors), Lecture Notes in Logic, vol. 19, AK Peters, to appear.

[12] B. Zilber, *A theory of a generic function with derivations*, ***Logic and algebra*** (Yi Zhang, editor), Contemporary Mathematics, vol. 302, American Mathematical Society, 2002, pp. 85–100.

LABORATOIRE DE L'INFORMATIQUE DU PARALLÉLISME
ECOLE NORMALE SUPÉRIEURE DE LYON - CNRS
46, ALLÉE D'ITALIE, 69364 LYON CEDEX 07, FRANCE
E-mail: Pascal.Koiran@ens-lyon.fr *URL:* http://perso.ens-lyon.fr/pascal.koiran

MOSCHOVAKIS'S NOTION OF MEANING AS APPLIED TO LINGUISTICS

MICHIEL VAN LAMBALGEN AND FRITZ HAMM

Wenn man nicht weiß was man selber will,
muß man zuerst wissen was die anderen wollen.
General Stumm von Bordwehr[1]

§1. Introduction: Moschovakis's approach to intensionality. G. Frege introduced two concepts which are central to modern formal approaches to natural language semantics; i.e., the notion of *reference* (denotation, extension, Bedeutung) and *sense* (intension, Sinn) of proper names[2]. The sense of a proper name is *wherein the mode of presentation (of the denotation) is contained.* For Frege proper names include not only expressions such as *Peter*, *Shakespeare* but also definite descriptions like *the point of intersection of line* l_1 *and* l_2 and furthermore sentences which are names for truth values. Sentences denote the *True* or the *False*. The sense of a sentence is the proposition (Gedanke) the sentence expresses. In the tradition of possible world semantics the proposition a sentence expresses is modelled via the set of worlds in which the sentence is true. This strategy leads to well known problems with propositional attitudes and other intensional constructions in natural languages since it predicts for example that the sentences in (1) are equivalent.

(1) a. Jacob knows that the square root of four equals two.
b. Jacob knows that any group G is isomorphic to a transformation group.

Even an example as simple as (1) shows that the standard concept of proposition in possible world semantics is not a faithful reconstruction of Frege's notion *sense*.

Frege developed his notion of sense for two related but conceptually different reasons. We already introduced the first one by considering propositional attitudes. The problem here is how to develop a general concept which can handle the semantics of Frege's *ungerade Rede*. The second problem is how

[1] In Robert Musil's novel 'Der Mann ohne Eigenschaften'.
[2] See especially [4] and the English translation [3].

Logic Colloquium '01
Edited by M. Baaz, S. Friedman, and J. Krajíček
Lecture Notes in Logic, 20

to distinguish a statement like $a = a$ which is rather uninformative from the informative statement $a = b$ or phrased differently how to account for the semantic difference between (2-a) and (2-b).

(2) a. Scott is Scott.
 b. Scott is the author of Waverly.

Frege's intuitive concept of sense therefore was meant both to model information and provide denotations for intensional constructions.

[12] develops a formal analysis of sense and denotation which is certainly closer to Frege's intentions than is possible world semantics. Moschovakis's motivations are (at least) twofold. The first motivation is to give a rigorous definition of the concept *algorithm* [13] and thereby provide the basics for a mathematical theory of algorithms. The second motivation is a philosophical and linguistic one. It consists in providing a more adequate formal reconstruction of the Fregean notions *sense* and *denotation* via the formalised concepts *algorithm* and *value of an algorithm*. Such a formalisation has a wide range of potential applications. We already mentioned propositional attitudes and other intensional constructions in natural language, but these formalised concepts could also contribute to a better understanding of such difficult notions like *synonymy* or *faithful translation* familiar from philosophical discussions.

Let us briefly explain the intuitive basis of Moschovakis's new approach to intensionality. Assume that a first-order (sorted) structure **A** for the *Formal Language of Recursion* (FLR) is given. The first step consists in recursively defining a *denotation* $den_{\mathbf{A}}(\overline{x}, \phi)$ with respect to **A**, where ϕ be an FLR-formula all of whose free variables are among $\overline{x}$.

In the second step an intension $int_{\mathbf{A}}(\overline{x}, \phi)$ is defined for ϕ. The intuition is that $int_{\mathbf{A}}(\overline{x}, \phi)$ is the algorithm which computes $den_{\mathbf{A}}(\overline{x}, \phi)$.

$$int_{\mathbf{A}}(\overline{x}, \phi) = \text{the algorithm which computes the truth value of } \phi.$$

Of course it is this definition of intension which is novel. It is developed in a rather indirect way via a syntactic reduction calculus which step by step transforms a given formula ϕ into the parts it consists of. The transformation process ends when only those parts of ϕ are left whose denotations are given by the structure **A**. In this case one says that the formula ϕ has been reduced to its *normal form*. It is the existence and uniqueness of normal forms for FLR-formulas which allows the recursive definition of *intension* in [11], and which justifies speaking of *the* intension in the definition of $int_{\mathbf{A}}(\overline{x}, \phi)$.

In the rest of this section we will explain the basic ideas of Moscchovakis's approach to intensionality. This will be done in a rather informal and sketchy way with emphasis on ideas rather than on technical details. The reader interested in a more thorough understanding of the formal theory is urged to consult Moschovakis's work, primarily [12], [10] and [11].

We will concentrate on the presentation in [12] which introduces the language LPCR (*Lower Predicate Calculus with Reflection*). LPCR is a simplified version of the formal language of recursion FLR developed in [10].

Apart from the logical symbols *eq* (equality), $\neg$, $\wedge$, $\vee$, $\rightarrow$, $\exists$ and $\forall$ LPCR contains k-ary *partial* relation variables P_1^k, P_2^k, ..., thought of as partial functions from the k-ary Cartesian product into the truth values. As a special case—when $k = 0$—we obtain the formula $P()$. The novel feature of LPCR is a new variable binding operator *rec* introduced in [10]. The syntax of this operator is given by (the types are omitted here):

DEFINITION 1. *If* $u_1, \ldots, u_n$ *are individual variables, if* $p_1, \ldots, p_n$ *are distinct partial function (pf) variables and if* $\phi_0, \phi_1, \ldots, \phi_n$ *are formulas, then*

$$\phi = rec(u_1, p_1, \ldots, u_n, p_n)[\phi_0, \phi_1, \ldots, \phi_n]$$

is a formula. The formulas $\phi_0, \ldots, \phi_n$ *are called the parts of the recursive formula* ϕ. *The formula* ϕ_0 *is the output or head part.*

In LPCR the following more intuitive notation is used (where the ϕ_is are formulas):

$$\phi = \phi_0 \text{ where } \{p_1(u_1) \simeq \phi_1, \ldots, p_n \simeq \phi_n\}^3$$

One may give a denotational semantics of LCPR relative to a many sorted functional structure **A** which contains *basic sets*, Cartesion products, domains for partial functions from Cartesian products of basic sets to basic sets and further domains for partial monotone functions from Cartesian products of basic and partial function spaces to basic sets. Moreover such a structure provides denotations for the logical constants *eq*, $\neg$, $\wedge$, $\vee$, $\rightarrow$, $\exists$ and $\forall$.

Given a functional structure **A**, it is possible to define by induction a partial monotone functional $den_{\mathbf{A}}(\overline{x}, \phi)\colon X \rightarrow TV$ which assigns to each formula ϕ and each $\overline{x}$ its truth value in **A**. Here X is the space of all n-tuples $(x_1, \ldots, x_n)$ of objects[4] in **A** corresponding to the variables in $\overline{x}$ and TV is the set of truth values. We will skip here the definitions which fix the standard denotations of terms and concentrate on the clause for the new operator *rec*. Consider

$$\phi = rec(u_1, p_1, \ldots, u_n, p_n)[\phi_0, \phi_1, \ldots, \phi_n]$$

For each $\overline{x}$, the induction hypothesis gives us a sequence of partial monotone functionals $f_{\overline{x},\phi_i}$ satisfying for $i = 0, \ldots, n$:

$$f_{\overline{x},\phi_i}(u_i, p_1, \ldots, p_n) \simeq den_{\mathbf{A}}(u_i, p_1, \ldots, p_n, \phi_i)(u_i, p_1, \ldots, p_n).$$

Since all $f_{\overline{x},\phi_i}$ are monotone it follows that the system of equations

$$p_i(u_i) = f_{\overline{x},\phi_i}(u_i, p_1, \ldots, p_n)$$

[3]Since the denotational semantics for FLR is a partial semantics $u \simeq v$ denotes identity for the case where both terms u and v are defined.

[4]We will use the same notation for variables and the objects which interpret the variables here.

has a (unique) sequence of least fixpoints $\overline{p}_{\overline{x},1}, \ldots, \overline{p}_{\overline{x},n}$ as solutions. The denotation of $rec(u_1, p_1, \ldots, u_n, p_n)[\phi_0, \phi_1, \ldots, \phi_n]$ is then obtained by substituting these fixpoints for the respective variables in ϕ_0, i.e., by setting

$$den(\overline{x}, \phi)(\overline{x}) \simeq f_{\overline{x},\phi_0}(\overline{p}_{\overline{x},1}, \ldots, \overline{p}_{\overline{x},n}).$$

This concludes the brief exposition of the denotational semantics of LPCR. What makes LPCR so interesting is that Moschovakis also provides an intensional semantics, that is, a semantics which assigns to each formula ϕ and each $\overline{x}$ an intension $int_{\mathbf{A}}(\overline{x}, \phi)$, a kind of abstract recursive algorithm which computes the denotation $den_{\mathbf{A}}(\overline{x}, \phi)$. The theory of (referential) intensions is developed in [11]. The central concept is that of a *recursor*.

DEFINITION 2. *A* recursor *on a set X to W is any tuple of partial monotone functionals* $\mathbf{f} = [f_0, f_1, \ldots, f_n]$ *such that the following equations are correctly typed*:

$$p_1(u_1) \simeq f_{x,1}(u_1, p_1, \ldots, p_n)$$

$$\vdots$$

$$p_n(u_n) \simeq f_{x,n}(u_n, p_1, \ldots, p_n)$$

Moreover it is supposed that the equation

$$\bar{\mathbf{f}} = f_{x,0}(\overline{p}_1^x, \ldots, \overline{p}_n^x)$$

makes sense too.

The functionals $f_0, \ldots, f_n$ are called the parts of $\mathbf{f}$ *and* $\bar{\mathbf{f}}$: $X \to W$ *is called the denotation of the recursor* $\mathbf{f}$.

One says that two recursors $\mathbf{f}$ and $\mathbf{g}$ are *equal* if they have the same number of parts n and for some permutation σ of $\{0, \ldots, n\}$ with $\sigma(0) = 0$ and inverse σ^{-1} the following holds:

$$f_{x,i}(u_i, p_1, \ldots, p_n) \simeq g_{x,\sigma^{-1}(i)}(u_i, p_{\sigma(1)}, \ldots, p_{\sigma(n)}).$$

The definition of equality for recursors thus says that changing the order of any parts of a recursor with the exception of the first part does not change the identity of the recursor.

The aim of the intensional semantics for LPCR is both to interpret the intension $int_{\mathbf{A}}(\overline{x}, \phi)$ as a recursor, and to let $int_{\mathbf{A}}(\overline{x}, \phi)$ compute $den_{\mathbf{A}}(\overline{x}, \phi)$. It is readily seen that algorithms which compute $den_{\mathbf{A}}(\overline{x}, \phi)$ will involve nested recursions, and one of Moschovakis's technical achievements is to supply a reduction calculus which reduces nested recursions to an unnested normal form. The reduction calculus axiomatises two relations between LPCR formulas. The first is $t \sim s$, with the intended meaning that t and s define the same abstract algorithm. The second relation is $t \to s$ which means that s defines the algorithm more directly (in terms of already given information) than t. Moreover, the uniqueness of normal forms of the reduction calculus allows one to assign to every expression of LPCR a recursor as intension in a unique way. We will illustrate these ideas by means of a few linguistic examples.

Assume first that LPCR is enriched by a construction $(\textit{the } x)\phi(x)$ which assigns to every formula a term. The denotation of this new construction with respect to a structure **A** is given by the functional $\mathbf{the}_{\mathbf{A}}$ which assigns to every partial function $f : A \rightharpoonup TV$ from the domain A to the set of truthvalues a unique object in A. We can then translate the sentences (2-a) and (2-b) repeated here for convenience as (3-a) and (3-b) into the formalism of LPCR.

(3) a. Scott is Scott.
b. Scott is the author of Waverly.

Possible LPCR representations are (4-a) and (4-b) where s is an individual constant and AW is a one-place predicate representing the expression *author of Waverly*. We assume that the interpretation of AW is given by a set of structure **A**.

(4) a. $eq(s, s)$
b. $eq(s, (\textit{the } x)\, AW(x))$

The reduction calculus allows to derive the following (rough) normal forms for formulas (4-a), (4-b).

(5) a. $eq(p(), q())$ *where* $\{p() \simeq s, q() \simeq s\}$
b. $eq(p(), q())$ *where* $\{p() \simeq s, q() \simeq (\textit{the } x)R(x), R(x) \simeq AW(x)\}$

It is clear that the normal forms (5-a) and (5-a) represent different recursors. Therefore the intensions assigned to the sentences (3-a) (3-b) are different although their denotation might well be the same; namely if in the structure **A** Scott is indeed the unique author of Waverly.

As a second application consider relations R and Q which are interpreted as converse relations on structure **A**. The reduction calculus allows to derive the following normal forms for $R(a, b)$ and $Q(b, a)$.

(6) a. $R(p(), q())$ *where* $\{p() \simeq a, q() \simeq b\}$
b. $Q(q(), p())$ *where* $\{q() \simeq b, p() \simeq a\}$

Given that R and Q are interpreted as given (intensionless) converse relations on the structure **A**, the intensions of $R(a, b)$ and $Q(b, a)$ are the same by the definition of equality for recursors (take a permutation σ with $\sigma(1) = 2$ and $\sigma(2) = 1$). Therefore if we choose to interpret the passive as the converse of the corresponding active relation we derive that the sentences *John hit Mary* and *Mary was hit by John* express the same intension.

We have seen one example where intensions are different, and one example where they are the same. In a letter to Husserl, Frege remarked on this topic

> It seems to me that we must have an objective criterion for recognising a thought as the same thought, since without such a criterion a logical analysis is not possible.

Moschovakis supplies a significant theoretical insight concerning the identity relation $\sim_{\mathbf{A}}$ between intensions on an appropriate structure **A**. The result says that the intensional identity relation is decidable. For a precise statement we need the following definition:

Definition 3. *For each structure* **A** *and arbitrary integers* n, m, *let*

$$S_{\mathbf{A}}(n,m) \iff n, m \textit{ are Gödel numbers of sentences } \theta_n, \theta_m \textit{ of } \mathrm{LPCR} \textit{ and } \theta_n \sim_{\mathbf{A}} \theta_m.$$

The precise formulation of the main theorem from [12] is:

Theorem 1. *For each recursor structure* **A** *of finite signature, the relation* $S_{\mathbf{A}}(n,m)$ *between Gödel numbers of formulas of* LPCR *interpreted on* **A** *is decidable.*

A corollary of theorem 1 is that the relation $S_{\mathbf{A}}(n,m)$ is definable in LPCR over each structure which allows the coding necessary for Gödelisation. (This requires just pairing and projection functions.)

This result brings us to our third linguistic application. Moschovakis gives an analysis of propositional attitudes which goes far beyond the usual accounts in possible world semantics. Consider the following example from [12].

(7) Othello believed that Cassio and Desdemona were lovers.

According to Frege the propositional attitude *believed* relates Othello not to the denotation of the sentence *Cassio and Desdemona were lovers* but to its sense. Technically this amounts to analyse example (7) as *Othello believed* m where m is the Gödel number of the sentence *Cassio and Desdemona were lovers*. It is now possible to formulate a principle which says within LPCR that *belief* is sensitive not to syntactic form but to sense.

(8) Othello believes $m \wedge S_{\mathbf{A}}(m,n) \Longrightarrow$ Othello believes n.

If we assume that conjunction is interpreted on the structure **A** by a commutative operation—which means that the formulas $\phi \wedge \psi$ and $\psi \wedge \phi$ express the same intension with respect to **A**—then sentence (9) follows from (7).

(9) Othello believed that Desdemona and Cassio were lovers.

This concludes our brief exposition of Moschovakis's notion of intension. We believe that this notion is highly relevant to the semantics of natural language, but that it has to be recast in a different formal framework to yield all that it promises. To put it roughly, whereas Moschovakis's emphasis is on how grammatical constructions affect intension, our emphasis will be on intension at the lexical level. The next section will explain the reasons for this shift of emphasis.

§2. **Meaning and logic programming.** For reasons that will become gradually clear, it is advantageous to represent the meaning of a set of natural language expressions by means of a constraint logic program involving those expressions. We will now explain how Moschovakis's ideas can be transferred to this context.[5]

At the most general level, the correspondence is this. Let P be some logic program, consisting of a number of clauses of the form $\theta_1 \wedge \cdots \wedge \theta_k \rightarrow \varphi$. For example, in the case of Prolog, the θ_i, φ are atomic; in the case of a *general* logic program, embodying full negation as failure, the θ_i may also be negations of atoms. Considering the latter case, it has turned out that a proper semantics for negation as failure involves Kleene's (strong) three valued logic applied to the *completion* of a program. The completion can be viewed as a set of simultaneous recursion equations; likewise, solutions to these equations are obtained as fixed points of suitable monotone operators (for general logic programs, three-valued consequence operators). Thus, the meaning of a natural language expression is given by a set of recursion equations (in a suitable logic programming language); denotations are obtained from fixed point solutions of these equations.

In formal semantics for natural language it is not common practice to associate algorithms to expressions. We have seen that it is usually assumed that all one needs is the intension of an expression, defined as a function which maps a possible world into an extension of the expression in that possible world. It seems to us that this picture of meaning is too static, and by and large cognitively irrelevant. A phenomenon that is very difficult to capture in terms of the traditional notion of meaning is nonmonotonicity. Recall that a consequence relation $\models$ is *nonmonotonic* if $\psi \models \varphi$ does not imply $\psi, \theta \models \varphi$. Natural language abounds with nonmonotonic phenomena. For instance, the distinction between

(10) John crossed the street

and

(11) John was crossing the street

[5]For information on constraint logic programming, see for example [6] and [16]. A mathematically sophisticated introduction to ordinary logic programming, including negation as failure, can be found in [1].

is that from (10) it follows that John reached the other side, whereas this is only implied by default by (11). One can add the clause 'when he was hit by a truck' to (11), but not to (10). When we think of the denotation of 'cross the street' as an event, then the temporal extent of the event described by (11) is cut short by the addition of the clause 'when he was hit by a truck'.[6] Native speakers of English do this automatically. This suggests an algorithmic interpretation process whereby the denotation of an expression is constantly re-computed on the basis of incoming data (linguistic or otherwise). The algorithm might work in such a way that it always computes a minimal model compatible with the present data; 'minimal' in the sense that nothing is assumed beyond what is given by the data. This point bears some elaboration. Both monotonic and nonmonotonic reasoning start from the maxim:

(M) 'assume only what is given in the premisses'

but they implement (M) in different ways. Nonmonotonic reasoning takes (M) to mean: all existence assumptions beyond those required by the premisses are false; instead, monotonic reasoning interprets (M) as: suspend judgement on statements which do not follow (and whose negations do not follow) from the premisses. In the interesting cases, these two interpretations of (M) can be reformulated as follows. In nonmonotonic reasoning, people construct a *minimal model* of the premisses (which is often unique); in monotonic reasoning, they must consider *all* models of the premisses. We believe that the intension or sense of an expression can be profitably identified with an algorithm constructing such minimal models. It remains to find the proper notion of algorithm for this context.

The general logic programs introduced above are not yet general enough for our purposes, mainly due to the fundamental role of time in natural language. For example, in the case of the progressive form ('John is writing') we want to be able to express that the activity *write* takes place more or less continuously during a certain period of time. We thus need a background structure of time, which for definiteness we take to be the structure $(\mathbb{R}, 0, 1, +, \cdot, <)$, and hence a version of logic programming which allows one to deal with the reals. In the case at hand, we would like to write a logic program for 'write', which should return as an answer the set of time points at which John is writing in a particular situation, also described in the program. The usual machinery of computed answer substitutions is not very helpful here, as it might present each point in that set as a separate solution. It is much more helpful to have the set represented by means of a definition; in the paradigmatic case, as a finite union of intervals. We have therefore opted for *constraint* logic programming, to which we now give a very brief introduction (cutting some corners in the process).

[6]This example will be treated in detail later, once we have introduced some technical machinery.

Constraint logic programming is in general concerned with the interplay of two languages. Let $\mathcal{L}$ be the language $\{0, 1, +, \cdot, <\}$, $\mathcal{T}$ the complete theory of $(\mathbb{R}, 0, 1, +, \cdot, <)$ in $\mathcal{L}$, i.e., the theory of real-closed fields. Let $\mathcal{K}$ be another language, consisting of programmed predicate symbols. The constraint programming language $CLP(\mathcal{T})$ consists on the one hand of *constraints*, which are first order formulas from the language $\mathcal{L}$, and on the other hand of formulas from $\mathcal{K}$, whose terms come from $\mathcal{L}$.[7] Constraint logic programs differ from logic programs by allowing constraints in the bodies of rules and in queries. For example, primitive constraints in $CLP(\mathcal{T})$ include formulas of the form $s < t$ and $s = t$, where s, t are terms from $\mathcal{L}$. *Definite* constraint logic programs have the form

$$B_1 \wedge \cdots \wedge B_m \wedge c \to A,$$

where the B_i, A are atoms in $\mathcal{K}$ and c is a constraint. Likewise, a query has the logical form

$$B_1 \wedge \cdots \wedge B_m \wedge c \to \bot.$$

We shall use the notation

$$?c, B_1 \ldots B_m$$

for queries, always with the convention that c denotes the constraint, and that the remaining formulas come from $\mathcal{K}$. The words 'query' and 'goal' will be used interchangeably.

The aim of a constraint computation is to express a programmed predicate symbol entirely in terms of constraints. Thus, unlike the case of ordinary logic programming, the last node of a successful branch in a computation tree contains a constraint instead of the empty clause.

For our purposes, even definite constraint logic programs are not yet expressive enough. We follow [16] in allowing (classical equivalents of) arbitrary first order formulas in the body of program clauses. More precisely, define a *complex subgoal* recursively to be

1. an atom in $\mathcal{K}$, or
2. $\neg\exists\overline{x}(B_1 \wedge \ldots B_m \wedge c)$, where c is a constraint and each B_i is a complex subgoal.

DEFINITION 4. *A* complex body *is a conjunction of complex subgoals. A* normal program *is a formula* $\psi \to \varphi$ *of* $CLP(\mathcal{T})$ *such that* ψ *is a complex body and* φ *is a complex subgoal.*

If, in the second clause of the above definition, we take $\overline{x}$ to be empty, we obtain ordinary goals (with constraints), which as indicated will be written as $?c, B_1 \ldots B_m$.

[7] We shall slightly relax this condition later.

The interpretation of negation most congenial to constraint logic programming is *constructive* negation [16]. In the customary negation as failure paradigm, negative queries differ from positive queries: the latter yield computed answer substitutions, the former only the answers 'true' or 'false'. Constructive negation tries to make the situation more symmetrical by also providing computed answer substitutions for negative queries. Applied to constraint logic programming, this means that both positive and negative queries can start successful computations ending in constraints. As in the case of negation as failure, the fundamental technical tool is the *completion* of a program:

DEFINITION 5. *Let $\mathcal{P}$ be a normal program, consisting of clauses*

$$\overline{B}^1 \wedge c_1 \rightarrow p^1(\overline{t}^1), \ldots, \overline{B}^n \wedge c_n \rightarrow p^n(\overline{t}^n),$$

where the p^i are atoms.[8] *The completion of $\mathcal{P}$, denoted by $comp(\mathcal{P})$, is computed by the following recipe:*

1. *choose a predicate p that occurs in the head of a clause of $\mathcal{P}$*
2. *choose a sequence of new variables $\overline{x}$ of length the arity of p*
3. *replace in the i-th clause of $\mathcal{P}$ all occurrences of a term in $\overline{t}_i$ by a corresponding variable in $\overline{x}$ and add the conjunct $\overline{x} = \overline{t}_i$ to the body; we thus obtain $\overline{B}^i \wedge c_i \wedge \overline{x} = \overline{t}_i \rightarrow p^i(\overline{x})$*
4. *for each i, let $\overline{z}_i$ be the set of free variables in $\overline{B}^i \wedge c_i \wedge \overline{x} = \overline{t}_i$ not in $\overline{x}$*
5. *given p, let $n_1, \ldots, n_k$ enumerate the clauses in which p occurs as head*
6. *define $Def(p)$ to be the formula*

$$\forall \overline{x}(p(\overline{x}) \leftrightarrow \exists \overline{z}_{n_1}(\overline{B}^{n_1} \wedge c_{n_1} \wedge \overline{x} = \overline{t}_{n_1}) \vee \cdots \vee \exists \overline{z}_{n_k}(\overline{B}^{n_k} \wedge c_{n_k} \wedge \overline{x} = \overline{t}_{n_k}).$$

7. *$comp(\mathcal{P})$ is then obtained as the formula $\bigwedge_p Def(p)$, where the conjunction ranges over predicates p occurring in the head of a clause of $\mathcal{P}$.*

As in the case of negation as failure, the proper logic for negation in constraint logic programming is Kleene's strong three-valued logic with truth values $\{\mathbf{t}, \mathbf{f}, \mathbf{u}\}$. Semantic consequence in this logic will be denoted by $\models_3$.

By the definition of complex subgoal, $\leftrightarrow$ does not occur in normal programs. The occurrence of $\leftrightarrow$ in the completion is interpreted in the manner of Łukasiewicz: $\varphi \leftrightarrow \psi$ is assigned $\mathbf{t}$ when φ, ψ are assigned the same truth value in $\{\mathbf{t}, \mathbf{f}, \mathbf{u}\}$, and $\mathbf{f}$ otherwise. Note the correspondence between this interpretation of $\leftrightarrow$ and the use of $\simeq$ in recursion theory.

DEFINITION 6. *A partial interpretation I is a function which maps ground atoms (in $\mathcal{P}$) to $\{\mathbf{t}, \mathbf{f}, \mathbf{u}\}$, and constraints to $\{\mathbf{t}, \mathbf{f}\}$*

DEFINITION 7. *Given a normal program $\mathcal{P}$, a real-closed field $\mathcal{A}$, a partial interpretation I and a ground atom A, the (immediate)* consequence operator *$\Phi_{\mathcal{P}}$ is defined as*

---

[8] This is not quite the general case, but it suffices for our purposes.

1. $\Phi_{\mathcal{P}}(I)(\mathcal{A}) = \mathbf{t}$ *if there exists a clause* $\overline{B} \wedge c \rightarrow p(\overline{s})$ *in* $\mathcal{P}$ *and an assignment* α *into* $\mathcal{A}$ *such that* $A = p(\overline{s})\alpha$ *and* $I(c\alpha) = I(\overline{B\alpha}) = \mathbf{t}$.
2. $\Phi_{\mathcal{P}}(I)(\mathcal{A}) = \mathbf{f}$ *if for each clause in* $\mathcal{P}$ *of the form* $\overline{B} \wedge c \rightarrow p(\overline{s})$ *and each assignment* α *into* $\mathcal{A}$ *such that* $A = p(\overline{s})\alpha$: $I(c\alpha) = \mathbf{f}$ *or* $I(B\alpha) = \mathbf{f}$.

$\Phi_{\mathcal{P}}$ is monotone, hence has a (least) fixed point, but in general itsclosure ordinal is not ω (as it would be for Horn clause programs) but can be anything below ω_1^{CK}. Nevertheless, there exist useful general soundness and completeness theorems, which in essence go back to a saturation argument due to Kunen [7]. The basic idea is that the closure ordinal is still ω if the underlying structure is recursively saturated. A fully accurate formulation of these theorems would require us to give a definition of the appropriate derivation trees. For this, we refer the reader to [16]. In terms of these trees, one has

THEOREM 2. *Let* $\mathcal{T}$ *be the theory of real-closed fields,* $\mathcal{P}$ *a normal program,* $?c, G$ *a query.*

1. $?c, G$ *succeeds iff*

$$\mathcal{T} \wedge comp(\mathcal{P}) \models_3 \overline{\forall}(c \rightarrow G)$$

2. $?c, G$ *fails finitely iff*

$$\mathcal{T} \wedge comp(\mathcal{P}) \models_3 \neg\overline{\exists}(c \wedge G).$$

Here, $\overline{\forall}(\overline{\exists})$ *denotes universal (existential) closure.*

At this stage it may seem far from clear why such an involved formalism is necessary, or even useful, to represent meaning in natural language. However, a glance at our main semantic tool, the event calculus (to be presented in section 3), will show that constraint logic programming is actually tailor-made for this application. We will motivate the introduction of the event calculus by means of a brief discussion of the semantics of verbs.

§3. Ontology for verb-semantics and the event calculus. Speaking intuitively, a verb denotes a kind of event. In slightly more detail,we may think of that event as being parametrised by the subject, direct object and indirect object of the verb (when appplicable). But what are events? Consider one of the most complex classes of verbs, the so-called *accomplishments*, of which examples are 'draw a circle', 'write a letter', 'cross the street'. Events representing such verbs have an elaborate internal structure. On the one hand there is an activity taking place (draw, write, cross), on the other hand an 'object' is being 'constructed': the circle, the letter, or the path across the street.

In this vein, Dowty (in [2]) analyses the progressivised accomplishment

(12) Mary is drawing a circle

as

(13) CAUSE[Mary draws something, a circle comes into existence],

that is, the sentence is decomposed into an *activity* ('Mary draws something') and a *partial, changing, object* ('circle'). Furthermore, an activity naturally comes with events marking the beginning or end of that activity.

We can see from this example that a rather baroque ontology is necessary to account for the semantics of verbs.

1. individuals
2. real numbers, both to represent time and to code 'stages' of partial objects
3. time-dependent properties, such as activities
4. changing partial objects
5. events, marking the beginning and end of time-dependent properties.

Both time-dependent properties and changing partial objects can be brought under the heading of a *fluent*.[9] A fluent is a function which may contain variables for individuals and reals, and which is interpreted in a model as a set of time points. Events will be taken in the sense of event types, from which tokens are obtained by anchoring the event type to a time point (sometimes also to an interval).

The literature contains several formalisms for reasoning about events, which have their roots in planning systems in artificial intelligence. It has been suggested several times that such formalisms might be useful for the semantics of natural language, although [5] seems to be the first paper where the actual computations are done. We borrow the basic format from [14], although the computational tools will be different.

Given this ontology, the following choice of basic predicates seems natural. We want to be able to say that fluents are initiated and terminated by events, or that a fluent was true at the beginning of time. If f is a variable over fluents, e a variable over events, and t a variable over time points, we may write the required predicates as

1. $Initially(f)$
2. $Happens(e, t)$
3. $Initiates(e, f, t)$
4. $Terminates(e, f, t)$

These predicates are to be interpreted in such a way, that if $Happens(e, t) \wedge Initiates(e, f, t)$, then f will begin to hold after (but not at) t; if $Happens(e, t) \wedge Terminates(e, f, t)$, then f will still hold at t. Strictly speaking this convention makes sense only for events which are not extended in time. For the general case one needs some axioms additional to those presented below; we will omit these for the sake of simplicity.

[9] The name is appropriated from Newton's treatise on the calculus, where all variables are assumed to depend implicitly on time.

The possibility of having changing partial objects requires its own special predicates, namely

5. *Trajectory*(f_1, t, f_2, d)
6. *Releases*(e, f, t)

We have seen above that an activity, while it is going on, may change a partial object. The first predicate in the above list embodies this. Here, one should think of f_1 as an activity, and of f_2 as a certain stage of a partial object. The predicate then expresses that if f_1 holds from t until $t + d$, then at $t + d$, f_2 holds. In applications, f_2 will have a real argument, and will be of the form $f_2(g(t + d))$ for some continuous function g. The predicate *Releases* is necessary to cancel the effect of those axioms of the event calculus which intend to express the so-called 'principle of inertia': 'normally, nothing changes'.

These axioms have the form: if there are no f-relevant events between t_1 and t_2, then the truth value of f at t_1 is the same as that at t_2. We introduce two special predicates for f-relevant events. The first predicate expresses that there is a terminating or releasing event between t_1 and t_2; the second predicate expresses that there is an initiating or releasing event between t_1 and t_2.

7. *Clipped*(t_1, f, t_2)
8. *Declipped*(t_1, f, t_2)

Lastly, we need the truth predicate

9. *HoldsAt*(f, t).

In the usual set-up of the event calculus, it is only *said* that *HoldsAt* is a truth predicate; the defining axioms for the truth predicate are lacking. In planning applications of the event calculus, fluents are typically derived from first order formulas in a language disjoint from that of the event calculus itself. 'Derive' here refers to an operation which transforms formulas into terms, for example, Gödelisation, or what in AI is termed reification. It is of course easy to declare a two-valued truth predicate applying to such fluents only. However, once the fluents are codes of formulas which may also contain predicates from the event calculus, this becomes problematic, and in that case one needs to add a logic program for the truth predicate to the event calculus. The proper logic for such a truth predicate is again Kleene's strong three-valued logic. We do not know of any planning application which needs this generality, but, as explained in [5], for a semantics of natural language this generality is essential. Gödelisation or reification is the formal counterpart of the syntactic operation of nominalisation, and when this procedure is iterated, as in

(14) My father objecting to my not going to church

the truth conditions for sentences involving this expression involve nested occurrences of *HoldsAt* as well. These observations notwithstanding, in the interest of brevity and legibility we shall not add a truth theory, and we shall assume some external interpreter available which relates a fluent to the formula (not containing predicates of the event calculus) from which it derives.

All in all, we then have

DEFINITION 8. *An EC-struct ure is a many-sorted structure of the form*[10] $(R \uplus D \uplus E \uplus F; 0, 1, +, \cdot; a_1 \ldots a_k, f_1 \ldots f_n, e_1 \ldots e_m, Basic)$, *where*

1. *R denotes a real closed field, D a set of individual objects, E a set of events, F a set of fluents,*
2. *the $a_1 \ldots a_k$ are elements of D and interpret individual constants,*
3. *the $f_1 \ldots f_n$ are the interpretations of fluent functions (with variables for reals and individuals),*
4. *the $e_1 \ldots e_m$ are the interpretations of event functions (with variables for reals and individuals),*
5. *Basic is an interpretation of the basic predicates introduced above.*

DEFINITION 9. *The predicates Initially, Happens, Initiates, Terminates, Trajectory and Releases are called the* primitive predicates *of the event calculus. (We consider HoldsAt to belong to the truth theory.)*

Here is an example of how the language may be put to work: the formalisation of accomplishments. Keeping in mind Dowty's analysis, one can see that we need at least the following

1. a cause-fluent f_1 (an activity, possibly containing variables for individuals),
2. a fluent representing a partial object $f_2(x)$, where $x \in \mathbb{R}^+$ (i.e., the set of nonnegative reals),
3. a function $g: \mathbb{R}^+ \longrightarrow \mathbb{R}^+$.

These objects should satisfy the following formula

$$HoldsAt(f_1, now) \wedge \forall t(HoldsAt(f_2(g(t)), t) \rightarrow \forall d > 0\, Trajectory(f_1, t, f_2(g(t + d)), d)).$$

In words: if the cause f_1 exerts its influence uninterruptedly from t until $t + d$, at $t + d$ the state of the partial object will be $f_2(g(t + d))$.

Apart from this we also need to include a culminating event, and a consequent state which follows upon the heels of the culminating event. These latter two parts will be illustrated in the detailed scenario for an accomplishment to be given below.

Formulas such as this capture part of the lexical meaning of a verb. In order to derive predictions, e.g., on when the drawing of the circle will be completed, one needs additional formulas, conveniently divided in axioms, holding for every situation, and a scenario, laying down properties of a particular situation.

---

[10] $\uplus$ denotes disjoint union.

The axioms of the event calculus given below are modified from [14], the difference being due to the fact that we prefer a constraint logic programming approach, whereas Shanahan uses circumscription. In the following, all variables are assumed to be universally quantified.

AXIOM 1. $Initially(f) \wedge \neg Clipped(0, f, t) \rightarrow HoldsAt(f, t)$.

AXIOM 2. $Happens(e, t) \wedge Initiates(e, f, t) \wedge t < t' \wedge \neg Clipped(t, f, t') \rightarrow HoldsAt(f, t')$.

AXIOM 3. $Happens(e, t) \wedge Terminates(e, f, t) \wedge t < t' \wedge \neg Declipped(t, f, t') \rightarrow \neg HoldsAt(f, t')$.

AXIOM 4. $Happens(e, t) \wedge Initiates(e, f_1, t) \wedge t < t' \wedge t = t + d \wedge Trajectory(f_1, t, f_2, d) \wedge \neg Clipped(t, f_1, t') \rightarrow HoldsAt(f_2, t')$.

AXIOM 5. $Happens(e, s) \wedge t < s < t' \wedge (Terminates(e, f, s) \vee Releases(e, f, s)) \rightarrow Clipped(t, f, t')$.

AXIOM 6. $Happens(e, s) \wedge t < s < t' \wedge (Initiates(e, f, s) \vee Releases(e, f, s)) \rightarrow Declipped(t, f, t')$.

The set of axioms of the event calculus will be abbreviated by EC. In the absence of further axioms, one can construct a model for EC by interpreting fluents as finite unions of halfopen intervals (of the form $(a, b]$), and assuming that each event either initiates or terminates a fluent, and that fluents are initiated or terminated by events only. The last section of the paper will show that such models are also available in the presence of further sentences, laying down preconditions on *Happens* and the other primitive predicates.

We see that the axioms form a normal logic program in the sense of $CLP(\mathcal{T})$, with the constraints of the form $s < t$.[11] In the logic program, the programmed predicates are *HoldsAt*, *Clipped* and *Declipped*. We also need programs for the remaining primitive predicates. This requires a preliminary definition.

DEFINITION 10. *A state $S(t)$ at time t is a complex body (cf. definition 4) involving only*

1. *literals of the form $(\neg) HoldsAt(f, t)$, for t fixed and possibly different f,*
2. *equalities between fluent terms, and between event terms,*
3. *formulas in the constraint language $\mathcal{L}$.*

DEFINITION 11. *A* scenario *is a conjunction of statements of the form*

1. $Initially(f)$, *or*
2. $\forall t(S(t) \rightarrow Initiates(e, f, t))$, *or*
3. $\forall t(S(t) \rightarrow Terminates(e, f, t))$, *or*
4. $\forall t(S(t) \rightarrow Releases(e, f, t))$, *or*
5. $\forall t(S(t) \rightarrow Happens(e, t))$,

---

[11] Axiom 3 may seem an exception but actually this axiom follows from the others using negation as failure.

where $S(t)$ is a state in the sense of definition 10. These formulas may contain additional constants for objects[12]*, reals or time points and can be prefixed by universal quantifiers over time points, reals and objects.*

Applying program-completion (cf. definition 5) to a scenario in the sense of definition 11 entails that the predicates *Happens*, *Initiates* and *Terminates* are definable in terms of *HoldsAt*. While this is computationally pleasant, it involves a simplification: it may seem more reasonable to also allow *Happens* to occur in the body of 5, to allow for one event causing another given a precondition. This however introduces an additional loop in the computation, thus possibly affecting termination. In section 5 we shall see that our main representation theorem indeed fails when the concept of scenario is thus liberalised.

An important part of the lexical meaning of a verb is enshrined in the dynamics:

DEFINITION 12. *A* dynamics *is a set of statements of the form*

$$S(t) \rightarrow \mathit{Trajectory}(f_1, t, f_2(g(t+d)), d),$$

where $S(t)$ is a state in the sense of definition 10, and g is a function definable in the constraint language.

§4. Computing with the event calculus. Given a scenario SCEN and a dynamics DYN, we now would like to know, preferably effectively, for which t, $\mathit{HoldsAt}(f,t)$, or, more generally, for which $(\overline{x},t)$, $\mathit{HoldsAt}(f(\overline{x}),t)$. Formulated in this way, the question will have mostly trivial answers, because the theory EC + SCEN + DYN will in general have many models, some of which may contain events and influences of events which are not part of SCEN. As we have explained above, we are only interested in *minimal* models, that is, models where intuitively speaking, nothing is true beyond what is given in SCEN and DYN, and what is forced to be true by EC.

Shanahan [14] uses *circumscription* (due to McCarthy; cf. [8] for an elaborate exposition) for this purpose.

DEFINITION 13. *Let P, Q be predicate symbols of the same arity. Put*

$$P = Q := \forall x(P(x) \leftrightarrow Q(x)),$$
$$P \leq Q := \forall x(P(x) \rightarrow Q(x)),$$
$$P < Q := P \leq Q \wedge \neg(P = Q).$$

DEFINITION 14. *Let $\varphi(P)$ be a sentence containing an occurrence of the predicate symbol P. The* circumscription of P in $\varphi(P)$ *is defined as the following*

[12]In order the simplify the exposition of the theorems in section 5 we do not allow functions on objects.

formula of second order logic:

$$\varphi(P) \wedge \neg\exists p[\varphi(p) \wedge p < P],$$

where p is a predicate variable of the same arity as P. The resulting formula will be denoted by $CIRC[\varphi; P]$.

Hence, if S is a finitely axiomatisable theory, $\overline{P}$ the list of predicates occurring in S, then $CIRC[S; \overline{P}]$ picks out the minimal model of S if there exists one.[13] From a computational point of view, this has the disadvantage that one still needs a mechanism to determine what follows from the circumscribed theory, which is in general second order. On the other hand, if one uses logic programming, one has a computational mechanism which computes the atomic formulae true in the minimal fixed point of the consequence operator (and perhaps more complex formulae as well). The downside of this is that the consequence operator associated to a logic program S need not have a unique fixed point, so that, in general, the completion of S (which is first order) does not define the minimal model, unlike $CIRC(S)$. But, still speaking generally, the least fixed point need not be recursive, although it is recursively enumerable, hence circumscription does not hold an advantage over logic programming here. More importantly, we shall see that in many cases of interest, the least fixed point is unique and computable, and hence that constraint logic programming delivers all that one could want.

4.1. An example. Consider the sentence

(15) 'John is crossing the street.'

The verb phrase 'cross the street' is an accomplishment, hence to specify its meaning one needs an activity, something that changes under the (causal) influence of that activity, a culminating event, and the resulting end-state. These are all introduced and connected in the following scenario, which also contains episodic information relating to the situation at hand.

1. $HoldsAt(distance(x), t) \rightarrow Trajectory(crossing, t, distance(x + d), d)$,
2. $HoldsAt(crossing, now)$,
3. $t_0 < now$,
4. $Initially(one\text{-}side)$,
5. $Initially(distance(0))$,
6. $Happens(start, t_0)$,
7. $HoldsAt(distance(m), t) \wedge HoldsAt(crossing, t) \rightarrow Happens(reach, t)$,
8. $Initiates(start, crossing, t)$,
9. $x < m \rightarrow Releases(start, distance(x), t)$,
10. $Initiates(reach, other\text{-}side, t)$,
11. $Terminates(reach, crossing, t)$,

[13]This is just an illustration; [14] uses a more subtle form of circumscription.

12. $HoldsAt(distance(x), t) \rightarrow Initiates(reach, distance(x), t)$.

The *scenario* SCEN consists of 4–12. The *dynamics* DYN consists of 1.

We now discuss the relation of this scenario to the meaning of the progressive form in English. The literature on the progressive is voluminous, so we have to be very brief; a more satisfactory discussion, including references, can be found in [5]. We basically follow Dowty's analysis in [2] which characterises the progressive by means of a default entailment: 'John is crossing the street' is true now if (a) John is now crossing, and (b) 'John will have crossed the street' is true in all *inertia* worlds, that is, in all possible worlds which are identical to the actual world up to the present time, and which in the future develop in ways 'most compatible with the present course of events', to quote Dowty. Thus, Dowty attempted to characterise the progressive by means of its entailments. The description of inertia worlds is only a gesture toward a definition, but a satisfactory definition can be obtained using the minimal models furnished by constraint logic programming. We will come back to the proposed characterisation of the progressive after some pertinent technical results.

If SCEN contains everything we know about the situation, we would like to be able to derive that John will get to the other side eventually. Let *RCF* denote the axioms for the theory of real-closed fields. We then have

THEOREM 3. *For* $s > t_0$:

$$RCF + comp(\text{SCEN} + \text{DYN} + \text{EC}) \models HoldsAt(other\text{-}side, s + m).$$

PROOF. The premisses *uniquely* determine a model to be of the form:

1. *crossing* holds in the interval $(t_0, t_0 + m]$, and is false outside this interval
2. $distance(0)$ holds on $[0, t_0]$, $distance(x)$ holds at $t_0 + x$, for $x \leq m$, and $distance(m)$ holds after $t_0 + m$
3. *start* happens at t_0, *reach* at $t_0 + m$
4. *one-side* holds before (and including) t_0, and is false thereafter
5. *other-side* holds at and after $t_0 + m$, and is false at other times. ⊣

Now consider the sentence

(16) John was crossing the street, when the truck hit him.

(16) leads to the following additions to the scenario

13. for some r, $t_0 < r < t_0 + m$, $Happens(hit, r)$
14. $Terminates(hit, crossing, t)$.

Now $Clipped(t_0, crossing, t_0 + m)$ becomes derivable, and theorem 3 is no longer true, although one can still compute an interpretation for *crossing*.

Let us now consider the question whether the interpretation for the progressive proposed here is in fact correct. The anonymous referee asked whether it is not more correct to say that 'John is crossing the street' leaves it entirely open whether he will make it, and pointed out that, also according to Dowty, the progressive of an impossibility is never true. This is also the case in our set-up, because scenarios must be consistent. Furthermore, the assumption is always that a scenario contains *all* of one's knowledge pertaining to a given situation. Thus, the default inference should be interpreted, as the referee indeed suggests, as an inference from 'John is F-ing' to '*For all one knows*, John will F'.

Coming back to our main theme, we see that what is constant across situations is the algorithmic content of 'cross the street', not its meaning in the sense of denotation.

We will now turn these considerations into a definition. In the example, one may distinguish between general properties of a situation of 'crossing the street' and formulas fixing parameters of the situation. In the original situation, all formulas except 2, 3 and 6 are general (one has to replace m by a variable in 7). Let us call this the *lexical* component of 'cross the street'. Formulas fixing parameters are 2, 3 and 6. In addition, we need as input a parameter for the width of the street. In the expanded situation, formulas 13 and 14 are added. Let us call this the *episodic* component of 'cross the street', or *episode* for short. An episode thus contains formulas and a substitution of ground terms for certain variables. We may then formulate

DEFINITION 15. *The* sense *of an expression is the constraint logic program (in the sense of $CLP(T)$) representing the lexical component, viewed as an algorithm which transforms an episode into the denotation of the expression in a model, using the axioms of the event calculus.*[14]

The question then arises, what these denotations look like. The definition is a bit vague on this, as it speaks about '*a* model'. We know however that models computed by a logic program are in a sense minimal. It turns out that the denotations in such a minimal model must always be of a very simple form, which matches well with intuition.

§5. **Structure of the models.** The main mathematical question raised by the preceding definition is thus to determine what is the structure of a denotation computed by the above recipe; or, more precisely, how the scenario, the dynamics and the axioms of the event calculus jointly affect the structure of the denotation. To our knowledge, this question has not been investigated in any detail in the relevant AI literature. In the example the scenario determines a unique model, i.e., there is a unique fixed point of the three-valued consequence operator (which is moreover reached in finitely many steps). The model obtained is actually a two-valued model, and the fluents without parameter, *crossing*, *one-side* and *other-side* are represented by finite sets of intervals.

[14]The full version would also need a logic program for the truth predicate.

To describe the structure of the fluent with a real parameter, *distance*(x), one needs the more general notion of a semialgebraic set:

DEFINITION 16. *A subset of* $\mathbb{R}^n$ *is* semialgebraic *if it is a finite union of sets of the form* $\{x \in \mathbb{R}^n \mid f_1 = \cdots = f_k = 0,\ g_1 > 0, \ldots, g_l > 0\}$, *where the* f_i, g_j *are polynomials.*

Observe also that the events (*start* and *reach*) mark the beginning and end of fluents. The structure of the model is thus very similar to that of the canonical model of the event calculus given in section 3.

The question is how far this generalises. Intuitively at least, fluents corresponding to natural language expressions (e.g., verbs) are semialgebraic, and we would like this to fall out of the set-up, without further stipulations. In natural language semantics it is a contested issue whether the fundamental temporal entities are points or intervals. The event calculus neatly sidesteps this issue, by taking the basic entities to be events and fluents, which are not explicit functions of time and which can be interpreted on structures with very different ontologies for time. Even if we take the structure underlying time to be $\mathbb{R}$, that does not constitute an ontological commitment to points. Ontological commitment is generated rather by representation theorems, which correlate the events and fluents with point sets in a given structure. It may then very well turn out that, even when time is taken to be $\mathbb{R}$, fluents and events can be represented as sets of intervals, so that points have no role of their own to play.

The situation is slightly more complicated in the case of fluents admitting real parameters, for example fluents representing possibly changing partial objects. Again speaking intuitively, one would expect change to be continuous, with at most a finite number of jumps. The kind of change it is possible to program depends on the one hand on the constraint language chosen, on the other hand on the constraint logic program. Now in *RCF* only semialgebraic sets can be defined; but would it be possible to extend the range of definable sets by a constraint logic program to *RCF*?

The next few theorems give some pertinent results. This material leans heavily on [16].[15] In this paper we deal only with the case where fluents do not to contain predicates from the event calculus, so that we need not use the program for the truth predicate.

DEFINITION 17. *A finite branch in a computation tree is* successful *if its last node contains a constraint only. A finite branch is* finitely failed *if its last node is of the form* $?c', G'$ *(*G' *nonempty) with no more resolution steps possible. A query* $?c, G$ *is* finitely evaluable *if all branches in a derivation tree starting from* $?c, G$ *end either in success or in finite failure.*

[15] Theorem 4 was announced in [5]. Since that time we have become aware of the relevance of Stuckey's work and we will not publish our own proof.

THEOREM 4. *Let RCF be the theory of real-closed fields. Let* $\mathcal{P} = \text{EC} + \text{SCEN} + \text{DYN}$ *and* $?G$ *a finitely evaluable query in the language of the event calculus. Then there exists a semialgebraic set c such that* $RCF + comp(\mathcal{P}) \models \forall(G \leftrightarrow c)$.

PROOF. We do not quantify over events or fluents, hence the predicates of the event calculus, even when they contain constants for objects, fluents or events, can be thought of as interpreted only on the constraint domain, in this case a real-closed field. Thus, the machinery of constraint logic programming is applicable. By hypothesis the computation tree whose top node is $?G$ is finite, hence those terminal nodes which are not marked as failures are marked by a constraint from the language of real-closed fields. It then follows from lemma 7.3 in [16] together with quantifier elimination for real-closed fields, that G represents a semialgebraic set. ⊣

COROLLARY 1. *Let* $\mathcal{P} = \text{EC} + \text{SCEN} + \text{DYN}$. *If for all fluents* $f(\overline{x})$ *in the scenario and dynamics, the query* $?\,HoldsAt(f(\overline{x}), t)$ *is finitely evaluable, the theory* $RCF + comp(\mathcal{P})$ *has a unique model (modulo the underlying real-closed field), which is obtained after finitely many iterations of the consequence operator. In this model the primitive predicates are also represented by semialgebraic sets.*

PROOF. Since the scenario is finite, it mentions only finitely many fluents (possibly containing parameters for reals or individuals). Every predicate is determined once the interpretation of *HoldsAt* is fixed. Since the computation tree for $?\,HoldsAt(f(\overline{x}), t)$ is finite, by the previous theorem $RCF + comp(\mathcal{P})$ implies that $HoldsAt(f(\overline{x}), t)$ is equivalent to a constraint. Theorem 7.1 in [16] shows that the interpretation of $HoldsAt(f(\overline{x}), t)$ is then determined by finitely many iterations of the consequence operator Φ_P. (This is nontrivial, since the consequence operator applies to ground atoms, and the domain is infinite.) By definition of scenario and dynamics, in $RCF + comp(\mathcal{P})$, every primitive predicate can be defined in terms of *HoldsAt* and relations and functions from the constraint language $\mathcal{L}$, and hence in terms of semialgebraic sets only. It then follows from Stuckey's completeness theorem 8.4 that the least fixed point of Φ_P is reached after finitely many iterations. Since given $RCF + comp(\mathcal{P})$ all predicates have semialgebraic definitions, the least fixed point is the only fixed point. A structure is a model of $RCF + comp(\mathcal{P})$ iff it is a fixed point of Φ_P, hence it also follows that the model determined by the answer to the queries $?\,HoldsAt(f(\overline{x}), t)$ (for each f), is a model of $RCF + comp(\mathcal{P})$, and the unique model. ⊣

COROLLARY 2. *Let* $\mathcal{P} = \text{EC} + \text{SCEN} + \text{DYN}$. *The following are equivalent:*

(a) *any model of* $RCF + comp(\mathcal{P})$ *is completely determined by its restriction to RCF;*
(b) $HoldsAt(f(\overline{x}), t)$ *is semialgebraic.*

PROOF. The direction from (a) to (b) follows from the previous corollary. The converse direction follows from Beth's definability theorem and quantifier elimination for real-closed fields. ⊣

The hypothesis of finite evaluability of $?HoldsAt(f(\overline{x}),t)$ is rather strong, although satisfied in the example given. In principle one can determine $HoldsAt(f(\overline{x}),t)$ completely by starting up derivations from both $?HoldsAt(f(\overline{x}),t)$ and $?\neg HoldsAt(f(\overline{x}),t)$ and stop when the collected answer constraint in one tree is the complement of that in the other; the trees may then still contain branches which are neither successful nor failed. Let us call this notion *essential evaluability*.

THEOREM 5. *Let RCF be the theory of real-closed fields. Let* $\mathcal{P} = \mathrm{EC} + \mathrm{SCEN} + \mathrm{DYN}$ *and* $?G$ *a query in the language of the event calculus such that* $?G$ *is essentially evaluable. Then there exists a semialgebraic set c such that* $RCF + comp(\mathcal{P}) \models \forall(G \leftrightarrow c)$.

PROOF. Along the same lines as that of theorem 4, replacing lemma 7.3 in [16] by theorem 7.4. ⊣

There is also a corresponding completeness result.

THEOREM 6. *Let* $\mathcal{P} = \mathrm{EC} + \mathrm{SCEN} + \mathrm{DYN}$. *The query* $?HoldsAt(f(\overline{x}),t)$ *is essentially evaluable if* $RCF + comp(\mathcal{P}) \models \forall t(HoldsAt(f,t) \leftrightarrow c(t))$ *for a constraint c.*

PROOF. The hypothesis implies that both $RCF + comp(\mathcal{P}) \models \forall t(c(t) \rightarrow HoldsAt(f,t))$ and $RCF + comp(\mathcal{P}) \models \forall t(\neg c(t) \rightarrow \neg HoldsAt(f,t))$. By theorem 8.4 in [16], the goals $?c, HoldsAt$ and $?\neg c, \neg HoldsAt$ succeed. By lemma 5.6(a) in [16] the goal $?HoldsAt$ has a successful derivation ending in a constraint implied by c, and analogously for $?\neg HoldsAt$. ⊣

We started out this section by asking how the structure of the denotation of an expression is affected by the structure of scenario, dynamics and EC. One may observe that these proofs would no longer go through if we were to allow in scenarios formulas of the form '$S(t,t') \wedge Happens(e',t') \rightarrow Happens(e,t)$. Consider a scenario containing

1. $Happens(e,0)$,
2. $Happens(e,t') \wedge t = t' + 1 \rightarrow Happens(e,t)$.

Happens will not be interpretable by a semialgebraic set in models of such a scenario, and it is easy to set up the scenario in such a way that a fluent f now consists of a *countable* set of halfopen intervals.

Next, consider a scenario containing

1. $Happens(e,0)$,
2. $Happens(e,1)$,
3. $Happens(e,t') \wedge Happens(e,t'') \wedge t = \dfrac{t'+t''}{2} \rightarrow Happens(e,t)$,
4. $\neg Happens(e,t) \rightarrow Happens(e',t)$,
5. $\neg Initially(f)$,

6. *Initiates*(e, f, t),
7. *Terminates*(e', f, t).

The event e occurs on a dense set of points. In this case we will have that *HoldsAt*(f, t) is assigned **u** outside $t = 0$, because $\neg$ *Clipped*(s, f, t) will always be assigned **u**. Thus a naive attempt to construct a fluent f which has to be interpreted by a nontrivial dense set of points fails, mainly because the axioms of the event calculus have a built-in 'inertia'.

These observations of course leave us with the bigger question, which scenarios make *HoldsAt* (essentially) evaluable. Not all scenarios in the sense of definition 11 force fluents to be interpreted by sets of intervals. Consider the scenario S for a fluent f determined by

1. $\neg$ *Initially*(f),
2. $\neg$ *HoldsAt*$(f, t) \rightarrow$ *Happens*(e, t),
3. *HoldsAt*$(f, t) \rightarrow$ *Happens*(e', t),
4. *Initiates*(e, f, t),
5. *Terminates*(e', f, t).

When t is taken to range over $\mathbb{R}$, the only models of S are those where $\{t > 0 \mid HoldsAt(f, t)\}$ and $\{t > 0 \mid \neg HoldsAt(f, t)\}$ are interpreted by fractal-like dense sets on $(0, \infty)$, while *HoldsAt*$(f, 0)$ is *false*. On the proof theoretical side, this is manifested by the fact that every derivation tree starting from a query $?t > 0, HoldsAt(f, t)$ is infinite in all its branches. The least fixed point of three-valued consequence operator assigns **t** to $\neg$ *Initially*(f), $\neg$ *HoldsAt*$(f, 0)$, *Happens*$(e, 0)$ and *Initiates*$(e, f, 0)$, and will assign **u** to all other ground instances of $(\neg)$ *HoldsAt*. S is of course blatantly circular, but all interesting programs involve some sort of cycle; cf. formulas 1 and 12 in the example of section 4.1. However, at present we have no idea how to isolate the scenarios which exhibit the completeness properties of corollary 2.

The minimal models of scenarios may thus be partial[16]. How bad this is, depends on the kind of question one wants the scenario to answer. When one wants to know whether some fluent f holds at a particular time t_0, it suffices to find a constraint c such that $c(t_0)$ and the query $?c, HoldsAt(f, t)$ is successful. This may be a fairly local question. On the other hand, asking whether some event will ever happen, may involve computing an entire model, and in that case it may be unfortunate not to know whether the answer to the query is just around the corner, or whether the scenario is essentially incomplete.

Although it would be pleasant to have a theorem indicating which programs of the type introduced here lead to terminating computations, in general partiality seems to be an inherent feature of natural language: there is no reason why the lexical and episodic information embodied in the scenario is sufficiently exhaustive to make the scenario coplete. In a sense it is gratifying

[16] Essentially so, when the program for the truth predicate is added.

that partiality is a consequence of the present approach to meaning, and does not have to be built in. It should be noted in this context that the use of Kleene's three-valued logic is not essential. There exist general techniques to transform three-valued structures into two-valued structures, essentially through replacing each predicate A by a pair $(A, \overline{A})$ consisting of the *positive* and *negative* part of A. For logic programming, this has been elaborated by Stärk in [15].

§6. **Conclusion.** The main conceptual difference between the approch to linguistic meaning advocated here and Moschovakis's notion of meaning concerns the rôle of the lexical component in the semantic architecture. In Moschovakis's approach the basic meanings are more or less given by the structure **A**. The phrase *more or less* means that there may very well be words with complex sense; the German word *Sonnenschutz* (sun protection)[17] may be an example of a lexical item with complex sense. Neglecting such examples the lexical component of semantics is given by structure **A**. This is in accordance with Montague's general program for semantics (see [9].). The task of structural semantics is to build new meanings on the basis of given (lexical) meanings. Semantic relations between between items which cannot be further analysed have to be specified in a rather ad hoc way by meaning postulates[18].

Moschovakis's major achievement from a linguistic point of view is his novel explication of the Fregean notion *sense* which emphasises the computational aspect of this concept. His account is certainly superior to any account based on possible worlds both for conceptual and empirical reasons.

To see how the approach to linguistic meaning advocated in this paper differs from Moschovakis's theory consider again the following formula.

$$\begin{aligned}&\mathit{HoldsAt}(f_1, \mathit{now}) \wedge \forall t(\mathit{HoldsAt}(f_2(g(t)), t)\\ &\qquad\qquad \rightarrow \forall d > 0\ \mathit{Trajectory}(f_1, t, f_2(g(t+d)), d)).\end{aligned}$$

This formula, which we took to be part of the dynamics, characterises the common semantic feature of a *class* of verbs, i.e., the class of accomplishments. It applies to both *crossing a street* and *drawing a circle* and distinguishes these expressions from activities like *pushing a cart*. But this is certainly not a sufficiently fine grained analysis of the meaning of accomplishments since *crossing a street* and *drawing a circle* clearly differ in meaning. The first involves information about say the width of a street the second information about the radius of a circle. This additional lexical information is provided by a specific *scenario*. Both the general characterisation of accomplishments given by the

[17]There may however be serious compositionality problems with examples like this. The word *Sonnenschutz* means protection from sun but certainly the word *Arbeitsschutz* (industrial safety) does not mean anything like protection from work but protection from losing work.

[18]For lots of quite intricate examples compare [2].

formula above and the specific scenario are involved in the computation of denotations in a prefered model. Although the lexical information expressed in scenarios may be rather idiosyncratic it nevertheless finds a systematic place in this computational process.

To summarise, the common feature of Moschovakis's approach to linguistic meaning and the approach advocated in this paper is their emphasis on the computational aspect of the notion *sense*. The difference is that here Moschovakis's approach is radicalised, and also taken to apply at the lexical level. Moschovakis also provides a reduction calculus and a decidable criterion for identity of intensions. This we have not yet been able to do. While the definition of scenario (11) provides a kind of normal form for intension, a full theory of identity of intensions in this context would require us to tackle the notoriously difficult problem of equivalence of logic programs.

Note added in proof. A book on this topic, entitled ***The proper treatment of events*** will be published by Blackwell (November 2004).

REFERENCES

[1] K. Doets, ***From logic to logic programming***, The MIT Press, Cambridge, MA, 1994.

[2] D. Dowty, ***Word meaning and Montague grammar***, Reidel, Dordrecht, 1979.

[3] G. Frege, *On sense and denotation*, ***Translations from the Philosophical Writings of Gottlob Frege*** (P. Geach and M. Black, editors), Blackwell, Oxford, 1952.

[4] ———, *Sinn und Bedeutung*, ***G. Frege: Funktion, Begriff, Bedeutung.*** Fünf logische Studien (G. Patzig, editor), Vandenhoeck, Göttingen, 1962.

[5] F. Hamm and M. van Lambalgen, *Event calculus, nominalisation and the progressive*, ***Linguistics and Philosophy***, vol. 26 (2003), pp. 381–458.

[6] J. Jaffar and M. Maher, *Semantics of constraint logic programs*, ***Journal of Logic Programming***, vol. 12 (1994).

[7] K. Kunen, *Negation in logic programming*, ***Journal of Logic Programming***, vol. 7 (1987), pp. 231–245.

[8] V. Lifschitz, *Circumscription*, ***Handbook of Logic in Artificial Intelligence and Logic Programming*** (D. Gabbay, C. Hogger, and J. Robinson, editors), vol. 3, Clarendon Press, Oxford, 1994.

[9] R. Montague, *The proper treatment of quantification in ordinary english*, ***Formal philosophy*** (R. Thomason, editor), Yale University Press, Yale, 1974.

[10] Y. Moschovakis, *The formal language of recursion*, ***The Journal of Symbolic Logic***, vol. 54 (1989), pp. 1216–1252.

[11] ———, *A mathematical modeling of pure recursive algorithms*, ***Logic at Botik '89*** (A. Meyer and M. Taitslin, editors), Lecture Notes in Computer Science, vol. 363, Springer Verlag, Berlin, 1989.

[12] ———, *Sense and denotation as algorithm and value*, ***Logic colloquium '90*** (J. Oikkonen and J. Väänänen, editors), Lecture Notes in Logic, vol. 2, Springer, Berlin, 1994, pp. 210–249.

[13] ———, *What is an algorithm*, ***Mathematics unlimited – 2001 and beyond*** (B. Engquist and W. Schmid, editors), Springer, Berlin, 2001.

[14] M. P. Shanahan, ***Solving the frame problem***, The MIT Press, Cambridge MA, 1997.

[15] R. F. Stärk, *From logic programs to inductive definitions*, ***Logic: from foundations to applications (European Logic Colloquium '93)*** (W. Hodges, M. Hyland, C. Steinhorn, and J. Truss,

editors), Oxford University Press, Oxford, 1996, pp. 453–481.

[16] P. J. Stuckey, *Negation and constraint logic programming*, ***Information and Computation***, vol. 118 (1995), pp. 12–33.

DEPARTMENT OF PHILOSOPHY
UNIVERSITY OF AMSTERDAM
NIEUWE DOELENSTRAAT 15
NL 1012 CP AMSTERDAM, THE NETHERLANDS
E-mail: M.vanLambalgen@uva.nl

SEMINAR FÜR SPRACHWISSENSCHAFT
UNIVERSITÄT TÜBINGEN
WILHELMSTRASSE 113
D-72074 TÜBINGEN, GERMANY
E-mail: friedrich.hamm@uni-tuebingen.de

TAMENESS IN EXPANSIONS OF THE REAL FIELD

CHRIS MILLER

Abstract. What might it mean for a first-order expansion of the field of real numbers to be tame or well behaved? In recent years, much attention has been paid by model theorists and real-analytic geometers to the o-minimal setting: expansions of the real field in which every definable set has finitely many connected components. But there are expansions of the real field that define sets with infinitely many connected components, yet are tame in some well-defined sense (*e.g.*, the topological closure of every definable set has finitely many connected components, or every definable set has countably many connected components). The analysis of such structures often requires a mixture of model-theoretic, analytic-geometric and descriptive set-theoretic techniques. An underlying idea is that first-order definability, in combination with the field structure, can be used as a tool for determining how complicated is a given set of real numbers.

Throughout, m and n range over $\mathbb{N}$ (the non-negative integers).

Given a first-order structure $\mathfrak{M}$ with underlying set M, "definable" (in $\mathfrak{M}$) means "definable in $\mathfrak{M}$ with parameters from M" unless otherwise noted. If no ambient space M^n is specified, then "definable set" means "definable subset of some M^n". I use "reduct" and "expansion" in the sense of definability, that is, given structures $\mathfrak{M}_1$ and $\mathfrak{M}_2$ with common underlying set M, I say that $\mathfrak{M}_1$ is a reduct of $\mathfrak{M}_2$—equivalently, $\mathfrak{M}_2$ is an expansion of $\mathfrak{M}_1$, or $\mathfrak{M}_2$ expands $\mathfrak{M}_1$—if every set definable in $\mathfrak{M}_1$ is definable in $\mathfrak{M}_2$. For the most part, we shall be concerned with the definable sets of a structure, so we identify $\mathfrak{M}_1$ and $\mathfrak{M}_2$ if they are interdefinable (that is, each is a reduct of the other). An expansion $\mathfrak{M}$ of a dense linear order $(M, <)$ is **o-minimal** if every definable subset of M is a finite union of points and open intervals.

From now on, $\mathfrak{R}$ denotes a fixed, but arbitrary, expansion of the real line $(\mathbb{R}, <)$. "Definable" means "definable in $\mathfrak{R}$" unless noted otherwise. The real field $(\mathbb{R}, +, \cdot)$ is denoted by $\overline{\mathbb{R}}$.

The sequel consists of two parts: Part 1 is mostly expository and somewhat informal; technical details and proofs are mostly deferred to Part 2. General references for background include: van den Dries [29] (a model-theoretic

2000 *Mathematics Subject Classification.* Primary 03C64; Secondary 28A05, 54H05.

This material is based upon work supported by the National Science Foundation under Grant No. 9988855, and is the basis for a lecture that I delivered at Logic Colloquium '01. I thank the organizers for inviting me to address the Colloquium.

Logic Colloquium '01
Edited by M. Baaz, S. Friedman, and J. Krajíček
Lecture Notes in Logic, 20

survey of o-minimality) and [32] (a text on topological o-minimality, with essentially no model theory); van den Dries and Miller [33] (focussing on the analytic geometry of o-minimal expansions of $\overline{\mathbb{R}}$); and anything along the lines of Hausdorff [7], Kechris [8], Kuratowski [9] and Oxtoby [17]. *Please note:* I attempt neither to cite only original sources nor to provide an historical survey.

Numbering. Lemmas, propositions, theorems and corollaries are not numbered explicitly if they appear singly within a section or subsection (or perhaps if they are not referred to later in the paper). For example, if there is one theorem in §5, then it will not be numbered, but referred to later as Theorem 5; if there are two propositions in §3.2, then they will be labelled as Propositions 3.2.1 and 3.2.2 respectively (and so on).

Part 1.

§1. Introduction. What might it mean for a first-order expansion of $\overline{\mathbb{R}}$ to be tame or well behaved? In recent years, much attention has been paid by model theorists and real-analytic geometers to the o-minimal setting: expansions of $\overline{\mathbb{R}}$ in which every definable subset of $\mathbb{R}$ is a finite union of points and open intervals; indeed, for any fixed $p \in \mathbb{N}$, every definable set is a finite disjoint union of connected embedded C^p-submanifolds, each of which is definable. But there are expansions of $\overline{\mathbb{R}}$ that define sets having infinitely many connected components (even locally), yet are tame in some well-defined sense.

1.1. Consider the following structures:

- $\overline{\mathbb{R}}$.
- $(\overline{\mathbb{R}}, 2^{\mathbb{Z}})$, where $2^{\mathbb{Z}} = \{2^k : k \in \mathbb{Z}\}$.
- $(\overline{\mathbb{R}}, \mathrm{Fib})$, where Fib $= \{$Fibonacci numbers$\}$.
- $(\overline{\mathbb{R}}, \psi)$, where $\psi : \mathbb{R}^2 \to \mathbb{R}$ is defined by

$$(x, y) \mapsto \begin{cases} 2^{(\log_2 x)(\log_2 y)} & \text{if } x, y \in 2^{\mathbb{Z}}, \\ 0 & \text{otherwise.} \end{cases}$$

- $(\overline{\mathbb{R}}, \mathcal{S})$, where $\mathcal{S} = \{(e^t \cos t, e^t \sin t) : t \in \mathbb{R}\}$.
- $(\overline{\mathbb{R}}, \mathbb{R}_{\mathrm{alg}})$, where $\mathbb{R}_{\mathrm{alg}}$ denotes the set of all real algebraic numbers.
- $(\overline{\mathbb{R}}, \mathbb{Z})$.

Now, $\overline{\mathbb{R}}$ is o-minimal, and is generally considered to be very well behaved. The study of its definable sets leads to the subject of real algebraic geometry (see Bochnak *et al.* [2]). At the other end, the structure $(\overline{\mathbb{R}}, \mathbb{Z})$ may be identified with the real projective hierarchy of classical descriptive set theory, that is, $A \subseteq \mathbb{R}^n$ is definable in $(\overline{\mathbb{R}}, \mathbb{Z})$ if and only A is projective (see [8, (37.6)] and [32, pg. 16]). From now on, we write PH instead of $(\overline{\mathbb{R}}, \mathbb{Z})$. PH is quite complicated. Many set-theoretic independence issues arise naturally in the study of its definable sets; for example, the statement that every real projective set is Lebesgue measurable is independent of ZFC. (Of course, compared to arbitrary sets in $\mathbb{R}^n$, projective sets might be regarded as rather tame). All of the structures listed above are reducts of PH.

It follows from [26, Theorem III] that every subset of $\mathbb{R}$ definable in $(\overline{\mathbb{R}}, 2^{\mathbb{Z}})$ is the union of an open (definable) set and finitely many (definable) discrete sets; indeed, this holds with $2^{\mathbb{Z}}$ replaced by $\alpha^{\mathbb{Z}}$ ($= \{\alpha^k : k \in \mathbb{Z}\}$) for any $\alpha > 0$. Hence, $\mathbb{Q}$ is not definable in $(\overline{\mathbb{R}}, 2^{\mathbb{Z}})$, so neither is $\mathbb{Z}$. In other words, $(\overline{\mathbb{R}}, 2^{\mathbb{Z}})$ is a proper reduct of PH. Moreover, $\mathrm{Th}(\overline{\mathbb{R}}, 2^{\mathbb{Z}})$ is decidable [26, Theorem I].

Every subset of $\mathbb{R}$ definable in $(\overline{\mathbb{R}}, \mathrm{Fib})$ is also the union of an open set and finitely many discrete sets, because Fib is definable in $(\overline{\mathbb{R}}, \varphi^{\mathbb{Z}})$, where $\varphi = (1+\sqrt{5})/2$. (Note that

$$\mathrm{Fib} = \{(\varphi^{2n} - \varphi^{-2n})/\sqrt{5} : n \geq 1\} \cup \{(\varphi^{2n+1} + \varphi^{-2n-1})/\sqrt{5} : n \geq 0\}.)$$

An isomorphic copy of $(\mathbb{Z}, +, \cdot)$ is definable in $(\overline{\mathbb{R}}, \psi)$, namely

$$(2^{\mathbb{Z}}, \cdot \restriction (2^{\mathbb{Z}} \times 2^{\mathbb{Z}}), \psi \restriction (2^{\mathbb{Z}} \times 2^{\mathbb{Z}})).$$

(Here and throughout, given a map $f : X \to Y$ and $A \subseteq X$, $f \restriction A$ denotes the restriction of f to A.) By comparison with PH, one might think $(\overline{\mathbb{R}}, \psi)$ would be quite complicated. Of course, $\mathrm{Th}(\overline{\mathbb{R}}, \psi)$ is undecidable, but what about the definable sets? We shall see that, like $(\overline{\mathbb{R}}, 2^{\mathbb{Z}})$, every subset of $\mathbb{R}$ definable in $(\overline{\mathbb{R}}, \psi)$ is the union of an open set and finitely many discrete sets.

The set $\mathcal{S}$ is an infinite spiral and a trajectory of the linear vector field

$$(x, y) \mapsto (x - y, x + y) : \mathbb{R}^2 \to \mathbb{R}^2.$$

The natural parameterization of $\mathcal{S}$ involves the exponential, sine and cosine functions, but none of these functions are definable in $(\overline{\mathbb{R}}, \mathcal{S})$ (as we shall see later). Indeed, again, every subset of $\mathbb{R}$ definable in $(\overline{\mathbb{R}}, \mathcal{S})$ is the union of an open set and finitely many discrete sets.

(More properties of the structures $(\overline{\mathbb{R}}, \alpha^{\mathbb{Z}})$, $(\overline{\mathbb{R}}, \psi)$ and $(\overline{\mathbb{R}}, \mathcal{S})$ are established in sections 3 and 4.)

Now, $\mathbb{R}_{\mathrm{alg}}$ is dense and co-dense (in $\mathbb{R}$), so certainly not the union of an open set and finitely many discrete sets. But $(\overline{\mathbb{R}}, \mathbb{R}_{\mathrm{alg}})$ is, loosely speaking, topologically almost o-minimal: By [31, Theorem 4], every *closed* definable subset of $\mathbb{R}$ is a finite union of points and open intervals. This has nice consequences for all definable sets (to be made precise in §5). In particular, $\mathbb{Z}$ is not definable in $(\overline{\mathbb{R}}, \mathbb{R}_{\mathrm{alg}})$.

1.2. The properties of the structures $(\overline{\mathbb{R}}, 2^{\mathbb{Z}})$ and $(\overline{\mathbb{R}}, \mathbb{R}_{\mathrm{alg}})$ suggest that we look beyond o-minimality for other kinds of tame behavior. First, let us consider some notions already in use by model theorists or analytic geometers.

An expansion $\mathfrak{M}$ of a dense linear order $(M, <)$ is **weakly o-minimal** if every definable subset of M is a finite union of convex definable sets, and is **locally o-minimal** if for every definable $A \subseteq M$ and $x \in M$, there exist $a, b \in M$ such that $a < x < b$ and $A \cap (a, b)$ is a finite union of points and open intervals. Of course, if $M = \mathbb{R}$, then $\mathfrak{M}$ is weakly o-minimal if and only if it is o-minimal, since a convex subset of the real line is just an interval of some sort. Now, there are expansions of the real line that that are locally o-minimal but not o-minimal—$(\mathbb{R}, <, +, \mathbb{Z})$ is one; see *e.g.*, Friedman and Miller [6]—but every locally o-minimal expansion of $\overline{\mathbb{R}}$ is o-minimal (the proof is an exercise). Consequently, for expansions of $\overline{\mathbb{R}}$, neither weak nor local o-minimality yields any generality beyond o-minimality. There are yet more exotic variations—for examples, see Belegradek *et al.* [1] and Macpherson [12]—but they either do not generalize o-minimality for expansions of $\overline{\mathbb{R}}$ or are not enough topologically based for present purposes (some do not even make sense over $\mathbb{R}$).

Based on differential- and analytic-geometric considerations, useful forays have been made beyond the realm of first-order structures—for example, geometric categories [33] and Shiota's $\mathfrak{X}$-systems [25]—but still, all sets dealt with have *locally* only finitely many connected components.

Another natural model-theoretic condition we might impose on $\mathfrak{R}$ is **uniform finiteness**: for each definable $A \subseteq \mathbb{R}^{m+n}$, there exists $N_A \in \mathbb{N}$ such that for every $x \in \mathbb{R}^m$, the set $\{y \in \mathbb{R}^n : (x, y) \in A\}$ is finite only if it contains at most N_A elements. Every o-minimal structure has the uniform finiteness property, but so does $(\overline{\mathbb{R}}, \mathbb{R}_{\text{alg}})$ [31, Corollary 4.5]. If $\mathfrak{R}$ expands $\overline{\mathbb{R}}$ and has the uniform finiteness property, then—like $(\overline{\mathbb{R}}, \mathbb{R}_{\text{alg}})$—every closed definable set is a finite union of points and open intervals (see Proposition 5.2 below); in particular, $\mathfrak{R}$ defines no infinite discrete sets.

1.3. The approach taken in this paper is to work with first-order structures on $(\mathbb{R}, <)$ under various topological, measure- or descriptive set-theoretic assumptions, with special emphasis on expansions of $\overline{\mathbb{R}}$ that define infinite discrete sets. Now, "topological definability theory" is certainly not new; see *e.g.*, A. Robinson [20] or Pillay [18] (and basic o-minimality as exposed in [32] may be regarded as topological definability theory). However, unlike most model-theoretic investigations, we do not shy away from making full use of special facts about the real numbers: uncountability, completeness, separability of the topology, measure theory, descriptive set theory, and so on (indeed, some proofs rely so heavily on the combination of separability and the Baire category theorem that I would not know how to avoid it). This approach is not particularly new in the study of o-minimal expansions of $\overline{\mathbb{R}}$ either, but here we focus on moving beyond o-minimality. (In addition to the above-mentioned [6, 26, 31], a first paper in this direction is Miller and Speissegger [16].)

I regard many of the results herein as preliminary or suggestive of further lines of inquiry. A number of questions are scattered throughout. Probably, some are relatively easy, but many appear to be quite hard. Possibly, some are independent of ZFC (but I have tried to avoid asking such questions). Many are posed for arbitrary expansions of the real line, but I am interested primarily in the answers for expansions of $\overline{\mathbb{R}}$.

Here is an outline of the remainder of Part 1. Some topological preliminaries are established in §2. In §3, we consider some conditions, more general than o-minimality, to impose on the definable subsets of $\mathbb{R}$, and investigate some corresponding consequences for all definable sets. Fundamental to the study of o-minimal structures are the notions of "cell" and "decomposition". By relaxing the definition of decomposition, we obtain tameness conditions that make sense for any definable set (as opposed to just the definable subsets of the line); see §4. We go further in §5 by relaxing the definition of cell, via the notion of the open core of a structure. Some structures that are interdefinable with PH are given in §6.

§2. Topological preliminaries. Let X be a topological space. Equip cartesian powers X^m with the product topology. ($X^0 = \{\emptyset\}$; regard a map $f: X^0 \to X^n$ as the corresponding constant $f(\emptyset)$.) If $A \subseteq X^m \times X^n$, then πA denotes the projection of A on the first m coordinates; for $x \in X^m$, put

$$A_x = \{y \in X^n : (x, y) \in A\},$$

the **fiber** of A over x. Whenever convenient, we identify $X^m \times X^n$ with X^{m+n}.

Let $A \subseteq X$.

Put:

$$\begin{aligned} \text{int}(A) &= \text{the interior of } A, \\ \text{cl}(A) &= \text{the closure of } A, \\ \text{bd}(A) &= \text{the boundary of } A \ (= \text{cl}(A) \setminus \text{int}(A)), \\ \text{fr}(A) &= \text{the frontier of } A \ (= \text{cl}(A) \setminus A), \\ \text{isol}(A) &= \text{the isolated points of } A. \end{aligned}$$

If $X = \mathbb{R}^n$ and $A \subseteq \mathbb{R}^n$ is definable, then so are each of the above sets. (This does not depend on working over $\mathbb{R}$, rather, only that the collection of all open boxes in $\mathbb{R}^n$ is a definable family.) We say that A:

- is **constructible** if it is a (finite) boolean combination of open sets.
- is **discrete** if $A = \text{isol}(A)$.
- **has no interior** if $\text{int}(A) = \emptyset$ and **has interior** if $\text{int}(A) \neq \emptyset$.
- is **dense** if $\text{cl}(A) = X$, **co-dense** if $\text{cl}(X \setminus A) = X$, **somewhere dense** if $\text{cl}(A)$ has interior, and **nowhere dense** if $\text{cl}(A)$ has no interior.
- is **meager** if it is countable union of nowhere dense sets.

For $x \in A$, A is **locally closed at** x if there is an open neighborhood U of x such that $A \cap U = \text{cl}(A) \cap U$; A is **locally closed** if A is locally closed at each $x \in A$. It is easy to check that the following are equivalent:

- A is locally closed.
- $A = \text{cl}(A) \cap U$ for some open U.
- $A = F \cap U$ for some open U and closed F.

- $\operatorname{fr}(A)$ is closed.
- $A \cap \operatorname{cl}(\operatorname{fr}(A)) = \emptyset$.

If $X = \mathbb{R}^n$ and $A \subseteq \mathbb{R}^n$ is definable and locally closed, then $A = \operatorname{cl}(A) \cap U$ for some definable open $U \subseteq \mathbb{R}^n$; see *e.g.*, [32, pg. 18] or [33, Appendix B].

Let $\operatorname{lc}(A)$ denote the set of all locally closed points of A, equivalently, $\operatorname{lc}(A) = A \setminus \operatorname{cl}(\operatorname{fr}(A))$. Note that $\operatorname{lc}(A)$ is locally closed and is the relative interior of A in $\operatorname{cl}(A)$. We say that A **has a locally closed point** if $\operatorname{lc}(A) \neq \emptyset$.

For ordinals λ, define sets $A^{(\lambda)}$ as follows:

$$\begin{aligned} A^{(0)} &= A, \\ A^{(\lambda+1)} &= A^{(\lambda)} \setminus \operatorname{lc}(A^{(\lambda)}), \\ A^{(\lambda)} &= A \setminus \bigcup_{\mu<\lambda} \operatorname{lc}(A^{(\mu)}) \quad \text{if } \lambda \text{ is a limit.} \end{aligned}$$

If $B \subseteq X^{m+n}$ and $x \in X^m$, then $B_x^{(\lambda)}$ denotes $(B_x)^{(\lambda)}$, not $(B^{(\lambda)})_x$. (There is an obvious notion of rank arising from this construction, but we shall not bother to define it formally.) If $X = \mathbb{R}^n$ and $A \subseteq \mathbb{R}^n$ is definable, then so is each $A^{(k)}$, $k \in \mathbb{N}$.

2.1. *The following are equivalent*:

1. *A is constructible.*
2. *There exists $k \in \mathbb{N}$ such that $A^{(k)} = \emptyset$.*
3. *There exists $k \in \mathbb{N}$ such that $A = \bigcup_{j=0}^{k} \operatorname{lc}(A^{(j)})$.*
4. *A is a finite disjoint union of locally closed sets.*

(For 1⇒2, see Dougherty and Miller [3]. Indeed, if $j \in \mathbb{N}$ is such that A is boolean combination of $2j$ open sets, then $A^{(j+1)} = \emptyset$.)

2.2 ([7, §30] or [9, §34, VI]). *If X is a Polish space, then $A \in \mathrm{F}_\sigma \cap \mathrm{G}_\delta$ if and only if there is a countable ordinal λ such that $A^{(\lambda)} = \emptyset$. In particular, if $\emptyset \neq A \in \mathrm{F}_\sigma \cap \mathrm{G}_\delta$, then $\operatorname{lc}(A) \neq \emptyset$.*

For ordinals λ, define sets $A^{[\lambda]}$ as follows:

$$\begin{aligned} A^{[0]} &= A, \\ A^{[\lambda+1]} &= A^{[\lambda]} \setminus \operatorname{isol}(A^{[\lambda]}), \\ A^{[\lambda]} &= A \setminus \bigcup_{\mu<\lambda} \operatorname{isol}(A^{[\mu]}) \quad \text{if } \lambda \text{ is a limit.} \end{aligned}$$

If $B \subseteq X^{m+n}$ and $x \in X^m$, then $B_x^{[\lambda]}$ denotes $(B_x)^{[\lambda]}$, not $(B^{[\lambda]})_x$. The Cantor-Bendixson rank (often defined only for closed sets) arises from this construction. Note that $\operatorname{isol}(A) = \operatorname{isol}(\operatorname{cl}(A))$, so $A^{[\lambda]} = \emptyset$ if $\operatorname{cl}(A)^{[\lambda]} = \emptyset$, but the converse need not hold. If $X = \mathbb{R}^n$ and $A \subseteq \mathbb{R}^n$ is definable, then so is each $A^{[k]}$ for $k \in \mathbb{N}$.

2.3. *The set A is the union of an open set and finitely many discrete sets if and only if there exists $k \in \mathbb{N}$ such that $(A \setminus \operatorname{int}(A))^{[k]} = \emptyset$.*

(The proof is an exercise.)

We make frequent (but often without explicit mention) use of the following consequences of 2.1 and 2.3.

2.4. *A definable set is:*

- *constructible if and only it is a finite disjoint union of locally closed definable sets.*
- *the union of an open set and finitely many discrete sets if and only if it is the disjoint union of an open definable set and finitely many discrete definable sets.*

The **dimension** of a nonempty set $A \subseteq X^n$, denoted by **dim** A, is the maximal integer d such that, after some permutation of coordinates, the projection of A on the first d coordinates has interior. (Put $\dim \emptyset = -\infty$ and $X^{-\infty} = \emptyset$.) Clearly, if $A \subseteq B \subseteq X^n$, then $\dim A \leq \dim B$.

Let $\Pi(n,m)$ denote the collection of all coordinate projection maps

$$(x_1, \dots, x_n) \mapsto (x_{\lambda(1)}, \dots, x_{\lambda(m)}) \colon X^n \to X^m,$$

where λ is a strictly increasing function from $\{1, \dots, m\}$ into $\{1, \dots, n\}$. If $d \in \mathbb{N}$ and $A \subseteq X^n$, then $\dim A \geq d$ if and only if there exists $\mu \in \Pi(n,d)$ such that μA has interior.

From now on, unless otherwise noted, "box" means "nonempty open box" and topological notions will be taken with respect to the box topologies induced by the order topology of the real line.

§3. Tameness conditions on definable subsets of the line. In this section, we consider various conditions to impose on the definable subsets of $\mathbb{R}$, and begin to investigate the corresponding consequences for all definable sets.

3.1. Many candidates for tameness conditions on $\mathfrak{R}$ imply that every definable subset of $\mathbb{R}$ either has interior or is nowhere dense.

CONDITIONS. Every definable subset of $\mathbb{R}$:

1. has interior or is finite.
2. has interior or is a finite union of discrete sets.
3. is constructible.
4. has interior or is countable.
5. has interior or an isolated point (or is empty).
6. has interior or is null (that is, has Lebesgue measure 0).
7. is F_σ.
8. has a locally closed point (or is empty).
9. has interior or is nowhere dense.

PROPOSITION. $1 \Rightarrow 2 \Rightarrow (3 \& 4)$. $3 \Rightarrow 7$. $4 \Rightarrow (5 \& 6 \& 7)$. $5 \Rightarrow 8$. $6 \Rightarrow 9$. $7 \Rightarrow 8 \Rightarrow 9$.

PROOF. Most of these implications are obvious. (Observe that if $A \subseteq \mathbb{R}$ is dense and co-dense in an open interval I, then $\mathrm{lc}(A \cap I) = \emptyset$ and: at most one of $A \cap I$, $A \setminus I$ is countable; at most one of $A \cap I$, $A \setminus I$ is meager; and at most one of $A \cap I$, $A \setminus I$ is null.) For $4 \Rightarrow 5$, recall that every nonempty perfect subset of $\mathbb{R}$ is uncountable. For $7 \Rightarrow 8$, use 2.2. ⊣

QUESTION. $(\overline{\mathbb{R}}, 2^{\mathbb{Z}})$ witnesses $2 \not\Rightarrow 1$. Do the other converse implications fail?

Lifting the conditions. Condition 1 holds if and only if $\mathfrak{R}$ is o-minimal, so an assumption on the definable subsets of $\mathbb{R}$ implies that all definable sets have certain nice properties. What might the other conditions listed above imply about all definable sets?

3.2. Constructibility. Rather than working sequentially through the list, we start with what is probably the best known of these "lifting questions":

QUESTION 3.2.1. If every definable subset of $\mathbb{R}$ is constructible, is every definable set constructible?

As far as I know, even the following weaker version is still open:

QUESTION 3.2.2. If for every $\mathfrak{M} \equiv \mathfrak{R}$, every subset of M definable in $\mathfrak{M}$ is constructible, is every set definable in $\mathfrak{R}$ constructible?

Why doesn't the stronger assumption seem to help? First, a routine compactness argument (using 2.1) shows that Question 3.2.2 is equivalent to:

QUESTION 3.2.3. Suppose that for every m and definable $A \subseteq \mathbb{R}^{m+1}$ there exists $N \in \mathbb{N}$ such $A_x^{(N)} = \emptyset$ for all $x \in \mathbb{R}^m$. Does it follow that for every m and definable $A \subseteq \mathbb{R}^{m+1}$ there exists $N \in \mathbb{N}$ such $(A^{(N)})_x = \emptyset$ for all $x \in \mathbb{R}^m$?

Now let $F \subseteq [0,1]^2$ be closed and σ be a permutation of $[0,1]$. Put $A = F \setminus \mathrm{graph}(\sigma)$; then every fiber A_x—as well as every set $\{x \in \mathbb{R} : (x,y) \in A\}$—is locally closed. Certainly, it can happen that $\mathrm{lc}(A) = \emptyset$, even if F is nowhere dense (indeed, take F to be an arbitrary Cantor subset of the unit cube). And, of course, one can remove from F much more complicated sets than the graph of a permutation of $[0,1]$. Such constructions appear to make unlikely a positive answer to Question 3.2.2, as well as to the obvious lifting question associated to Condition 8. However, if we rule out wild behavior by the dimension 0 definable sets, then we begin to obtain some positive results.

PROPOSITION. *If every dimension 0 definable set has a locally closed point, then $A^{(1)}$ is nowhere dense in A, for every definable set A.*

(See §8.1 for the proof.)

Hence, if every dimension 0 definable set has a locally closed point, then for any definable set A we have:

$$\begin{aligned} A &= \mathrm{lc}(A) \cup A^{(1)} \\ &= \mathrm{lc}(A) \cup \mathrm{lc}(A^{(1)}) \cup A^{(2)} \\ &\ \ \vdots \\ &= \mathrm{lc}(A) \cup \cdots \cup \mathrm{lc}(A^{(k)}) \cup A^{(k+1)} \\ &\ \ \vdots \end{aligned}$$

where $\mathrm{lc}(A^{(k)}) \neq \emptyset$ or $A^{(k+1)} = \emptyset$. Of course, in general, there is no reason to believe that this process terminates after finitely many iterations—that A is constructible—even if $\dim A = 0$. This suggests trying a weaker formulation of Question 3.2.2, and finally we have a reasonable result:

THEOREM. *Suppose that for every $\mathfrak{M} \equiv \mathfrak{R}$, every dimension 0 set definable in $\mathfrak{M}$ is constructible. Then every set definable in $\mathfrak{R}$ is constructible.*

(See §8.2 for the proof.) The converse also holds. The assumption is equivalent to: For every m, n and definable $A \subseteq \mathbb{R}^{m+n}$ there exists $N \in \mathbb{N}$ such that for all $x \in \mathbb{R}^m$ either $\dim A_x > 0$ or $A_x^{(N)} = \emptyset$.

3.3. The proof of Theorem 3.2 uses special facts about $\mathbb{R}$. Moreover, the most natural way to construct a proof is via results that are established first for the "interior or nowhere dense" condition. Hence, we jump to the end of the list.

CONVENTION. From now on, p ranges over $\mathbb{N}$.

THEOREM. *Suppose every definable subset of $\mathbb{R}$ has interior or is nowhere dense. Then every definable set has interior or is nowhere dense. If $U \subseteq \mathbb{R}^m$ is open and $f: U \to \mathbb{R}$ is definable, then there is an open definable $V \subseteq U$ such that $U \setminus V$ is nowhere dense and the restriction $f \restriction V: V \to \mathbb{R}$ is continuous. If $m = 1$, then V may be chosen so that for each connected component I of V, $f \restriction I$ is either constant or strictly monotone. If moreover $\mathfrak{R}$ expands $\overline{\mathbb{R}}$, then the above holds with "C^p" in place of "continuous" (so*

$$\bigcap_{p \in \mathbb{N}} \{x \in U : f \text{ is } C^p \text{ on an open ball about } x\}$$

is dense-G_δ in U).

(See §8.3 for the proof.)

3.4. Smoothness and d-minimality. "Submanifold" always means "embedded submanifold, everywhere of the same dimension, but not necessarily connected". For $A \subseteq \mathbb{R}^n$, let $\mathrm{reg}^p(A)$ denote the set of $a \in A$ such that for some $d \in \mathbb{N}$, $\mu \in \Pi(n, d)$ and box U about a, $\mu \restriction (A \cap U)$ maps $A \cap U$ C^p-diffeomorphically onto an open subset of $\mathbb{R}^d$. (A C^0-diffeomorphism is just a homeomorphism.) Note that $\mathrm{reg}^p(A)$ is open in A and is a finite disjoint union of C^p-submanifolds of $\mathbb{R}^n$. If $\dim A = 0$, then $\operatorname{reg}^p(A) = \operatorname{isol}(A)$. If A is open, then $\operatorname{reg}^p(A) = A$. If A is definable, then $\operatorname{reg}^0(A)$ is definable; if moreover $\mathfrak{R}$ expands $\overline{\mathbb{R}}$, then each $\operatorname{reg}^p(A)$ is definable (see *e.g.*, [33, Appendix B]).

PROPOSITION. *If every dimension* 0 *definable subset of* $\mathbb{R}$ *has an isolated point, then* $A \setminus \operatorname{reg}^0(A)$ *is nowhere dense in* A, *for every definable set* A. *If moreover* $\mathfrak{R}$ *expands* $\overline{\mathbb{R}}$, *then this holds for each* p (*so* $\bigcap_{p\in\mathbb{N}} \operatorname{reg}^p(A)$ *is dense*-G_δ *in* A).

(See §8.4 for the proof.)

Informally, the above says that if $\mathfrak{R}$ expands $\overline{\mathbb{R}}$ and defines no Cantor subsets of the line, then every definable set is something like a countable union of C^∞-submanifolds: For every fixed p and definable set A, we have:

$$\begin{aligned} A &= \operatorname{reg}^p(A) \cup (A \setminus \operatorname{reg}^p(A)) \\ &= \operatorname{reg}^p(A) \cup \operatorname{reg}^p(A \setminus \operatorname{reg}^p(A)) \cup (A \setminus \operatorname{reg}^p(A \setminus \operatorname{reg}^p(A))) \\ &\;\;\vdots \end{aligned}$$

Again, there is no apparent reason to believe that this process stabilizes after finitely many iterations (consider $(\overline{\mathbb{R}}, E)$, where E is a countable closed subset of $\mathbb{R}$ with infinite Cantor-Bendixson rank). Naturally, we might prefer that it does.

Now, Condition 2 may be rephrased as: Every definable subset of $\mathbb{R}$ is a finite disjoint union of particularly nice C^p-submanifolds, each of which is definable. This condition does lift to all definable sets if we assume some extra uniformity (but I don't know if the extra assumption is necessary). We need some definitions before we can make this precise.

I say that $\mathfrak{R}$ is **d-minimal** (short for "discrete-minimal") if for every $\mathfrak{M} \equiv \mathfrak{R}$, every subset of M definable in $\mathfrak{M}$ is the union of an open set and finitely many discrete sets. Equivalently (by a routine compactness argument), $\mathfrak{R}$ is d-minimal if for every m and definable $A \subseteq \mathbb{R}^{m+1}$ there exists $N \in \mathbb{N}$ such that for all $x \in \mathbb{R}^m$, A_x either has interior or is a union of N discrete sets. Note that if $\mathfrak{R}$ is d-minimal, then every reduct of $\mathfrak{R}$ over $(\mathbb{R}, <)$ is d-minimal.

Aside. The notion of d-minimality makes sense as stated for expansions of arbitrary first-order topological structures (as defined in [18]) but it is unclear to me how useful it would be (even over arbitrary real closed fields). In this paper, every interesting fact established about d-minimality seemingly depends on working over $\mathbb{R}$. In any case, here I only scratch the surface of the subject; another paper is in preparation.

Let us say that a d-dimensional C^p-submanifold M of $\mathbb{R}^n$ is **special** if there exists $\mu \in \Pi(n,d)$ such that for each $y \in \mu M$ there is an open box B about y such that each connected component X of $M \cap \mu^{-1}(B)$ projects C^p-diffeomorphically (via $\mu{\restriction}X$) onto B, *i.e.*, $\mu{\restriction}M : M \to \mu M$ is a C^p-smooth covering map. (For $d = 0$ or $d = n$, every dimension d submanifold of $\mathbb{R}^n$ is special: a dimension 0 submanifold of $\mathbb{R}^n$ is just a discrete set; a dimension n submanifold of $\mathbb{R}^n$ is just an open set.) If we wish to keep track of the projection μ, then say that M is **μ-special**. Note that if M is μ-special and $S \subseteq \mu M$ is simply connected, then (after some permutation of coordinates) $M \cap \mu^{-1}(S)$ is a countable disjoint of graphs of C^p maps $S \to \mathbb{R}^{n-d}$.

A collection $\mathcal{A}$ of subsets of $\mathbb{R}^n$ is **compatible** with a collection $\mathcal{B}$ of subsets of $\mathbb{R}^n$ if for every $A \in \mathcal{A}$ and $B \in \mathcal{B}$, either A is contained in B or A is disjoint from B. A set $A \subseteq \mathbb{R}^n$ is compatible with $\mathcal{B}$ if $\{A\}$ is compatible with $\mathcal{B}$, and similarly for $\mathcal{A}$ being compatible with a set $B \subseteq \mathbb{R}^n$.

THEOREM 3.4.1. *Assume $\mathfrak{R}$ is d-minimal. Let $\mathcal{A}$ be a finite collection of definable subsets of $\mathbb{R}^n$. Then there is a finite partition of $\mathbb{R}^n$ into special C^0-submanifolds, each of which is definable and compatible with $\mathcal{A}$. If moreover $\mathfrak{R}$ expands $\overline{\mathbb{R}}$, then this holds with "C^p" in place of "C^0".*

(See §8.5 for the proof.) The converse holds as well. More can be said if $\mathfrak{R}$ expands $(\mathbb{R}, <, +)$, as we shall see in the next section. But first we consider some examples.

Clearly, every o-minimal structure is d-minimal, but more is true: every locally o-minimal expansion of $(\mathbb{R}, <)$ is d-minimal (since every definable subset of $\mathbb{R}$ has interior or is discrete).

The **field of exponents** of an expansion of $\overline{\mathbb{R}}$ is the set of all $r \in \mathbb{R}$ such that the power function $t \mapsto t^r : (0, \infty) \to \mathbb{R}$ is definable.

THEOREM 3.4.2. *Suppose $\alpha > 0$ and $\mathfrak{R}$ is o-minimal, expands $\overline{\mathbb{R}}$ and has field of exponents $\mathbb{Q}$. Then $(\mathfrak{R}, \alpha^{\mathbb{Z}})$ is d-minimal.*

(See §8.6 for the proof.) $\overline{\mathbb{R}}$ is o-minimal and has field of exponents $\mathbb{Q}$—this is an easy consequence of quantifier elimination (in the language of ordered rings)—but there are far more exotic examples; see van den Dries and Speissegger [34, 35] and Rolin *et al.* [23].

NOTE. There is a converse of sorts: Every proper subgroup of $(\mathbb{R}^{>0}, \cdot)$ is either cyclic or both dense and co-dense in $(0, \infty)$, so if $\alpha > 1$ and r is irrational, then the set $\{xy^r : x, y \in \alpha^{\mathbb{Z}}\}$ is dense and co-dense in the positive real line. Hence, if $\mathfrak{R}$ is an expansion of $(\mathbb{R}, +, \cdot, \alpha^{\mathbb{Z}})$ such that every definable set has interior or is nowhere dense, then $\mathfrak{R}$ has field of exponents $\mathbb{Q}$ and every proper definable subgroup of $(\mathbb{R}^{>0}, \cdot)$ is of the form $\alpha^{q\mathbb{Z}}$ for some $q \in \mathbb{Q}$.

COROLLARY (joint with P. Speissegger). *$(\overline{\mathbb{R}}, S)$ is d-minimal, where S is the infinite spiral defined in* §1.1.

PROOF. The structure $\mathbb{R}^{\mathrm{RE}} := (\overline{\mathbb{R}}, \exp\restriction[0, 2\pi], \sin\restriction[0, 2\pi])$ is o-minimal and has field of exponents $\mathbb{Q}$; see [27]. (Here and throughout, exp denotes the function $t \mapsto e^t : \mathbb{R} \to \mathbb{R}$.) Note that $\cos\restriction[0, 2\pi]$ is definable. Put $\alpha = e^{2\pi}$. Then

$$(x, y) \in S \Leftrightarrow \exists g \in \alpha^{\mathbb{Z}},\ \exists t \in [0, 2\pi),\ x = ge^t \cos t \ \&\ y = ge^t \sin t.$$

Hence, $\mathcal{S}$ is definable in $(\mathbb{R}^{\mathrm{RE}}, \alpha^{\mathbb{Z}})$. Apply Theorem 3.4.2. ⊣

REMARKS.

- The restriction of exp to any bounded interval is definable in $\mathbb{R}^{\mathrm{RE}}$, so the argument goes through for any spiral $\{(e^{at}\cos t, e^{at}\sin t) : t \in \mathbb{R}\}$, $a \neq 0$.
- No restriction to an *unbounded* interval I of any of exp, sin or exp · sin is definable in any d-minimal expansion of $(\overline{\mathbb{R}}, \mathcal{S})$: We have

$$(\overline{\mathbb{R}}, \sin\restriction I) = (\overline{\mathbb{R}}, \exp\cdot\sin\restriction I) = \mathrm{PH}$$

(since $I \cap \pi\mathbb{Z} = \{t \in I : \sin t = 0\} = \{t \in I : e^t \sin t = 0\}$). The group $e^{2\pi\mathbb{Z}}$ is definable in $(\overline{\mathbb{R}}, \mathcal{S})$ and $(\overline{\mathbb{R}}, e^{2\pi\mathbb{Z}}, \exp\restriction I) = \mathrm{PH}$ (exp is definable over $\overline{\mathbb{R}}$ from $\exp\restriction I$, hence so is $\log : (0,\infty) \to \mathbb{R}$).
- $\mathbb{R}^{\mathrm{RE}}$ is of interest in its own right; see [28].

The next result produces some rather exotic examples built on some of those that we have obtained so far. Let $E \subseteq \mathbb{R}$. Put $\mathcal{S}_0 := \{\mathbb{R}^0, \emptyset\}$. Let $\mathcal{S}_{n+1}$ be the collection of all subsets of $\mathbb{R}^{n+1}$ of the form

$$A = \bigcup_{\alpha\in I}\bigcap_{u\in P_\alpha} Y_u$$

where $m \in \mathbb{N}$, $(P_\alpha)_{\alpha\in I}$ is an indexed family of subsets of the cartesian power E^m, and Y is of one of the following forms:

$$X \times \mathbb{R}$$

$$\{(x,t) \in \mathbb{R}^{n+1} : x \in X \,\&\, f(x) = t\}$$

$$\{(x,t) \in \mathbb{R}^{n+1} : x \in X \,\&\, f(x) < t\}$$

$$\{(x,t) \in \mathbb{R}^{n+1} : x \in X \,\&\, t < g(x)\}$$

$$\{(x,t) \in \mathbb{R}^{n+1} : x \in X \,\&\, f(x) < t < g(x)\}$$

where $X \subseteq \mathbb{R}^n$ is definable in $\mathfrak{R}$ and $f, g : \mathbb{R}^n \to \mathbb{R}$ are functions definable in $\mathfrak{R}$. There are no conditions on the functions f, g other than definability, and the index sets I are allowed to be completely arbitrary. Let $(\mathfrak{R}, E)^\infty$ denote the expansion of $\mathfrak{R}$ by all elements of each $\mathcal{S}_k$ $(k \geq 1)$. Every set definable in $\mathfrak{R}$, as well as every subset of any E^k $(k \geq 1)$, is definable in $(\mathfrak{R}, E)^\infty$. Of course, if E is finite, then the construction is of no interest—we just wind up with $\mathfrak{R}$—so we take E to be infinite. (On the other hand, it's not hard to see that if $\mathfrak{R}$ defines a function $f : \mathbb{R}^n \to \mathbb{R}$ such that $f(E^n)$ is dense, then each $\mathcal{S}_k$ is equal to the power set of $\mathbb{R}^k$, so again the construction is of no further interest.)

THEOREM (joint with H. Friedman, [6]). *Suppose $\mathfrak{R}$ is an o-minimal expansion of $(\mathbb{R}, <, +)$. Let $A \subseteq \mathbb{R}^{m+1}$ be definable in $(\mathfrak{R}, E)^\infty$. Then there exist $l \in \mathbb{N}$ and $f : \mathbb{R}^{l+m} \to \mathbb{R}$ definable in $\mathfrak{R}$ such that for every $x \in \mathbb{R}^m$ either A_x has interior or $A_x \subseteq \mathrm{cl}\{f(u,x) : u \in E^l\}$.*

(The above is not stated explicitly in [6], but follows from the claim there on page 62.)

COROLLARY. *Suppose* $\mathfrak{R}$ *is an o-minimal expansion of* $(\mathbb{R}, <, +)$, E *has no interior and* $(\mathfrak{R}, E)$ *is d-minimal. Then* $(\mathfrak{R}, E)^\infty$ *is d-minimal.*

PROOF. Since E has no interior and $(\mathfrak{R}, E)$ is d-minimal, E is a finite union of discrete sets, hence countable. For any $l, m \in \mathbb{N}$, $f : \mathbb{R}^{l+m} \to \mathbb{R}$ definable in $\mathfrak{R}$ and $x \in \mathbb{R}^m$, the set $\{f(u, x) : u \in E^l\}$ is countable and definable in $(\mathfrak{R}, E)$, so it is a finite union of discrete sets. Then the same is true of $\text{cl}\{f(u, x) : u \in E^l\}$. By d-minimality, there exists $N \in \mathbb{N}$ independent of x such that $\text{cl}\{f(u, x) : u \in E^l\}$ is a union of N discrete sets. Apply the theorem. ⊣

By combining with Theorem 3.4.2:

COROLLARY. *Suppose* $\alpha > 0$ *and* $\mathfrak{R}$ *is o-minimal, expands* $\overline{\mathbb{R}}$ *and has field of exponents* $\mathbb{Q}$. *Then* $(\mathfrak{R}, \alpha^{\mathbb{Z}})^\infty$ *is d-minimal.*

COROLLARY. $(\overline{\mathbb{R}}, \psi)$ *(as defined in* §1.2) *is d-minimal.*

PROOF. $(\overline{\mathbb{R}}, 2^{\mathbb{Z}})^\infty$ is d-minimal and $(\overline{\mathbb{R}}, \psi)$ is a reduct of $(\overline{\mathbb{R}}, 2^{\mathbb{Z}})^\infty$. ⊣

QUESTION. Suppose every definable subset of $\mathbb{R}$ either has interior is a finite union of discrete sets. Is $\mathfrak{R}$ d-minimal? What if $\mathfrak{R}$ expands $\overline{\mathbb{R}}$?

3.5. Countability. C^0-submanifolds of $\mathbb{R}^n$ have only countably many connected components, so if $\mathfrak{R}$ is d-minimal, then every definable set has only countably many connected components.

QUESTION. If every definable subset of $\mathbb{R}$ has interior or is countable, does every definable set have only countably many connected components?

It might seem reasonable, at first thought, that the answer should be "Yes"; after all, this is certainly true for every open definable set, and is easily seen to be true for every dimension 0 definable set. But complications similar to those associated to Question 3.2.2 arise; perhaps further assumptions on $\mathfrak{R}$ are needed.

3.6. Lebesgue measure. Condition 6 almost lifts:

PROPOSITION. *Suppose every definable subset of* $\mathbb{R}$ *has interior or is null. Then every definable set has interior or is null if and only if every definable set is Lebesgue measurable.*

By Fubini's theorem and its converse, the above follows easily from Proposition 3.1 and Theorem 3.3 (the details are left to the reader). Unfortunately, knowing that a subset of $\mathbb{R}^n$ is null doesn't really say much (especially for $n > 1$) and one might hope for stronger results, *e.g.*, if every dimension 0 definable set has Hausdorff dimension 0, does every definable set have integer-valued Hausdorff dimension? But probably we would need yet further assumptions on $\mathfrak{R}$.

3.7. Borel structures. Here is the most general lifting result associated to Condition 7 that I know of:

PROPOSITION. *Suppose* $\mathfrak{R}$ *is an expansion* in the syntactic sense *of* $(\mathbb{R}, <)$ *by Borel relations and functions. If every definable subset of* $\mathbb{R}$ *is* F_σ, *then* $\mathfrak{R}$ *is Borel.*

(I say that $\mathfrak{R}$ is **Borel** if every definable set is Borel.)

PROOF. Since boolean combinations of Borel sets are Borel, it suffices to show that if $A \subseteq \mathbb{R}^{n+1}$ is definable and Borel, then $\pi A \subseteq \mathbb{R}^n$ is Borel. By assumption, each fiber A_x $(x \in \mathbb{R}^n)$ is F_σ. Apply Arsenin and Kunugui [8, (35.46)]. ⊣

REMARK. The above holds without the assumption that $\mathfrak{R}$ defines <.

Every Borel set is projective, but not every projective set is Borel. Hence, every Borel expansion of $\overline{\mathbb{R}}$ is a proper reduct of PH. The structure $(\overline{\mathbb{R}}, \mathbb{R}_{\mathrm{alg}})$ is Borel; indeed, every definable set is a boolean combination of definable F_σ sets (this follows from [31, Theorem 1]). Hence, all particular proper reducts of PH that we have examined so far are Borel.

A set $A \subseteq \mathbb{R}^n$ is **analytic** (also called "Souslin" or "Suslin" in the literature) if it is the continuous image of a Borel subset of $\mathbb{R}$. By Souslin's Theorem [8, (14.11)], A is Borel if and only if both A and its complement are analytic. Hence, if $\mathfrak{R}$ is not Borel, then $\mathfrak{R}$ defines a non-analytic set.

Powerful tools from geometric measure theory can be used to analyze the definable sets in Borel expansions of $\overline{\mathbb{R}}$. I will not go into details in this paper (but see Edgar and Miller [4, 5] for some related material).

§4. Cells and decompositions. Another way to generalize the notion of o-minimality is to relax one of the fundamental definitions in the subject; for convenience, we review it (but see [32, Chapter 3] for a thorough treatment).

Cells (**$\mathfrak{R}$-cells**, if more precision is needed) are defined by induction on n:

- $\mathbb{R}^0$ is the unique cell contained in $\mathbb{R}^0$.
- Let $D \subseteq \mathbb{R}^n$ be a cell. Then $D \times \mathbb{R}$ is cell. Let $f: D \to \mathbb{R}$ be continuous and definable; then

$$\mathrm{graph}(f)$$
$$\{(x,t) \in D \times \mathbb{R} : f(x) < t\}$$
$$\{(x,t) \in D \times \mathbb{R} : t < f(x)\}$$

  are cells. If $g: D \to \mathbb{R}$ is continuous and definable and $f(x) < g(x)$ for all $x \in D$, then $\{(x,t) \in D \times \mathbb{R} : f(x) < t < g(x)\}$ is a cell.

Cells are connected and special (as defined in §3.4) C^0-submanifolds. Indeed, a cell is definably homeomorphic to an open cell in $\mathbb{R}^{\dim A}$ via some $\mu \in \Pi(n, \dim A)$.

NOTE. Every cell is a PH-cell.

A **(finite) decomposition** of $\mathbb{R}^n$ is defined by induction on n:

- $\{\mathbb{R}^0\}$ is the unique decomposition of $\mathbb{R}^0$.
- A decomposition of $\mathbb{R}^{n+1}$ is a finite partition $\mathcal{D}$ of $\mathbb{R}^{n+1}$ into cells such that the collection of projections $\pi\mathcal{D} := \{\pi D \colon D \in \mathcal{D}\}$ is a decomposition of $\mathbb{R}^n$.

Different kinds of cells and decompositions are defined by imposing extra conditions. In particular, C^p-cells are defined by requiring that the functions f and g (in the definition of "cell") be C^p. Again, C^p-cells are connected special C^p-submanifolds, each definably C^p-diffeomorphic to an open C^p-cell in $\mathbb{R}^{\dim A}$ via some coordinate projection.

Arguably, the most fundamental result in o-minimality is the following:

CELL DECOMPOSITION (Pillay and Steinhorn, [19]). *$\mathfrak{R}$ is o-minimal if and only if for every n and finite collection $\mathcal{A}$ of definable subsets of $\mathbb{R}^n$ there is a finite decomposition of $\mathbb{R}^n$ compatible with $\mathcal{A}$.*

If $\mathfrak{R}$ expands $\overline{\mathbb{R}}$, the theorem holds using C^p-cells and finite C^p-decompositions; see [32, Chapter 7]. (These results do *not* rely on working over $\mathbb{R}$.)

It is natural to consider relaxing the definitions of cells, decompositions, or both. In this section, we retain the definition of cell and relax the definition of decomposition.

First, let us consider what might be the weakest acceptable notion of a cell decomposition. Define a **weak decomposition** of $\mathbb{R}^n$ by induction on n:

- $\{\mathbb{R}^0\}$ is the unique weak decomposition of $\mathbb{R}^0$.
- A weak decomposition of $\mathbb{R}^{n+1}$ is a partition $\mathcal{D}$ of $\mathbb{R}^{n+1}$ into cells such that $\pi\mathcal{D}$ is a weak decomposition of $\mathbb{R}^n$ and $\bigcup\{D \in \mathcal{D} \colon \operatorname{int}(D) = \emptyset\}$ has no interior.

Let us say that $\mathfrak{R}$ **admits weak decomposition** if for every n and finite collection $\mathcal{A}$ of definable subsets of $\mathbb{R}^n$ there is a weak decomposition of $\mathbb{R}^n$ compatible with $\mathcal{A}$. It is immediate from the definitions that if $\mathfrak{R}$ admits weak decomposition, then every definable set has interior or is nowhere dense. I doubt if the converse holds, but a counterexample is needed. (Of course, it could be that the notion of "weak decomposition" is *too* weak to be useful.)

Define a **countable decomposition** of $\mathbb{R}^n$ by induction on n:

- $\{\mathbb{R}^0\}$ is the unique countable decomposition of $\mathbb{R}^0$.
- A countable decomposition of $\mathbb{R}^{n+1}$ is a countable partition $\mathcal{D}$ of $\mathbb{R}^{n+1}$ into cells such that $\pi\mathcal{D}$ is a countable decomposition of $\mathbb{R}^n$.

$\mathfrak{R}$ **admits countable decomposition** if, for every n and finite collection $\mathcal{A}$ of definable subsets of $\mathbb{R}^n$, there is a countable decomposition of $\mathbb{R}^n$ compatible with $\mathcal{A}$.

As with finite decompositions, different kinds of weak or countable decompositions are obtained by imposing extra conditions on the cells; for example, a countable C^p-decomposition of $\mathbb{R}^n$ is a countable decomposition of $\mathbb{R}^n$ by C^p cells.

THEOREM. *Every d-minimal expansion of* $(\mathbb{R}, <, +)$ *admits countable decomposition. Every d-minimal expansion of* $\overline{\mathbb{R}}$ *admits countable* C^p*-decomposition.*

(See §8.7 for the proof.)

QUESTION. If $\mathfrak{R}$ expands the field and admits countable C^p-decomposition, is $\mathfrak{R}$ d-minimal? (I doubt it.)

Of course, $(\overline{\mathbb{R}}, \mathbb{R}_{\text{alg}})$ does not admit even weak decomposition. But there is a connection to cell decomposition, as we shall see in the next section.

REMARK. For a survey of notions of cell decompositions for structures other than expansions of dense linear orders, see Mathews [13].

§5. Open cores. Finally we consider an option—a pseudo-cell decomposition condition—for dealing with the case that $\mathfrak{R}$ does not satisfy the "interior or nowhere dense" condition. (The material in this section is based in part on joint work with P. Speissegger; see [16] for more detailed information, examples and applications.)

The **open core** of $\mathfrak{R}$, denoted by $\mathfrak{R}^\circ$, is the reduct of $\mathfrak{R}$ generated by the collection of all open sets definable in $\mathfrak{R}$. Note that $\mathfrak{R}^\circ$ expands $(\mathbb{R}, <)$ and is a reduct of PH. If every set definable in $\mathfrak{R}$ is constructible, then $\mathfrak{R}^\circ = \mathfrak{R}$. If $\mathfrak{R}$ expands $\overline{\mathbb{R}}$ and defines $\mathbb{Z}$, then $\mathfrak{R}^\circ =$ PH (and conversely). Fortunately, we have less trivial examples. In particular, suppose $\mathfrak{R}$ expands $\overline{\mathbb{R}}$ and is o-minimal. Let M be the underlying set of a proper elementary substructure of $\mathfrak{R}$. By [31, Theorem 5], the open core of $(\mathfrak{R}, M)$ is $\mathfrak{R}$. A canonical example is $\mathfrak{R} = \overline{\mathbb{R}}$ and $M = \mathbb{R}_{\text{alg}}$, so $(\overline{\mathbb{R}}, \mathbb{R}_{\text{alg}})^\circ = \overline{\mathbb{R}}$; indeed, $(\overline{\mathbb{R}}, K)^\circ = \overline{\mathbb{R}}$ for any real closed subfield K of $\mathbb{R}$. (*Aside:* There are projective, but non-Borel, real closed subfields of $\overline{\mathbb{R}}$, so dense pairs provide examples of proper reducts of PH that are not Borel. But I do not know of any expansions of $\overline{\mathbb{R}}$ *by constructible sets* that are not Borel, other than PH.)

Every constructible set definable in $\mathfrak{R}$ is definable in $\mathfrak{R}^\circ$ (so a set in $\mathbb{R}^n$ is an $\mathfrak{R}$-cell if and only if it is an $\mathfrak{R}^\circ$-cell). In particular, if A is definable in $\mathfrak{R}$, then all of $\text{int}(A)$, $\text{cl}(A)$, $\text{lc}(A)$ and $\text{isol}(A)$ are definable in $\mathfrak{R}^\circ$. This suggests that if the sets definable in $\mathfrak{R}^\circ$ are suitably well behaved, then the behavior of the sets definable in $\mathfrak{R}$ should not be too much worse. This loose notion can be made more precise:

PROPOSITION 5.1. $\mathfrak{R}^\circ$ *is o-minimal if and only if for every* m *and finite collection* $\mathcal{A}$ *of subsets of* $\mathbb{R}^m$ *definable in* $\mathfrak{R}$, *there is a finite decomposition* $\mathcal{D}$ *of* $\mathbb{R}^m$ *such that for each* $A \in \mathcal{A}$ *and* $D \in \mathcal{D}$, *either* A *is disjoint from* D, *or* A *contains* D, *or* A *is dense and co-dense in* D.

(As usual, if $\mathfrak{R}$ expands $\overline{\mathbb{R}}$, then the above holds using C^p-cells and decompositions.)

PROOF. See [16, pg. 203] for the forward implication.

Conversely, assume that such a pseudo-decomposition property holds for $\mathfrak{R}$. Then every constructible definable subset of $\mathbb{R}$ is a finite union of points and open intervals, so it suffices to show that every set definable in $\mathfrak{R}^\circ$ is constructible. Now, every set that is quantifier-free definable in $\mathfrak{R}^\circ$ (regarded in its natural language) is constructible, so it suffices to show that if $A \subseteq \mathbb{R}^{m+1}$ is constructible and definable, then πA is constructible, and for this it suffices to show that A is a finite union of cells. Put $\mathcal{A} = \{A\}$ and let $\mathcal{D}$ be a decomposition of $\mathbb{R}^{m+1}$ as described. If A is dense in some some cell $D \in \mathcal{D}$, then A is not co-dense in D—since both A and D are constructible—so A contains D. Hence, A is a disjoint union of cells in $\mathcal{D}$. ⊣

Note an easy consequence:

COROLLARY. *$\mathfrak{R}$ is o-minimal if and only if $\mathfrak{R}^\circ$ is o-minimal and every subset of $\mathbb{R}$ definable in $\mathfrak{R}$ has interior or is nowhere dense.*

As the reader might imagine, one can formulate various results of the above kind, based on whatever nice properties $\mathfrak{R}^\circ$ might have (*e.g.*, d-minimality). Of course, knowing that $\mathfrak{R}^\circ$ = PH would not very useful: PH is the open core of the expansion of $(\mathbb{R}, <)$ by *all* subsets of each $\mathbb{R}^n$ ($n \geq 1$), so it is difficult to see how any interesting conclusions could be drawn about the sets definable in $\mathfrak{R}$.

There is nothing special in the preceding two results about working over the real numbers; they both hold for expansions of arbitrary dense linear orders. The next result is quite a different matter.

THEOREM (joint with P. Speissegger, [16]). *If every definable subset of $\mathbb{R}$ is finite or uncountable, then $\mathfrak{R}^\circ$ is o-minimal. If $\mathfrak{R}$ expands $\overline{\mathbb{R}}$, then $\mathfrak{R}^\circ$ is o-minimal if and only if every discrete definable subset of $\mathbb{R}$ is finite.*

The techniques used in the proof rely, in a seemingly crucial way, upon that $\mathbb{R}$ (with its natural metric) is a Polish space.

INFORMAL COROLLARY. *The study of expansions of $\overline{\mathbb{R}}$ breaks down to the study of those with o-minimal open core and those that define an infinite discrete subset of $\mathbb{R}$.*

There is an important (but logically trivial) dichotomy in the case that $\mathfrak{R}^\circ$ is o-minimal: Either $\mathfrak{R}$ defines a function whose graph is somewhere dense, or it does not. We shall not pursue this matter in this paper (but see [16, pg. 204]). Also, there is an *a priori* subdivision of the case that $\mathfrak{R}$ defines an infinite discrete set: Either $\mathfrak{R}$ defines an infinite discrete *closed* subset of $\mathbb{R}$ or it does not (I don't know if the latter can happen, even if $\mathfrak{R}$ expands $\overline{\mathbb{R}}$). The next result (another consequence of the preceding theorem; see [16, pg. 201]) illustrates why we are interested in this distinction.

COROLLARY. *Suppose $\mathfrak{R}$ expands $\overline{\mathbb{R}}$. Let $A \subseteq \mathbb{R}^{m+n}$ be definable and constructible. Then there is a closed definable $B \subseteq \mathbb{R}^{m+2}$ such that the projection* *of B on the first m coordinates is equal to the projection of A on the first m coordinates. If moreover $\mathfrak{R}$ defines an infinite discrete closed subset of $\mathbb{R}$, then B may be taken in $\mathbb{R}^{m+1}$.*

COROLLARY. *Suppose $\mathfrak{R}$ expands $\overline{\mathbb{R}}$, defines an infinite discrete closed subset of $\mathbb{R}$, and πA is constructible for every n and closed definable $A \subseteq \mathbb{R}^{n+1}$. Then every definable set is constructible.*

QUESTION. If $A \subseteq \mathbb{R}$ is infinite and discrete, does $(\overline{\mathbb{R}}, A)$ define an infinite discrete closed set? (For example, let A be the set of midpoints of the complementary intervals of an arbitrary Cantor subset of $\mathbb{R}$.)

LEMMA. *If $\mathfrak{R}$ expands $(\mathbb{R}, <, +)$ and has the uniform finiteness property (recall the definition from §1.2) then every discrete definable subset of $\mathbb{R}$ is finite.*

PROOF. Let $A \subseteq \mathbb{R}$ be discrete and definable. Assume, toward a contradiction, that A is infinite.

Suppose A is closed. Fix $a \in A$. Then at least one of $A \cap [a, \infty)$ or $A \cap (-\infty, a]$ is infinite; say the former. Put $B = \{(x, y) : x, y \in A \;\&\; a \leq y \leq x\}$. Define $\sigma : A \to \mathbb{R}$ by $\sigma(x) = \min(A \cap (x, \infty))$, that is, $\sigma(x)$ is the successor of x in A. Then

$$B_a = \{a\}, \quad B_{\sigma(a)} = \{a, \sigma(a)\}, \quad B_{\sigma(\sigma(a))} = \{a, \sigma(a), \sigma(\sigma(a))\}$$

and so on, contradicting uniform finiteness.

Now suppose A is not closed; then $\operatorname{fr}(A)$ is nonempty and closed (since A is locally closed). The distance function $x \mapsto d(x, \operatorname{fr}(A)) : \mathbb{R} \to \mathbb{R}$ (taken with respect with the sup norm) is continuous and definable. Hence, for each $r > 0$, $\{x \in A : d(x, \operatorname{fr}(A)) \geq r\}$ is discrete and closed. By the previous case (and uniform finiteness) there exists $N \in \mathbb{N}$ such that

$$\operatorname{card}\{x \in A : d(x, \operatorname{fr}(A)) \geq r\} \leq N$$

for all $r > 0$. But then $\{x \in A : d(x, \operatorname{fr}(A)) < \varepsilon\} = \emptyset$ for some $\varepsilon > 0$, contradicting that $\operatorname{fr}(A) \neq \emptyset$. ⊣

REMARK. If $\mathfrak{M}$ is an expansion of a densely ordered group $(M, <, +)$ and the least upper bound property holds for definable subsets of M (see [15] for more information) then the lemma holds for $\mathfrak{M}$.

Taken together with [16, Theorem] we have:

PROPOSITION 5.2. *If $\mathfrak{R}$ expands $\overline{\mathbb{R}}$ and has the uniform finiteness property, then $\mathfrak{R}^\circ$ is o-minimal.*

QUESTION. If $\mathfrak{R}$ expands $\overline{\mathbb{R}}$ and $\mathfrak{R}^\circ$ is o-minimal, does $\mathfrak{R}$ have the uniform finiteness property?

QUESTIONS. Let $E \subseteq \mathbb{R}^m$ be any "natural" mathematical object. Identify and describe the open core of $(\overline{\mathbb{R}}, E)$. (Of course, if E is constructible, then this is asking to describe the definable sets of $(\overline{\mathbb{R}}, E)$ itself.) Note that any finite sequence $E_1 \subseteq \mathbb{R}^{m(1)}, \dots, E_l \subseteq \mathbb{R}^{m(l)}$ may be definably identified with $E_1 \times \cdots \times E_l$.

Some interesting candidates for E include (finite sequences of): infinitely generated proper subgroups of $(\mathbb{R}, +)$; noncyclic proper subgroups of $(\mathbb{R}^{>0}, \cdot)$; subrings; subfields; the torsion points of the circle group $S^1 \subseteq \mathbb{R}^2$; rational points of an irreducible algebraic variety; fractal subsets of the plane; trajectories of vector fields (and so on).

Let us consider some concrete cases.

If E is a subfield of $\mathbb{R}$, is $(\overline{\mathbb{R}}, E)^\circ$ equal to either $\overline{\mathbb{R}}$ or PH? We have already noted that if E is real closed, then $(\overline{\mathbb{R}}, E)^\circ = \overline{\mathbb{R}}$. On the other hand, if E is either a finite degree algebraic extension of $\mathbb{Q}$, or of the form $K(\alpha)$ with α transcendental over a subfield K, then $\mathbb{Z}$ is definable in $(E, +, \cdot)$—see J. Robinson [21] for the former and R. Robinson [22] for the latter—so $(\overline{\mathbb{R}}, E) = \text{PH}$.

We know (by Theorems 3.4.1 and 3.4.2) that if $\alpha > 0$ and $E = \alpha^{\mathbb{Z}}$, then $(\overline{\mathbb{R}}, E)$ is d-minimal and its own open core. What can be said if $E = \alpha^{\mathbb{Z}} \cdot \beta^{\mathbb{Z}}$ $(= \{\alpha^j \beta^k : j, k \in \mathbb{Z}\})$ or $E = \alpha^{\mathbb{Z}} \times \beta^{\mathbb{Z}}$? Note that if $\beta \notin \alpha^{\mathbb{Q}}$, then $\alpha^{\mathbb{Z}} \cdot \beta^{\mathbb{Z}}$ is dense and co-dense in $(0, \infty)$, and is definable in $(\overline{\mathbb{R}}, \alpha^{\mathbb{Z}}, \beta^{\mathbb{Z}})$. To be fair, I should point out that even the number theory of the set $\{3^m \pm 2^n : m, n \in \mathbb{N}\}$ is not well understood.

Is $(\overline{\mathbb{R}}, 2^{\mathbb{Z}}, \mathbb{R}_{\text{alg}})^\circ = (\overline{\mathbb{R}}, 2^{\mathbb{Z}})$? (I think this is probably true: just amalgamate all relevant proofs in [26] and [31].)

§6. Interdefinability with PH. We have considered several examples of proper reducts of PH. In this section, we consider a few sets that do generate PH over the field.

6.1. Of course, $(\overline{\mathbb{R}}, \mathbb{Q}) = \text{PH}$, since $\mathbb{Z}$ is definable in $(\mathbb{Q}, +, \cdot)$.

PROPOSITION. *Let R be a subring of $\mathbb{R}$ and $(G, +)$ be a finitely generated subgroup of $(R, +)$ with $1 \in G$. Then $\mathbb{Z}$ is definable in $(R, +, \cdot, G)$.*

PROOF. The definable set $D := \{r \in R : rG \subseteq G\}$ is a subring of R contained in G. Since G is finitely generated, D is the ring of integers of a number field. By [21], $\mathbb{Z}$ is definable in $(D, +, \cdot)$, hence also in $(R, +, \cdot, G)$. ⊣

COROLLARY. *If $(G, +)$ is a nontrivial finitely generated subgroup of $(\mathbb{R}, +)$, then $(\overline{\mathbb{R}}, G) = \text{PH}$.*

6.2. "Natural" subsets of natural numbers. We have seen that for certain familiar $E \subseteq \mathbb{N}$ (*e.g.*, Fib and $\{\alpha^n : n \in \mathbb{N}\}$ for α a fixed positive integer), we have $(\overline{\mathbb{R}}, E) \neq \text{PH}$. Let us examine some other cases.

Obviously, $(\overline{\mathbb{R}}, \mathbb{N}) = \text{PH}$.

If $f : \mathbb{R} \to \mathbb{R}$ is semialgebraic (equivalently, definable in $\overline{\mathbb{R}}$), then either f is ultimately constant or ultimately strictly monotone. Hence, if

$E = \{f(n) : n \in \mathbb{N}\}$ and is infinite—say, f is an ultimately positive non-constant polynomial with integer coefficients—then every sufficiently large natural number is definable in $(\overline{\mathbb{R}}, E)$, hence $\mathbb{N}$ is as well.

If $E = \{n! : n \in \mathbb{N}\}$, then $(\overline{\mathbb{R}}, E) = \mathrm{PH}$. (Note that for any $A \subseteq \mathbb{R}$ having order type ω, the successor function σ on A is definable in $(\mathbb{R}, <, A)$ and the set $\{\sigma(x)/x : x \in A \setminus \{0\}\}$ is definable in $(\overline{\mathbb{R}}, A)$.)

By Vinogradov [36], every sufficiently large odd integer is a sum of three prime numbers. It follows easily that $(\overline{\mathbb{R}}, E) = \mathrm{PH}$ if E is the set of all primes.

QUESTION. Let $E \subseteq \mathbb{N}$ and suppose that $(\overline{\mathbb{R}}, E)$ is not Borel. Is $(\overline{\mathbb{R}}, E) = \mathrm{PH}$?

6.3. $(\overline{\mathbb{R}}, \alpha^{\mathbb{Z}}, \exp) = \mathrm{PH}$ for any $\alpha > 1$, since $x \mapsto \log_\alpha : (0, \infty) \to \mathbb{R}$ is definable in $(\overline{\mathbb{R}}, \exp)$. More interesting and less trivial:

PROPOSITION. $(\overline{\mathbb{R}}, \mathbb{R}_{\mathrm{alg}}, \exp) = \mathrm{PH}$.

PROOF. It suffices to show that $\mathbb{Q}$ is definable in $(\overline{\mathbb{R}}, \mathbb{R}_{\mathrm{alg}}, \exp)$. The function $t \mapsto 2^t : \mathbb{R} \to \mathbb{R}$ is definable in $(\overline{\mathbb{R}}, \exp)$. By the Gelfond-Schneider theorem (see *e.g.*, Lang [10, pg. 682]), $t \in \mathbb{R}$ is rational if and only if both t and 2^t are algebraic. ⊣

6.4. PH is even obtained as the amalgamation of two o-minimal expansions of $\mathbb{R}$:

PROPOSITION. [23] *There exist functions $f, g : \mathbb{R} \to \mathbb{R}$ such that both $(\overline{\mathbb{R}}, f)$ and $(\overline{\mathbb{R}}, g)$ admit (finite) C^∞-cell decomposition and have field of exponents $\mathbb{Q}$, but $(\overline{\mathbb{R}}, f, g) = \mathrm{PH}$.*

6.5. Consider the vector field

$$(x, y, z) \mapsto (-x^2, xy - z, xz + y) : \mathbb{R}^3 \to \mathbb{R}^3.$$

The set $T := \{(1/t, t\cos t, t\sin t) : t > 0\}$ is a trajectory. By intersecting T with the xy-plane and then projecting on the x-axis, we obtain the set

$$\{1/(\pi k) : 0 < k \in \mathbb{Z}\}.$$

So $(\overline{\mathbb{R}}, T) = \mathrm{PH}$.

REMARK. If $\mathfrak{R}$ expands $(\mathbb{R}, <, +)$, defines $\mathbb{Z}$, and 1 is $\emptyset$-definable, then $\mathbb{Z}$ is $\emptyset$-definable, since $\mathbb{Z}$ is the unique $S \subseteq \mathbb{R}$ such that $(0, 1] \cap S = \{1\}$ and $x - y \in S$ for all $x, y \in S$.

Part 2.

We now proceed to proofs and further technical details. Results may be stated in a preliminary form or in greater generality than is needed for this paper.

Recall that if $A \subseteq \mathbb{R}^{m+n} \cong \mathbb{R}^m \times \mathbb{R}^n$ and the base space $\mathbb{R}^m$ is clear from context, then πA denotes the projection of A on the first m coordinates.

BCT is an abbreviation for the Baire category theorem.

§7. Lemmas.

FIBER LEMMA. *Let $A \subseteq \mathbb{R}^{m+n}$ be definable. Then the definable set*

$$B := \{x \in \mathbb{R}^m \colon \operatorname{cl}(A)_x \neq \operatorname{cl}(A_x)\}$$

is a countable union of definable subsets of $\mathbb{R}^m$, each having no interior. If moreover $\dim A = m$, *then the definable set*

$$C := \{x \in \mathbb{R}^m \colon \dim A_x > 0\}$$

is a countable union of definable subsets of $\mathbb{R}^m$, each having no interior.

PROOF. Let $\mathcal{V}$ be the collection of all boxes in $\mathbb{R}^n$ with vertices in $\mathbb{Q}^n$; then

$$B = \bigcup_{V \in \mathcal{V}} \{x \in \mathbb{R}^m \colon V \cap \operatorname{cl}(A)_x \neq \emptyset \,\&\, V \cap \operatorname{cl}(A_x) = \emptyset\}.$$

For each $V \in \mathcal{V}$, we have

$$\{x \in \mathbb{R}^m \colon V \cap \operatorname{cl}(A)_x \neq \emptyset \,\&\, V \cap \operatorname{cl}(A_x) = \emptyset\} \subseteq \operatorname{fr}(\pi((\mathbb{R}^m \times V) \cap A)).$$

Frontiers of sets have no interior.

The set C is the union of all sets of the form $\{x \in \mathbb{R}^m : I \subseteq \mu A_x\}$ where $\mu \in \Pi(n, 1)$ and $I \subseteq \mathbb{R}$ is an open interval with endpoints in $\mathbb{Q}$. Clearly, such a set has no interior (otherwise, $\dim A > m$). ⊣

For each n, let $\mathfrak{R}_\sigma(n)$ denote the collection of all countable unions of definable subsets of $\mathbb{R}^n$. I might say "A is $\mathfrak{R}_\sigma(n)$", or even just "A is $\mathfrak{R}_\sigma$", instead of "$A \in \mathfrak{R}_\sigma(n)$".

EASY OBSERVATIONS.

- Every open subset of $\mathbb{R}^n$ is $\mathfrak{R}_\sigma$.
- Every countable subset of $\mathbb{R}^n$ is $\mathfrak{R}_\sigma$.
- Every element of $\mathfrak{R}_\sigma(n)$ is a countable increasing union of bounded definable subsets of $\mathbb{R}^n$.
- $\mathfrak{R}_\sigma(n)$ is closed under taking countable unions and finite intersections.
- If $A, B \in \mathfrak{R}_\sigma$ and B is closed in A, then $A \setminus B \in \mathfrak{R}_\sigma$.
- Let $f : \mathbb{R}^m \to \mathbb{R}^n$ be definable. If $A \in \mathfrak{R}_\sigma(m)$ then $f(A) \in \mathfrak{R}_\sigma(n)$. If $B \in \mathfrak{R}_\sigma(n)$ then $f^{-1}(B) \in \mathfrak{R}_\sigma(m)$. In particular, coordinate projections, as well as the associated fibers, of $\mathfrak{R}_\sigma$ sets are $\mathfrak{R}_\sigma$.

MAIN LEMMA. *The following are equivalent:*

1. *For all definable $A \subseteq \mathbb{R}$,* $\dim \operatorname{cl}(A) = \dim A$.
2. *Every definable subset of $\mathbb{R}$ has interior or is nowhere dense.*
3. *Every definable set has interior or is nowhere dense.*
4. *For all definable A,* $\dim \operatorname{cl}(A) = \dim A$.
5. *For all m, n and definable $A \subseteq \mathbb{R}^{m+n}$,* $\{x \in \mathbb{R}^m \colon \operatorname{cl}(A)_x \neq \operatorname{cl}(A_x)\}$ *is nowhere dense.*
6. *For all m, n and definable $A \subseteq \mathbb{R}^{m+n}$,* $\{x \in \mathbb{R}^m \colon \operatorname{fr}(A)_x \neq \operatorname{fr}(A_x)\}$ *is nowhere dense.*

7. *Every definable subset of* $\mathbb{R}$ *has interior or is meager.*
8. *Every definable set has interior or is meager.*
9. *For all definable* A, $\{x \in \mathbb{R}^{\dim A} \colon \dim A_x > 0\}$ *is nowhere dense.*
10. *For all definable* $A, B \subseteq \mathbb{R}$, $\dim(A \cup B) = \max\{\dim A, \dim B\}$.
11. *For all* n *and definable* $A, B \subseteq \mathbb{R}^n$, $\dim(A \cup B) = \max\{\dim A, \dim B\}$.
12. *Every* $\mathfrak{R}_\sigma(1)$ *has interior or is meager.*
13. *Every* $\mathfrak{R}_\sigma$ *has interior or is meager.*
14. *For all* m, n *and* $A \in \mathfrak{R}_\sigma(m+n)$, A *has interior if and only if*
$$\{x \in \mathbb{R}^m \colon A_x \text{ has interior}\}$$
has interior.
15. *For all* $A \in \mathfrak{R}_\sigma$, $\{x \in \mathbb{R}^{\dim A} \colon \dim A_x > 0\}$ *is meager.*
16. *For all* n *and* $\{A_k \colon k \in \mathbb{N}\} \subseteq \mathfrak{R}_\sigma(n)$,
$$\dim \bigcup_{k \in \mathbb{N}} A_k = \max\{\dim A_k \colon k \in \mathbb{N}\}.$$

Proof. Many of the various implications are obvious, or become so after seeing the key tricks (and there is a marked resemblance to some basic results about F_σ sets; see *e.g.*, [16, 1.5]). Hence, I do only a few of these implications and leave the rest to the reader.

2⇒3. Let $n \geq 1$ and assume inductively that every definable subset of $\mathbb{R}^n$ has interior or is nowhere dense. Let $A \subseteq \mathbb{R}^{n+1}$ be definable. Suppose that A is somewhere dense, that is, $\operatorname{cl}(A)$ has interior. By the Fiber Lemma and the inductive assumptions, $\{x \in \mathbb{R}^n \colon A_x \text{ has interior}\}$ has interior. By BCT, there is an open interval $I \subseteq \mathbb{R}$ such that $\{x \in \mathbb{R}^n \colon I \subseteq A_x\}$ is nonmeager (and thus has interior). Then A has interior.

3⇒4. Let $A \subseteq \mathbb{R}^n$ be definable. Put $d = \dim \operatorname{cl}(A)$. The result is clear if $d = 0$. If $d = n$, then A is somewhere dense, and thus has interior. Suppose now that $0 < d < n$. Without loss of generality, assume that the projection of $\operatorname{cl}(A)$ on the first d coordinates contains a box U. By the Fiber Lemma, $\{x \in U \colon \operatorname{cl}(A)_x \neq \operatorname{cl}(A_x)\}$ is nowhere dense, so $\{x \in U \colon A_x \neq \emptyset\}$ has interior.

5⇔3 follows easily from BCT and the Fiber Lemma.

5⇔6 is just symbol chasing.

Next do 7⇒2⇒3⇒8⇒7.

(And so on.) ⊣

Definition. Items 5, 6, 9, 14 and 15 above will be referred to, collectively, as the **fiber properties.**

By the Main Lemma, every definable subset of $\mathbb{R}$ has interior or is nowhere dense if and only if the same is true for every definable set. We use this observation in the sequel without further mention.

DEFINITION. The **full dimension** of $A \subseteq \mathbb{R}^n$, denoted by fdim A, is the pair (d, k), ordered lexicographically, where $d = \dim A$ and

$$k = \operatorname{card}\{\mu \in \Pi(n, \dim A) \colon \operatorname{int}(\mu A) \neq \emptyset\}.$$

Note that k is independent of d if $d \in \{-\infty, 0, n\}$, so we identify dim and fdim in these cases.

DEFINITION. A set $A \subseteq \mathbb{R}^{m+n}$ is **π-good** (relative to $\mathfrak{R}$) if:

- A is definable;
- $\dim A = m$;
- πA is open;
- $\pi(A \cap U)$ has interior for every $a \in A$ and open neighborhood U of a;
- For all $x \in \pi A$, $\dim A_x = 0$ and $\operatorname{cl}(A_x) = \operatorname{cl}(A)_x$.

More generally: A is μ-good ($\mu \in \Pi(m+n, m)$) if there is a permutation σ of coordinates such that $\mu = \pi \circ \sigma$ and $\sigma(A)$ is π-good. Finally, $A \subseteq \mathbb{R}^n$ is **Π-good** if it is μ-good for some $\mu \in \Pi(n, \dim A)$. A collection $\mathcal{P}$ of subsets of $\mathbb{R}^n$ is Π-good if $\mathcal{P}$ is a finite collection of Π-good subsets of $\mathbb{R}^n$.

Every nonempty open definable subset of $\mathbb{R}^n$ is Π-good (but $\emptyset$ is not Π-good). Every dimension 0 definable subset of $\mathbb{R}^n$ is Π-good.

PARTITION LEMMA. *Suppose every definable set has interior or is nowhere dense. Let $\mathcal{A}$ be a finite collection of definable subsets of $\mathbb{R}^n$. Then there is a Π-good partition $\mathcal{P}$ of $\mathbb{R}^n$ compatible with $\mathcal{A}$.*

PROOF. We proceed by induction on $(d, k) = \max\{\operatorname{fdim} A \colon A \in \mathcal{A}\}$, where $d \geq 0$ (the result is trivial if $d = -\infty$). It suffices to deal with the case that the elements of $\mathcal{A}$ are pairwise disjoint.

Suppose $d = 0$. Put $\mathcal{A}' = \mathcal{A} \cup \{\operatorname{fr}(A) \colon A \in \mathcal{A}\}$. By the usual tricks, there is a finite partition $\mathcal{P}_0$ of $\bigcup \mathcal{A}'$, compatible with $\mathcal{A}'$, with each $P \in \mathcal{P}_0$ definable. Put $\mathcal{P} = \mathcal{P}_0 \cup \{\mathbb{R}^n \setminus \bigcup \mathcal{A}'\}$.

Suppose $d = n$. Put $\mathcal{P} = \{\operatorname{int}(A) \colon A \in \mathcal{A} \,\&\, \operatorname{int}(A) \neq \emptyset\} \cup \mathcal{P}'$, where $\mathcal{P}'$ is obtained by applying the inductive assumption to $\{A \setminus \operatorname{int}(A) \colon A \in \mathcal{A}\}$.

Suppose $0 < d < n$. Let $A \subseteq \mathbb{R}^n$ be definable such that $\dim A = d$ and πA has interior. Let Y be the set of all $a \in A$ such that $\pi(A \cap U)$ has interior for every box U containing a. Note that Y is definable. Let $a \in A \setminus Y$; then there is a box U with rational vertices such that $a \in U$ and $\pi(A \cap U)$ has no interior, and thus is nowhere dense. Hence, $\pi(A \setminus Y)$ is meager, so $\operatorname{fdim}(A \setminus Y) < \operatorname{fdim} A$. By the fiber properties, the set

$$S := \{x \in \mathbb{R}^d \colon \dim Y_x > 0 \text{ or } \operatorname{cl}(Y_x) \neq \operatorname{cl}(Y)_x\}$$

is nowhere dense. Then $P := Y \cap \pi^{-1}(\operatorname{int}(\pi A) \setminus \operatorname{cl}(S))$ is π-good and $\operatorname{fdim}(A \setminus P) < \operatorname{fdim} A$. (The rest of the proof is routine.) ⊣

NOTE. The partition $\mathcal{P}$ above is obtained canonically if we fix an ordering of the elements of $\Pi(n, d)$ for each $n \in \mathbb{N}$ and $d \in \{0, \ldots, n\}$. For example, if no $A \in \mathcal{A}$ has interior, then we may always deal first with any $A \in \mathcal{A}$ such that $\dim A = n-1$ and the projection of A on the first $n-1$ coordinates has interior. (This observation is used later in the proof of Theorem 4.)

§8. Proofs.

8.1. Proof of Proposition 3.2.

LEMMA. *Suppose every definable set has interior or is nowhere dense. Let $A \subseteq \mathbb{R}^{m+n}$ be definable. Then:*

1. $\{x \in \mathbb{R}^m : \operatorname{lc}(A_x) \neq \operatorname{lc}(A)_x\}$ *is nowhere dense.*
2. *For each $k \in \mathbb{N}$, $\{x \in \mathbb{R}^m : (A^{(k)})_x \neq A_x^{(k)}\}$ is nowhere dense.*
3. *If $\{x \in \mathbb{R}^m : \operatorname{lc}(A_x) \neq \emptyset\}$ is somewhere dense, then $\operatorname{lc}(A) \neq \emptyset$.*

PROOF. By the fiber properties, $\{x \in \mathbb{R}^m : \operatorname{fr}(A)_x \neq \operatorname{fr}(A_x)\}$ is nowhere dense, which in turn yields that $\{x \in \mathbb{R}^m : \operatorname{cl}(\operatorname{fr}(A))_x \neq \operatorname{cl}(\operatorname{fr}(A_x))\}$ is nowhere dense. For all $x \in \mathbb{R}^m$, we have $\operatorname{lc}(A)_x = A_x \setminus \operatorname{cl}(\operatorname{fr}(A))_x$ and $\operatorname{lc}(A_x) = A_x \setminus \operatorname{cl}(\operatorname{fr}(A_x))$.

Item 2 follows from 1 by an easy induction on k.

Item 3 is immediate from 1. ⊣

We are now ready to finish the proof of Proposition 3.2. Suppose every dimension 0 definable set has a locally closed point. Then every nonempty definable subset of $\mathbb{R}$ has a locally closed closed point, so every definable subset of $\mathbb{R}$ has interior or is nowhere dense, which in turn yields that every definable set has interior or is nowhere dense.

Let $\emptyset \neq A \subseteq \mathbb{R}^n$ be definable; we must show that $A^{(1)}$ is nowhere dense in A. Since $\operatorname{lc}(A)$ is open in A, it suffices to show that $\operatorname{lc}(A)$ is dense in A. Since $\operatorname{lc}(U \cap A) = U \cap \operatorname{lc}(A)$ for every open $U \subseteq \mathbb{R}^n$, it suffices to show $\operatorname{lc}(A) \neq \emptyset$. The case $\dim A = 0$ holds by assumption and the result is obvious if A has interior, so suppose $0 < \dim A < n$. Since $\sigma(\operatorname{lc}(A)) = \operatorname{lc}(\sigma(A))$ for any permutation σ of coordinates, we may reduce (by the Partition Lemma) to the case that πA is open (where π denotes projecting on the first d coordinates) and $\dim A_x = 0$ for all $x \in \pi A$. Then $\operatorname{lc}(A_x) \neq \emptyset$ for all $x \in \pi A$. Apply the lemma. ⊣

8.2. Proof of Theorem 3.2. It suffices to show that the following are equivalent:

1. For every $\mathfrak{M} \equiv \mathfrak{R}$, every dimension 0 set definable in $\mathfrak{M}$ is constructible.
2. For every $\emptyset$-definable $A \subseteq \mathbb{R}^{m+n}$ there exists $N \in \mathbb{N}$ such that for all $x \in \mathbb{R}^m$, if $\dim A_x = 0$, then $(A_x)^{(N)} = \emptyset$.
3. Every $\emptyset$-definable set is a finite union of $\emptyset$-definable locally closed sets.
4. Every definable set is constructible.

1⇒2 is a routine compactness argument (using 2.1).

Assume 2. Note that every definable set has interior or is nowhere dense. We proceed by induction on full dimension. Let $\emptyset \neq A \subseteq \mathbb{R}^n$ be $\emptyset$-definable and $(d, k) = \operatorname{fdim} A$. The case $d = 0$ holds by assumption. If $d = n$, then $A \setminus \operatorname{int}(A)$ is $\emptyset$-definable and $\dim(A \setminus \operatorname{int}(A)) < \dim A$. Suppose $0 < d < n$. By the Partition Lemma, we may reduce to the case that A is π-good (where π denotes projecting on the first d coordinates). Let N be as guaranteed by the hypothesis. By Lemma 8.1, $\{x \in \mathbb{R}^d : (A^{(N)})_x \neq \emptyset\}$ is nowhere dense. Put

$$Y = A \cap \pi^{-1}(\operatorname{int}\{x \in \mathbb{R}^d : (A^{(N)})_x = \emptyset\}).$$

Then $Y^{(N)} = \emptyset$ (so Y is constructible) and $\operatorname{fdim}(A \setminus Y) < \operatorname{fdim} A$.

Assume 3. Let $A \subseteq \mathbb{R}^n$ be definable. Then $A = Y_x$ for some $m \in \mathbb{N}$, $\emptyset$-definable $Y \subseteq \mathbb{R}^{m+n}$ and $x \in \mathbb{R}^m$. Since Y is constructible, and fibers of constructible sets are constructible, A is constructible.

(The converse implications are all easy.) ⊣

8.3. Proof of Theorem 3.3. Suppose every definable subset of $\mathbb{R}$ has interior or is nowhere dense.

We already know that every definable set has interior or is nowhere dense.

Let $U \subseteq \mathbb{R}^m$ be open and $f : U \to \mathbb{R}$ be definable.

Almost continuity. Let V be the set of points in U such that f is continuous on a box about x. We must show that V is dense in U. Let $B \subseteq U$ be a box. Now, B is the union of the definable sets $\{x \in B : |f(x)| \leq k\}$, $k \in \mathbb{N}$; by BCT, there exists $N \in \mathbb{N}$ such that $\{x \in B : |f(x)| \leq N\}$ is somewhere dense, and thus has interior. So we may assume that f is bounded on B. By the fiber properties, the set

$$\{x \in B : \operatorname{cl}(\operatorname{graph}(f))_x = \{f(x)\}\}$$

contains a box B'. Then $f \restriction B'$ is continuous.

Monotonicity. Suppose $m = 1$. We must show there is an open definable $V \subseteq U$ such that $U \setminus V$ is nowhere dense, $f \restriction V$ is continuous, and f is either constant or strictly monotone on each connected component of V. (The proof resembles that of the monotonicity theorem for o-minimal structures, but the setting is different enough to warrant giving some details.) By almost continuity, we may reduce to the case that f is continuous. The sets

$$V_1 = \{x \in U : f \text{ is constant on an open interval about } x\},$$
$$V_2 = \{x \in U : f \text{ is strictly increasing on an open interval about } x\},$$
$$V_3 = \{x \in U : f \text{ is strictly decreasing on an open interval about } x\}$$

are each open and definable, and a routine argument shows that f is either constant or strictly monotone on each connected component of each of these sets, so it suffices to show that $U \setminus (V_1 \cup V_2 \cup V_3)$ has no interior. Suppose otherwise; then $U \setminus (V_1 \cup V_2 \cup V_3)$ contains a compact interval I. Since $f \restriction I$ is continuous and nonconstant, $f(I)$ contains an open interval J. Define $g : J \to \mathbb{R}$ by $g(r) = \min(I \cap f^{-1}\{r\})$. Now, g is injective, so by almost continuity there is an open interval $J' \subseteq J$ such that g maps J' homeomorphically onto an open interval $I' \subseteq I$. But then $f \restriction I'$ is strictly monotone; contradiction.

Almost C^p-smoothness. Assume $\mathfrak{R}$ expands $\overline{\mathbb{R}}$. We must show there is an open definable $V \subseteq U$ such that $U \setminus V$ has no interior and $f \restriction V$ is C^p. We proceed by induction on $p \geq 0$. We have already established the case $p = 0$. Assume the result holds for a certain $p \geq 0$; we show it holds for $p+1$. By the inductive assumption, we reduce to the case that f is C^p.

Suppose $m = 1$. By monotonicity, we may assume that $f^{(p)}$ is monotone on each connected component of U. By the Lebesgue differentiability theorem (*e.g.*, Royden [24, Ch. 5]), the set

$$\{x \in U \colon f^{(p)} \text{ is not differentiable at } x\}$$

is null, hence nowhere dense. Now apply almost continuity.

Suppose $m > 1$. It suffices to show that the set of all $x \in U$ such that f is C^{p+1} on a neighborhood of x has interior. Let g be some partial derivative of f of order p; we need only show that the set of all $x \in U$ such that g is C^1 on a box about x has interior. Fix some $i \in \{1, \dots, m\}$; for convenience, say $i = m$. Let πU be the projection of U on the first $m-1$ coordinates and μU be the projection on the last coordinate. By the case $m = p = 1$, for each $u \in \pi U$ there is an open interval $I(u) \subseteq \mu U$ such that $x_m \mapsto g(u, x_m) \colon I(u) \to \mathbb{R}$ is differentiable. By BCT, there is an open interval $I \subseteq \mu U$ such that the set

$$\{u \in \pi U \colon x_m \mapsto g(u, x_m) \colon I \to \mathbb{R} \text{ is differentiable}\}$$

is nonmeager, so there is a box $B \subseteq \pi U$ such that $\partial g / \partial x_m$ exists at all $x \in B \times I$. By repeating the argument for $i = 1, \dots, m-1$, we obtain a box $B' \subseteq B$ such that the gradient of g exists at all $x \in B'$. By almost continuity (and standard facts), g is C^1 on some box contained in B'. $\dashv$

REMARK. Perhaps arguments of Laskowski and Steinhorn [11] could be modified to show that almost C^1-smoothness holds without the assumption that $\mathfrak{R}$ defines multiplication.

8.4. A stronger version of Proposition 3.4.

DEFINITION. For $A \subseteq \mathbb{R}^n$, $\mu \in \Pi(n, d)$ and $p \in \mathbb{N}$, let $\operatorname{reg}^p_\mu(A)$ denote the set of all $a \in A$ such that, for some open neighborhood U about a, $\mu \restriction (A \cap U)$ maps $A \cap U$ C^p-diffeomorphically onto some open $V \subseteq \mathbb{R}^d$. For $\mu = \pi$, this just means that $A \cap U = \operatorname{graph}(f)$ for some C^p map $f \colon V \to \mathbb{R}^{n-d}$. Note that $\operatorname{reg}^0_\mu(A)$ is definable, open in A, and a C^0-submanifold of $\mathbb{R}^n$ of dimension d; similarly for $\operatorname{reg}^p_\mu(A)$ if $\mathfrak{R}$ expands $\overline{\mathbb{R}}$.

LEMMA. *Suppose every definable subset of $\mathbb{R}$ has interior or is nowhere dense. Let $A \subseteq \mathbb{R}^{m+n}$ be definable such that $\{x \in \mathbb{R}^m \colon \operatorname{isol}(A_x) \neq \emptyset\}$ has interior. Then $\operatorname{reg}^0_\pi(A) \neq \emptyset$. If $\mathfrak{R}$ expands $\overline{\mathbb{R}}$, then this holds for each $\operatorname{reg}^p_\pi(A)$.*

PROOF. For each $x \in \mathbb{R}^m$ such that $\operatorname{isol}(A_x) \neq \emptyset$, there exist $y \in A_x$ and a box V (with rational vertices) about y such that $V \cap A_x = \{y\}$. By BCT, there is a box $V \subseteq \mathbb{R}^n$ such that $\{x \in \mathbb{R}^m\colon \operatorname{card}(V \cap A_x) = 1\}$ is somewhere dense, and thus contains a box U. Define $f\colon U \to \mathbb{R}^n$ by letting $f(x)$ be the unique element of $V \cap A_x$. By Theorem 3.3, there is a box $B \subseteq U$ such that $f\restriction B$ is continuous (C^p if $\mathfrak{R}$ expands $\overline{\mathbb{R}}$). Then $\operatorname{graph}(f\restriction B)$ is contained in $\operatorname{reg}^0_\pi(A)$ (and in $\operatorname{reg}^p_\pi(A)$ if $\mathfrak{R}$ expands $\overline{\mathbb{R}}$). ⊣

PROPOSITION. *Suppose every dimension* 0 *definable subset of* $\mathbb{R}$ *has an isolated point. Let* $\mathcal{A}$ *be a finite collection of definable subsets of* $\mathbb{R}^n$. *Then there is a* Π-*good partition* $\mathcal{P}$ *of* $\mathbb{R}^n$, *compatible with* $\mathcal{A}$, *such that* $P \setminus \operatorname{reg}^0_\mu(P)$ *is nowhere dense in* P *for every projection* μ *and* $P \in \mathcal{P}$ *such that* P *is* μ-*good. If moreover* $\mathfrak{R}$ *expands* $\overline{\mathbb{R}}$, *then this holds with* "$\operatorname{reg}^p_\mu(P)$" *in place of* "$\operatorname{reg}^0_\mu(P)$".

PROOF. First, note that every definable set has interior or is nowhere dense. By the Partition Lemma, it suffices to show that if $P \subseteq \mathbb{R}^n$ is π-good, then $P \setminus \operatorname{reg}^0_\pi(P)$ is nowhere dense in P. The result is trivial if $\dim P = n$ (since P is open). An easy induction on n handles the case $\dim P = 0$ (note that $\operatorname{reg}^p_\pi(P) = \operatorname{isol}(P)$ if $\dim P = 0$). For $0 < \dim P < n$, apply the lemma. ⊣

8.5. Proof of Theorem 3.4.1.

LEMMA. *The following are equivalent*

1. $\mathfrak{R}$ *is d-minimal.*
2. *For every* m *and definable* $A \subseteq \mathbb{R}^{m+1}$ *there exists* $N \in \mathbb{N}$ *such that for every* $x \in \mathbb{R}^m$, A_x *either has interior or is a union of* N *discrete sets.*
3. *For every* m, n *and definable* $A \subseteq \mathbb{R}^{m+n}$ *there exists* $N \in \mathbb{N}$ *such that for every* $x \in \mathbb{R}^m$, *either* $\dim A_x > 0$ *or* A_x *is a union of* N *discrete sets.*

PROOF. (1)⇒(2) is a routine compactness argument (using 2.3).

(2)⇒(3) is an easy induction on n.

Assume (3). Let $\mathfrak{M} \equiv \mathfrak{R}$ and $S \subseteq M$ be definable in $\mathfrak{M}$. Then there exist $m \in \mathbb{N}$, $x \in M^m$ and $A \subseteq M^{m+1}$, $\emptyset$-definable in $\mathfrak{M}$, such that $S = A_x$. Since $\mathfrak{M} \equiv \mathfrak{R}$, S either has interior or is a finite union of discrete sets. Hence, $\mathfrak{R}$ is d-minimal. ⊣

We now begin the proof proper. Let $n \in \mathbb{N}$ and $\mathcal{A}$ be a finite collection of definable subsets of $\mathbb{R}^n$. We show that there is a finite partition of $\mathbb{R}^n$ into special C^0-submanifolds, each of which is definable and compatible with $\mathcal{A}$. By the Partition Lemma and Proposition 8.4, it suffices to consider the case that $\mathcal{A}$ is a Π-good partition of $\mathbb{R}^n$ such that $A \setminus \operatorname{reg}^0_\mu(A)$ is nowhere dense in A for every projection μ and $A \in \mathcal{A}$ such that A is μ-good. For the open $A \in \mathcal{A}$, there is nothing further to do. Every dimension 0 definable set is a finite disjoint union of discrete definable sets, so the dimension 0 sets in $\mathcal{A}$ are disposed of as well. It suffices now to show that if $A \in \mathcal{A}$ with $0 < d := \dim A < n$, then there is a definable $M \subseteq A$ such that M is a C^0-submanifold and $\operatorname{fdim}(A \setminus M) < \operatorname{fdim}(A)$. By permuting coordinates, it suffices to consider the case that A is π-good, where π is the projection on the first d coordinates. Since there exists $N \in \mathbb{N}$ such that A is the disjoint union of the sets $\{(x, y) \in A : y \in \operatorname{isol}(A_x^{[k]})\}$ ($k = 0, \dots, N$), we may reduce to the case that, in addition to the above data, each A_x is discrete.

Let S be the (definable) open set of $x \in \pi A$ such that, for some box U about x, for every $y \in A_x$ there is a bounded box $V \subseteq \mathbb{R}^{n-d}$ about y with $\operatorname{card}(A_z \cap V) = 1$ for every $z \in U$. Note that $A \cap \pi^{-1}(U) \subseteq \operatorname{reg}_\pi^0(A)$—since $\operatorname{cl}(A_x) = \operatorname{cl}(A)_x$ for all $x \in \pi A$; compare with the proof of almost continuity in Theorem 3.3—so $A \cap \pi^{-1}(S)$ is a special C^0-submanifold. It suffices now to show that S is dense in $\operatorname{int}(\pi A)$ (since then $\operatorname{fdim}(A \setminus \pi^{-1}(S)) < \operatorname{fdim} A$). Let $B \subseteq \pi A$ be a box. Each A_x is discrete, so (by BCT) there exist $J \subseteq \mathbb{N}$ and a pairwise disjoint collection $(V_j)_{j \in J}$ of bounded boxes in $\mathbb{R}^{n-d}$ such that the (not necessarily definable) set

$$S' := \{x \in B : A_x \subseteq \bigcup_{j \in J} V_j \;\&\; \forall j \in J,\ \operatorname{card}(A_x \cap V_j) = 1\}$$

is nonmeager. By shrinking B, we reduce to the case that S' is dense in B. By further shrinking B, we may assume that $\operatorname{cl}(\operatorname{reg}_\pi^0(A))_x = \operatorname{cl}(\operatorname{reg}_\pi^0(A)_x)$ for all $x \in B$. Since $\operatorname{reg}_\pi^0(A)$ is dense in A, we have $\operatorname{cl}(A_x) = \operatorname{cl}(\operatorname{reg}_\pi^0(A)_x)$ for all $x \in B$; then $(x, y) \in \operatorname{reg}_\pi^0(A)$ for every $x \in B$ and $y \in A_x$ (since each A_x is discrete). By density of S' in B, we have $\operatorname{card}(A_x \cap V_j) = 1$ for all $j \in J$ and $x \in B$. Hence, $B \subseteq S$.

(We have established the C^0 version.)

Suppose that $\mathfrak{R}$ expands $\overline{\mathbb{R}}$ and let $p \in \mathbb{N}$. In order to obtain the C^p statement, replace "reg_π^0" by "reg_π^p" in the above proof prior to the point of defining the set S. Let T be the set of all $x \in \pi A$ such that $A \cap \pi^{-1}(U) \subseteq \operatorname{reg}_\pi^p(A)$ for some box U about x. We need only show that T is dense in S. Let $B \subseteq S$ be a box. After shrinking B, there is a countable set J and continuous definable maps $\phi_j : B \to \mathbb{R}$ such that $A \cap \pi^{-1}(B) = \bigcup_{j \in J} \operatorname{graph}(\phi_j)$. By Theorem 3.3, each ϕ_j is C^p off a nowhere dense definable subset of B. Then $B \setminus T$ is meager, hence nowhere dense. ⊣

8.6. Proof of Theorem 3.4.2. Let $\alpha > 1$ and $\mathfrak{R}$ be an o-minimal expansion of $\overline{\mathbb{R}}$ having field of exponents $\mathbb{Q}$. We show that $(\mathfrak{R}, \alpha^{\mathbb{Z}})$ is d-minimal.

First, we have:

PROPOSITION (*cf.* [26, Theorem I]). $\operatorname{Th}(\mathfrak{R}, \alpha^{\mathbb{Z}})$ *is axiomatized over* $\operatorname{Th}(\mathfrak{R})$ *by axioms expressing:*

- *the cut of* α *in* $\mathbb{Q}$
- $\alpha^{\mathbb{Z}}$ *is a multiplicative subgroup of* $(0, \infty)$
- $\alpha^{\mathbb{Z}} \cap (1, \alpha] = \{\alpha\}$
- *for every* $t > 0$ *there exists* $g \in \alpha^{\mathbb{Z}}$ *such that* $g \leq t < \alpha g$.

REMARK. If α is $\emptyset$-definable in $\mathfrak{R}$ then the axioms for the cut of α are unnecessary.

Outline of the proof of the Proposition. Let $L \supseteq \{<, +, -, \cdot, 0, 1\}$ be a first-order language such that $\mathfrak{R}$ is an L-structure. By adding a constant, we may assume that α is $\emptyset$-definable. By expanding $\mathfrak{R}$ by all $\emptyset$-definable functions, we reduce to the case that $\mathrm{Th}(\mathfrak{R})$ admits QE and has a universal axiomatization as an L-structure, and that L has no relation symbols other than $<$.

For $t > 0$, let $\lfloor t \rfloor = \max((0, t] \cap \alpha^{\mathbb{Z}})$. For $t \leq 0$, put $\lfloor t \rfloor = 0$. Note that $\lfloor\ \rfloor$ is $\emptyset$-definable in $(\mathfrak{R}, \alpha^{\mathbb{Z}})$ and $\alpha^{\mathbb{Z}}$ is $\emptyset$-definable in $(\mathfrak{R}, \lfloor\ \rfloor)$.

Let L^* be the result of extending L by a new unary function symbol for $\lfloor\ \rfloor$. Let T be the L^*-theory $\mathrm{Th}(\mathfrak{R})$ together with axioms expressing:

- $x \leq 0 \to \lfloor x \rfloor = 0$
- $1 \leq x < \alpha \to \lfloor x \rfloor = 1$
- $\lfloor \alpha \rfloor = \alpha$
- $\lfloor \lfloor x \rfloor y \rfloor = \lfloor x \rfloor \lfloor y \rfloor$
- $x > 0 \to \lfloor x \rfloor \leq x < \alpha \lfloor x \rfloor$

It suffices now to show $T = \mathrm{Th}(\mathfrak{R}, \lfloor\ \rfloor)$. Since $(\mathfrak{P}, \lfloor\ \rfloor)$ embeds into every model of T, where $\mathfrak{P}$ is the prime submodel of $\mathfrak{R}$, it suffices to show T admits QE; the proof is a routine modification of known results, so I provide only a brief sketch. Note that T is universal and [30, Theorem C] generalizes [14, 1.2]. Combine the technique of [14, 2.2] with the exponential image (so to speak) of the argument in [15, Appendix]. ⊣

We are now ready to finish the proof of the theorem. We work with $(\mathfrak{R}, \lfloor\ \rfloor)$ instead of $(\mathfrak{R}, \alpha^{\mathbb{Z}})$. Let $(\mathfrak{M}, \lfloor\ \rfloor) \equiv (\mathfrak{R}, \lfloor\ \rfloor)$ and $A \subseteq M$ be definable in $(\mathfrak{M}, \lfloor\ \rfloor)$. We show that A has interior or is a finite union of discrete sets.

Let L^*_M be the expansion of the language L^* by constants for elements of M. By QE, it suffices to consider the case that

$$A = \{t \in M : \tau_0(t) = 0, \tau_1(t) < 0, \dots, \tau_k(t) < 0\}$$

where $\tau_0, \dots, \tau_k$ are unary L^*_M-terms. By an easy induction on complexity (*cf.* [26, Theorem III]) there exist $m \in \mathbb{N}$, an $(m+1)$-ary function $f : M^{m+1} \to M$ definable in $\mathfrak{M}$, and $E \subseteq M$ such that:

- E is definable and a finite union of discrete sets;
- If $1 \leq i \leq k$ and $(a, b) \subseteq M \setminus E$, then there exists $x \in M^m$ such that $\tau_i(t) = f(x, t)$ for all $t \in (a, b)$.

Now, $\mathfrak{M}$ is o-minimal (since $\mathfrak{M} \equiv \mathfrak{R}$), so for any interval $(a, b) \subseteq M \cup \{\pm\infty\}$ and $x_0, \dots, x_k \in M^m$, the set

$$\{t \in (a, b) : f(x_0, t) = 0, f(x_1, t) < 0, \dots, f(x_k, t) < 0\}$$

either has interior or is finite. Then either A has interior or $A \setminus E$ is discrete. Hence, either A has interior or is a finite union of discrete sets. ⊣

Remark. By Theorems 3.4.2 and 4, if $\mathfrak{R}$ is an o-minimal expansion of $\overline{\mathbb{R}}$ having field of exponents $\mathbb{Q}$, then $(\mathfrak{R}, \alpha^{\mathbb{Z}})$ admits countable C^p-decomposition. But more can be said: Since T is universal and admits QE, every function

$f: \mathbb{R}^n \to \mathbb{R}$ definable in $(\mathfrak{R}, \alpha^{\mathbb{Z}})$ is given piecewise by $L^*_{\mathfrak{R}}$-terms. With a little more work, the proof (below) of Theorem 4 can be modified to show that the cells of the decompositions can be taken to be $\mathfrak{R}$-cells. Moreover, if $\mathfrak{R}$ admits C^∞ (or analytic) decomposition, then the cells can be taken to be C^∞ (or analytic) cells. In particular, every set definable in $(\overline{\mathbb{R}}, \alpha^{\mathbb{Z}})$ is a countable disjoint union of analytic semialgebraic cells, and every set definable in $(\mathbb{R}_{\text{an}}, \alpha^{\mathbb{Z}})$ is a countable disjoint union of analytic, globally subanalytic cells (see *e.g.*, [33] for a definition of $\mathbb{R}_{\text{an}}$). The details are left to a possible future paper (or to the reader as an exercise).

8.7. Proof of Theorem 4. First, we dispose of some preliminaries.

For $\varepsilon > 0$ and $x \in \mathbb{R}^n$, $B(x, \varepsilon)$ denotes the cube centered at x with side length 2ε.

For $A, B \subseteq \mathbb{R}$, write $A \cong B$ if A and B are order-isomorphic. For $X \subseteq \mathbb{R}$ we have:

- $X \cong \mathbb{N}$ iff X is discrete, $\min X$ exists, $\max X$ does not exist, and X is closed in the interval $(\min X, \sup X)$.
- $X \cong -\mathbb{N}$ iff X is discrete, $\max X$ exists, $\min X$ does not exist, and X is closed in $(\inf X, \max X)$.
- $X \cong \mathbb{Z}$ iff X is discrete, neither $\min X$ nor $\max X$ exist, and X is closed in $(\inf X, \sup X)$.
- X is finite iff X is discrete, closed and bounded.

Hence, if $A \subseteq \mathbb{R}^{n+1}$, then the following sets are definable in $(\mathbb{R}, <, A)$:

$$\{x \in \mathbb{R}^n : A_x \cong \mathbb{N}\},$$
$$\{x \in \mathbb{R}^n : A_x \cong -\mathbb{N}\},$$
$$\{x \in \mathbb{R}^n : A_x \cong \mathbb{Z}\},$$
$$\{x \in \mathbb{R}^n : A_x \text{ is finite}\}.$$

For $-\infty \leq a < b \leq +\infty$, put

$$\operatorname{midpt}(a,b) = \begin{cases} (a+b)/2 & \text{if } a, b \in \mathbb{R}, \\ 0 & \text{if } a = -\infty \text{ and } b = +\infty, \\ a+1 & \text{if } a \in \mathbb{R} \text{ and } b = +\infty, \\ b-1 & \text{if } a = -\infty \text{ and } b \in \mathbb{R}. \end{cases}$$

For $U \subseteq \mathbb{R}$ open, let $\operatorname{midpts}(U)$ be the set of all points $\operatorname{midpt}(a,b)$, where (a,b) is a connected component of U. Note that if $A \subseteq \mathbb{R}$, then $\operatorname{bd}(A) \cup \operatorname{midpts}(\operatorname{int}(A))$ is closed, has no interior, and is definable in $(\mathbb{R}, <, +, A)$.

We now begin the proof proper. Suppose $\mathfrak{R}$ expands $(\mathbb{R}, <, +)$ and is d-minimal. We show that $\mathfrak{R}$ admits countable decomposition (and that the C^p version holds if $\mathfrak{R}$ expands $\overline{\mathbb{R}}$). We proceed by induction on $n \geq 1$ and $(d, e) := \max\{\operatorname{fdim}(\pi A) : A \in \mathcal{A}\}$ to show that if $\mathcal{A}$ is a finite collection of definable subsets of $\mathbb{R}^n$, then there is a countable decomposition of $\mathbb{R}^n$ compatible with $\mathcal{A}$.

Suppose $n = 1$. Each $A \in \mathcal{A}$ has countable boundary (since $\mathrm{bd}(A)$ is a finite union of discrete sets), so the collection of the connected components of the sets

$$\bigcup_{A\in\mathcal{A}} \mathrm{bd}(A), \quad \mathbb{R} \setminus \bigcup_{A\in\mathcal{A}} \mathrm{bd}(A)$$

is a countable decomposition of $\mathbb{R}$ compatible with $\mathcal{A}$.

Let $n > 1$ and assume the result holds for all $m \leq n$; we show the result holds for $n+1$. Suppose $d = 0$. Inductively, there is a countable decomposition $\mathcal{C}$ of $\mathbb{R}^n$ compatible with $\pi\mathcal{A} := \{\pi A\colon A \in \mathcal{A}\}$. Then $\{C\times\mathbb{R}\colon C \in \mathcal{C},\ C \not\subseteq \bigcup \pi\mathcal{A}\}$, together with the connected components of the sets

$$\{x\} \times \bigcup_{A\in\mathcal{A}} \mathrm{bd}(A_x), \quad \mathbb{R} \setminus (\{x\} \times \bigcup_{A\in\mathcal{A}} \mathrm{bd}(A_x))$$

$(x \in \bigcup \pi\mathcal{A})$ is a countable decomposition of $\mathbb{R}^{n+1}$ compatible with $\mathcal{A}$.

Suppose $d > 0$. Put

$$Y = \bigcup_{A\in\mathcal{A}} \{(x,t) \in \mathbb{R}^{n+1}\colon t \in \mathrm{bd}(A_x) \cup \mathrm{midpts}(\mathrm{int}(A_x))\}.$$

We need only find a countable decomposition of $\mathbb{R}^{n+1}$ compatible with Y (since it will be compatible with $\mathcal{A}$). Each Y_x is closed and has no interior. By d-minimality, there exists $N \in \mathbb{N}$ such that $Y_x^{[N]} = \emptyset$ for all $x \in \mathbb{R}^n$. We proceed now by induction on $N \geq 1$.

Suppose $N = 1$. Then each Y_x is closed and discrete, so πY is equal the union of the following definable sets:

$$\begin{aligned}
S_1 &:= \{x \in \mathbb{R}^n\colon Y_x \cong \mathbb{N}\},\\
S_2 &:= \{x \in \mathbb{R}^n\colon Y_x \cong -\mathbb{N}\},\\
S_3 &:= \{x \in \mathbb{R}^n\colon 0 \in Y_x \cong \mathbb{Z}\},\\
S_4 &:= \{x \in \mathbb{R}^n\colon 0 \notin Y_x \cong \mathbb{Z}\},\\
S_5 &:= \{x \in \mathbb{R}^n\colon Y_x \text{ is finite and nonempty}\}.
\end{aligned}$$

For each $l = 1, \ldots, 4$, $Y \cap \pi^{-1}(S_l)$ is a countable union of graphs of definable functions $S_l \to \mathbb{R}$, as we now show. Define $(f_{1,j}\colon S_1 \to \mathbb{R})_{j\in\mathbb{N}}$ (by induction):

$$\begin{aligned}
f_{1,0}(x) &= \min Y_x,\\
f_{1,j+1}(x) &= \min(Y_x \cap (f_{1,j}(x), \infty)).
\end{aligned}$$

Define $(f_{2,j}\colon S_2 \to \mathbb{R})_{j\in-\mathbb{N}}$ by:

$$\begin{aligned}
f_{2,0}(x) &= \max Y_x,\\
f_{2,j-1}(x) &= \max(Y_x \cap (-\infty, f_{2,j}(x))).
\end{aligned}$$

Define $(f_{3,j}\colon S_3 \to \mathbb{R})_{j\in\mathbb{Z}}$ by letting $f_{3,0}$ be the 0 map on S_3 and then combining the previous arguments. Define $(f_{4,j}\colon S_4 \to \mathbb{R})_{j\in\mathbb{Z}}$ by $f_{4,0}(x) = \min(Y_x \cap (0,\infty))$; again finish by combining previous arguments.

We now consider the special case $d = n$ (that is, πY has interior). By arguing as in §8.5, there is a definable open $S \subseteq \mathbb{R}^n$ such that $\pi Y \setminus S$ has no interior and $Y \cap \pi^{-1}(S)$ is a π-special C^0-submanifold. Inductively, there is a countable decomposition $\mathcal{C}$ of $\mathbb{R}^n$ compatible with $\{S_1 \cap S, \dots, S_5 \cap S\}$. Note that if $C \in \mathcal{C}$ is contained in $S_5 \cap S$, then $Y \cap \pi^{-1}(C)$ is a finite disjoint union of graphs of continuous functions $C \to \mathbb{R}$, each of which is definable (by the constant finite cardinality of the fibers). Hence, if $C \in \mathcal{C}$ is contained in $(S_1 \cup \cdots \cup S_5) \cap S$, then every connected component of either $\pi^{-1}(C) \cap Y$ or $\pi^{-1}(C) \setminus Y$ is a cell, compatible with Y, that projects onto C. Then there is a countable decomposition $\mathcal{D}_1$ of $\mathbb{R}^{n+1}$ compatible with $Y \cap \pi^{-1}(S)$ such that if $D \in \mathcal{D}_1$ and $\pi D \cap (\mathbb{R}^n \setminus S) \neq \emptyset$, then $D = \pi D \times \mathbb{R}$. Since $\pi Y \setminus S$ has no interior, there is (inductively) a countable decomposition $\mathcal{D}_2$ of $\mathbb{R}^{n+1}$, compatible with $Y \setminus \pi^{-1}(S)$, such that if $D \in \mathcal{D}_2$ and $\pi D \cap S \neq \emptyset$, then $D = \pi D \times \mathbb{R}$. Hence,

$$\mathcal{D} := \{D \in \mathcal{D}_1 : \pi D \subseteq S\} \cup \{D \in \mathcal{D}_2 : \pi D \subseteq \mathbb{R}^n \setminus S\}$$

is a countable decomposition of $\mathbb{R}^{n+1}$ compatible with Y (hence also with $\mathcal{A}$).

The proof for the case $0 < d < n$ is a minor modification. Let $\mu \in \Pi(n,d)$ be such that $(\mu \circ \pi)(Y)$ has interior. By arguing as in §8.5, there is a definable open $S \subseteq \mathbb{R}^d$ such that $(\mu \circ \pi)(Y) \setminus S$ has no interior and $Y \cap (\mu \circ \pi)^{-1}(S)$ is a $(\mu \circ \pi)$-special C^0-submanifold of $\mathbb{R}^{n+1}$. Inductively, there is a countable decomposition $\mathcal{C}$ of $\mathbb{R}^n$ compatible with $\{S_1 \cap \mu^{-1}(S), \dots, S_5 \cap \mu^{-1}(S)\}$. If $C \in \mathcal{C}$ is contained in any of $S_1 \cap \mu^{-1}(S), \dots, S_5 \cap \mu^{-1}(S)$, then every connected component of either $\pi^{-1}(C) \cap Y$ or $\pi^{-1}(C) \setminus Y$ is a cell, compatible with Y, that projects onto C. Then there is a countable decomposition $\mathcal{D}_1$ of $\mathbb{R}^{n+1}$ compatible with $Y \cap (\mu \circ \pi)^{-1}(S))$ such that if $D \in \mathcal{D}_1$ and $\pi D \cap (\mathbb{R}^n \setminus \mu^{-1}(S)) \neq \emptyset$, then $D = \pi D \times \mathbb{R}$. Since $\operatorname{fdim}(\pi Y \setminus \mu^{-1}(S)) < \operatorname{fdim}(\pi Y)$, there is (inductively) a countable decomposition $\mathcal{D}_2$ of $\mathbb{R}^{n+1}$, compatible with $Y \setminus (\mu \circ \pi)^{-1}(S)$, such that if $D \in \mathcal{D}_2$ and $\pi D \cap \mu^{-1}(S) \neq \emptyset$, then $D = \pi D \times \mathbb{R}$. Hence,

$$\mathcal{D} := \{D \in \mathcal{D}_1 : \pi D \subseteq \mu^{-1}(S)\} \cup \{D \in \mathcal{D}_2 : \pi D \subseteq \mathbb{R}^n \setminus \mu^{-1}(S)\}$$

is a countable decomposition of $\mathbb{R}^{n+1}$ compatible with Y.

We have finished the case $N = 1$. Having done it in detail, we now concentrate on the remaining main ideas of the proof, leaving more routine details to the reader.

For $A \subseteq \mathbb{R}^n$ and $f, g\colon A \to \mathbb{R} \cup \{\pm\infty\}$, put

$$(f, g) = \{(x,t) \in A \times \mathbb{R} : f(x) < t < g(x)\}.$$

Suppose $N > 0$ and $Y_x^{[N+1]} = \emptyset$ for all $x \in \mathbb{R}^n$. We do the details only of the case $d = n$. (As before, the case $0 < d < n$ is a minor modification.) Put

$$W = \{(x,t) \in \mathbb{R}^{n+1} : t \in \operatorname{isol}(Y_x)\},$$
$$Z = \{(x,t) \in \mathbb{R}^{n+1} : t \in Y_x^{[1]}\}.$$

Then Y is the disjoint union of W and Z. For every $x \in \mathbb{R}^n$: $W_x = \operatorname{isol}(Y_x)$; $Z_x \subseteq \operatorname{fr}(W_x)$; Z_x is closed; and $Z_x^{[N]} = \emptyset$. Put

$$W' = \{(x,t) \in W : t \in \operatorname{midpts}(\mathbb{R} \setminus Z_x)\},$$
$$S_0 = \{x \in \pi W' : \exists \varepsilon > 0,\ W \cap \pi^{-1}(B(x,\varepsilon)) \text{ is a } \pi\text{-special } C^0\text{-submanifold}\},$$
$$T_0 = \{x \in \pi W : \exists \varepsilon > 0,\ W \cap \pi^{-1}(B(x,\varepsilon)) \text{ is a } \pi\text{-special } C^0\text{-submanifold}\}.$$

There exist definable S, T such that S is dense-open in S_0 and T is dense-open in T_0. By the inductive assumption on N, there is a countable decomposition $\mathcal{D}$ of $\mathbb{R}^{n+1}$ compatible with $\{Z, Z \cap \pi^{-1}(S), Z \cap \pi^{-1}(T)\}$. Let $C \in \pi\mathcal{D}$ be a cell contained in T. Note that either $C \subseteq S$ or $C \subseteq T \setminus S$. Every connected component of $Z \cap \pi^{-1}(C)$ is a non-open cell that projects onto C. Since $W \cap \pi^{-1}(T)$ is a π-special C^0-submanifold and C is simply connected, there is a countable family $(\phi_j : C \to \mathbb{R})_{j \in J}$ of continuous functions (J some index set) such that $W \cap (C \times \mathbb{R})$ is the disjoint union of the graphs of the ϕ_j; the only remaining non-routine work is to show that these functions are definable. Fix one ϕ. Since $Z \cap \pi^{-1}(C)$ is a disjoint union of graphs of continuous functions $C \to \mathbb{R}$, we have exactly three cases to consider:

1. $\phi(x) > t$ for all $x \in C$ and $t \in Z_x$.
2. $\phi(x) < t$ for all $x \in C$ and $t \in Z_x$.
3. For every $x \in C$, both $\max(Z_x \cap (-\infty, \phi(x)))$ and $\min(Z_x \cap (\phi(x), \infty))$ exist (recall that Z_x is closed).

Suppose Case 1 holds. Put $h(x) = \max Z_x$ for $x \in C$. Note that h is definable and $h(x) + 1 = \operatorname{midpt}(h(x), \infty)$ for all $x \in C$. There exists $K \subseteq J$ such that $W \cap (h, \infty) = \bigcup_{j \in K} \operatorname{graph}(\phi_j)$. We have five subcases:

(i) K is finite (then certainly each ϕ_j with $j \in K$ is definable, and we are done).
(ii) $K \cong \mathbb{N}$.
(iii) $K \cong -\mathbb{N}$.
(iv) $K \cong \mathbb{Z}$ and $h(x) + 1 \in W_x$ for all $x \in C$.
(v) $K \cong \mathbb{Z}$ and $h(x) + 1 \notin W_x$ for all $x \in C$.

If (ii) holds, then $\min(W_x \cap (h(x), \infty))$ exists for all $x \in C$. Hence, each ϕ_j lying strictly above h is definable by induction (as in the case $N = 1$). Subcase (iii) is similar (but note also that $h(x) = \inf W_x$ for all $x \in C$). Assume (iv) or (v) holds. For ease of notation, take $K = \mathbb{Z}$ and, if $j, k \in \mathbb{Z}$ with $j < k$, then $\phi_j < \phi_k$. After re-indexing, we have either $\phi_0 = h + 1$ or

$\phi_0 < h + 1 < \phi_1$ (definitely the latter if $C \subseteq T \setminus S$). In either case, ϕ_0 is definable. Again, we define all ϕ_j with $j \in K$ by induction.

Case 2 is handled by an easy modification.

For Case 3, define $g, h\colon C \to \mathbb{R}$ by:

$$g(x) = \max(Z_x \cap (-\infty, \phi(x))),$$
$$h(x) = \min(Z_x \cap (\phi(x), \infty)).$$

It is not clear from their definitions that g and h are definable—we do not yet know if ϕ is definable—but they are: their graphs are contained in Z and thus are cells. Note that $\mathrm{midpt}(g(x), h(x)) = (g(x) + h(x))/2$ for all $x \in C$. The rest of the argument is a routine modification of that for Case 1.

(We have now finished the proof of the C^0 version of the theorem.)

If $\mathfrak{R}$ expands $\overline{\mathbb{R}}$, then countable C^p-decomposition is obtained just by replacing "C^0" (or "continuous", as the case may be) with "C^p". ⊣

Added in proof. In regard to §3.4: For more examples of d-minimal structures, see Friedman and Miller, *Expansions of o-minimal structures by fast sequences*, to appear in ***The Journal of Symbolic Logic***; Miller and Tyne, *Expansions of o-minimal structures by iteration sequences*, to appear in ***Notre Dame Journal of Formal Logic***.

In contrast to §6: See Miller, *Avoiding the projective hierarchy in expansions of the real field by sequences*, to appear in ***Proceedings of the American Mathematical Society***.

In regard to the last sentence of the second paragraph of §5: Friedman and Miller have produced compact $E \subseteq \mathbb{R}$ such that $(\overline{\mathbb{R}}, E)$ is non-Borel but does not define $\mathbb{Z}$. A paper is in preparation.

Miller and Steinhorn have produced a generalization of Proposition 5.2 that does not require working over $\mathbb{R}$: If $\mathfrak{R}$ is an o-minimal expansion of a densely ordered group whose underlying set is definably connected in $\mathfrak{R}$, and $\mathfrak{R}$ has the uniform finiteness property, then the open core of $\mathfrak{R}$ is o-minimal. A paper is in preparation.

REFERENCES

[1] O. Belegradek, F. Wagner, and Y. Peterzil, *Quasi-o-minimal structures*, ***The Journal of Symbolic Logic***, vol. 65 (2000), pp. 1115–1132.

[2] J. Bochnak, M. Coste, and M.-F. Roy, ***Real algebraic geometry***, Ergebnisse der Mathematik und ihrer Grenzgebiete (3. Folge), vol. 36, Springer-Verlag, 1998.

[3] R. Dougherty and C. Miller, *Definable boolean combinations of open sets are boolean combinations of open definable sets*, ***Illinois Journal of Mathematics***, vol. 45 (2001), pp. 1347–1350.

[4] G. Edgar and C. Miller, *Hausdorff dimension, analytic sets and transcendence*, ***Real Analysis Exchange***, vol. 27, 2001/2, pp. 335–339.

[5] ———, *Borel subrings of the reals*, ***Proceedings of the American Mathematical Society***, vol. 131 (2003), pp. 1121–1129.

[6] H. Friedman and C. Miller, *Expansions of o-minimal structures by sparse sets*, ***Fundamenta Mathematicae***, vol. 167 (2001), pp. 55–64.

[7] F. Hausdorff, ***Set theory***, 4th English ed., Chelsea, 1991.

[8] A. Kechris, ***Classical descriptive set theory***, Graduate Texts in Mathematics, vol. 156, Springer-Verlag, 1995.

[9] K. Kuratowski, ***Topology***, vol. 1, Academic Press, 1966.

[10] S. Lang, ***Algebra***, 2nd ed., Addison-Wesley, 1984.

[11] M. Laskowski and C. Steinhorn, *On o-minimal expansions of Archimedean ordered groups*, ***The Journal of Symbolic Logic***, vol. 60 (1995), pp. 817–831.

[12] D. Macpherson, *Notes on o-minimality and variations*, ***Model theory, algebra, and geometry***, Mathematical Sciences Research Institute Publications, vol. 39, Cambridge University Press, 2000, pp. 97–130.

[13] L. Mathews, *Cell decomposition and dimension functions in first-order topological structures*, ***Proceedings of the London Mathematical Society (3)***, vol. 70 (1995), pp. 1–32.

[14] C. Miller, *Expansions of the real field with power functions*, ***Annals of Pure and Applied Logic***, vol. 68 (1994), pp. 79–94.

[15] ———, *Expansions of dense linear orders with the intermediate value property*, ***The Journal of Symbolic Logic***, vol. 66 (2001), pp. 1783–1790.

[16] C. Miller and P. Speissegger, *Expansions of the real line by open sets: o-minimality and open cores*, ***Fundamenta Mathematicae***, vol. 162 (1999), pp. 193–208.

[17] J. Oxtoby, ***Measure and category***, 2nd ed., Graduate Texts in Mathematics, vol. 2, Springer-Verlag, 1980.

[18] A. Pillay, *First order topological structures and theories*, ***The Journal of Symbolic Logic***, vol. 52 (1987), pp. 763–778.

[19] A. Pillay and C. Steinhorn, *Definable sets in ordered structures*. I, ***Transactions of the American Mathematical Society***, vol. 295 (1986), pp. 565–592.

[20] A. Robinson, *A note on topological model theory*, ***Fundamenta Mathematicae***, vol. 81 (1974), pp. 159–171.

[21] J. Robinson, *The undecidability of algebraic rings and fields*, ***Proceedings of the American Mathematical Society***, vol. 10 (1959), pp. 950–957.

[22] R. Robinson, *The undecidability of pure transcendental extensions of real fields*, ***Zeitschrift für Mathematische Logik und Grundlagen der Mathematik***, vol. 10 (1964), pp. 275–282.

[23] J.-P. Rolin, P. Speissegger, and A. Wilkie, *Quasianalytic Denjoy-Carleman classes and o-minimality*, ***Journal of the American Mathematical Society***, vol. 16 (2003), pp. 751–777.

[24] H. Royden, ***Real analysis***, 3rd ed., Macmillan, 1988.

[25] M. Shiota, ***Geometry of subanalytic and semialgebraic sets***, Progress in Mathematics, vol. 150, Birkhäuser, 1997.

[26] L. van den Dries, *The field of reals with a predicate for the powers of two*, ***Manuscripta Mathematica***, vol. 54 (1985), pp. 187–195.

[27] ———, *A generalization of the Tarski-Seidenberg theorem, and some nondefinability results*, ***Bulletin of the American Mathematical Society (New Series)***, vol. 15 (1986), pp. 189–193.

[28] ———, *On the elementary theory of restricted elementary functions*, ***The Journal of Symbolic Logic***, vol. 53 (1988), pp. 796–808.

[29] ———, *o-Minimal structures*, ***Logic: From foundations to applications***, Oxford Science Publications, Oxford University Press, 1996, pp. 137–185.

[30] ———, *T-convexity and tame extensions*. II, ***The Journal of Symbolic Logic***, vol. 62 (1997), pp. 14–34.

[31] ———, *Dense pairs of o-minimal structures*, ***Fundamenta Mathematicae***, vol. 157 (1998), pp. 61–78.

[32] ———, *Tame topology and o-minimal structures*, ***London Mathematical Society Lecture Note Series***, vol. 248, Cambridge University Press, 1998.

[33] L. van den Dries and C. Miller, *Geometric categories and o-minimal structures*, ***Duke Mathematical Journal***, vol. 84 (1996), pp. 497–540.

[34] L. van den Dries and P. Speissegger, *The real field with convergent generalized power series*, ***Transactions of the American Mathematical Society***, vol. 350 (1998), pp. 4377–4421.

[35] ———, *The field of reals with multisummable series and the exponential function*, ***Proceedings of the London Mathematical Society (3)***, vol. 81 (2000), pp. 513–565.

[36] I. Vinogradov, *The representation of an odd number as a sum of three primes*, ***Doklady Akademii Nauk SSSR***, vol. 16 (1937), pp. 139–142, (in Russian).

DEPARTMENT OF MATHEMATICS
THE OHIO STATE UNIVERSITY
231 WEST 18TH AVENUE
COLUMBUS, OHIO 43210, USA
E-mail: miller@math.ohio-state.edu

THE MODEL THEORY OF COMPACT COMPLEX SPACES

RAHIM N. MOOSA†

The usual model-theoretic approach to complex algebraic geometry is to view complex algebraic varieties as living definably in the structure $(\mathbb{C}, +, \times)$. Various model-theoretic properties of algebraically closed fields (such as quantifier elimination and strong minimality) are then used to obtain geometric information about the varieties. This approach extends to other geometric contexts by considering expansions of algebraically closed fields to which the methods of stability or simplicity apply. For example, differential algebraic varieties live in differentially closed fields, and difference algebraic varieties in algebraically closed fields equipped with a generic automorphism. Another approach would be to consider the variety as a structure in its own right, equipped with the algebraic (respectively differential or difference algebraic) subsets of its cartesian powers. This point of view is compatible with the theory of Zariski-type structures, developed by Hrushovski and Zilber (see [21] and [41]). While the two approaches are equivalent (i.e., bi-interpretable) in the case of complex algebraic varieties, the latter point of view extends to certain fragments of complex *analytic* geometry in a manner that does not seem accessible by the former.

Zilber showed in [41] that a compact complex analytic space with the structure induced by the analytic subsets of its cartesian powers is of finite Morley rank and admits quantifier elimination. Since then, there have been a number of papers investigating various aspects of the model theory of such structures, as well as possible applications to complex analytic geometry. The most notable achievement has been a kind of Chevalley Theorem due to Pillay and Scanlon [35] that classifies meromorphic groups in terms of complex tori and linear algebraic groups. It has become clear that compact complex spaces serve as a particularly rich setting in which many of the more advanced phenomena of stability theory are witnessed. For example, the full trichotomy for strongly minimal sets (trivial, not trivial but locally modular, and not locally modular) occurs in this category. It also seems reasonable to expect that a greater model-theoretic understanding of compact complex spaces can contribute to complex analytic geometry, particularly around issues concerning the bimeromorphic classification of Kähler-type spaces.

†Supported by the Natural Sciences and Engineering Research Council of Canada.

Logic Colloquium '01
Edited by M. Baaz, S. Friedman, and J. Krajíček
Lecture Notes in Logic, 20

In this article, I hope to give both an introduction to, and a survey of, the model theory of compact complex spaces. Notions from complex analytic geometry that are of particular interest will be explained in some detail, and this is partially responsible for the length of the article. Formal proofs will generally be omitted, but I will make some attempt to motivate the results and provide information about the techniques employed in obtaining them. While most of the material I have included has already appeared (or is to appear) elsewhere, I have also taken this oppurtunity to describe some of the results obtained in my thesis [25] which is still being prepared for publication.[1]

Here are some of the topics that are covered. I begin with a brief introduction to the basic objects of study in Section 1, followed by a summary of the first results in the model theory of compact complex spaces (Section 2). In Section 3, I discuss how the work of Hrushovski and Zilber on Zariski geometries applies to this category in order to establish a dichotomy for strongly minimal sets in terms of their relationship to projective space.[2] The notion of a meromorphic group and the classification theorem of Pillay and Scanlon are presented in Section 4, together with the results of Kowalski and Pillay on the socle of a commutative meromorphic group. A somewhat detailed discussion of Douady spaces and their relationship to issues of saturation for compact complex spaces, as it appears in my thesis, is given in Section 5. In Section 6, I explain how the model theory of elementary extensions of a compact complex space can be related to various notions about families of analytic sets from complex analytic geometry. Finally, in Section 7, I give an analogue of the Riemann Existence Theorem for elementary extensions (also from my thesis). Several examples and open questions are dicussed along the way.

The emphasis given to the various topics in this article may very well have more to do with my own knowledge of them rather than their relative worth. Moreover, this is an active area of research, and I have probably not discussed all the work that has been done. One obvious omission is the work of Peterzil and Starchenko on complex analysis in o-minimal structures ([27] and [28]), whereby—despite my comments at the beginning of this introduction—complex analytic spaces can be viewed as living in an ambient enriched field ($\mathbb{R}_{an}$), though outside the stable/simple context.

Acknowledgements. I am grateful to Anand Pillay for his guidance during my tenure as a graduate student of the University of Illinois at Urbana-Champaign. I also thank Thomas Scanlon with whom I have had several discussions that were helpful in preparing this paper. Finally, I am indebted to Matthias Aschenbrenner, Anand Pillay, and the referee for their comments on an earlier draft.

---

[1]Please see *Note added in proof* at the end of the article.

[2]Recent work of Pillay [32], based on results of Campana [2] and Fujiki [11], gives a direct proof of the dichotomy that does not involve Zariski geometries. I have included a brief appendix on these methods.

§1. Complex analytic spaces. In this section I will introduce complex analytic spaces and some of the basic notions associated with them. While I will sometimes use the language of sheaves and ringed spaces, only a rudimentary understanding of these objects will be assumed. More detailed introductions to the theory of complex analytic spaces can be found in [16] (for a classical treatment) and [38] (for a modern treatment).

Just as algebraic geometry is essentially concerned with the zero sets of polynomials, the fundamental objects of study in complex analytic geometry are controlled by the zero sets of holomorphic functions. Let D be a domain in $\mathbb{C}^n$ (some $n \geq 1$), and suppose that $f_1, f_2, \ldots, f_m$ are holomorphic functions on D. Then the set of common zeros, $V = V(f_1, \ldots, f_m)$, is called an *analytic set in D*. If we let $\mathcal{O}_D$ denote the sheaf of germs of holomorphic functions on D, then the quotient of $\mathcal{O}_D$ by the ideal sheaf generated by $f_1, \ldots, f_m$ equips V with a structure sheaf. Note that this sheaf may not be reduced—it may contain nilpotent elements. However, if we let $\mathcal{I}_V \subset \mathcal{O}_D$ denote the ideal sheaf of *all* holomorphic functions vanishing on V, then the quotient, $\mathcal{O}_V = \mathcal{O}_D/\mathcal{I}_V$, gives rise to a reduced sheaf structure on V. Each section of $\mathcal{O}_V$ can be naturally identified with a unique continuous $\mathbb{C}$-valued function on V, and is called a *holomorphic function* on V. If W is another analytic set in some domain, then a *holomorphic map* (respectively *biholomorphic map*) from V to W is a morphism (respectively isomorphism) between the ringed spaces $(V, \mathcal{O}_V)$ and $(W, \mathcal{O}_W)$.

Complex analytic spaces are ringed spaces that are locally modeled after analytic sets in domains of $\mathbb{C}^n$ (various n). This can be expressed as follows: a second countable Hausdorff topological space X is a (reduced) *complex analytic space* if there exists an open cover $\{X_\alpha\}$ of X, such that for each α there is a homeomorphism $\phi_\alpha : X_\alpha \to V_\alpha$, where V_α is an analytic set in a domain of $\mathbb{C}^{n_\alpha}$; and such that for all α and β for which $X_\alpha \cap X_\beta \neq \emptyset$, the induced transition function

$$\phi_\beta \phi_\alpha^{-1} : \phi_\alpha(X_\alpha \cap X_\beta) \to \phi_\beta(X_\alpha \cap X_\beta)$$

is a biholomorphic map between analytic sets in domains of $\mathbb{C}^{n_\alpha}$ and $\mathbb{C}^{n_\beta}$ respectively. We obtain a reduced sheaf structure on X by pulling back the $\mathcal{O}_{V_\alpha}$'s and using the transition functions to glue them together. This structure sheaf is denoted by $\mathcal{O}_X$, and its sections are the *holomorphic functions* on X. If we allowed nonreduced sheaf structures on the V_α's (as described in the above paragraph), then we would obtain the general notion of a complex analytic space (i.e., not necessarily reduced). However, since the model-theoretic perspective is essentially set-theoretic, and not sheaf-theoretic, I will usually only consider reduced complex analytic spaces and deal explicitly with nonreduced spaces as they appear. Again, if X and Y are complex analytic spaces, then a *holomorphic map* (respectively *biholomorphic map* or *isomorphism*) between X and Y is a morphism (respectively isomorphism) of the ringed spaces $(X, \mathcal{O}_X)$ and $(Y, \mathcal{O}_Y)$.

Suppose X is a complex analytic space, and $x \in X$. The (complex) *dimension of X at x*, denoted $\dim_x X$, is the least $d \geq 0$ such that there exists a finite-to-one holomorphic map $f: U \to D$, where U is a neighbourhood around x in X, and D is a domain in $\mathbb{C}^d$. If X is connected then $\dim_x X$ is constant for all $x \in X$, and we let the *dimension of X* (denoted by $\dim X$) be this quantity. We say that X is *smooth at x* if there is a neighbourhood about x that is biholomorphic with a domain in $\mathbb{C}^d$, in which case $d = \dim_x X$. A *complex manifold* can then be described as a complex analytic space that is smooth at every point.

Given a complex analytic space X, we will be interested in those subsets of X and its cartesian powers (which are again complex analytic spaces) that are given locally by holomorphic functions. A subset $A \subset X$ is *analytic* if for all $x \in X$ there is a neighbourhood U of x in X, and finitely many holomorphic functions on U, $f_1, \ldots, f_m$, such that $A \cap U$ is the set of common zeros of $f_1, \ldots, f_m$ in U. Note that an analytic subset of X is closed and inherits from X the structure of a complex analytic space in its own right. The analytic subsets form the closed sets of another topology on X that is coarser than the underlying complex topolgy. I will refer to this as the *Zariski topology*, and use the terms "Zariski closed set" and "analytic set" interchangeably. We say that X is *irreducible* to mean that it is irreducible in the Zariski topology; it cannot be written as the union of two proper analytic subsets. In fact, if X is irreducible and $A \subset X$ is a proper analytic subset, then A is nowhere dense. If X is irreducible and P is a property of points in X, then I will say that P *holds for general $x \in X$* if it holds in some nonempty Zariski open subset.

Suppose X and Y are irreducible complex analytic spaces. A holomorphic map, $f: X \to Y$, is a *modification* if it is proper, surjective, and there exist proper analytic subsets $A \subset X$ and $B \subset Y$ such that f restricts to a biholomorphic map from $(X \setminus A)$ to $(Y \setminus B)$. By a *meromorphic map* from X to Y (written $g: X \to Y$) I will mean a multivalued map whose graph, $\Gamma(g) \subset X \times Y$ is an irreducible analytic set, such that the first coordinate projection map $\Gamma(g) \to X$ is a modification. Off a proper analytic subset of X, $\Gamma(g)$ is the graph of a well-defined holomorphic map to Y. Note that a meromorphic map $g: X \to Y$ is a holomorphic map (everywhere) exactly when $\Gamma(g) \to X$ is an isomorphism. For any $y \in Y$, X_y will denote the set-theoretic fibre of g above y—that is, the analytic set given by $\{x \in X: (x, y) \in \Gamma(g)\}$. In the case when g is holomorphic, this is just the pre-image $g^{-1}(y)$. If the second coordinate projection $\Gamma(g) \to Y$ is surjective, then the meromorphic map g is called *surjective*. If $\Gamma(g) \to Y$ is also a modification, then $g: X \to Y$ is called a *bimeromorphic map*. (Bi)meromorphic maps are the analogue in complex analytic geometry of (bi)rational maps in algebraic geometry.

For ease of notation, by a *complex variety* I will mean a reduced irreducible complex analytic space.[3] For much of this article I will restrict my attention to *compact* complex varieties. Part of what makes compact complex varieties accessible to model-theoretic analysis is that in this case the Zariski topology is rather well behaved. If X is a compact complex variety then every Zariski closed set in X can be written uniquely as an irredundant union of finitely many irreducible Zariski closed subsets, which are called its *irreducible components*. We can then define the *dimension* of a Zariski closed set in X to be the maximum of the dimensions of its irreducible components. Moreover, if $A \subset X$ is a complex subvariety and $B \subset A$ is a Zariski closed set, then $\dim B = \dim A$ if and only if $B = A$. This yields a descending chain condition on Zariski closed sets, showing that the Zariski topology on X is noetherian. It would not be too inaccurate to describe the model-theoretic point of view as being that of ignoring the underlying complex topology on X and considering only the structure induced by the Zariski topology.

I will conclude this section with a few familiar (and not so familiar) examples of compact complex varieties. First of all, consider the algebraic case. I will use $\mathbb{P}_n(\mathbb{C})$, or just $\mathbb{P}_n$, to denote projective n-space viewed as a compact complex variety. Chow's Theorem states that every analytic set in projective space is algebraic (i.e., given by homogeneous polynomials). Hence, for projective space, the analytic Zariski topology defined above coincides with the usual algebraic Zariski topology. A *projective variety* is a compact complex variety that is biholomorphic to a closed subvariety of $\mathbb{P}_n$, for some $n \geq 0$.

More generally, a *Moishezon variety* is a compact complex variety that is bimeromorphic to a projective variety. In classifying compact complex varieties one is often only interested in bimeromorphic equivalence classes. From this point of view, Moishezon varieties are part of the "algebraic universe". It is a nontrivial fact that a Moishezon variety can also be characterised as a compact complex variety that is the holomorphic image of a projective variety.

Recall that a real $2n$-dimensional lattice of $\mathbb{C}^n$ is an additive subgroup of the form $\Delta = \{m_1\alpha_1 + \cdots + m_{2n}\alpha_{2n} \mid m_1, \ldots, m_{2n} \in \mathbb{Z}\}$, where $(\alpha_1, \ldots, \alpha_{2n})$ is an $\mathbb{R}$-basis for $\mathbb{C}^n$. An n-dimensional *complex torus* is a quotient of $\mathbb{C}^n$ by a real $2n$-dimensional lattice, equipped with the induced analytic structure. Complex tori are compact complex manifolds, and, depending on the choice of the lattice, may give rise to non-Moishezon varieties. The group structure induced on a complex torus from $\mathbb{C}^n$ is holomorphic.

Finally, let me briefly mention a class of compact complex varieties that will eventually play an important role in this article. A *Kähler manifold* is a compact complex manifold that admits a Hermitian metric whose associated

[3]This terminology differs from the one I used in my thesis [25], where a "complex variety" was assumed to also be compact. In doing so I was following Ueno [40], but now agree with those who objected to that convention.

differential 2-form of type $(1,1)$ is closed. A compact complex variety is said to be of *Kähler type* if it is the holomorphic image of a Kahler manifold. While these definitions may not be particularly enlightening, certain relevant properties of the class of all Kähler-type varieties, referred to as the class $\mathcal{C}$, will become clear later. For now, I only mention that the class $\mathcal{C}$ contains all Moishezon varieties and complex tori, and is closed under taking cartesian products and bimeromorphic equivalence. Kähler-type varieties have been extensively studied (see, for example, Fujiki [12]), mainly because many methods from algebraic geometry are applicable to them.

§2. Basic model-theoretic properties. A compact complex variety can be viewed as a structure in the sense of model theory by taking the Zariski closed subsets of all its (finite) cartesian powers as the basic relations. In order to deal with several compact complex varieties at once, it is convenient to consider the many-sorted structure $\mathcal{A}$ where there is a sort for each compact complex variety, and where the relations are now the Zariski closed subsets of the cartesian products of the sorts. I let $\mathcal{L}$ denote the corresponding many-sorted language consisting of predicates for these relations. In this section I will survey some of the first results on the model theory of $\mathcal{A}$.

Clearly, a quantifier-free definable set in $\mathcal{A}$ will be a (finite) boolean combination of Zariski closed sets. *A priori*, arbitrary definable sets are obtained from Zariski closed sets using the usual boolean operations together with the coordinate projection maps. That projection maps turn out to be unnecessary is the first indication that $\mathrm{Th}(\mathcal{A})$ is well-behaved:

THEOREM 2.1 (Łojasiewicz [24], Zilber [41]). *$\mathrm{Th}(\mathcal{A})$ admits quantifier elimination; every definable set is a boolean combination of Zariski closed sets.*

Two classical results of Remmert are responsible for quantifier elimination. The first is Remmert's Proper Mapping Theorem which states that the image of an analytic set under a proper holomorphic map is again analytic.[4] In the compact case properness comes for free, and so we have that the holomorphic image of a Zariski closed set is Zariski closed. Quantifier elimination says something slightly different, that the holomorphic image of a Zariski constructible set is Zariski constructible. Nevertheless, Theorem 2.1 can be deduced from the Proper Mapping Theorem using an inductive argument that involves the following fact (also due to Remmert) about how the dimension of an analytic set varies in families:[5]

FACT 2.2 (Dimension Formula). Suppose X and Y are compact complex varieties and $f: X \to Y$ is a surjective holomorphic map. Then there exists a nonempty Zariski open set $U \subset Y$, such that for all $y \in U$ and $x \in X_y$,

[4]A proof of this can be found on page 162 of [16].

[5]See Chapter 3 of [7]

$\dim_x X_y = \dim X - \dim Y$. In particular, each irreducible component of a general fibre of f is of dimension $\dim X - \dim Y$. Moreover, this is the least dimension of any fibre of f.

The above fact is of independent interest and implies the definability of dimension; if S is a definable set and $\{F_s : s \in S\}$ is a definable family of Zariski closed sets parametrised by S, then for each $n \geq 0$, the set

$$U(n) = \{s \in S: \dim F_s = n\}$$

is a definable subset of S. As we will see later, it is the definability of dimension in $\mathcal{A}$ that allows for a well-behaved notion of dimension for definable sets in elementary extensions of $\mathcal{A}$.

Suppose F is a definable subset of some compact complex variety X. By quantifier elimination, F has a unique irredundant expression of the form:

$$F = (S_1 \setminus T_1) \cup (S_2 \setminus T_2) \cup \cdots \cup (S_k \setminus T_k)$$

where each S_i is an irreducible Zariski closed subset of X and T_i is a proper Zariski closed subset of S_i. The Zariski closure of F, denoted by $\overline{F}$, is then $S_1 \cup \cdots \cup S_k$. I say that F is *irreducible* if $k = 1$, which is equivalent to $\overline{F}$ being irreducible. For each i, $S_i \setminus T_i$ in the above expression is called an *irreducible component of* F. By the *dimension of* F I mean the dimension of $\overline{F}$, and denote it by $\dim F$. Note that the dimension of a definable set is equal to the maximum of the dimensions of its irreducible components.

A natural consequence of quantifier elimination is that every definable map is "piecewise meromorphic":

COROLLARY 2.3. *Suppose $f: A \to B$ is a definable map between definable sets. Then $A = A_1 \cup \cdots \cup A_m$, where each A_i is a Zariski open subset of a compact complex variety X_i, and such that the restriction of f to each A_i is holomorphic on A_i and meromorphic on X_i.*

This may require some explanation. Suppose X and Y are compact complex varieties, U is a nonempty Zariski open subset of X, and $f: U \to Y$ is a holomorphic map. I will say that f is *meromorphic on* X to mean that there exists a meromorphic map from X to Y that agrees with f on U. It should be clear that in this case f is definable. The above corollary says that all definable maps are of this form, after a finite decomposition of the domain.

Quantifier elimination also yields a correspondence between complete types and irreducible Zariski closed sets. Suppose x is a (finite) tuple of variables corresponding to a certain cartesian product of sorts from $\mathcal{A}$, say X, and $p(x)$ is a complete type over the empty set in these variables.[6] By quantifier elimination and the noetherianity of the Zariski topology, there exists a unique irreducible Zariski closed subset $F \subset X$, such that $p(x)$ is determined by the

[6]Notice that this is equivalently to $p(x)$ being over $\mathcal{A}$, since every point of a compact complex variety, being itself a Zariski closed set, is named in $\mathcal{L}$.

formulae stating that "$x \in F$ and $x \notin G$ for any proper Zariski closed subset $G \subset F$". Conversely, if F is any irreducible Zariski closed subset of X, then the collection of formulae stating that "$x \in F$ and $x \notin G$ for any proper Zariski closed subset $G \subset F$" is consistent and determines a complete type. This type is often referred to as the *generic type of F over the empty set*. If F is an irreducible definable set, then by the *generic type of F* I will mean the generic type of its Zariski closure $\overline{F}$.

Every element of $\mathcal{A}$ is named in $\mathcal{L}$. In particular, $\mathcal{A}$ is not even ω-saturated. Hence in order to evaluate complete types, one is forced to pass to elementary extensions of $\mathcal{A}$. However, using quantifier elimination and the fact that no complex variety can be written as a countable union of proper analytic subsets (this is a consequence of the Baire Category Theorem), the following weakening of saturation can be obtained:

THEOREM 2.4 (Zilber [41]). *$\mathcal{A}$ is ω_1-compact; the intersection of any countable collection of definable sets is nonempty as long as all of its finite subcollections have nonempty intersection.*

For an ω_1-compact structure it is not necessary to consider elementary extensions in order to decide whether the structure is of finite Morley rank. Moreover, if this is the case, the Morley rank of the definable sets can be computed directly inside the given model.[7] That is, the following definitions are valid for an ω_1-compact structure $\mathcal{M}$:

(a) For F a definable set and $n \in \omega$, $\mathrm{RM}(F) \geq n$ can be defined inductiveley as follows:
  - $\mathrm{RM}(F) \geq 0$ iff F is nonempty;
  - $\mathrm{RM}(F) \geq n+1$ iff there are disjoint definable subsets $(F_i)_{i\in\omega}$ of F with $\mathrm{RM}(F_i) \geq n$ for all $i \in \omega$;

(b) $\mathcal{M}$ is of *finite Morley rank* if for any definable set F, there exists $n \in \omega$, such that $\mathrm{RM}(F)$ is not greater than or equal to $n+1$; in which case the minimal such number is the *Morley rank* of F and is denoted by $\mathrm{RM}(F)$.

Using quantifier elimination and Theorem 2.4, Zilber shows:

THEOREM 2.5 (Zilber [41]). $\mathrm{Th}(\mathcal{A})$ *is of finite Morley rank. Moreover, for definable sets in $\mathcal{A}$, Morley rank is bounded by dimension.*

The methods of stability theory are therefore applicable to the structure $\mathcal{A}$. One useful consequence is *uniform definability of types*.[8] This allows us to assume, in a uniform manner, that the parameters for a given definable set come from the sort in which the definable set itself lives. More precisely, suppose $\phi(x, y)$ is an $\mathcal{L}$-formula where x is an n-tuple of variables belonging to a sort X and y is an m-tuple of variables belonging to a sort Y. Then there exists an $\mathcal{L}$-formula $\psi(x, z)$, where z is now an m'-tuple of variables also

[7]For example, see Proposition 0.16 in [25].

[8]See III.1.24 of [1].

belonging to the sort X, such that for all $c \in Y^m$ there is a $d \in X^{m'}$, such that $\phi(x, c)$ and $\psi(x, d)$ define the same subset of X^n.

REMARK 2.6. The notion of a *Zariski-type* structure, as described by Zilber in [41], provides a framework in which to study compact complex varieties. Indeed, all the basic model-theoretic properties that I have discussed thus far can be deduced from the fact that the sorts of $\mathcal{A}$ are Zariski-type structures. As I will not be adopting this point of view, I omit a discussion of these axioms here, and instead refer the interested reader to the aforementioned article.

Every compact complex variety can be obtained as the holomorphic image of a compact complex manifold. Indeed, by resolution of singularities, every complex variety is the image of a complex manifold under a modification. It follows that a compact complex variety is interpretable in a compact complex manifold; it can be obtained as the quotient of a compact complex manifold by a definable equivalence relation. Hence, even if we had begun by only considering the smooth case, arbitrary compact complex varieties would have appeared naturally as imaginary sorts. Moreover, having allowed all compact complex varieties as sorts of $\mathcal{A}$, the process of taking quotients of definable sets by definable equivalence relations does not produce anything more:

THEOREM 2.7. $\text{Th}(\mathcal{A})$ *admits elimination of imaginaries.*

In [31] Pillay describes how a result of Grauert can, in principle, be used to obtain Theorem 2.7. Grauert's result concerns definable equivalence relations that satisfy certain geometric conditions—what are called "meromorphic equivalence relations". In [14] Grauert shows that the quotient of a compact complex manifold by a meromorphic equivalence relation exists (at least generically) in the category of compact complex varieties. In my thesis [25], I have provided the details of how quantifier elimination and the dimension formula can be used to show that all definable equivalence relations are essentially meromorphic, and then to conclude from this that $\text{Th}(\mathcal{A})$ admits elimination of imaginaries.

The projective line over $\mathbb{C}$ is a sort in $\mathcal{A}$. I will use $\mathbb{P}^n(\mathbb{C})$, or just $\mathbb{P}^n$, to denote the nth cartesian power of $\mathbb{P}(\mathbb{C})$ (to be distinguished from projective n-space which is denoted by $\mathbb{P}_n$). Note that $\mathbb{P}^n$ is a projective variety (this is witnessed by the Segre embedding). By quantifier elimination for $\mathcal{A}$ together with Chow's theorem, every definable subset of $\mathbb{P}^n$ is given by a boolean combination of algebraic subsets. Since every algebraic set is naturally interpretable in $(\mathbb{C}, +, \times)$, the full structure induced on $\mathbb{P}(\mathbb{C})$ by $\mathcal{A}$, is interpretable in $(\mathbb{C}, +, \times)$. On other hand, after fixing an identification of $\mathbb{C}$ with a Zariski open subset of $\mathbb{P}(\mathbb{C})$, the complex field $(\mathbb{C}, +, \times)$ is definable in $\mathbb{P}(\mathbb{C})$. Hence the full structure induced on the sort $\mathbb{P}(\mathbb{C})$ by $\mathcal{A}$ is bi-interpretable with the pure algebraically closed field $(\mathbb{C}, +, \times)$.

§3. Strongly minimal sets and Zariski geometries. Recall that in an ω_1-compact structure, a definable set is *strongly minimal* if all of its definable subsets are either finite or cofinite.[9] In $\mathcal{A}$, the first examples of strongly minimal sets are the compact complex curves (i.e., irreducible 1-dimensional Zariski closed sets). The Riemann Existence Theorem says that these are exactly the projective curves. However, part of what makes the model theory of $\mathcal{A}$ interesting is that there are higher dimensional examples of strongly minimal sets. Recall that a complex torus of dimension $n \geq 0$ is a compact complex manifold that can be obtained as the quotient of $\mathbb{C}^n$ by a $2n$-dimensional latttice $\Lambda \subset \mathbb{C}^n$. I will say that the complex torus $M = \mathbb{C}^n/\Lambda$ is *generic*, if Λ is generated by complex numbers $(a_1 + b_1 i), \ldots, (a_{2n} + b_{2n} i)$, where $\{a_1, \ldots, a_{2n}, b_1, \ldots, b_{2n}\} \subset \mathbb{R}$ are algebraically independent over $\mathbb{Q}$. Generic complex tori have no proper infinite analytic subsets, and are therefore strongly minimal in $\mathcal{A}$. Note that this also yields a family of examples where Morley rank and dimension differ—the Morley rank of a strongly minimal set is 1 while the dimension of M above is n.

Recall the notion of (model-theoretic) algebraic closure: a tuple b is in the *algebraic closure* of a set A (denoted $b \in \operatorname{acl} A$) if b belongs to a finite definable set with parameters from A. The algebraic closure relation on a strongly minimal set gives rise to a pregeometry with an associated notion of independence and dimension, that I will refer to here as *acl-independence* and *acl-dimension*, respectively. Strongly minimal sets can then be studied in terms of the behaviour of this relation. For example, there is a natural dividing line among the strongly minimal sets that serves to isolate those on which acl-independence behaves in an essentially "linear" fashion. More precisely, recall that a strongly minimal set X is *locally modular* if the following condition holds after passing to a sufficiently saturated elementary extension of the ambient structure, and possibly adding names for finitely many parameters: for all algebraically closed sets $A, B \subset X$, A is acl-independent from B over $A \cap B$. Equivalently,

$$\text{acl-dim}(A \cup B) = \text{acl-dim}(A) + \text{acl-dim}(B) - \text{acl-dim}(A \cap B).$$

In [29], Pillay gives a direct proof that strongly minimal complex tori (for example, generic complex tori) of dimension greater than 1 are locally modular. On the other hand, it is not very difficult to see that a projective curve is not locally modular.

In [31], Pillay points out that every strongly minimal set in $\mathcal{A}$ that is not locally modular is essentially algebraic. This follows from the deep results of Hrushovski and Zilber [21] on Zariski geometries, and it is worth the digression to give a brief description of this abstract setting. Recall that a

---

[9]In the absence of ω_1-compactness one must reformulate this condition to hold uniformly in the parameters.

topological space is *noetherian* if it satisfies the descending chain condition on closed sets. In a noetherian topological space, the *noetherian dimension* of a closed set, denoted by Ndim, is the maximal length of a chain of irreducible closed subsets. A *Zariski geometry* on a set X is a noetherian topology on X^n for each $n \geq 1$, such that the following conditions hold:

Z0. *If* $f: X^n \to X^m$ *is given by* $f(x) = (f_1(x), \dots, f_m(x))$, *where each* $f_i: X^n \to X$ *is either a coordinate projection or a constant map, then* f *is continuous. Moreover, the diagonals* $x_i = x_j$ *are closed in* X^n.

Z1. *Suppose* $\pi: X^n \to X^k$ *is the projection map onto the first* k *coordinates, and* $C \subset X^n$ *is a closed set. There is a proper closed subset* $F \subset \mathrm{cl}(\pi C)$ *(where* cl *denotes topological closure), such that* $\mathrm{cl}(\pi C) \setminus F \subset \pi C$.

Z2. *Suppose* $C \subset X^n \times X$ *is closed, and denote by* $C_a = \{x \in X : (a, x) \in C\}$ *the fibre of* C *above* $a \in X^n$. *Then there exists* $m \geq 0$, *such that for all* $a \in X^n$, *either* C_a *or* $X \setminus C_a$ *is of size at most* m.

Z3. *Suppose* $F \subset X^n$ *is an irreducible closed set and* Δ_{ij} *is the diagonal* $x_i = x_j$ *in* X^n. *Then every irreducible component of* $F \cap \Delta_{ij}$ *has noetherian dimension at least* $\mathrm{Ndim}(F) - 1$.

For example, if K is an algebraically closed field and C is a smooth algebraic curve over K, then C equipped with the usual Zariski topology on its cartesian powers is a Zariski geometry. Indeed, the main purpose of [21] is to isolate topological and geometric conditions that characterise this example.

Suppose X is a Zariski geometry. Then we can view X as a structure by taking the closed subsets of X^n as the basic relations. Equipped with this structure, X admits quantifier elimination and is strongly minimal (this follows from Z1 and Z2). One of the main theorems in [21] is that if X is in addition not locally modular, then there is an algebraically closed field K interpretable in X. Moreover, there is a finite-to-one surjective map $f: X \to \mathbb{P}(K)$, such that for each $n \geq 1$, $f^n: X^n \to \mathbb{P}(K)^n$ is continuous and maps constructible sets to constructible sets (where the topology on $\mathbb{P}(K)^n$ is taken to be the usual Zariski topology). This says that X is rather close to being a smooth algebraic curve over K.

How does this theory apply to the structure $\mathcal{A}$? Suppose X is a strongly minimal set in $\mathcal{A}$ that is given as a nonempty Zariski open subset of a compact complex variety $\overline{X}$. For each $n \geq 1$, we have a noetherian topology on X^n coming from the relatively Zariski closed subsets (intersections of Zariski closed sets in $\overline{X}^n$ with X^n). Notice that X equipped with these subsets of its powers is exactly the induced structure on X from $\mathcal{A}$. Now assume, moreover, that X is smooth. That is, the analytic structure on X inherited from $\overline{X}$ is smooth. It is pointed out in [20] that X together with this Zariski topology on each of its cartesian powers, is a Zariski geometry. It follows that if X is not locally modular, then there is an algebraically closed field interpretable in X. By elimination of imaginaries, this field is definable in $\mathcal{A}$. However, the only infinite definable field in $\mathcal{A}$, up to definable isomorphism, is the complex field $(\mathbb{C}, +, \times)$ living on the sort $\mathbb{P}(\mathbb{C})$.[10] The results on Zariski geometries mentioned above, thus imply that if X is not locally modular then there is a definable, finite-to-one, surjection from X to $\mathbb{P}(\mathbb{C})$. It follows that there is a finite-to-one meromorphic surjection from $\overline{X}$ to $\mathbb{P}(\mathbb{C})$. In particular, $\dim \overline{X} = 1$, and by the Riemann Existence Theorem, $\overline{X}$ is a projective curve. To summarise, if X is not locally modular then it is a Zariski open subset of a projective curve.

It was assumed in the previous paragraph that X was given as a Zariski open subset of a compact complex variety, and that it was smooth. These conditions are somewhat superfluous. By quantifier elimination, after possibly removing finitely many points, every strongly minimal set X in $\mathcal{A}$ is a Zariski open subset of a compact complex variety. Also, the non-smooth locus of a compact complex variety is a proper analytic subset[11]—and hence definable. By strong minimality, after possibly removing another finite set of points, X is smooth. Hence, up to finitely many points, every strongly minimal set in $\mathcal{A}$ is a Zariski geometry, and the results described above apply. That is, the following dichotomy holds:[12]

THEOREM 3.1. *Every strongly minimal set in $\mathcal{A}$ is either locally modular or a projective curve up to finitely many points.*

§4. Definable groups.

Recall that a *definable group* is a definable set G with a definable map from $G \times G$ to G that satisfies the conditions of a group operation. Definable groups are an intrinsic part of pure model theory, in the sense that they arise naturally when studying the structural properties of a stable theory. *Binding groups*, which are infinitely definable groups of automorphisms that play a role rather similar to that of Galois groups, arise when two definable sets (or types) are related (i.e., nonorthogonal) but the interaction requires additional parameters.[13] Definable groups also appear when one tries to classify strongly minimal sets according to the behaviour of the algebraic closure relation. For example, among strongly minimal sets there is a further dividing line coming from the notion of *triviality*—when the algebraic closure of the union of sets is the union of their algebraic closures. For Zariski geometries, nontriviality is witnessed by the interpretability of an inifinite definable group, in the way that not being locally modular is witnessed by the interpretability of an infinite definable field.

[10] An argument for this is sketched in [31] (the discussion following Remark 3.10), and uses the classification of locally compact fields as well as the Riemann Existence Theorem.

[11] For example, see 2.14 in [7].

[12] See the appendix for a recent proof of this theorem that does not appeal to the theory of Zariski geometries.

[13] See Poizat [37] Section 2.5 for a discussion of binding groups in stable theories.

Recall that a definable group is *definably connected* if it has no definable subgroups of finite index, and *definably simple* if it has no proper infinite normal definable subgroups. For example, any strongly minimal group is definably connected and definably simple. In a theory of finite Morley rank, an understanding of the definable groups can often be reduced to an understanding of those groups that are definably connected and definably simple.

Complex algebraic groups, being definable in $(\mathbb{C}, +, \times)$, are the first examples of groups definable in $\mathcal{A}$. Indeed, every definable group from the sort $\mathbb{P}(\mathbb{C})$ is definably isomorphic to a complex algebraic group.[14] However, there are also non-algebraic groups definable in $\mathcal{A}$. For example, a complex torus is equipped with a holomorphic (and hence definable) group structure, and I have mentioned that generic complex tori of dimension greater than 1 are non-algebraic. Indeed, they are strongly minimal and locally modular. As it turns out, in the definably connected and definably simple case, all non-algebraic groups are strongly minimal and locally modular:

THEOREM 4.1. *A definably connected and definably simple group in $\mathcal{A}$ is either strongly minimal and locally modular or is definably isomorphic to a complex algebriac group.*

I will describe how the above theorem follows from the dichotomy for strongly minimal sets given by Theorem 3.1, and techniques from the model theory of groups of finite Morley rank. Suppose G is a definably connected and definably simple group in $\mathcal{A}$. Using Zilber's Indecomposability Theorem, one obtains a strongly minimal set $X \subset G$, such that G is the image of some cartesian power of X under a definable surjective map, $f : X^n \to G$.[15] Suppose X is not locally modular. By Theorem 3.1, up to finitely many points, X is a projective curve. In particular it is definably isomorphic to a subset of some cartesian power of $\mathbb{P}(\mathbb{C})$. Hence G is interpretable in the sort $\mathbb{P}(\mathbb{C})$ and so definably isomorphic to an algebraic group. Now suppose that X is locally modular. It follows that G, being the image of some cartesian power of X under a definable map, is 1-*based.* 1-basedness is a generalisation of the phenomenon of local modularity from strongly minimal sets to arbitrary definable sets in a stable theory. I will not define 1-basedness here, but instead refer the reader to Section 4.4 of [30]. Hrushovski and Pillay [19] gave a rather complete description of the structure of 1-based groups. For example any definably connected 1-based group is abelian. Applying this to G, which is both definably connected and definably simple, we obtain that G contains no infinite proper definable subgroups. Moreover, every definable subset of a 1-based group is a (finite) boolean combination of cosets of definable subgroups. It follows that every definable subset of G is either finite or cofinite. That is,

---

[14]This follows from the Weil-vdDries-Hrushovski Theorem, see [37].

[15]See Chapter 2 of [37].

G is strongly minimal and locally modular (local modularity and 1-basedness agree on strongly minimal sets).

In [35], Pillay and Scanlon show that every strongly minimal locally modular group is definably isomorphic to a complex torus. Hence, strongly minimal complex tori of dimension greater than 1 are the only examples of non-algebraic definably connected and definably simple groups. The main result in [35] is much stronger; it classifies all definable groups in $\mathcal{A}$. It is convenient to state this result in the more geometric (though equivalent) category of "meromorphic groups" (Definition 2.2 from [35]).

DEFINITION 4.2. A *meromorphic group* is a complex Lie group G for which there exists a finite open cover $\{W_i\}$, and isomorphisms $\phi_i: W_i \to U_i$, where U_i is a Zariski open subset of some compact complex variety X_i, and such that the following hold:

1 The transition maps are meromorphic. That is, for each $i \neq j$,

$$\phi_i\phi_j^{-1}: \phi_j(W_i \cap W_j) \to \phi_i(W_i \cap W_j)$$

extends to a meromorphic map from X_i to X_j.

2 The group operation is meromorphic. That is, for all i, j, k,

$$\{(x, y) \in U_i \times U_j : \phi_i^{-1}(x)\phi_j^{-1}(y) \in W_k\}$$

is Zariski open in $X_i \times X_j$ and the map $U_i \times U_j \to U_k$ induced by the group operation extends to a meromorphic map $X_i \times X_j \to X_k$.

A *meromorphic subgroup* of G is a closed subgroup H of G, such that for each i, $\phi_i(H \cap W_i)$ is the intersection of a Zariski closed subset of X_i with U_i. A *morphism* (respectively *isomorphism*) of meromorphic groups is a holomorphic (respectively biholomorphic) homomorphism which when restricted to the charts is meromorphic.

Every complex algebraic group is meromorphic. Indeed, the definition of a meromorphic group is designed to extend the notion of a complex algebraic group to the category of compact complex varieties. The relationship between meromorphic groups and groups definable in $\mathcal{A}$ is analogous to the relationship between complex algebraic groups and groups definable in the complex field. By elimination of imaginaries in $\mathcal{A}$, every meromorphic group can be identified with a definable group in $\mathcal{A}$. Conversely, the Weil-vdDries-Hrushovski Theorem for the algebraic case extends to this context: every definably connected group in $\mathcal{A}$ has the structure of a connected meromorphic group that is unique up to isomorphism. This equivalence means that we can move freely from definable groups to meromorphic groups and back. For example, one can see that quotient objects exist in the category of meromorphic groups by applying elimination of imaginaries to groups definable in $\mathcal{A}$.

Here is the classification of meromorphic groups alluded to above:

THEOREM 4.3 (Pillay, Scanlon [35]). *Suppose G is a connected meromorphic group. Then G has a unique normal connected meromorphic subgroup L, such that L is isomorphic to a linear complex algebraic group and G/L is isomorphic to a complex torus.*

Fujiki [9] had proven this result in certain special cases, all of which assumed that G had a "nice" global compactification. This turned out to be a useful notion, and I give the definition here. Let G be a meromorphic group and $\mu: G \times G \to G$ the group operation. A *Fujiki-compactification* of G is a compact complex analytic space, G^*, which contains G as a dense Zariski open subset, and a meromorphic map $\mu^*: G^* \times G^* \to G^*$ that is holomorphic on $(G \times G^*) \cup (G^* \times G)$ and that agrees with μ on $G \times G$. Essentailly this says that the group operation on G extends meromorphically to the compactification G^*, while the action of G on itself (in both the right and left senses) extends *holomorphically* to an action of G on all of G^*. For example, a complex torus, being already compact, has itself as a Fujiki-compactification. Also, every complex algebraic group has a Fujiki-compactification.[16] Fujiki proved Theorem 4.3 in the case when G is commutative and has a Fujiki-compactification. He also showed that a meromorphic group has a Fujiki-compactification of Kähler-type if and only if it satisfies the conclusion of Theorem 4.3.

Pillay and Scanlon first find Fujiki-compactifications for commutative meromorphic groups that are either strongly minimal or arise as extensions of a 1-dimensional complex algebraic group by a simple complex torus. Once such a compactification has been found, Theorem 4.3 for these cases follows from Fujiki's results. The general case is then deduced using, among other things, techniques from the model theory of groups of finite Morley rank. Notice that as a consequence of Theorem 4.3, and the results of Fujiki mentioned above, we obtain:

COROLLARY 4.4 (Pillay, Scanlon [35]). *Every connected meromorphic group has a Fujiki compactification of Kähler-type.*

At the beginning of this section, I mentioned that there was a further division among strongly minimal sets, triviality and non-triviality, that in the case of Zariski geometries was connected with the interpretability of an infinite group. Using the classification of meromorphic groups, Pillay and Scanlon obtain the following trichotomy for strongly minimal compact complex manifolds (this was already observed by Scanlon in [39]):

THEOREM 4.5. *Let X be a non-trivial strongly minimal compact complex manifold. Then X is either a projective curve or a complex torus.*

[16]This is pointed out in Remark 2.3 of [9], and uses the fact that Theroem 4.3 is true for algebraic groups.

The full trichotomy for strongly minimal compact complex manifolds is witnessed in $\mathcal{A}$. Indeed, smooth projective curves are strongly minimal compact complex manifolds that are not locally modular, and we have seen that generic complex tori of dimension greater than 1 yield examples of strongly minimal compact complex manifolds that are locally modular but non-trivial. In [22], Kowalski and Pillay point out that certain K3 surfaces are trivial.

I will end this section with a brief discussion of some results obtained by Kowalski and Pillay [22] on the structure of a Zariski closed subset of a commutative connected meromorphic group. Suppose G is a connected meromorphic group with a fixed Fujiki-compactification G^*. Then G is naturally equipped with a Zariski topology induced from G^*; the Zariski closed subsets of G being the intersections of G with Zariski closed subsets of the compact complex variety G^*. If $X \subset G$ is a Zariski closed set, then the *stabiliser of* X is the meromorphic subgroup of G given by $\mathrm{stab}(X) = \{g \in G : g + X = X\}$.

THEOREM 4.6 (Kowalski, Pillay [22]). *Suppose G is a commutative and connected meromorphic group, and $X \subset G$ is an irreducible Zariski closed set with finite stabiliser. Then X is contained in a translate of an algebraic subgroup.*

Since we can always quotient out by $\mathrm{stab}(X)$, the above theorem implies that every irreducible Zariski closed subset of G is algebraic modulo its stabiliser. This allows one to lift certain results about algebraic groups to meromorphic groups. For example, the following Mordell-Lang type theorem for cyclic subgroups of commutative meromorphic groups follows from the algebraic case:

COROLLARY 4.7 (Kowalski, Pillay [22]). *Suppose G is a commutative connected meromorphic group, X is an irreducible Zariski closed subset of G, and $\Gamma \subset G$ is a cyclic subgroup. If $X \cap \Gamma$ is Zariski dense in X, then X is a translate of a meromorphic subgroup of G.*

Indeed, X is a translate of a meromorphic subgroup if and only if it is a translate of its stabiliser. Equivalently, X', the image of X in $G/\mathrm{stab}(X)$, is a single point. Applying Theorem 4.6 to $X' \subset G/\mathrm{stab}(X)$, we obtain that some translate of X' is contained in an algebraic subgroup. By the truth of Corollary 4.7 for commutative complex algebraic groups, X' must be a translate of $\mathrm{stab}(X')$. Since $\mathrm{stab}(X')$ is trivial, X' is a singleton, and X is a translate of $\mathrm{stab}(X)$ as desired.

The proof of Theorem 4.6 is connected to the "socle argument" which appears in Hrushovski's proof of the Mordell-Lang conjecture for function fields [18]. Suppose G is any commuatative group of finite Morley rank. A definable subgroup $H \subset G$ is *almost strongly minimal* if there is a strongly minimal set $X \subset G$ such that , after passing to a saturated elementary extension of G, $H \subset \mathrm{acl}(F \cup X)$, where F is a finite set of parameters. The *socle of*

G was introduced by Hrushovski in [18], and can be described as the sum of all definably connected almost strongly minimal subgroups of G.[17] Zilber's Indecomposability Theorem implies that socle(G) is itself a definably connected subgroup of G, and is in fact a finite sum of definably connected almost strongly minimal subgroups. For example, using the dichotomy for strongly minimal sets in $\mathcal{A}$ from Theorem 3.1, one can show that if G is a connected meromorphic group, then socle(G) $= A + T$, where A is the maximal connected algebraic subgroup of G and T is a sum of strongly minimal locally modular complex tori.

In [18], Hrushovski shows that for commutative groups of finite Morley rank with certain "rigidity" conditions, every definable subset with finite stabiliser is contained in a translate of the socle, up to sets of smaller Morley rank. The rigidity hypothesis need not be satisfied by a commutative meromorphic group. Nevertheless, using the classification of meromorphic groups, together with some arguments from complex analytic geometry, Kowalski and Pillay are able to show that if G is a commutative connected meromorphic group, and $X \subset G$ is an irreducible Zariski closed set with finite stabiliser, then X is contained in some translate of socle(G). Moreover, following [18], they are then able to conclude that X must be contained in a translate of A, where socle(G) $= A + T$ as above.[18] Theorem 4.6 follows.[19]

§5. Douady spaces and saturation. One obstacle to applying model-theoretic techniques directly to $\mathcal{A}$ is that $\mathcal{A}$ is not saturated. Since ω_1-saturation and ω_1-compactness are equivalent for countable languages, the reduct of $\mathcal{A}$ to any countable sublanguage of $\mathcal{L}$ is ω_1-saturated. Of course, in reducing the language we may lose some of the structure on $\mathcal{A}$ that we are interested in. It may be the case, nevertheless, that for certain sorts of $\mathcal{A}$ the full structure (allowing parameters) is induced by some countable sublanguage of $\mathcal{L}$. In other words, there may be sorts of $\mathcal{A}$ in which saturation fails for only syntactic reasons. In my thesis [25], I discuss such sorts and give a characterisation of them in terms of their Douady spaces. In this section I will first give a rather detailed exposition of the ideas from complex analytic geometry that are involved, and then discuss their implication on the model theory of $\mathcal{A}$.

Suppose that $f: X \to S$ and $g: T \to S$ are surjective holomorphic maps on compact complex varieties. Then $(T \times_S X)_{\mathrm{red}}$, the set-theoretic fibred

[17]This was not the original definition in [18], however the above characterisation follows from arguments appearing there. See [22] for a more detailed discussion of the socle.

[18]This follows from the fact that A and T are fully orthogonal, T is 1-based, and X has finite stabiliser.

[19]Theorem 4.6 can also be obtained, more directly, from Theorem A.2 of the appendix—see Pillay [32] for details.

product of X and T over S, will be denoted by X_T:

$$X_T = (T \times_S X)_{\text{red}} = \{(t,x) \mid f(x) = g(t)\} \subset T \times X.$$

The first coordinate projection map restricted to X_T will be denoted by $f_T\colon X_T \to T$. The fibres of f_T can be identified (by the second projection map) with analytic sets in X. Hence, $X_T \subset T \times X$ yields a family of analytic sets in X parametrised by T. Under this identification, $(X_T)_t = f^{-1}(g(t))$ for all $t \in T$. So $f_T\colon X_T \to T$ is just the lifting by g of the original family given by $f\colon X \to S$ to a new set of parameters. The following diagram is illustrative:

$$\begin{array}{ccc} X & & X_T \subset T \times X \\ \downarrow{\scriptstyle f} & & \downarrow{\scriptstyle f_T} \\ S & \xleftarrow[g]{} & T \end{array}$$

This process is referred to as *base change*. Note that if $T = S$ and g is the identity map, then X_S is just the graph of f. Also, if $T = \{s\}$ is a point in S and $g\colon \{s\} \hookrightarrow S$ is the identity embedding of this point, then X_s is just the fibre of f above s, which agrees with earlier notation.

One complication is that X_T need no longer be irreducible. However, if the general fibres of $f\colon X \to S$ are irreducible (that is, f is a *fibre space*), then there is a unique irreducible component of X_T that projects onto T. I will call this compact complex variety the *strict pull back* of X in X_T and denote it by $X_{(T)}$. The general fibres of $X_T \to T$ and $X_{(T)} \to T$ are the same.

Definable families of analytic sets, though natural from the model-theoretic point of view, are not sufficiently well-behaved. The correct geometric notion of a family of analytic sets, involves a flatness condition. Let $f\colon X \to S$ be a holomorphic map between complex analytic spaces, and suppose $\mathcal{F}$ is a coherent analytic sheaf on X. Then $\mathcal{F}$ is said to be f*-flat* if for all $x \in X$, $\mathcal{F}_x$ is a flat $\mathcal{O}_{S,f(x)}$-module. The map f itself is *flat* if the structure sheaf $\mathcal{O}_X$ is f-flat.[20] Essentially, flatness of a holomorphic surjection between compact complex varieties means that the family of analytic sets it defines varies "nicely" from the geometric perspective. For example, the fibres of a flat map are pure dimensional and of constant dimension.

Fortunately, if we allow bimeromorphic changes, every definable family can be made flat. Suppose X and S are compact complex varieties, and $G \subset S \times X$ is an irreducible Zariski closed set such that $G \to S$ is a not necessarily flat fibre space. Hironaka's Flattening Theorem [17] says that after changing $G \to S$ bimeromorphically, we can obtain a *flat* family of analytic

[20] I suggest Section 6 of [38] as a reference to the theory of coherent analytic sheaves, and Section 2 of [26] for more details on the concept of flatness in complex analytic geometry.

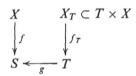

sets in X. Indeed, there exists a modification $g\colon T \to S$ such that the strict pull back $G_{(T)} \to T$ is now flat. Again a diagram is illustrative:

$$\begin{array}{ccccc} S \times X \supset G & \xleftarrow{g \times id} & G_{(T)} \subset T \times_S (S \times X) = T \times X \\ \downarrow & & \downarrow \\ S & \xleftarrow{g} & T \end{array}$$

Since g is a modification, $G_{(T)} \to T$ is a bimeromorphic copy of $G \to S$, that now defines a flat family of analytic sets in X parametrised by T.

A notion from complex analytic geometry that is particularly relevant to saturation issues in $\mathcal{A}$, is the universal flat family of analytic subsets of a complex analytic space, constructed by Douady in [6]:

Fact 5.1 (Existence of Douady Spaces). Let X be a complex analytic space. Then there exists a possibly nonreduced complex analytic space $\mathcal{D} = \mathcal{D}(X)$ and an analytic subspace $\mathcal{Z} = \mathcal{Z}(X) \subset \mathcal{D} \times X$ such that:

(a) The projection $\mathcal{Z} \to \mathcal{D}$ is a flat and proper surjection.
(b) If S is a complex analytic space and G is an analytic set in $S \times X$ that is flat and proper over S, then there exists a unique holomorphic map $g\colon S \to \mathcal{D}$ such that $G \simeq S \times_{\mathcal{D}} \mathcal{Z}$ canonically.

$\mathcal{D}(X)$ is called the *Douady Space of* X, $\mathcal{Z}(X)$ is called the *universal family of* X, and $g\colon S \to \mathcal{D}(X)$ as in (b) is called the *Douady map associated to* $G \subset S \times X$.

What does this say in the cases that we are interested in? Suppose X and S are compact complex varieties and $G \subset S \times X$ is an irreducible Zariski closed subset such that $G \to S$ is flat. So G is a flat family of analytic sets in X parametrised by S. Then Fact 5.1 says that there exists a unique holomorphic map, the Douady map, $g\colon S \to \mathcal{D}(X)$, such that $G \simeq S \times_{\mathcal{D}(X)} \mathcal{Z}(X)$ in a canonical fashion. Set-theoretically speaking, this means that g extends by identity to a holomorphic map from G to $\mathcal{Z}(X)$,

$$\begin{array}{ccc} S \times X \supset G & \xrightarrow{g \times id} & \mathcal{Z}(X) \subset D(X) \times X \\ \downarrow & & \downarrow \\ S & \xrightarrow{g} & \mathcal{D}(X) \end{array}$$

such that $G_s = \mathcal{Z}(X)_{g(s)}$ for all $s \in S$. Essentially, every flat family of analytic sets in X lives in $\mathcal{Z}(X) \to \mathcal{D}(X)$. By Hironaka's Flattening Theorem, the assumption of flatness will not be very restrictive in applications.

I point out a particular case. Suppose $A \subset X$ is any given Zariski closed subset. Take S to be a fixed point $\{s\}$, and G to be the product $\{s\} \times A$ viewed as a Zariski closed subset of $\{s\} \times X$. G is a one-member family of analytic sets in X, and is trivially flat over $\{s\}$. Hence there is a Douady map $g : \{s\} \to \mathcal{D}(X)$ such that the sheaf theoretic fibre of $\mathcal{Z}(X)$ over $g(s)$, $\{g(s)\} \times_{\mathcal{D}(X)} \mathcal{Z}(X)$, is A. Since A is reduced, so is $\{g(s)\} \times_{\mathcal{D}(X)} \mathcal{Z}(X)$. In other words, $\mathcal{Z}(X)_{g(s)} = A$. As A was arbitrary, we have that *every* Zariski closed set in X occurs (uniquely) as a reduced fibre of $\mathcal{Z}(X) \to \mathcal{D}(X)$.

It is already becoming clear that when dealing with Douady spaces, nonreduced spaces naturally enter the picture. I have pointed out that the reduced fibres of the Douady space are in bijective correspondence to the Zariski closed subsets of X. The nonreduced fibres, correspond to subspaces of X equipped with possibly nonreduced structure sheaves. I will avoid most of these nonreduced fibres by only considering a certain subspace of the Douady space (following Fujiki in [8]):

DEFINITION 5.2. Suppose X is a compact complex variety. Let $D(X)$ be the subspace of $\mathcal{D}(X)$ that is obtained by taking the union of all the irreducible components, D_α, of $\mathcal{D}(X)_{\text{red}}$ such that for some $d \in D_\alpha$, the (sheaf-theoretic) fibre $\{d\} \times_{\mathcal{D}(X)} \mathcal{Z}(X)$ is reduced and pure dimensional. Let $Z(X)$ be the subspace of $\mathcal{Z}(X)$ obtained by restricting to $D(X)$. We call $D(X)$ the *restricted Douady space of* X, and $Z(X)$ the *restricted universal family of* X.

By passing from $\mathcal{Z}(X) \to \mathcal{D}(X)$ to $Z(X) \to D(X)$, we are focusing on only certain components of the Douady space. Our choice of which components (those that have at least one reduced and pure dimensional fibre) is justified by the following fact:[21] Suppose D_α is an irreducible component of $\mathcal{D}(X)_{\text{red}}$ and Z_α is the restriction of $\mathcal{Z}(X)$ to D_α. Then the following are equivalent

- D_α is a component of the restricted Douady space, $D(X)$,
- Z_α is reduced and pure dimensional,
- There is a dense Zariski open subset $U \subset D_\alpha$ such that for all $u \in U$, the (sheaf-theoretic) fibre of Z_α over u is reduced and pure dimensional.

A consequence is that the collection of pure dimensional Zariski closed subsets of X are in bijective correspondence with a dense Zariski open subset of $D(X)$. Moreover, if S is a compact complex variety and $G \subset S \times X$ is an irreducible Zariski closed subset that is flat and surjective over S, then the Douady map associated to $G \to S$, will map S to $D(X)$.[22] The restricted Douady space and the corresponding restricted universal family are sufficiently universal for our purposes.

A crucial property of Douady spaces is that it has only countably many irreducible components (due to Fujiki [10]). Notice that we have potentially moved out of the structure $\mathcal{A}$. Even when X is a compact complex variety, it is not in general the case that the components of $D(X)$ are compact. Indeed,

---

[21] See Lemma 1.4 of [8] and Lemma 3 of [11].

[22] This is because a general (sheaf-theoretic) fibre of $G \to S$ will be reduced and pure dimensional.

the question of when the components of the Douady space are compact is very much related to how "saturated" a sort of $\mathcal{A}$ is.

Before making this notion precise, consider the sort $\mathbb{P}(\mathbb{C})$, which is a typical example of the phenomenon described at the beginning of this section. The full structure induced by $\mathcal{A}$ on $\mathbb{P}(\mathbb{C})$ comes from the complex field $(\mathbb{C}, +, \times)$. In particular, there is a finite language, $\mathcal{L}_0 \subset \mathcal{L}$, such that every $\mathcal{L}$-definable set in a cartesian power of $\mathbb{P}(\mathbb{C})$ is $\mathcal{L}_0$-definable (with parameters). Thus $\mathbb{P}(\mathbb{C})$ both retains all of its structure and becomes ω_1-saturated, when it is considered as an $\mathcal{L}_0$-structure. The following definition is from [25]:

DEFINITION 5.3. Let X be a compact complex variety. A *full countable language for* X is a countable sublanguage $\mathcal{L}_0$ of $\mathcal{L}$, such that every $\mathcal{L}$-definable subset of a cartesian power of X is $\mathcal{L}_0$-definable with parameters from $\mathcal{A}$.

One potential defect in the above definition is that a full countable language, for a compact complex variety X, may involve other sorts from $\mathcal{A}$. That is, the parameters involved in the $\mathcal{L}_0$-formulae may come from outside X. However, by uniform definability of types, we can always pull these parameters into X. It follows that X has a full countable language if and only if *there exists a countable language $\mathcal{L}(X)$ and an $\mathcal{L}(X)$-structure on X such that for all $F \subset X^n$, F is $\mathcal{L}$-definable if and only if F is $\mathcal{L}(X)$-definable with parameters from X.* Since $\mathcal{L}$ and $\mathcal{L}(X)$ induce the same definable sets on X, X is still of finite Morley rank as an $\mathcal{L}(X)$-structure, and every $\mathcal{L}(X)$-definable set in X is still a boolean combination of Zariski closed sets. Also, and this is the point of the definition, ω_1-compactness of $\mathcal{A}$ and the countability of $\mathcal{L}(X)$ imply that X is ω_1-saturated as an $\mathcal{L}(X)$-structure. An argument using finite U-rank considerations (due to Pillay), together with the Baire Category Theorem, yields that X is 2^ω-saturated.[23] That is, X is saturated as an $\mathcal{L}(X)$-structure (it is $|X|$-saturated and strongly $|X|$-homogeneous). For this reason, I will often say that X is *saturated*, to mean that it has a full countable language.

Here is the promised connection with Douady spaces. Let X be a compact complex variety, and suppose that for all $n > 0$, each component C of $D(X^n)$ is compact. Then C is a sort in $\mathcal{A}$. Let Z_C denote the restriction of $Z(X^n)$ to C. Then Z_C is a Zariski closed subset of $C \times X^n$, and hence a basic relation in $\mathcal{A}$. Let $\mathcal{L}_0$ be the sublanguage of $\mathcal{L}$ consisting of all such relations Z_C, as C and n vary. Since $D(X^n)$ has only countably many components for each n, $\mathcal{L}_0$ is countable. I have pointed out that every irreducible Zariski closed subset of X^n occurs as a fibre of $Z(X^n) \to D(X^n)$. Hence, for all n, every irreducible Zariski closed subset of X^n is $\mathcal{L}_0$-definable with parameters (where the parameters come from the components of $D(X^n)$). By quantifier elimination, $\mathcal{L}_0$ is a full countable language for X. In fact, the converse of this is also true:

[23]This would of course follow from the Continuum Hypothesis. Fortunately such an assumption is not necessary, see Proposition 0.24 in [25].

THEOREM 5.4 (Moosa [25]). *A compact complex variety X has a full countable language if and only if for all $n > 0$, every irreducible component of the restricted Douady space of X^n is compact.*

Given a full countable language, one uses quantifier elmination and Hironaka's Flattening Theorem to cover each component of $D(X^n)$ by countably many holomorphic images of compact complex varieties. It follows that every such component is in fact the holomorphic image of a single compact complex variety, and hence is itself compact.

EXAMPLE 5.5 (The class $\mathcal{C}$). Fujiki has shown that the components of the Douady space of a Kähler-type variety are compact, and even of Kähler-type themselves ([8] and [11]). Moreover the class of Kähler-type varieties, $\mathcal{C}$, is closed under taking cartesian products. So by Theorem 5.4 they are saturated. In particular, all complex tori and Moishezon varieties are saturated. Moreover, since every meromorphic group has a Fujiki-compactification that is of Kähler-type (Corollary 4.4), every definable group in $\mathcal{A}$ is saturated (or rather, lives on a saturated sort).

Some of the methods used in complex analytic geometry to study the classification problem for Kähler-type varieties have a very model-theoretic flavour. For example, I show in my thesis [25], that the *relative Moishezon reduction* of a fibre space living in a saturated compact complex variety exists.[24] This was shown by Campana [3] and Fujiki [13] (independently) for fibre spaces in the class $\mathcal{C}$, and is used to carry out what is essentially a $\mathbb{P}(\mathbb{C})$-analysis (in the model theoretic sense) of a Kähler-type variety.

There are a number of open questions about compact complex varieties with full countable languages. For example, is every saturated compact complex variety of Kähler-type? In fact, it is not even known whether saturation is closed under bimeromorphism. I will end this section with an example of a compact complex variety that does *not* have a full countable language.

EXAMPLE 5.6 (Hopf manifolds). Consider the original Hopf surface, H, defined as follows: let W be the analytic space $\mathbb{C}^2 - \{(0,0)\}$ and $g: W \to W$ the analytic automorphism of W given by $g(z_1, z_2) = (\frac{1}{2}z_1, \frac{1}{2}z_2)$. Then the Hopf manifold $H = W/ < g >$ is a compact complex surface. In [9] Fujiki points out that $D(H \times H)$ has a noncompact component (by analysing the group of analytic automorphisms of H). By Theorem 5.4, this implies that H is not a saturated complex variety. In particular, H is not of Kähler-type.

§6. The universal domain. As $\mathcal{A}$ is not a saturated model of $\mathrm{Th}(\mathcal{A})$, and not every sort of $\mathcal{A}$ has a full countable language, it is often necessary to consider elementary extensions of $\mathcal{A}$. The sorts of such an extension no longer possess

[24]See [25] for a definition and discussion of what these are.

the structure of a complex analytic space. All that is retained from the standard model is a formal "Zariski topology".

In classical algebraic geometry the notion of a universal domain in which all the objects live, and where "generic" points can be found, is (or at least was) familiar. There does not seem to be an analogue of this in complex analytic geometry. Passing to elementary extensions of the standard universe is characteristic of the model-theoretic approach, and in the case of compact complex varieties it is one of the "new" techniques that model theory brings to complex analytic geometry.

Let $\kappa > |\mathcal{L}|$ be a fixed cardinal, and let $\mathcal{A}'$ be a κ-saturated elementary extension of $\mathcal{A}$ of cardinality κ.[25] That is, $\mathcal{A}'$ is a saturated model of $\mathrm{Th}(\mathcal{A})$; all types over sets of size less than κ are realised in $\mathcal{A}'$ (κ-saturation), and all elementary bijections between sets of size less than κ extend to automorphisms of $\mathcal{A}'$ (strong κ-homogeneity). I will treat $\mathcal{A}'$ as a universal domain for $\mathrm{Th}(\mathcal{A})$. All parameters sets are from now on asumed to be of size less than κ. As opposed to the situation in $\mathcal{A}$, one is forced to consider parameters when dealing with definable sets in $\mathcal{A}'$. A $\emptyset$-definable set in $\mathcal{A}'$ is just the interpretation in $\mathcal{A}'$ of a definable set in $\mathcal{A}$. If X and Y are compact complex varieties (hence sorts of $\mathcal{A}$), and $G \subset Y^n \times X^m$ is a definable subset, then I will sometimes use the notation $G(\mathcal{A})$ and $G(\mathcal{A}')$ to distinguish between G and its interpretation in $\mathcal{A}'$—or, stated differently, between the $\mathcal{A}$-points and the $\mathcal{A}'$-points of G. An arbitrary definable set in $\mathcal{A}'$ is then obtained as the fibre of a $\emptyset$-definable set. They are of the form

$$G(\mathcal{A}')_s = \{x \in X(\mathcal{A}')^m : (s, x) \in G(\mathcal{A}')\}$$

where s is an n-tuple from $Y(\mathcal{A}')$. In particular, $G(\mathcal{A}')_s$ is called *Zariski closed* if G can be chosen to be a Zariski closed set. By quantifier elimination, every definable set in $\mathcal{A}'$ is a finite boolean combination of Zariski closed sets.

There is a more canonical description of a Zariski closed set in $\mathcal{A}'$. Suppose $F = G(\mathcal{A}')_s$ is a Zariski closed set, where G and s are as above. Let $S \subset Y^n$ be the smallest Zariski closed set such that $s \in S(\mathcal{A}')$. That is, s realises the generic type of S over the empty set. Recall that this type is determined by the formulae stating that it is contained in S but not in any proper Zariski closed subset of S. We say that s is a *generic point of S* (or $S(\mathcal{A}')$) *over the empty set*, and that S is the *$\emptyset$-locus of s*. Notice that if P is a $\emptyset$-definable property of points in S, then P holds for a generic point of S if and only if P holds for general $x \in S$. Letting $H = G \cap (S \times X^m)$, the restriction of G to S, we obtain F as a *generic fibre* of the family of Zariski closed sets $H \subset S \times X^m$ under the projection map $H \to S$. Such a description is somewhat more stable in the following sense: if F is also given as a generic fibre of some other family $K \subset S \times X^m$, then K and H have the same general fibres. That is, there is a

[25] By total transcendentality, such models exist.

nonempty Zariski open subset $U \subset S$, such that for every $u \in U$, $K_u = H_u$. It follows that K and H share those irreducible components that project onto S. The description of F as a generic fibre of a Zariski closed subset of $S \times X^m$ over S is therefore unique up to irreducible components whose projections are proper subsets of S (at least for a fixed parameter set).

The above description also gives us a canonical way of going from one set of parameters to another. Suppse that t is an $n+k$ tuple from $Y(\mathcal{A}')$ that extends s, and that $T \subset Y^{n+k}$ is the $\emptyset$-locus of t. Then F can be viewed as being definable over t as well, and as such is obtained as a generic fibre of some Zariski closed subset of $T \times X^m$. How do these two descriptions of F compare? Let π be the coordinate projection map $Y(\mathcal{A}')^{m+k} \to Y(\mathcal{A}')^m$ which takes t to s. Transferring back to the standard model, we get a surjective holomorphic map $\pi\colon T \to S$. We can lift $G \to S$ to T by base change:

$$\begin{array}{ccc} S \times X^m \supset G & & G_T = (T \times_S G)_{\mathrm{red}} \subset T \times X^m \\ \downarrow & & \downarrow \\ S & \xleftarrow{\ \pi\ } & T \end{array}$$

Recall that $G_T = \{(y,x) \in T \times X^m\colon (\pi(y),x) \in G\}$. Note that for any $v \in T$, the fibre of G_T above v is exactly $G_{\pi(v)}$. The fibres of $G_T \to T$ and $G \to S$ are the same. In particular, F is also obtained as a generic fibre of $G_T \to T$. Working with additional parameters in $\mathcal{A}'$ corresponds to base change in the standard model.

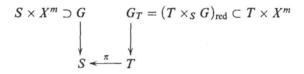

Using this description of Zariski closed sets with parameters, it is not hard to deduce that they also form a noetherian topology on the sorts of $\mathcal{A}'$. A consequence of the descending chain condition is that every Zariski closed set in $\mathcal{A}'$ has a unique expression as the union of finitely many irreducible Zariski closed sets. I will use the term *absolutely irreducible* (instead of just irreducible) Zariski closed set in order to distinguish it from the following relative notion: Suppose X is a compact complex variety, A is a set of parameters, and $F \subset X(\mathcal{A}')$ is a Zariski closed set definable over A. Then F is said to be *irreducible over A* if it cannot be written as the union of two proper A-definable Zariski closed sets. Note that F is irreducible over A if and only if for any tuple s from A over which F is defined, there is an irreducible Zariski closed set $G \subset S \times X$ where S is the $\emptyset$-locus of s and $F = G(\mathcal{A}')_s$. On the other hand, F is absolutely irreducible if and only if $G \to S$ can be chosen with general fibres irreducible (that is, $G \to S$ is a *fibre space*). Suppose F is an A-definable Zariski closed set in $\mathcal{A}'$. Then its absolutely irreducible components are defined over $\mathrm{acl}(A)$. Indeed, since there are only finitely many such components, and automorphisms of $\mathcal{A}'$ that fix A pointwise must permute them, each absolutely irreducible component of F has only finitely many A-conjugates. By elimination of imaginaries and saturation, they are definable over $\mathrm{acl}(A)$. Irreducibility and absolute irreducibility agree over algebraically closed sets of parameters.

One also obtains, in this way, a notion of dimension. Suppose F is a Zariski closed set in $\mathcal{A}'$ obtained as a generic fibre of a holomorphic surjection $G \to S$. The *dimension* of F is the dimension of the general fibres of $G \to S$. This definition makes sense because of the definability of dimension in $\mathcal{A}$. Indeed, by the dimension formula (Fact 2.2) the general fibres of each irreducible component of G that projects onto S are of constant dimension, and hence the general fibres of $G \to S$ are also of constant dimension. Moreover, if F is obtained as a generic fibre of another analytic family $H \to T$, then by base change, we see that the general fibres of $G \to S$ and $H \to T$ have the same dimension. The dimension of F is well-defined.

Suppose A is a set of parameters, c is a tuple of elements from $\mathcal{A}'$, and F is an A-definable Zariski closed in $\mathcal{A}'$. I will say that F is the *A-locus* of c, and that *c is a generic point in F over A*, if F is the smallest Zariski closed set over A that contains c. By quantifier elimination, c is generic in F over A if and only if $c \in F$, and $c \notin G$ for any proper A-definable Zariski closed subset of F. By saturation, every A-definable A-irreducible Zariski closed set has a generic point. Moreover, the generic type over A is unique. The *dimension of c over A* (or of $\mathrm{tp}(c/A)$), denoted by $\dim(c/A)$, is then the dimension of the A-locus of c.

The structure of definable sets in $\mathcal{A}'$ is still rather mysterious. For example, does the classification of definable groups in $\mathcal{A}$ given by Theorem 4.3 extend to $\mathcal{A}'$? On the other hand, it is known that Corollary 4.4 is *not* true in $\mathcal{A}'$; there are definable groups in $\mathcal{A}'$ that are not definably isomorphic to groups living inside the $\mathcal{A}'$-points of a Kähler-type variety. Pillay and Scanlon give a counterexample that also shows that Morley rank and U-rank differ in $\mathcal{A}'$. I will briefly describe their construction (taken from Lieberman [23]).

EXAMPLE 6.1 (Pillay, Scanlon [34]). Fix an elliptic curve $(E, 0, +)$, and let $\sigma\colon E \times E \to E \times E$ be the automorphism taking (a, b) to $(a + b, a + 2b)$. Fix a complex number τ whose imaginary part is positive. Then $\mathbb{Z} \times \mathbb{Z}$ acts on $E \times E \times \mathbb{C}$ by

$$(m, n)((a, b), s) = (\sigma^n(a, b), s + m + n\tau).$$

Let X be the the quotient of $E \times E \times \mathbb{C}$ by this action, equipped with the induced structure of a compact complex manifold. The coordinate projection $E \times E \times \mathbb{C} \to \mathbb{C}$ induces a holomorphic surjection $q\colon X \to S$, where S is the elliptic curve given as the quotient of $\mathbb{C}$ by the lattice generated by 1 and τ. It is not difficult to see that each fibre of q is isomorphic to the product of elliptic curves $E \times E$. In fact, $q\colon X \to S$ is locally trivial. That is, for every $p \in S$, there is a neighbourhood about p, U, such that $X|_U = q^{-1}(U)$ is isomorphic to $E \times E \times U$ over U. Moreover, the local trivialisations induce a fibrewise group structure on $X \to S$. That is, there is a holomorphic map $X \times_S X \to X$ such that for all $p \in S$, the induced map $X_p \times X_p \to X_p$ is the group operation on X_p that is obtained as the pull back of addition on $E \times E$. Hence, if we let $s \in S(\mathcal{A}')$ be generic in S over $\emptyset$, then $G = X(\mathcal{A}')_s$ is a definable group.

Since $\dim X = 3$, and the fibres of $X \to S$ are of Morley rank 2, the Morley rank of X is also 3. On the other hand, Campana's analysis of $q: X \to S$ in [4] shows that if τ is chosen to be sufficiently general, then X has no 2-dimensional subvarieties that project onto S. It follows that the generic fibre $G = X(\mathcal{A}')_s$ is strongly minimal. As S is a curve, finite U-rank computations imply that X (or rather its unique generic type) is of U-rank 2. This yields an example where Morley rank and U-rank differ.

Now let Z be a Kähler-type variety, and suppose that G is definably isomorphic to a group $H \subset Z(\mathcal{A}')$. Hence H is a strongly minimal group. Using uniform definability of types, we obtain H as a generic fibre of a family of definable groups $Y \to T$, where Y and T are definable sets living in the sort Z. That is, $H = Y(\mathcal{A}')_t$, for $t \in T(\mathcal{A}')$ generic over $\emptyset$. Since H is definably isomorphic to $G = X(\mathcal{A}')_s$, there exists a nonempty definable set $U \subset S$, which is Zariski dense and open in $\overline{S}$, such that for each $u \in U$, Y_u is definably isomorphic to X_p for some $p \in S$. On the other hand, since Z is of Kähler-type, it has a full countable language $\mathcal{L}(Z)$. By saturation of Z with respect to $\mathcal{L}(Z)$, there exists $q \in U$, in the standard model, such that the $\mathcal{L}(Z)$-type of q is the same as that of t. In particular, Y_q is strongly minimal. But as Y_q is definably isomorphic to some X_p, it is definably isomorphic to the product of elliptic curves $E \times E$. This contradiction shows that G cannot be definably isomorphic to a group that lives on the $\mathcal{A}'$-points of a Kähler-type variety.

§7. Nonstandard algebraicity. The Riemann Existence Theorem states that every 1-dimensional compact complex variety is algebraic—it can be biholomorphically embedded into some complex projective space. An extension of this to the universal domain $\mathcal{A}'$ was asked for by Pillay in [31] and obtained in my thesis [25]. In order to state this theorem, I need to make precise what it means for a Zariski closed set in $\mathcal{A}'$ to be algebraic. It should, of course, have something to do with being embeddable into projective space over the non-standard algebraically closed field in $\mathcal{A}'$ that extends $\mathbb{C}$, which I will denote by $\mathbb{C}^{\mathcal{A}'}$.[26] One key point is that we should allow such an embedding to be defined with additional patramaters from $\mathcal{A}'$. Another issue is to make precise what kind of embedding is sought. Here are two possible notions that correspond to Moishezon and projective in the standard model:

Definition 7.1. Suppose X and Y are compact complex varieties. A holomorphic surjection $X \to Y$ is *trivially Moishezon* if for some $n \geq 1$, there is

[26]Note that $\mathbb{P}(\mathbb{C}^{\mathcal{A}'})$ is just the interpretation of the sort $\mathbb{P}(\mathbb{C})$ in $\mathcal{A}'$.

a meromorphic map, $g: X \to Y \times \mathbb{P}_n(\mathbb{C})$, which is bimeromorphic with its image and which commutes with the projection map $Y \times \mathbb{P}_n(\mathbb{C}) \to Y$:

$$\begin{array}{ccc} X & \xrightarrow{\quad g \quad} & Y \times \mathbb{P}_n(\mathbb{C}) \\ & \searrow \quad \swarrow & \\ & Y & \end{array}$$

If in addition, g and n can be chosen such that for general $y \in Y$, the fibrewise map $g_y: X_y \to \mathbb{P}_n(\mathbb{C})$ is biholomorphic with its image, then $X \to Y$ is called *trivially projective*. An irreducible Zariski closed set F in $\mathcal{A}'$ is *Moishezon* (respectively *projective*) if it can be obtained as a generic fibre of a trivially Moishezon (respectively trivially projective) surjection between compact complex varieties.

The first thing to notice about these definitions is that for a fibre space to be trivially Moishezon or trivially projective is stable under base change (and strict pull backs). Hence for an absolutely irreducible Zariski closed set in $\mathcal{A}'$ to be Moishezon or projective does not depend on a choice of parameters. This can be expressed as follows. Suppose that F is of the form $G(\mathcal{A}')_s$, where $G \subset S \times X$ is an irreducible Zariski closed set, X and S are compact complex varieties, the projection $G \to S$ is a fibre space, and $s \in S(\mathcal{A}')$ is a generic point of S over the empty set. Then F is Moishezon (respectively projective) if and only if for some compact complex variety T and holomorphic surjection $T \to S$, $G_{(T)} \to T$ (the strict pull back of G in the fibred product $S \times_T G$) is trivially Moishezon (respectively trivially projective). In diagrams:

$$\begin{array}{cccc} G & G_{(T)} & \xrightarrow{\ g\ } & T \times \mathbb{P}_n(\mathbb{C}) \\ \downarrow & \downarrow & \swarrow & \\ S & \longleftarrow \ T & & \end{array}$$

where g satisifies the appropriate conditions of Definition 7.1.

The notion of Moishezon in $\mathcal{A}'$, though weaker than projective, is more stable under model-theoretic manipulations. Suppose F is given as above. Using quantifier elimination together with the dictionary set up in the previous section that allows one to translate from Zariski closed sets and additional parameters in $\mathcal{A}'$, to families of Zariski closed sets and base change in $\mathcal{A}$, it is not hard to see that F is Moishezon if and only if:

- *There is a definable embedding of a nonempty Zariski open subset of F into some cartesian power of the sort $\mathbb{P}(\mathbb{C}^{\mathcal{A}'})$.*

Moreover, using saturation in $\mathcal{A}'$, this is equivalent to:

- *There is a tuple of parameters t extending s and a generic point of F over t that is interdefinable with a tuple from $\mathbb{P}(\mathbb{C}^{\mathcal{A}'})$ over t.*

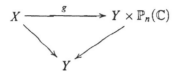

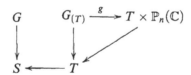

Finally, from the definition of internality together with the fact that the induced structure on the sort $\mathbb{P}(\mathbb{C})$ eliminates imaginaries (as it is that of a pure algebraically closed field), we can conclude that F is Moishezon if and only if:

- *The generic type of* F *over* s *is internal to* $\mathbb{P}(\mathbb{C}^{\mathcal{A}'})$.

It should now be clear in what sense being Moishezon in $\mathcal{A}'$ corresponds to being "generically" algebraic.

THEOREM 7.2 (Moosa [25]). *Every irreducible* 1*-dimensional Zariski closed set in* $\mathcal{A}'$ *is Moishezon.*

The above Theorem says that every fibre space of curves in $\mathcal{A}$, possibly after base change, is trivially Moishezon. Indeed, this follows from a result of Campana together with some observations regarding projective linear spaces. Since these spaces also appear in other aspects of the model theory of compact complex varieties, I will give a rather detailed description of them here.

Fix a compact complex variety S and consider the category of compact complex varieties over S. In complex analytic geometry, the analogue of projective space in this relative category is the notion of a projective linear space over S. Let $\mathcal{F}$ be a coherent analytic sheaf on S. I will sketch a (local) construction of the *projective linear space over* S *associated to* $\mathcal{F}$, denoted by $\pi\colon \mathbb{P}(\mathcal{F}) \to S$. Let $U \subset S$ be a small open subset for which there exists a resolution of $\mathcal{F}_U$ as follows:

$$\mathcal{O}_U^p \xrightarrow{\alpha} \mathcal{O}_U^q \longrightarrow \mathcal{F}_U \longrightarrow 0$$

As an $\mathcal{O}_U$-linear homomorphism, α can be represented by a $q \times p$ matrix $M = (m_{ij})$, where each $m_{ij} \in \mathcal{O}_U$. Letting X be coordinates for U and $(Y_1 : \cdots : Y_q)$ homogeneous coordinates for $\mathbb{P}_{q-1}$, $\mathbb{P}(\mathcal{F})_U$ is the analytic subset of $U \times \mathbb{P}_{q-1}$ defined by the equations:

$$m_{1i}(X)Y_1 + \cdots + m_{qi}(X)Y_q = 0$$

for $i = 1, \ldots, p$. One checks that $\mathbb{P}(\mathcal{F})_U$ depends only on the coherent sheaf $\mathcal{F}_U$, and not on the particular resolution chosen above. We then patch the $\mathbb{P}(\mathcal{F})_U$ to obtain $\mathbb{P}(\mathcal{F})$. The structure of a fibre space over S is induced on $\mathbb{P}(\mathcal{F})$ by the coordinate projection maps $U \times \mathbb{P}_{q-1} \to U$.

For each $s \in S$, the fibre $\mathbb{P}(\mathcal{F})_s$ is isomorphic to $\mathbb{P}_r$, where $r+1$ is the rank of $\mathcal{F}$ at s. In the special case when $\mathcal{F}$ is locally free, this rank is constant and $\pi\colon \mathbb{P}(\mathcal{F}) \to S$ is called a *projective bundle over* S. Projective bundles are locally trivial in the following strong sense. There is an open cover $\{U_i\}$ of S and local trivialisations

$$\begin{array}{ccc} \mathbb{P}(\mathcal{F})_{U_i} & \xrightarrow[\simeq]{h_i} & U_i \times \mathbb{P}_r \\ & {\scriptstyle\pi}\searrow \quad \swarrow & \\ & U_i & \end{array}$$

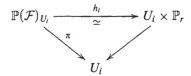

such that the induced transition functions

$$\begin{array}{ccc} U_i \cap U_j \times \mathbb{P}_r & \xrightarrow[\simeq]{h_i h_j^{-1}} & U_i \cap U_j \times \mathbb{P}_r \\ & \searrow \quad \swarrow & \\ & U_i \cap U_j & \end{array}$$

are of the form $h_i h_j^{-1}(x, p) = (x, g_{ij}(x)(p))$, where $g_{ij}\colon U_i \cap U_j \to PGL(r, \mathbb{C})$ is holomorphic. In other words, a projective bundle over S of rank r is locally a product of the base with projective r-space, where the transition functions fix the base points while permuting the fibres as elements of the projective general linear group. A *trivial* projective bundle over S is one of the form $S \times \mathbb{P}_r(\mathbb{C}) \to S$. Equivalently, $\mathcal{F} = \mathcal{O}_S^{r+1}$.

A *projective morphism*, $f\colon X \to S$, is a holomorphic map that factors through an embedding into a projective linear space over S. That is, there is a coherent analytic sheaf $\mathcal{F}$ on S and an embedding of X into $\mathbb{P}(\mathcal{F})$ over S. A *Moishezon morphism* is a holomorphic map, $f\colon X \to S$, that is bimeromorphic over S to a projective morphism.

REMARK 7.3. One open question about saturated compact complex varieties is whether they are closed under preimages of projective morphisms. Indeed, the question of whether they are stable under bimeromorphic equivalence can be reduced to this.

Notice that Definition 7.1 (of trivially projective/Moishezon morphisms), is derived from the above notions by replacing "projective linear space" by "trivial projective bundle". Indeed, from the model-theoretic point of view, trivial projective bundles, and not projective linear spaces, seem to be the more natural relative form of projectivity. As it turns out, this discrepancy does not pose too many difficulties. To begin with, while a projective morphism need not embed into a product of the base with projective space, it does embed bimeromorphically into an object that is at least locally of that form. That is, every projective linear space is, after a modification of the base, bimeromorphic to a projective bundle. This is a consequence of Hironaka's Flattening Theorem applied to coherent analytic sheafs (a coherent analytic sheaf is locally free if and only if it is flat).

Moreover, and this turns out to be quite useful, every projective bundle is bimeromorphic after base change to a trivial projective bundle. Following ideas of Fujiki in [12], and using his results on relative Douady spaces, I give a proof of this fact in Proposition 2.37 of [25]. Putting these together, we get that after base change every Moishezon morphism is trivially Moishezon.

Returning to the proof of 7.2, we show that every fibre space of curves in $\mathcal{A}$, $f\colon X \to Y$, possibly after base change, is trivially Moishezon. By the above remarks, Moishezon is enough. Campana has shown (see Lemma 3.10

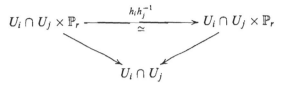

in [5]) that if f admits a holomorphic section then it is Moishezon. Since we are allowing for additional parameters, we can always obtain a holomorphic section by base change with the fibre space itself and considering the diagonal map. Theorem 7.2 follows.

I will conclude with an application to fields definable in $\mathcal{A}'$. Recall that every infinite field definable in $\mathcal{A}$ is definably isomorphic to $(\mathbb{C}, +, \times)$ (see the discussion after Remark 3.10 in [31]). The only known argument for extending this to $\mathcal{A}'$ uses Theorem 7.2. Indeed, one uses the result for $\mathcal{A}$ to conclude that an infinite field in $\mathcal{A}'$ is of dimension 1, and hence can be definably embedded into a cartesian power of the sort $\mathbb{P}(\mathbb{C}^{\mathcal{A}'})$. Since there is a unique definable field in any pure algebraically closed field, one obtains:[27]

COROLLARY 7.4. *Every infinite field definable in $\mathcal{A}'$ is definably isomorphic to $(\mathbb{C}^{\mathcal{A}'}, +, \times)$.*

Appendix: The dichotomy theorem revisited. During preparation of this article, a new development has emerged in the model theory of compact complex spaces that fits in well with the particular approach taken here. Interpreting a result due independently to Campana [2] and Fujiki [11], Pillay has obtained a direct proof of Theorem 3.1 (the dichotomy for strongly minimal sets) that does not use the theory of Zariski Geometries. I will give a brief sketch of the ideas involved, now assuming greater familiarity with notions from stability theory. To begin with, here is the result referred to above, on which Pillay's observations are based:

FACT A.1 (Campana [2], Fujiki [11]). *Suppose X is a compact complex variety, $\mathcal{D}(X)$ is the Douady space of X, $\mathcal{Z}(X) \subset \mathcal{D}(X) \times X$ is the universal family of X, and $B \subset \mathcal{D}(X)$ is a reduced and irreducible compact analytic subspace such that for general $b \in B$, $\mathcal{Z}(X)_b$ is reduced and irreducible. Let $Z \subset B \times X$ denote the restriction of $\mathcal{Z}(X)$ to B, and $\pi: Z \to X$ the second coordinate projection. Then π is Moishezon.*[28]

Since every Moishezon morphism is trivially Moishezon after base change, Fact A.1 says that the generic fibres of $\pi: Z \to X$ in $\mathcal{A}'$ are Moishezon. (See the the discussion on these issues in Section 7.) Equivalently, the generic type of a generic fibre of $\pi: Z \to X$ in $\mathcal{A}'$ is internal to $\mathbb{P}(\mathbb{C}^{\mathcal{A}'})$. Pillay translates this fact into the following model-theoretic statement:

THEOREM A.2 (Pillay [32]). *Suppose b and c are finite tuples from $\mathcal{A}'$ such that $\operatorname{tp}(c/b)$ is stationary, and b is the canonical base of $\operatorname{tp}(c/b)$. Then $\operatorname{tp}(b/c)$ is internal to $\mathbb{P}(\mathbb{C}^{\mathcal{A}'})$.*

Indeed, let $X = \operatorname{locus}(c)$, $B = \operatorname{locus}(b)$, and $Z = \operatorname{locus}(b, c)$. Using Hironaka's Flattening Thoerem we may assume that $Z \to B$ is flat. Moreover,

[27]See Corollary 2.41 in [25] for a more detailed proof.

[28]Campana's version uses cycle spaces instead of Douady spaces.

as $b = \mathrm{Cb}(c/b)$, the general fibres of $Z \to B$ are distinct as subspaces of X. It follows that the Douady map associated to $Z \to B$ is a meromorphic embedding. We may therefore assume that $B \subset \mathcal{D}(X)$, and Z is the restriction of $\mathcal{Z}(X)$ to B. Note that $\mathrm{tp}(b/c)$ is the generic type of a generic fibre of $Z \to X$. Theorem A.2 now follows from Fact A.1.

The following corollary implies that every strongly minimal set in $\mathcal{A}$ is either locally modular or a projective curve up to finitely many points—that is, we obtain a new and direct proof of Theorem 3.1.

COROLLARY A.3 (Pillay [32]). *Every stationary U-rank 1 type in $\mathcal{A}'$ is either locally modular or nonorthogonal to the generic type of $\mathbb{P}(\mathbb{C}^{\mathcal{A}'})$.*

To see how this follows from Theorem A.2, let $p(x)$ be a stationary U-rank 1 type in $\mathcal{A}'$. For ease of dicussion, we suppress the parameters of p. If $p(x)$ is not locally modular, then there is a tuple of realisations of $p(x)$, c, and a tuple b from $\mathcal{A}'$, such that $b = \mathrm{Cb}(c/b)$ and $b \notin \mathrm{acl}(c)$. By the theorem, $\mathrm{tp}(b/c)$ is internal to $\mathbb{P}(\mathbb{C}^{\mathcal{A}'})$. Since $\mathrm{tp}(b/c)$ is not (model-theoretically) algebraic and $\mathbb{P}(\mathbb{C}^{\mathcal{A}'})$ is minimal, this implies that $\mathrm{tp}(b/c)$ is nonorthogonal to the generic type of $\mathbb{P}(\mathbb{C}^{\mathcal{A}'})$. As c is a tuple of realisations of $p(x)$ and $b = \mathrm{Cb}(c/b)$, it follows that $p(x)$ is nonorthogonal to the generic type of $\mathbb{P}(\mathbb{C}^{\mathcal{A}'})$, as desired.

In [32], using Theorem A.2, Pillay also obtains a generalisation of Theorem 4.6 on subvarieties of meromorphic groups, without reference to the theory of the socle.

The ideas involved in this section and in the proof of Fact A.1 have been useful in obtaining similar results in other algebraic/model-theoretic contexts; namely, for differential and difference fields (see [36] and [33]).[29]

REFERENCES

[1] J. BALDWIN, ***Fundamentals of stability theory***, Springer-Verlag, Berlin, 1988.

[2] F. CAMPANA, *Algébricité et compacité dans l'espace des cycles d'un espace analytique complexe*, ***Mathematische Annalen***, vol. 251 (1980), no. 1, pp. 7–18.

[3] ———, *Réduction algébrique d'un morphisme faiblement Kählerien propre et applications*, ***Mathematische Annalen***, vol. 256 (1981), no. 2, pp. 157–189.

[4] ———, *Exemples de sous-espaces maximaux isolés de codimension deux d'un espace analytique compact*, ***Institut Élie Cartan***, vol. 6 (1982), pp. 106–127.

[5] F. CAMPANA and T. PETERNELL, *Cycle spaces*, In Grauert et al. [15], pp. 319–349.

[6] A. DOUADY, *Le problème des modules pour les sous espaces analytiques compacts d'un espace analytique donné*, ***Annales de l'Institut Fourier (Grenoble)***, vol. 16 (1966), pp. 1–95.

[7] G. FISCHER, ***Complex analytic geometry***, Springer-Verlag, Berlin, 1976.

---

[29]***Note added in proof***: The results from my thesis discussed in this survey have now been written-up in two papers: "A nonstandard Riemann existence theorem" (*Transactions of the American Mathematical Society*, vol. 356 (2004), no. 5, pp. 1781–1797) and "On saturation and the model theory of compact Kähler manifolds" (to appear in Crelle's Journal). Moreover, my notes entitled "Jet spaces in complex analytic geometry: an exposition", which describe the techniques used to prove Fact A.1, is now available at `http://arxiv.org/abs/math.LO/0405563`.

[8] A. FUJIKI, *Closedness of the Douady spaces of compact Kähler spaces*, ***Publication of the Research Institute for Mathematical Sciences***, vol. 14 (1978), no. 1, pp. 1–52.

[9] ———, *On automorphism groups of compact Kähler manifolds*, ***Inventiones Mathematicae***, vol. 44 (1978), no. 3, pp. 225–228.

[10] ———, *Countability of the Douady space of a complex space*, ***Japanese Journal of Mathematics (New Series)***, vol. 5 (1979), pp. 431–447.

[11] ———, *On the Douady space of a compact complex space in the category C*, ***Nagoya Mathematical Journal***, vol. 85 (1982), pp. 189–211.

[12] ———, *On the structure of compact complex manifolds in C*, ***Algebraic varieties and analytic varieties***, Advanced Studies in Pure Mathematics, vol. 1, North-Holland, Amsterdam, 1983, pp. 231–302.

[13] ———, *Relative algebraic reduction and relative Albanese map for a fibre space in C*, ***Publications of the Research Institute for Mathematical Sciences***, vol. 19 (1983), no. 1, pp. 207–236.

[14] H. GRAUERT, *On meromorphic equivalence relations*, ***Contributions to several complex variables*** (A. Howard and P. Wong, editors), Aspects of Mathematics, Vieweg, Wiesbaden, 1986, pp. 115–147.

[15] H. Grauert, T. Peternell, and R. Remmert (editors), ***Several complex variables VII***, Encyclopedia of Mathematical Sciences, vol. 74, Springer-Verlag, Berlin, 1994.

[16] R. GUNNING and H. ROSSI, ***Analytic functions of several complex variables***, Prentice-Hall, Edgewood Cliffs, 1965.

[17] H. HIRONAKA, *Flattening theorem in complex-analytic geometry*, ***American Journal of Mathematics***, vol. 97 (1975), pp. 503–547.

[18] E. HRUSHOVSKI, *The Mordell-Lang conjecture for function fields*, ***Journal of the American Mathematical Society***, vol. 9 (1996), no. 3, pp. 667–690.

[19] E. HRUSHOVSKI and A. PILLAY, *Weakly normal groups*, ***Logic Colloquium 1985*** (Orsay) (Paris Logic Group, editor), North-Holland, 1987, pp. 233–44.

[20] E. HRUSHOVSKI and B. ZILBER, *Zariski geometries*, ***Bulletin of the American Mathematical Society***, vol. 28 (1993), no. 2, pp. 315–322.

[21] ———, *Zariski geometries*, ***Journal of the American Mathematical Society***, vol. 9 (1996), pp. 1–56.

[22] P. KOWALSKI and A. PILLAY, *Subvarieties of commutative meromorphic groups*, preprint.

[23] D. LIEBERMAN, *Compactness of the Chow scheme: applications to automorphisms and deformations of Kähler manifolds*, ***Fonctions de plusieurs variables complexes, III (Séminaire François Norguet, 1975–1977)***, Lecture Notes in Mathematics, vol. 670, Springer-Verlag, Berlin, 1978, pp. 140–186.

[24] S. ŁOJASIEWICZ, ***Introduction to complex analytic geometry***, Birkhäuser Verlag, Basel, 1991, English edition.

[25] R. MOOSA, *Contributions to the model theory of fields and compact complex spaces*, ***Ph.D. thesis***, University of Illinois, Urbana-Champaign, 2001.

[26] T. PETERNELL and R. REMMERT, *Differential calculus, holomorphic maps and linear structures on complex spaces*, In Grauert et al. [15], pp. 97–144.

[27] Y. PETERZIL and S. STARCHENKO, *Expansions of algebraically closed fields in o-minimal structures*, ***Selecta Mathematica. New Series***, vol. 7 (2001), no. 3, pp. 409–455.

[28] ———, *Expansions of algebraically closed fields II. Functions of several variables*, ***Journal of Mathematical Logic***, vol. 3 (2003), no. 1, pp. 1–35.

[29] A. PILLAY, *Definable sets in generic complex tori*, ***Annals of Pure and Applied Logic***, vol. 77 (1996), no. 1, pp. 75–80.

[30] ———, ***Geometric stability theory***, Oxford Logic Guides, vol. 32, Oxford Science Publications, Oxford, 1996.

[31] ———, *Some model theory of compact complex spaces*, ***Workshop on Hilbert's tenth problem: relations with arithmetic and algebraic geometry*** (Ghent), vol. 270, Contemporary Mathematics, 2000.

[32] ———, *Model-theoretic consequence of a theorem of Campana and Fujiki*, ***Fundamenta Mathematicae***, vol. 174 (2002), no. 2, pp. 187–192.

[33] ———, *Mordell-Lang conjecture for function fields in characteristic zero, revisited*, ***Compositio Mathematica***, vol. 140 (2004), no. 1, pp. 64–68.

[34] A. Pillay and T. Scanlon, *Compact complex manifolds with the DOP and other properties*, ***The Journal of Symbolic Logic***, vol. 67 (2002), no. 2, pp. 737–743.

[35] ———, *Meromorphic groups*, ***Transactions of the American Mathematical Society***, vol. 355 (2003), no. 10, pp. 3843–3859.

[36] A. Pillay and M. Ziegler, *Jet spaces of varieties over differential and difference fields*, ***Selecta Mathematica. New Series***, vol. 9 (2003), no. 4, pp. 579–599.

[37] B. Poizat, ***Stable groups***, Mathematical Surveys and Monographs, vol. 87, American Mathematical Society, 2001, Translation from the French, ***Groupes Stables***, Nur Al-Mantiq Wal-Ma'rifah, 1987.

[38] R. Remmert, *Local theory of complex spaces*, In Grauert et al. [15], pp. 10–96.

[39] T. Scanlon, *Locally modular groups in compact complex manifolds*, preprint.

[40] K. Ueno, ***Classification theory of algebraic varieties and compact complex spaces***, Lecture Notes in Mathematics, vol. 439, Springer-Verlag, Berlin, 1975.

[41] B. Zilber, *Model theory and algebraic geometry*, ***Proceedings of the 10th Easter conference on model theory*** (Berlin), 1993.

DEPARTMENT OF PURE MATHEMATICS
UNIVERSITY OF WATERLOO
200 UNIVERSITY AVE. WEST
WATERLOO, ONTARIO, N2L 3G1, CANADA
E-mail: rmoosa@math.uwaterloo.ca

"NATURAL" REPRESENTATIONS AND EXTENSIONS OF GÖDEL'S SECOND THEOREM

KARL-GEORG NIEBERGALL

Abstract. Building on [9], several types of examples for consistent extensions of PA proving their own consistency, employing only what may be regarded as natural formalizations of their consistency-assertions, are presented. I also suggest and discuss explications of "natural formalization" and analyze the results given before in the light of this discussion.

§1. Introduction. In [9], I have presented a method for obtainig extensions of, e.g., Peano Arithmetic (PA), which are consistent, but nonetheless prove their own consistency. For these results, it is of crucial importance that the theories considered are provided with representations that are "natural". That is, "T proves its own consistency" is explained as "$T \vdash \mathrm{Con}_\tau$", where "$\mathrm{Con}_\tau$" is defined in the usual way from the usual arithmetized proof-predicate, and τ is a "natural" representation of a "natural" *axiom-set* of T.[1] How to understand "*natural* representation" is, of course, a decisive question. In [9], I made some suggestions on this topic and, moreover, embedded the metamathematical results given there in some broader philosophical perspective.

This paper is a continuation of [9].[2] Its first part (section 2.1) lists several examples of theories which prove their own consistency employing only natural formalizations of consistency-assertions—in short: *theories proving their own consistency naturally*—and methods to obtain such theories. Whereas in that part, the expression "natural representation" is merely *used*, but not defined, section 3 attempts to provide for the missing definition: I propose and discuss in it several explications of "α is a natural representation of A".

Cordial thanks to the organizers of LC 2001, in particular Prof. Isaacson and Prof. Visser, for inviting me to that conference. I also would like to thank Prof. H. Friedman and Prof. Willard for their comments on the talk, the referees for their critical remarks and suggestions to improve this paper, and M. Barrios and N. Silich.

[1]In his classical [1], Feferman has shown that—in some sense—even PA can prove its own consistency by supplying a representation pa* of an *axiom-set* of PA such that PA $\vdash \mathrm{Con}_{\mathrm{pa}^*}$ (cf. section 3); but $\mathrm{Pr}_{\mathrm{pa}^*}$ will hardly be regarded as a "natural" representation of PA.

[2]As such, it deals more deeply (see especially sections 3 and 4) with some of the issues I had treated only sketchily or left open in [9], but is rather brief with respect to others: this holds, in particular, for the philosophical background, which is elaborated in [9].

Logic Colloquium '01
Edited by M. Baaz, S. Friedman, and J. Krajíček
Lecture Notes in Logic, 20

The paper closes with some "negative" metamathematical results (section 4): if all conditions for "α is a natural representation of A" taken into consideration are conjoined, it seems to be hard to find theories proving their own consistency naturally.

§2. Theories proving their own consistency. For simplicity, only theories S in L[PA] (or definitional extensions of it) which extend $I\Sigma_1$ will be considered (Robinson Arithmetic (Q) is an exception; see [12], [4]). Their underlying logic is assumed to be classical first-order logic with identity; it is convenient to axiomatize it such that only *modus ponens* is a primitive rule of inference (see [1] for more details). As usual, a theory S is understood as a set of sentences which is deductively closed; i.e., using "$\overline{A}$" for the set of sentences from L[A] derivable from A, S is a theory $\iff S = \overline{S}$. Let me note that, as such, theories need not be recursively enumerable (r.e.) or even, say, arithmetically definable: think at the *theory of* $\mathcal{N}$, i.e., the set of all sentences from L[PA] being true in $\langle \mathbb{N}, S, +, \cdot, 0\rangle$, as an example for a non arithmetical theory. In fact, non r.e. theories will be the main object of investigation in this paper.[3]

If S fails to be r.e., it is not axiomatizable; for lack of a better term, nonetheless even in this case I will say that S has an *axiom-set*.[4] That is,

$$A \text{ is an } \textit{axiom-set} \text{ for } S \iff S = \overline{A}.$$

Let me emphasize here that for sets of formulas to be theories, they have to be closed only under the usual rules of inference of classical first order logic.[5] Thus, even if a theory S is non r.e., it is r.e. in each of its *axiom-sets*. Moreover, if A is an *axiom-set* for S, S is certainly the set of sentences *derivable* (where the meaning of "derivable" has not been stretched) from A.[6]

Some *axiom-sets* will play a distinguished role for the following investigations: Ax(PA), Ax(Q), Ax($I\Sigma_n$) ($n \geq 1$) are the usually chosen *axiom-sets* of PA, Q, $I\Sigma_n$, Ax(PA) being recursive, the others being finite. $\mathrm{Tr}_{\Pi^0_k}$ and $\mathrm{Tr}_{\Sigma^0_k}$ are the sets of true Π^0_k-sentences and the set of true Σ^0_k-sentences ($k \geq 1$).[7]

[3] Besides, it will not be assumed that theories need to be arithmetically sound.

[4] Alternative terminological choices are, e.g., "set of postulates" or "set of pseudoaxioms" (cf. [7]). But see also [1], in which the expression "axiom-system" is used where I prefer "*axiom-set*".

[5] In particular, no infinitary rules (like, e.g., the ω-rule) or rules defined by semantical conditions are employed.

[6] There are further questions, some of which are philosophically more important, like: Do non r.e. theories exist? Are they worthy of studying? Can they be used as *means* of investigation? These deserve a close investigation, which I will not even attempt to carry out here, however (personally, I think that whatever the answers may be, those who assert that the analogous questions concerning r.e. theories should be answered differently are in need of providing reasons for their claims).

[7] See [4] and [5] for the theories and the background on recursion theory and the arithmetical hierarchy.

For the topic "Gödel's incompleteness theorems", the arithmetization of metamathematical formulas is of importance.[8] Thus, let $A \subseteq \mathbb{N}^k$ and α a k-place arithmetical formula ($k \geq 1$); in general, $\overline{n}$ will be the numeral denoting the natural number n; then

$$\alpha \text{ is a representation of } A :\Longleftrightarrow$$
$$\forall n_1, \ldots n_k \in \mathbb{N}\ (\langle n_1, \ldots, n_k \rangle \in A \Longleftrightarrow \mathcal{N} \models \alpha(\overline{n_1}, \ldots, \overline{n_k})).$$

I assume that the usual "syntactic" metamathematical vocabulary—like "x is a sequence", "the length (of sequence) x", "x is the conditional of y and z", "x is a formula", "x is a Π^0_n-formula"—is represented by Σ^0_0-formulas—like "$Seq(x)$", "$lh(x)$", "$x = y \dot{\rightarrow} z$", "$Fml(x)$", "$\Pi^0_n(x)$"—in such a way that conditions characteristic for these notions are provable in the theories considered here (with the exception of Q).[9] For a finite set $\{n_1, \ldots, n_k\}$ of natural numbers, the formula "$x = \overline{n_1} \vee \cdots \vee x = \overline{n_k}$" is its so called *canonical representation*. In particular, I write "$[q](x)$" and "$[\mathrm{i}\sigma_n](x)$" for the canonical representations of Ax(Q) and Ax($\mathrm{I}\Sigma_n$) ($n \geq 1$). Furthermore, I use "LogAx(x)" for the representation of a recursive set of axioms of first-order logic (to be specific, take the axiomatization from [1]), "pa(x)" for the representations of Ax(PA) and "$\mathrm{Tr}_{\Pi^0_k}(x)$" and "$\mathrm{Tr}_{\Sigma^0_k}(x)$" for the (natural) representations of $\mathrm{Tr}_{\Pi^0_k}$ and $\mathrm{Tr}_{\Sigma^0_k}$ ($k \geq 1$) (see [5] for details). With the exception of "$\mathrm{Tr}_{\Pi^0_k}(x)$" and "$\mathrm{Tr}_{\Sigma^0_k}(x)$", which are Π^0_k and Σ^0_k, all representations are Σ^0_0.

Given a representation τ of an *axiom-set* T, arithmetical formulas representing the relations or metatheoretical formulas *proof in* τ, *provability in* τ and τ *is consistent* are defined as usual:[10]

DEFINITION 1.

$$\mathrm{Proof}_\tau(x, y) :\Longleftrightarrow Seq(x) \wedge y = x_{lh(x) \dot{-} 1} \wedge$$
$$\forall v < lh(x)(\mathrm{LogAx}(x_v) \vee \tau(x_v) \vee \exists uw < v(x_w = x_u \dot{\rightarrow} x_v)),$$
$$\mathrm{Pr}_\tau(y) :\Longleftrightarrow \exists x\, \mathrm{Proof}_\tau(x, y),$$
$$\mathrm{Con}_\tau :\Longleftrightarrow \neg \mathrm{Pr}_\tau(\ulcorner \bot \urcorner).$$

Furthermore, RFN[σ] is the uniform reflection principle for S (σ, to be more precise), i.e., the set of all formulas "$\forall x(\mathrm{Pr}_\sigma(\overline{\ulcorner \psi(\dot{x}) \urcorner}) \rightarrow \psi(x))$" for arbitrary formulas ψ in L[PA]; $\mathrm{RFN}_{\Sigma^0_k}[\sigma]$ is the restriction of RFN[σ] to the

[8] With respect to the notation employed, I follow [1]; cf. also [11]; in particular, I employ the *dot-notation* presented there. For questions of provability in theories weaker than PA, see [4].

[9] Some gödelization is presupposed here: for an expression t from L[PA] (or some if its extensions), $\ulcorner t \urcorner$ is the Gödel-number and $\overline{\ulcorner t \urcorner}$ the Gödel-numeral of t. Given this, α is called a representation of Σ, too, if $A = \{\ulcorner \psi \urcorner \mid \psi \in \Sigma\}$ and α represents A.

[10] Usually, the phrases "proof in T", "provability in T" and "T is consistent" are used here. Besides, one should perhaps say "derivation" and "derivable" instead of "proof" and "provable"; but the use of the latter is more common in metamathematics.

Σ_k^0-formulas; and Rfn[σ] is the local reflection principle for S, i.e., the set of all sentences "$\mathrm{Pr}_\sigma(\ulcorner\psi\urcorner) \to \psi$" for arbitrary sentences ψ in L[PA].

In this context,[11] Gödel's Second Incompleteness Theorem can be stated quite generally as follows:

GÖDEL'S SECOND THEOREM. For each consistent r.e. extension T of PA and for each representation τ of an *axiom-set* of T which is Σ_1^0, $T \nvdash \mathrm{Con}_\tau$.

Thus, theories extending PA which prove their own consistency naturally can probably not be found under the r.e. ones (if the interpretation of "naturally" is not stretched too much).

2.1. Mutual consistency proofs. The examples of theories proving their own consistency naturally given in [9] rest on the fact (taken from [8]) that there exist theories S, T such that S proves the consistency of T and T proves the consistency of S (while Gödel's incompleteness theorems hold for both).

FIRST EXAMPLE for mutual consistency proofs:

Let $S_1 := \mathrm{PA} + \mathrm{Tr}_{\Pi_1^0}$. S_1 has the natural *axiom-set* $\mathrm{Ax(PA)} \cup \mathrm{Tr}_{\Pi_1^0}$, which is represented in a natural way by the formula σ_1, where

$$\sigma_1(x) :\longleftrightarrow \mathrm{pa}(x) \vee \mathrm{Tr}_{\Pi_1^0}(x).$$

Now, let $T_1 := \mathrm{PA} + \mathrm{Con}_{\sigma_1}$ (i.e., $T_1 := \mathrm{PA} + \mathrm{Con}_{\mathrm{pa}+\mathrm{Tr}_{\Pi_1^0}}$). T_1 has the natural *axiom-set* $\mathrm{Ax(PA)} \cup \{\mathrm{Con}_{\sigma_1}\}$, which is represented in a natural way by the formula τ_1, where

$$\tau_1(x) :\longleftrightarrow \mathrm{pa}(x) \vee x = \ulcorner \mathrm{Con}_{\sigma_1} \urcorner.$$

By definition of T_1, $T_1 \vdash \mathrm{Con}_{\sigma_1}$. And since T_1 is r.e. and consistent, its consistency assertion Con_{τ_1} is a true Π_1^0-sentence, whence provable in S_1.

SECOND EXAMPLE for mutual consistency proofs:

Let $S_{2,k} := \mathrm{I}\Sigma_1 + \mathrm{Tr}_{\Pi_{k+1}^0}$ $(k \in \mathbb{N})$, $T_2 := \mathrm{PA}$. $S_{2,k}$ and T_2 have natural *axiom-sets* $\mathrm{Ax}(\mathrm{I}\Sigma_1) \cup \mathrm{Tr}_{\Pi_{k+1}^0}$ and Ax(PA). These *axiom-sets* are represented in a natural way by the formulas $\sigma_{2,k}$ and τ_2, where

$$\sigma_{2,k}(x) :\longleftrightarrow [\mathrm{i}\sigma_1](x) \vee \mathrm{Tr}_{\Pi_{k+1}^0}(x), \quad \text{and} \quad \tau_2(x) :\longleftrightarrow \mathrm{pa}(x).$$

In [8], it is shown that $T_2 \vdash \mathrm{RFN}[\sigma_{2,k}]$ for each $k \in \mathbb{N}$.[12] Since $\mathrm{RFN}_{\Sigma_k^0}[\tau_2]$ is a set of true Π_{k+1}^0-sentences, $S_{2,k} \vdash \mathrm{RFN}_{\Sigma_k^0}[\tau_2]$.

[11] It may be that very weak theories formulated in L[PA] or languages with similar or less expressive richness prove their own consistency; see [14] for work going in this direction and further remarks on the relevant literature.

[12] Provability in $S_{2,k}$ is usually arithmetized in a way different from $\mathrm{Pr}_{\sigma_{2,k}}$ (see e.g., [11]), making it easier to formalize metamathematical arguments about $S_{2,k}$. But $\sigma_{2,k}$ is surely the representation which is intuitively more natural.

Thus, it is not only possible to obtain mutual consistency proofs, but $S_{2,k}$ and T_2 provide for examples of the mutual provability of partial soundness (up to an arbitrarily given extent).[13]

2.2. System-internal consistency proofs. Having theories proving mutually their consistency at hand, it is easy to obtain theories proving their own consistency (naturally): one simply has to take intersections of theories proving their mutual consistency.

In order to carry out this idea, one should have control over representations of intersections of theories. Thus, let S and T be arithmetically definable theories and $\sigma(x)$ and $\tau(x)$ be representations of *axiom-sets* of S and T. Then "$\mathrm{Pr}_\sigma(x)$" and "$\mathrm{Pr}_\tau(x)$" are representations of the theories S and T, and "$\mathrm{Pr}_\sigma(x) \wedge \mathrm{Pr}_\tau(x)$" is a representation of $S \cap T$. Moreover, it seems plausible to me that this formula is a *natural* representation of $S \cap T$, if the representations σ and τ are natural (see section 3 for more on this). Thus, I define

$$(\sigma \wedge \tau)(x) :\Longleftrightarrow \mathrm{Pr}_\sigma(x) \wedge \mathrm{Pr}_\tau(x).$$

Now, by what has been said so far, it also seems that "$\mathrm{Pr}_{\sigma\wedge\tau}$" is a natural representation of $\overline{S \cap T}$. But $\overline{S \cap T} = S \cap T$—whence we have two "natural" representations ("$\sigma \wedge \tau$" and "$\mathrm{Pr}_{\sigma\wedge\tau}$") for just one object (the theory $S \cap T$). In general, I do not think that the mere circumstance that we have several representations of the same theory presents a problem (see again section 3). Besides, the case considered here seems to me particularly harmless: the reason is that these two representations are PA-provably equivalent.

Lemma 1. (a) $\mathrm{PA} \vdash \forall x\,((\sigma \wedge \tau)(x) \longleftrightarrow \mathrm{Pr}_{\sigma\wedge\tau}(x))$.
(b) $\mathrm{PA} \vdash \mathrm{Con}_\sigma \vee \mathrm{Con}_\tau \longleftrightarrow \mathrm{Con}_{\sigma\wedge\tau}$.

Example 1. Take S_1, T_1 and σ_1, τ_1 from the first example.

Since $S_1 \vdash \mathrm{Con}_{\tau_1}$, it follows that $S_1 \vdash \mathrm{Con}_{\sigma_1\wedge\tau_1}$.
Since $T_1 \vdash \mathrm{Con}_{\sigma_1}$, it follows that $T_1 \vdash \mathrm{Con}_{\sigma_1\wedge\tau_1}$.

Therefore, $S_1 \cap T_1 \vdash \mathrm{Con}_{\sigma_1\wedge\tau_1}$.

Example 2. Take $S_{2,k}, T_2$ and $\sigma_{2,k}, \tau_2$ ($k \in \mathbb{N}$) from the second example.

Since $T_2 \vdash \mathrm{RFN}[\sigma_{2,k}]$, it follows that $T_2 \vdash \mathrm{RFN}[\sigma_{2,k} \wedge \tau_2]$.
Since $S_{2,k} \vdash \mathrm{RFN}_{\Sigma^0_k}[\tau_2]$, it follows that $S_{2,k} \vdash \mathrm{RFN}_{\Sigma^0_k}[\sigma_{2,k} \wedge \tau_2]$.

Therefore, $S_{2,k} \cap T_2 \vdash \mathrm{RFN}_{\Sigma^0_k}[\sigma_{2,k} \wedge \tau_2]$.

Thus, consistent extensions of PA can also prove their own soundness naturally up to an arbitrarily preassigned degree.

The foregoing considerations can be found (more or less) already in [9]. Coming to the new work in what follows, let me first give a further example.

---

[13]Evidently, the second example is obtained as a special case of a general procedure for building theories which mutually prove their consistency. In fact, there are many theories which mutually prove their consistency or partial soundness.

EXAMPLE 3. Let $(\mathrm{PA}_n)_{n\in\mathbb{N}}$ be a sequence (as discovered by Feferman, Friedman, Solovay and possibly others; see [11]) of consistent r.e. extensions of PA such $\forall n \in \mathbb{N}\, \mathrm{PA}_n \vdash \mathrm{Con}_{\mathrm{pa}_{n+1}}$. To be more specific, I follow the version presented in [6].

Here (for each $n \in \mathbb{N}$), $\mathrm{PA}_n := \mathrm{PA} + \beta(\overline{n})$, where each $\beta(\overline{n})$ is a certain Π^0_1-sentence satisfying

$$\mathrm{PA}_n \vdash \beta(\overline{n+1}) \wedge \mathrm{Con}_{\mathrm{pa}+\beta(\overline{n+1})}\,.^{14}$$

Now define $(S_3 :=)\ \mathrm{PA}_\omega := \bigcap\{\mathrm{PA}_n \mid n \in \mathbb{N}\}$, and

$$\alpha(x, y) :\longleftrightarrow \mathrm{pa}(x) \vee x = \overline{\ulcorner\beta(\dot{y})\urcorner},$$
$$\mathrm{pa}_\omega(x) :\longleftrightarrow \forall y\ \mathrm{Pr}_{\alpha(,y)}(x).^{15}$$

Then,

$$\mathcal{N} \models \alpha(\overline{\ulcorner\psi\urcorner}, \overline{n}) \iff \mathcal{N} \models \mathrm{pa}(\overline{\ulcorner\psi\urcorner}) \vee \overline{\ulcorner\psi\urcorner} = \overline{\ulcorner\beta(\overline{n})\urcorner}$$
$$\iff \psi \in \mathrm{Ax}(\mathrm{PA}) \cup \{\beta(\overline{n})\},$$

and

$$\mathcal{N} \models \mathrm{pa}_\omega(\overline{\ulcorner\psi\urcorner}) \iff \forall n\, (\mathrm{Ax}(\mathrm{PA}) \cup \{\beta(\overline{n})\}) \vdash \psi \iff \mathrm{PA}_\omega \vdash \psi.$$

That is, pa_ω is a—I think, natural—representation of the theory PA_ω. But there is, like before, a further representation of the latter which deserves to be called "natural": $\mathrm{Pr}_{\mathrm{pa}_\omega}$. Yet, again, these two representations are PA-provably equivalent with each other:

CLAIM. $\mathrm{PA} \vdash \forall x\, (\mathrm{pa}_\omega(x) \longleftrightarrow \mathrm{Pr}_{\mathrm{pa}_\omega}(x))$.

PROOF. Only "$\longleftarrow$" has to be proved. But since $\mathrm{PA} \vdash \forall xy\ (\mathrm{pa}_\omega(x) \wedge \mathrm{pa}_\omega(x \dot{\rightarrow} y) \longrightarrow \mathrm{pa}_\omega(y))$, this direction clearly holds, too (cf. [1]).

CLAIM. $\mathrm{PA}_\omega \vdash \mathrm{Con}_{\mathrm{pa}_\omega}$.

PROOF. Surely, for each $n \in \mathbb{N}$, $\mathrm{PA} \vdash \forall x\, (\forall y\ \mathrm{Pr}_{\alpha(,y)}(x) \longrightarrow \mathrm{Pr}_{\mathrm{pa}+\beta(\overline{n})}(x))$, i.e.,

$$\mathrm{PA} \vdash \forall x\, (\mathrm{pa}_\omega(x) \longrightarrow \mathrm{Pr}_{\mathrm{pa}_n}(x)),$$

whence (by the first claim) $\mathrm{PA} \vdash \mathrm{Con}_{\mathrm{pa}_n} \longrightarrow \mathrm{Con}_{\mathrm{pa}_\omega}$.

Since $\mathrm{PA}_n \vdash \mathrm{Con}_{\mathrm{pa}_{n+1}}$, $\mathrm{PA}_n \vdash \mathrm{Con}_{\mathrm{pa}_\omega}$ is obtained. As n was arbitrary, it follows that $\mathrm{PA}_\omega \vdash \mathrm{Con}_{\mathrm{pa}_\omega}$.

As a next step, let me present some methods of proceeding from theories proving their own consistency to other types of theories which also prove their own consistency. For this, let σ, τ be representations of *axiom-sets* of consistent arithmetically definable theories S, T extending $\mathrm{I}\Sigma_1$.

---

14 Therefore, each PA_n is a sound r.e. extension of $\mathrm{PA} + \mathrm{Con}_{\mathrm{pa}}$.

15 This notation is taken from [2].

METHOD 1. Add a consistent Σ_1^0-sentence.

LEMMA 2. If $S \vdash \mathrm{Con}_\sigma$ and ψ is Σ_1^0, and $S \vdash \forall x\, (\mathrm{Pr}_{\mathrm{q}}(x) \rightarrow \mathrm{Pr}_\sigma(x))$,[16] then $S + \psi \vdash \mathrm{Con}_{\sigma+\psi}$.

PROOF. Since $S \vdash \forall x\, (\mathrm{Pr}_{\mathrm{q}}(x) \rightarrow \mathrm{Pr}_\sigma(x))$ and Q is S-provably Σ_1^0-complete (S being an extension of $\mathrm{I}\Sigma_1$), Con_σ implies $\mathrm{Rfn}_{\Pi_1^0}[\sigma]$ over S. Therefore, $S \vdash \mathrm{Pr}_{\sigma+\psi}(\ulcorner \bot \urcorner) \longrightarrow \mathrm{Pr}_\sigma(\ulcorner \psi \rightarrow \bot \urcorner) \longrightarrow (\psi \rightarrow \bot)$, for $S \vdash \mathrm{Con}_\sigma$ and $\psi \rightarrow \bot$ is Π_1^0. It follows that $S + \psi \vdash \neg \mathrm{Pr}_{\sigma+\psi}(\ulcorner \bot \urcorner)$.

METHOD 2. Intersect with a theory proving its own consistency.

LEMMA 3. If $S \vdash \mathrm{Con}_\sigma$ and $T \vdash \mathrm{Con}_\sigma$, then $S \cap T \vdash \mathrm{Con}_{\sigma \wedge \tau}$.

PROOF. Since $S \cap T \vdash \mathrm{Con}_\sigma$, the conclusion follows by Lemma 1.

METHOD 3. Maximize in a provable way.

Each consistent arithmetically definable theory has arithmetically definable completions, i.e., maximally consistent extensions; moreover, this statement can be formalized in $\mathrm{I}\Sigma_1$ (see [1], [6] and especially [13] for more detailed presentations of this "formalized completeness theorem"). A weakened version of this metatheorem which suffices here is:

For each representation σ of some *axiom-set* there is an arithmetical formula $v(\sigma)$ representing an *axiom-set* of a complete extension $V(S)$ of S such that

$$\mathrm{I}\Sigma_1 + \mathrm{Con}_\sigma \vdash \mathrm{Con}_{v(\sigma)}\,.$$

In addition, $v(\sigma)$ can be found in Δ_{k+1}^0 if σ is Σ_k^0 ($k \geq 1$).

Now, let S be an extension of $\mathrm{I}\Sigma_1$ with representation σ such that $S \vdash \mathrm{Con}_\sigma$. Then $V(S) \vdash \mathrm{Con}_{v(\sigma)}$; i.e., $V(S)$ is a complete extension of PA proving its own consistency.

METHOD 4. Intersect provable extensions.

LEMMA 4. Assume $S \cap T \vdash \mathrm{Con}_{\sigma \wedge \tau}$; and let α, β be arithmetical formulas such that $S \cap T \vdash \mathrm{Con}_\sigma \rightarrow \mathrm{Con}_\alpha$ and $S \cap T \vdash \mathrm{Con}_\tau \rightarrow \mathrm{Con}_\beta$. Then $S \cap T \vdash \mathrm{Con}_{\alpha \wedge \beta}$.

PROOF. Since $\mathrm{PA} \vdash \mathrm{Con}_{\sigma \wedge \tau} \longrightarrow \mathrm{Con}_\sigma \vee \mathrm{Con}_\tau$ and $\mathrm{PA} \vdash \mathrm{Con}_\alpha \vee \mathrm{Con}_\beta \longrightarrow \mathrm{Con}_{\alpha \wedge \beta}$ (by Lemma 1), the assumptions yield the claim.

I give some applications of these methods.

EXAMPLE 4 (by method 1). Let $S_4 := \mathrm{PA}_\omega + \{\neg \mathrm{Con}_{\mathrm{pa}_1}\}$, $\sigma_4(x) := \mathrm{pa}_\omega(x) \vee x = \overline{\ulcorner \neg \mathrm{Con}_{\mathrm{pa}_1} \urcorner}$. S_4 is consistent, and $S_4 \vdash \mathrm{Con}_{\sigma_4}$ by Lemma 2.

For the following examples, let $T_5 := S_{2,0} \cap T_2$ with representation $\tau_5(x) :\longleftrightarrow \sigma_{2,0}(x) \wedge \tau_2(x)$.

---

[16]By adding "$[q](x)$" to "$\sigma(x)$", this can always be achieved.

EXAMPLE 5 (by method 2). Take $I\Sigma_1 + \mathrm{Con}_{\tau_5}$ with natural representation "$[i\sigma_1](x) \vee x = \ulcorner \mathrm{Con}_{\tau_5} \urcorner$". Since $T_5 \vdash \mathrm{Con}_{\tau_5}$, by method 2, $(I\Sigma_1 + \mathrm{Con}_{\iota_5}) \cap T_5$ proves its own consistency with representation $([i\sigma_1] + \mathrm{Con}_{\tau_5}) \wedge \tau_5$.

EXAMPLE 6 (by method 3). Since $T_5 \vdash \mathrm{Con}_{\tau_5}$, there is a completion $V(T_5)$ with representation $v(\tau_5)$ in Δ_3^0 such that $V(T_5) \vdash \mathrm{Con}_{v(\tau_5)}$.

EXAMPLE 7 (by method 4). Let $V(S_{2,0})$ be a completion in Δ_3^0 of $S_{2,0}$ with an *axiom-set* represented by $v(\sigma_{2,0})$ and $V(T_2)$ be a completion in Δ_2^0 of T_2 with an *axiom-set* represented by $v(\tau_2)$ such that

$$T_5 + \mathrm{Con}_{\sigma_{2,0}} \vdash \mathrm{Con}_{v(\sigma_{2,0})} \text{ and } T_5 + \mathrm{Con}_{\tau_2} \vdash \mathrm{Con}_{v(\tau_2)}\,.$$

Then $T_5 \vdash \mathrm{Con}_{v(\sigma_{2,0}) \wedge v(\tau_2)}$ by Lemma 4, and $V(S_{2,0}) \cap V(T_2)$ proves its own consistency with representation $v(\sigma_{2,0}) \wedge v(\tau_2)$.

§3. **Natural representations.** In the previous section, I have presented some methods which allow the construction of many theories which prove their own consistency. Moreover, I have repeatedly asserted that the representations of those theories considered there are natural, but not supplied any arguments for these claims. In fact, without a conceptual analysis of "natural representation", it seems hardly possible to do so. In this section, I will attempt to close that gap.[17]

When it comes to the task of formulating an explication of "α is a natural representation of A", there is no sense in denying that we share linguistic intuitions about which representations are natural and which ones fail to be natural.—On the one hand, there should be "universal" agreement on several paradigmatic cases—where we could just *stipulate* which representations are natural and unnatural (or artificial). Here, we should find (or so do I believe) the usual representations "$\mathrm{pa}(x)$" of Ax(PA) and the the usual representations of the recursive syntactic notions—like "$x = y \dot{\rightarrow} z$", "$Fml(x)$", "$\Pi_n^0(x)$"—but also such formulas as "$x + y = z$" for the addition function. Furthermore, finite sets of formulas $\{\psi_1, \dots, \psi_k\}$ are naturally represented by their *canonical representation* (see section 2.1).—On the other hand, we have clear intuitions about how to obtain natural representations of B from natural representations of $A_1, \dots, A_n$, provided B is constructed by simple methods from $A_1, \dots, A_n$. Let me give two examples:

(a) if α is a natural representation of A, then "Pr_α" is a natural representation of $\overline{A}$;
(b) if α is a natural representation of A and ψ is a sentence, then "$\alpha(x) \vee x = \ulcorner \psi \urcorner$" is a natural representation of $A \cup \{\psi\}$.

What I will certainly not assume is that each set of natural numbers (e.g., Gödel numbers) has only one natural representation. Quite the opposite: I

[17] For further considerations on this topic, see [9].

think each one has infinitely many natural representations.[18] If, e.g., $\alpha(x)$ is a natural representation of A, what about "$\alpha(x) \wedge x = x$"? More interesting is the following example: for each theorem ψ of A, "$\text{Pr}_{\alpha+\psi}(x)$" is a natural representation of $\overline{A+\psi}$ (by (a) and (b)); but this set is just $\overline{A}$ (but see section 3.4).[19]

Surely, these examples and considerations do not automatically deliver a precise and generally adequate explication of "α is a natural representation of A". But they also should make it plausible that the search for such an explication is not futile. Thus, in the remaining part of this section, I will propose three ways of giving sufficient or necessary conditions for "α is a natural representation of A" which should be faithful to what I said in these introductory reflections.[20]

3.1. The inductive approach. The idea that natural representations are built from natural representations by using procedures which lead from natural to natural representations is an *inductive conception of naturality*. Rather than using it as a mere means for testing the adequacy of definitions of "α is a natural representation of A", I will employ it to *obtain* such a definition.[21]

BASE.

(B1) (by stipulation[22]): "pa(x)" is a natural representation of Ax(PA) and "$[\text{i}\sigma_1](x)$" is a natural representation of Ax$(\text{I}\Sigma_1)$. "$\Pi^0_k(x)$" and "$\Sigma^0_k(x)$" (see [5]) are natural representations of the sets (of Gödel numbers) of Π^0_k- and Σ^0_k-formulas ($k \geq 1$). The usual function signs for some primitive recursive functions (like successor, addition, multiplication) are natural representations of those functions.

(B2) (by a general principle): if Σ is a finite set of formulas $\psi_1, \ldots, \psi_n$ and $A_\Sigma = \{\ulcorner\psi\urcorner \mid \psi \in \Sigma\}$, then

$$x = \overline{\ulcorner\psi_1\urcorner} \vee \cdots \vee x = \overline{\ulcorner\psi_n\urcorner}$$

is a natural representation of A_Σ (and Σ).

---

[18]I agree that it would also be interesting to have a more fine-grained conception of natural representation.

[19]Sometimes, different axiom systems are common for one theory: Zermelo-Fraenkel set theory (ZF) is an interesting example, with Ax_1(ZF) being its recursive *axiom-set* containing the axiom scheme of replacement and Ax_2(ZF) being its recursive *axiom-set* containing the axiom scheme of collection. This leads to two natural representations, "$\text{zf}_1(x)$" and "$\text{zf}_2(x)$" of these *axiom-sets* and to two natural representations of ZF itself.

[20]It may be asked "What about Löb's derivibality conditions for theories which prove their own consistency?" Provable closure under *modus ponens* is, of course, for free. Moreover, not all of the derivability conditions can hold for such theories, because this would imply the unprovability of consistency for them. Of the two remaining ones, closure under necessitation is the least one could hope for. See section 4 for more on this.

[21]There is nothing new to the following clauses; it just seems that they have not been put to work as part of an explication of "α is a natural representation of A".

[22]The list given here is certainly not supposed to be exhaustive. I think this has to remain like this: for one, it is open which theories and, therefore, representations could one day be taken to be worthy of investigation.

STEP. Let $\alpha(x)$, $\beta(x)$ be a $k+1$-place arithmetical formulas and $A, B \subseteq \mathbb{N}^{k+1}$; then

(I1) If $\alpha(x)$ is a natural representation of A and $\beta(x)$ is a natural representation of B, then $\alpha(x) \vee \beta(x)$ is a natural representation of $A \cup B$;

(I2) If $\alpha(x)$ is a natural representation of A and $\beta(x)$ is a natural representation of B, then $\alpha(x) \wedge \beta(x)$ is a natural representation of $A \cap B$;

(I3) If $\alpha(x)$ is a natural representation of A, then $\neg\alpha(x)$ is a natural representation of $\mathbb{N}^{k+1} \setminus A$;

(I4) If $\alpha(x)$ is a natural representation of A, then $\mathrm{Pr}_\alpha(x)$ is a natural representation of $\overline{A}$;

(I5) If for each $n \in \mathbb{N}$, $\alpha(x, \overline{n})$ is a natural representation of $\{k \in \mathbb{N} \mid \langle k, n\rangle \in A\}$, then $\forall y\ \alpha(x, y)$ is a natural representation of $\{k \in \mathbb{N} \mid \forall n\ (\langle k, n\rangle \in A)\}$.

Surely, if no clauses are added to the ones given here for the inductive definition of "α is a natural representation of A", then α is a representation of A if it is a natural representation of A. But this set of conditions is supposed to be merely a first suggestion: it may not contain *all* relevant conditions.

3.2. The complexity approach. When a theory T is r.e. (that is, Σ^0_1), it does not only have a r.e. set of axioms (which is trivial), but even a recursive one (Craig's theorem). These sets of axioms can be represented by a Σ^0_1- and a Σ^0_0-formula, resp. An analogous result holds if T is Σ^0_{k+1} ($k \in \mathbb{N}$): in this case, T has an *axiom-set* of complexity Π^0_k (see [3]). Of course, a r.e. theory T may also have an *axiom-set* A_T which has a higher complexity than Σ^0_1. But it seems that such an *axiom-set* must be quite unusual, if not odd.

This suggests that for a representation α of A to be natural, it should not be unnecessarily complex. More formally put, let α be a k-place arithmetical formula and $B \subset \mathbb{N}^k$ be arithmetically definable. In comparing the complexity of these objects, I will take as relevant only what is expressible in the arithmetical hierachy. Thus, with "x" and "y" standing for "α" or "B", I define:

$x <_{Co} y :\Longleftrightarrow$ for some $n \in \mathbb{N}$, x is Σ^0_n and y fails to be Σ^0_n, or x is Δ^0_n and y fails to be Δ^0_n.[23]

My second proposal for (a necessary condition for) "α is a natural representation of A" is now as follows (for arithmetical α representing A):

$$\text{If } \alpha \text{ is a natural representation of } A\text{, then } \alpha \leq_{Co} \overline{A}.^{24}$$

[23]For a theory T, let's explain "α is $\Sigma^{0,T}_n$ / $\Pi^{0,T}_n$ / $\Delta^{0,T}_n$" as "There is a Σ^0_n-formula φ / Π^0_n-formula ψ / Σ^0_n-formula χ_1 and Π^0_n-formula χ_2 such that $T \vdash \alpha \leftrightarrow \varphi$ / $T \vdash \alpha \leftrightarrow \psi$ / $T \vdash \alpha \leftrightarrow \chi_1$ and $T \vdash \alpha \leftrightarrow \chi_2$" and deal similarly with A instead of α. Now, since it may be preferable to regard bounded quantifiers as not introducing further complexity, one should opt rather for the following definition: "$x <_{Co} y :\Longleftrightarrow$ for some $n \in \mathbb{N}$, x is $\Sigma^{0,\mathrm{PA}}_n$ and y fails to be $\Sigma^{0,\mathrm{PA}}_n$, or x is $\Delta^{0,\mathrm{PA}}_n$ and y fails to be $\Delta^{0,\mathrm{PA}}_n$."

[24]"$x \leq_{Co} y$" is defined as "$\neg(y <_{Co} x)$".

3.3. The numeration approach. If τ is a representation of (an *axiom-set* of) T, but T "does not notice it", then "strange" results about T-provable sentences in which τ occurs may not be as strange after all. For example, τ may be chosen in such a way that, in T, "Pr_τ" behaves like "$\mathrm{Pr}_{[E]}$" for some finitely axiomatizable subtheory E of T. If, now, T is reflexive, it may prove "Con_τ"; but this is nothing to wonder about.

The anthropomorphic "T does not notice that τ is a representation of (an *axiom-set*) of T" may be explained in two ways. One is: there are certain conditions which actually hold for formulas containing τ, but it is not T-provable that they do.[25] The second is: τ does not numerate (an *axiom-set*) of T in T. Here,

α numerates A in S $:\Longleftrightarrow$

$$\forall n_1,\dots n_k \in \mathbb{N}\big(\langle n_1,\dots,n_k\rangle \in A \Longleftrightarrow S \vdash \alpha(\overline{n_1},\dots,\overline{n_k})\big).$$

I will only deal with the second idea. This, then, is my third suggestion for (a necessary condition for) "α is a natural representation of A" (for arithmetical α representing A):

If α is a natural representation of A, then α numerates A in $\overline{A}$.

3.4. Comments on the approaches: positive instances and open ends. To begin with, let's have a look at the examples presented in section 2 from the perspective of the definitions given in section 3.

Relative to the *inductive approach*—and apart from the cases where complete theories are taken into account[26]—all representations both of the theories and their *axiom-sets* discussed above are natural. Let me show this for PA_ω and pa_ω:

By the base clauses, "$\mathrm{pa}(x)$" and "$x = \overline{\ulcorner\beta(\overline{n})\urcorner}$" are natural representation of $\mathrm{Ax(PA)}$ and $\{\beta(\overline{n})\}$. Thus, by the induction step for "$\vee$", "$\alpha(x,\overline{n})$" is a natural representation of $\mathrm{Ax(PA)}\cup\{\beta(\overline{n})\}$ for each $n \in \mathbb{N}$. Therefore, for each $n \in \mathbb{N}$, "$\mathrm{Pr}_{\alpha(,\overline{n})}(x)$" is a natural representation of $\{\psi \mid \mathrm{Ax(PA)}\cup\{\beta(\overline{n})\}\vdash\psi\}$. Finally, by the induction step for the universal quantifier, "$\mathrm{pa}_\omega(x)$" is a natural representation of $\{\psi \mid \forall n\ (\mathrm{Ax(PA)}\cup\{\beta(\overline{n})\})\vdash\psi\}$, i.e., of PA_ω.

Although these considerations are an indication that the set of natural representations of arithmetically definable theories is wide enough, there are examples which suggest that this may not be so: there just seems to be too much dependency on notational features built into it. Take, e.g., the formula "$\neg\forall y\neg\,\mathrm{Proof}_{\mathrm{pa}}(y,x)$"; the inductive analysis does not put it into the set of natural representations of PA—it simply does not have the right form. Yet,

[25]This has to do with a certain understanding of "intensional" to be found in proof-theory; see [1].

[26]The inductive definition of "α is a natural representation of A" does not contain any clause dealing with complete theories. At the moment, I do not have strong intuitions on how to include one.

the fact that this formula is provably equivalent to "$\mathrm{Pr}_{\mathrm{pa}}(x)$" may motivate putting the following principle on top of the inductive definition:

- If $\alpha(x)$ is a natural representation of A and $\vdash \forall x\ (\alpha(x) \leftrightarrow \beta(x))$, then $\beta(x)$ is a natural representation of A.

But perhaps logical equivalence is too narrow. In fact, consider the formula pa'_ω defined by

$$\mathrm{pa}'_\omega(x) :\longleftrightarrow \forall y\ (\mathrm{Pr}_{\mathrm{pa}}(\overline{\ulcorner \beta(\dot{y}) \urcorner \dot{\rightarrow} x})).$$

pa'_ω is a representation of PA_ω which seems to be as natural as pa_ω; and, again, it is not declared to be a natural representation of PA_ω by the inductive definition. But now, "$\forall x\ (\mathrm{pa}_\omega(x) \longleftrightarrow \mathrm{pa}'_\omega(x))$" is not provable in first-order predicate logic alone. As a reply, the previous extension of the inductive approach could be strengthened to

- If $\alpha(x)$ is a natural representation of A and $T \vdash \forall x\ (\alpha(x) \leftrightarrow \beta(x))$, then $\beta(x)$ is a natural representation of A.

The problem with this principle, however, is its dependency on a theory T: which one should we choose? T should not be too weak for its purpose of obtaining sufficiently many "new" natural representations of A. From this viewpoint, PA is a good choice.[27] Or should we take some theory α is about for T (i.e., PA_ω in our example)? Furthermore, it would be possible to leave T just as it is—as an additional parameter. Yet, in this case we would get a notion of *T-natural representation* instead of the originally sought *representation*.—Nevertheless, I think that *some* version of the principle should be added to the inductive definition.[28]

In case of the *complexity approach*, I consider the situation as somewhat problematic in the sense that for most of the *theories* dealt with in section 2—i.e., for $S_1 \cap T_1$, $S_{2,k} \cap T_2$, S_3, S_4 and $V(T_5)$—their arithmetical complexity is not known to me.[29] In example 5, we have a theory with an *axiom-set* which is Σ^0_2; but that theory is just $\mathrm{I}\Sigma_1 + \mathrm{Con}_{\tau_5}$, which is r.e.. Thus, the representation "$([\mathrm{i}\sigma_1] + \mathrm{Con}_{\tau_5}) \wedge \tau_5$" is not a natural one.

---

[27]See also [9]; in fact, PA proves "$\forall x\ (\mathrm{pa}_\omega(x) \longleftrightarrow \mathrm{pa}'_\omega(x))$".

[28]As sort of a converse, it may be plausible to conjecture that if Pr_τ and $\mathrm{Pr}_{\tau'}$ are natural representations of one theory T, they are T- (or PA-) provably equivalent. But I think there are counterexamples: take, e.g., $S_{2,0}$. One *axiom-set* of it is $\mathrm{Ax(PA)} \cup \mathrm{Tr}_{\Pi^0_1}$. Since $S_{2,0}$ proves (each instance of the scheme of) *local ω-consistency* for PA [10] (in short: ω-Con[pa]), $\mathrm{Ax(PA)} \cup \mathrm{Tr}_{\Pi^0_1} \cup \omega$-Con[pa] is an *axiom-set* for $S_{2,0}$, too. Now, take a natural representation $\omega Con_{\mathrm{pa}}(x)$ of ω-Con[pa] and the natural representations $\sigma_{2,0}$ and $\sigma_{2,0}(x) \vee \omega Con_{\mathrm{pa}}(x)$ of the *axiom-sets*. It turns out that not even $S_{2,0}$ can prove "$\forall x\ (\mathrm{Pr}_{\sigma_{2,0}}(x) \leftrightarrow \mathrm{Pr}_{\sigma_{2,0}+\omega Con_{\mathrm{pa}}}(x))$"; see [8].

[29]If, e.g., PA_ω is Π^0_2, the complexity condition is satisfied and pa_ω may well be taken to be a natural representation of PA_ω. But if this theory is r.e., it's representation pa_ω is distinguished from Feferman's pa* only in that *intuitively* (what I would still claim to hold), pa_ω is a natural representation of PA_ω, whereas $\mathrm{Pr}_{\mathrm{pa}^*}$ is no natural representation of PA.

Coming to example 7, the theory $V(S_{2,0}) \cap V(T_2)$ is Δ^0_3 (being an intersection of a Δ^0_3- and a Δ^0_2-set). Moreover, it has no lower complexity: for assume it were Σ^0_2; since $V(S_{2,0})$ and $V(T_2)$ are both complete, they are inconsistent with each other; thus, for some sentence φ, $(V(S_{2,0}) \cap V(T_2)) + \varphi = V(S_{2,0})$. Yet, following from the assumption, $(V(S_{2,0}) \cap V(T_2)) + \varphi$ is Σ^0_2, whence $V(S_{2,0})$ would turn out to be Σ^0_2, too; contradiction.[30] But what is the complexity of "$v(\sigma_{2,0}) \wedge v(\tau_2)$"? Notationally, it is Σ^0_3; as evaluated in $\mathcal{N}$, it is Δ^0_3; and relative to PA (see footnote 23), it is Δ^0_3, too. For (as a closer look at the formalized completeness theorem would show) the formula "$v(\sigma_{2,0})$" is Σ^0_3, but $T_5 + \mathrm{Con}_{\sigma_{2,0}}$-provably equivalent to a Π^0_3-formula and equivalent to "$\mathrm{Pr}_{v(\sigma_{2,0})}$". Now, since that theory is a subtheory of PA, we have

$$\begin{aligned} \mathrm{PA} \vdash (v(\sigma_{2,0}) \wedge v(\tau_2))(x) &\longleftrightarrow \mathrm{Pr}_{v(\sigma_{2,0})}(x) \wedge \mathrm{Pr}_{v(\tau_2)}(x) \\ &\longleftrightarrow v(\sigma_{2,0})(x) \wedge \mathrm{Pr}_{v(\tau_2)}(x) \end{aligned}$$

—and by what has just been noted, "$v(\sigma_{2,0})(x) \wedge \mathrm{Pr}_{v(\tau_2)}(x)$" is Δ^0_3 in PA. Thus, "$v(\sigma_{2,0}) \wedge v(\tau_2)$" may be taken to be a natural representation of an *axiom-set* of $V(S_{2,0}) \cap V(T_2)$.

After this discussion of theories which were presented as examples for natural theory-internal consistency proofs, let's consider, as sort of a different test case, Feferman's well known representation pa* (see [1]) of an *axiom-set* of PA more thoroughly. pa* is defined thus:

$$\mathrm{pa}^*(x) :\longleftrightarrow \mathrm{pa}(x) \wedge \forall z \leq x \ \mathrm{Con}_{\mathrm{pa}\lceil z}\,.$$

As far as I know, there is agreement that "$\mathrm{Pr}_{\mathrm{pa}^*}$" is not a natural representation of the theory PA and that "pa*" is not a natural representation of Ax(PA). But it is perhaps not as clear whether "pa*" may not be regarded as a natural representation of the set $\{\psi \mid \mathcal{N} \models \mathrm{pa}^*(\ulcorner\psi\urcorner)\}$.

Relative to the complexity analysis, "pa*" and "$\mathrm{Pr}_{\mathrm{pa}^*}$" are no natural representations of the sets $\{\psi \mid \mathcal{N} \models \mathrm{pa}^*(\ulcorner\psi\urcorner)\}$ and PA: for PA is Σ^0_1, whence "pa*" and "$\mathrm{Pr}_{\mathrm{pa}^*}$" needed to be Σ^0_1, too—but they are (Π^0_1 and) Σ^0_2. Given the inductive approach, we have the same result: pa* is certainly not put into the set of natural representations of an *axiom-set* of PA by a base-clause; moreover, there is also no induction step which is applicable to "pa*" in order to make it a natural representation of an *axiom-set* of PA (for "$\mathrm{Pr}_{\mathrm{pa}^*}$", the reasoning is similar).

Yet, what if "pa*" *is* accepted as a natural representation of $\{\psi \mid \mathcal{N} \models \mathrm{pa}^*(\ulcorner\psi\urcorner)\}$? Besides "refuting" the complexity analysis, this would present a problem for clause (I4) of the inductive approach. That is, the inductive analysis would turn out to be too coarse in the step from an *axiom-set* of

[30]See [9] for a further example where this type of reasoning, employing that the theories intersected are inconsistent with each other, is applied.

a theory to the theory itself. But note now that—since PA *is* consistent—$\{\psi \mid \mathcal{N} \models \mathrm{pa}^*(\ulcorner\psi\urcorner)\}$ simply is the same set as Ax(PA). Thus we would have a violation of the following scheme:

If α is a natural representation of A and $A = B$,
then α is a natural representation of B.

I admit that the rejection of that scheme has to be taken into consideration seriously. Yet, without further arguments, I am reluctant to accept the *intensionality* of "α is a natural representation of ... ". Therefore, at this point, I am satisfied with the result that both "pa*" and "$\mathrm{Pr}_{\mathrm{pa}^*}$" are no natural representations of the sets of sentences represented by these formulas.

Nevertheless, the fact that "pa*" is no natural representation of $\{\psi \mid \mathcal{N} \models \mathrm{pa}^*(\ulcorner\psi\urcorner)\}$ in the sense of the inductive approach points to a weakness of the latter: "α is a natural representation of A" it is not explained for sentences α. And a clause of the form

If $\alpha(x)$ is a natural representation of A and $\mathcal{N} \models \psi$,
then "$\alpha(x) \wedge \psi$" is a natural representation of $\{\psi \mid \mathcal{N} \models \alpha(\ulcorner\psi\urcorner) \wedge \psi\}$

is certainly inacceptable, as long as "natural representation" is not construed intensionally.[31]

§4. **Numerations and closure under necessitation.** So far, it seems that the inductive and the complexity approach support each other quite well: *ideally*—with some exceptions noted—the first one generates enough, but probably too many representations as being natural—whence it is the task of the second analysis to eliminate the inacceptable ones. Continuing along this line of thought, I would propose to explain "α is a natural representation of A" by conjoining its three *explicanta* suggested above. Under this assumption, let me consider the question

Are there examples of theories proving their own consistency naturally?

anew. To answer it, I will investigate more deeply into the numeration approach. In fact, it can be shown for several types of theories T that, if α is a representation of an *axiom-set* of T, it does not numerate that *axiom-set* in T.

To start with, let me present a lemma which is closely related to the main theorem from [9].[32] For its formulation note that, if α numerates A in $\overline{A}$, then

[31] In principle, one could also try to answer these problems by distinguishing between "natural" and "unnatural" *axiom-sets* of theories. In fact, in section 2 I have already called several *axiom-sets* "natural" and, moreover, I think that there are theories, like PA, which are so often presented with just one axiom-system that one may decide to *individuate* them by employing these axiom-systems. But, in general, I have no idea of how to distinguish natural from unnatural *axiom-sets* in a formal way, and I am sceptical if there is any convincing informal motivation for this distinction.

[32] This is: Let $T_1, \ldots, T_k$ be arithmetically definable extensions of PA, having *axiom-sets* with representations $\tau_1, \ldots, \tau_k$ of complexity $\Sigma^0_{i_1}, \ldots, \Sigma^0_{i_k}$ (w.l.o.g. $i_1 \leq i_2 \leq \cdots \leq i_k$), such that PA $\vdash \forall x(\mathrm{Pr}_q(x) \to \mathrm{Pr}_{\tau_i}(x))$ $(1 \leq i \leq k)$ and $\forall\varphi$ (φ is a $\Sigma^0_{i_j}$-sentence $\Longrightarrow$ PA $\vdash \varphi \to \mathrm{Pr}_{\tau_j}(\ulcorner\varphi\urcorner)$); let γ

A is closed under $\Pr_\alpha$-necessitation: i.e.,

$$\forall\psi \in L[A]\,(A \vdash \psi \Longrightarrow A \vdash \Pr_\alpha(\ulcorner\psi\urcorner)).$$

LEMMA 5. If T is a sound extension of S such that $T <_{Co} S$, and if σ is a representation of an *axiom-set* of S, then S is not closed under necessitation with respect to $\Pr_\sigma$.

PROOF. Assume that S is closed under necessitation with respect to $\Pr_\sigma$, and let ψ in $L[S]$ be arbitrary. Then $S \vdash \psi \Longrightarrow S \vdash \Pr_\sigma(\ulcorner\psi\urcorner) \Longrightarrow T \vdash \Pr_\sigma(\ulcorner\psi\urcorner)$. Since T is assumed to be sound, $\mathcal{N} \models \Pr_\sigma(\ulcorner\psi\urcorner)$ follows, whence $S \vdash \psi$.

Now, the function f mapping (the Gödel number of) ψ to (the Gödel number of) "$\Pr_\sigma(\ulcorner\psi\urcorner)$" is recursive and satisfies

$$S \vdash \psi \iff T \vdash f(\psi).$$

for all $\psi \in L[S]$. That is, S is many-one reducible to T. But then, $S \leq_{Co} T$ must hold; contradiction.

In [9], I gave a theorem to the effect that completions of PA *never* numerate themselves with formulas also representing *axiom-sets* of them. Here, I present two approximations to a generalization of this result to the case of intersections of theories such that at least one of them is complete (the second is stated (without proof) as Theorem C in [9]).[33]

THEOREM 1, VERSION 1. Assume that $I\Sigma_1 \subseteq S$, S is complete, T is inconsistent with S and φ is a sentence such that $(S \cap T) + \varphi = S$.[34] If α is a representation of an *axiom-set* A of $S \cap T$ and $S \nvdash \Pr_{\alpha+\varphi}(\ulcorner\bot\urcorner)$, then α does not numerate A in $S \cap T$.

PROOF. Assume that it does, and let ψ be an arbitrary sentence from $L[S]$ such that $S \vdash \psi$; then $(S \cap T) + \varphi \vdash \psi$, whence $(S \cap T) \vdash \varphi \rightarrow \psi$, which implies $(S \cap T) \vdash \Pr_\alpha(\ulcorner\varphi \rightarrow \psi\urcorner)$ by the assumption on α. Therefore, $S \vdash \Pr_{\alpha+\varphi}(\ulcorner\psi\urcorner)$.

This shows that S is closed under $\Pr_{\alpha+\varphi}$-necessitation. Now, let by an application of the diagonalization lemma G be the usual *Gödel*-fixed point of "$\Pr_{\alpha+\varphi}$", i.e.,

$$S \vdash G \longleftrightarrow \neg\Pr_{\alpha+\varphi}(\ulcorner G\urcorner).$$

If $S \vdash G$, then, being closed under $\Pr_{\alpha+\varphi}$-necessitation, S is inconsistent. Thus, $S \vdash \neg G$, since S is complete. But then, by $\Pr_{\alpha+\varphi}$-necessitation, $S \vdash$

be a sentence from $L[PA]$ which is consistent with $(T_1 \cap T_2 \cap \cdots \cap T_k)$; and let $(T_1 \cap T_2 \cap \cdots \cap T_k)$ be closed under $\Pr_{\tau_1 \wedge \cdots \wedge \tau_k}$-necessitation, then: $(T_1 \cap T_2 \cap \cdots \cap T_k) + \gamma \nvdash \mathrm{Con}_{\tau_1 \wedge \cdots \wedge \tau_k + \gamma}$.

[33] In case of the second version, it would be interesting to know for which extensions S of PA and for which representations α of *axiom-sets* of S, $S \vdash \mathrm{Rfn}[\alpha\lceil n]$ holds.

[34] Since T inconsistent with S, such sentences exist.

$\mathrm{Pr}_{\alpha+\varphi}(\ulcorner\neg G\urcorner)$; and because of the fixed point, $S \vdash \mathrm{Pr}_{\alpha+\varphi}(\ulcorner G\urcorner)$. But then $S \vdash \mathrm{Pr}_{\alpha+\varphi}(\ulcorner\bot\urcorner)$; contradiction.

For the second version, I need an additional concept and two technical lemmata. The new notion is *relative interpretability*, for which I write "$\preceq$"; that is, "$S \preceq T$" is short for "S is relatively interpretable in T". Since it is essentially only one, though central, result of the *theory of relative interpretability* that is employed here, I will omit giving the definition of "$\preceq$" and refer the reader to [1], [4], [6] instead.

LEMMA 6. Let $\mathrm{Q} \subseteq S$, α be a numeration of an *axiom-set* A of a theory U in S, and let ψ be a formula; then $(\alpha+\psi)(x)$, i.e., $\alpha(x) \vee x = \overline{\ulcorner\psi\urcorner}$, is a numeration of the *axiom-set* $A \cup \{\psi\}$ of $U + \psi$ in S.

PROOF. Because of the assumption on A and α, and since "$\overline{n} = \overline{\ulcorner\psi\urcorner}$" is a Σ^0_0-sentence we have

$$\forall n \in \mathbb{N}\,(n \in A \iff S \vdash \alpha(\overline{n})) \text{ and } \forall n \in \mathbb{N}\,(n = \ulcorner\psi\urcorner \iff S \vdash \overline{n} = \overline{\ulcorner\psi\urcorner}).$$

This implies for all $n \in \mathbb{N}$

$$n \in A \cup \{\ulcorner\psi\urcorner\} \Longrightarrow S \vdash \alpha(\overline{n}) \vee S \vdash \overline{n} = \overline{\ulcorner\psi\urcorner} \Longrightarrow S \vdash \alpha(\overline{n}) \vee \overline{n} = \overline{\ulcorner\psi\urcorner}.$$

On the other hand, assume (*) $S \vdash \alpha(\overline{n}) \vee \overline{n} = \overline{\ulcorner\psi\urcorner}$, but also $n \notin A \cup \{\ulcorner\psi\urcorner\}$. Then (**) $n \notin A$ and (***) $\mathcal{N} \models \overline{n} \neq \overline{\ulcorner\psi\urcorner}$.

(**) implies $S \nvdash \alpha(\overline{n})$, for α is assumed to be a numeration of A in S, and (***) implies $S \vdash \overline{n} \neq \overline{\ulcorner\psi\urcorner}$. But this contradicts (*).

LEMMA 7. Assume that PA $\subseteq S$; then for all α, A, φ, U:

if α is a representation of A,
A is an *axiom-set* of U such that $\forall n\ (U \vdash \mathrm{Rfn}[\alpha\lceil n])$, and
φ is a sentence such that $U + \varphi$ is a complete theory,

then α is not a numeration of A in U.

PROOF. Applying the diagonalization lemma, let O be the *Orey-sentence* for $\alpha + \varphi$; i.e.,

$$\mathrm{PA} \vdash O \longleftrightarrow \forall x(\mathrm{Con}_{(\alpha+\varphi+O)\lceil x} \to \mathrm{Con}_{(\alpha+\varphi+\neg O)\lceil x}).^{35}$$

Then, since $\forall n\ (U \vdash \mathrm{Rfn}[\alpha\lceil n])$,

$$\forall n\ U + \varphi + O \vdash \mathrm{Con}_{(\alpha+\varphi+O)\lceil n},$$

Therefore,

$$\forall n\ U + \varphi + O \vdash \mathrm{Con}_{(\alpha+\varphi+\neg O)\lceil n}\,. \tag{*}$$

[35] "$\sigma\lceil z(x)$" is defined as "$\sigma(x) \wedge x \leq z$".

Following [1], let $(\alpha + \varphi + \neg O)^*$ be defined by

$$(\alpha + \varphi + \neg O)^*(x) :\longleftrightarrow \left(\alpha(x) \vee x = \ulcorner\varphi\urcorner \vee x = \ulcorner\neg O\urcorner\right) \wedge \forall z \leq x \ \mathrm{Con}_{(\alpha+\varphi+\neg O)\lceil z}.$$

Assume that α *is a numeration of* A *in* U. Then:

(i) $U + \varphi + O \vdash \mathrm{Con}_{(\alpha+\varphi+\neg O)^*}$,

(ii) for all $\psi \in \mathrm{L}[U]$
$((U + \varphi + \neg O) \vdash \psi \Longrightarrow (U + \varphi + O) \vdash \mathrm{Pr}_{(\alpha+\varphi+\neg O)^*}(\ulcorner\psi\urcorner))$.

For (i), see the proof of Theorem 5.9 in [1] (which does not presuppose that $U + \varphi + \neg O$ is consistent or r.e.).

For (ii), one first shows that (**) for all $\psi \in \mathrm{L}[U]$,

$$\psi \in A \cup \{\varphi, \neg O\} \Longrightarrow (U + \varphi + O) \vdash \mathrm{Pr}_{(\alpha+\varphi+\neg O)^*}(\ulcorner\psi\urcorner).$$

Thus, let $\psi \in A \cup \{\varphi, \neg O\}$; then $U \vdash (\alpha+\varphi+\neg O)(\ulcorner\psi\urcorner)$ follows by Lemma 6. With (*), this yields $(U + \varphi + O) \vdash (\alpha + \varphi + \neg O)^*(\ulcorner\psi\urcorner)$, and therefore

$$(U + \varphi + O) \vdash \mathrm{Pr}_{(\alpha+\varphi+\neg O)^*}(\ulcorner\psi\urcorner).$$

Now, (ii) can be proved from (**) by an induction on the length of proof in $(U + \varphi + \neg O)$.

From (i) and (ii) it follows by the proof of Theorem 6.1 from [1] that

$$(U + \varphi + \neg O) \preceq (U + \varphi + O). \tag{1}$$

Furthermore, the fixed-point yields

$$\mathrm{PA} \vdash \neg O \longrightarrow \forall x (\mathrm{Con}_{(\alpha+\varphi+\neg O)\lceil x} \rightarrow \mathrm{Con}_{(\alpha+\varphi+O)\lceil x}).$$

Just like before, this implies

$$\forall n \ U + \varphi + \neg O \vdash \mathrm{Con}_{(\alpha+\varphi+O)\lceil n},$$

and therefore, employing $(\alpha + \varphi + O)^*$, it can be shown that

$$(U + \varphi + O) \preceq (U + \varphi + \neg O). \tag{2}$$

Now $U + \varphi$ is consistent, whence $(U + \varphi + O)$ is consistent or $(U + \varphi + \neg O)$ is consistent. Since relative interpretability implies relative consistency, we have in both cases by (1) and (2) that $(U + \varphi + O)$ and $(U + \varphi + \neg O)$ are consistent; but this contradicts the completeness of $U + \varphi$.

THEOREM 1, VERSION 2. Assume that $\mathrm{PA} \subseteq S$, S is complete and T not a proper subtheory of S; then for all α, A:

(+) if α is a representation of A, A is an *axiom-set* of $S \cap T$ such that $\forall n\, (S \cap T \vdash \mathrm{Rfn}[\alpha\lceil n])$,

then α is not a numeration of A in $S \cap T$.

PROOF. (a) Assume $S = T$; then $S \cap T$ is complete. In this case, the claim follows from Theorem 1 in [9].

(b) Assume $S \neq T$; since S is complete, it is inconsistent with T. Thus there is a sentence φ in L[PA] satisfying $S \vdash \varphi$, $T \vdash \neg\varphi$. Let $U := S \cap T$: then, by (+) A is an *axiom-set* of U with representation α such that $\forall n$ $(U \vdash \mathrm{Rfn}[\alpha\lceil n])$, and such that $U + \varphi (= (S \cap T) + \varphi = S)$ is complete. Thus, Lemma 7 yields that α is not a numeration of A in $S \cap T$.

To close, let me summarize this paper's results for the question whether there are counterexamples to analogues of Gödel's second theorem to non r.e. theories extending (theories like) PA. In section 2, I have presented a wide range of theories T which actually do this; moreover, the arithmetized consistency assertions for these T are formulated with representations of *axiom-sets* of T which are intuitively natural.[36] All of these representations were also natural in the light of the inductive analysis, but only some of them from the perspective of the complexity analysis of "natural representation". Now that we have some results concerning the numeration approach, the picture is as follows:

By Theorem 2 of [9] (in principle), the representations given for $S_1 \cap T_1$ and $S_{2,k} \cap T_2$ do not numerate these theories in themselves. If S_3 is r.e., it's representation pa_ω is too complex; the same type of result holds for S_4. If S_3 fails to be r.e., Lemma 5 applies and pa_ω is no numeration of an *axiom-set* of PA_ω in PA_ω. The representation from example 5 violates the complexity condition, and the one from example 7 is excluded as being natural by Theorem 1, Version 1.[37] Finally, $v(\tau_5)$ (from example 6) is no numeration of $V(T_5)$ in $V(T_5)$ by Theorem 1 from [9].—From the examples discussed in this text, S_4 remains as a serious candidate for a theory which proves its own consistency naturally while fulfilling all adequacy conditions presented here.

REFERENCES

[1] S. FEFERMAN, *Arithmetization of metamathematics in a general setting*, ***Fundamenta Mathematicae***, vol. 49 (1960), pp. 35–92.

[2] ———, *Transfinite recursive progressions of axiomatic theories*, ***The Journal of Symbolic Logic***, vol. 27 (1962), pp. 259–316.

[3] A. GRZEGORCZYK, A. MOSTOWSKI, and C. RYLL-NARDZEWSKI, *The classical and the ω-complete arithmetic*, ***The Journal of Symbolic Logic***, vol. 23 (1958), pp. 188–206.

[4] P. HÁJEK and P. PUDLAK, ***Metamathematics of First-Order Arithmetic***, Springer, Berlin, 1993.

[5] R. KAYE, ***Models of Peano Arithmetic***, Clarendon Press, Oxford, 1991.

---

[36] Probably with the exception of the representation from example 5.

[37] This holds under the assumption that not both $V(S_{2,0}) \vdash \mathrm{Pr}_{(v(\sigma_{2,0}) \wedge v(\tau_2)) + \varphi}(\ulcorner \bot \urcorner)$ and $V(T_2) \vdash \mathrm{Pr}_{(v(\sigma_{2,0}) \wedge v(\tau_2)) + \neg\varphi}(\ulcorner \bot \urcorner)$ (φ being a sentence such that $(V(S_{2,0}) \cap V(T_2)) + \varphi = V(S_{2,0})$ and $(V(S_{2,0}) \cap V(T_2)) + \neg\varphi = V(T_2)$). At present, I cannot exclude this possibility.

[6] P. LINDSTRÖM, ***Aspects of Incompleteness***, Springer, Berlin, 1997.

[7] G. H. MÜLLER, *Über die unendliche Induktion*, ***Infinitistic Methods***, Pergamon Press, Oxford, London, NY, Paris, 1961, pp. 75–95.

[8] K. G. NIEBERGALL, ***Zur Metamathematik nichtaxiomatisierbarer Theorien***, CIS, München, 1996.

[9] ———, *On the limits of Gödel's second incompleteness theorem*, ***Argument und Analyse. Proceedings of GAP4*** (C. U. Moulines and K. G. Niebergall, editors), Mentis, 2002, pp. 109–136.

[10] C. SMORYNSKI, *The incompleteness theorems*, ***Handbook of mathematical logic*** (J. Barwise, editor), North-Holland, 1977, pp. 821–865.

[11] ———, ***Self-reference and Modal Logic***, Springer, Berlin, 1985.

[12] A. TARSKI, A. MOSTOWSKI, and R. M. ROBINSON, ***Undecidable Theories***, North-Holland, Amsterdam, 1953.

[13] A. VISSER, *The formalization of interpretability*, ***Studia Logica***, vol. 50 (1991), pp. 81–105.

[14] D. WILLARD, *Self-verifying axiom systems, the incompleteness theorem and related reflection principles*, ***The Journal of Symbolic Logic***, vol. 66 (2001), pp. 536–596.

SEMINAR FÜR PHILOSOPHIE, LOGIK UND WISSENSCHAFTSTHEORIE
PHILOSOPHIE-DEPARTMENT
LUDWIG-MAXIMILIANS-UNIVERSITÄT MÜNCHEN
LUDWIGSTR. 31, D-80539 MÜNCHEN, GERMANY
E-mail: kgn@lrz.uni-muenchen.de

EFFECTIVE HAUSDORFF DIMENSION

JAN REIMANN AND FRANK STEPHAN†

Abstract. We continue the study of effective Hausdorff dimension as it was initiated by LUTZ. Whereas he uses a generalization of martingales on the Cantor space to introduce this notion we give a characterization in terms of effective s-dimensional Hausdorff measures, similar to the effectivization of Lebesgue measure by MARTIN-LÖF. It turns out that effective Hausdorff dimension allows to classify sequences according to their 'degree' of algorithmic randomness, i.e., their algorithmic density of information. Earlier the works of STAIGER and RYABKO showed a deep connection between Kolmogorov complexity and Hausdorff dimension. We further develop this relationship and use it to give effective versions of some important properties of (classical) Hausdorff dimension. Finally, we determine the effective dimension of some objects arising in the context of computability theory, such as degrees and spans.

§1. **Introduction.** Generally speaking, the concepts of Hausdorff measure and dimension are a generalization of Lebesgue measure theory. In the early 20th century, HAUSDORFF [9] used CARATHEODORY's construction of measures to define a whole family of outer measures. For examining a set of a peculiar topological or geometrical nature Lebesgue measure often is too coarse to investigate the features of the set, so one may 'pick' a measure from this family of outer measures that is suited to study this particular set. This is one reason why Hausdorff measure and dimension became a prominent tool in fractal geometry.

Hausdorff dimension is extensively studied in the context of dynamical systems, too. On the Cantor space, the space of all infinite binary sequences, the interplay between dimension and concepts from dynamical systems such as entropy becomes really close. Results of BESICOVITCH [3] and EGGLESTON [7] early brought forth a correspondence between the Hausdorff dimension of frequency sets (i.e., sets of sequences in which every symbol occurs with a certain frequency) and the entropy of a process creating such sequences as typical outcomes. Besides, under certain conditions the Hausdorff dimension of a set in the Cantor space equals the topological entropy of this set, viewed as a shift space.

An effective version of measure and entropy has been developed since the middle of the 20th century. MARTIN-LÖF [14] effectivized the notion of a

†Supported by the Deutsche Forschungsgemeinschaft (DFG) Heisenberg grant Ste 967/1-1.

Logic Colloquium '01
Edited by M. Baaz, S. Friedman, and J. Krajíček
Lecture Notes in Logic, 20

Lebesgue nullset in order to characterize objects (sequences) that are algorithmically random (namely those that do not have effective measure 0). The theory of *Kolmogorov complexity* (see [11] for a thorough introduction), on the other hand, can be regarded as an effective version of entropy, which makes it possible to determine the entropy of individual objects, just as Martin-Löf randomness can declare individual sequences as random. And indeed, ways emerged how to characterize randomness in terms of Kolmogorov complexity.

Nevertheless, the border between randomness and non-randomness is quite stiff. The theory of Martin-Löf randomness offers no possibility to distinguish between different "degrees" of randomness. It might be that one sequence behaves "close to random" (for instance by satisfying a lot of statistical laws) although it is not, whereas other nonrandom sequences are very regular. In contrast to this, the notion of entropy can be interpreted to describe the degree of randomness of a dynamical system. Kolmogorov complexity, as an algorithmic version of entropy, does the same for finite binary sequences.

By developing an effective version of Hausdorff measure and dimension, one may hope to obtain a tool for classifying sets and sequences according to their degree of algorithmic randomness. Besides, as Hausdorff dimension is also a geometrical notion (it is invariant under bi-Lipschitz transformations), effectivizing the theory might point new techniques coming from fractal geometry for use in algorithmic measure and information theory.

These ideas, of course, are not entirely new. Brudno [4] and White [22] studied the relationship between the entropy of a symbolic dynamical system and the Kolmogorov complexity of an individual trajectory of a system. Ryabko [19], Staiger [20] and Cai and Hartmanis [5] observed close links between the Hausdorff dimension of a set and the Kolmogorov complexity of its members. Lutz [13] was the first to explicitly define an effective notion of Hausdorff dimension. He also introduced a resource bounded version ([12], see also [1]).

In this article, we further develop effective Hausdorff measure and dimension along the line of Lutz [12], [13]. However, we do not follow his martingale approach and develop a Martin-Löf style definition instead. The outline of the paper is as follows. In Section 2 we give a short overview over classical Hausdorff dimension on the Cantor space. In Section 3 we present effective Hausdorff measure and dimension, along with effective versions of some of their important properties. To achieve this, the close connection of effective dimension to Kolmogorov complexity will be used. In Section 4 we determine the effective dimension of some objects arising in the context of computability theory, such as degrees or spans.

Notation. Our notation is fairly standard. $\{0,1\}^\infty$ denotes the *Cantor space*, the set of all infinite binary sequences. The greek letters ζ, η, ξ and ω denote elements of the Cantor space. We write $\xi(n)$, $n \in \mathbb{N}$, to denote the n-th bit of the sequence ξ, and $\xi|_n$ denotes the n-bit initial segment of ξ, that is $\xi|_n = \xi(0)\xi(1)\dots\xi(n-1)$. We identify subsets of the natural numbers with their characteristic sequences, so sometimes we will regard them as elements of the Cantor space, too. Therefore, subsets of the Cantor space are also called *classes*. The lower case roman letters i, j, k, m, n denote natural numbers, whereas v, w, x, y, z usually denote finite binary strings; $l(x)$ denotes the length of a string, so $x = x(0)\dots x(l(x)-1)$, and $\{0,1\}^*$ denotes the set of all finite binary strings. We write $x \prec y$ if x is an initial segment of y, i.e., $l(x) < l(y)$ and $\forall i < l(x)$: $x(i) = y(i)$. $x \prec \xi$, $\xi \in \{0,1\}^\infty$, is defined analogously. We also say that ξ and y *extend* x.

Furthermore, we assume some familiarity with the basic concepts of computability theory such as recursive and recursively enumerable sets, reducibilities, degrees and spans. For an extensive treatment, see the textbook by ODIFREDDI [17].

§2. Classical Hausdorff dimension. The basic idea behind Hausdorff dimension is to generalize the process of measuring a set by approximating (covering) it with sets whose measure is already known. Especially, the size of the sets used in the measurement process will be manipulated by certain transformations, thus making it harder or easier to approximate a set with a covering of small accumulated measure. This gives rise to the notion of *Hausdorff measure*.

We will introduce this notion on the Cantor space $\{0,1\}^\infty$ directly, where we can make use of some of its special features in order to simplify some definitions. For a general treatment of Hausdorff dimension and measure on metric or measure spaces, see the textbooks by EDGAR [6], FALCONER [8] or MATTILA [15].

We endow the Cantor space with the usual metric d for sequences. For two sequences $\xi, \omega \in \{0,1\}^\infty$, define $c(\xi,\omega)$ to be their *maximal common initial segment*. Now let

$$d(\xi,\omega) = 2^{-l(c(\xi,\omega))}.$$

We write $l(\xi,\omega)$ for $l(c(\xi,\omega))$. The *diameter* $d(X)$ of a class $X \subseteq \{0,1\}^\infty$ is defined by

$$d(X) = \sup\{d(\xi,\omega) \colon \xi,\omega \in X\}.$$

The standard topology of $\{0,1\}^\infty$ is generated by the *basic open cylinders*

$$C_w = \{\xi \in \{0,1\}^\infty \colon w \prec \xi\},\ w \in \{0,1\}^*.$$

Assigning each of these cylinders the measure

$$\lambda(C_w) = 2^{-l(w)} = d(C_w)$$

induces the Lebesgue measure λ on $\{0,1\}^\infty$, which is measure-theoretically isomorphic to the standard Lebesgue measure on $\{r : 0 \le r < 1\}$, the unit interval.

Now we can introduce Hausdorff measures. Let $X \subseteq \{0,1\}^\infty$, $\delta > 0$. A (countable) family $\{C_{w_i}\}_{i \in \mathbb{N}}$ is a *δ-cover* of X, if $(\forall i)\ d(C_{w_i}) \le \delta$ and $X \subseteq \bigcup C_{w_i}$. For $s \ge 0$, define

$$\mathcal{H}^s_\delta(X) = \inf\Big\{\sum_{i \in \mathbb{N}} d(C_{w_i})^s : \{C_{w_i}\}_{i\in\mathbb{N}} \text{ is a } \delta\text{-cover of } X\Big\}$$

As δ decreases, there are fewer δ-covers available, hence $\mathcal{H}^s_\delta$ is non-decreasing. Consequently, the value

$$\mathcal{H}^s(X) = \lim_{\delta \to 0} \mathcal{H}^s_\delta(X)$$

is well defined, but may be infinite.

$\mathcal{H}^s(X)$ is called the *s-dimensional Hausdorff measure* of X. It can be shown that $\mathcal{H}^s$ is an outer measure and that the Borel sets of $\{0,1\}^\infty$ are $\mathcal{H}^s$-measurable. Of course, for $s = 1$ we get the usual Lebesgue outer measure.

The outer measures $\mathcal{H}^s$ have an important property.

Proposition 2.1. *Let $X \subseteq \{0,1\}^\infty$. If, for some $s \ge 0$, $\mathcal{H}^s(X) < \infty$, then $\mathcal{H}^t(X) = 0$ for all $t > s$.*

Proof. Let $\mathcal{H}^s(X) < \infty$, $t > s$. If $\{C_{w_i}\}$ is a δ-cover of X, $\delta > 0$, we have

$$\sum_{i\in\mathbb{N}} d(C_{w_i})^t \le \delta^{t-s} \sum_{i\in\mathbb{N}} d(C_{w_i})^s$$

so, taking infima, $\mathcal{H}^t_\delta(X) \le \delta^{t-s}\mathcal{H}^s_\delta(X)$. As $\delta \to 0$, the result follows. ⊣

This means that there can exist only one point $s \ge 0$ where a given class might have finite positive s-dimensional Hausdorff measure. This point is the *Hausdorff dimension* of the class.

Definition 2.2. For a class $X \subseteq \{0,1\}^\infty$, define the *Hausdorff dimension* of X as

$$\dim_{\mathrm{H}}(X) = \inf\{s \ge 0 : \mathcal{H}^s(X) = 0\}.$$

In the following, we list some characteristic properties of Hausdorff dimension.

Refinement of measure 0. If $\lambda(X) \ne 0$ then $\dim_{\mathrm{H}}(X) = 1$. This follows from the fact that $\mathcal{H}^1$ is the Lebesgue outer measure. In particular, $\mathcal{H}^1(\{0,1\}^\infty) = \lambda(\{0,1\}^\infty) = 1$. On the other hand, no $X \subseteq \{0,1\}^\infty$ can have Hausdorff dimension greater than 1, as $\mathcal{H}^s(X) = 0$ for all $s > 1$. (This can be seen by taking the 'trivial' covering consisting for every $\delta = 1/2^n$ of all cylinders C_w with $l(w) = n$.)

Monotonicity. If $X \subseteq Y$ then $\dim_{\mathrm{H}}(X) \le \dim_{\mathrm{H}}(Y)$.

Stability. For $X, Y \subseteq \{0,1\}^\infty$ we have

$$\dim_H(X \cup Y) = \max\{\dim_H(X), \dim_H(Y)\}.$$

This can be generalized to the case of countable unions.

Countable Stability. Let $\{X_i\}_{i\in\mathbb{N}}$ be a countable family of classes. Then

$$\dim_H\Big(\bigcup_{i\in\mathbb{N}} X_i\Big) = \sup_{i\in\mathbb{N}}\{\dim_H(X_i)\}.$$

Geometric Invariance. Let $X \subseteq \{0,1\}^\infty$ and $f: X \to \{0,1\}^\infty$ be a bi-Lipschitz transformation, i.e., there exists $c_1, c_2 > 0$ such that $c_1 d(\xi,\omega) \leq d(f(\xi), f(\omega)) \leq c_2 d(\xi,\omega)$ for all $\xi,\omega \in X$. Then $\dim_H(f(X)) = \dim_H(X)$. This property follows easily from the behaviour of Hausdorff measure/dimension under *Hölder transformations*.

Proposition 2.3. *Let $X \subseteq \{0,1\}^\infty$. Suppose $f: X \to \{0,1\}^\infty$ satisfies a Hölder condition: There exist $c, \alpha > 0$ such that $d(f(\xi), f(\omega)) \leq c d(\xi,\omega)^\alpha$ for all $\xi, \omega \in X$. Then*

$$\dim_H(f(X)) \leq \tfrac{1}{\alpha}\dim_H(X).$$

The primary interest, of course, now lies in determining the Hausdorff dimension of classes and in exposing the structure of sets having non-integral dimension. However, the main obstacle for a direct applicability of Hausdorff dimension in the area of computability theory is the property of countable stability, which easily implies that all countable classes have dimension 0. Therefore, as in the case of effective measure theory, the notion of Hausdorff dimension first has to be effectivized, which is done in the next section.

§3. Effective Hausdorff dimension. The effectivization of Hausdorff dimension resembles the effectivization of Lebesgue measure on $\{0,1\}^\infty$, as it was done by Martin-Löf [14] in order to characterize algorithmic randomness. As measure 0 is defined via coverings the crucial step lies in allowing effective, that is recursively enumerable coverings only. Hausdorff measure is an outer measure, defined via coverings as well. Therefore, the same strategy may be applied in effectivizing Hausdorff dimension, using the following alternative characterization of s-dimensional measure 0 (a special version of Theorem 32 in [18]).

Proposition 3.1. *A class $X \subseteq \{0,1\}^\infty$ has s-dimensional Hausdorff measure 0 if and only if there exists a set $C \subseteq \{0,1\}^*$ such that*

$$(1) \qquad \sum_{w\in C} 2^{-l(w)s} < \infty \quad \textit{and} \quad (\forall \xi \in X)\,(\exists^\infty w \in C)\,[w \prec \xi].$$

Proof. Suppose $\mathcal{H}^s(X) = 0$. Then, for all n, there is a set $C_n \subseteq \{0,1\}^*$ that is a 2^{-n}-covering of X and for which $\sum_{w\in C_n} 2^{-l(w)s} < 2^{-n}$ holds (since

$\mathcal{H}^s_{2^{-n}}(X) = 0$). Set $C = \bigcup C_n$. Then we have $\sum_{w\in C} 2^{-l(w)s} < 1$ (absolute convergence) and each $\xi \in X$ extends infinitely many $w \in C$.

Now suppose that there is a set $C = \{w_0, w_1, w_2, \dots\}$ with $\sum_{w\in C} 2^{-l(w)s} < \infty$ and for all $\xi \in X$ there are infinitely many n such that $w_n \prec \xi$. Let m be a natural number. It suffices to show that for any $\varepsilon > 0$ there exists a 2^{-m}-cover W of X such that $\sum_{w\in W} 2^{-l(w)s} < \varepsilon$.

Choose N so large that

$$\sum_{n\geq N} 2^{-l(w)s} < \varepsilon \text{ and for all } n \geq N,\ l(w_n) \geq m.$$

As each $\xi \in X$ extends infinitely many w_n, for each $\xi \in X$ there exists some $n \geq N$ (in fact, infinitely many n) such that $w_n \prec \xi$. Thus $\{w_n : n \geq N\}$ is a 2^{-m}-cover of X with the desired properties. ⊣

We call a set C that satisfies (1) an *s-covering set*.

It is quite obvious now how to define effective Hausdorff measure 0.

DEFINITION 3.2. Let $s \geq 0$. A class $X \subseteq \{0,1\}^\infty$ has *effective s-dimensional Hausdorff measure* 0, $\mathcal{H}^{1,s}(X) = 0$, if there exists a recursively enumerable set C that satisfies (1).

In this definition it is not presupposed that s is computable. However, as we go on to investigate effective versions of Hausdorff measure and dimension, often it will be necessary that s is in somer form given effectively, too. Therefore (for sake of simplicity), from now on we concentrate on rational s, which is sufficient, as $\mathbb{Q}$ lies dense in $\mathbb{R}$.

Furthermore, note that this approach to effective measure is not the usual one, as it was first proposed by MARTIN-LÖF. Definition 3.2 is a generalization of SOLOVAY's approach to randomness. Nevertheless, the following lemma ensures that, as in the case of effective Lebesgue measure, the two approaches lead to the same concept.

LEMMA 3.3. *Let $X \subseteq \{0,1\}^\infty$, $s \in \mathbb{Q}$. Then $\mathcal{H}^{1,s}(X) = 0$ if and only if there is a recursive sequence $C_1, C_2, C_3, \dots$ of r.e. sets of strings such that for all n*

$$(\forall \xi \in X)\,(\exists w \in C_n)\, w \prec \xi \quad \textit{and} \quad \sum_{w\in C_n} 2^{-l(w)s} \leq 2^{-n}. \tag{2}$$

So, in particular, $\mathcal{H}^{1,1}(X) = 0$ means that X is an effective nullclass in the sense of MARTIN-LÖF. We denote this special case by $\lambda^1(X) = 0$.

It is easy to see that the following 'effective' version of Proposition 2.1 holds.

PROPOSITION 3.4. *Let $X \subseteq \{0,1\}^\infty$. If, for some $s \geq 0$, $\mathcal{H}^{1,s}(X) = 0$, then $\mathcal{H}^t(X) = 0$ for all $t > s$, too.*

The definition of effective Hausdorff dimension follows in a straightforward way.

DEFINITION 3.5. The *effective Hausdorff dimension* of a class $X \subseteq \{0,1\}^\infty$ is defined as

$$\dim_H^1(X) = \inf\{s \geq 0 \colon \mathcal{H}^{1,s}(X) = 0\}.$$

We check some basic properties of effective dimension.

Dimension Conservation. We have $\dim_H^1(\{0,1\}^\infty) = 1$. Obviously, the trivial cover $C = \{0,1\}^*$ is r.e. and $\sum_{w \in \{0,1\}^*} 2^{-l(w)s} < \infty$ if $s > 1$.

Monotonicity. $\dim_H^1(X) \leq \dim_H^1(Y)$ for $X \subseteq Y$ follows just as in the non-effective case.

Refinement of effective Lebesgue measure 0. $\lambda^1(X) \neq 0 \Rightarrow \dim_H^1(X) = 1$. This is another straightforward analogy to the classical case.

Classical and effective Hausdorff dimension. $\dim_H^1(X) \geq \dim_H(X)$ follows directly from the definition.

The other important properties of Hausdorff dimension, countable stability and invariance under bi-Lipschitz transformations, require more careful treatment.

One of the great advantages of effective measure is the existence of a maximal effective nullclass, i.e., one that contains all other effective nullclasses. This has as an easy corollary the closure of effective nullclasses under countable unions. In order to prove a similar property for $\mathcal{H}^{1,s}$-measure, one has to take into account that s is a real number, i.e., might not be computable. However, for rational s, we do not face any problems. So, using Lemma 3.3 and adapting the usual construction of a maximal nullclass, we can prove the following.

PROPOSITION 3.6. *Let $s \geq 0$ be rational. There exists a maximal $\mathcal{H}^{1,s}$-nullclass, i.e., some U_s such that for all $X \subseteq \{0,1\}^\infty$,*

$$\mathcal{H}^{1,s}(X) = 0 \Leftrightarrow X \subseteq U_s.$$

Obviously, Proposition 3.6 yields the countable stability of effective dimension.

Besides, it now makes sense to consider the effective dimension of an individual sequences (viewed as a singleton class), as these have not automatically effective dimension 0. (In the following, we write $\dim_H^1(\xi)$ for $\dim_H^1(\{\xi\})$, $\xi \in \{0,1\}^\infty$.)

An example are the *Martin-Löf random sequences.* These are precisely the sequences not contained in the maximal λ^1-nullclass. Every single Martin-Löf random sequence has effective dimension 1.

Furthermore, the effective dimension of a class can be characterized in terms of the effective dimension of its members. The following theorem has first been proved by LUTZ [13].

THEOREM 3.7 (LUTZ). *For any class $X \subseteq \{0,1\}^\infty$,*

$$\dim_H^1(X) = \sup_{\xi \in X} \dim_H^1(\xi).$$

As regards an effective version of bi-Lipschitz invariance, the problem, of course, lies in the fact that Hölder transformations of the Cantor space need not to be effective, which means that an effective covering of some class does not automatically yield an effective covering of its image. Easy objects could be transformed in to very complicated ones from a computability point of view.

In order to get results of a similar flavour as those of Proposition 2.3 we have to take into account the computational behaviour as well as the geometrical behaviour of a mapping. In this setting, Kolmogorov complexity will prove quite useful.

There exist different versions of Kolmogorov complexity, 'plain' complexity C and prefix complexity K. (Sometimes, these are also referred to as K and H, respectively.) Since we are interested in their asymptotical behaviour, which is equal (i.e., $C(x) \leq K(x) \leq C(x) + 2\log C(x)$), it does not matter which concept we use. As effective Lebesgue measure and randomness can be approached via prefix complexity K, we stick to this notion.

Let us fix a universal prefix-free Turing machine U, that is, a machine for which no two halting inputs are prefixes of one another and that can simulate all other prefix-free Turing machines. We define

$$K(x \mid y) = K_U(x \mid y) = \min\{l(p) \colon U(p, y) = x\}$$

and

$$K(x) = K(x \mid \epsilon)$$

where ϵ denotes the empty string. We do not go into details of this theory here, the interested reader may consult the comprehensive book by Li and Vitanyi [11].

In the context of Hausdorff dimension, an important feature of prefix complexity is its characterization as a universal recursively enumerable semimeasure (see [23]). This characterization might be interpreted in the following way: The (prefix) Kolmogorov complexity of a finite initial segment of a sequence tells us how well this sequence at this level can be described by an recursively enumerable process, or, to put it another way, individuated from the other possible strings of the same length. This leads to a nice characterization of effective dimension. For $\xi \in \{0,1\}^\infty$, define $\underline{K}(\xi) = \liminf_{n\to\infty} K(\xi|_n)/n$.

Theorem 3.8. *For any* $\xi \in \{0,1\}^\infty$,

$$\dim^1_{\mathrm{H}}(\xi) = \underline{K}(\xi). \tag{3}$$

The proof of Theorem 3.8 was first given by Lutz [13] and Mayordomo [16], but much of it is inherent in the works of Ryabko [19], Staiger [20] and Cai and Hartmanis [5].

Lutz showed that $\underline{K}(\xi) \leq \dim^1_{\mathrm{H}}(\xi) \leq \overline{K}(\xi) \stackrel{\mathrm{def}}{=} \limsup K(\xi|_n)/n$, using a martingale characterization of effective dimension (which he introduced as *constructive dimension*). The first author has given an alternative proof using an effectivized *mass distribution principle* (see [8] for this technique).

We give a direct proof of the inequality $\dim^1_{\mathrm{H}}(\xi) \leq \underline{K}(\xi)$.

PROOF (OF THEOREM 3.8). Let $s > \underline{K}(\xi)$ be rational. We show that for such s, $\mathcal{H}^{1,s}(\xi) = 0$. Note that

$$C = \{w \in \{0,1\}^*: K(w) < sl(w)\}$$

is a recursively enumerable set. We claim that it is also an s-covering sequence for ξ, i.e., satisfies (1) for $\{\xi\}$. First, there are infinitely many n such that $\xi|_n \in C$, for $s > \underline{K}(\xi)$. Second, it is easy to see that

$$\sum_{w \in C} 2^{-l(w)s} < \infty,$$

for $w \in C$ implies $K(w) < sl(w)$ and therefore $2^{-K(w)} > 2^{-l(w)s}$, and we know that $\sum_{w \in \{0,1\}^*} 2^{-K(w)}$ converges, due to the Kraft-Chaitin inequality (see [11]). ⊣

REMARK 3.9. It is possible to characterize Hausdorff dimension in terms of martingales. This is the approach chosen by LUTZ[13]. A *martingale* m is a function mapping binary strings to positive rationals that satisfies $m(w0) + m(w1) = 2m(w)$ for all $w \in \{0,1\}^*$. One can compute a weighted sum over all martingales for which the set $\{(w,c): w \in \{0,1\}^*,\ c \in \mathbb{Q},\ m(w) < c\}$ is recursively enumerable and obtain a maximal martingale m_0 (see , for instance, [13] for details). Now the effective Hausdorff dimension of a class C is

$$\dim^1_{\mathrm{H}}(C) = \inf\{s: (\forall \xi \in C)(\exists^\infty n) m_0(\xi|_n) > 2^{n(1-s)}\}.$$

Now we can start investigating the behaviour of effective dimension under transformations. Let $f: \{0,1\}^\infty \to \{0,1\}^\infty$ be a Hölder transformation on the Cantor space, i.e.,

$$\forall \xi, \omega \in \{0,1\}^\infty\ d(f(\xi), f(\omega)) \leq c d(\xi, \omega)^\alpha \tag{4}$$

for some $\alpha, c > 0$. This implies (recall the definition of metric d)

$$l(f(\xi), f(\omega)) \geq \alpha l(\xi, \omega) - \log c.$$

Indeed, one can exploit the special structure of the Cantor space to study a more general class of mappings based on functions operating on strings. Call a mapping $\varphi: \{0,1\}^* \to \{0,1\}^*$ *α-expansive*, $\alpha > 0$, if

(I) $x \preceq y \Rightarrow \varphi(x) \preceq \varphi(y)$ (φ is *monotone*),
(II) $\forall \omega \in \{0,1\}^\infty \liminf_{n\to\infty} l(\varphi(\omega|_n))/n \geq \alpha$.

Obviously, property (II) implies that $\forall \omega \in \{0,1\}^\infty \lim_{n\to\infty} l(\varphi(\omega|_n)) = \infty$, so (I) and (II) induce a mapping $\hat{\varphi}: \{0,1\}^\infty \to \{0,1\}^\infty$. One can show that for each Hölder transformation $f: \{0,1\}^\infty \to \{0,1\}^\infty$ there exists an

α-expansive φ such that $\hat{\varphi} = f$. On the other hand, not every function $\hat{\varphi}$ induced by a mapping φ satisfying (I) and (II) is necessarily Hölder.

Theorem 3.8 enables us to analyze the effective dimension of $\hat{\varphi}(\xi)$ by investigating the asymptotical behaviour of the Kolmogorov complexity $K(\varphi(\xi|_n))$ of its initial segments. Note that, if $\varphi\colon \{0,1\}^* \to \{0,1\}^*$ is recursive, then $K(\varphi(x)) \leq K(x)$. (CAI and HARTMANIS [5] gave a similar result for classical Hausdorff dimension.)

THEOREM 3.10. *Let $\varphi\colon \{0,1\}^* \to \{0,1\}^*$ be a recursive α-expansive mapping. Then it holds that, for any $X \subseteq \{0,1\}^\infty$,*

$$\dim^1_{\mathrm{H}}(\hat{\varphi}(X)) \leq \tfrac{1}{\alpha} \dim^1_{\mathrm{H}}(X). \tag{5}$$

PROOF. It suffices to show that, for any $\xi \in X$,

$$\dim^1_{\mathrm{H}}(\hat{\varphi}(\xi)) \leq \tfrac{1}{\alpha} \dim^1_{\mathrm{H}}(\xi).$$

As $\varphi\colon \{0,1\}^* \to \{0,1\}^*$ is α-expansive, for every $\varepsilon > 0$ there exists some n_0 such that $l(\varphi(\xi|_n)) \geq (\alpha - \varepsilon)n$ for all $n \geq n_0$. Hence, for every $\varepsilon > 0$,

$$\liminf_{n\to\infty} \frac{K(\hat{\varphi}(\xi)|_n)}{n} \leq \liminf_{n\to\infty} \frac{K(\xi|_n)}{(\alpha - \varepsilon)n}.$$

It follows with Theorem 3.8 that

$$\dim^1_{\mathrm{H}}(\hat{\varphi}(\xi)) = \liminf_{n\to\infty} \frac{K(\hat{\varphi}(\xi)|_n)}{n} \leq \liminf_{n\to\infty} \frac{1}{\alpha}\frac{K(\xi|_n)}{n} = \dim^1_{\mathrm{H}}(\xi).$$

This completes the proof. ⊣

Now suppose f satisfies a Hölder condition from below:

$$\exists \alpha, c > 0\, \forall \xi, \omega \in \{0,1\}^\infty\; c\,d(\xi,\omega)^\alpha \leq d(f(\xi), f(\omega)). \tag{6}$$

Note that this implies that f is injective. Suppose further that f has a recursive monotone representation $\varphi\colon \{0,1\}^* \to \{0,1\}^*$ with $\hat{\varphi} = f$ that is injective, too. This means that $K(x \mid \varphi(x), K(\varphi(x))) = O(1)$, since we can always simply scan through all possible strings for a preimage of a given $\varphi(x)$.

Therefore, we get $K(\varphi(x)) = K(x) + O(1)$, and using the lower bound on the length of $\varphi(x)$, we see that, for any ξ,

$$\underline{K}(\varphi(\xi|_n)) \geq \tfrac{1}{\alpha}\underline{K}(\xi|_n).$$

Combining this observation with Theorem 3.10, the invariance of effective dimension under recursive bi-Lipschitz mappings follows.

COROLLARY 3.11. *Let $f\colon \{0,1\}^\infty \to \{0,1\}^\infty$ be a bi-Lipschitz transformation such that there exists a recursive, 1-expansive, injective mapping $\varphi\colon \{0,1\}^* \to \{0,1\}^*$ with $\hat{\varphi} = f$. Then, for any $X \subseteq \{0,1\}^\infty$,*

$$\dim^1_{\mathrm{H}}(f(X)) = \dim^1_{\mathrm{H}}(X).$$

After having introduced effective dimension, we can now start to determine the dimension of some classes occurring in computability theory such as degrees, spans, etc., especially of those known to have effective measure 0. In particular, we may try to answer the following questions:

- What are interesting examples of classes of non-integral dimension such as the *middle-third Cantor set*
$$C_{1/3} = \left\{ \sum_{i=1}^{\infty} \zeta(i)3^{-i} : \zeta \in \{0,2\}^{\infty} \right\}$$
(which has dimension $\log 2/\log 3$) in the classical setting?
- Is there a class of effective measure 0 but effective dimension 1?
- Which (nontrivial) classes have effective dimension 0?

§4. **Some examples of effective dimension.** We now present some results on effective Hausdorff dimension. We consider classes arising in the context of computability theory.

4.1. A small class of maximal dimension. First, we further employ the Kolmogorov complexity characterization of effective dimension (Theorem 3.8) to get invariance results for other, not necessarily recursive mappings. This will allow us to show that the effective dimension of a degree of a set and its lower span coincide.

Note that the *symmetry of algorithmic information* for prefix complexity says that there exists some constant c such that for any $x, y \in \{0,1\}^*$

$$K(x,y) = K(x) + K(y \mid x, K(x)) + c. \tag{7}$$

A proof of this identity can be found in [11]. Rewriting this in two different ways and replacing y by $\varphi(x)$, where φ maps strings to strings, we get

$$K(\varphi(x)) = K(x) + K(\varphi(x) \mid x, K(x)) - K(x \mid \varphi(x), K(\varphi(x))) + c. \tag{8}$$

Next, we have to generalize the notion of a *join* of two sets (sequences). Let Z be an infinite recursive subset of $\mathbb{N}$ with infinite complement. We shall call Z simply a *recursive partition*. Define the *Z-join* of two sequences $\xi, \omega \in \{0,1\}^{\infty}$, $\xi \oplus_Z \omega$, to be the unique sequence ζ which satisfies

$$\zeta|_Z = \xi \quad \text{and} \quad \zeta|_{\overline{Z}} = \omega.$$

If $\lim_{n\to\infty} \frac{1}{n}|Z \cap \{0,\ldots,n-1\}|$ exists, call this number the *density* δ of Z and consider, for given ω, the "insertion mapping" $g\colon \xi \to \xi \oplus_Z \omega$. This mapping g satisfies a Hölder condition: Define $\varphi\colon \{0,1\}^* \to \{0,1\}^*$ by

$$\varphi(\xi|_n) = (\xi \oplus_Z \omega)|_{\pi_Z(n)},$$

where $\pi_Z(n)$ denotes the nth element of Z. Note that for each $\varepsilon > 0$ there exist constants $c_1, c_2 > 0$ such that

$$\tfrac{n}{\delta+\varepsilon} - c_1 \le \pi_Z(n) \le \tfrac{n}{\delta-\varepsilon} + c_1$$

for all n. Furthermore, $\hat{\varphi} = g$ and both φ and g are injective. Now we can use (8) to determine the complexity of φ. Obviously, since φ is injective and x is contained in $\varphi(x)$, i.e., can be identified in $\varphi(x)$ since Z is recursive, $K(x \mid \varphi(x), K(\varphi(x)))$ is bounded by a constant. On the other hand, in order to compute $\varphi(x)$ given x, it suffices to specify the bits that are "inserted" into x (at the $\overline{Z}$-positions). These are at most $\pi_Z(n) - n$ bits, and it follows that, for every $\varepsilon > 0$, $\varepsilon < \delta$,

$$K(\varphi(x) \mid x, K(x)) \leq l(x)(\tfrac{1}{\delta - \epsilon} - 1) + O(1).$$

Hence, if $\delta = 1$, we can conclude that, for every $X \subseteq \{0,1\}^\infty$,

$$\dim^1_{\mathrm{H}}(X) = \dim^1_{\mathrm{H}}(\hat{\varphi}(X)).$$

We now use this invariance property to code information into classes of sequences without changing the effective dimension of these classes.

Let r be a standard reducibility in computability theory, i.e., one of 1, m, $btt(1)$, btt, tt, wtt, T. For a set $A \subseteq \mathbb{N}$, let $\deg_r(A)$ and $\mathrm{span}_r(A)$ be its r-degree and lower r-span, respectively.

THEOREM 4.1. *For any set $A \subseteq \mathbb{N}$, it holds that*

$$\dim^1_{\mathrm{H}}(\deg_r(A)) = \dim^1_{\mathrm{H}}(\mathrm{span}_r(A)).$$

PROOF. Let $A \subseteq \mathbb{N}$. Obviously, $\dim^1_{\mathrm{H}}(\deg_r(A)) \leq \dim^1_{\mathrm{H}}(\mathrm{span}_r(A))$. To obtain the reverse inequality, choose some recursive partition Z with

$$\lim_{n\to\infty} \tfrac{1}{n}|Z \cap \{0,\dots,n-1\}| = 1.$$

Define the mapping $f\colon \{0,1\}^\infty \to \{0,1\}^\infty$ by $f(\xi) = \xi \oplus_Z A$. (As already mentioned, we identify subsets of the natural numbers with their characteristic sequences.) Then, obviously, $f(\mathrm{span}_r(A)) \subseteq \deg_r(A)$, and by the preceding remarks,

$$\dim^1_{\mathrm{H}}(\mathrm{span}_r(A)) = \dim^1_{\mathrm{H}}(f(\mathrm{span}_r(A))) \leq \dim^1_{\mathrm{H}}(\deg_r(A)).$$

Equality follows from $\deg_r(A) \subseteq \mathrm{span}_r(A)$. ⊣

Theorem 4.1 allows us to exhibit an interesting example of an effective nullclass that nevertheless has effective dimension 1.

It is a known fact that the lower tt-span of the halting problem K contains a Martin-Löf random sequence, hence it does not have effective measure 0 $\lambda^1(\mathrm{span}_{tt}(K)) \neq 0$. On the other hand, K does not tt-reduce to a Martin-Löf random sequence, which implies $\lambda^1(\deg_{tt}(K)) = 0$. (For details on this results refer to TERWIJN [21].) Therefore, we have the following corollary.

COROLLARY 4.2. $\lambda^1(\deg_{tt}(K)) = 0$ *but* $\dim^1_{\mathrm{H}}(\deg_{tt}(K)) = 1$.

Note that Corollary 4.2 holds for truth-table reducibility only. The lower btt-span of K is known to effective measure 0, and we can strengthen this result.

THEOREM 4.3. $\dim^1_H(\deg_{btt}(K)) = 0$ *and* $\dim^1_H(\mathrm{span}_{btt}(K)) = 0$.

PROOF. It is sufficient to prove $\dim^1_H(\mathrm{span}_{btt}(K)) = 0$ as the degree is a subset of the lower span. Using the stability theorem, let ξ be any set btt-reducible to K. There is a constant c such that every $\xi(n)$ depends only on c places of K, and these places can be computed without querying K. Therefore, one can compute for given n the up to cn places which are necessary to compute $\xi|_n$ from a code for n. Furthermore, one can enumerate K at the queried places until all elements have shown up provided one knows how many will eventually do so. These two numbers can be codes with $(2c+2)\log(n)$ many bits and so one has that the overall number of bits needed to compute $\xi|_n$ is in $O(\log(n))$. It follows that $\dim^1_H(\xi) = 0$ and thus $\dim^1_H(\mathrm{span}_{btt}(K)) = 0$. ⊣

4.2. Another example of a class of dimension 0. KOLMOGOROV has made an easy but fundamental observation.

THEOREM 4.4. *Let* $A \subseteq \mathbb{N} \times \{0,1\}^*$ *be recursively enumerable. Suppose* $A_m = \{x : (m,x) \in A\}$ *is finite. Then, for some constant* c *and for all* $x \in A_m$,

$$C(x \mid m) \leq \log |A_m| + c.$$

Here C is the 'plain' (non-prefix) version of Kolmogorov complexity. As plain complexity and prefix complexity are asymptotically equal, we can immediately deduce the following *covering principle*:

PROPOSITION 4.5. *Suppose for some* $\xi \in \{0,1\}^\infty$ *there is an r.e. set* $A \subseteq \{0,1\}^*$ *such that for infinitely many* n *it holds that* $\xi|_n \in A \cap \{0,1\}^n$. *Then*

$$\dim^1_H(\xi) \leq \liminf_{n\to\infty} \frac{\log|A_n|}{n}.$$

(Compare this with Theorem 3.8.) We give an application of this principle:

A set $A \subseteq \mathbb{N}$ is *semirecursive*, if there is a recursive function $f : \mathbb{N} \times \mathbb{N} \to \mathbb{N}$ such that

(I) $f(m,n) \in \{m,n\}$ for all $m, n \in \mathbb{N}$,
(II) $m \in A \vee n \in A$ implies $f(m,n) \in A$.

Semirecursive sets were introduced by JOCKUSCH [10] and have been used to give a structural solution to *Post's Problem* (for details refer to ODIFREDDI's monograph [17]). Note that there are semirecursive sets which are not recursively enumerable (and hence not recursive). In fact, every tt-degree contains a semirecursive set (JOCKUSCH). Nevertheless, from the point of view of effective dimension, semirecursive sets are not very complex.

THEOREM 4.6. *The class of all semirecursive sets has effective dimension* 0.

PROOF. It suffices to show that every semirecursive set (that is, its characteristic sequence) has effective dimension 0.

Let ξ be the characteristic sequence of a semirecursive set. Note that, for any two numbers m, n, we can recursively exclude one of four possible values of the two-bit string $\xi(m)\xi(n)$. This observation was used by BEIGEL, KUMMER and STEPHAN [2], who studied semirecursive sets in the broader context of *frequency computability*, to prove the following lemma.

LEMMA 4.7. *If $\xi \in \{0,1\}^\infty$ is the characteristic sequence of a semirecursive set, then there exists a recursive set $C \subseteq \{0,1\}^*$ such that, for any n, $|C \cap \{0,1\}^n| \leq n+1$ and $\xi|_n \in C \cap \{0,1\}^n$.*

The proof of the lemma, which occurs in [2] in a much more general form, uses the SAUER-PERLES-SHELAH-Lemma from extremal combinatorics. Lemma 4.7 allows us to immediately deduce that $\dim^1_{\mathrm{H}}(\xi) = 0$, because of Proposition 4.5. ⊣

4.3. Lower spans of non-integral dimension. Of course, when developing a notion of effective Hausdorff dimension, a central interest lies in exhibiting classes of non-integral dimension. In the Cantor space, the most important example was given by BESICOVITCH [3] and EGGLESTON [7], who showed that the class of sequences in which ones occur with a given limiting frequency β has Hausdorff dimension $H(\beta)$, where H denotes the binary entropy function. LUTZ [13] has shown that these results carry over to the effective case. In this paper we further extend this to a special family of lower spans. For this purpose we have to introduce (generalized) Bernoulli measures on the Cantor space.

Let $p_0, p_1, p_2, \ldots$ be a sequence of real numbers such that $0 \leq p_i \leq 1$ for all i. This sequence induces a measure on $\{0,1\}^\infty$ in the following way: Let C_w, $|w| = n$, be an open cylinder. For $i \geq 0$, let $\mu_i(1) = p_i$, $\mu_i(0) = 1 - p_i$ and

$$\mu(C_w) = \prod_{i=0}^{n-1} \mu_i(w(i)). \tag{9}$$

The resulting measure (as can be easily checked) is a probability measure on $\{0,1\}^\infty$, called a *generalized Bernoulli measure*. It is a *Bernoulli measure*, if all the p_i are identical, i.e., $p_i = p$ for all i.

The (generalized) Bernoulli measures can be seen as an infinite product of measures on $\{0,1\}$, endowed with the obvious σ-algebra and a probability measure given by μ_i as defined above. For $p_i = 1/2$ for all $i \geq 0$ we get the usual Lebesgue measure λ on $\{0,1\}^\infty$.

Bernoulli measures can be effectivized in the usual way, even as Hausdorff measures.

DEFINITION 4.8. Let μ be a generalized Bernoulli measure and $s \geq 0$. A class $X \subseteq \{0,1\}^\infty$ has *effective s-dimensional μ-measure* 0, $\mu^{1,s}(X) = 0$, if there exists a recursively enumerable set C such that

$$\sum_{w \in C} \mu(C_w)^s < \infty \quad \text{and} \quad (\forall \xi \in X)\,(\exists^\infty w \in C)\,[w \prec \xi]. \tag{10}$$

It is possible to define a concept of effective μ-dimension as well (the first author has done so), but here we restrict our attention to *μ-random sequences.*

DEFINITION 4.9. Let μ be a generalized Bernoulli measure. A sequence $\xi \in \{0,1\}^\infty$ is *1-μ-random*, if it does not have effective 1-dimensional μ-measure 0, i.e., $\mu^{1,1}(\{\xi\}) \neq 0$.

1-μ-random sequences can be regarded as typical outcomes of an infinite sequence of coin tosses, where in each round the bias of the coin is chosen according to μ_i. The notion of a 1-μ-random sequence with respect to a generalized Bernoulli measure was already considered by MARTIN-LÖF, when he gave his definition of algorithmic randomness.

Now we turn to a special class of generalized Bernoulli measures. Let μ be such a measure with bias sequence $p_0, p_1, p_2, \ldots$ converging to a real number $\beta \in (0,1)$. The law of large numbers implies that μ-almost every sequence in $\{0,1\}^\infty$ must have limiting frequency β, that is

$$\lim_{n\to\infty} \frac{\sum_{i=0}^{n-1} \xi(i)}{n} = \beta \ \mu\text{-almost surely.} \tag{11}$$

In particular, every 1-μ-random sequence satisfies (11).

Building on results by EGGLESTON [7], LUTZ [13] showed the following:

THEOREM 4.10. *Let μ be a generalized Bernoulli measure with computable bias sequence $p_0, p_1, p_2, \ldots$ of rational numbers converging to a real number $\beta \in (0,1)$. Then, for every 1-μ-random sequence ξ,*

$$\dim^1_{\mathrm{H}}(\xi) = H(\beta),$$

where $H(\beta) = -[\beta \log \beta + (1-\beta)\log(1-\beta)]$ is the binary entropy function.

Theorem 4.10 is also a straightforward consequence of the characterization of 1-μ-random sequences via Kolmogorov complexity (using sequential tests, see [11]) due to LEVIN and SCHNORR, together with the law of large numbers. Employing this characterization, one can easily extend this result as follows (see [1] for the somewhat more complicated resource-bounded version).

THEOREM 4.11. *Let μ be as in Theorem 4.10. Then, for every 1-μ-random sequence ξ,*

$$\dim^1_{\mathrm{H}}(\mathrm{span}_m(\xi)) = H(\beta).$$

Note that Theorem 4.11 implies the existence of lower m-spans of arbitrary Δ^0_2-computable dimension (since H is a continuous surjective function on $[0,1]$).

Acknowledgements. The authors would like to thank an anonymous referee for many helpful comments and suggestions.

REFERENCES

[1] KLAUS AMBOS-SPIES, WOLFGANG MERKLE, JAN REIMANN, and FRANK STEPHAN, *Hausdorff dimension in exponential time*, ***Proceedings of the Sixteenth Annual IEEE Conference on Computational Complexity***, IEEE Computer Society, 2001, pp. 210–217.

[2] RICHARD BEIGEL, MARTIN KUMMER, and FRANK STEPHAN, *Quantifying the amount of verboseness*, ***Information and Computation***, vol. 118 (1995), no. 1, pp. 73–90.

[3] ABRAM S. BESICOVITCH, *On the sum of digits of real numbers represented in the dyadic system*, ***Mathematische Annalen***, vol. 110 (1934), pp. 321–330.

[4] A. A. BRUDNO, *Entropy and the complexity of the trajectories of a dynamical system*, ***Transactions of the Moscow Mathematical Society***, vol. 1983 (1983), no. 2, pp. 127–151.

[5] JIN-YI CAI and JURIS HARTMANIS, *On Hausdorff and topological dimensions of the Kolmogorov complexity of the real line*, ***Journal of Computer and System Sciences***, vol. 49 (1994), no. 3, pp. 605–619.

[6] GERALD A. EDGAR, ***Measure, topology, and fractal geometry***, Springer-Verlag, 1990.

[7] H. G. EGGLESTON, *Sets of fractional dimensions which occur in some problems of number theory*, ***Proceedings of the London Mathematical Society, II. Ser.***, vol. 54 (1951), pp. 42–93.

[8] KENNETH FALCONER, ***Fractal geometry***, John Wiley & Sons Ltd., 1990.

[9] FELIX HAUSDORFF, *Dimension und äußeres Maß*, ***Mathematische Annalen***, vol. 79 (1919), pp. 157–179.

[10] CARL JOCKUSCH, *Semirecursive sets and positive reducibility*, ***Transactions of the American Mathematical Society***, vol. 131 (1968), pp. 420–436.

[11] MING LI and PAUL VITÁNYI, ***An introduction to Kolmogorov complexity and its applications***, 2nd ed., Springer-Verlag, 1997.

[12] JACK H. LUTZ, *Dimension in complexity classes*, ***Proceedings of the Fifteenth Annual IEEE Conference on Computational Complexity***, IEEE Computer Society, 2000, pp. 158–169.

[13] ———, *Gales and the constructive dimension of individual sequences*, ***Proceedings of the Twenty-Seventh International Colloquium on Automata, Languages, and Programming*** (Ugo Montanari, José D. P. Rolim, and Emo Welzl, editors), Springer-Verlag, 2000, pp. 902–913.

[14] PER MARTIN-LÖF, *The definition of random sequences*, ***Information and Control***, vol. 9 (1967), pp. 602–619.

[15] PERTTI MATTILA, ***Geometry of sets and measures in Euclidean spaces***, Cambridge University Press, 1995.

[16] ELVIRA MAYORDOMO, *A Kolmogorov complexity characterization of constructive Hausdorff dimension*, ***Technical Report TR01-059***, Electronic Colloquium on Computational Complexity, 2001.

[17] PIERGIORGIO ODIFREDDI, ***Classical Recursion Theory***, North-Holland, 1989.

[18] C. A. ROGERS, ***Hausdorff measures***, Cambridge University Press, 1970.

[19] BORIS YA. RYABKO, *The complexity and effectiveness of prediction algorithms*, ***Journal of Complexity***, vol. 10 (1994), no. 3, pp. 281–295.

[20] LUDWIG STAIGER, *Kolmogorov complexity and Hausdorff dimension*, ***Information and Computation***, vol. 103 (1993), no. 2, pp. 159–194.

[21] SEBASTIAAN A. TERWIJN, *Computability and measure*, ***Ph.D. thesis***, Universiteit van Amsterdam, 1998.

[22] HOMER S. WHITE, *Algorithmic complexity of points in dynamical systems*, ***Ergodic Theory and Dynamical Systems***, vol. 13 (1992), no. 4, pp. 807–830.

[23] ALEXANDER K. ZVONKIN and LEONID A. LEVIN, *The complexity of finite objects and the development of the concepts of information and randomness by means of the theory of algorithms*, ***Russian Mathematical Surveys***, vol. 25 (1970), no. 6, pp. 83–124.

MATHEMATISCHES INSTITUT
RUPRECHT-KARLS-UNIVERSITÄT HEIDELBERG
HEIDELBERG, GERMANY
E-mail: reimann@math.uni-heidelberg.de

SCHOOL OF COMPUTING AND DEPARTMENT OF MATHEMATICS
NATIONAL UNIVERSITY OF SINGAPORE
3 SCIENCE DRIVE 2
SINGAPORE 117543, SINGAPORE
E-mail: fstephan@comp.nus.edu.sg

MUTUAL STATIONARITY IN THE CORE MODEL

RALF SCHINDLER

Abstract. Foreman and Magidor in [3] study mutual stationarity in Gödel's constructible universe L. We shall extend their analysis to the core model. We include a discussion of what it is that turns an extender model into a core model.

The present paper links mutual stationarity with core model theory. The concept of mutual stationarity found a powerful application in recent work of Foreman and Magidor (cf. [3]). The paper [3] also studies mutual stationarity in Gödel's universe L. We shall extend Foreman and Magidor's analysis to higher core models.

Section 1 will recall basic information about mutually stationary sequences. Section 2 contains a general discussion of the concept of a core model. We want to emphasize that only core models, rather than arbitrary extender models, will be amenable to our analysis; we therefore think that this paper provides a reasonable place for a discussion of what it is that turns an extender model into a core model. Section 2 is thus of independent interest. Section 3 will prove a new condensation result for our core model below $0^\P$ (cf. [14]), and in section 4 we shall prove or main result, Theorem 4.6. Section 5 contains a list of open problems.

The author would like to thank Ernest Schimmerling, Hugh Woodin, and Martin Zeman for their very helpful comments on this paper.

§1. Mutual stationarity. Let A be a non-empty set of regular uncountable cardinals, and let $(S_\kappa : \kappa \in A)$ be such that $S_\kappa \subset \kappa$ for all $\kappa \in A$. Recall that $(S_\kappa : \kappa \in A)$ is called *mutually stationary* [3, Definition 22] if and only if for each large enough regular cardinal θ and for each model $\mathfrak{A}$ with universe H_θ there is some $X \prec \mathfrak{A}$ such that for all $\kappa \in X \cap A$, $sup(X \cap \kappa) \in S_\kappa$. We refer the reader to [3] (in particular, to [3, Section 7]) for key results and background information on mutual stationarity.

For a regular cardinal θ let $cf(\theta)$ denote the class of all ordinals of cofinality θ. Foreman and Magidor proved (cf. [3, Theorem 7]) that if A is a non-empty set of regular uncountable cardinals, and if $(S_\kappa : \kappa \in A)$ is such that for each $\kappa \in A$, $S_\kappa \subset \kappa \cap cf(\omega)$ and S_κ is stationary in κ, then $(S_\kappa : \kappa \in A)$

Logic Colloquium '01
Edited by M. Baaz, S. Friedman, and J. Krajíček
Lecture Notes in Logic, 20

is mutually stationary. This immediately gives that if λ is a singular cardinal with cofinality θ then the non-stationary ideal on $\mathcal{P}_{\omega_1}(\lambda)$ is never λ^θ-saturated (cf. [3, Corollary 8]).[1] Despite of this application the authors of [3] regard the concept of mutual stationarity to be also of independent interest; it is this approach that will be adopted here. We shall link mutual stationarity with core model theory.

It is known that, provided $2^{\aleph_0} < \aleph_\omega$, $\aleph_\omega$ is Jonsson if and only if there is a sequence $(k_n : n < \omega)$ of integers tending to ω such that $(\omega_n \cap cf(\omega_{k_n}) : n < \omega)$ is mutually stationary. ($\Leftarrow$ is trivial, $\Rightarrow$ is due to Silver.) We shall therefore focus on ordinals of bounded cofinality; inspired by [3, §7.2] we'll in fact focus on ordinals of a fixed uncountable cofinality.

It is an open problem to decide whether there is a model of set theory in which $(S_n : n < \omega)$ must be mutually stationary provided each individual $S_n \subset \aleph_n$ is stationary and consists of points of an uncountable cofinality fixed in advance (cf. the last paragraph of [3, §7.1]). On the other hand, Foreman and Magidor have shown that in Gödel's constructible universe L for each $k < \omega$ there is some $(S_n : n < \omega)$ such that each $S_n \subset \aleph_n \cap cf(\omega_k)$ is stationary, but $(S_n : n < \omega)$ is not mutually stationary (cf. [3, Theorems 24 and 27]). The purpose of this paper is to extend their result [3, Theorem 27] to higher core models.

We shall need the following lemma, which is essentially due to Baumgartner [2].

LEMMA 1.1. *Let $m < \omega$, and let $A_0, A_1, \ldots, A_m$ be non-empty sets of successor cardinals such that $sup(A_l) < MIN(A_{l+1})$ for all $l < m$. Let, for each $l \leq m$, $(S_\kappa : \kappa \in A_l)$ be mutually stationary and such that for all $\kappa \in A_l$ and for all $\alpha \in S_\kappa$, $cf(\alpha) < MIN(A_0)$. Then $(S_\kappa : \kappa \in \bigcup_{l \leq m} A_l)$ is mutually stationary as well.*

PROOF. It certainly suffices to prove 1.1 for $m = 1$. Let ν be the cardinal predecessor of $MIN(A_0)$. Let $\mathfrak{A}$ be a model expanding $(H_\theta; \in)$ for some large regular θ. As $(S_\kappa : \kappa \in A_1)$ is mutually stationary we may pick some $X \prec \mathfrak{A}$ such that $sup(A_0) \subset X$ and for all $\kappa \in X \cap A_1$, $sup(\kappa \cap X) \in S_\kappa$. Pick $F : (X \cap A_1) \times \nu \to X$ such that for all $\kappa \in X \cap A_1$, $F(\kappa, -)$ is cofinal in $sup(\kappa \cap X)$. Expand X by F to get (X, F). As $(S_\kappa : \kappa \in A_0)$ is mutually stationary we may pick some $Y \prec (X, F)$ such that $\nu \subset Y$ and for all $\kappa \in Y \cap A_0$, $sup(\kappa \cap Y) \in S_\kappa$. However, due to the presence of F, we'll also have that for all $\kappa \in Y \cap A_1$, $sup(\kappa \cap Y) = sup(\kappa \cap X) \in S_\kappa$. As $Y \prec \mathfrak{A}$, we are done. □

§2. Extender models and the core model. This section intends to discuss and stress the difference between *extender models* and *core models*, a difference

[1]This in turn subsumes under the striking more general result of [3] that the non-stationary ideal on $\mathcal{P}_\kappa(\lambda)$ is never λ^+-staturated unless $\kappa = \lambda = \omega_1$.

which is crucial in this paper and elsewhere. We shall also state a technical lemma which will be used later on.

An *extender model*[2] is a premouse $\mathcal{M}$ which is a proper class (equivalently, such that $OR \subset \mathcal{M}$). Unfortunately, the literature knows a handful of formal definitions of what a premouse is. Currently, the two most common ones are [10, Definition 3.5.1] (US style "Mitchell-Steel premice"; see also [16]) and [4, §4 p. 2] (European style "Friedman-Jensen premice"). The conceptual differences arise from how the respective authors choose to index the extenders on the extender sequences of premice.[3] The current paper is based on Friedman-Jensen premice, so that we here take [4, §4 p. 2] to officially define what a premouse is. In fact, the construction to follow will be based on [14], which in turn builds upon [4]; the reader may find a publisized definition of a (Friedman-Jensen) premouse as well as many other background informations in [14, §1].

Roughly speaking, a premouse is a transitive model of the form

$$J_\alpha[\vec{E}] = (J_\alpha[\vec{E}]; \in, \vec{E}, E_\alpha)$$

for some $\alpha \leq OR$ where $\vec{E}^\frown E_\alpha$ is a coherent sequence of extenders witnessing that certain ordinals are "measurable to a certain extent." We call a premouse $\mathcal{M}$ *weakly iterable* (cf. [4, §11 p. 5]) if $\mathcal{P}$ is $\omega_1 + 1$ iterable for all countable transitive $\mathcal{P}$ which (sufficiently) elementarily embed into $\mathcal{M}$. It is an empirical fact that all the extender models constructed so far are weakly iterable. Examples include the models built in [10, §11] or (more recently) [1] and [11].

For premice transcending a fixed smallness condition weak iterability no longer implies full iterability. We call a premouse $\mathcal{M}$ *fully iterable*, or just *iterable* (cf. [4, §4 p. 26]), if $\mathcal{M}$ is α iterable for every $\alpha \in OR$ (we urge the reader to consult [14, §1] for an exact statement). The iterability condition imposed on extender models is one of the key ingredients which creates a core model.

The ω_1+1 iterability of a given premouse $\mathcal{M}$ allows to prove facts which can be expressed by first order statements over $\mathcal{M}$. (Prototype examples include the solidity and universality of the standard parameter.) On the other hand, in order to compare two premice $\mathcal{M}$ and $\mathcal{N}$ one needs to have (in general) that both $\mathcal{M}$ and $\mathcal{N}$ are $(max\{Card(\mathcal{M}), Card(\mathcal{N})\})^+ + 1$ iterable. In order to prove significant covering properties of a given extender model $L[E]$ at arbitrary ordinals we seem to need the full iterability of $L[E]$.

---

[2]The term "extender model" has been suggested by S. Friedman and will be adopted here. Other terms in use to denote an object we shall refer to as an extender model are "fine structural inner model," "Mitchell-Steel model," "$L[E]$-model," and "weasel." Some authors even use "core model" to refer to an extender model, a habit which might be found at least a bit confusing in the light of the discussion to follow.

[3]We here suppress that there are good reasons for choosing either of the indexing systems depending on the purpose in mind.

A (fully) iterable extender model is called *universal* if it does not loose the coiteration against any other coiterable premouse (cf. [14, Definition 4.3]).

DEFINITION 2.1. *An extender model $L[E]$ is called a core model provided the following conditions are met.*[4]

(1) *$L[E]$ is fully iterable,*
(2) *$L[E]$ is universal and it elementarily embeds into any other universal weasel,*
(3) *$L[E]$ is rigid, i.e., there is no non-trivial elementary embedding $\pi\colon L[E] \to L[E]$,*
(4) *$L[E]$ satisfies weak covering in that $cf^V(\kappa^{+L[E]}) \geq Card^V(\kappa)$ for all $\kappa \geq \aleph_2$,*
(5) *$L[E]$ has a forcing absolute definition, i.e., there is a formula $\Phi(-)$ in the language of set theory such that for every poset $\mathbb{P} \in V$ we have that* (a) *for every premouse $\mathcal{M}$,*

$$\mathcal{M} \lhd L[E] \Leftrightarrow \mathbb{P} \Vdash \Phi(\check{\mathcal{M}}),$$

and (b) *$\mathbb{P}$ forces that for all $\mathcal{M}$ and $\mathcal{N}$ with $\Phi(\mathcal{M}) \wedge \Phi(\mathcal{N})$, $\mathcal{M} \trianglelefteq \mathcal{N} \vee \mathcal{N} \trianglelefteq \mathcal{M}$,*[5] *and*
(6) *$L[E]$ has a uniform local definition, i.e., there is a formula $\Phi(-)$ in the language of set theory*[6] *such that for every large enough*[7] *cardinal κ we have that*

$$\mathcal{M} \lhd L_\kappa[E] \Leftrightarrow H_\kappa \models \Phi(\mathcal{M}).$$

Notice that by (2) and (3) there can be at most one core model. Woodin has shown that if $L[E]$ is 1-small and $L[E] \models$ "there is a Woodin cardinal," then $L[E]$ satisfies "I am not iterable." In particular, in this situation $L[E]$ thinks that there can be no core model![8] On the other hand, the informed guess is that if there is no inner model with a Woodin cardinal then the core model exists. This view is supported by the seminal [17]. The present author has shown in [14] that the core model exists, starting from an assumption being somewhat stronger than "there is no inner model with a Woodin cardinal."

We say that $0^{\P}$ ("zero hand grenade") does not exist (cf. [14, Definition 2.3]) if there is no premouse $\mathcal{M}$ with a measurable cardinal κ such that

$$\mathcal{M} \,\|\, \kappa \models \text{"there is a proper class of strong cardinals."}^{9}$$

The paper [14] proves the existence of the core model in the theory $ZFC + 0^{\P}$ does not exist.

---

[4]Hugh Woodin informs us that his work on $AD_{\mathbb{R}}$ shows that the present conditions might be too demanding in general to yield an appropriate definition of the concept of "core model." However, our conditions (1) through (6) hold for any known core model in a universe in which all premice are tame, i.e., don't have extenders overlapping Woodin cardinals.

[5]$\trianglelefteq$ means "is an initial segment of," and $\lhd$ means "is a proper initial segment of."

[6]We may allow set parameters.

[7]$\kappa > 2^{\aleph_0}$ will do in general.

[8]Cf. also the discussion in [5, §4].

[9]$\mathcal{N} \,\|\, \alpha$ is $\mathcal{N}$ cut off at α with top extender $E_\alpha^{\mathcal{N}}$ (provided $E_\alpha^{\mathcal{N}} \neq \emptyset$). We'll also confuse $\mathcal{N} \,\|\, \alpha$ with its underlying universe.

If $0^\P$ does not exist then we let K denote the core model. We aim to briefly discuss how (1) through (6) in Definition 2.1 materialize for K, if $0^\P$ does not exist. (1) is given by [14, Lemma 3.3] (which garantees the existence and iterability of K^c, a preliminary version of K) and the fact that, in [14, §8], K is constructed as the collapse of a hull of K^c. (2) and (3) follow from [14, Corollary 8.17] and (the proofs of) [17, Theorem 8.10] and [17, Theorem 8.8]. (4) is [14, Theorem 8.18] and in fact is given by the proofs of [9] and [8]. (5) is given by the proof of [17, Theorem 5.18 (3)]. We now aim to prove (6) by a lemma which is of independent interest and which generalizes a lemma Jensen has shown to hold under the stronger assumption that $0^\P$ does not exist. We shall actually use Lemma 2.2 in the proof of Theorem 4.6.

By $\neg 0^\P$ we abbreviate the statement that $0^\P$ does not exist.

LEMMA 2.2. ($\neg 0^\P$) *Let $\kappa \geq \aleph_2$ be a cardinal in K. Let $\mathcal{M} \trianglerighteq K \parallel \kappa$ be an iterable sound premouse with $\rho_\omega(\mathcal{M}) \leq \kappa$. Then $\mathcal{M} \lhd K$.*

PROOF. By standard methods, it suffices to prove that the phalanx $((K, \mathcal{M}), \kappa)$ is iterable (cf. [17, §6]). As $\neg 0^\P$, [14, Lemma 2.7] shows that for this in turn it suffices to prove that if $\mathcal{N} \trianglerighteq K \parallel \kappa$ is an iterable premouse and $F = E_\nu^{\mathcal{N}} \neq \emptyset$ is such that $\mu = c.p.(F) < \kappa \leq \nu$ then $Ult(K; F)$ is well-founded (and therefore iterable). In what follows we may and shall assume that F is the top extender of $\mathcal{N}$.

Set $\tau = \mu^{+K}$ if $cf(\mu^{+K}) > \omega$, and $\tau = \aleph_2$ otherwise. Notice that $\mu < \tau \leq \kappa$ by [14, Theorem 8.18] and $\kappa \geq \aleph_2$.

Let θ be a large enough regular cardinal, and let $\pi \colon \bar{H} \to H_{\theta^+}$ be elementary and such that $\bar{H}$ is countable and transitive and $ran(\pi)$ contains all the sets of current interest. Let

$$\bar{K} = Ult(\pi^{-1}(K \parallel \theta); \pi \restriction \pi^{-1}(K \parallel \tau))$$

be the ultrapower of $\pi^{-1}(K \parallel \theta)$ by the (long) extender $\pi \restriction \pi^{-1}(K \parallel \tau)$. Also, let

$$\bar{\mathcal{N}} = Ult(\pi^{-1}(\mathcal{N}); \pi \restriction \pi^{-1}(K \parallel \tau))$$

be the ultrapower of $\pi^{-1}(\mathcal{N})$ by the (long) extender $\pi \restriction \pi^{-1}(K \parallel \tau)$, and let

$$k \colon \bar{\mathcal{N}} \to \mathcal{N}$$

be the canonical embedding. Set $\bar{\tau} = k^{-1}(\tau)$, and notice that $\bar{\tau} = c.p.(k)$. Let $\bar{F}$ be the top extender of $\bar{\mathcal{N}}$. It is important to notice that $\bar{\mathcal{N}}$ is a premouse (rather than a proto-mouse), as the ultrapower map producing $\bar{\mathcal{N}}$ is continuous at $\pi^{-1}(\mu^{+K})$.

It now suffices to prove that $Ult(\bar{K}; \bar{F})$ is well-founded, as we might have thrown in arbitrary coordinates witnessing the alleged ill-foundedness of $Ult(K; F)$ into the range of π.

Let $(\mathcal{T}, \mathcal{U})$ denote the coiteration of $((\mathcal{N}, \bar{\mathcal{N}}), \bar{\tau})$ with $\mathcal{N}$. [4, § 8 Lemma 1] shows that the last model $\mathcal{M}_\infty^\mathcal{T}$ of $\mathcal{T}$ will sit above $\bar{\mathcal{N}}$, that there will be no drop along the main branch of $\mathcal{T}$, and that $\mathcal{M}^{\mathcal{T}}_{\infty} \trianglelefteq \mathcal{M}^{\mathcal{U}}_{\infty}$. Set $\varphi = \pi^{\mathcal{T}}_{0\infty}$, and $\mathcal{N}^* = \mathcal{M}^{\mathcal{T}}_{\infty}$. We thus have

$$\varphi: \bar{\mathcal{N}} \to \mathcal{N}^*,$$

where $\mathcal{N}^*$ is an iterate of $\mathcal{N}$ and $\varphi \restriction \bar{\tau} = id$. As $\bar{\mathcal{N}}$ has a top extender, namely $\bar{F}$, with critical point $\mu < \bar{\tau}$, $\mathcal{N}^*$ will also have a top extender, call it F^*, with critical point μ. As $\neg\, 0^{\P}$, we'll therefore have that $\mathcal{U}$ can only use extenders with critical point $> \mu$ (cf. Claim 2 in the proof of [14, Lemma 2.4]).

Another application of [4, § 8 Lemma 1] shows that the coiteration of $((K, \bar{K}), \bar{\tau})$ with K produces an embedding

$$\chi: \bar{K} \to K^*$$

where K^* is an iterate of K and $\chi \restriction \bar{\tau} = id$. In order to finish the proof of Lemma 2.2 we now split the argument into two cases according to whether $\tau = \mu^{+K}$ or $\tau = \omega_2$.

Let us now first assume that $\tau = \mu^{+K}$. Then $\bar{\tau} = \mu^{+\bar{\mathcal{N}}} = \mu^{+\mathcal{N}^*} < \tau = \mu^{+\mathcal{N}} = \mu^{+K}$. This and the fact that $\mathcal{U}$ only uses extenders with critical point $> \mu$ implies that we may actually construe $\mathcal{U}$ as an iteration of $\mathcal{N} \parallel \tau = K \parallel \tau$. Especially, $\mathcal{N}^*$ is an iterate of K. Therefore (cf. [9, Fact 3.19.1]), $Ult(\mathcal{P}; F^*)$ is well-founded whenever $\mathcal{P}$ is an iterate of K with $\mathcal{P} \parallel \bar{\tau} = K \parallel \bar{\tau}$ and $\bar{\tau} = \mu^{+\mathcal{P}}$. Thus, using φ, $Ult(\mathcal{P}; \bar{F})$ is well-founded whenever $\mathcal{P}$ is an iterate of K with $\mathcal{P} \parallel \bar{\tau} = K \parallel \bar{\tau}$ and $\bar{\tau} = \mu^{+\mathcal{P}}$. We therefore in particular know that $Ult(K^*; \bar{F})$ is well-founded. Hence $Ult(\bar{K}; \bar{F})$ is well-founded, too, using χ.

Let us finally suppose that $\tau = \omega_2$. By [14, Lemma 2.7], $\mathcal{T}$ is really an iteration of $\bar{\mathcal{N}}$. Moreover, $\bar{\mathcal{N}} \cap OR < \omega_2$, so that by [12, Theorem 3.4] $K \parallel \tau = \mathcal{N} \parallel \tau$ wins the coiteration against $\bar{\mathcal{N}}$. In other words, we may again construe $\mathcal{U}$ as an iteration of $\mathcal{N} \parallel \tau = K \parallel \tau$. We may now continue exactly as in the previous case. □

Lemma 2.2 readily implies the following which is now easy to verify.[10]

THEOREM 2.3 ($\neg\, 0^{\P}$). *$K \parallel \kappa$ is uniformly $\Sigma_3^{H_\kappa}$ in the parameter $K \parallel \omega_2$ for all cardinals $\kappa \geq \aleph_3$, i.e., there is a Σ_3 formula $\Phi(v_0, v_1)$ such that for all such κ, $\mathcal{M} \lhd K \parallel \kappa$ if and only if $H_\kappa \models \Phi(\mathcal{M}, K \parallel \omega_2)$.*

As a matter of fact, if $\neg\, 0^{\P}$ then K satisfies that "$0^{\P}$ does not exist and I am the core model."[11] The proof of our main result, theorem 4.6, will be run in the theory ZFC + $0^{\P}$ does not exist + $V = K$. We want to emphasize that we do not know whether Theorem 4.6 remains true if we replace "$0^{\P}$ does not exist" by a weaker anti large cardinal assumption. *A fortiori*, we don't know

[10]Steel and the author (independently from each other) have shown for the K of [17] that $K \parallel \kappa$ is uniformly lightface $\Sigma_3^{H_\kappa}$ for all cardinals $\kappa > 2^{\aleph_0}$. On the other hand, it is open whether Lemma 2.2 holds for the K of [17].

[11]We shall let $V = K$ abbreviate the statement "I am the core model." It is in fact true that if $\neg\, 0^{\P}$ then any weakly iterable extender model satisfies "$0^{\P}$ does not exist and $V = K$."

whether 4.6 remains true if we moreover replace $V = K$ by "V is a weakly iterable extender model."

We are now going to state a lemma which is implicit in [13]. We formulate it under the assumption that $0^{\P}$ does not exist, although it is known to hold under much weaker circumstances. The reader should consult [9, §2.3] or [13, Section 2] on proto-mice.[12]

LEMMA 2.4 ($\neg 0^{\P}$). *Let $\mathcal{P}$ be an iterable sound premouse with top extender G, and let $\mathcal{Q}$ be a proto-mouse with top extender fragment F. Let $k\colon \mathcal{Q} \to_{\Sigma_0} \mathcal{P}$. Suppose that $\lambda = max\{\xi\colon \sigma \restriction \xi = id\} = \kappa^{+\mathcal{Q}}$, $\rho_1(\mathcal{P}) \leq \kappa$, and $\mu = c.p.(F) < \kappa$. Suppose also that $\rho_1(\mathcal{Q}) \leq \kappa$ and $\mathcal{Q}$ is sound and solid above κ. Let ρ with $\lambda < \rho < \kappa^{+\mathcal{P}}$ be largest such that F measures the subsets of μ which exist in $\mathcal{Q} \parallel \rho$. Let*

$$\pi\colon \mathcal{Q} \parallel \rho \to_F \mathcal{R} = Ult_n(\mathcal{Q} \parallel \rho; F),$$

where $n < \omega$ is such that $\rho_{n+1}(\mathcal{Q} \parallel \rho) \leq \mu < \rho_n(\mathcal{Q} \parallel \rho)$. Then $\rho_{n+1}(\mathcal{R}) \leq \kappa$, $\mathcal{R}$ is sound above κ, and in fact $\mathcal{R} \lhd \mathcal{P}$. Moreover, $\mathcal{Q} \in \mathcal{P}$.

PROOF SKETCH. We may define $l\colon \mathcal{R} \to \mathcal{P} \parallel i_G(\rho)$ by setting

$$[a, f]_F^{\mathcal{Q}\parallel\rho} \mapsto [k(a), f]_G^{\mathcal{P}\parallel\mu^{+\mathcal{P}}},$$

where a and f are appropriate. It can be verified that $\rho_{n+1}(\mathcal{R}) = \rho_1(\mathcal{Q}) \leq \kappa$ and that $\mathcal{R}$ is sound above κ. We may then apply the condensation lemma, [4, § 8 Lemma 4], to get that $\mathcal{R} \lhd \mathcal{P}$. We shall also have that F is definable over $\mathcal{R}$, and therefore $\mathcal{Q} \in \mathcal{P}$. The reader may find a full proof of Lemma 2.4 along the lines of the proof of [13, Lemma 2.19]. □

§3. Condensation and collapsing structures.

Let κ be a cardinal, and let $\pi\colon \bar{K} \to K \parallel \kappa$ be elementary. We may ask: under which circumstances is $\bar{K}$ an iterate of K? This section will provide a sufficient critierion for when the answer is "yes." This criterion will be used in the next section.

THEOREM 3.1. *Suppose that $0^{\P}$ does not exist and that $V = K$. Let κ be a cardinal, and let $\pi\colon \bar{K} \to K \parallel \kappa$ be elementary. Suppose that*

$$cf(sup(\pi''\alpha^{+\bar{K}})) > \omega$$

whenever α is an infinite cardinal in $\bar{K}$ (we understand that $\alpha^{+\bar{K}} = \bar{K} \cap OR$ if α is the largest cardinal in $\bar{K}$). Then $\bar{K}$ is a normal iterate of K, i.e., there is a normal iteration tree $\mathcal{T}$ on K with a last model $\mathcal{M}_\infty^{\mathcal{T}}$ such that $\bar{K} \trianglelefteq \mathcal{M}_\infty^{\mathcal{T}}$.

PROOF. We may assume without loss of generality that $\pi \neq id$. Let $\delta = c.p.(\pi)$, and let $\eta \leq \delta$ be least such that $(\mathcal{P}(\eta) \cap K) \setminus \bar{K} \neq \emptyset$. Note that $(\mathcal{P}(\delta) \cap K) \setminus \bar{K} \neq \emptyset$, because otherwise (as there are no superstrong extenders) $\pi \restriction \mathcal{P}(\delta) \cap K$ would be an extender which collapses the cardinal $\pi(\delta)$.

[12] Proto-mice are called fragments in [13].

Let $(\mathcal{U}, \mathcal{T})$ denote the (padded) coiteration of $\bar{K}$ with K. We'll have that $[0, \infty)_U \cap \mathcal{D}^{\mathcal{U}} = \emptyset$ by the universality of K (alternatively, by Dodd-Jensen). By $\neg 0^\P$ we shall also have that $\pi^{\mathcal{U}}_{0\infty} \restriction \delta = id$ (cf. the proof of [14, Lemma 5.2]).

We want to show that $\mathcal{U}$ is trivial (i.e., that $\bar{K}$ doesn't move in the comparison with K), and in fact that $\mathcal{M}^{\mathcal{T}}_\infty$ is set-sized and $\rho_\omega(\mathcal{M}^{\mathcal{T}}_\infty) < \bar{K} \cap OR$.

CLAIM 1. Let μ be such that $\pi^{\mathcal{U}}_{0\infty} \restriction \mu = id$. Then for no $F = E^{\mathcal{T}}_\nu$ do we have that $\bar{\mu} = c.p.(F) < \eta$ and $F(\bar{\mu}) \leq \mu$.

PROOF. Suppose otherwise. Notice first that $F(\bar{\mu})$ is a cardinal in $\mathcal{M}^{\mathcal{T}}_\infty$, hence in $\mathcal{M}^{\mathcal{U}}_\infty$, and hence in $\bar{K}$. Therefore, $\pi \circ F(\bar{\mu})$ is a cardinal in K $(= V)$. However, $\pi \circ F$ would be an extender which collapses the cardinal $\pi \circ F(\bar{\mu})$, again as there are no superstrong extenders. Contradiction! □

CLAIM 2. $\mathcal{U}$ is trivial.

PROOF. Assume not. Let $F = E^{\mathcal{U}}_\varepsilon$, where $\varepsilon + 1$ is least in $(0, \infty]_U$, and let $\mu = c.p.(F) \geq \delta$. (Remember that $\pi^{\mathcal{U}}_{0\infty} \restriction \delta = id$.) Let $\beta < lh(\mathcal{U}) = lh(\mathcal{T})$ be minimal with $\mathcal{M}^{\mathcal{T}}_\beta \trianglerighteq \bar{K} \parallel \mu^{+\bar{K}}$. We let $(\kappa_\gamma : \gamma \leq \theta)$ enumerate the cardinals of $\bar{K}$ in the closed interval $[\eta, \mu]$, and we let $\lambda_\gamma = \kappa_\gamma^{+\bar{K}}$ for $\gamma \leq \theta$. For each $\gamma \leq \theta$ we let $\delta(\gamma) \leq \beta$ be the least δ such that $\mathcal{M}^{\mathcal{T}}_\delta \parallel \lambda_\gamma = \bar{K} \parallel \lambda_\gamma$, and we let $\mathcal{P}_\gamma$ be the largest initial segment $\mathcal{P}$ of $\mathcal{M}^{\mathcal{T}}_{\delta(\gamma)}$ such that all bounded subsets of λ_γ which are in $\mathcal{P}$ are in $\bar{K}$ as well. In particular, $\lambda_\gamma = \kappa_\gamma^{+\mathcal{P}_\gamma}$. By Claim 1 we shall have that $\rho_\omega(\mathcal{P}_\gamma) \leq \kappa_\gamma$ for all $\gamma \leq \theta$. Moreover, $\mathcal{P}_\gamma$ is sound above κ_γ.

We let $\vec{\mathcal{P}}$ denote the phalanx

$$((K^\frown(\mathcal{P}_\gamma : \gamma \leq \theta)^\frown \mathcal{M}^{\mathcal{U}}_\varepsilon), \eta^\frown(\lambda_\gamma : \gamma \leq \theta)).$$

In the language of [9, Definition 2.4.5], $\vec{\mathcal{P}}$ is a special phalanx of premice.

Let $\mathcal{U}^*$ and $\mathcal{V}$ be the padded iteration trees arising from the comparison of $\vec{\mathcal{P}}$ with K. We understand that $\mathcal{U}^*$ either has a last ill-founded model, or else that $\mathcal{M}^{\mathcal{U}^*}_\infty$ and $\mathcal{M}^{\mathcal{V}}_\infty$ are lined up. The following says that it is the latter which will hold.

SUBCLAIM. $\vec{\mathcal{P}}$ is coiterable with K.

PROOF. Suppose that $\mathcal{U}^*$ has a last ill-founded model. Let $\sigma : \bar{H} \to H_\Omega$ be such that Ω is regular and large enough, $\bar{H}$ is countable and transitive, and $\{\vec{\mathcal{P}}, \mathcal{U}^*\} \subset ran(\sigma)$. Then $\sigma^{-1}(\mathcal{U}^*)$ witnesses that $\sigma^{-1}(\vec{\mathcal{P}})$ is not iterable (in a special respect).

Fix $\gamma \in \theta + 1 \cap ran(\sigma)$ for a moment. Let

$$\mathcal{Q}_\gamma = Ult_n(\sigma^{-1}(\mathcal{P}_\gamma); \sigma \restriction \sigma^{-1}(\mathcal{P}_\gamma \parallel \lambda_\gamma)),$$

where $n < \omega$ is such that $\rho_{n+1}(\mathcal{P}_\gamma) \leq \kappa_\gamma < \rho_n(\mathcal{P}_\gamma)$. Let $\sigma_\gamma \supset \sigma \restriction \sigma^{-1}(\mathcal{P}_\gamma \parallel \lambda_\gamma)$ denote the canonical embedding from $\sigma^{-1}(\mathcal{P}_\gamma)$ into $\mathcal{Q}_\gamma$. Notice that by $cf(\lambda_\gamma) > \omega$ it is clear that $sup\ \sigma''\sigma^{-1}(\lambda_\gamma)$ is not cofinal in λ_γ. Hence if

k_γ denotes the canonical embedding from $\mathcal{Q}_\gamma$ into $\mathcal{P}_\gamma$ then $sup\ \sigma''\sigma^{-1}(\lambda_\gamma) = \kappa_\gamma^{+\mathcal{Q}_\gamma} = k_\gamma^{-1}(\lambda_\gamma)$ is the critical point of k_γ.

Unfortunately, $\mathcal{Q}_\gamma$ might not be premouse but rather a proto-mouse; this will in fact be the case if $\mathcal{P}_\gamma$ has a top extender with critical point $\bar{\mu} < \kappa_\gamma$ and $\rho_1(\mathcal{P}_\gamma) \leq \kappa_\gamma$, as then σ_γ is discontinuous at $\sigma_\gamma^{-1}(\bar{\mu}^{+\mathcal{P}_\gamma})$ and the top extender fragment of $\mathcal{Q}_\gamma$ will not measure all the subsets of $\bar{\mu}$ which exist in $\mathcal{Q}_\gamma$. Let us therefore define an object $\mathcal{R}_\gamma$ as follows. We set $\mathcal{R}_\gamma = \mathcal{Q}_\gamma$ if $\mathcal{Q}_\gamma$ is a premouse. Otherwise, if $\bar{\mu} < \kappa_\gamma$ is the critical point of the top extender of $\mathcal{P}_\gamma$, we let

$$\mathcal{R}_\gamma = Ult_n(\mathcal{Q}_\gamma \parallel \rho; G),$$

where G is the top extender fragment of $\mathcal{Q}_\gamma$, ρ is maximal with

$$\bar{\mu}^{+\mathcal{Q}_\gamma \| \rho} = sup\ \sigma_\gamma''\sigma^{-1}(\bar{\mu})^{+\sigma^{-1}(\mathcal{P}_\gamma)},$$

and $n < \omega$ is such that $\rho_{n+1}(\mathcal{Q}_\gamma) \leq \bar{\mu} < \rho_n(\mathcal{Q}_\gamma)$.

We can apply [4, § 8 Lemma 4] (if $\mathcal{Q}_\gamma$ is a premouse) and Lemma 2.4 (if $\mathcal{Q}_\gamma$ is not a premouse) and deduce that $\rho_\omega(\mathcal{R}_\gamma) \leq \kappa_\gamma$ and $\mathcal{R}_\gamma \lhd \mathcal{P}_\gamma$. But as $\kappa_\gamma^{+\mathcal{R}_\gamma} < \lambda_\gamma = \kappa_\gamma^{+\mathcal{P}_\gamma}$, we'll of course have that $\mathcal{R}_\gamma \lhd \mathcal{P}_\gamma \parallel \lambda_\gamma = \mathcal{M}_\varepsilon^{\mathcal{U}} \parallel \lambda_\gamma$. Therefore $\mathcal{R}_\gamma$ is an initial segment of $\mathcal{M}_\varepsilon^{\mathcal{U}}$.

Now let $\vec{\mathcal{Q}}$ denote the special phalanx of proto-mice

$$(K^\frown(\mathcal{Q}_\gamma : \gamma \in \theta + 1 \cap ran(\sigma))^\frown \mathcal{M}_\varepsilon^{\mathcal{U}}, \eta^\frown(\lambda_\gamma : \gamma \in \theta + 1 \cap ran(\sigma))),$$

and let $\vec{\mathcal{R}}$ denote the special phalanx of premice

$$(K^\frown(\mathcal{R}_\gamma : \gamma \in \theta + 1 \cap ran(\sigma))^\frown \mathcal{M}_\varepsilon^{\mathcal{U}}, \eta^\frown(\lambda_\gamma : \gamma \in \theta + 1 \cap ran(\sigma))).$$

Due to the existence of the family of maps

$$\sigma \restriction \sigma^{-1}(K \parallel \kappa), (\sigma_\gamma : \gamma \in \theta + 1 \cap ran(\sigma)), \sigma \restriction \sigma^{-1}(\mathcal{M}_\varepsilon^{\mathcal{U}})$$

we know that $\vec{\mathcal{Q}}$ cannot be iterable, as $\sigma^{-1}(\vec{\mathcal{P}})$ is not iterable. But then, arguing axactly as for [9, Lemma 3.18], $\vec{\mathcal{R}}$ cannot be iterable. However, any iteration of $\vec{\mathcal{R}}$ can be construed as an iteration of $((K, \mathcal{M}_\varepsilon^{\mathcal{U}}), \delta)$, and thus in turn of $((K, \bar{K}), \delta)$. But $((K, \bar{K}), \delta)$ is iterable. Contradiction! Subclaim □

Now notice that $\mathcal{U}^* \restriction \varepsilon$ is trivial, $\mathcal{V} \restriction \varepsilon = \mathcal{T} \restriction \varepsilon$, and that $F = E_\varepsilon^{\mathcal{U}} = E_\varepsilon^{\mathcal{U}^*}$ will be the first extender used in $\mathcal{U}^*$. By [14, Lemma 2.7], we'll then[13] in fact have that no extender from $\mathcal{U}^*$ will be applied to (an inital segment of) the last model of the phalanx $\vec{\mathcal{P}}$, $\mathcal{M}_\varepsilon^{\mathcal{U}}$. We may therefore finally argue as in the proof of [17, Theorem 8.6] to derive a contradiction. Claim 2 □

We have shown Theorem 3.1 Theorem 3.1 □

---

[13]This is the only place in this proof where we really use the assumption that K is below $0^\P$ in a way which does not seem to be avoidable. If the assumption that K is below $0^\P$ is dropped then at the time of writing I don't see how to prove the iterability of the relevant phalanx (which would then have to be longer than the phalanx $\vec{\mathcal{P}}$ defined above) needed to verify Claim 2.

We do not know how to prove Theorem 3.1 if the assumption that $V = K$ is removed from its statement.

We aim to continue the discussion which was begun in the proof of Theorem 3.1. Specifically, we want to see how the argument leads to collapsing structures.

DEFINITION 3.2 ($\neg 0^{\P}$). *Let $\alpha \leq \gamma \in OR$. We say that $K \parallel \gamma$ is a collapsing structure for α provided the following holds true. For all $\bar{\gamma} \leq \gamma$, $\bar{\gamma} = \gamma$ if and only if there are $n < \omega$, $\delta < \alpha$, and $\vec{p} \in {}^{<\omega}\bar{\gamma}$ such that*

$$Hull_n^{K\parallel\bar{\gamma}}(\delta \cup \{\vec{p}\}) \cap \alpha$$

is cofinal in α.

Of course, collapsing structures are unique so that we may and shall talk about *the* collapsing structure for a given α.

Now let $\pi\colon \bar{K} \to K \parallel \kappa$ and everything else be as in the proof of Theorem 3.1. We wish to isolate collapsing structures for $sup(\pi''\alpha^{+\bar{K}})$ whenever $\alpha \geq \eta$.

Let $(\kappa_\gamma\colon \gamma < \theta)$ enumerate the cardinals of $\bar{K}$ in the half-open interval $[\eta, \bar{K} \cap OR)$, and let $\lambda_\gamma = \kappa_\gamma^{+\bar{K}}$ for $\gamma < \theta$. For each $\gamma < \theta$ we let $\delta(\gamma) < lh(\mathcal{T})$ be the least δ such that $\mathcal{M}_\delta^{\mathcal{T}} \parallel \lambda_\gamma = \bar{K} \parallel \lambda_\gamma$, we let $\mathcal{P}_\gamma$ be the largest initial segment $\mathcal{P}$ of $\mathcal{M}_{\delta(\gamma)}^{\mathcal{T}}$ such that all bounded subsets of λ_γ which are in $\mathcal{P}$ are in $\bar{K}$ as well, and we let $n(\gamma)$ be the $n < \omega$ such that

$$\rho_{n+1}(\mathcal{P}_\gamma) \leq \kappa_\gamma < \rho_n(\mathcal{P}_\gamma).$$

Notice that $n(\gamma)$ will always be defined by Claims 1 and 2. Let for each $\gamma < \theta$,

$$\mathcal{Q}_\gamma = Ult_{n(\gamma)}(\mathcal{P}_\gamma; \pi \upharpoonright \bar{K} \parallel \lambda_\gamma).$$

Let σ_γ denote the canonical embedding from $\mathcal{P}_\gamma$ into $\mathcal{Q}_\gamma$. We also set $\tilde{\lambda}_\gamma = sup\ \pi''\lambda_\gamma$. Notice that if $\mathcal{Q}_\gamma$ is well-founded then $\tilde{\lambda}_\gamma = \sigma_\gamma(\lambda_\gamma) = \pi(\kappa_\gamma)^{+\mathcal{Q}_\gamma} < \pi(\lambda_\gamma)$, $\rho_{n+1}(\mathcal{Q}_\gamma) \leq \pi(\kappa_\gamma)$, and $\mathcal{Q}_\gamma$ is sound above $\pi(\kappa_\gamma)$. Unfortunately, again even if it is well-founded, $\mathcal{Q}_\gamma$ might not be premouse but rather a proto-mouse, namely if $\mathcal{P}_\gamma$ has a top extender with critical point $\bar{\mu} < \kappa_\gamma$ and $n(\gamma) = 0$.

Let us therefore define, inductively for $\gamma < \theta$, objects $\mathcal{R}_\gamma$ as follows. We understand that we let the construction break down as soon as one of the models defined is ill-founded. It will be clear from the construction that if $\mathcal{R}_\gamma$ is well-defined then $\mathcal{R}_\gamma$ is a premouse, $\mathcal{R}_\gamma \trianglerighteq K \parallel \tilde{\lambda}_\gamma$, $\tilde{\lambda}_\gamma = \pi(\kappa_\gamma)^{+\mathcal{R}_\gamma}$, and $\rho_\omega(\mathcal{R}_\gamma) \leq \kappa_\gamma$; we shall then let $n^*(\gamma)$ be the least $n < \omega$ such that $\rho_{n+1}(\mathcal{R}_\gamma) \leq \kappa_\gamma < \rho_n(\mathcal{R}_\gamma)$.

Fix $\gamma < \theta$, and suppose $\mathcal{R}_{\bar{\gamma}}$ to be given for all $\bar{\gamma} < \gamma$. We set $\mathcal{R}_\gamma = \mathcal{Q}_\gamma$ if $\mathcal{Q}_\gamma$ is a premouse. Otherwise, if $\kappa_{\bar{\gamma}}$ is the critical point of the top extender of $\mathcal{P}_\gamma$, we let

$$\mathcal{R}_\gamma = Ult_{n^*(\bar{\gamma})}(\mathcal{R}_{\bar{\gamma}}; G),$$

where G is the top extender fragment of $\mathcal{Q}_\gamma$.

LEMMA 3.3. *For each $\gamma < \theta$ we have that $\mathcal{R}_\gamma$ is well-defined premouse. In fact, $\mathcal{R}_\gamma \lhd K$, and $\mathcal{R}_\gamma$ is the collapsing structure for $\pi(\kappa_\gamma)$.*

PROOF. We shall verify that $\mathcal{R}_\gamma$ is an iterable premouse. This will suffice via Lemma 2.2.

Let $\sigma : \bar{H} \to H_\Omega$ be such that Ω is regular and large enough, $\bar{H}$ is countable and transitive, and $ran(\sigma)$ contains all the sets of current interest. For $\gamma \in \theta \cap ran(\sigma)$ we shall inductively choose $\mathcal{Q}^*_\gamma \in \bar{K} \parallel \lambda_\gamma$ and $\mathcal{R}^*_\gamma \lhd \bar{K} \parallel \lambda_\gamma$ together with embeddings

$$\bar{\varphi}_\gamma : \mathcal{Q}_\gamma \cap ran(\sigma) \to \pi(\mathcal{Q}^*_\gamma) \quad \text{and} \quad \varphi_\gamma : \mathcal{R}_\gamma \cap ran(\sigma) \to \pi(\mathcal{R}^*_\gamma)$$

such that $\bar{\varphi}_\gamma \restriction (\tilde{\lambda}_\gamma \cap ran(\sigma)) = \varphi_\gamma \restriction (\tilde{\lambda}_\gamma \cap ran(\sigma)) = id$. We shall also inductively maintain that $\mathcal{Q}^*_\gamma$ is a premouse or a proto-mouse with $\rho_{n(\gamma)+1}(\mathcal{Q}^*_\gamma) \leq \kappa_\gamma < \rho_{n(\gamma)}(\mathcal{Q}^*_\gamma)$ which is sound and solid above κ_γ, that $\mathcal{R}^*_\gamma$ is a premouse with $\rho_{n^*(\gamma)+1}(\mathcal{R}^*_\gamma) \leq \kappa_\gamma < \rho_{n^*(\gamma)}(\mathcal{R}^*_\gamma)$ which is sound and solid above κ_γ, and that there is some $\lambda'_\gamma < \lambda_\gamma$ such that $\mathcal{Q}^*_\gamma \trianglerighteq K \parallel \lambda'_\gamma$, $\mathcal{R}^*_\gamma \trianglerighteq K \parallel \lambda'_\gamma$, and $\lambda'_\gamma = \kappa_\gamma^{+\mathcal{Q}^*_\gamma} = \kappa_\gamma^{+\mathcal{R}^*_\gamma}$. As we might have thrown in potential witnesses to $\mathcal{R}_\gamma$ not being iterable (for some fixed γ) into the range of σ, it will be clear from the construction that this does the job.

Fix $\gamma \in \theta \cap ran(\sigma)$, and let us suppose that $\mathcal{Q}^*_{\bar{\gamma}}$, $\mathcal{R}^*_{\bar{\gamma}}$, $\bar{\varphi}_{\bar{\gamma}}$, and $\varphi_{\bar{\gamma}}$ have already been defined for all $\bar{\gamma} \in \gamma \cap ran(\sigma)$. We let

$$\mathcal{Q}^*_\gamma = Ult_{n(\gamma)}(\sigma^{-1}(\mathcal{P}_\gamma); \sigma \restriction \sigma^{-1}(\mathcal{P}_\gamma \parallel \lambda_\gamma)),$$

and we let k_γ be the canonical embedding from $\mathcal{Q}^*_\gamma$ into $\mathcal{P}_\gamma$. As $cf(\lambda_\gamma) > \omega$, the critical point of k_γ will be $k_\gamma^{-1}(\lambda_\gamma) = \lambda'_\gamma$. If $\mathcal{Q}^*_\gamma$ is a premouse (which will be the case if and only if $\mathcal{Q}_\gamma$ is a premouse, which in turn is the case if and only if $\mathcal{P}_\gamma$ does not have a top extender with critical point $\bar{\mu} < \kappa_\gamma$ or else $n(\gamma) > 0$) then we set $\mathcal{R}^*_\gamma = \mathcal{Q}^*_\gamma$. Otherwise, let $\kappa_{\bar{\gamma}}$ be the critical point of the top extender of $\mathcal{P}_\gamma$, $\bar{\gamma} < \gamma$, and let G^* be the top extender fragment of $\mathcal{P}^*_\gamma$; then set

$$\mathcal{R}^*_\gamma = Ult_{n^*(\bar{\gamma})}(\mathcal{R}^*_{\bar{\gamma}}; G^*)$$

(notice that by our inductive hypotheses $\mathcal{R}^*_{\bar{\gamma}}$ is also the longest initial segment of $\mathcal{R}^*_\gamma$ to which G^* could be applied).

By using [4, § 8, Lemma 4] and Lemma 2.4 we get that $\mathcal{Q}^*_\gamma \in \mathcal{P}_\gamma \parallel \lambda_\gamma = \bar{K} \parallel \lambda_\gamma$ and $\mathcal{R}^*_\gamma \lhd \mathcal{P}_\gamma \parallel \lambda_\gamma = \bar{K} \parallel \lambda_\gamma$. It is easy to see that

$$[a, f]^{\mathcal{P}_\gamma}_{\pi \restriction \bar{K} \parallel \lambda_\gamma} \mapsto \pi \circ k_\gamma^{-1}(f)(a),$$

where a and f are appropriate, defines an embedding from $\mathcal{Q}_\gamma \cap ran(\sigma)$ into $\pi(\mathcal{Q}^*_\gamma)$. This is our embedding $\bar{\varphi}_\gamma$.

If $\mathcal{Q}^*_\gamma$ is a premouse then we let $\varphi_\gamma = \bar{\varphi}_\gamma$. We are left with having to define $\bar{\varphi}_\gamma$ in the case that $\mathcal{Q}^*_\gamma$ is not a premouse. Let G be the top extender fragment of $\mathcal{Q}_\gamma$, and let $\tilde{G}$ be the top extender fragment of $\pi(\mathcal{Q}^*_\gamma)$. Notice that

$$\pi(\mathcal{R}^*_\gamma) = Ult_{n^*(\bar{\gamma})}(\pi(\mathcal{R}^*_{\bar{\gamma}}); \tilde{G}).$$

It is thus straightforward to verify that

$$[a, f]^{\mathcal{R}_{\bar{\gamma}}}_G \mapsto [\bar{\varphi}_\gamma(a), \varphi_{\bar{\gamma}}(f)]^{\pi(\mathcal{R}^*_{\bar{\gamma}})}_{\tilde{G}} = i_{\tilde{G}} \circ \varphi_{\bar{\gamma}}(f)(\bar{\varphi}_\gamma(a)),$$

for appropriate a and f, defines an embedding from $\mathcal{R}_\gamma \cap ran(\sigma)$ into $\pi(\mathcal{R}^*)$. This is our embedding φ_γ.

It is easy to check our inductive hypotheses for γ. □

We remark that in order for Lemma 3.3 to hold true it is not important how the models $\mathcal{P}_\gamma$ were actually obtained to begin with. Lemma 3.3 remains true if the models $\mathcal{R}_\gamma$ are defined starting from any sequence $(\mathcal{P}_\gamma : \gamma < \theta)$ of iterable premice such that $\mathcal{P}_\gamma \trianglerighteq \bar{K} \parallel \lambda_\gamma$, $\mathcal{P}_\gamma$ is sound and solid above κ_γ, and $\rho_\omega(\mathcal{P}_\gamma) \leq \kappa_\gamma$. It is this observation which we shall make use of in the next section.

§4. Mutual stationarity in the core model. We shall need yet another condensation result in the proof of Theorem 4.6. This result, however, only holds under an additional assumption.

Definition 4.1 (cf. [14, Definition 1.2]). *Let $\mathcal{M}$ be a premouse, and let $\kappa < \kappa^{+\mathcal{M}} < \tau < \mathcal{M} \cap OR$ be such that κ and τ are cardinals of $\mathcal{M}$. Then κ is said to be $< \tau$-strong in $\mathcal{M}$ if for all $\alpha < \tau$ there is some $E^{\mathcal{M}}_\beta \neq \emptyset$ with critical point κ and such that $\alpha \leq \beta < \tau$.*

Definition 4.2. *Let $\mathcal{M}$ be a premouse. We say that $\mathcal{M}$ does not reach $o(\kappa) = \kappa^{++}$ if there is no $\lambda \in \mathcal{M}$ such that λ is $< \lambda^{++\mathcal{M}}$-strong in $\mathcal{M}$.*

William Mitchell, in [6] and [7], had shown that K exists if no extender model reaches $o(\kappa) = \kappa^{++}$. Of course, if no extender model reaches $o(\kappa) = \kappa^{++}$ then $\neg\, 0^\P$ holds. However, an extender model W which does not reach $o(\kappa) = \kappa^{++}$ can be such that $0^\P \in W$, or $0^\P \in W$, or even $M_1^\# \in W$, etc.

Lemma 4.3 ($\neg\, 0^\P$; cf. [14, Corollary 1.3]). *Let $\mathcal{M}$ be a 0-iterable premouse which has a top extender F with critical point κ. Let $\kappa^{+\mathcal{M}} < \tau \leq \rho_1(\mathcal{M})$, where τ is a cardinal in $\mathcal{M}$. Then κ is $< \tau$-strong in $\mathcal{M}$.*

Corollary 4.4. *Let $\mathcal{M}$ be a 0-iterable premouse which does not reach $o(\kappa) = \kappa^{++}$ and which has a top extender F with critical point κ. Then $\rho_1(\mathcal{M}) \leq \kappa^{+\mathcal{M}}$.*

It can be true, though, that the top extender F as in Corollary 4.4 has more than one generator. The following lemma is part of the folklore.

Lemma 4.5. *Let $\mathcal{M}$ be an iterable premouse which does not reach $o(\kappa) = \kappa^{++}$. Let $n < \omega$, and let λ be a cardinal in $\mathcal{M}$ with $\rho_{n+1}(\mathcal{M}) \leq \lambda < \rho_n(\mathcal{M})$. Let*

$\vec{p} \in {}^{<\omega}(\mathcal{M} \cap OR)$ be the standard parameter of $\mathcal{M}$. Let

$$\pi\colon \bar{\mathcal{M}} \cong Hull^{\mathcal{M}}_{n+1}(\lambda \cup \{\vec{p}\}) \prec \mathcal{M},$$

where $\bar{\mathcal{M}}$ is transitive. Then $\lambda^{+\bar{\mathcal{M}}} = \lambda^{+\mathcal{M}}$, and $\bar{\mathcal{M}} \parallel \lambda^{+\mathcal{M}} = \mathcal{M} \parallel \lambda^{+\mathcal{M}}$.

Proof sketch. By [4, § 8 Lemma 4] we know that $\bar{\mathcal{M}} \parallel \lambda^{+\bar{\mathcal{M}}} = \mathcal{M} \parallel \lambda^{+\bar{\mathcal{M}}}$, and that if $(\mathcal{U}, \mathcal{T})$ denotes the coiteration of $((\mathcal{M}, \bar{\mathcal{M}}), \lambda)$ with $\mathcal{M}$ then $\mathcal{M}^{\mathcal{U}}_\infty$ sits above $\bar{\mathcal{M}}$, neither $[0,\infty]_U$ nor $[0,\infty]_T$ contains any drop, and $\mathcal{M}^{\mathcal{U}}_\infty = \mathcal{M}^{\mathcal{T}}_\infty$. We have that $\mathcal{P}(\lambda) \cap \bar{\mathcal{M}} = \mathcal{P}(\lambda) \cap \mathcal{M}^{\mathcal{U}}_\infty$. Let us now assume that $\lambda^{+\bar{\mathcal{M}}} < \lambda^{+\mathcal{M}}$.

It is easy to see that there can be then no $E^{\mathcal{M}}_\nu \neq \emptyset$ with $c.p.(E^{\mathcal{M}}_\nu) < \lambda$ and $\lambda^{+\bar{\mathcal{M}}} \leq \nu < \lambda^{+\mathcal{M}}$. The reason is that otherwise we'd have that $c.p.(E^{\mathcal{M}}_\nu)$ would be $< \lambda^{+\bar{\mathcal{M}}}$-strong in $\bar{\mathcal{M}}$ and hence $< \lambda^{+\mathcal{M}}$-strong in $\mathcal{M}$, contradicting the assumption that $\mathcal{M}$ does not reach $o(\kappa) = \kappa^{++}$. It is also easy to see that this fact is inherited by all iterates of $\mathcal{M} \parallel \lambda^{+\mathcal{M}}$; more precisely, if $\mathcal{P}$ is an iterate of $\mathcal{M} \parallel \lambda^{+\mathcal{M}}$ via an iteration which only uses extenders with indices $\geq \lambda^{+\bar{\mathcal{M}}}$ then there can be no $E^{\mathcal{P}}_\nu \neq \emptyset$ with $c.p.(E^{\mathcal{P}}_\nu) < \lambda$ and $\lambda^{+\bar{\mathcal{M}}} \leq \nu \leq \mathcal{P} \cap OR$.

Now $\mathcal{U}$ certainly only uses extenders with indices $\geq \lambda^{+\bar{\mathcal{M}}}$. By the preceding paragraph, we thus know that $\lambda^{+\bar{\mathcal{M}}} < \lambda^{+\mathcal{M}}$ implies that $\mathcal{U}$ only uses extenders with critical points $\geq \lambda$, and therefore that there must be a drop along the main branch of $\mathcal{U}$. Contradiction! □

We can now state and prove our main result. Our proof will closely follow [3, Section 7.2] to a certain extent. By an interval (of ordinals) we mean a set of ordinals of the form $[\alpha, \beta)$.

Theorem 4.6. *Suppose that $0^{\P}$ does not exist and that $V = K$. Suppose that K does not reach $o(\kappa) = \kappa^{++}$. Let $k \in OR\setminus 1$. There is then a sequence $(S^n_i : i \in OR\setminus k, 0 < n < \omega)$ such that*

- *for all $i \in OR\setminus k$ and $n < \omega \setminus 1$ we have that $S^n_i \subset \aleph_{i+1}$ is stationary in $\aleph_{i+1}$ and $\alpha \in S^n_i \Rightarrow cf(\alpha) = \aleph_k$, and*
- *for all limit ordinals λ and for all $f\colon \lambda \to \omega \setminus 1$ we have that $(S^{f(i)}_i : i \in \lambda \setminus k)$ is mutually stationary if and only if we can split the domain $\lambda \setminus k$ of f into a finite partition $D_1 \cup \cdots \cup D_m$ of intervals such that $f \restriction D_l$ is constant whenever $1 \leq l \leq m$.*

Proof. We commence by defining the sequence $(S^n_i : i \in OR\setminus k, 0 < n < \omega)$. If α is a singular ordinal, then we let $(\gamma(\alpha), n(\alpha), \delta(\alpha), \vec{p}(\alpha))$ be the lexicographically least tuple $(\gamma, n, \delta, \vec{p})$ such that

$$Hull^{K\parallel\gamma}_n(\delta \cup \vec{p}) \cap \alpha$$

is cofinal in α. In particular, $K \parallel \gamma(\alpha)$ is the collapsing structure for α. For an ordinal $i \geq k$ and a natural number n we define

$$S^n_i = \{\alpha < \aleph_{i+1} \colon cf(\alpha) = \aleph_k \wedge n(\alpha) = n\}.$$

We are now going to show that $(S^n_i : i \in OR\setminus k, 0 < n < \omega)$ witnesses the truth of Theorem 4.6. Let λ be a limit ordinal.

We shall first prove that if $f : \lambda \setminus k \to \omega \setminus 1$ is such that we can split the domain $\lambda \setminus k$ of f into a finite partition $D_1 \cup \cdots \cup D_m$ of intervals with $Card(f''D_l) = 1$ for all l then $(S_i^{f(i)} : i \in \lambda \setminus k)$ is mutually stationary. By Lemma 1.1 it will be enough if we prove this under the assumption that $ran(f) = \{n\}$ for some $n \in \omega \setminus 1$.

Let $\mathfrak{A}$ be a model with universe $K \parallel \kappa$, where κ is a large regular cardinal. Let γ_0 be least such that $\mathfrak{A} \in K \parallel \gamma_0$, and let γ be the $\aleph_k^{th}$ ordinal β such that $\beta > \gamma_0$ and $K \parallel \beta \prec_{\Sigma_{n-1}} K \parallel \kappa^+$. Let

$$X = Hull_n^{K \parallel \gamma}(\aleph_k \cup \{\mathfrak{A}\}).$$

One can then argue exactly as for [3, Lemma 25] that for every $i \in X \cap \lambda$, $sup(X \cap \aleph_{i+1}) \in S_i^n$. The only thing to notice is that the use of the condensation lemma for L can be replaced by [4, § 8, Lemma 4] in a straightforward way.

Let us now fix some $f : \lambda \setminus k \to \omega \setminus 1$ such that $(S_i^{f(i)} : i \in \lambda \setminus k)$ is mutually stationary. We aim to prove that we can split the domain $\lambda \setminus k$ of f into a finite partition $D_1 \cup \cdots \cup D_m$ of intervals with $Card(f''D_l) = 1$ for all l. We shall exploit the covering argument of section 3.

Let $\kappa = \aleph_\lambda$, and let

$$N \prec (K \parallel \kappa; \in, \dots)$$

be such that for all $i \in N \cap \lambda$ we have that $sup(N \cap \aleph_{i+1}) \in S_i^{f(i)}$. In particular, $cf(sup(N \cap \aleph_{i+1})) = \aleph_k > \omega$ for each such i. Let

$$\pi : \bar{K} \cong N \prec K \parallel \kappa$$

be such that $\bar{K}$ is transitive.

We shall now apply the results of section 3. Let us adopt the notation from there. In particular, by the proof of Theorem 3.1, $\mathcal{T}$ is a normal iteration tree on K such that $\mathcal{M}_\infty^\mathcal{T} \trianglerighteq \bar{K}$.

Let $\mathcal{D} = \mathcal{D}^\mathcal{T} \cap (0, \infty]_T$, and let $\mathcal{D} = \{\alpha_0 + 1 < \cdots < \alpha_N + 1\}$ where $N < \omega$. Let α_i^* be the T-predecessor of $\alpha_i + 1$ for $i \leq N$. Notice that $\alpha_0^* = 0$. Let $\kappa^{-1} = \eta$, let $\kappa^i = c.p.(E_{\alpha_i}^\mathcal{T})$ for $0 < i \leq N$, and let $\kappa^{N+1} = \bar{K} \cap OR$. For any $\gamma < \theta$, let $i(\gamma)$ be the unique $i \leq N$ such that $\kappa_\gamma \in [\kappa^i, \kappa^{i+1})$. Notice that $\mathcal{M}_{\alpha_{i+1}^*}^\mathcal{T} \trianglerighteq \bar{K} \parallel \kappa^{i+1}$ and, by the proof of Theorem 3.1, $\rho_\omega(\mathcal{M}_{\alpha_{i+1}^*}^\mathcal{T}) \leq \kappa^i$. For any $\gamma < \theta$, let $\eta(\gamma)$ be the largest $\eta \leq \mathcal{M}_{\alpha_{i+1}^*}^\mathcal{T} \cap OR$ such that

$$\mathcal{P}(\kappa_\gamma) \cap \bar{K} = \mathcal{P}(\kappa_\gamma) \cap \mathcal{M}_{\alpha_{i+1}^*}^\mathcal{T} \parallel \eta,$$

and let $m(\gamma)$ be the least $m < \omega$ with $\rho_{m+1}(\mathcal{M}_{\alpha_{i+1}^*}^\mathcal{T} \parallel \eta(\gamma)) \leq \kappa_\gamma$. If $\kappa^i \leq \kappa_\gamma \leq \kappa_{\gamma'} < \kappa^{i+1}$ then $(\eta(\gamma'), m(\gamma')) \leq_{\text{lex}} (\eta(\gamma), m(\gamma))$, where $\leq_{\text{lex}}$ denotes the lexicographical ordering. This shows:

CLAIM 1. We can split θ into a finite partition $I_0, \dots, I_p$ of intervals such that whenever $\{\gamma, \gamma'\} \subset I_i$ then $i(\gamma) = i(\gamma')$, $\eta(\gamma) = \eta(\gamma')$, and $m(\gamma) = m(\gamma')$.

For $\gamma < \theta$ we may let $\mathcal{P}'_\gamma$ be the unique transitive $\mathcal{P}$ such that

$$\mathcal{P} \cong Hull_{m(\gamma)+1}^{\mathcal{M}^{\mathcal{T}}_{\alpha^*_{i+1}} \| \eta(\gamma)}(\kappa_\gamma \cup \{\vec{p}\}) \prec \mathcal{M}^{\mathcal{T}}_{\alpha^*_{i+1}} \| \eta(\gamma),$$

where $\vec{p}$ is the standard parameter of $\mathcal{M}^{\mathcal{T}}_{\alpha^*_{i+1}} \| \eta(\gamma)$. As K does not reach $o(\kappa) = \kappa^{++}$, by Lemma 4.5 we shall have that

$$\mathcal{P}(\kappa_\gamma) \cap \mathcal{P}'_\gamma = \mathcal{P}(\kappa_\gamma) \cap \mathcal{M}^{\mathcal{T}}_{\alpha^*_{i+1}} \| \eta(\gamma) = \mathcal{P}(\kappa_\gamma) \cap \bar{K}.$$

Moreover, we'll clearly have that $\rho_{m(\gamma)+1}(\mathcal{P}'_\gamma) \leq \kappa_\gamma < \rho_{m(\gamma)}(\mathcal{P}'_\gamma)$.

We now use the models $\mathcal{P}'_\gamma$ to define $\mathcal{Q}'_\gamma$, $\mathcal{R}'_\gamma$, and $m^*(\gamma)$ in exactly the same way as we had defined $\mathcal{Q}_\gamma$, $\mathcal{R}_\gamma$, and $n^*(\gamma)$ using the models $\mathcal{P}_\gamma$ in the previous section.[14] We set

$$\mathcal{Q}'_\gamma = Ult_{m(\gamma)}(\mathcal{P}'_\gamma; \pi \restriction \bar{K} \| \lambda_\gamma).$$

We set $\mathcal{R}'_\gamma = \mathcal{Q}'_\gamma$ and $m^*(\gamma) = n^*(\gamma)$ if $\mathcal{Q}'_\gamma$ is a premouse. If $\mathcal{Q}'_\gamma$ is a proto-mouse rather than a premouse then we set

$$\mathcal{R}'_\gamma = Ult_{m^*(\bar{\gamma})}(\mathcal{R}'_{\bar{\gamma}}; G),$$

where $\kappa_{\bar{\gamma}}$ is the critical point of the top extender of $\mathcal{P}'_\gamma$ and G is the top extender fragment of $\mathcal{Q}'_\gamma$.

Suppose that $\{\gamma, \gamma'\} \subset I_i$, where I_i is an interval as in Claim 1. Then $\mathcal{P}'_\gamma$ has a top extender if and only if $\mathcal{P}'_{\gamma'}$ has a top extender. Suppose this is the case, and let us further suppose that $\kappa_{\bar{\gamma}} < \kappa_\gamma \leq \kappa_{\gamma'}$, where $\kappa_{\bar{\kappa}}$ is the critical point of the top extender of $\mathcal{P}'_\gamma$ as well as of $\mathcal{P}'_{\gamma'}$. Then $\mathcal{Q}'_\gamma$ is a proto-mouse if and only if $\mathcal{Q}'_{\gamma'}$ is a proto-mouse, and in fact $m^*(\gamma) = m^*(\gamma')$. This shows:

Claim 2. We can split θ into a finite partition $I'_0, \ldots, I'_{p'}$ of intervals such that whenever $\{\gamma, \gamma'\} \subset I'_i$ then and $m^*(\gamma) = m^*(\gamma')$.

By the remark right after the proof of Lemma 3.3 we shall now have the following.

Claim 3. For each $\gamma < \theta$ we have that $\mathcal{R}'_\gamma$ is a well-defined premouse. In fact, $\mathcal{R}'_\gamma \lhd K$, and $\mathcal{R}'_\gamma$ is the collapsing structure for $\pi(\kappa_\gamma)$. We have that $\rho_{m^*(\gamma)+1}(\mathcal{R}'_\gamma) \leq \kappa_\kappa < \rho_{m^*(\gamma)}(\mathcal{R}'_\gamma)$.

By Claim 3,

$$n(sup(\pi''\lambda_\gamma)) = m^*(\gamma) + 1$$

whenever $\gamma < \theta$. Now suppose that we couldn't split the domain $\lambda \setminus k$ of f into a finite partition $D_1 \cup \cdots \cup D_m$ with $Card(f''D_l) = 1$ for all l. We could then have expanded $(K \| \kappa; \in, \ldots)$ so as to make sure that there is a sequence $(\gamma_q : q < \omega) \in {}^\omega\lambda$ such that $f \restriction \{\pi(\gamma_q) : q < \omega\}$ is not eventually constant. This would contradict Claim 2. $\square$

[14]It is well possible that $\mathcal{P}'_\gamma \neq \mathcal{P}_\gamma$. We'll have to have that $\mathcal{R}'_\gamma = \mathcal{R}_\gamma$, though.

§5. **Open problems.** Many questions remain open. Can Theorem 4.6 be extended to the core model of [14], or to the one of [17] (cf. [15, Problem # 10])? Or can Theorem 4.6 even be extended to extender models which do not know how to fully iterate themselves? Finally: Can one use methods provided by the current paper to get a reasonable lower bound for the consistency strength of the assumption that $(S_n : n < \omega)$ must be mutually stationary provided every individual $S_n \subset \aleph_n$ is stationary?

REFERENCES

[1] A. Andretta, I. Neeman, and J. R. Steel, *The domestic levels of K^c are iterable*, ***Israel Journal of Mathematics***, vol. 125 (2001), pp. 157–201.

[2] J. Baumgartner, *On the size of closed unbounded sets*, ***Annals of Pure and Applied Logic***, vol. 54 (1991), pp. 195–227.

[3] M. Foreman and M. Magidor, *Mutually stationary sequences of sets and the non-saturation of the non-stationary ideal on $P_\kappa(\lambda)$*, ***Acta Mathematica***, vol. 186 (2001), pp. 271–300.

[4] R. Jensen, *A new fine structure for higher core models*, handwritten notes, 1997, available at http://www.mathematik.hu-berlin.de/.

[5] B. Löwe and J. R. Steel, *An introduction to core model theory*, ***Sets and proofs*** (Cooper and Truss, editors), Cambridge University Press, 1999.

[6] W. J. Mitchell, *The core model for sequences of measures*, ***Mathematical Proceedings of the Cambridge Philosophical Society***, vol. 95 (1984), pp. 228–260.

[7] ———, *The core model for sequences of measures II*, typescript.

[8] W. J. Mitchell and E. Schimmerling, *Covering without countable closure*, ***Mathematical Research Letters***, vol. 2 (1995), pp. 595–609.

[9] W. J. Mitchell, E. Schimmerling, and J. R. Steel, *The covering lemma up to a Woodin cardinal*, ***Annals of Pure and Applied Logic***, vol. 84 (1997), pp. 219–255.

[10] W. J. Mitchell and J. R. Steel, ***Fine structure and iteration trees***, Lecture Notes in Logic, vol. 3, Springer-Verlag, 1994.

[11] I. Neeman, *Inner models in the region of a woodin limit of Woodin cardinals*, preprint.

[12] E. Schimmerling and J. R. Steel, *The maximality of the core model*, ***Transactions of the American Mathematical Society***, vol. 351 (1999), no. 8, pp. 3119–3141.

[13] E. Schimmerling and M. Zeman, *A characterization of $\Box_\kappa$ in core models*, ***Journal of Mathematical Logic***, vol. 4 (2004), pp. 1–72.

[14] R. Schindler, *The core model for almost linear iterations*, ***Annals of Pure and Applied Logic***, vol. 116 (2002), pp. 207–274.

[15] R. Schindler and J. R. Steel, *List of open problems in inner model theory*, available at http://wwwmath.uni-muenster.de/math/inst/logik/org/staff/rds/list.html.

[16] R. Schindler, J. R. Steel, and M. Zeman, *Deconstructing inner model theory*, ***The Journal of Symbolic Logic***, vol. 67 (2002), pp. 721–736.

[17] J. Steel, ***The core model iterability problem***, Lecture Notes in Logic, vol. 8, Springer-Verlag, 1996.

INSTITUT FÜR MATHEMATISCHE LOGIK UND GRUNDLAGENFORSCHUNG
UNIVERSITÄT MÜNSTER
48149 MÜNSTER, GERMANY
E-mail: rds@math.uni-muenster.de
URL: http://wwwmath.uni-muenster.de/math/inst/logik/org/staff/rds

THE PAIR $(\aleph_n, \aleph_0)$ MAY FAIL $\aleph_0$-COMPACTNESS

SAHARON SHELAH

Abstract. Let P be a distinguished unary predicate and $\mathbf{K} = \{M : M$ a model of cardinality $\aleph_n$ with P^M of cardinality $\aleph_0\}$. We prove that consistently for $n = 4$, that for some countable first order theory T we have: T has no model in $\mathbf{K}$ whereas every finite subset of T has a model in $\mathbf{K}$. We then show how we prove it also for $n = 2$.

ANNOTATED CONTENTS

0. Introduction 2

1. Relevant identities 7

We deal with the 2-identities we shall use.

2. Definition of the forcing 9

We define (historically) our forcing notion, which depends on Γ, a set of 2-identities and on a model M^* with universe λ and $\aleph_0$ functions.

The program is to force with (the finite support product) $\prod_n \mathbb{P}_{\Gamma_n}$ where the forcing $\mathbb{P}_{\Gamma_n}$ adds a colouring (= a function) $\underset{\sim}{c}_n : [\lambda]^2 \to \aleph_0$ satisfying $ID_2(\underset{\sim}{c}_n) \cap ID^* = \Gamma_n$, but no $\underset{\sim}{c} : [\lambda]^2 \to \aleph_0$ has $ID_2(\underset{\sim}{c})$ too small.

3. Why does the forcing work 14

We state the partition result in the original universe which we shall use (in 3.1). Then we prove that if, e.g., Γ contains only identities which restricted to $\leq m(*)$ elements are trivial, then this holds for the colouring in any $p \in \mathbb{P}_\Gamma$ (see 3.2).

We prove that $\mathbb{P}_\Gamma$ preserves identities from $ID_2(\lambda, \mu)$ which are in Γ (because we allow in the definition of the forcing appropriate amalgamations (see 3.3(1))). We have weaker results for $\prod_n \mathbb{P}_{\Gamma_n}$, (see 3.3(2)).

Key words and phrases. Model theory, two cardinal theorems, compactness, partition theorems.

I would like to thank Alice Leonhardt for the beautiful typing.

Research supported by the United States-Israel Binational Science Foundation. Publication 604.

Logic Colloquium '01
Edited by M. Baaz, S. Friedman, and J. Krajíček
Lecture Notes in Logic, 20

On the other hand, forcing with $\mathbb{P}_\Gamma$ gives a colouring showing relevant 2-identities are not in $ID_2(\lambda, \mu)$. Lastly, we derive the main theorem; e.g., incompactness for $(\aleph_4, \aleph_0)$, (see (3.5)).
4. Improvements and additions 22
We show that we can deal with the pair $(\aleph_2, \aleph_0)$ (see 4.1–4.6).
5. Open problems and concluding remarks 28
We list some open problems, and note a property of $ID(\aleph_n, \aleph_0)$ under the assumption MA $+2^{\aleph_0} > \aleph_n$. We note on when k-simple identities suffice and an alternative proof of $(\aleph_\omega, \aleph_2) \to (2^{\aleph_0}, \aleph_0)$.

§0. **Introduction.** Interest in two cardinal models comes from the early days of model theory, as generalizations of the Lowenheim-Skolem theorem. Already Mostowski [Mo57] considered a related problem concerning generalized quantifiers. Let us introduce the problem. Throughout the paper λ, μ and κ stand for infinite cardinals and n, k for natural numbers.

We consider a countable language = vocabulary τ with a distinguished unary relation symbol P and models M for τ; i.e., τ-models.

0.1. Notation. We let

$$K_{(\lambda,\mu)} =: \{M : \|M\| = \lambda \ \& \ |P^M| = \mu\}.$$

0.2. Definition. (1) We say that $K_{(\lambda,\mu)}$ is $(< \kappa)$-compact when every first order theory T in the vocabulary τ (i.e., in the first order logic $\mathbb{L}(\tau)$) with $|T| < \kappa$, satisfies:

if every finite $t \subseteq T$ has a model in $K_{(\lambda,\mu)}$, then T has a model in $K_{(\lambda,\mu)}$.

We similarly give the meaning to $(\leq \kappa)$-compactness. We say that (λ, μ) is $(< \kappa)$-compact if $K_{(\lambda,\mu)}$ is.

(2) We say that

$$(\lambda, \mu) \to'_\kappa (\lambda', \mu')$$

when for every first order theory T in $\mathbb{L}(\tau)$ with $|T| < \kappa$, if every finite $t \subseteq T$ has a model in $K_{(\lambda,\mu)}$, then T has a model in $K_{(\lambda',\mu')}$. Instead "κ^+" we may write "$\leq \kappa$".

(3) We say that

$$(\lambda, \mu) \to_\kappa (\lambda', \mu')$$

when for every first order theory T of L with $|T| < \kappa$, if T has a model in $K_{(\lambda,\mu)}$, then T has a model in $K_{(\lambda',\mu')}$.

(4) In both $\to'_\kappa$ and $\to_\kappa$ we omit κ if $\kappa = \aleph_0$.

Note. Note that $\to_\kappa$ is transitive and $\to'_\kappa$ is as well. Also note that $\to_{\aleph_0}$ and $\to'_{\aleph_0}$ are equivalent.

We consider the problem of $K_{(\lambda,\mu)}$ being compact. Before we start, we review the history of the problem. Note that a related problem is the one of completeness, i.e., if

$$\{\psi : \psi \text{ has a model in } K_{(\lambda,\mu)}\}$$

is recursively enumerable and other related problems, see in the end. We do not concentrate on those problems here.

We review some of the history of the problem, in an order which is not necessarily chronological.

Some early results on the compactness are due to Furkhen [Fu65]. He showed that

(A) if $\mu^\kappa = \mu$, then $K_{(\lambda,\mu)}$ is $(\le \kappa)$-compact.

The proof is by using ultraproducts over regular ultrafilters on κ, generalizing the well known proof of compactness by ultrapowers. Morley related result is

(B) (Morley [Mo68]) If $\mu^{\aleph_0} \le \mu' \le \lambda' \le \lambda$, then $(\lambda,\mu) \to_{\le\lambda} (\lambda',\mu')$.

Next result we mention is one of Silver [Si71] concerning Kurepa trees,

(C) (Silver [Si71]) From the existence of a strongly inaccessible cardinal, it follows that the following is consistent with ZFC:

$$GCH + (\aleph_3,\aleph_1) \not\to_{\aleph_0} (\aleph_2,\aleph_0).$$

Using special Aronszajn trees Mitchell showed

(D) (Mitchell [Mi72]) From the existence of a Mahlo cardinal, it follows that it is consistent with ZFC to have

$$(\aleph_1,\aleph_0) \not\to_{\aleph_2} (\aleph_2,\aleph_1).$$

A later negative consistency result is the one of Schmerl in [Sc74]

(E) (Schmerl [Sc74]) Con(if $n < m$ then $(\aleph_n,\aleph_{n+1}) \not\to (\aleph_m,\aleph_{m+1})$).

Earlier, Vaught proved two positive results

(F) (Vaught [MV62]) $(\lambda^+,\lambda) \to'_{\aleph_1} (\aleph_1,\aleph_0)$.

Keisler [Ke66] and [Ke66a] has obtained more results in this direction.

(G) (Vaught [Va65]) If $\lambda \ge \beth_\omega(\mu)$ and $\lambda' > \mu'$, then $(\lambda,\mu) \to'_{\le\mu'} (\lambda',\mu')$.

In [Mo68] Morley gives another proof of this result, using Erdős-Rado Theorem and indiscernibles.

Another early positive result is the one of Chang:

(H) (Chang [Ch]) If $\mu = \mu^{<\mu}$ then $(\lambda^+,\lambda) \to'_{\le\mu} (\mu^+,\mu)$.

Jensen in [Jn] uses $\Box_\mu$ to show

(I) (Jensen [Jn]) If $\mathbf{V} = \mathbf{L}$, then $(\lambda^+,\lambda) \to'_{\le\mu} (\mu^+,\mu)$. (The fact that $0^\#$ does not exist suffices.)

Hence, Jensen's result deals with the case of μ is singular, which was left open after the result of Chang. For other early consistency results concerning gap-1 two cardinal theorems, including consistency, see [Sh:269], Cummings, Foreman and Magidor [CFM].

In [Jn] there is actually a simplified proof of (I) due to Silver. A further result of Jensen, using morasses, is:

(J) (Jensen, see [De73] for $n = 2$) If $\mathbf{V} = \mathbf{L}$, then $(\lambda^{+n}, \lambda) \to'_{\leq\mu} (\mu^{+n}, \mu)$ for all $n < \omega$.

Note that by Vaught's result [MV62] stated in (F) we have: the statement in (I), in the result of Chang etc., (λ^+, λ) can be without loss of generality replaced by $(\aleph_1, \aleph_0)$.

Finally, there are many more related results, for example the ones concerning Chang's conjecture. A survey article on the topic was written by Schmerl in [Sc74].

Note that many of the positive rsults above (F)–(J), their proof also gives compactness of the pair, e.g., $(\aleph_0, \aleph_1)$ by [MV62].

We now mention some results of the author which will have a bearing to the present paper.

(α) (Shelah [Sh:8] and the abstract [Sh:E17]). If $K_{(\lambda,\mu)}$ is $(\leq \aleph_0)$-compact, then $K_{(\lambda,\mu)}$ is $(\leq \mu)$-compact and $(\lambda, \mu) \to_{\leq\mu'} (\lambda', \mu')$ when $\lambda \leq \lambda' \leq \mu' \leq \mu$.

More than $(\leq \mu)$-compactness cannot hold for trivial reasons. In the same work we have the analogous result on $\to'$ and:

(β) (Shelah [Sh:8] and the abstract [Sh:E17]) $(\lambda, \mu) \to'_{\aleph_1} (\lambda', \mu')$ is actually a problem on partition relations, (see below), also it implies $(\lambda, \mu) \to'_{\leq\mu'} (\lambda', \mu')$ see 0.4(1) below.

We state a definition from [Sh:8] that will be used here too. We do not consider the full generality of [Sh:8], there problems like considering K with several λ_ℓ-like $(P^2_\ell, <_\ell)$ and $|P^1_\ell| = \mu_\ell$ were addressed.

(We can use below only ordered a and increase h, it does not matter much.)

0.3. Definition. (1) An *identity*[1] is a pair (a, e) where a is a finite set and e is an equivalence relation on the finite subsets of a, having the property

$$b \; e \; c \Rightarrow |b| = |c|.$$

The equivalence class of b with respect to e will be denoted b/e.

(2) We say that $\lambda \to (a, e)_\mu$, if for every $f : [\lambda]^{<\aleph_0} \to \mu$, there is $h : a \xrightarrow{1\text{-}1} \lambda$ such that

$$b \; e \; c \Rightarrow f(h''(b)) = f(h''(c)).$$

[1] identification in the terminology of [Sh:8].

(3) We define

$$ID(\lambda, \mu) =: \{(n, e) \colon n < \omega \ \& \ (n, e) \text{ is an identity and } \lambda \to (n, e)_\mu\}$$

and for $f : [\lambda]^{<\aleph_0} \to X$ we let

$$ID(f) =: \{(n, e) \colon (n, e) \text{ is an identity such that for some one-to-one function } h \text{ from } n = \{0, \dots, n-1\} \text{ to } \lambda \text{ we have } (\forall b, c \subseteq n)(b\, e\, c \Rightarrow f(h''(b)) = f(h''(c)))\}$$

0.4. Claim (Shelah [Sh:8] and the abstract [Sh:E17]). *$(\lambda, \mu) \to'_{\aleph_1} (\lambda', \mu')$ is equivalent to the existence of a function $f : [\lambda']^{<\aleph_0} \to \mu'$ such that*

$$ID(f) \subseteq ID(\lambda, \mu)$$

(*more on this see* [Sh:74, Theorem 3] *statement there on* $\to'_{\aleph_1}$, *see details in* [Sh:E28]).

0.5. Remark. The identities of $(\beth_\omega, \aleph_0)$ are clearly characterized by Morley's proof of Vaught's theorem (see [Mo68]). The identities of $(\aleph_\omega, \aleph_0)$ are stated explicitly in [Sh:37] and [Sh:49], when $\aleph_\omega \le 2^{\aleph_0}$ where it is also shown that $(\aleph_\omega, \aleph_0) \to' (2^{\aleph_0}, \aleph_0)$. For $(\aleph_1, \aleph_0)$, the identities are characterized in [Sh:74] (for some details see [Sh:E28]). The identities for λ-like models, λ strongly ω-Mahlo are clear, see Schmerl and Shelah [ScSh:20] (for strongly n-Mahlo this gives positive results, subsequently sharpened (replacing $n+2$ by n) and the negative results proved by Schmerl, see [Sc85]).

By the referee request we indicate the proof for $(\aleph_\omega, \aleph_0)$ in 5.12.

We generally neglect here three cardinal theorems and λ-like models (and combinations, see [Sh:8], [Sh:18]), the positive results (like 0.4) are similar. Recently Shelah and Vaananen deal with recursiveness, completeness, and identities [ShVa:790] and see [ShVa:E47].

In Gilchrist, Shelah [GcSh:491] and [GcSh:583], we dealt with 2-identities.

0.6. Definition. (1) A two-identity or 2-identity[2] is a pair (a, e) where a is a finite set and e is an equivalence relation on $[a]^2$. Let $\lambda \to (a, e)_\mu$ mean $\lambda \to (a, e^+)_\mu$ where $be^+c \leftrightarrow (bec) \vee (b = c \subseteq a)$ for any $b, c \subseteq a$.

(2) We defined

$$ID_2(\lambda, \mu) =: \{(n, e) \colon (n, e) \text{ is a 2-identity and } \lambda \to (n, e)_\mu\}$$

we define $ID_2(f)$ when $f : [\lambda]^2 \to X$ as

[2]It is not an identity as e is an equivalence relation on too small set.

$$\{(n,e)\colon (n,e) \text{ is a two-identity such that for some } h, \text{ a one-to-one function from } \{0,\ldots,n-1\} \text{ into } \lambda \text{ we have } \{\ell_1,\ell_2\}e\{k_1,k_2\} \text{ implies that } \ell_1 \neq \ell_2 \in \{0,\ldots,n-1\}, k_1 \neq k_2 \in \{0,\ldots,n-1\} \text{ and } f(\{h(\ell_1),h(\ell_2)\}) = f(\{h(k_1),h(k_2)\})\}.$$

(3) Let us define

$$ID_2^{\circledast} =: \{({}^n2,e)\colon ({}^n2,e) \text{ is a two-identity and if } \{\eta_1,\eta_2\} \neq \{\nu_1,\nu_2\} \text{ are } \subseteq {}^n2, \text{ then } \{\eta_1,\eta_2\}e\{\nu_1,\nu_2\} \Rightarrow \eta_1 \cap \eta_2 = \nu_1 \cap \nu_2\}.$$

By [Sh:49], under the assumption $\aleph_\omega < 2^{\aleph_0}$, the families $ID_2(\aleph_\omega,\aleph_0)$ and $ID_2^{\circledast}$ coincide (up to an isomorphism of identities). In Gilchrist and Shelah [GcSh:491] and [GcSh:583] we considered the question of the equality between these $ID_2(2^{\aleph_0},\aleph_0)$ and $ID_2^{\circledast}$ under the assumption $2^{\aleph_0} = \aleph_2$. We showed that consistently the answer may be "yes" and may be "no".

Note that $(\aleph_n,\aleph_0) \not\twoheadrightarrow (\aleph_\omega,\aleph_0)$ so $ID(\aleph_2,\aleph_0) \neq ID(\aleph_\omega,\aleph_0)$, but for identities for pairs (i.e., ID_2) the question is meaningful.

The history of the problem suggested to me that there should be a model where $K_{(\lambda,\mu)}$ is not $\aleph_0$-compact for some λ,μ; I do not know about the opinion of others and it was not easy for me as I thought a priori. As mathematicians do not feel that a strong expectation makes a proof, I was quite happy to be able to prove the existence of such a model. This was part of my lectures in a 1995 seminar in Jerusalem and notes of the lecture were taken by Mirna Džamonja and I thank her for this, but because the proof was not complete, it was delayed.

The following is the main result of this paper (proved in 3.5):

0.7. Main Theorem. Con(the pair $(\aleph_n,\aleph_0)$ is not $\aleph_0$-compact $+\, 2^{\aleph_0} \geq \aleph_n$) *for* $n \geq 4$.

Later in the paper we deal with the case $n = 2$ which is somewhat more involved. This is the simplest case by a reasonable measure: if you do not like to use large cardinals then assuming that there is no inaccessible in **L**, all pairs (μ^+,μ) are known to be $\aleph_0$-compact and if $\mathbf{V} = \mathbf{L}$ also all logic $L(\exists^{\geq\lambda})$, $\lambda > \aleph_0$ are (by putting together already known results; $\mathbf{V} = \mathbf{L}$ is used just to imply that there is no limit, uncountable not strong limit cardinal).

How much this consistency result will mean to a model theorist, let us not elaborate, but instead say an anecdote about Jensen. He is reputed to have said: "When I started working on the two-cardinal problem, I was told it was the heart of model theory. Once I succeeded to prove something, they told me what I did was pure set theory, and were not very interested"; also, mathematics is not immune to fashion changes.

My feeling is that there are probably more positive theorems in this subject waiting to be discovered. Anyway, let us state the following

THESIS. Independence results help us clear away the waste, so the possible treasures can stand out.

Of course, I have to admit that, having spent quite some time on the independence results, I sometimes look for the negative of the picture given by this thesis.

The strategy of our proof is as follows. It seems natural to consider the simplest case, i.e., that of two-place functions, and try to get the incompactness by constructing a sequence $\langle f_k : k < \omega \rangle$ of functions from $[\aleph_n]^2$ into $\aleph_0$ such that for all n we have $ID_2(f_k) \supseteq ID_2(f_{k+1})$, yet for no $f : [\aleph_n]^2 \to \aleph_0$ do we have $ID_2(f) \subseteq \bigcap_{k<\omega} ID_2(f_k)$. This suffices. Related proofs to our main results were [Sh:522].

Note that another interpretation of 0.7 is that if we add to first order logic the cardinality quantifiers $(\exists^{\geq \lambda} x)$ for $\lambda = \aleph_1, \aleph_2, \aleph_3, \aleph_4$ we get a noncompact logic.

We thank the referee for many helpful comments and the reader should thank him also for urging the inclusion of several proofs.

This work is continued in [ShVa:790] and [Sh:824].

§1. Relevant identities. We commence by several definitions. For simplicity, for us all identities, colorings etc. will be 2-place.

1.1. DEFINITION. (1) For $m, \ell < \omega$ let

$$\text{dom}_{\ell,m} = \{\eta \in {}^{\ell+1}\omega : \eta \restriction \ell \in {}^{\ell}2 \text{ and } \eta(\ell) < m\}$$

$$\begin{aligned} ID^1_{\ell,m} = \{(\text{dom}_{\ell,m}, e) : \ & e \text{ is an equivalence relation on } [\text{dom}_{\ell,n}]^2 \\ & \text{such that } \{\eta_1, \eta_2\} e \{\nu_1, \nu_2\} \ \& \ \{\eta_1, \eta_2\} \neq \{\nu_1, \nu_2\} \\ & \Rightarrow \eta_1 \cap \eta_2 = \nu_1 \cap \nu_2 \wedge lg(\eta, m_2) < \ell\}. \end{aligned}$$

(2) Let

$$\begin{aligned} ID^1_\ell &= \cup\{ID^1_{\ell,m} : m < \omega\}, \\ ID^1 &= \cup\{ID^1_\ell : \ell < \omega\}. \end{aligned}$$

(3) For $\mathbf{s} = (\text{dom}_{\ell,m}, e) \in ID^1_{\ell,m}$ and $\nu \in {}^{\ell \geq}2$ let

$$\text{dom}^{[\nu]}_{\ell,m} = \{\rho \in \text{dom}_{\ell,m} : \nu \trianglelefteq \rho\}$$

and if $\nu \in {}^{\ell >}2$ we let

$$e_{\langle \nu \rangle}(\mathbf{s}) = e \restriction \{\{\eta_0, \eta_1\} : \nu^\frown\langle i \rangle \triangleleft \eta_i \text{ for } i = 0, 1\}.$$

We use $\mathbf{s}$ to denote identities so $\mathbf{s} = (\text{dom}_{\mathbf{s}}, e(\mathbf{s}))$; and if $\mathbf{s} \in ID^1$ then let $\mathbf{s} = (\text{dom}_{\ell(\mathbf{s}), m(\mathbf{s})}, e(\mathbf{s}))$.

(4) An equivalence class is nontrivial if it is not a singleton.

Note that it follows that every e-equivalence class is an $e_{\langle \nu \rangle}$-equivalence class for some ν. We restrict ourselves to

1.2. DEFINITION. (1) Let $ID^2_{\ell,m}$ be the set of $\mathbf{s} \in ID^1_{\ell,m}$ such that for every $\nu \in {}^{\ell>}2$ the equivalence relation $e_{\langle \nu \rangle}(\mathbf{s})$ has at most one non-singleton equivalence class, which we call $e_{[\nu]} = e_{[\nu]}(\mathbf{s})$.

So we also allow $e_{\langle \nu \rangle} =$ empty, in which case we choose a representative equivalence class $e_{[\nu]}$ as the first one under, say, lexicographical ordering.

(2) $ID^2_\ell = \bigcup\{ID^2_{\ell,m} : m < \omega\}$.

1.3. DEFINITION. (1) We define for $k < \omega$ when $\mathbf{s} = (\mathrm{dom}_{\ell,m}, e)$ is k-nice: the demands are

(a) $\mathbf{s} \in ID^1_{\ell,m}$,

(b) if $\nu \in {}^\ell 2$ and $(\nu \restriction i)^\frown\langle 1 - \nu(i)\rangle \lhd \rho_i \in \mathrm{dom}_{\ell,m}$ for each $i < \ell$ then $\{\eta : \nu \lhd \eta \in \mathrm{dom}_{\ell,m}$ and for each $i < \ell$ the set $\{\rho_i, \eta\}/e$ is not a singleton$\}$ has at least two members,

(c) the graph $H[e]$, see below, has no cycle $\leq k$ (for $k \leq 2$ this holds trivially),

(d) the graph $H[e]$ has a cycle.

(2) We can interpret $\mathbf{s} = (\mathrm{dom}_{\ell,m}, e)$ as the graph $H[\mathbf{s}]$ with set of nodes $\mathrm{dom}_{\ell,m}$ and set of edges $\{\{\eta, \nu\} : \{\eta, \nu\}/e$ not a singleton (and of course $\eta \neq \nu$ are from $\mathrm{dom}_{\ell,m}$)$\}$.

(3) We may write $e(\mathbf{s})$ instead of $\mathbf{s}$ if $\mathrm{dom}_{\ell,m}$ can be reconstructed from e (e.g., if the graph has no isolated point (e.g., if it is 0-nice, see clause (b) of part (1)). Saying nice we mean $[\log_2(m)]$-nice.

1.4. CLAIM. (1) *If (λ, μ) is $\aleph_0$-compact and $c_n : [\lambda]^{<\aleph_0} \to \mu$ and $\Gamma_n = ID(c_n)$ for $n < \omega$, then for some $c : [\lambda]^{<\aleph_0} \to \mu$ we have $ID(c) \subseteq \bigcap_{n<\omega} \Gamma_n$ (in fact equality holds).*

(2) *Similarly using ID_2.*

REMARK. By the same proof, if we just assume $(\lambda_1, \mu_1) \to'_{\aleph_1} (\lambda_2, \mu_2)$ and $c_n : [\lambda_1]^{<\aleph_0} \to \mu_1$, then we can deduce that there is $c : [\lambda_2]^{<\aleph_0} \to \mu_2$ satisfying $ID(c) \subseteq \bigcap_{n<\omega} ID(c_n)$.

PROOF. Straightforward.

(1) In details, let F_m be an m-place function symbol and P the distinguished unary predicate and let $T = \{\psi_n : n < \omega\} \cup \{\neg\psi_{\mathbf{s}} : c$ is an identity of the form (n, e) not from $\bigcap_{n<\omega} ID(c_n)\}$ where

(a) $\psi_n = (\forall x_0)(\forall x_1)\ldots(\forall x_{n-1})(P(F_n(x_0, \ldots, x_{n-1}))$ &
$\bigwedge\{(\forall x_0)\ldots(\forall x_n)F_n(x_0, \ldots, x_{n-1}) = F_n(x_{\pi(0)}, \ldots, x_{\pi(n-1)}) :$
π a permutation of $\{0, \ldots, n-1\}\}$,

(b) if $\mathrm{s} = (n, e)$ is an identity then $\psi_{\mathrm{s}} = (\exists x_0) \ldots (\exists x_{n-1})[\bigwedge_{\ell<m<n} x_\ell \neq x_m$ & $\bigwedge_{b,c \subseteq n, bec} F_{|b|}(\ldots, x_\ell, \ldots)_{\ell \in b} = F_{|b|}(\ldots, x_\ell, \ldots,)_{\ell \in c}]$.

Clearly T is a (first order) countable theory so it suffices to prove the following two statements $\boxtimes_1, \boxtimes_2$.

$\boxtimes_1$ if $M \in K_{(\lambda,\mu)}$ is a model of T, then there is $c\colon [\lambda]^{<\aleph_0} \to \mu$ such that $ID(c) \subseteq \bigcap_{n<\omega} \Gamma_n$.
[Why does $\boxtimes_1$ hold? There is $N \cong M$ such that N has universe $|N| = \lambda$ and $P^N = \mu$. Now we define c: if $u \in [\lambda]^{<\aleph_0}$, let $\{\alpha_\ell^u : \ell < |u|\}$ enumerate u in increasing order and let $c(u) = F_{|u|}^N(\alpha_0^u, \alpha_1^u, \ldots, \alpha_{|u|-1}^u)$. Note that because $N \models \psi_n$ for $n < \omega$ clearly c is a function from $[\lambda]^{<\aleph_0}$ into μ. Also because $N \models \psi_n$, if $n < \omega$ and $\alpha_0, \ldots, \alpha_{n-1} < \lambda$ are with no repetitions then $F_n^N(\alpha_0, \ldots, \alpha_{n-1}) = c\{\alpha_0, \ldots, \alpha_{n-1}\}$. Now if $\mathrm{s} \in ID(c)$ let $\mathrm{s} = (n, e)$ and let $u = \{\alpha_0, \ldots, \alpha_{n-1}\} \in [\lambda]^n \subseteq [\lambda]^{<\aleph_0}$ exemplify that $\mathrm{s} \in ID(c)$, hence easily $N \models \psi_{\mathrm{s}}$ so necessarily $\neg\psi_{\mathrm{s}} \notin T$ hence $\mathrm{s} \in \bigcap_{n<\omega} \Gamma_n$. This implies that $ID(c) \subseteq \bigcap_{n<\omega} \Gamma_n$ is as required.]

$\boxtimes_2$ if $T' \subseteq T$ is finite then T' has a model in $K_{(\lambda,\mu)}$.

[Why? So T' is included in $\{\psi_m : m < m^*\} \cup \{\neg\psi_{\mathrm{s}_k} : k < k^*\}$ for some $m^*, k^* < \omega$, $\mathrm{s}_k = (n_k, e_k)$ an identity not from $\bigcap_{\ell<\omega} ID(c_\ell)$, so we can find $\ell(k) < \omega$ such that $\mathrm{s}_k \notin ID(c_{\ell(k)})$. Let H be a one-to-one function from ${}^{k^*}\mu$ into μ. We define a model M:

(a) its universe $|M|$ is λ,
(b) $P^M = \mu$,
(c) if $n < \omega, \{\alpha_0, \ldots, \alpha_{n-1}\} \in [\lambda]^n$ then
$$F_n^M(\alpha_0, \ldots, \alpha_{n-1}) = H(c_{\ell(0)}\{\alpha_0, \ldots, \alpha_{n-1}\}, c_{\ell(1)}\{\alpha_0, \ldots, \alpha_{n-1}\}, \ldots, c_{\ell(k^*-1)}\{\alpha_0, \ldots, \alpha_{n-1}\}).$$

If $n < \omega$ and $\alpha_0, \ldots, \alpha_{n-1} < \lambda$ is with repetitions we let $F_n^M(\alpha_0, \ldots, \alpha_{n-1}) = 0$. Clearly M is a model from $K_{(\lambda,\mu)}$ of the vocabulary of T. Also M satisfies each sentence ψ_m by the way we have defined F_m^M. Lastly, for $k < k^*, M \models \neg\psi_{\mathrm{s}_k}$ because $(n_k, e_k) \notin ID(c_{\ell(k)})$ by the choice of the F_n's as H is a one-to-one function.] $\square_{1.4}$

Of course

1.5. Observation. (1) *For every* $\ell < \omega, k < \omega$ *for some* m *there is a* k*-nice* $\mathrm{s} = (\mathrm{dom}_{\ell,m}, e)$.
(2) *If* s *is* k*-nie and* $m \leq k$, *then* s *is* m*-nice.*

§2. Definition of the forcing. We have outlined the intended end of the proof at the end of the introductory section. It is to construct a sequence of functions $\langle f_n : n < \omega \rangle$ with certain properties. As we have adopted the decision of dealing only with 2-identities from ID^1_ℓ, all our functions will be colorings of pairs, and we shall generally use the letter c for them.

Our present Theorem 0.7 deals with $\aleph_4$, but we may as well be talking about some $\aleph_{n(*)}$ for a fixed natural number $n(*) \geq 2$. Of course, the set of identities will depend on $n(*)$. We shall henceforth work with $n(*)$, keeping in mind that the relevant case for Theorem 0.7 is $n(*) = 4$. Also we fix $\ell(*) = n(*) + 1$ on which the identities depend (but vary m). Another observation about the proof is that we can replace $\aleph_0$ with an uncountable cardinal κ such that $\kappa = \kappa^{<\kappa}$ replacing $\aleph_n$ by κ^{+n}. Of course, the pair (κ^{+n}, κ) is compact because $[\kappa = \kappa^{\aleph_0} < \lambda \Rightarrow (\kappa, \lambda)$ is $\leq \kappa$-compact], however, much of the analysis holds.

We may replace $(\aleph_n, \aleph_0)$ by $(\kappa^{+n(*)}, \kappa)$ if $\kappa^{+n(*)} \leq 2^{\aleph_0}$; we hope to return to this elsewhere.

To consider (κ^+, κ) we need large cardinals; even more so for considering $(\mu^+, \mu), \mu$ strong limit singular of cofinality $\aleph_0$, and even $(\kappa^{+n}, \kappa), \mu \leq \kappa < \kappa^{+n} \leq \mu^{\aleph_0}$.

We now describe the idea behind the definition of the forcing notion we shall be concerned with. Each "component" of the forcing notion is supposed to add a coloring

$$c \colon [\lambda]^2 \to \mu$$

preserving some of the possible 2-identities, while "killing" all those which were not preserved, in other words it is concerned with adding f_n; specifically we concentrate on the case $\lambda = \aleph_{n(*)}, \mu = \aleph_0$. Hence, at first glance the forcing will be defined so that to preserve an identity we have to work hard proving some kind of amalgamation for the forcing notion, while killing an identity is a consequence of adding a colouring exemplifying it. By preserving a set Γ of identities, we mean that $\Gamma \subseteq ID(c)$, and more seriously $\Gamma \subseteq ID_2(\lambda, \mu)$; we restrict ourselves to some ID^*, an infinite set of 2-identity.

We shall choose $ID^* \subseteq ID^{\circledast}_2$ below small enough such we can handle the identities in it..

We define the forcing by putting in its definition, for each identity that we want to preserve, a clause specifically assuring this. Naturally this implies that not only the desired identities are preserved, but also some others so making an identity be not in $ID(\lambda, \mu)$ becomes now the hard part. So, we lower our sights and simply hope that, if $\Gamma \subseteq ID^*$ is the set of identities that we want to preserve, than no identity $(a, e) \in ID^* \setminus \Gamma$ is preserved; this may depend on Γ.

How does this control over the set of identities help to obtain the non-compactness? We shall choose sets $\Gamma_n \subseteq ID^*$ of possible identities for $n < \omega$. The forcing we referred to above, let us call it $\mathbb{P}^{\Gamma_n}$, add a colouring $\underset{\sim}{c}_n \colon [\lambda]^2 \to \omega$ such that $ID_2(\underset{\sim}{c}_n)$ includes Γ_n and is disjoint to $ID^* \setminus \Gamma_n$; also it will turn out to have a strong form of the ccc. We shall force with $\mathbb{P} =: \prod_{n \in \omega} \mathbb{P}^{\Gamma_n}$, where the product is taken with finite support. Because of the strong version of ccc possessed by each $\mathbb{P}^{\Gamma_n}$, also $\mathbb{P}$ will have ccc. Now, in $\mathbf{V}^{\mathbb{P}}$ we have for every n

a colouring $c_n\colon [\lambda]^2 \to \omega$ which preserves the identities in Γ_n, moreover $\mathbf{V}^{\mathbb{P}} \models \Gamma_n \subseteq ID(c_n) \cap ID^*$.

We shall in fact obtain that

$$ID^* = \Gamma_0 \supseteq \Gamma_1 \ \&\ \Gamma_1 \supseteq \Gamma_2 \ \&\ \cdots \ \&\ \bigcap_{n<\omega} \Gamma_n = \emptyset \ \&\ ID(c_n) \cap \Gamma_0 = \Gamma_n.$$

If we have $\aleph_0$-compactness for $(\lambda, \aleph_0)$, then by 1.4(2) there must be a colouring $c\colon [\lambda]^2 \to \omega$ in $\mathbf{V}^{\mathbb{P}}$ such that

$$ID_2(c) \cap \Gamma_0 \subseteq \bigcap_{n<\omega} \Gamma_n = \emptyset.$$

We can find a name $\underset{\sim}{c}$ in $\mathbf{V}$ for such c, so by ccc, for every $\{\alpha, \beta\} \in [\lambda]^2$, the name $\underset{\sim}{c}(\{\alpha, \beta\})$ depends only on $\aleph_0$ "coordinates". At this point a first approximation to what we do is to apply a relative of Erdős-Rado theorem to prove that there are an n, a large enough $w \subseteq \lambda$ and for every $\{\alpha, \beta\} \in [w]^2$ a condition $p_{\{\alpha,\beta\}} \in \prod_{\ell<n} \mathbb{P}^{\Gamma_\ell}$, such that $p_{\{\alpha,\beta\}}$ forces a value to $\underset{\sim}{c}(\{\alpha, \beta\})$ in a "uniform" enough way. We shall be able to extend enough of the conditions $p_{\{\alpha,\beta\}}$ by a single condition p^* in $\prod_{\ell<n} \mathbb{P}^{\Gamma_\ell}$, which gives an identity in $ID_2(\underset{\sim}{c})$ which belongs to $\bigcap_{\ell<n} \Gamma_\ell \setminus \Gamma_n$, contradiction.

Before we give the definition of the forcing, we need to introduce a notion of closure. The properties of the closure operation are the ones possible to obtain for $(\lambda, \aleph_0)$, but not for $(\aleph_\omega, \aleph_0)$. We of course need to use somewhere such a property, as we know in ZFC that $(\aleph_\omega, \aleph_0)$ has all those identities, i.e., $ID_2^{\circledast} = ID_2(\lambda, \aleph_0)$.

On a similar proof see [Sh:424] (for ω-place functions) and also (2-place functions), [Sh:522]. The definition of the closure in [GcSh:491] is close to ours, but note that the hard clause from [GcSh:491] is not needed here.

2.1. Definition. Let $ID^*_{\ell(*)} =: \{\mathrm{s} \in ID^2_{\ell(*)}\colon \mathrm{s} \text{ is 0-nice}\}$.

Remark. We can consider $\{\mathrm{s}_n\colon n < \omega\}$, which hopefully will be independent, i.e., for every $X \subseteq \omega$ for some c.c.c. forcing notion $\mathbb{P}$, in $\mathbf{V}^{\mathbb{P}}$ we have $\lambda \to (\mathrm{s}_n)_\mu$ iff $n \in X$. It is natural to try $\{\mathrm{s}_n\colon n < \omega\}$ where $\mathrm{s}_n = (\mathrm{dom}_{\ell(*),m_n}, e_n)$ where $m_n = n$ (or 2^{2^n} may be more convenient) and e_n is $[\log\log(n)]$-nice.

2.2. Definition (λ is our fixed cardinal). (1) Let M^* (or M^*_λ) be a model with universe λ, countable vocabulary, and its relations and functions are exactly those defined in $(\mathscr{H}(\chi), \in, <^*_\chi)$ for $\chi = \lambda^+$ (and some choice of $<^*_\chi$, a well ordering of $\mathscr{H}(\chi)$).

(2) For $\bar{a} \in {}^{\omega>}(M^*_\lambda)$ let $c\ell_\ell(\bar{a}) = \{\beta < \lambda\colon$ for some first order $\varphi(y, \bar{x})$ we have $M^*_\lambda \models \varphi[\beta, \bar{a}]$ & $(\exists^{\leq \aleph_\ell} x)\varphi(x, \bar{a})\}$ and $c\ell(\bar{a}) = \{\beta < \lambda\colon$ for some first order $\varphi(y, \bar{x})$ we have $M^*_\lambda \models \varphi[\beta, \bar{a}]$ & $(\exists^{\leq \aleph_0} x)\varphi(x, \bar{a})\}$.

(3) For a model M and $A \subseteq M$ let $c\ell_M(A)$ be the smallest set of elements of M including A and closed under the functions of M (so including the individual constants).

Note that

2.3. FACT. If $\beta_0, \beta_1 \in c\ell_{\ell+1}(\bar{\alpha})$ then for some $i \in \{0,1\}$ we have $\beta_i \in c\ell_\ell(\bar{\alpha}\hat{}\langle\beta_{1-i}\rangle)$.

PROOF. Easy.

The idea of our forcing notion is to do historical forcing (see [RoSh:733] for more on historical forcing and its history). That is, we put in only those conditions which we have to put in order to meet our demands, so every condition in the forcing has a definite rule of creation. In particular, (see below), in the definition of our partial colourings, we avoid giving the same color to any pairs for which we can afford this, if the rule of creation is to be respected. We note that the situation here is not as involved as the one of [RoSh:733], and we do not in fact need the actual history of every condition.

We proceed to the formal definition of our forcing.

Clearly Case 0 for $k \geq 0$ is not necessary from a historical point of view but it simplifies our treatment later; also Case 1 is used in clause (η) of Case 3.

Note that in Case 2 below we do not require that the conditions are isomorphic over their common part (which is natural for historic forcing) as the present choice simplifies clause (ζ)(iv) in Case 3.

2.4. MAIN DEFINITION. Let $n(*) \geq 2, n(*) \leq \ell(*) < \omega, \lambda = \aleph_{n(*)}, \mu = \aleph_0$ be fixed. All closure operations we shall use are understood to refer to $M^*_{\aleph_{n(*)}}$ from 2.2(2). Let $\Gamma \subseteq ID^*_{\ell(*)}$ be given. For two sets u and v of ordinals with $|u| = |v|$, we let $OP_{v,u}$ stand for the unique order preserving 1–1 function from u to v.

We shall define $\mathbb{P} =: \mathbb{P}_\Gamma = \mathbb{P}^\lambda_\Gamma$, it is $\subseteq \mathbb{P}^*_\lambda$.

Members of $\mathbb{P}^*_\lambda$ are the pairs of the form $p = (u, c) =: (u^p, c^p)$ with

$$u \in [\lambda]^{<\aleph_0} \text{ and } c : [u]^2 \rightarrow \omega.$$

The order in $\mathbb{P}^*_\lambda$ is defined by

$$(u_1, c_1) \leq (u_2, c_2) \Leftrightarrow (u_1 \subseteq u_2 \ \& \ c_1 = c_2 \restriction [u_1]^2).$$

For $p \in \mathbb{P}^*_\lambda$ let $n(p) = \sup(\mathrm{Rang}(c^p)) + 1$; this is $< \omega$.

We now say which pairs (u, c) of the above form (i.e., $(u, c) \in \mathbb{P}^*_\lambda$) will enter $\mathbb{P}$. We shall have $\mathbb{P} = \bigcup_{k<\omega} \mathbb{P}_k$ where $\mathbb{P}_k =: \mathbb{P}^{\lambda,\Gamma}_k$ are defined by induction on $k < \omega$, as follows.

CASE 0. $k = 4\ell$. If $k = 0$ let $\mathbb{P}_0 =: \{(\emptyset, \emptyset)\}$.

If $k = 4l > 0$, a pair $(u, c) \in \mathbb{P}_k$ iff for some $(u', c') \in \bigcup_{m<k} \mathbb{P}_m$ we have $u \subseteq u'$ and $c = c' \restriction [u]^2$; we write $(u, c) = (u', c') \restriction u$.

CASE 1. $k = 4\ell+1$. (This rule of creation is needed for density arguments.)

A pair (u, c) is in $\mathbb{P}_k$ iff (it belongs to $\mathbb{P}^*_\lambda$ and) there is a $(u_1, c_1) \in \bigcup_{m<k}\mathbb{P}_m$ and $\alpha < \lambda$ satisfying $\alpha \notin u_1$ such that:

(a) $u = u_1 \cup \{\alpha\}$,
(b) $c \restriction [u_1]^2 = c_1$ and
(c) For every $\{\beta, \gamma\}$ and $\{\beta', \gamma'\}$ in $[u]^2$ which are not equal, if $c(\{\beta, \gamma\})$ and $c(\{\beta', \gamma'\})$ are equal, then $\{\beta, \gamma\}, \{\beta', \gamma'\} \in [u_1]^2$. (Hence, c does not add any new equalities except for those already given by c_1.)

CASE 2. $k = 4\ell+2$. (This rule of creation is needed for free amalgamation, used in the Δ-system arguments for the proof of the c.c.c..)

A pair (u, c) is in $\mathbb{P}_k$ iff (it belongs to $\mathbb{P}^*_\lambda$ and) there are $(u_1, c_1), (u_2, c_2) \in \bigcup_{m<k}\mathbb{P}_m$ for which we have

(a) $u = u_1 \cup u_2$.
(b) $c \restriction [u_1]^2 = c_1$ and $c \restriction [u_2]^2 = c_2$.
(c) c does not add any unnecessary equalities, i.e., if $\{\beta, \gamma\}$ and $\{\beta', \gamma'\}$ are distinct and in $[u]^2$ and $c(\{\beta, \gamma\}) = c(\{\beta', \gamma'\})$, then $\{\{\beta, \gamma\}, \{\beta', \gamma'\}\} \subseteq [u_1]^2 \cup [u_2]^2$.
Note that $[u_1]^2 \cap [u_2]^2 = [u_1 \cap u_2]^2$
(d) $c\ell_0(u_1 \cap u_2) \cap (u_1 \cup u_2) \subseteq u_1$ (usually $c\ell_0(u_1 \cap u_2) \cap (u_1 \cup u_2) \subseteq u_1 \cap u_2$) is O.K. too for present §2, §3 but not, it seems, in 4.6).

MAIN RULE.

CASE 3. $k = 4\ell + 3$. (This rule[3] is like the previous one, but the amalgamation is taken over a graph $\mathbf{s} = (\mathrm{dom}_{\ell(*),m}, e) \in \Gamma$).

A pair $(u, c) \in \mathbb{P}_k$ iff there are $\mathbf{s} = (\mathrm{dom}_{\ell(*),m(*)}, e) \in \Gamma$ and a sequence of conditions

$$\bar{p} = \langle p_y \colon y \in Y \rangle \text{ where } Y = \{y \in [\mathrm{dom}_\mathbf{s}]^2 \colon |y/e| > 1\}$$

from $\bigcup_{m<k}\mathbb{P}_m$ AND we have a sequence of finite sets $\bar{v} = \langle v_t \colon t \in Y^+ \rangle$ where

$$Y^+ = \{t \colon t \in Y \text{ or } t = \emptyset \text{ or } t = \{\eta\}, \text{ where } \eta \in \mathrm{dom}_\mathbf{s}\}$$

such that

(a) $u = \bigcup\{u^{p_y} \colon y \in Y\}$,
(b) $(u, c) \in \mathbb{P}^*_\lambda$ and $c \restriction [u^{p_y}]^2 = c^{p_y}$ for all $y \in Y$,
(c) if $\alpha_1 \neq \alpha_2, \beta_1 \neq \beta_2$ are from u and $\{\alpha_1, \alpha_2\} \neq \{\beta_1, \beta_2\}$ and $c\{\alpha_1, \alpha_2\} = c\{\beta_1, \beta_2\}$ then $(\exists y)[\{\alpha_1, \alpha_2\} \subseteq u^{p_y}]$ and $(\exists y)[\{\beta_1, \beta_2\} \subseteq u^{p_y}]$,
(d) $v_t \cap v_s \subseteq v_{t \cap s}$ for $t, s, \in Y^+$,
(e) $c\ell_0(v_t) \cap u^{p_y} \subseteq v_t$ for all $y \in Y$ and $t \in \{\emptyset\} \cup \{\{\eta\} \colon \eta \in \mathrm{dom}_\mathbf{s}\}$,
(f) $u^{p_y} \subseteq v_y$ for all $y \in Y$,

[3]You may understand it better seeing how it is used in the proof of 3.3.

(g) if $y_1, y_2 \in Y$ and $t \in \{\emptyset\} \cup \{\{\eta\} : \eta \in \mathrm{dom}_s\}$ and $t = y_1 \cap y_2$, then $p_{y_1} \restriction v_t = p_{y_2} \restriction v_t$; equivalently: $\{p_\eta : \eta \in Y\}$ has a common upper bound in $\mathbb{P}^*_\lambda$.

2.5. CLAIM. (1) *$\mathbb{P}^\lambda_\Gamma$ satisfies the c.c.c. and even the Knaster condition.*
(2) *For each $\alpha < \lambda$ the set $\mathcal{J}_\alpha = \{p \in \mathbb{P}^\lambda_\Gamma : \alpha \in u^p\}$ is dense open.*
(3) $\Vdash_{\mathbb{P}^\lambda_\Gamma}$ *"$\underset{\sim}{c} = \bigcup\{c^p : p \in \underset{\sim}{G}\}$ is a function from $[\lambda]^2$ to ω".*

PROOF. (1) By Case 2.
In detail, assume that $p_\varepsilon \in \mathbb{P}^\lambda_\Gamma$ for $\varepsilon < \omega_1$ and let $p_\varepsilon = (u_\varepsilon, c_\varepsilon)$. As each u_ε is a finite subset of λ, by the Δ-system lemma without loss of generality for some finite $u^* \subseteq \lambda$ we have: if $\varepsilon < \zeta < \omega_1$ then $u_\varepsilon \cap u_\zeta = u^*$. By further shrinking, without loss of generality $\alpha \in u^* \Rightarrow \langle |u_\varepsilon \cap \alpha| : \varepsilon < \omega_1\rangle$ is constant and $\varepsilon < \zeta < \omega_1 \Rightarrow |u_\varepsilon| = |u_\zeta|$. Also without loss of generality the set $\{(\ell, m, k)$: for some $\alpha \in u_\varepsilon$ and $\beta \in u_\varepsilon$ we have $\ell = |\alpha \cap u_\varepsilon|, m = |\beta \cap u_\varepsilon|$ and $k = c_\varepsilon\{\alpha, \beta\}\}$ does not depend on ε. We can conclude that $\varepsilon < \zeta < \omega_1 \Rightarrow OP_{u_\zeta, u_\varepsilon}$ maps p_ε to p_ζ over u^*. Clearly for $\varepsilon < \omega_1$, the set $c\ell(u_\varepsilon)$ is countable hence for every $\zeta < \omega_1$ large enough we have $u_\zeta \cap c\ell_0(u_\varepsilon) = u^*$ so restricting $\langle p_\varepsilon : \varepsilon < \omega_1\rangle$ to a club we get that $\varepsilon < \zeta < \omega_1 \Rightarrow c\ell_0(u_\varepsilon) \cap u_\zeta = u^*$ (this is much more than needed). Now for any $\varepsilon < \zeta < \omega_1$ we can define $q_{\varepsilon,\zeta} = (u_{\varepsilon,\zeta}, c_{\varepsilon,\zeta})$ with $u_{\varepsilon,\zeta} = u_\varepsilon \cup u_\zeta$ and $c_{\varepsilon,\zeta} : [u_{\varepsilon,\zeta}]^2 \to \omega$ is defined as follows: for $\alpha < \beta$ in $u_{\varepsilon,\zeta}$ let $c_{\varepsilon,\zeta}\{\alpha, \beta\}$ be $c_\varepsilon\{\alpha, \beta\}$ if defined, $c_\zeta\{\alpha, \beta\}$ if defined, and otherwise $\sup(\mathrm{Rang}(c_\varepsilon)) + 1 + (|u_{\varepsilon,\zeta} \cap \alpha| + |u_{\varepsilon,\zeta} \cap \beta|)^2 + |u_{\varepsilon,\zeta} \cap \alpha|$. Now $q \in \mathbb{P}^\lambda_\Gamma$ by Case 2, and $p_\varepsilon \le q_\varepsilon, p_\zeta \le q_{\varepsilon,\zeta}$ by the definition of order.
(2) By Case 1.
In detail, let $p \in \mathbb{P}^\lambda_\Gamma$ and $\alpha < \lambda$ and we shall find q such that $p \le q \in \mathcal{J}_\alpha$. If $\alpha \in u^p$ let $q = p$, otherwise define $q = (u^q, c^q)$ as follows $u^q = u^p \cup \{\alpha\}$ and for $\beta < \gamma \in u^q$ we let $c^q\{\beta, \gamma\}$ be: $c^p\{\beta, \gamma\}$ when it is well defined and $\sup(\mathrm{Rang}(c^p)) + 1 + (|\beta \cap u^q| + |\gamma \cap u^q|)^2 + |\beta \cap u^q|$ when otherwise. Now $q \in \mathbb{P}^\lambda_\Gamma$ by Case 1 of Definition 2.4, $p \le q$ by the order's definition and $q \in \mathcal{J}_\sigma$ trivially.
(3) Follows from part (2). $\square_{2.5}$

§3. Why does the forcing work.

We shall use the following claim for $\mu = \aleph_0$.

3.1. CLAIM. (1) *If $f : [\lambda]^2 \to \mu$ and M is an algebra with universe λ, $|\tau_M| \le \mu$ and $w_t \subseteq [\lambda]$, $|w_t| < \aleph_0$ for $t \in [\lambda]^2$ and $\lambda \ge \beth_2(\mu^+)^+$, then for some $\langle v_t : t \in [W]^{\le 2}\rangle$ we have:*
(a) *$W \subseteq \lambda$ is infinite in fact $|W| = \mu^+$,*
(b) *$f \restriction [W]^2$ is constant,*
(c) *$t \cup w_t \subseteq v_t \in [\lambda]^{<\aleph_0}$ for $t \in [W]^2$,*
(d) *$v_{t_1} \cap v_{t_2} \subseteq v_{t_1 \cap t_2}$ when $t_1, t_2 \in [W]^{\le 2}$ but for no $\alpha < \beta < \gamma$ do we have $\{t_1, t_2\} = \{\{\alpha, \beta\}, \{\beta, \gamma\}\}$,*

(e) *if* $t_1, t_2 \in [W]^i$, *where* $i \in \{1, 2\}$ *then* $|v_{t_1}| = |v_{t_2}|$ *and* $OP_{v_{t_2}, v_{t_1}}$ *maps* t_1 *onto* t_2 *and* w_{t_1} *onto* w_{t_2} *and* v_{t_1} *onto* v_{t_2}, $w_{\{Min(t_1)\}}$ *onto* $w_{\{Min(t_2)\}}$ *and* $w_{\{Max(t_1)\}}$ *onto* $w_{\{Max(t_2)\}}$, $v_{\{Min(t_1)\}}$ *onto* $v_{\{Min(t_2)\}}$, *and* $v_{\{Max(t_1)\}}$ *onto* $v_{\{Max(t_2)\}}$,

(f) $v_{\{\alpha,\beta\}} \cap c\ell_M(v_{\{\gamma\}}) \subseteq v_\gamma$ *for* $\alpha, \beta, \gamma \in W$.

(2) *If* $[u \in [\lambda]^{<\aleph_0} \Rightarrow c\ell_M(u) \in [M]^{<\mu}$, *then* $\lambda = (\beth_2(\mu))^+$ *is enough.*

REMARK. See more in [Sh:289]; this is done for completeness.

PROOF. (1) Let $w_t \cup t = \{\zeta_{t,\ell} \colon \ell < n_t\}$ with no repetitions and we define the function c, c_0, c_1 with domain $[\lambda]^3$ as follows: if $\alpha < \beta < \gamma < \lambda$ then

$$c_0\{\alpha, \beta, \gamma\} = \{(\ell_1, \ell_2) \colon \ell_1 < n_{\{\alpha,\beta\}}, \ell_2 < n_{\{\alpha,\gamma\}} \text{ and } \zeta_{\{\alpha,\beta\},\ell_1} = \zeta_{\{\alpha,\gamma\},\ell_2}\},$$
$$c_1\{\alpha, \beta, \gamma\} = \{(\ell_1, \ell_2) \colon \ell_1 < n_{\{\alpha,\gamma\}}, \ell_2 < n_{\{\beta,\gamma\}} \text{ and } \zeta_{\{\alpha,\gamma\},\ell_1} = \zeta_{\{\beta,\gamma\},\ell_2}\},$$
$$c\{\alpha, \beta, \gamma\} = (c_0\{\alpha, \beta, \gamma\}, c_1\{\alpha, \beta, \gamma\}, f\{\alpha, \beta\}).$$

By Erdős-Rado theorem for some $W_1 \subseteq \lambda$ of cardinality and even order type μ^{++} for part (1), μ^+ for part (2) such that $c \restriction [W_0]^3$ is constant. Let $\{\alpha_\varepsilon \colon \varepsilon < \mu^{++}\}$ list W_0 in increasing order. If $2 < i < \mu^{++}$, let

$$v_{\{\alpha_i\}} =: \{\zeta_{\{\alpha_i,\alpha_{i+1}\},\ell_1} \colon \text{for some } \ell_2 \text{ we have } (\ell_1, \ell_2) \in c_0\{\alpha_i, \alpha_{i+1}, \alpha_{i+2}\}\}$$
$$\cup \{\zeta_{\{\alpha_0,\alpha_i\},\ell_1} \colon \text{for some } \ell_2 \text{ we have } (\ell_1, \ell_2) \in c_1\{\alpha_0, \alpha_1, \alpha_i\}\}$$

(clearly $\alpha_i \in v_{\{\alpha_i\}}$).

For $i < j$ in $(2, \mu^{++})$ let $v_{\{\alpha_i,\alpha_j\}} = v_{\{\alpha_i\}} \cup v_{\{\alpha_j\}} \cup w_{\{\alpha_i,\alpha_j\}}$. Now for some unbounded $W_2 \subseteq W_1 \setminus \{\alpha_0, \alpha_1\}$ and $Y \in [\lambda]^{\leq\mu}$ we have:

$$\text{if } \alpha \neq \beta \in W_2 \text{ then } c\ell_M(v_{\{\alpha\}}) \cap c\ell_M(v_{\{\beta\}}) \subseteq Y.$$

Now by induction on $\varepsilon < \mu^+$ we can choose $\gamma_\varepsilon \in W_2$ strictly increasing with ε, γ_ε large enough. It is easy to check that $W = \{\gamma_\varepsilon \colon \varepsilon < \mu^+\}$ is as required.

(2) The same proof. $\square_{3.1}$

3.2. CLAIM. *Let* $n(*), \ell(*), \lambda$ *be as in Definition* 2.4, *and see Definition* 2.1. *Assume that* $\Gamma_1, \Gamma_2, m^*, p^*$ *satisfies*:

(a) $\Gamma_1, \Gamma_2 \subseteq ID^*_{\ell(*)}$,

(b) *if* $(\mathrm{dom}_{\ell(*),m}, e) \in ID^*_{\ell(*)}$ *and* $(\mathrm{dom}_{\ell(*),m}, e)$ *is not* m^**-nice then* $(\mathrm{dom}_{\ell(*),m}, e) \in \Gamma_1 \Leftrightarrow (\mathrm{dom}_{\ell(*),m}, e) \in \Gamma_2$,

(c) $p^* \in \mathbb{P}^*_\lambda$ *and* $|u^{p^*}| < m^*$.

Then $p^* \in \mathbb{P}^\lambda_{\Gamma_1} \Leftrightarrow p^* \in \mathbb{P}^\lambda_{\Gamma_2}$.

PROOF. We prove by induction on $k < \omega$ that

$(*)_k$ if $r' \in \mathbb{P}^{\lambda,\Gamma_1}_k$ (see Definition 2.4 before Case 0) and $r \leq r'$ and $|u^r| < m^*$, then $r \in \mathbb{P}^\lambda_{\Gamma_2}$.

This is enough by the symmetry in our assumptions.

For a fixed k we prove this by induction on $|u^r|$. The proof splits according to the Case in Definition 2.4 which hold for r'.

CASE 0. Trivial.

CASE 1. Easy.

CASE 2. Should be clear but let us check, so $r' = (u', c')$ is gotten from $(u'_1, c'_1), (u'_2, c'_2)$ as in clauses (a)–(d) of Case 2, and let $r = (u, c) \leq r'$.

Let $u_\ell = u'_\ell \cap u$, $c_\ell = c'_\ell \restriction [u_\ell]^2$. By the induction hypothesis $(u_\ell, c_\ell) \in \mathbb{P}^\lambda_{\Gamma_2}$ and it is enough to check that $(u, c), (u_1, c_1), (u_2, c_2)$ are in Case (2) of Definition 2.4 which is easy, e.g., in clause (d) we use monotonicity of $c\ell_0$.

CASE 3. So let r' be gotten from $\mathrm{s} = (\mathrm{dom}_{\ell(*),m}, e)$, $\langle p_y : y \in Y \rangle$, $\langle v_t : t \in Y^+ \rangle$ as there. Of course, we have $(\mathrm{dom}_{\ell(*),m}, e) \in \Gamma_1$ and $p_y \in \bigcup_{\ell<k} \mathbb{P}^{\lambda,\Gamma_1}_\ell$ so by the induction hypothesis clearly $p_y \restriction u^r \in \mathbb{P}^\lambda_{\Gamma_2}$.

SUBCASE 3A: $\mathrm{nice}(\mathrm{dom}_{\ell(*),m}, e) < m^*$ (see Definition 1.3(1)).
Hence $(\mathrm{dom}_{\ell(*),m}, e) \in \Gamma_2$ and the desired conclusion easily holds.
[Why? We can find $p^*_y = p_y \restriction u^r = p_y \restriction (u^{p_y} \cap u^r) = r' \restriction (u^{p_y} \cap u^r)$ hence $|u^{p^*_y}| < m^*$.

By the induction hypothesis p^*_y belongs to $\mathbb{P}^\lambda_{\Gamma_2}$ for each $y \in Y$. Now r, $\langle p^*_y : y \in Y \rangle$ and $\langle v_t : t \in Y^+ \rangle$ satisfies clauses (a)–(g) of Case 3 of Definition 2.4. Hence by Case 3 of Definition 2.4 $r'' =: r' \restriction (\cup\{u^{p^*_y} : y \in Y\})$ belong to $\mathbb{P}^\lambda_{\Gamma_2}$ but $r = r''$ so $r \in \mathbb{P}^\lambda_{\Gamma_2}$.]

SUBCASE 3B: Not subcase 3A.

So $\mathrm{nice}(\mathrm{dom}_{\ell(*),m}, e) \geq m^* > |u^r|$. For $a \subseteq \mathrm{dom}_{\ell(*),m}$ let $u_a = \{\alpha \in u^r : \alpha \in u^{p_y}$ for some $y \in Y$ satisfies $y \subseteq a$ or $\alpha \in v_{\{\eta\}}, \eta \in a$ or $\alpha \in v_\emptyset\}$. Now

$(*)_0$ if $u^r \subseteq v_{\{\eta\}}$ for some $\eta \in \mathrm{dom}_{\ell(*),m}$ then $r \in \mathbb{P}^\lambda_{\Gamma_2}$.
[Why? By applying Case 2 (and 0) of Definition 2.4.]

$(*)_1$ if for some $\eta \in \mathrm{dom}_{\ell(*),m}$ we have $[(\{\alpha_1, \beta_1\} \neq \{\alpha_2, \beta_2\} \in [u]^2)$ & $c^r\{\alpha_1, \beta_1\} = c^r\{\alpha_2, \beta_2\} \Rightarrow \{\alpha_1, \beta_1, \alpha_2, \beta_2\} \subseteq v_{\{\eta\}}]$ then $r \in \mathbb{P}^\lambda_{\Gamma_2}$.
[Why? By $(*)_0$ and uses of Case 1 of Definition 2.4.]

$(*)_2$ if $y \in Y$ and $u^r \subseteq v_y$ then $r \in \mathbb{P}^\lambda_{\Gamma_2}$.
[Why? Similarly.]

Now

$(*)_3$ It is enough to find $a, b \subseteq \mathrm{dom}_{\ell(*),m}$ such that:

$(*)^3_{a,b}$ $u_a \neq u_a \cap u_b, u_b \neq u_a \cap u_b, u^r \subseteq u_a \cup u_b, u^r \nsubseteq u_a, u^r \nsubseteq u_b$ and $[\eta_1 \in u_a \backslash u_b$ & $\eta_2 \in u_b \backslash u_a \Rightarrow (\{\eta_1, \eta_2\}/e)$ is a singleton] and $|a \cap b| \leq 1$.

[Why is this enough? As then r is gotten by Case 2 of Definition 2.4 from $(u_a, c^p \restriction [u_a]^2), (u_b, c^b \restriction [u_b]^2)$. The main point is why clause (d) of this case holds; now we shall prove more $c\ell_0(u_a \cap u_b) \cap (u_a \cup u_b) \subseteq u_a \cap u_b$; now by clause (e) of Case 3 of Definition 2.4 letting $t = a \cap b$

(it $\in \{\emptyset\} \cup \{\{\eta\} : \eta \in \mathrm{dom}_{\ell(*),m}\}$ by the last statement in $(*)^3_{a,b}$) we have $u_a \cap u_b = u_t$ (see (d), (f) Definition 2.4, Case 3), hence $c\ell_0(u_a \cap u_b) = c\ell_0(u_t), u_t \subseteq v_t$ hence $c\ell_0(u_a \cap u_b) \subseteq c\ell_0(v_{a\cap b})$ which is disjoint to $u_a \backslash u_b$ and to $u_b \backslash u_a$ by clause (e) in Case 3 of Definition 2.4 as $u_a \cap v_t = u_t$ and $u_b \cap v_t = u_t$.]

So now why can we find such a, b?

We try to choose $a_i \subseteq \mathrm{dom}_{\ell(*),m}$ for $i = 2, 3, \ldots$ or for $i = 1, 2, \ldots,$ such that $|a_i| = i, a_i \subseteq a_{i+1}$ and $i \leq |u_{a_i}|$. First assume that we cannot find neither a_2 nor a_1, then $y \in Y \Rightarrow |u^{p_y} \cap u^r| \leq 1$ and $\eta \in \mathrm{dom}_{\ell(*),m} \Rightarrow |v_{\{\eta\}} \cap u^r| = 0$. If $(*)_2$ applies we are done, so there are $\langle (y_\ell, \gamma_\ell) : \ell < k \rangle$ satisfying $y_\ell = \{\eta_{1,\ell}, \eta_{2,\ell}\} \in Y$ such that $u^r \cap u^{p_{y_\ell}} \backslash v_{\{\eta_{1,\ell}\}} \backslash v_{\{\eta_{2,\ell}\}} = \{\gamma_\ell\}$ and $k \geq 2$ so $u^r \backslash v_\emptyset = \{\gamma_\ell : \ell < k\}$. Let $u_1 = (u^r \cap v_\emptyset) \cup \{\gamma_0\}, u_2 = u^r \backslash \{\gamma_0\}$, clearly r is gotten from $r \restriction u_1, r \restriction u_2$ as in Case 2 of Definition 2.4.

Second, assume a_1 or a_2 is defined. So we are stuck in $a_{i(*)}$ for some $i(*)$, i.e., $a_{i(*)}$ is chosen but we cannot choose $a_{i(*)+1}$. If $u_{a_{i(*)}} \neq u^r$, let $a = a_{i(*)}, b = \mathrm{dom}_{\ell(*),m} \backslash a_{i(*)}$, so we get $(*)^3_{a,b}$ and we are done. So $u^r = u_{a_{i(*)}}$, hence $i(*) = |a_{i(*)}| = |u^r| < m^*$ and we can assume that $(*)_2$ does not apply. By the niceness of $(\mathrm{dom}_{\ell(*),m}, e)$ the graph $H[e] \restriction a_{i(*)}$ has no cycle so is a tree in the graph theoretic sense and so for some $c, b \subseteq a_{i(*)}$ we have $c \cap b = \{\eta\}, c \cup b = a_{i(*)}, b \neq \{\eta\}, c \neq \{\eta\}$ and $[\eta' \in b \backslash \{\eta\}\ \&\ \eta'' \in c \backslash \{\eta\} \Rightarrow \{\eta', \eta''\}$ not an $H[e]$-edge]; so we get $(*)^3_{a,b}$ and we are done. (So if we change slightly the claim demanding only $2|u^r| < m^*$, the proof is simpler). $\square_{3.2}$

3.3. THE PRESERVATION CLAIM. *Let $n(*), \ell(*), \lambda, \mu = \aleph_0$ be as in Definition 2.4 and assume $\lambda > \beth_2(\mu^+)$.*

(1) *If $\mathbb{P} = \mathbb{P}^\lambda_\Gamma$ and $(\mathrm{dom}_{\ell(*),m}, e) \in \Gamma \subseteq ID^*_{\ell(*)}$ then in $\mathbf{V}^{\mathbb{P}}$ we have $(\mathrm{dom}_{\ell(*),m}, e) \in ID_2(\lambda, \aleph_0)$.*

(2) *Assume that $\mathbb{P} = \prod_{n<\gamma} \mathbb{P}^\lambda_{\Gamma_n}$ where $\Gamma_n \subseteq ID^*_{\ell(*)}$ and $\gamma \leq \omega$ and $p^* \in \mathbb{P}$ forces that $\underset{\sim}{c}$ is a function from $[\lambda]^2$ to ω. Then for some finite $d \subseteq \gamma$ for any $\mathbf{s} \in \bigcap_{n\in d} \Gamma_n$ we have $p^* \nVdash_{\mathbb{P}}$ "$\mathbf{s} \notin ID_2(\underset{\sim}{c})$".*

PROOF. (1) Follows from (2), letting $\gamma = 1, \Gamma_0 = \Gamma$.

(2) Assume $p^* \in \mathbb{P}$ and $p^* \Vdash_{\mathbb{P}}$ "$\underset{\sim}{c}$ is a function from $[\lambda]^2$ to ω". Let $k(*) = 2^{\ell(*)} - 1$ and let $k(\nu) = |\{\rho \in {}^{\ell(*)>}2 : \rho <_{\mathrm{lex}} \nu\}|$ for $\nu \in {}^{\ell(*)>}2$. For $p \in \mathbb{P}$ let $u[p] = \bigcup\{u^{p(n)} : n \in \mathrm{Dom}(p)\}$, so $u[p] \in [\lambda]^{<\aleph_0}$ and for any $q \in \mathbb{P}$ we let $n[q] = \sup(\bigcup\{\mathrm{Rang}(c^{q(n)}) : n \in \mathrm{Dom}(q)\})$. For any $\alpha < \beta < \lambda$ letting $t = \{\alpha, \beta\}$ we define, by induction on $k \leq k(*)$ the triple $(n_{t,k}, w_{t,k}, d_{t,k})$ such that:

(*) $n_{t,k} < \omega, w_{t,k} \in [\lambda]^{<\aleph_0}$ and $d_{t,k} \subseteq \gamma$ is finite.

CASE 1. $k = 0$: $n_{t,k} = n[p^*] + 2$ and $w_{t,k} = \{\alpha, \beta\} \cup u^{p^*}$ and $d_{t,k} = \mathrm{Dom}(p^*)$.

CASE 2. $k+1$:

Let $\mathscr{P}_{t,k} = \{q \in \mathbb{P}\colon p^* \leq q, u[q] \subseteq w_{t,k}$ and $n[q] \leq n_{t,k}$ and $\mathrm{Dom}(q) \subseteq d_{t,k}\}$; clearly it is a finite set, and for every $q \in \mathscr{P}_{t,k}$ we choose $p_{t,q}$ such that $q \leq p_{t,q} \in \mathbb{P}$ and $p_{t,q}$ forces a value, say $\zeta_{t,q}$ to $\underset{\sim}{c}(t)$. Now we let

$$w_{t,k+1} = w_{t,k} \cup \bigcup\{u[p_{t,k}]\colon q \in \mathscr{P}_{t,k}\},$$
$$d_{t,k+1} = d_{t,k} \cup \{\mathrm{Dom}(q_{t,p})\colon p \in \mathscr{P}_{t,k}\},$$
$$n_{t,k+1} = \mathrm{Max}\{|w_{t,k+1}|^2, n_{t,k}+1, n[p_q]+1\colon q \in \mathscr{P}_{t,k}\}.$$

We next define an equivalence relation E on $[\lambda]^2$: $t_1 E t_2$ iff letting $t_1 = \{\alpha_1, \beta_1\}$, $t_2 = \{\alpha_2, \beta_2\}$, $\alpha_1 < \beta_1$, $\alpha_2 < \beta_2$ and letting $h = OP_{w_{\{\alpha_2,\beta_2\},k(*)}, w_{\{\alpha_1,\beta_1\},k(*)}}$, we have

(i) $w_{t_1,k(*)}, w_{t_2,k(*)}$ has the same number of elements,

(ii) h maps α_1 to α_2 and β_1 to β_2 and $w_{t_1,k}$ onto $w_{t_2,k}$ for $k \leq k(*)$ (so h is onto),

(iii) $d_{t_1,k} = d_{t_2,k}$ for $k \leq k(*)$ (hence h maps $\mathscr{P}_{t_1,k}$ onto $\mathscr{P}_{t_2,k}$),

(iv) if $q_1 \in \mathscr{P}_{t_1,k}, k < k(*)$ then h maps q_1 to some $q_2 \in \mathscr{P}_{t_2,k}$ and it maps p_{t,q_1} to p_{t,q_2} and we have $\zeta_{t_1,q_1} = \zeta_{t_2,q_2}$.

Clearly E has $\leq \aleph_0$ equivalence classes. So let $c\colon [\lambda]^2 \to \aleph_0$ be such that $c(t_1) = c(t_2) \Leftrightarrow t_2 E t_2$ and let $w_t = w_{t,k(*)}$.

By Claim 3.1, recalling that we have assumed $\lambda > \beth_2(\aleph_1)$ we can find $W \subseteq \lambda$ of cardinality μ^+ and $\bar{v} = \langle v_t\colon t \in [W]^{\leq 2}\rangle$ as there; i.e., we apply it to an expansion of M^*_λ such that $c\ell_0(-) = c\ell_M(-)$.

Let $d^*_k = d_{t,k} \subseteq \omega$ for $t \in [W]^2$ and $k \leq k(*)$, now we choose $d = d^*_{k(*)} \subseteq \gamma$, and we shall show that it is as required in the claim. Let $\mathbf{s} = (\mathrm{dom}_{\ell(*),m(*)}, e) \in \bigcap_{\ell \in d} \Gamma_\ell$ and let $Y_v = Y_{e,v} = \{\{\eta_0, \eta_1\}\colon v\hat{\ }\langle i\rangle \trianglelefteq \eta_i \in \mathrm{dom}_{\mathbf{s}}$ for $i = 0, 1$ and $\{\eta_0, \eta_1\}/e$ is not a singleton$\}$ for $v \in {}^{\ell(*)>}2$ and let $Y = \bigcup\{Y_{e,v}\colon v \in {}^{\ell(*)>}2\}$.

We now choose $\alpha_\eta \in W$ for $\eta \in \mathrm{dom}_{\mathbf{s}}$ such that $\eta_1 <_{\mathrm{lex}} \eta_2 \Rightarrow \alpha_{\eta_1} < \alpha_{\eta_2}$. Let $S = \{\alpha_\eta\colon \eta \in \mathrm{dom}_{\mathbf{s}}\}$. For $y \in Y$ let $t(y) = \{\alpha_\eta\colon \eta \in y\}$. Let $\langle v^*_\ell\colon \ell < k(*) = 2^{\ell(*)} - 1\rangle$ list ${}^{\ell(*)>}2$ in increasing order by $\leq_{\mathrm{lex}}$.

We now define q_ℓ and $q_{\eta,\ell}$ for $\eta \in \mathrm{dom}_{\ell(*),m(*)}$ and $p_{y,\ell}$ for $y \in Y$ by induction on $\ell \leq k(*)$ such that

(a) $p_{y,\ell} \in \mathscr{P}_{t(y),\ell}$ hence $u^{p_{y,\ell}} \subseteq w_{t,\ell}$ for every $y \in Y$,

(b) $\mathbb{P} \models$ "$p_{y,m} \leq p_{y,\ell}$" for $m \leq \ell$,

(c) if $y \in Y$ and $\eta \in Y$ then $\mathrm{Dom}(q_{\eta,\ell}) = \mathrm{Dom}(p_{y,\ell})$ and for each $\beta \in \mathrm{Dom}(q_{\eta,\ell})$ we have $q_{\eta,\ell}(\beta) = p_{y,\ell}(\beta) \restriction v_{\{\eta\}}$ hence $m \leq \ell \Rightarrow q_{\eta,m} \leq q_{\eta,\ell}$ and $\mathrm{Dom}(q_\ell) = \mathrm{Dom}(q_{\eta,\ell}) \cap v_\emptyset$ and for each $\beta \in \mathrm{Dom}(q_\ell)$ we have $(q_{\eta,\ell}(\beta) \restriction v_\emptyset = q_\ell(\beta)$ so $m < \ell \Rightarrow q_m \leq q_\ell$,

(d) if $v^*_\ell\hat{\ }\langle i\rangle \trianglelefteq \eta_i \in \mathrm{dom}_{\ell(*),m(*)}$ for $i = 0, 1$ then $p_{y,\ell+1}$ forces a value to $\underset{\sim}{c}\{\alpha_\eta\colon \eta \in y\}$.

For $\ell = 0$ there is no problem. For $\ell + 1$ choose η_0^ℓ, η_1^ℓ such that $\nu_\ell^{*\frown}\langle i\rangle \trianglelefteq \eta_i^\ell \in \mathrm{dom}_{\ell(*),m(*)}$ for $i = 0, 1$ and $\{\eta_0^\ell, \eta_1^\ell\}/e^{\mathbf{s}}$ is not a singleton and let $y_\ell = \{\eta_0, \eta_1\}$. As $p_{y_\ell,\ell} \in \mathscr{P}_{t(y_\ell),\ell}$ by the choice of $\mathscr{P}_{t(y),\ell+1}$ there is $p_{y_\ell}^\ell \in \mathscr{P}_{t(y_\ell),\ell+1}$ above $p_{y_\ell,\ell}$, which forces a value to $\underset{\sim}{c}(t(y_\ell))$. For $p \in \mathbb{P}$ and $u \subseteq \lambda$ let $q = p \upharpoonright u$ means $\mathrm{Dom}(p) = \mathrm{Dom}(q)$ and $\beta \in \mathrm{Dom}(p) \Rightarrow q(p) = (p(\beta)) \upharpoonright u$.

Now we define $\langle p_{y,\ell+1} : y \in Y_{\nu_i}\rangle$:

$$\text{if } y \in Y_{\nu_i} \text{ then } p_{y,\ell+1} = OP_{\upsilon_y,\upsilon_{y_\ell}}(p_{y_\ell}^\ell).$$

So necessarily

$(*)_1$ if $y'' \neq y'' \in Y_{\nu_\ell}$ then $p_{y',\ell+1} \upharpoonright \upsilon_\emptyset = p_{y'',\ell+1} \upharpoonright \upsilon_\emptyset$ is above (by $\leq_{\mathbb{P}}$) q_ℓ,

$(*)_2$ if $y' \neq y'' \in Y_{\nu_\ell}$ and $y' \cap y'' \neq \emptyset$ then for some $\eta \in \mathrm{dom}_{\mathbf{s}}$ we have $y' \cap y'' = \{\eta\}$ and $p_{y',\ell+1} \upharpoonright \upsilon_{\{\eta\}} = p_{y'',\ell+1} \upharpoonright \upsilon_{\{\eta\}}$ is above $q_{\eta,\ell}$.
[Why? As if let $y' = \{\eta_0', \eta_1'\}, y'' = \{\eta_0'', \eta_1''\}, \nu_\ell{}^\frown\langle i\rangle \trianglelefteq \eta_i', \eta_i''$ for $i = 0, 1$ then either $\eta_0' = \eta_0'', y' \cap y'' = \{\eta_0'\}$ or $\eta_1' = \eta_1'', y' \cap y'' = \{\eta_1'\}$ but $\eta_0' \neq \eta_1''$ & $\eta_1' \neq \eta_0''$. Now use the properties from 3.1 and clause (iv) above.]

Let $q_{\emptyset,\ell+1} = p_{y_\ell}^{\ell+1} \upharpoonright \upsilon_\emptyset$. The $q_{\eta_i',\ell}$ is defined as $q_{\eta_i',\ell+1} = p_{\{\eta_0',\eta_1'\},\ell+1} \upharpoonright \upsilon_{\{\eta_i'\}}$ for $i = 0, 1$ if $\nu_\ell{}^\frown\langle i\rangle \trianglelefteq \eta_i'$ & $\{\eta_0', \eta_1'\} \in Y_{\nu_\ell}$.

Let $q_{\eta,\ell+1}$ be the result of free amalgamation (i.e., Case 2 of Definition 2.4) in each coordinate β of $q_{\eta,\ell}$ and $q_{\emptyset,\ell+1}$ if $\eta \in \mathrm{dom}_{\mathbf{s}} \wedge \neg(\nu_\ell \trianglelefteq \eta)$ and $\eta \in \mathrm{dom}_{\mathbf{s}}$.

Let $p_{y,\ell+1}$ be the result of free amalgamation (i.e., Case 2 of Definition 2.4) in each coordinate (twice) of $p_{y,\ell}, q_{\{\eta_0'\},\ell+1}, q_{\{\eta_1\},\ell+1}$ if $y = \{\eta_0, \eta_1\} \in Y \setminus Y_{\nu_\ell}$.

Of course, putting two conditions together using Case 2 of Definition 2.4, not repeating colours except when absolutely necessary.

Lastly, let p^+ be such that $\mathrm{Dom}(p^+) = d_{k(*)}^*$ and for each $\beta \in d_{k(*)}^*$

$$u^{p^+}(\beta) = \bigcup\{u^{p_{y,k(*)}} : y \in Y\};$$

if $u^{p_{y,k(*)}(\beta)}$ is not defined, it means $\emptyset$

$$c^{p^+}(\beta) \text{ extend each } c^{p_{y,k(*)}(\beta)} \text{ otherwise is 1-to-1 with new values.}$$

So $p^+ \geq p^*$ forces that $\{\alpha_\eta : \eta \in \mathrm{dom}_{\ell(*),m(*)}\}$ exemplify $\mathbf{s} = (\mathrm{dom}_{\ell(*),m(*)}, e) \in ID(\underset{\sim}{f})$, a contradiction. $\square_{3.3}$

3.4. The Example Claim. *Let $n(*), \ell(*), \lambda$ be as in Definition 2.4. Assume*

(a) $(\mathrm{dom}_{\ell(*),m(*)}, e^*) \in ID_{\ell(*)}^*$,

(b) $\Gamma \subseteq ID_{\ell(*)}^*$,

(c) *if* $\mathbf{s} \in \Gamma$ *then* $\mathbf{s}$ *is* $(2^{\ell(*)}m(*))$*-nice,*

(d) $\mathbb{P} = \mathbb{P}_\Gamma^\lambda$,

(e) $\underset{\sim}{c}$ *is the* $\mathbb{P}$*-name* $\bigcup\{c^p : p \in \underset{\sim}{G}_{\mathbb{P}}\}$,

(f) $\ell(*) \geq n(*)$.

Then $\Vdash_{\mathbb{P}}$ *"$\underset{\sim}{c}$ is a function from $[\lambda]^2$ to μ exemplifying* $(\mathrm{dom}_{\ell(*),m(*)}, e^*)$ *does not belong to* $ID_2(\lambda, \mu)$*".*

REMARK. The proof is similar to [GcSh:491].

PROOF. So assume toward contradiction that $p \in \mathbb{P}$ and $\alpha_\eta < \lambda$ for $\eta \in \mathrm{dom}_{\ell(*),m(*)}$ are such that p forces that $\eta \mapsto \alpha_\eta$ is a counterexample, i.e., $\langle \alpha_\eta : \eta \in \mathrm{dom}_{\ell(*),m(*)} \rangle$ is with no repetitions and p forces that $t_1 e^* t_2 \Rightarrow \underset{\sim}{c}(\{\alpha_\eta : \eta \in t_1\}) = \underset{\sim}{c}(\{\alpha_\eta : \eta \in t_2\})$. By 2.5(2) without loss of generality $\{\alpha_\eta : \eta \in \mathrm{dom}_{\ell(*),m(*)}\} \subseteq u^p$.

Let $Y = Y_{e^*} = \{y : y \in \mathrm{Dom}(e)$ and y/e is not a singleton$\}$ and for $\nu \in {}^{\ell(*)>}2$ let $Y_\nu = Y_{\nu,e^*} = \{\{\eta_0, \eta_1\} \in Y_{e^*} : \nu\hat{}\langle i \rangle \trianglelefteq \eta_i$ for $i = 0, 1\}$ as in the previous proof. We now choose by induction on $\ell \leq n(*)$ the objects $\eta_\ell, \nu_\ell, Z_\ell$ and first order formulas $\varphi_\ell(x, y_0, \ldots, y_{\ell-1}), <^\ell_{y_0,\ldots,y_{\ell-1}}(x, y)$ in the vocabulary of M^*_λ such that:

⊠ (a) $\nu_\ell \in {}^\ell 2, \eta_\ell \in \mathrm{dom}_{\ell(*),m(*)}$ and $M^*_\lambda \models (\exists^{\leq \aleph_{n(*)-\ell}} x)\varphi_\ell(x, \alpha_{\eta_0}, \ldots, \alpha_{\eta_{\ell-1}})$,
(b) $<^\ell_{\alpha_{\eta_0}}, \ldots, \alpha_{\eta_{\ell-1}}$ is a well ordering of $\{x : M^*_\lambda \models \varphi_\ell[x, \alpha_{\eta_0}, \ldots, \alpha_{\eta_{\ell-1}}]\}$ of order type a cardinal $\leq \aleph_{n(*)-\ell}$,
(c) $\nu_0 = \langle\rangle, \varphi_0 = [x = x]$,
(d) $\nu_{\ell+1} = (\eta_\ell \restriction \ell)\hat{}\langle 1 - \eta_\ell(\ell)\rangle$ and $\nu_\ell \triangleleft \eta_\ell$,
(e) $Z_\ell = \{\eta : \nu_\ell \triangleleft \eta \in \mathrm{dom}_{\ell(*),m(*)}$ and $\{\eta_s, \eta\} \in e_{\nu \restriction s}$ for $s = 0, 1, \ldots, \ell - 1\}$,
(f) $\eta \in Z_\ell \Rightarrow \alpha_\eta \in \{\beta : M^*_\lambda \models \varphi_\ell[\beta, \alpha_{\eta_0}, \ldots, \alpha_{\eta_{\ell-1}}]\}$,
(g) η_ℓ is such that:
(α) $\nu_\ell \triangleleft \eta_\ell \in Z_\ell$,
(β) if $\nu_\ell \trianglelefteq \eta \in Z_\ell$ then $\alpha_\eta \leq^\ell_{\alpha_{\eta_0},\ldots,\alpha_{\eta_{\ell-1}}} \alpha_{\eta_\ell}$.

(See similar proof with more details in 4.3).

Let $\nu^* = \nu_{n(*)}, Z = Z_{n(*)}, Z^+ = \{\eta_\ell : \ell < n(*)\} \cup Z$; note that by Definition 1.3(1), clause (b) and Definition 2.1 we have $|Z| \geq 2$, i.e., this is part of $(\mathrm{dom}_{\ell(*),m(*)}, e^*)$ being 0-nice. For $\nu \in \{\nu_\ell : \ell < n(*)\}$ let s_ν be such that: $\rho_1 \cap \rho_2 = \nu$ & $\rho_1, \rho_2 \in \{\eta_\ell : \ell < n(*)\} \cup Z \Rightarrow s_\nu = c\{\alpha_{\rho_1}, \alpha_{\rho_2}\}$ (clearly exists). By Case 0 in Definition 2.4, without loss of generality

$$u^p = \{\alpha_\eta : \eta \in Z^+\},$$

that is, we may forget the other $\alpha \in u^p$; by claim 3.2 we have $p \in \mathbb{P}^\lambda_\emptyset$ so for some k we have $r \in \mathbb{P}^{\lambda,\emptyset}_k$.

So we have

⊞ $\langle \eta_\ell : \ell < n(*)\rangle, Z, Z^+, \langle \nu_\ell : \ell \leq n(*)\rangle, \langle s_{\eta_\ell \restriction \ell} : \ell < n(*)\rangle$ and p are as above, that is
(i) $\eta_\ell \in \mathrm{dom}_{\ell(*),m(*)}, \nu_0 = \langle\rangle, \nu_{\ell+1} = (\eta_\ell \restriction \ell)\hat{}\langle(1 - \eta_\ell(\ell))\rangle, Z = \{\rho \in \mathrm{dom}_{\ell(*),m(*)} : \nu_{n(*)} \triangleleft \rho$ and $\{\eta_\ell, \rho\}/e$ is not a singleton for each $\ell < n(*)\}$ hence $|Z| \geq 2$ and $Z^+ = Z \cup \{\eta_\ell : \ell < n(*)\}$,
(ii) $p \in \mathbb{P}^{\lambda,\emptyset}_k$,
(iii) $\alpha_\eta \in u^p$ for $\eta \in Z^+$,

(iv) $\langle \alpha_\eta : \eta \in Z^+ \rangle$ is with no repetitions,
(v) $c^p \restriction \{\alpha_\eta : \eta \in Z^+\}$ satisfies:
if $\ell < n(*)$ and $\nu \in Z \cup \{\eta_t : \ell < t < n(*)\}$ so $\eta_\ell \cap \nu = \eta_\ell \restriction \ell$ then $(\alpha_\nu \neq \alpha_{\eta_\ell}$ and) $c(\{\alpha_\nu, \alpha_{\eta_\ell}\} = s_{\eta_\ell \restriction \ell}$,
(vi) $\{\alpha_\eta : \eta \in Z\} \subseteq c\ell_0\{\alpha_{\eta_\ell} : \ell < n(*)\}$,
(vii) Z has at least two members.

Among all such examples choose one with $k < \omega$ minimal. The proof now splits according to the cases in Definition 2.4.

CASE 0. $k = 0$.

Trivial.

CASE 1. Let p_1, α be as there, so recall that $\{\alpha, \beta\} e^{p_1} \{\alpha', \beta'\} \Rightarrow \{\alpha, \beta\} = \{\alpha', \beta'\}$. Hence obviously, by clauses (v) and (vii) above, $\eta \in Z^+ \Rightarrow \alpha_\eta \neq \alpha$, so $\{\alpha_\eta : \eta \in Z^+\} \subseteq u^{p_1}$, contradicting the minimality of k.

CASE 2. Let $p_i = (u_i, c_i) \in \bigcup_{\ell < k} \mathbb{P}_\ell^{\lambda, \emptyset}$ for $i = 1, 2$ be as there. We now prove by induction on $\ell < n(*)$ that $\alpha_{\eta_\ell} \in u_0 \cap u_1$. If $\ell < n(*)$ and it is true for every $\ell' < \ell$, but (for some $i \in \{1, 2\}$), $\alpha_{\eta_\ell} \in u_i \backslash u_{3-i}$, it follows by clause (v) of $\boxplus$ that the sequence $\langle c(\{\eta_\ell, \nu\}) : \nu \in Z_\ell^* \rangle$ is constant where we let $Z_\ell^* = \{\eta_{\ell+1}, \eta_{\ell+2}, \ldots, \eta_{n(*)-1}\} \cup Z$, hence $\{\alpha_\nu : \nu \in Z_\ell^*\}$ is disjoint to $u_{3-i} \backslash u_i$, so $\{\alpha_\nu : \nu \in Z^+\} \subseteq u_i$, so we get contradiction to the minimality of k.

As $\{\alpha_{\eta_\ell} : \ell < n(*)\} \subseteq u_2 \cap u_1$ necessarily (by clause (vi) of $\boxplus$) we have $\{\alpha_\nu : \nu \in Z_{n(*)}^*\} \subseteq c\ell_0\{\alpha_{\eta_\ell} : \ell < n(*)\} \subseteq c\ell_0(u_2 \cap u_1)$. But $\{\alpha_\nu : \nu \in Z_{n(*)}^*\} \subseteq u_2 \cup u_1$ by $\boxplus$(iii), and we know that $c\ell_0(u_2 \cap u_1) \cap (u_2 \cap u_1) \subseteq u_1$ by clause (d) of Definition 2.4, Case 2 hence $\{\alpha_\nu : \nu \in Z_0^*\} \subseteq u_1$ contradiction to "k minimal".

CASE 3. This case never occurs as $p \in \mathbb{P}_k^{\lambda, \emptyset}$. $\square_{3.4}$

3.5. THEOREM. (1) *Let* $n(*) = 4$ (*or just* $n(*) \geq 4$), $\lambda = \aleph_{n(*)}, \ell(*) = n(*) + 1$ *and* $2^{\aleph_\ell} = \aleph_{\ell+1}$ *for* $\ell < n(*)$.

For some c.c.c. forcing $\mathbb{P}$ *of cardinality* λ *in* $\mathbf{V}^{\mathbb{P}}$ *the pair* $(\lambda, \aleph_0)$ *is not* $\aleph_0$-*compact.*

(2) *For given* $\chi = \chi^{\aleph_0} \geq \lambda$ *we can add* $\mathbf{V}^{\mathbb{P}} \models$ "$2^{\aleph_0} = \chi$".

PROOF. (1) Let $\Gamma_n = \{\mathbf{s} \in ID_{\ell(*)}^* : \mathbf{s}$ is n-nice$\}$, see Definition 2.1, clearly $\Gamma_{n+1} \subseteq \Gamma_n$ and $\Gamma_n \neq \emptyset$ (see 1.5) for $n < \omega$ and $\emptyset = \bigcap_{n<\omega} \Gamma_n$ and let $\mathbb{P}_n = \mathbb{P}_{\Gamma_n}^\lambda$ and let $\underset{\sim}{c}_n = \cup\{c^p : p \in \underset{\sim}{G}_{\mathbb{P}_n}\}$, it is a $\mathbb{P}_n$-name and $\mathbb{P}$ is $\prod_{n<\omega} \mathbb{P}_n$ with finite support. Now the forcing notion $\mathbb{P}$ satisfies the c.c.c. as $\mathbb{P}_n$ satisfies the Knaster condition (by 2.5(1)). By 3.4 we know that $\Vdash$ "$ID_2(\underset{\sim}{c}_n) \cap ID_{\ell(*)}^* \subseteq \Gamma_n$" for $\mathbb{P}_n$ hence for $\mathbb{P}$, in fact it is not hard to check that equality holds. If $\aleph_0$-compactness holds then in $\mathbf{V}^{\mathbb{P}}$ for some $c : [\lambda]^2 \to \omega$ we have $ID_2(c) \cap ID_{\ell(*)}^* \subseteq \bigcap_n \Gamma_n = \emptyset$ by claim 1.4.

But $\mathbf{V}^{\mathbb{P}}$, if $c : [\lambda]^2 \to \omega$ then by 3.3(2) it realizes some $\mathbf{s} \in \cup\{\Gamma_n : n < \omega\} \subseteq ID_{\ell(*)}^*$ (even k-nice one for every $k < \omega$).

Together we get that the pair $(\lambda, \aleph_0)$ is not $\aleph_0$-compact.

(2) We let $\mathbb{Q}$ be adding χ Cohen reals, i.e., $\{h\colon h$ a finite function from χ to $\{0,1\}\}$ ordered by inclusion. Let $\mathbb{P}$ be as above and force with $\mathbb{P}^+ = \mathbb{P} \times \mathbb{Q}$, now it is easy to check that $\mathbb{P}^+$ is as required. $\square_{3.5}$

§4. Improvements and additions. Though our original intention was to deal with the possible incompactness of the pair $(\aleph_2, \aleph_0)$, we have so far dealt with $(\lambda, \aleph_0)$ where $2^{\aleph_0} \geq \lambda = \aleph_{n(*)}$ & $n(*) \geq 4$. For dealing with $(\aleph_3, \aleph_0)$, $(\aleph_2, \aleph_0)$, that is $n(*) = 3, 2$ we need to choose M^*_λ more carefully.

What is the problem in §3 concerning $n(*) = 2$?

On the one hand in the proof of 3.4 we need that there are many dependencies among ordinals $< \lambda$ by M^*_λ; so if λ is smaller this is easier, but really just make us use larger $\ell(*)$ help.

On the other hand, in the proof of 3.3 we use 3.1, a partition theorem, so here if λ is bigger it is easier; but instead we can use demands specifically on M^*_λ. Along those lines we may succeed for $n(*) = 3$ using 3.1(2) rather than 3.1(1) but we still have problems for the pair $(\aleph_2, \aleph_0)$; here we change the main definition 2.4, in Case 3 changes $\langle v_y\colon y \in Y^+\rangle$, i.e., for $\eta \in \mathrm{dom}_s$ we have $v^+_{\{\eta\}}, v^-_{\{\eta\}}$ instead $v_{\{\eta\}}$. For this we have to carefully reconsider 3.3, but the parallel of 3.1 is easier. Note that in §2, §3 we could have used a nontransitive version of $c\ell_M(-)$.

4.1. Definition. We say that M^* is $(\lambda, <\mu, n(*), \ell(*))$-suitable if:

(a) M^* is a model of cardinality λ,

(b) λ is $> \mu, \leq \mu^{+n(*)}$ and $n(*) < \ell(*) < \omega$,

(c) τ_{M^*}, the vocabulary of M^*, is of cardinality $\leq \mu$,

(d) for every subset A of M^* of cardinality $< \mu$,
the set $c\ell_{M^*}(A)$ has cardinality $< \mu$,

(e) for some $m^* < \omega$ we have:
if $s = (\mathrm{dom}_{\ell(*),m}, e) \in ID^*_{\ell(*)}$ and $a_\eta \in M^*$ for $\eta \in \mathrm{dom}_{\ell(*),m}$ and s is m^*-nice, $m > m^*$,
then we can find $\langle \eta_\ell\colon \ell < n(*)\rangle$ and $\langle \nu_\ell\colon \ell \leq n(*)\rangle$ such that

(α) $\eta_\ell \in \mathrm{dom}_{\ell(*),m}$,

(β) $\nu_0 = \langle\rangle, \nu_{\ell+1} = (\eta_\ell \restriction \ell)\char`^\langle 1 - \eta_\ell(\ell)\rangle$,

(γ) $\nu_\ell \lhd \eta_\ell$,

(δ) $Z = \{\rho \in \mathrm{dom}_{\ell(*),m}\colon \nu_{n(*)} \lhd \rho$ and in the graph $H[e]$, ρ is connected to η_ℓ for $\ell = 0, \dots, n(*) - 1\}$,

(ε) $\{\alpha_\rho\colon \rho \in Z\} \subseteq c\ell_{M^*}\{\alpha_{\eta_\ell}\colon \ell < n(*)\}$.

4.2. Definition. (1) We say that M is explicitly[1] $(\lambda, <\mu, n(*))$-suitable if:

(a) M^* is a model of cardinality λ,

(b) $\lambda = \mu^{+n(*)}$,

(c) τ_{M^*}, the vocabulary of M^*, is of cardinality $\leq \mu$,

(d) for $A \subseteq M^*$ of cardinality $< \mu$, the set $c\ell_{M^*}(A)$ has cardinality $< \mu$ and $A \neq \emptyset \wedge \mu > \aleph_0 \Rightarrow \omega \subseteq c\ell_{M^*}(A)$,

(e) for some $\langle R_\ell : \ell \leq n(*)\rangle$ we have

(α) R_ℓ is an $(\ell+2)$-place predicate in τ_{M^*}; we may write $R_\ell(x, y, z_0, \dots, z_{\ell-1})$ as $x <_{z_0,\dots,z_{\ell-1}} y$ or $x <_{\langle z_0,\dots,z_{\ell-1}\rangle} y$,

(β) for any $c_0, \dots, c_{\ell-1} \in M^*$, the two place relation $<_{c_0,\dots,c_{\ell-1}}$ (i.e., $\{(a,b) : \langle a, b, c_0, \dots, c_{\ell-1}\rangle \in R^{M^*}\}$) is a well ordering of $A_{c_0,\dots,c_{\ell-1}} =: A_{\langle c_0,\dots,c_{\ell-1}\rangle} =: \{b : (\exists x)(x <_{c_0,\dots,c_{\ell-1}} b \vee b <_{c_0,\dots,c_{\ell-1}} x)\}$ of order-type a cardinal,

(γ) $R_0^{M^*}$ is a well ordering of M^* of order type λ,

(δ) if $\bar c = \langle c_\ell : \ell < k\rangle$ and $<_{\bar c}$ is a well ordering of $A_{\bar c}$ of order type μ^{+m} then for every $c_k \in M^*_{\bar c}$ we have $A_{\bar c \frown \langle c_k\rangle} = \{a \in A_{\bar c} : a <_{\bar c} c_k\}$ so is empty if $c_k \notin A_{\bar c}$, so if $lg(\bar c) = n(*)$ this is a definition of $A_{\bar c \frown \langle c_k\rangle}$ as it is not covered by clause (β)

(ε) if $\bar c = \langle c_\ell : \ell < k\rangle \in {}^k(M^*)$ and $|A_{\bar c}| < \mu$ then $A_{\bar c} \subseteq c\ell_{M^*}(\bar c)$.

(2) We say that M is explicitly[2] $(\lambda, <\mu, n(*))$-suitable if:

(a)–(d) as in part (1),

(e) for some $\langle R_\ell : \ell \leq n(*)\rangle$ we have (like (e) but we each time add z's and see clause (δ))

(α) R_ℓ is a $(2\ell+2)$-place predicate in τ_μ; we may write $R_\ell(x, y, z_0, \dots, z_{2\ell-1})$ or $x <_{z_0,\dots,z_{2\ell-1}} y$ or $x <_{\langle z_0,\dots,z_{2\ell-1}\rangle} y$,

(β) for any $c_0, \dots, c_{2\ell-1} \in M^*$ the two-place relation $<_{c_0,\dots,c_{2\ell-1}}$ (i.e., $\{(a,b) : \langle a, b, c_0, \dots, c_{2\ell-1}\rangle \in R_\ell^{M^*}\}$) is a well ordering of $A_{c_0,\dots,c_{2\ell-1}} = A_{\langle c_0,\dots,c_{2\ell-1}\rangle} = \{b : \text{for some } a, \langle a, b, c_0, \dots, c_{2\ell-1}\rangle \in R_\ell^{M^*} \text{ or } \langle b, a, c_0, \dots, c_{2\ell-1}\rangle \in R_\ell^{M^*}\}$,

(γ) $R_0^{M^*}$ is a well ordering of M^* of order type λ; for simplicity $R_0^{M^*} = c \restriction \lambda$,

(δ) if $\bar c = \langle c_\ell : \ell < 2k\rangle$ and $<_{\bar c}$ is a well ordering of $A_{\bar c}$ of order type μ^{+m} then for any $c_{2k}, c_{2k+1} \in M^*$ we have $A_{\bar c \frown \langle c_{2k}, c_{2k+1}\rangle}$ is empty if $\{c_{2k}, c_{2k+1}\} \nsubseteq A_{\bar c}$ and otherwise is $\{a \in A_{\bar c} : a <_{\bar c} c_{2k}$ and $a < c_{2k+1}\}$. If $k = n(*)$ this is a definition of $A_{\bar c^* \langle c_{2k}, c_{2k+1}\rangle}$.

4.3. Observation. (1) *If M is an explicitly[1] $(\lambda, <\mu, n(*))$-suitable model, then M is a $(\lambda, <\mu, n(*)+1, \ell(*))$-suitable model if $\ell(*) > n(*)+1$.*

(2) *If M is an explicitly[2] $(\lambda, <\mu, n(*))$-suitable model, then M is a $(\lambda, <\mu, 2n(*)+2, 2n(*)+3)$-suitable model.*

Proof. (1) Straightforward, similar to inside the proof of 3.4 and as we shall use part (2) only and the proof of (1) is similar but simpler, we do not elaborate.

(2) Clearly clauses (a)–(d) of Definition 4.1 holds, so we deal with clause (e). So assume $\ell(*) \geq 2n(*)$ and $\mathbf{s} = (\mathrm{dom}_{\ell(*),m}, e) \in ID^*_{\ell(*)}$ and $\alpha_\eta \in M$ for $\eta \in \mathrm{dom}_{\ell(*),m}$ are pairwise distinct. We choose by induction on $\ell \leq n(*)$ the

objects $\eta_{2\ell}, \nu_{2\ell+1}, Z_{2\ell}, \eta_{2\ell+1}, \nu_{2\ell+2}, Z_{2\ell+1}$ such that node $\nu_\ell = \langle\,\rangle$ and $\nu_{2\ell+2}$ is chosen in stage ℓ

⊠ (a) $\nu_\ell \in {}^\ell 2, \eta_\ell \in \mathrm{dom}_{\ell(*),m(*)}$ and $M \models (\exists^{\leq \aleph_{n(*)-\ell}} x)\varphi_\ell(x, \alpha_{\eta_0}, \ldots, \alpha_{\eta_{2\ell-1}})$,

(b) $<^\ell_{\alpha_{\eta_0},\ldots,\alpha_{\eta_{2\ell-1}}}$ is a well ordering of $A_{\langle \alpha_{\eta_0},\ldots,\alpha_{2\ell-1}\rangle} =: \{x\colon M \models \varphi_\ell[x, \alpha_{\eta_0}, \ldots, \alpha_{\eta_{\ell-1}}]\}$ of order type a cardinal $\leq \aleph_{n(*)-\ell}$,

(c) $\nu_0 = \langle\,\rangle, \varphi_0 = [x = x]$,

(d) $\nu_{\ell+1} = (\eta_\ell \restriction \ell)\hat{}\langle 1 - \eta_\ell(\ell)\rangle$,

(e) $Z_\ell = \{\eta\colon \nu_{2\ell} \lhd \eta \in \mathrm{dom}_{\ell(*),m(*)}$ and $\{\eta_s, \eta\} \in e_{\nu\restriction s}$ for $s = 0, 1, \ldots, \ell - 1\}$,

(f)′ $\eta \in Z_\ell \Rightarrow \alpha_\eta \in A_{\langle a_{\eta_k}\colon k<2\ell\rangle}$,

(g) η_ℓ is such that:

(α) $\nu_\ell \lhd \eta_\ell \in Z_\ell$,

(β) if $\nu_\ell \trianglelefteq \eta \in Z_\ell$ then $[\ell$ even $\Rightarrow \alpha_\eta \leq_{\alpha_{\eta_0},\ldots,\alpha_{\eta_{\ell-1}}} \alpha_{\eta_\ell}]$ and $[\ell$ odd $\Rightarrow \alpha_\eta \leq \alpha_{\eta_\ell}]$.

How do we do the induction step? Arriving to ℓ we have already defined $\langle \nu_k\colon k \leq 2\ell\rangle, \langle \eta_k\colon k < 2\ell\rangle$ and $\langle Z_k\colon k < 2\ell\rangle$, recalling $\nu_0 = \langle\,\rangle$. So by the definition of Z_k also $Z_{2\ell}$ is well defined and $\{\alpha_\eta\colon \eta \in Z_{2\ell}\}$ is included in $A_{\langle a_{\eta_k}\colon k<2\ell\rangle}$ and let $\eta_{2\ell} \in Z_{2\ell}$ be such that $\eta \in Z_{2\ell} \Rightarrow a_\eta \leq_{\langle a_{\eta_k}\colon k<2\ell\rangle} a_{\eta_{2\ell}}$ and $\nu_{2\ell+1} = \nu_{2\ell}\hat{}\langle 1 - \eta_{2\ell}(2\ell)\rangle = (\eta_{2\ell} \restriction (2\ell))\hat{}\langle 1 - \eta_{2\ell}(2\ell)\rangle$ so $Z_{2\ell+1}$ is well defined. Let $\eta_{2\ell+1} \in Z_{2\ell+1}$ be such that $\eta \in Z_{2\ell} \Rightarrow a_\eta \leq \alpha_{\eta_{2\ell+1}}$ and $\nu_{2\ell+2} = \nu_{2\ell+1}\hat{}\langle 1 - \eta_{2\ell+1}(2\ell + 1)\rangle$ and we have carried the induction. $\square_{4.3}$

Are there such models? We shall use 4.4(2), the others are for completeness (i.e., part (3) is needed for $\lambda = \aleph_3$ and part (4) says concerning $\lambda = \aleph_2$ it suffices to use ID_3^*):

4.4. OBSERVATION. (1) *For μ regular uncountable, there is an explicitly*[1] *$(\mu^{+2}, < \mu, 2)$-suitable model.*

(2) *If $\mu = \aleph_0$, then there is an explictly*[2] *$(\mu^{+2}, < \mu, 2)$-suitable model.*

(3) *If μ is regular uncountable, $t = 1$ or $\mu = \aleph_0$ & $t = 2$ and $n \in [3, \omega)$, then there is an explicitly*[t] *$(\mu^{+n}, < \mu, n)$-suitable model.*

(4) *If $2^{\aleph_0} = \aleph_1, \mu = \aleph_0$ then for some $\aleph_2$-c.c., $\aleph_1$-complete forcing notion $\mathbb{Q}$ of cardinality $\aleph_2$ in $\mathbf{V}^{\mathbb{Q}}$ there is an explicitly $(\aleph_2, < \aleph_0, 2)$-suitable model.*

4.5. REMARK. (1) It should be clear that if $\mathbf{V} = \mathbf{L}$ (or just $\neg 0^\#$), then this works also for singular μ *but* more reasonable is to use nontransitive closure.

PROOF (1), (2). Let $t = 1$ for part (1) and $t = 2$ for part (2). Let $n(*) = 2$ and $\lambda = \mu^{+2}$. We choose M_α by induction on $\alpha \leq \lambda$ such that:

(α) M_α is a τ^--model where $\tau^- = \{R_0, R_1, R_2\}$ with R_ℓ is $(t\ell + 2)$-predicate and $x <_{\bar z} y$ means $R_\ell(x, y, \bar z)$,

(β) M_α is increasing with α and has universe $1 + \alpha$,

(γ) $R_0^{M_\alpha}$ is $<\restriction \alpha$ (and $A_{\langle\,\rangle}^{M_\alpha} = \alpha$),

(δ) for $\bar{c} \in {}^{tk}(M_\alpha), k = 0, 1, 2$ we have $<_{\bar{c}}$ is a well ordering of $A_{\bar{c}}^{M_\alpha} =: \{a\colon M_\alpha \models (\exists x)(a <_{\bar{c}} x \vee x <_{\bar{c}} a)\}$ of order type a cardinal $< \mu^{+(n(*)+1-k)}$,

(ε) (i) if $t = 1, \bar{c} \in {}^k(M_\alpha), k = 0, 1, 2$ and $d \in A_{\bar{c}}^{M_\alpha}$ then $A_{\bar{c}\hat{}<d>}^{M_\alpha} = \{a \in A_{\bar{c}}^{M_\alpha}\colon M_\alpha \models a <_{\bar{c}} d\}$,

(ii) if $t = 2, \bar{c} \in {}^{2k}(M_\alpha), k = 0, 1, 2$ and $d_0, d_1 \in A_{\bar{c}}^{M_\alpha}$ then $A_{\bar{c}\hat{}\langle d_0, d_1\rangle}^{M_\alpha} = \{a \in A_{\bar{c}}^{M_\alpha}\colon M_\alpha \models \text{“}a <_{\bar{c}} d_0 \ \&\ a < d_1\text{”}\}$,

(ζ) if A is a subset of M_α of cardinality $< \mu$ then $c\ell^*_{M_\alpha}(A)$ is of cardinality $< \mu$ and $c\ell^*_{M_\alpha}(c\ell^*_{M_\alpha}(A)) = c\ell^*_{M_\alpha}(A)$ where

⊠ for $A \subseteq M_{\alpha'}, c\ell^*_{M_\alpha}(A)$ is the minimal set B such that: $A \subseteq B$ and $(\forall \bar{c} \in {}^{3t}B)(|A_{\bar{c}}^{M_\alpha}| < \mu \rightarrow A_{\bar{c}} \subseteq B)$; clearly B exists and $c\ell^*_{M_\alpha}(\emptyset) = \emptyset$,

(η) for every $\beta < \alpha, k = 1, 2$ and $\bar{c} \in {}^k(M_\beta)$ we have $A_{\bar{c}}^{M_\alpha} = A_{\bar{c}}^{M_\beta}$,

(θ) if $A \subseteq \beta < \alpha$ then $c\ell^*_{M_\beta}(A) = c\ell^*_{M_\alpha}(A)$,

(ι) if $t = 2$ and $\mu = \aleph_0$ and $A \subseteq \alpha$ is finite, β is the last element in A, then for some finite $B \subseteq \beta$ we have $c\ell^*_{M_\alpha}(A) = \{\beta\} \cup c\ell^*_{M_\beta}(B)$.

We leave the cases $\alpha < \mu$ and α a limit ordinal to the reader (for (ζ) we use (θ)) and assume $\alpha = \beta + 1$ and M_γ for $\gamma \leq \beta$ are defined. We can choose $\langle B_{\beta,i}\colon i < \mu^+\rangle$, a (not necessarily strictly) increasing sequence of subsets of β, each of cardinality $\leq \mu$, $B_{\beta,0} = \emptyset$ and $\bigcup\{B_{\beta,i}\colon i < \mu^+\} = \beta$ and $c\ell^*_{M_\beta}(B_{\beta,i}) = B_{\alpha,i}$.

For each $i < \mu^+$ let $\langle B_{\beta,i,\varepsilon}\colon \varepsilon < \mu\rangle$ be (not necessarily strictly) increasing sequence of subsets of $B_{\beta,i}$ with union $B_{\beta,i}$ such that $c\ell^*_{M_\beta}(B_{\beta,i,\varepsilon}) = B_{\beta,i,\varepsilon}$, $B_{\beta,0} = \emptyset$. Let $<^*_\beta$ be a well ordering of $\{\gamma\colon \gamma < \beta\}$ such that each $B_{\beta,i}$ is an initial segment so it has order type μ^+. For $\gamma \in B_{\beta,i+1} \backslash B_{\beta,i}$ let $<^*_{\beta,\gamma}$ be a well ordering of $A^*_{(\beta,\gamma)} = \{\xi\colon \xi <^*_\beta \gamma\}$ of order type $\leq \mu$ such that $(\forall \varepsilon < \mu)(B_{\beta,i+1,\varepsilon} \cap A^*_{(\beta,\gamma)}$ is an initial segment of $A^*_{(\beta,\gamma)}$ by $<^*_{\beta,\gamma}$.

Now we define M_α:

$$\text{universe is } \alpha$$
$$R_0^{M_\alpha} = <\restriction \alpha$$

CASE 1. $t = 1$.

$R_1^{M_\alpha} = R_1^{M_\beta} \cup \{(a, b, \beta)\colon a <^*_\beta b\}$

$R_2^{M_\alpha} = R_2^{M_\beta} \cup \{(a, b, \beta, \gamma)\colon \gamma < \beta$, and $a <^*_{\beta,\gamma} b$ hence $a <^*_\beta \gamma \ \&\ b <^*_\beta \gamma$ and $a, b \in B_{\beta,i+1}$ for the unique i such that $\gamma \in B_{\beta,i+1} \backslash B_{\beta,i}\}$.

CASE 2. $t = 2$.

$R_1^{M_\alpha} = R_1^{M_\beta} \cup \{(a, b, \beta, \gamma)\colon$
$a <^*_\beta b$ and $a < \gamma, b < \gamma$ and, of course, $a, b, \beta \in \alpha\}$.

$$R_2^{M_\alpha} = R_2^{M_\beta} \cup \{(a,b,\beta,\gamma_0,\beta_1,\gamma_1)\colon a,b,\gamma_0,\beta_1 \in \alpha \text{ and } a<\beta,\ b<\beta, a<\gamma_0, b<\gamma_0,\ a,b,\beta_1,\gamma_1 \in A_{<\beta,\gamma_0>} \text{ and }\ a <^*_{\beta,\gamma_0} b \text{ and } a<\gamma_1, b<\gamma_1\}.$$

To check for clause (ζ) is easy if $\mu = \mathrm{cf}(\mu) > \aleph_0$ and follows by clause (ι) if $\mu = \aleph_0$.

Having carried the induction we define M: it is M_λ expanded by $\langle F_i^M\colon i<\mu\rangle$ such that: if $\bar{c} \in {}^{3\iota}\lambda = {}^{3\iota}(M_\lambda)$ and $A_{\bar{c}}$ is a non empty well defined and of cardinality $<\mu$ (which follows) then $\{F_i^M(\bar{c})\colon i<\mu\}$ list $A_{<c_0,c_1,c_2>} \cup \{0\}$ otherwise $\{F_i^M(\bar{c})\colon i<\mu\}$ is $\{0\}$.

(3) Similar and used only for $(\aleph_3, \aleph_0)$ so we do not elaborate.

(4) Let $\mathbb{Q}$ be defined as follows:

$p \in \mathbb{Q}$ iff

(α) p is a τ^--model, as in (α) of the proof of part (1),

(β) the universe $\mathrm{univ}(p)$ of p is a countable subset of λ, we let $A^p_{\langle\rangle} = \mathrm{univ}(p)$,

(γ) $R_0^p = < \restriction \mathrm{univ}(p)$ and $<_{\langle\rangle} = R_0^p$,

(δ) if $\bar{c} \in {}^{\iota k}(\mathrm{univ}(p)), k = 1,2$ then $<_{\bar{c}} = <^p_{\bar{c}}$ is a well ordering of $A^p_{\bar{c}} = \{a \in p\colon p \models (\exists x)(a <_{\bar{c}} x \vee x <_{\bar{c}} a)\}$ and for $d \in A^p_{\bar{c}}$ we let $A^p_{\bar{c}^\frown<d>} = \{a \in A^p_{\bar{c}}\colon a <^p_{\bar{c}} d\}$,

(ε) $(A^p_{\bar{c}}, <_{\bar{c}})$ has order type ω if $k=2$,

(ζ) if $A \subseteq \mathrm{univ}(p)$ is finite, then $c\ell^*_p(A)$ is finite (is defined as in(2)).

the order:

$\mathbb{Q} \models p \le q$ iff

(i) p is a submodel of q,

(ii) if $\bar{c} \in {}^2(\mathrm{univ}(p))$ then $A^p_{\bar{c}} = A^q_{\bar{c}}$,

(iii) if $\bar{c} \in {}^1(\mathrm{univ}(p))$ then $A^p_{\bar{c}}$ is an initial segment of $A^q_{\bar{c}}$ by $<_{\bar{c}}$.

The rest should be clear. $\square_{4.4}$

4.6. Claim. *Assume (main case is $n(*) = 2$)*

$(*)$ $2 \le n(*) < \omega, \lambda = \aleph_{n(*)}, \ell(*) = 2n(*)+3$ *and* $\lambda \le \chi = \chi^{\aleph_0}$.

Then for some $\mathbb{P}^*$ we have

(a) $\mathbb{P}^*$ *is a forcing notion of cardinality* χ,

(b) $\mathbb{P}^*$ *satisfies the c.c.c.*,

(c) *in* $\mathbf{V}^{\mathbb{P}^*}$ *the pair* $(\aleph_{n(*)}, \aleph_0)$ *is not compact*,

(d) *in* $\mathbf{V}^{\mathbb{P}^*}$ *we have* $2^{\aleph_0} = \chi$.

Remark. We intend to prepare a full version.

Proof. We repeat §2, §3 with the following changes.

If $n(*) \geq 3$, we need change (A) below and using 3.1(2) instead of 3.1(1). For $n(*) = 2$ we need all the changes below

(A) we replace M^*_λ by any model as in 4.4(2) if $n(*) = 2$, 4.4(3) if $n(*) \geq 3$,

(B) in Definition 2.4, Case 3: we add $\langle v^+_{\{\eta\}}, v^-_{\{\eta\}} : \eta \in \mathrm{dom}_{\ell(*),m}\rangle, v^+_\emptyset, v^-_\emptyset$

(d)′ (i) $v_y \supseteq u^{p_y}$ for $y \in Y$,

(ii) if $\eta_1 <_{\text{lex}} \eta_2 <_{\text{lex}} \eta_3$ are from $\mathrm{dom}_{\ell(*),m}$ and $\{\eta_1, \eta_2\}, \{\eta_1, \eta_3\} \in Y$, then

$$v_{\{\eta_1,\eta_2\}} \cap v_{\{\eta_1,\eta_3\}} = v^+_\eta,$$

(iii) if $\eta_1 <_{\text{lex}} \eta_2 <_{\text{lex}} \eta_3$ are from $\mathrm{dom}_{\ell(*),m}$ and $\{\eta_1, \eta_3\}, \{\eta_2, \eta_3\} \in Y$, then

$$v_{\{\eta_1,\eta_3\}} \cap v_{\{\eta_2,\eta_3\}} = v^-_\eta$$

$$p_{\{\eta_1,\eta_3\}} \restriction v^-_\eta = p_{\{\eta_2,\eta_3\}} \restriction v^-_\eta,$$

(iv) $v_{\{\eta\}} = v^+_\eta \cup v^-_\eta$,

(v) if $\eta_1 \neq \eta_2$ then $v^+_{\eta_1} \cap v^+_{\eta_2} = v^+_\emptyset$ and $v^-_{\eta_2} \cap v^-_{\eta_2} = v^-_\emptyset$,

(vi) if $\eta_1 <_{\text{lex}} \eta_2 <_{\text{lex}} \eta_2$ are from $\mathrm{dom}_{\ell(*),m}$, then $p \restriction v_\eta \in \bigcup_{r<k} \mathbb{P}_r$,

(e) if $\eta_1 <_{\text{lex}} \eta_2$ are from $\mathrm{dom}_{\ell(*),m}$ and $t = \{\eta_1, \eta_2\} \in Y$ then $c\ell(v^+_{\eta_1}) \cap v_{\{\eta_1,\eta_2\}} = v^+_{\eta_1}$, $c\ell(v^-_{\eta_2}) \cap v_{\{\eta_1,\eta_2\}} = v^-_{\eta_2}$ if $\{\eta_1, \eta_2\} \in Y, \eta \in \mathrm{dom}_{\ell(*),m} \setminus \{1, \eta_1\}$ then $c\ell(v^\pm_\eta) \cap v_{\{\eta_1,\eta_2\}} \subseteq v^\pm_{\eta_1}$,

(f) the functions $\langle c^{p_\eta} : \eta \in y\rangle$ are pairwise compatible,

(C) in 3.1

(a) $\lambda \geq (2^\mu)^+, \mu = \mu^{<\mu}, (\forall A \in [M]^{<\mu})(|c\ell_M(A)| < \mu)$,

(b) the conclusion: change as in Definition 2.4, Case 3,

(c) proof:

Case 1. $\mu = \aleph_0$: let $g : [\lambda]^2 \to \omega$ be $g(t) = |c\ell_M(\alpha, t \cup w_t)| < \omega$.

Let $W_1 \in [\lambda]^{\mu^+}$ be such that $g \restriction [W]^2$ is constant say $k(*)$ and $f \restriction [W]^2$ is constantly γ. Let $c\ell_M(t) = \{\zeta_{t,\ell} : \ell < g(t)\}$. By Ramsey theorem, there is an infinite $W \subseteq W_1$ such that:

⊛ the truth value on $\zeta_{\{\alpha_1,\beta_2\},\ell_1} = \zeta_{\{\alpha_2,\beta_2\},\ell_2}$ depend just on ℓ_1, ℓ_2, T.V.(α_i, β_j), T.V.$(\beta_j < \alpha_i)$ for $i, j \in \{1, 2\}$.

The conclusion should be clear.

(D) p in the proof of 3.2: only Case 3B need care, assuming $m(*) > 2|u^{p^*}|$, the relevant subgraph has no cycle by clause (e) of Case 3 of 2.4 we are done,

(E) in the proof of 3.3, we will have $q^+_{\eta,\ell}, q^-_{\eta,\ell}$ with domain $\subseteq v^-_\eta, v^+_\eta$ respectively and q^+_ℓ, q^-_ℓ such that if $\eta_1 <_{\text{lex}} \eta_2$ and $\{\eta_1, \eta_2\} \in Y$ then $p_{\{\eta_1,\eta_2\},\ell} \restriction v^+_{\eta_1} = q^+_{\eta_1,\ell}$ and $p_{\{\eta_1,\eta_2\},\ell} \restriction v^-_{\eta_2} = q^-_{\eta_2,\ell}$ and $q^+_{\eta,\ell}, q^-_{\eta,\ell}$ are compatible, $q^+_{\eta,\ell} \restriction v^+_\emptyset = q^+_\ell, q^-_{\eta,\ell} \restriction v_\emptyset = q^-_\ell$,

(F) 3.4: part of the work has already been done in 4.1–4.3.

§5. Open problems and concluding remarks. We finish the paper by listing some problems (some are old, see [CK]).

5.1. QUESTION. Suppose that λ is strongly inaccessible, $\mu > \aleph_0$ is regular not Mahlo and $\Box_\mu$. Then $\lambda \to \mu$ in the λ-like model sense, i.e., if a first order ψ has a λ-like model then it has a μ-like model.

If λ is ω-Mahlo, the answer is yes, see [ScSh:20] by appropriate partition theorems. The assumption that μ is Mahlo is necessary by Schmerl, see [Sc85].

5.2. QUESTION. (Maybe under $\mathbf{V} = \mathbf{L}$.) Suppose that $\lambda^{\beth_\omega(\kappa)} = \lambda$ and $\lambda_1^{<\lambda_1} = \lambda_1 > \kappa_1$. Then $(\lambda^+, \lambda, \kappa) \to (\lambda_1^+, \lambda_1, \kappa_1)$.

5.3. QUESTION. (GCH) If λ and μ are strong limit singulars and λ is a limit of supercompacts, then $(\lambda^+, \lambda) \to (\mu^+, \mu)$.

5.4. QUESTION. Find a universe with $(\beth_2(\aleph_0), \aleph_0) \to (2^{2^\lambda}, \lambda)$ for every λ.

(The author has a written sketch of a result which is close to this one. He starts with $\aleph_0 = \kappa_0 < \kappa_1 < \cdots < \kappa_m$ which are supercompacts and let $\mathbb{P}_n$ be the forcing which adds κ_{n+1} Cohen subsets to κ_n in $V^{\mathbb{P}_0 * \mathbb{P}_1 \ldots \mathbb{P}_{n-1}}$ for $n < m$. The idea is using the partition on trees from [Sh:288, §4]).

5.5. QUESTION. Are all pairs in the set

$$\{(\lambda, \mu)\colon 2^\mu = \mu^+ \ \&\ \mu = \mu^{<\mu} \ \&\ \mu^{+\omega} \leq \lambda \leq 2^{\mu^+}\}$$

such that there is μ^+-tree with $\geq \mu^+$ branches, equivalent for the two cardinal problem? More related to this particular work are

5.6. QUESTION. (1) Can we find $n < \omega$ and an infinite set Γ^* of identities (or 2-identities) such that for any $\Gamma \subseteq \Gamma^*$ for some forcing notion $\mathbb{P}$ in $\mathbf{V}^{\mathbb{P}}$ we have $\Gamma = \Gamma^* \cap ID(\aleph_n, \aleph_0)$.

(2) In (1) we can consider (λ, μ) with $\mu = \mu^{\aleph_0}, \lambda = \mu^{+n}$, so we ask: can we find a forcing notion $\mathbb{P}$ not adding reals such that for every $\Gamma \subseteq \Gamma^*$ for some $\mu = \mu^{<\mu}$ we have $\Gamma = \Gamma^* \cap ID(\mu^{+n}, \mu)$.

5.7. QUESTION. (1) Can we get results parallel to 3.5 for $(\aleph_2, \aleph_1) + 2^{\aleph_0} \geq \aleph_2$ (so we should start with a large cardinal, at least a Mahlo).

(2) The parallel to 5.6(1),(2).

5.8. QUESTION. (1) Can we get results parallel to 3.5 for $(\aleph_{\omega+1}, \aleph_\omega)$ + G.C.H. (or (μ^+, μ), μ strong limit singular + G.C.H.

(2) The parallel to 5.6(1),(2).

5.9. QUESTION. How does assuming MA + $2^{\aleph_0} > \aleph_n$ influence $ID(\aleph_n, \aleph_0)$? (see below).

We end with some comments:

5.10. DEFINITION. (1) For $k \leq \aleph_0$, we say (λ, μ) has k-simple identities when $(a, e) \subseteq ID(\lambda, \mu) \Rightarrow (a, e') \in ID(\lambda, \mu)$ whenever:

$(*)_k$ $a \subseteq \omega, (a,e)$ is an identity of (λ,μ) and e' is defined by $be'c$ iff $|b| = |c|$ & $(\forall b', c')[b' \subseteq b$ & $|b'| \leq k$ & $c' = OP_{c,b}(b') \rightarrow b'ec']$ recalling
$OP_{A,B}(\alpha) = \beta$ iff $\alpha \in A$ & $\beta \in B$ & $\text{otp}(\alpha \cap A) = \text{otp}(\beta \cap B)$.

5.11. Claim. (1) *If* (λ_1, μ_1) *has* k*-simple identities and there is* $f : [\lambda_2]^{\leq k} \rightarrow \mu_2$ *such that* $ID_{\leq k}(f) \subseteq ID_{\leq k}(\lambda_1, \mu_1)$, *then* $(\lambda_1, \mu_1) \rightarrow (\lambda_1', \mu_1')$.
(2) *If* $cf(\lambda_1) > \mu$, *then we can use* f *with domain* $[\lambda_1']^{\leq k} \backslash [\lambda_1']^{\leq 1}$.

Proof. Should be easy.

5.12. Claim. (1) $[\text{MA} + 2^{\aleph_0} > \aleph_n]$. *The*[4] *pair* $(\aleph_n, \aleph_0)$ *has 2-simple identities.*
(2) *If* $\mu = \mu^{<\mu}$ *and* $\gamma \leq \omega$ *then for some* μ^+*-c.c.,* $(< \mu)$*-complete forcing notions,* $\mathbb{P}$ *in* $\mathbf{V}^{\mathbb{P}}$ *we have* $2^\mu \geq \mu^{+\gamma}$ *and* $n \leq \gamma$ & $n < \omega \Rightarrow (\mu^{+n}, \mu)$ *has 2-simple identities.*
(3) *If* $m < n < \omega, \mu = \mu^{<\mu}$, *then* $[\mu^{+n}, \mu^{+m})$ *has* $(m+2)$*-simple identities in* $\mathbf{V}^{\mathbb{P}}$ *for appropriate* μ^+*-c.c.* $(< \mu)$*-complete forcing notion.*

Proof. (1) For any $c : [\aleph_n]^{<\aleph_0} \rightarrow \omega$ we define a forcing notion $\mathbb{P} = \mathbb{P}_c$ as follows:
$p \in \mathbb{P}$ iff:
(a) $p = (u, f) = (u^p, f^p)$,
(b) u is a finite subset of $\aleph_n$,
(c) f is a function from $[u]^2$ to ω,
(d) if $k < \omega, k \geq 2$ and $\alpha_0 < \cdots < \alpha_{k-1}$ are from u, $\beta_0 < \cdots < \beta_{k-1}$ are from u and $[\ell(1) < \ell(2) < k \Rightarrow f(\{\alpha_{\ell(1)}, \alpha_{\ell(2)}\}) = f(\{\beta_{\ell(1)}, \beta_{\ell(2)}\})]$, then $c(\{\alpha_0, \ldots, \alpha_{k-1}\}) = c(\{\beta_0, \ldots, \beta_{k-1}\})$.

The rest should be clear.
(2), (3) Similar (use e.g., [Sh:546]). $\square_{5.12}$

We can give an alternative proof of [Sh:49], note that by absoluteness the assumption MA is not a real one; it can be eliminated and $(\mu^{+\omega}, \mu) \rightarrow' (2^{\aleph_0}, \aleph_0)$ can be deduced.

5.13. Claim. *Assume* $\text{MA} + 2^{\aleph_0} > \aleph_\omega$.
Then $(\aleph_\omega, \aleph_0) \rightarrow (2^{\aleph_0}, \aleph_0)$.

Proof. Let $\{\eta_\alpha : \alpha < 2^{\aleph_0}\}$ list ${}^\omega 2$, and define $f : [2^{\aleph_0}]^2 \rightarrow {}^{\omega >}2$ by:
$(*)$ $f\{\alpha_0, \alpha_1\} = \eta_{\alpha_0} \cap \eta_{\alpha_1} \in {}^{\omega >}2$ for $\alpha_0 \neq \alpha_1$.

So by 5.11, 5.12 it is enough to prove that $ID_2(f) \subseteq ID_2(\aleph_\omega, \aleph_0)$.
Clearly
$(*)_1$ if $\lambda \leq 2^{\aleph_0}, (u, e) \in ID_2(\lambda, \aleph_0)$ then $(u, e) \in ID_2(f \restriction \lambda)$ hence $(u, e) \in ID_2(f)$ hence for some n, (u, e) can be embedded (in the natural sense) into $({}^n 2, e_n^*)$ where $(\{\eta_1, \eta_2\} e_n^* \{\nu_1, \nu_2\}) \equiv (\eta_1 \cap \eta_2 = \nu_1 \cap \nu_2)$.

[4]Of course the needed version of MA is quite weak; going more deeply in [Sh:522].

So it is enough to prove

$(*)_2$ $({}^n2, e^*_n) \in ID_2(\mu^{+n}, \mu)$.

We prove this by induction on n.

$n = 0$: Trivial.

$n+1$: Let $c\colon [\mu^{+n+1}]^2 \to \mu$, choose $M \prec (\mathscr{H}(\mu^{+n+2}), \in)$ of cardinality μ^{+n} such that $\mu^{+n} + 1 \subseteq M, c \in M$, so let $\delta = M \cap \mu^{+n}$.

Define $c_n\colon \mu^{+n} \to \mu$ by $c_n\{\alpha, \beta\} = (c\{\alpha,\beta\}, c\{\delta,\alpha\}, c\{\delta,\beta\})$ for $\alpha < \beta < \mu^{+n}$. By the induction hypothesis there is a sequence $\langle \beta_\eta\colon \eta \in {}^n2\rangle$ of distinct ordinal $< \mu^{+n}$ such that $\{\eta_1,\eta_2\}\, e^*_n\, \{\nu_1,\nu_2\} \Rightarrow c_n\{\beta_{\eta_1}, \beta_{\eta_2}\} = c_n\{\beta_{\nu_1}, \beta_{\nu_2}\}$.

Let

$$A = \{\gamma < \mu^{+n+1}\colon \gamma \notin \{\beta_\eta\colon \eta \in {}^n2\} \text{ and for every } \eta \in {}^n2 \text{ we have } c\{\beta_\eta, \gamma\} = c\{\beta_\eta, \delta\}\}.$$

Clearly $A \in M$ and $\delta \in A$ so $A \not\subseteq M$, hence necessarily $|A| = \mu^{+n+1}$. So by the induction hypothesis we can find a sequence $\langle \gamma_\eta\colon \eta \in {}^n2\rangle$ of distinct members of $A \backslash \delta$ such that

$$\{\eta_1,\eta_2\} e^*_n \{\nu_1,\nu_2\} \Rightarrow c\{\gamma_{\eta_1,\eta_2}, \gamma_{\eta_2}\} = c\{\gamma_{\nu_1}, \gamma_{\nu_2}\}.$$

Now we define

$$\alpha_\eta = \begin{cases} \beta_{\langle \eta(1+\ell)\colon \ell<n\rangle} & \text{if } \eta(0) = 0, \\ \gamma_{\langle \eta(1+\ell)\colon \ell<n\rangle} & \text{if } \eta(0) = 1. \end{cases}$$

It is easy to check that $\langle \alpha_\eta\colon \eta \in {}^{n+1}2\rangle$ is as required. $\square_{5.13}$

We further can ask:

5.14. QUESTION. Assume $\Gamma_i \subseteq ID^*$ for $i < i^*, \mathbb{P}$ is $\Pi\{\mathbb{P}^\lambda_{\Gamma_i}\colon i < i^*\}$ with finite support, $c\colon [\aleph_{n(*)}]^2 \to \omega$ in $\mathbf{V}^{\mathbb{P}}$ then $ID(c)$ is not too far from some $\bigcup_{i\in w}\Gamma_i, w \subseteq i^*$ finite.

5.15. DISCUSSION. We can look more at ordered identities (recall)

$(*)_1$ for $\mathbf{c}_i\colon [\lambda]^{<\aleph_0} < \mu$ let $OID(c) = \{(a,e)\colon a$ a set of ordinals and there is an ordered preserving $f\colon a \to \lambda\}$ such that $b_1 e b_2 \Rightarrow \mathbf{c}\{f''(b_1)) = \mathbf{c}(f''(b_2))$ and $OID(\lambda,\mu) = \{(n,e)\colon (n,e) \in OID(\mathbf{c})$ for every $\mathbf{c}\colon [\lambda]^{<\aleph_0} \to \mu$, and similarly OID_2, OID_k.

Of course,

$(*)_2$ $ID(\lambda,\mu)$ can be computed from $OID(\lambda,\mu)$.

REFERENCES

[Ch] CHEN C. CHANG, *A note on the two cardinal problem*, ***Proceedings of the American Mathematical Society***, vol. 16 (1965), pp. 1148–1155.

[CK] CHEN C. CHANG and JEROME H. KEISLER, *Model Theory*, Studies in Logic and the Foundation of Mathematics, vol. 73, North Holland Publishing Co., Amsterdam, 1973.

[CFM] JAMES CUMMINGS, MATTHEW FOREMAN, and MENACHEM MAGIDOR, *Squares, scales and stationary reflection*, ***Journal of Mathematical Logic***, vol. 1 (2001), pp. 35–98.

[De73] KEITH J. DEVLIN, ***Aspects of constructibility***, Lecture Notes in Mathematics, vol. 354, Springer-Verlag, 1973.

[Fu65] E. G. FURKHEN, *Languages with added quantifier "there exist at least $\aleph_\alpha$"*, ***The Theory of Models*** (J. V. Addison, L. A. Henkin, and A. Tarski, editors), North-Holland Publishing Company, 1965, pp. 121–131.

[GcSh:491] MARTIN GILCHRIST and SAHARON SHELAH, *Identities on cardinals less than $\aleph_\omega$*, ***The Journal of Symbolic Logic***, vol. 61 (1996), pp. 780–787, math.LO/9505215[5].

[GcSh:583] ———, *The Consistency of ZFC + $2^{\aleph_0} > \aleph_\omega + \mathrm{I}(\aleph_2) = \mathrm{I}(\aleph_\omega)$*, ***The Journal of Symbolic Logic***, vol. 62 (1997), pp. 1151–1160, math.LO/9603219.

[Jn] RONALD B. JENSEN, *The fine structure of the constructible hierarchy*, ***Annals of Mathematical Logic***, vol. 4 (1972), pp. 229–308.

[Ke66] JEROME H. KEISLER, *First order properties of pairs of cardinals*, ***Bulletin of the American Mathematical Society***, vol. 72 (1966), pp. 141–144.

[Ke66a] ———, *Some model theoretic results for ω-logic*, ***Israel Journal of Mathematics***, vol. 4 (1966), pp. 249–261.

[Mi72] WILLIAM MITCHELL, *Aronszajn trees and the independence of the transfer property*, ***Annals of Mathematical Logic***, vol. 5 (1972), no. 3, pp. 21–46.

[Mo68] M. D. MORLEY, *Partitions and models*, ***Proceedings of the Summer School in Logic, Leeds, 1967***, Lecture Notes in Mathematics, vol. 70, Springer-Verlag, 1968, pp. 109–158.

[Mo57] ANDRZEJ MOSTOWSKI, *On a generalization of quantifiers*, ***Fundamenta Mathematicae***, vol. 44 (1957), pp. 12–36.

[MV62] M. D. MORLEY and R. L. VAUGHT, *Homogeneous and universal models*, ***Mathematica Scandinavica***, vol. 11 (1962), pp. 37–57.

[RoSh:733] ANDRZEJ ROSŁANOWSKI and SAHARON SHELAH, *Historic forcing for Depth*, ***Colloquium Mathematicum***, vol. 89 (2001), pp. 99–115, math.LO/0006219.

[Sc74] JAMES H. SCHMERL, *Generalizing special Aronszajn trees*, ***The Journal of Symbolic Logic***, vol. 39 (1974), pp. 732–740.

[Sc85] ———, *Transfer theorems and their application to logics*, ***Model Theoretic Logics*** (J. Barwise and S. Feferman, editors), Springer-Verlag, 1985, pp. 177–209.

[ScSh:20] JAMES H. SCHMERL and SAHARON SHELAH, *On power-like models for hyperinaccessible cardinals*, ***The Journal of Symbolic Logic***, vol. 37 (1972), pp. 531–537.

[Sh:8] SAHARON SHELAH, *Two cardinal compactness*, ***Israel Journal of Mathematics***, vol. 9 (1971), pp. 193–198.

[Sh:18] ———, *On models with power-like orderings*, ***The Journal of Symbolic Logic***, vol. 37 (1972), pp. 247–267.

[Sh:37] ———, *A two-cardinal theorem*, ***Proceedings of the American Mathematical Society***, vol. 48 (1975), pp. 207–213.

[Sh:49] ———, *A two-cardinal theorem and a combinatorial theorem*, ***Proceedings of the American Mathematical Society***, vol. 62 (1976), pp. 134–136.

[Sh:74] ———, *Appendix to: "Models with second-order properties. II. Trees with no undefined branches"* (***Annals of Mathematical Logic*** vol. 14 (1978), no. 1, pp. 73–87), ***Annals of Mathematical Logic***, vol. 14 (1978), pp. 223–226.

[Sh:289] ———, *Consistency of positive partition theorems for graphs and models*, ***Set theory and its applications (Toronto, ON, 1987)*** (J. Steprans and S. Watson, editors), Lecture Notes in Mathematics, vol. 1401, Springer, Berlin-New York, 1989, pp. 167–193.

[Sh:269] ———, *"Gap 1" two-cardinal principles and the omitting types theorem for $\mathscr{L}(Q)$*, ***Israel Journal of Mathematics***, vol. 65 (1989), pp. 133–152.

[Sh:288] ———, *Strong partition relations below the power set: Consistency, Was Sierpiński right, II?*, ***Proceedings of the Conference on Set Theory and its Applications in honor of A. Hajnal and V. T. Sos, Budapest, 1/91***, Colloquia Mathematica Societatis Janos Bolyai. Sets, Graphs, and Numbers, vol. 60, 1991, math.LO/9201244, pp. 637–638.

[Sh:424] ———, *On CH + $2^{\aleph_1} \to (\alpha)^2_2$ for $\alpha < \omega_2$*, ***Logic Colloquium '90. ASL Summer Meeting in Helsinki***, Lecture Notes in Logic, vol. 2, Springer Verlag, 1993, math.LO/9308212, pp. 281–289.

[Sh:522] ———, *Borel sets with large squares*, ***Fundamenta Mathematicae***, vol. 159 (1999), pp. 1–50, math.LO/9802134.

[Sh:546] ———, *Was Sierpiński right? IV*, ***The Journal of Symbolic Logic***, vol. 65 (2000), pp. 1031–1054, math.LO/9712282.

[Sh:824] ———, *Two cardinals models with gap one revisited*, ***Mathematical Logic Quarterly***, to appear.

[Sh:E17] ———, *Two cardinal and power like models: compactness and large group of automorphisms*, ***Notices of the American Mathematical Society***, vol. 18 (1968), p. 425.

[Sh:E28] ———, *Details on* [Sh:74].

[ShVa:790] Saharon Shelah and Jouko Väänänen, *Recursive logic frames*, Preprint.

[Si71] Jack Silver, *Some applications of model theory in set theory*, ***Annals of Mathematical Logic***, vol. 3 (1971), pp. 45–110.

[Va65] R. L. Vaught, *A Löwenheim-Skolem theorem for cardinals far apart*, ***The Theory of Models*** (J. V. Addison, L. A. Henkin, and A. Tarski, editors), North-Holland Publishing Company, 1965, pp. 81–89.

INSTITUTE OF MATHEMATICS
THE HEBREW UNIVERSITY
JERUSALEM, ISRAEL
and
RUTGERS UNIVERSITY
MATHEMATICS DEPARTMENT
NEW BRUNSWICK, NJ, USA
E-mail: shelah@math.huji.ac.il

INCOMPLETENESS THEOREM AND ITS FRONTIER

GAISI TAKEUTI

As you know well, we celebrate 70th year of Gödel's "Incompleteness theorem" which he proved at the age of 25. This revolutionary theorem changed the way mathematicians think of mathematics drastically. Today, I would like to describe this theorem, its effects and expound on this subject.

Gödel's incompleteness theorem can be stated as follows.

1. Let T be a consistent axiomatic theory like set theory or the theory of analysis where Peano's arithmetic (denoted simply as arithmetic from now on) is included. Then T does not prove its consistency.
2. The consistency of T can be expressed as a sentence in arithmetic, therefore there exists an arithmetical sentence which cannot be proved in T.

This seemingly simple theorem changed our view of mathematics completely. Before this theorem, mathematicians believed that every problem in arithmetic could be solved by some stronger theory, e.g., set theory. But the Incompleteness theorem tells us that whatever stronger theory we use, there exists a true arithmetical sentence which cannot be proved in the theory.

After the Incompleteness theorem, the consistency statement becomes a landmark for the boundary of provable statements in the theory. Whenever we wish to show that the theory T' is strictly stronger than the theory T, we first try to show that the theory T' proves the consistency statement of T. In this way we can show that the theory of analysis is strictly stronger than the theory of arithmetic and that set theory is strictly stronger than the theory of analysis.

I would now like to speak of the impact that this theorem had on Hilbert's program. Hilbert was a genius to think of problems in a general setting, to find the essence of the problem and solve it. So, he was a leader of the movement of abstract systematic development and axiomatization in the 20th century mathematics. Cantor's set theory gave the ideal framework for this movement. Hilbert believed Cantor had created a new paradise for mathematicians. So, paradoxes of Cantor's set theory came as a great shock to him. Hilbert tried to save the modern mathematics and proposed the following program.

Logic Colloquium '01
Edited by M. Baaz, S. Friedman, and J. Krajíček
Lecture Notes in Logic, 20

1. To formalize mathematics by introducing symbols for basic relations, symbols for variables and symbols for logical notions and formalizing mathematical inferences and proofs.
2. Use only the clearest arguments on finitely many symbols. I.e., make concrete all the arguments. He called this method "finite standpoint".
3. The foundation of the formalized mathematics will be accomplished by proving the consistency of the system in the finite standpoint.

Gödel's Incompleteness theorem dealt a blow on Hilbert's program. It shows that Hilbert's program is "almost" impossible, since the finite standpoint allows us to use only elementary, clear, intuitive arguments on finite concrete figures but that consistency of set theory cannot be proved even in set theory which is far beyond the elementary finite standpoint. I said "almost" impossible in place of "absolutely" impossible only because there is no precise definition of the "finite standpoint". It is rather curious that Gödel's arithmetization somehow resembles Hilbert's formalization. I believe that this idea of arithmetization was inspired by Hilbert's method. It is an irony that as a result of Gödel carrying out Hilbert's idea methodically, he proved exactly the opposite of what Hilbert was trying to achieve. Bernays, who was Hilbert's primary advocate, once told me "Gödel accomplished through his diligence what we, Hilbert's disciples, were too lazy to pursue."

However, recently, there is an important new area in which the original Incompleteness theorem does not work as is. It is for the separations of theories. First I must explain about the polynomial hierarchy in the computational complexity.

PH (polynomial hierarchy):

$$
\begin{array}{ccccc}
 & | & & & | \\
 & \Delta_3^p & & --- \quad --- & S_2^3 \\
/ & & \backslash & & | \\
\Pi_2^p & & \Sigma_2^p & & | \\
\backslash & & / & & | \\
 & \Delta_2^p & & --- \quad --- & S_2^2 \\
/ & & \backslash & & | \\
\Pi_1^p = coNP & & \Sigma_1^p = NP & & | \\
\backslash & & / & & | \\
 & \Delta_1^p = P & & --- \quad --- & S_2^1
\end{array}
$$

P is the class of predicates or functions which can be computable in polynomial time by Turing machine, i.e., a computer. Σ_1^p, or NP equivalently, is the class of predicates which can be computed in polynomial time by a non-deterministic Turing machine. Δ_2^p is the class of predicates which can be computable in polynomial time by an oracle Turing machine with an oracle from Σ_1^p. I.e., the polynomial hierarchy is the class of predicates which is generated in polynomial time by Turing machines, non-deterministic Turing machines and oracle Turing machines. Whether $P = NP$ or $P \neq NP$ is the famous problem in this area. In general, it is conjectured that two classes in polynomial hierarchy are different. But this is a very difficult problem.

S. Buss introduced systems of bounded arithmetic S_2^i ($i = 0, 1, 2, \dots$) which corresponds to Δ_i^p. S_2^i is a very weak system of arithmetic. The separation of S_2^1 and S_2^2 is closely related to the separation of P and NP. If one proves the separation of S_2^i and S_2^j by a proof-theoretic method, then one can show $P \neq NP$.

S. Buss proved that Gödel's Incompleteness theorem holds for S_2^i, i.e., S_2^i does not prove its own consistency. On the other hand Wilkie and Paris proved that $S_2 + Exp$ does not prove the consistency of Robinson's system Q. $S_2 + Exp$ is stronger than any type of system of bounded arithmetic and Robinson's Q is the weakest system of bounded arithmetic. Therefore we cannot separate two systems of bounded arithmetic by original Gödel's Incompleteness theorem. For these cases let us see how far Gödel's idea works.

We consider the separation of S_2^1 and S_2^2. If we introduce finitely many functions $x + y$, $x \cdot y$, $|x|$, $\lfloor \frac{x}{2} \rfloor$, $x \# y$, ... from P, then a formula in Σ_1^p can be expressed in the form

$$\exists x \leq s(a) \forall y \leq |t(a)|\, A(a)$$

where $A(a)$ has no quantifiers and consists of $=$, $\leq$, the finitely many functions listed above and logical symbols $\neg$, $\wedge$, $\vee$. $|x|$ is the length of the binary representation of x. $|x|$ is very close to $\log_2(x)$. ($|x|$ is the smallest integer satisfying $\log_2(x+1) \leq |x|$). If we live in the world of polynomial, then exponential is an infinite and $\log_2(x)$ and $|x|$ are infinitesimal. We define $|x|_0 := x$ and $|x|_{i+1} := ||x|_i|$.

The main axiom of S_2^1 is PIND (polynomial time mathematical induction) on Σ_1^p-formula where PIND is weaker than the usual mathematical induction.

We denote the proof predicate of S_2^1 by PRF. Then Gödel's argument is as follows. There exists a Gödel sentence Φ satisfying

$$S_2^1 \vdash \Phi \leftrightarrow \forall x \neg PRF(x, \ulcorner \Phi \urcorner)$$

where $\ulcorner \Phi \urcorner$ is the Gödel number of Φ. Then $S_2^1 \nvdash \Phi$ and

$$S_2^1 \vdash \Phi \leftrightarrow \forall x \neg PRF(x, \ulcorner \rightarrow \urcorner)$$

where $\ulcorner \rightarrow \urcorner$ expresses the Gödel number of contradiction, i.e., $\forall x \neg PRF(x, \ulcorner \rightarrow \urcorner)$ expresses the consistency of S_2^1. Here I am using Gentzen system. A proof in S_2^1 is said to be normal if every cut formula in the proof is Σ_1^p. Then we have a theorem that every provable sequent (consisting of Σ_1^p formulas) in S_2^1 has a normal proof. Let us denote the normal proof predicate of S_2^1 by *Prf*. Then by Gödel's argument there exists φ_i for every $i = 0, 1, 2, \ldots$ satisfying

$$S_2^1 \vdash \varphi_i \leftrightarrow \forall x \neg Prf(|x|_i, \ulcorner \varphi_i \urcorner).$$

So there exists infinitely many Gödel sentences φ_i for the proof predicate *Prf*. Gödel's argument shows the following.

1. $S_2^1 \nvdash \varphi_i, i = 0, 1, 2, \ldots.$
2. $S_2^1 \vdash \varphi_0 \leftrightarrow \forall x \neg Prf(x, \ulcorner \rightarrow \urcorner)$.
3. $S_2^1 \vdash \varphi_0 \rightarrow \varphi_i$.

Now we can show $S_3^2 \vdash \varphi_1$, where S_3^2 is a system of bounded arithmetic which is stronger than S_2^2. Therefore this is not the result we wish to have. Nevertheless this shows that Wilkie-Paris' theorem does not hold for *Prf*. I believe that if we know the nature of φ_i or $\forall x \neg Prf(x, \ulcorner \rightarrow \urcorner)$ then we can separate S_2^1 and S_2^2 and very likely prove $P \neq NP$. We list several conjectures which imply the separation of S_2^1 and S_2^2.

1. If $S_2^2 \vdash \varphi_i$ for some $i \geq 0$, then we can separate S_2^1 and S_2^2.
2. We can show $S_2^2 \vdash \forall x \neg Prf(|x|, \ulcorner \rightarrow \urcorner)$ and conjecture $S_2^1 \nvdash \forall x \neg Prf(|x|, \ulcorner \rightarrow \urcorner)$.
   The conjecture implies the separation of S_2^1 and S_2^2.
3. We can show $S_2^2 \vdash \forall x \neg Prf(|x|_{i+1}, \ulcorner \varphi_i \urcorner)$ for every $i = 0, 1, 2, \ldots$ and conjecture $S_2^1 \nvdash \forall x \neg Prf(|x|_{i+1}, \ulcorner \varphi_i \urcorner)$ for some i. This conjecture also implies the separation of S_2^1 and S_2^2.

Now I would like to tell you some of my personal experiences with Gödel. In the 1950's, I was working on Hilbert's program, more precisely I proposed a higher order cut elimination theorem as my conjecture, which implies the consistency of analysis. I wished to contribute to Hilbert's program by working on a definite technical problem. Gödel was the first person to be interested in my work and invited me to the Institute for Advanced Study. In the beginning Gödel thought my conjecture false and thought that a counter-example can be found in impredicative cases. I told him that in many impredicative cases the cut elimination is trivially true. He was very surprised and changed his opinion on my conjecture. Gödel thought that it would work very well for my conjecture if Schütte and I were together. He invited Schütte to the Institute for this purpose. When I went to the Institute at the end of the summer, a stranger approached me and asked, "Do you know Takeuti?" I answered, "I am Takeuti." He introduced himself as Schütte and told me "I just saw Gödel. He is intensely interested in your conjecture." I think that Gödel showed him the relation between my conjecture and what Schütte was working on, and suggested a line of research. Consequently, Schütte became very interested in my work. The solution to my conjecture by Takahashi and Prawitz was reached by utilizing Schütte's method developed during that time. Gödel's insight was correct! In 1959, when I met Gödel for the first time, I did not know much logic. Until then, I had only read Gentzen's and Gödel's papers. So Gödel ended up teaching me logic. Since my English was terrible, he once took me to the library to show me a specific theorem in a book. Gödel's teaching of nonstandard arithmetic was unique. Let T be the theory of arithmetic. Then by his Incompleteness theorem, T and "T is inconsistent" are consistent. Therefore by Completeness Theorem there exists a model M of $T +$ "T is inconsistent". In M, there must exist a number n which is the Gödel number of the proof of the contradiction from T. Therefore n must be an infinite number.

One can see from this episode how much the Incompleteness theorem was at the root of his thought process.

One time, after he explained large cardinal axioms, he told me to go and ask Bernays, who was visiting the Institute at the time, for details. Gödel and Bernays spent many hours talking in Gödel's office. One day my curiosity got the better of me and I asked Bernays what they were discussing. Bernays answered, smiling broadly with pleasure: "We are chatting about a lot of things. This morning, we were reminiscing about a coffee house in Vienna." But I knew then that they were also discussing mathematics. Since Gödel suggested I ask Bernays set theory questions, I went to his office for this purpose. Instead of answering my questions, he started to talk about "Bernays' set theory". He had proved that Bernays' set theory is a conservative extension of ZF set theory. I told him that that is a direct corollary of Gentzen's cut elimination theorem. In the beginning, he seemed to think that I was mistaken. Soon after, he understood my explanation. Later I discovered that Gödel knew of this particular discourse. Thanks to Gödel, there were many logicians that year: Bernays, Schütte, Feferman, Putnam, Smullyan and many others. Logic seminar was held every week in jovial and optimistic atmosphere. Without Gödel at the center, this would not have been possible.

We had many discussions on my work and Hilbert's program. Once he told me that Gentzen's consistency proof is very close to the ideal finite standpoint. Since I had my own justification of Gentzen's transfinite induction, we agreed immediately. I regret that I do not remember the detail of his arguments though I think it closely related to his Dialectica paper. I was surprised to find that he had studied Gentzen's work at such considerable length. I felt that Gödel was genuinely interested in Hilbert's program. Later, I was shocked to read his sharp criticism of Hilbert's program in Hao Wang's book.

I stayed 3 times at the Institute as a member. The second time, Gödel and I had many contacts over the phone. He once complained that often he could not reach me at my office. I told him I preferred working at home since the chair in my office was hard and uncomfortable. His response was "Ask the secretary to put a sofa in your office." Of course, I could not act on this since I knew that I was not in a position to ask such favors. When I could not be reached at the office, he would call me at home. When my eight year old daughter answered the phone in my absence, his gentleness toward children was apparent. He gave the spelling of his name "K" "U" "R" "T" very slowly, letter by letter, and asked to have me call him back. My daughter still remembers and can imitate his special tone of voice.

Some describe Gödel as a strange, remote, and even crazy mathematician. Perhaps his shyness contributed to such an inaccurate impression. I will always remember him to be a kind, thoughtful and extremely perceptive mathematician. He was a truly great mathematician who changed the frontier of mathematics.

1420 LOCUST STREET #35R
PHILADELPHIA, PA 19102, USA
E-mail: takeuti@math.tsukuba.ac.jp

GROUPS IN SIMPLE THEORIES

FRANK WAGNER

Abstract. Groups definable in simple theories retain the chain conditions and decomposition properties known from stable groups, up to commensurability. In the small case, if a generic type of G is not foreign to some type q, there is a q-internal quotient. In the supersimple case, the Berline-Lascar decomposition works, and type-definable groups are the intersection of definable ones. One-based groups with simple theory are finite-by-abelian-by-finite.

§1. **Introduction.** In connection with the work by Hrushovski on pseudo-algebraically closed structures [8], Hrushovski and Chatzidakis [3] on difference fields, and Cherlin and Hrushovski [4] on smoothly approximated structures, simple theories of groups have recently begun to attract attention. While the above concentrate on those particular important examples, all of which have finite rank, simple theories in general have been studied by Shelah [19], [20], Kim [12], [13], [11], Kim and Pillay [14], Hart, Kim and Pillay [6], and in the group case by Pillay [15].

In this paper, we shall reconstitute the basic theory (chain conditions, internal quotients, components, definability, Zil'ber Indecomposability, one-based groups) for groups with simple theories, some under the additional condition of supersimplicity or smallness. Many of our results were proved by Hrushovski and Pillay [8], [9], [10] for S_1-theories. In a subsequent paper [24], many results have been further extended to the hyperimaginary context.

As usual, we shall be working in a κ-saturated κ-homogeneous monster model, for some big cardinal κ. Subsets and elements will come from the monster model, models will be elementary substructures of the monster model, all of cardinality $< \kappa$, and automorphisms will be automorphisms of the monster model, unless otherwise stated. A set is *bounded*, or *small*, if it is of cardinality less than κ. When dealing with groups we shall usually omit the

2000 *Mathematics Subject Classification*. Primary: 03C45; Secondary: 20F22.

Key words and phrases. simple theory, definable group.

Research done while visiting the Fields Institute, Toronto, Canada, whose hospitality and support are gratefully acknowledged. The author would like to thank Ambar Chowdhury, Bradd Hart, Bjunghan Kim and Anand Pillay for stimulating discussions.

Logic Colloquium '01
Edited by M. Baaz, S. Friedman, and J. Krajíček
Lecture Notes in Logic, 20

product symbol, so gh denotes the group product of g and h. Consequently, parameters will be separated by commas.

Note. In the meantime, the monograph [23] has appeared, which gives a detailed development of simplicity theory.

§2. Chain conditions. A family $\mathfrak{H}$ of subgroups of a (definable) group G is said to be *uniformly definable* if every element H of $\mathfrak{H}$ is of the form $\{g \in G: \models \varphi(g, \bar{a})\}$, where the formula φ remains fixed and only the parameters $\bar{a}$ vary. The most basic chain condition satisfied by a stable group is the chain condition on uniformly definable subgroups:

> Any chain of uniformly definable subgroups has finite length, bounded by some n depending only on the defining formula φ.

Clearly, the formula $\forall x\ [\varphi(x, \bar{y}) \to \varphi(x, \bar{z})]$ defines a partial order; as a simple theory does not have the strict order property, this chain condition must hold in any simple theory as well. (Note, however, that a simple theory may have the order property.)

The matter is different, however, when we consider intersections of uniformly definable subgroups. In a stable group, chains of such intersections still have finite length. This is no longer true in the simple case:

Example 1. Let V be an infinite-dimensional vector space over a finite field $\mathbf{F}_p$, and $\langle .,. \rangle$ a non-degenerate bilinear form on V. Then $(V, 0, +, \langle .,. \rangle)$ is simple [4], and has an infinite descending chain of intersections of uniformly definable subgroups $\{v \in V: \langle v, a \rangle = 0\}$.

Definition 1.

1. A group is *type-definable* if it is given as the set of realizations of a partial type in finitely many variables (in any model of the theory). It is $\bigcap$-definable if it is given as the intersection of definable groups.
2. If G is a group (not necessarily definable), then a subgroup H of G is *definable relative to G*, or simply *relatively definable*, if there is some formula φ such that $H = \{g \in G: \models \varphi(g)\}$. A family $\mathfrak{H}$ of subgroups of G is *uniformly definable relative to G* if every group in the family is definable relative to G by an instance of the same formula $\varphi(x, \bar{y})$. Note that $|\mathfrak{H}|$ need not be small.
3. A family $\mathfrak{H}$ of (relatively) uniformly definable subgroups of some group G in a model $\mathfrak{M}$ is *type-definable* if it consists of all groups $\{g \in G: \mathfrak{M} \models \varphi(g, \bar{a})\}$, where $\bar{a}$ runs through all realizations in $\mathfrak{M}$ of some partial type π.
4. If $H \leq G$ are two type-definable groups, then H has *bounded index* in G if $|G: H| < \kappa$ (or equivalently, if for any model $\mathfrak{M}$ of the ambient theory containing the relevant parameters the index $|G^{\mathfrak{M}}: H^{\mathfrak{M}}|$ remains bounded).

REMARK 1. 1. Clearly a $\bigcap$-definable group is type-definable. As for the converse, it was proved for the stable case by Hrushovski [7]; for a supersimple theory, this is our Theorem 5.6. The simple case is still open. Note that in any ω-saturated elementary extension of $\langle \mathbb{Q}, 0, +, < \rangle$ the infinitesimals form a type-definable subgroup which is not $\bigcap$-definable (but this structure is not simple).

2. We usually fix a model with an ambient group G, and view a family $\mathfrak{H}$ of subgroups as just a collection of subgroups of G. However, when G and $\mathfrak{H}$ are both type-definable, for any model $\mathfrak{M}$ we obtain a family $\mathfrak{H}^{\mathfrak{M}}$ of subgroups of $G^{\mathfrak{M}}$; we can then ask about properties independent of the particular choice of model.
3. If H is $\bigcap$-definable as $\bigcap_{i<\lambda} H_i$ and has bounded index in a type-definable group G, then by compactness every H_i intersects G in a subgroup of finite index. So for infinite λ the index of H in G is bounded by 2^λ. This bound still holds if H is type-defined by a partial type π of size λ: for every $\varphi(x) \in \pi$, there is some definable superset X of G and some $n < \omega$ such that for any n elements $g_1, \ldots, g_n \in X$ we have $\models \varphi(g_i g_j^{-1})$ for at least one pair $i \neq j$, as otherwise by compactness we could find models in which H has arbitrarily large index in G. Now suppose $\{g_i : i \in (2^\lambda)^+\}$ is a set of representatives for different cosets of H in G. Then to any pair $\{i, j\}$ we could associate a formula $\varphi \in \pi$ such that $\models \neg\varphi(g_i g_j^{-1})$. By the Erdös-Rado Theorem [5, Theorem 4(i)] there is a subset $J \subset (2^\lambda)^+$ of cardinality λ^+ such that all pairs from J are associated to the same formula. This contradicts the fact that we even have a finite bound on such sets.
4. One could also consider groups given by a partial type in infinitely many variables. Many results of this paper remain valid for that case.

We shall now review the local ranks, generic types, and stabilizers in a type-definable group G in a simple theory, as developed by Pillay [15]. We shall view G as a partial type over $\emptyset$, so $g \in G$ means $g \models G(x)$. Parameter sets A, B, etc. will come from the monster model and need not consist of group elements.

DEFINITION 2. For any $\mathcal{L}$-formula $\varphi(x, \bar{y})$ and $k < \omega$ the *local stratified rank* $D(., \varphi*, k)$ is defined for a partial type π with $\pi \vdash G(x)$ via:

1. $D(\pi(x), \varphi*, k) \geq 0$ if $\pi(x)$ is consistent.
2. $D(\pi(x), \varphi*, k) \geq n+1$ if there are elements $g_i \in G$ and $\bar{a}_i \in \mathfrak{M}$ for $i < \omega$, such that $\{\varphi(g_i x, \bar{a}_i) : i < \omega\}$ is k-inconsistent and $D(\pi(x) \wedge \varphi(g_i x, \bar{a}_i), \varphi*, k) \geq n$ for all $i < \omega$.

As usual, we write $D(a/A, \varphi*, k)$ for $D(\mathrm{tp}(a/A), \varphi*, k)$.

An element $g \in G$ is *generic* over A if for any $h \in G$ with $h \mathop{\smile\!\!\!\!\!\!\mid}_A g$ we have $hg \mathop{\smile\!\!\!\!\!\!\mid}_A h$. A type $\mathrm{tp}(g/A)$ is *generic* if g is generic over A. A partial type is *generic* if it can be extended to a generic type.

The *connected component* G_A^0 of G over A is the intersection of all subgroups of G of bounded index which are type-definable over A.

The *stabilizer* of a type $p \in S(\mathfrak{M})$ (for some model $\mathfrak{M}$) with $p \vdash (x \in G)$ is defined as follows: $S(p) = \{g \in G : gp \cup p$ does not fork over $\mathfrak{M}\}$, and $\mathrm{stab}(p) = S(p)S(p)$.

FACT 2.1. [15] *Let G be a type-definable group in a simple theory.*

1. *$D(\pi, \varphi*, k) < \omega$ for all partial types π, $\mathcal{L}$-formulas φ and $k < \omega$. For any partial type $\pi(x, \bar{y})$, $\mathcal{L}$-formula φ and $k, n < \omega$ the condition $D(\pi(x, \bar{y}), \varphi*, k) \geq n$ is type-definable in $\bar{y}$. If $D(\pi, \varphi*, k) \leq n$, then there is a formula $\psi \in \pi$ with $D(\psi, \varphi*, k) \leq n$.*
2. *If $g \in G$ and $h \in G \cap \mathrm{dcl}(A)$, then $D(g/A, \varphi*, k) = D(hg/A, \varphi*, k)$. Furthermore, for $g \in G$ and arbitrary A and B we have $g \mathop{\smile\!\!\!\!\!\!|}_A B$ if and only if $D(g/AB, \varphi*, k) = D(g/A, \varphi*, k)$ for all $\mathcal{L}$-formulas φ and all $k < \omega$.*
3. *An element $g \in G$ is generic over A if and only if $D(g/A, \varphi*, k) = D(G, \varphi*, k)$ for all $\mathcal{L}$-formulas φ and all $k < \omega$. A type-definable subgroup $H \leq G$ has bounded index in G if and only if $D(H, \varphi*, k) = D(G, \varphi*, k)$ for all $\mathcal{L}$-formulas φ and all $k < \omega$. In that case, generic types for H are also generic types for G (and H is generic for G); if H is type-definable over A, then $H \geq G_A^0$.*
4. *If $g \in G$ is generic over A, then g^{-1} and hg are generic over A for any $h \in G \cap \mathrm{dcl}(A)$.*
5. *G_A^0 is a normal subgroup of G of bounded index.*
6. *Let $\mathfrak{M}$ be a model and p imply $x \in G$. Then $S(p)$ is type-definable and* $\mathrm{stab}(p)$ *is a type-definable subgroup of G. All generic types of* $\mathrm{stab}(p)$ *are in $S(p)$. Furthermore, p is generic if and only if* $\mathrm{stab}(p)$ *has bounded index in G; in that case* $\mathrm{stab}(p) = G_{\mathfrak{M}}^0$. *For any generic type p of G and any generic type q of $G_{\mathfrak{M}}^0$, both over $\mathfrak{M}$, there are independent realizations $g \models p$ and $h \models q$ with $gh \models p$.*

REMARK 2. By symmetry, there also are local ranks $D(., \varphi*', k)$ for all $\mathcal{L}$-formulas φ and $k < \omega$, where we use multiplication by elements from G on the right rather than on the left; we can similarly define the stabilizer on the right $\mathrm{stab}'(p)$. Clearly these satisfy an analogous version of Fact 2.1.

PROPOSITION 2.2. *Let $(H_i : i \leq \alpha)$ be a descending chain of type-definable subgroups of a type-definable group in a simple theory, continuous at limits, such that each successor group has unbounded index its predecessor. Then $\alpha < |T|^+$.*

PROOF. For any $i < \alpha$ let (φ_i, k_i) be a pair such that $D(H_i, \varphi_i*, k_i) > D(H_{i+1}, \varphi_i*, k_i)$. If the chain had length $|T|^+$, then we could find a subchain J of length $|T|^+$ such that $(\varphi_i, k_i) = (\varphi, k)$ is constant for $i \in J$. As $D(H_j, \varphi*, k)$ is strictly descending for $j \in J$, this contradicts finiteness of $D(T, \varphi*, k)$. ⊣

For $\bigcap$-definable groups we get stronger results.

Definition 3. Two subgroups of some group are *commensurable* if their intersection has finite index in both of them. A group G is *uniformly commensurable* to a family $\mathfrak{H}$ of groups if the index of $G \cap H$ in G and in H is finite and bounded independently of $H \in \mathfrak{H}$.

Two type-definable subgroups are *commensurate* if their intersection has bounded index in both of them.

By compactness, if a relatively definable group has bounded index in a type-definable group, then the index must be finite. In particular, two commensurate definable groups are commensurable.

We shall need a theorem due essentially to Schlichting [18] and Bergman and Lenstra [1]:

Fact 2.3. *Let G be a group and $\mathfrak{H}$ a family of subgroups of G such that there is some $n < \omega$ bounding the index $|H : H \cap H'|$ for any $H, H' \in \mathfrak{H}$. Then there is a subgroup N which is uniformly commensurable to $\mathfrak{H}$ and invariant under all automorphisms of G which stabilize $\mathfrak{H}$ setwise. We have $\bigcap \mathfrak{H} \leq N \leq \langle \mathfrak{H} \rangle$; in fact N is a finite extension of a finite intersection of subgroups in $\mathfrak{H}$. In particular, if $\mathfrak{H}$ consists of relatively definable subgroups of a type-definable group, then N is relatively definable.*

Proof. Let $G_1 = G \rtimes \mathrm{Aut}(G)$, and K be the subgroup of elements of G_1 which stabilize $\mathfrak{H}$ setwise. Then K contains all automorphisms of G stabilizing $\mathfrak{H}$ setwise. By [1, Theorem 6] there is a K-invariant subgroup N_1 of G_1 which is uniformly commensurable with $\mathfrak{H}$; we put $N := (N_1 \cdot \bigcap \mathfrak{H}) \cap \langle \mathfrak{H} \rangle$ (which is easily seen to be a group commensurable with N_1).

Since $\bigcap \mathfrak{H} \leq N \cap H \leq H$ and N is commensurable with H for any $H \in \mathfrak{H}$, there is a finite intersection N_2 of groups in $\mathfrak{H}$ such that N_2 is a subgroup of N of finite index. So if $\mathfrak{H}$ consists of relatively definable subgroups, then N is relatively definable as a finite extension of N_2. ⊣

Lemma 2.4. *Let G be a type-definable group in a simple theory. Suppose the formula $\varphi(x, \bar{a})$ defines a subgroup $H(\bar{a}) \leq G$ relative to G for any $\bar{a}$ satisfying some partial type π. Then there is some integer $n < \omega$ and a definable superset X of G such that whenever $H(\bar{a}')$ is a subgroup of finite index in G for some $\bar{a}' \models \pi$, the index is bounded by n and X is covered by n translates of $\varphi(x, \bar{a}')$.*

Proof. By compactness, there is a definable superset Y of G such that on Y multiplication is defined and associative (but may go outside of Y), every $y \in Y$ has an inverse $y^{-1} \in Y$, and for any $x, y \in Y$ and $\bar{a} \models \pi$ with $\models \varphi(x, \bar{a}) \wedge \varphi(y, \bar{a})$ also $\models \varphi(x^{-1}y, \bar{a})$. Consider $\{g \in G : \models \varphi(xg, \bar{a})\}$, for some x with $xG \subseteq Y$. If g and h are in this set, then $g^{-1}h \in G$ and $\models \varphi(g^{-1}h, \bar{a})$, so this set must be a left translate of $H(\bar{a})$ in Y. Therefore, if there is a definable superset X of G and $x_0 = 1, x_1, \ldots, x_n$ with $\bigcup_{i \leq n} x_i X \subseteq Y$ and $\models \forall x \in X \bigvee_{i \leq n} \varphi(x_i x, \bar{a})$, then $H(\bar{a})$ has index at most $n + 1$ in G.

Conversely, if $H(\bar{a})$ has index at most $n+1$ in G, by compactness there must be such elements $x_0, \ldots, x_n$ (which we may even choose to come from G) and such a set X (which we may take definable over the parameters used to type-define G). It follows that the condition on $\bar{a}'$ that $H(\bar{a}')$ intersects G in a subgroup of finite index can be expressed as an infinite disjunction. However, it is also type-definable, since it is equivalent to $D(H(\bar{a}'), \varphi*, k) = D(G, \varphi*, k)$ for all $\mathcal{L}$-formulas φ and all $k < \omega$. By compactness, it is definable, and there is a bound n and a definable $X \supseteq G$ as required. ⊣

PROPOSITION 2.5. *Let G be a group type-definable over $\emptyset$ in a simple structure $\mathfrak{M}$, and $\mathfrak{H}$ a family of uniformly relatively $\mathfrak{M}$-definable subgroups of G. Then there is a finite intersection $N_0 := \bigcap_{i<n} H_i$ of elements in $\mathfrak{H}$, and a finite extension N of N_0 (which is thus also relatively definable) such that N is uniformly commensurable with $H \cap N_0$ for all $H \in \mathfrak{H}$, and N is invariant under all automorphisms of $\mathfrak{M}$ stabilizing $\mathfrak{H}$ setwise.*

In particular, any intersection of a family of uniformly relatively definable subgroups is commensurate with a finite subintersection.

PROOF. By Proposition 2.2 there is an intersection K of at most T groups in $\mathfrak{H}$ such that any bigger intersection of this form yields a subgroup of bounded index. Hence for any $H \in \mathfrak{H}$ the index $|K : K \cap H|$ is finite, and by Lemma 2.4 there is $k < \omega$ and a finite subintersection $N_1 = \bigcap_{i<n} H_i$ of K such that $|N_1 : N_1 \cap H| \leq k$ for all $H \in \mathfrak{H}$.

Now consider the family of conjugates of N_1 under automorphisms of $\mathfrak{M}$ stabilizing $\mathfrak{H}$ setwise. For any such conjugate N_1', the index $|N_1 : N_1 \cap N_1'|$ is bounded by k^n. By Fact 2.3, there is a finite intersection N_0 of these conjugates (which is thus a finite intersection of groups in $\mathfrak{H}$), and a finite extension N of N_0, such that N is invariant under all these automorphisms. It is clear that N is uniformly commensurable with $N_0 \cap H$ for all $H \in \mathfrak{H}$. ⊣

REMARK 3. If $\mathfrak{H}$ is invariant under all automorphisms fixing some parameter set A in a $|T \cup A|^+$-saturated and -homogeneous model $\mathfrak{M}$, then N is A-definable. This happens in particular if the family $\mathfrak{H}$ is type-definable over A, say of the form $\{\varphi(x, \bar{a}) : \models \pi(\bar{a})\}$ where π is a partial type over A. In this case, for every model $\mathfrak{M}$ containing A, we obtain a family $\mathfrak{H}^{\mathfrak{M}}$ of uniformly $\mathfrak{M}$-definable subgroups of $G^{\mathfrak{M}}$. Choosing $\mathfrak{M}$ sufficiently saturated, we obtain a relatively A-definable subgroup N such that $N^{\mathfrak{M}}$ is uniformly commensurable to $\mathfrak{H}^{\mathfrak{M}}$; as this is a first-order property, N is uniformly commensurable to $\mathfrak{H}$ in any model.

DEFINITION 4. Let G be a type-definable group over $\emptyset$. We call a type-definable subgroup $H \leq G$ *locally connected* if every commensurate G-conjugate or $\emptyset$-automorphic image of H is equal to H.

This definition is similar to the stable case; note that it depends on the ambient group G and its parameters. However, while a connected group in a stable theory is locally connected, a connected (over some parameters) group in a simple theory need not be locally connected. For instance, in Example 1 only V itself is locally connected, while over sufficiently many parameters A the connected component V_A^0 is a proper subgroup of V.

Corollary 2.6. *Let G be a group type-definable over $\emptyset$ in a simple theory, and H a relatively definable subgroup of G. If $\mathfrak{H}$ denotes the family of G-conjugates of images of H under $\emptyset$-automorphisms, then the equivalence relation on $\mathfrak{H}$ given by commensurability is definable. Furthermore, there is a locally connected relatively definable subgroup H^c of G commensurable with H.*

Note that H^c need not be unique.

Proof. Let $\mathfrak{H}_0$ be the subfamily of those G-conjugates of automorphic images of H which are commensurable with H. By Lemma 2.4 commensurability is type-definable and $\mathfrak{H}_0$ is a type-definable family of uniformly relatively definable subgroups of G. Fact 2.3 yields a relatively definable subgroup H^c commensurable with H, such that H^c is stabilized by any automorphism of G stabilizing $\mathfrak{H}_0$. These include in particular inner and $\emptyset$-automorphisms which preserve the commensurability class of H, so H^c is locally connected. ⊣

Remark 4. Let H' be a group-theoretic conjugate of an automorphic image of H^c. Then $x \in G$ and $D(H \cap H', \varphi*, k) = D(H, \varphi*, k)$ for all $\mathcal{L}$-formulas φ and all $k < \omega$ together imply $x \in H^c \leftrightarrow x \in H'$. By compactness, there is a definable superset X of H^c (namely the intersection of a formula relatively defining H^c with a definable superset of G) such that X relatively defines H^c, and whenever X' is a conjugate of an automorphic image of X such that $X \cap X'$ relatively defines a commensurable subgroup of G (which must be H^c itself), then $X = X'$. It follows that any automorphism stabilizes H^c setwise if and only if it stabilizes X setwise, i.e., if and only if it fixes the canonical parameter for X. Hence H^c, and more generally any relatively definable locally connected group, has a canonical parameter (over the parameters needed to relatively define the ambient group).

As a consequence, we recover the chain condition on intersections up to commensurability.

Theorem 2.7. *Let G be a group type-definable over $\emptyset$ in a simple theory, and $\mathfrak{H}$ a family of uniformly relatively definable subgroups of G. Then there are integers $k, k' < \omega$ such that there is no sequence $\{H_i : i \leq k\} \subset \mathfrak{H}$ with $|\bigcap_{i<j} H_i : \bigcap_{i \leq j} H_i| > k'$ for all $j \leq k$. In particular, any intersection N of elements in $\mathfrak{H}$ is commensurate with a subintersection N_0 of size at most k and $|N_0 : N_0 \cap H| \leq k'$ for any $H \in \mathfrak{H}$ with $H \geq N$. Furthermore, any chain of intersections of elements of $\mathfrak{H}$, each of unbounded index in its predecessor, has length at most k.*

PROOF. By enlarging $\mathfrak{H}$, we may assume that $\mathfrak{H}$ is a type-definable family. If the first assertion does not hold, then by compactness the condition

$$\{H_i \in \mathfrak{H}: i \in \omega\} \cup \{|\bigcap_{i<k} H_i : \bigcap_{i\leq k} H_i| > k' : k, k' \in \omega\}$$

is a consistent first-order condition on the parameters needed to define the groups $(H_i : i < \omega)$. However, a realization of it yields a family $\{H_i : i \in \omega\} \subset \mathfrak{H}$ such that no finite subintersection N_0 will be commensurable with $N_0 \cap H_i$ for all $i \in \omega$: if $N_0 = H_{i_0} \cap \cdots \cap H_{i_n}$, then $|N_0 : N_0 \cap H_i|$ is infinite for any $i > \max\{i_0, \ldots, i_n\}$. This contradicts Proposition 2.5.

Now consider an intersection N of elements of $\mathfrak{H}$, and the subfamily $\mathfrak{H}_0 := \{H \in \mathfrak{H}: H \geq N\}$. Choose successively $H_0, H_1, \cdots \in \mathfrak{H}_0$ such that $|\bigcap_{i<j} H_i : \bigcap_{i\leq j} H_i| > k'$ for all j. By the first assertion this sequence must stop after k such choices, and $N_0 := H_0 \cap \cdots \cap H_{k-1}$ will do.

Finally, consider a chain $C_0 > C_1 > \cdots > C_k$ of intersections of groups in $\mathfrak{H}$ such that every C_i has unbounded index in its predecessor. As the index of C_{i+1} in C_i is unbounded, there must be some element H_{i+1} in the intersection which forms C_{i+1}, such that $C_i \cap H_{i+1}$ has infinite index in C_i. Pick arbitrary $H_0 \in \mathfrak{H}$ with $H_0 \geq C_0$. Then the indices $|\bigcap_{i<j} H_i : \bigcap_{i\leq j} H_i|$ are infinite for all $j \leq k$, contradicting the first assertion. ⊣

We finish with three general lemmas.

LEMMA 2.8. *Suppose G is a type-definable group in a simple theory, and X is a non-empty type-definable subset of G such that for independent $x, x' \in X$ we have $x^{-1}x' \in X$. Then $X \cdot X =: Y$ is a type-definable subgroup of G, and X is generic in Y. In fact, X contains all generic types for Y.*

REMARK 5. Pillay proves this in [15, Lemma 4.4] for the special case $X = S(p)$ for some type p over a model $\mathfrak{M}$. Note also that we do not actually require G to be a group; it is enough if $x \cdot y \cdot z$ is defined and associative for all $x, y, z \in X$, every element in X has a two-sided inverse in X, and there is a unit $1 \in X$. In particular, those conditions are sufficient for the existence of a stabilizer, since the Independence Theorem below implies that $S(p)$ satisfies the assumptions of Lemma 2.8.

PROOF. As $X' = \{x \in X : x^{-1} \in X\}$ also satisfies the assumptions of the lemma and generates the same group (if a, b, c are three independent elements of X, then $a^{-1}b \in X'$ and $(b^{-1}a)c \in X'$, whence $c \in X' \cdot X'$), we may assume that X is closed under inversion.

Enumerate all pairs of formulas and natural numbers as (φ_i, k_i), for $i < \lambda$. By Fact 2.1.1 and compactness there is a type p extending the partial type $x \in X$, such that $D(p, \varphi_i*, k_i)$ takes a maximal value, say n_i, subject to $D(p, \varphi_j*, k_j) = n_j$ for all $j < i$. For any three elements a, b, c of X, choose

$d \models p$ with $d \mathop{\smile\!\!\!\!\!\!\mid}\; a, b, c$. Then $bd \in X$. Since for any $i < \omega$

$$n_i = D(d/b, \varphi_i *, k_i) = D(bd/b, \varphi_i *, k_i) \leq D(bd, \varphi_i *, k_i) \leq n_i$$

by the maximal choice of n_i, we have equality all the way and thus $bd \mathop{\smile\!\!\!\!\!\!\mid}\; b$ by Fact 2.1.2, whence $bd \mathop{\smile\!\!\!\!\!\!\mid}\; a, b$ and $abd \in X$. Finally $d \mathop{\smile\!\!\!\!\!\!\mid}\; c$, so $d^{-1}c \in X$, and $abc \in Y$. This shows that Y is a subgroup; it is clearly type-definable.

Now let a be a generic element of Y, say $a = bc$ for some $b, c \in X$. Let $d \models p$ independently from b, c. Then $cd \in X$, and by maximality of n_i again $D(cd/c, \varphi_i *, k_i) = D(cd, \varphi_i *, k_i) = n_i$ for all $i < \omega$, whence $cd \mathop{\smile\!\!\!\!\!\!\mid}\; c$. Therefore $b \mathop{\smile\!\!\!\!\!\!\mid}\; cd$, and $bcd = ad \in X$. But since a is generic independent from d, we have $ad \mathop{\smile\!\!\!\!\!\!\mid}\; d$, so $add^{-1} = a \in X$. ⊣

This will yield most of the type-definable groups we shall encounter. In order to normalize them (as for type-definable groups we cannot use Fact 2.3 and the subsequent results), we have to work some more.

Recall from [14] that the *Lascar strong type* of a over A, denoted $\text{lstp}(a/A)$, is given by $\text{tp}(a/A)$ together with the class of a modulo all type-definable equivalence relations over A with only boundedly many classes. In particular, for a model $\mathfrak{M}$ we have $\text{tp}(a/\mathfrak{M}) \vdash \text{lstp}(a/\mathfrak{M})$. The most important fact about Lascar strong types is the Independence Theorem:

FACT 2.9. [14, Theorem 5.8] *If* $A_1 \mathop{\smile\!\!\!\!\!\!\mid}_{A_0} A_2$, $a_1 \mathop{\smile\!\!\!\!\!\!\mid}_{A_0} A_1$, $a_2 \mathop{\smile\!\!\!\!\!\!\mid}_{A_0} A_2$ *and* $\text{lstp}(a_1/A_0) = \text{lstp}(a_2/A_0)$, *then there is a with* $a \mathop{\smile\!\!\!\!\!\!\mid}_{A_0} A_1A_2$, $\text{tp}(a/A_0A_1) = \text{tp}(a_1/A_0A_1)$, $\text{tp}(a/A_0A_2) = \text{tp}(a_2/A_0A_2)$ *and* $\text{lstp}(a/A_0) = \text{lstp}(a_1/A_0)$.

LEMMA 2.10. *Let G be a group type-definable over $\emptyset$ in a simple theory, and H a type-definable subgroup of G. Then there is a group N type-definable over $\emptyset$, and an intersection N_0 of automorphic images of H commensurate with N, such that any intersection of automorphic images of H with N_0 is commensurate with N_0.*

PROOF. By Proposition 2.2 there is an intersection $N(A)$ of automorphic images of H, type-definable over some parameters A, such that any further intersection has bounded index in $N(A)$. Put

$$X := \{g \in G : \exists Z\, [\text{lstp}(Z) = \text{lstp}(A) \wedge g \mathop{\smile\!\!\!\!\!\!\mid}\; Z \wedge g \in N(Z)\}.$$

Then $\text{lstp}(Z) = \text{lstp}(A)$ is type-definable over A, and $g \mathop{\smile\!\!\!\!\!\!\mid}\; Z$ is type-definable as $D(\bar{z}/g, \varphi, k) = D(\bar{a}, \varphi, k)$ for every $\mathcal{L}$-formula φ, every $k < \omega$, and every finite tuple $\bar{a} \subset A$ and corresponding tuple $\bar{z} \subset Z$. (See [14, Section 6] for D-rank and its type-definability.) As the existential quantifier commutes with infinite disjunctions (by saturation of the monster model), X is type-definable.

If $g, g' \in X$ and $g \mathop{\smile\!\!\!\!\!\!\mid}\; g'$, then by the Independence Theorem we may choose A' of the same Lascar strong type as A and independent of g, g' such that both g and g' are in $N(A')$, whence $g^{-1}g' \in N(A')$; since $g^{-1}g' \mathop{\smile\!\!\!\!\!\!\mid}\; A'$, we get $g^{-1}g' \in X$. By Lemma 2.8 the type-definable set X^2 forms a subgroup N_1.

Clearly, N_1 is stabilized setwise by all automorphisms which fix the Lascar strong type of A.

CLAIM. *N_1 is commensurate with $N(A)$.*

PROOF OF CLAIM. Consider A' of the same Lascar strong type as A and independent of A. By our choice of $N(A)$, the index of $N(A') \cap N(A)$ in $N(A)$ is bounded, so $N(A) \cap N(A')$ is generic for $N(A)$ by Fact 2.1.3. Hence there is an element $g \in N(A)$ generic over AA' which lies in $N(A')$. But then $A' \mathop{\smile\!\!\!\!\!\mid}_A g$ and $A' \mathop{\smile\!\!\!\!\!\mid} A$ yields $A' \mathop{\smile\!\!\!\!\!\mid} g$, whence $g \in X$ and $g \in N_1$. So $N_1 \cap N(A)$ is generic in $N(A)$.

Conversely, let g be generic in N_1 over A. Then $g \in X$, and there is A' of the same Lascar strong type as A with $g \mathop{\smile\!\!\!\!\!\mid} A'$ and $g \in N(A')$. But then for every φ and k

$$\begin{aligned} D(N_1, \varphi*, k) &= D(g/\emptyset, \varphi*, k) = D(g/A', \varphi*, k) \leq D(N(A'), \varphi*, k) \\ &= D(N(A), \varphi*, k) = D(N(A) \cap N_1, \varphi*, k) \leq D(N_1, \varphi*, k); \end{aligned}$$

equality of the ranks and commensurativity follow. ⊣

Now let N be the intersection of all automorphic images of N_1. Since N_1 is stabilized by all automorphisms which fix the Lascar strong type of A, it only has boundedly many automorphic images, and N is a type-definable subgroup of G. Clearly, N is type-definable over $\emptyset$. Furthermore, N_1 is commensurate with $N(A)$, so N is commensurate with a bounded intersection N_0 of automorphic images of $N(A)$. But N_0 is commensurate with $N(A)$ by the choice of $N(A)$. The assertion follows. ⊣

LEMMA 2.11. *Let G be a group type-definable over $\emptyset$ and H a type-definable subgroup of G. Then there is a type-definable normal subgroup N of G commensurate with a bounded intersection of G-conjugates of H.*

PROOF. By Proposition 2.2 there is a type-definable intersection H_1 of G-conjugates of H such that any further intersection has bounded index in H_1. Let $\mathfrak{M}$ be a model such that H_1 is type-defined over $\mathfrak{M}$. Choose a generic type p for G over $\mathfrak{M}$, and put

$$X := \{g \in G : \exists z\, [z \models p \wedge z \mathop{\smile\!\!\!\!\!\mid}_{\mathfrak{M}} g \wedge g \in H_1^z\}.$$

Then X is type-definable as in the proof of Lemma 2.10. By the Independence Theorem, X is closed under generic multiplication, and X^2 is a type-definable group N_1 by Lemma 2.8. Let h realize a generic type for $G^0_{\mathfrak{M}}$, and consider $g \in X$ with $g \mathop{\smile\!\!\!\!\!\mid}_{\mathfrak{M}} h$. Then there is some $z \models p$ with $z \mathop{\smile\!\!\!\!\!\mid}_{\mathfrak{M}} g$ and $g \in H_1^z$; by Fact 2.1.6 there also is a realization $z' \models p$ with $z' \mathop{\smile\!\!\!\!\!\mid}_{\mathfrak{M}} h$ and $z'h \models p$. By the Independence Theorem, we may assume $z = z'$ and $z \mathop{\smile\!\!\!\!\!\mid}_{\mathfrak{M}} g, h$. Since $zh \mathop{\smile\!\!\!\!\!\mid}_{\mathfrak{M}} h$, we get $g^h \mathop{\smile\!\!\!\!\!\mid}_{\mathfrak{M}} zh$ and $g^h \in H_1^{zh}$, whence $g^h \in X$. Now for any $n \in N_1$ and h generic for $G^0_{\mathfrak{M}}$ there are $g, g' \in X$ with $n = gg'$ and

$g \mathop{\smile\!\!\!\!\!\!\,|}_{\mathfrak{M}} h$ and $g' \mathop{\smile\!\!\!\!\!\!\,|}_{\mathfrak{M}} h$, whence $n^h = g^h g'^h \in X^2 = N_1$. It follows that $G^0_{\mathfrak{M}}$ normalizes N_1. So the intersection of all G-conjugates of N_1 is in fact a bounded intersection, and thus a type-definable normal subgroup of G.

Finally, H_1 is commensurable with N_1, and with N, for the same reasons as in the proof of Lemma 2.10. ⊣

Combining these two Lemmas, we can actually find a locally connected group commensurate with H.

§3. Small groups. Recall that a theory is *small* if it has only countably many pure types. Small stable groups are well-behaved; we shall see that this is similar in the simple case.

FACT 3.1. [13, Theorem 22] *In a small theory (not necessarily simple) every type-definable equivalence relation on finite tuples over a finite set is the conjunction of definable equivalence relations.*

An equivalence class modulo a type-definable equivalence relation is called a *hyperimaginary*. Fact 3.1 tells us that small theories eliminate finitary hyperimaginaries (i.e., classes of finite tuples modulo equivalence relations definable over finite sets): every such hyperimaginary is really just a set of imaginary elements.

REMARK 6. It has been shown in [2] that a supersimple theory eliminates all hyperimaginaries.

FACT 3.2. [6, Proposition 2.8] *In a simple theory, for every Lascar strong type p over a set A there is a type-definable equivalence relation E over $\emptyset$ such that p does not fork over A/E, and whenever p and some Lascar strong type q over some set B have a common non-forking extension, then A/E is definable over B.*

A/E is called the *canonical base* of p, and denoted $\mathrm{Cb}(p)$. We shall write $\mathrm{Cb}(\bar{b}/A)$ for $\mathrm{Cb}(\mathrm{lstp}(\bar{b}/A))$; note that $\mathrm{lstp}(\bar{b}/A)$ is a partial type over $A\bar{b}$. We may thus apply Fact 3.1 to see that $\mathrm{Cb}(\bar{b}/\bar{a})$ is an imaginary set for any finite tuples $\bar{a}, \bar{b}$; since $\mathrm{Cb}(B/\bar{a}) = \mathrm{dcl}\{\mathrm{Cb}(\bar{b}/\bar{a}) \colon \bar{b} \in B \text{ finite}\}$, this is also true for the canonical base of any Lascar strong type over finitely many parameters.

LEMMA 3.3. *Let G be a definable group in a small (not necessarily simple) theory. Suppose H is a subgroup of G which is type-definable over a finite set. Then H is $\bigcap$-definable.*

PROOF. The relation $xy^{-1} \in H$ is type-definable, and clearly an equivalence relation on G. By Fact 3.1, it is the intersection of definable equivalence relations E_i. Put

$$H_i = \{g \in G \colon \forall x \in G\ xE_i gx\}.$$

If $h \in H$ and $x \in G$, then $(hx)x^{-1} \in H$, whence $xE_i hx$ for all i, and $H \subseteq H_i$. On the other hand, if $h \in H_i$ for all i, then $1E_i h$ for all E_i (put $x = 1$), whence

$h \in H$. Furthermore, if $g, h \in H_i$, then $h^{-1}xE_ihh^{-1}x$, and xE_ihxE_ighx. It follows that H_i is a definable subgroup of G, and $H = \bigcap_i H_i$. ⊣

DEFINITION 5. Let Q be an A-invariant family of types. A type $p \in S(A)$ is *Q-internal* if for every realization a of p there is $B \mathop{\smile\!\!\!\!\!\!|}_A a$, types $\bar{q}$ from Q based on B, and realizations $\bar{c}$ of $\bar{q}$, such that $a \in \mathrm{dcl}(B\bar{c})$.

Note that in a simple theory B may well depend on the particular realization of p, contrary to the stable case where we may choose a B which works for all realizations of p. This arises from the fact that in a simple theory a type (even over a model) may have many different non-forking extensions.

PROPOSITION 3.4. *Suppose* $\mathrm{tp}(a)$ *is non-orthogonal to a type* q *in a small simple theory, and let* Q *be the family of* $\emptyset$*-conjugates of* q. *Then there is* $a_0 \in \mathrm{dcl}(a) - \mathrm{acl}(\emptyset)$ *such that* $\mathrm{tp}(a_0)$ *is* Q*-internal.*

PROOF. This is similar to the stable case. Let $\mathrm{tp}(c/A)$ be a non-forking extension of q such that $a \mathop{\smile\!\!\!\!\!\!|} A$ and $a \not\mathop{\smile\!\!\!\!\!\!|}_A c$, and consider a Morley sequence $(c_iA_i : i \leq \omega)$ in $\mathrm{lstp}(cA/a)$. Since the canonical base $\mathrm{Cb}(cA/a)$ consists of imaginary elements, it is in $\mathrm{acl}(a)$. On the other hand, $p_\omega := \mathrm{tp}(c_\omega A_\omega / c_iA_i : i < \omega)$ is a Lascar strong type: if $E(xX, yY)$ is an equivalence relation type-definable over $(c_iA_i : i < \omega)$ with boundedly many classes, then for any formula $\varphi(xX, yY)$ in E with parameters in $(c_iA_i : i < n)$ we must have $\varphi(c_iA_i, c_jA_j)$ for all $i, j > n$ by indiscernibility, and hence $\varphi(c_iA_i, c_\omega A_\omega)$. Furthermore $c_\omega A_\omega \mathop{\smile\!\!\!\!\!\!|}_{(c_iA_i :\, i<\omega)} a$, as any formula $\varphi(xX, a)$ with additional parameters in $(c_iA_i : i \leq n)$ (for any $n < \omega$) which is satisfied by $c_\omega A_\omega$ is satisfied by c_jA_j for any $j > n$, and hence cannot divide over $(c_iA_i : i < \omega)$. So $\mathrm{lstp}(cA/a)$ and p_ω have a common non-forking extension (realized by $c_\omega A_\omega$); by Fact 3.2 we get $\mathrm{Cb}(cA/a) \in \mathrm{dcl}(c_iA_i : i < \omega)$.

However, $\mathrm{Cb}(cA/a) \notin \mathrm{acl}(\emptyset)$, as $a \not\mathop{\smile\!\!\!\!\!\!|} cA$. So there is some element $a_1 \in \mathrm{acl}(a) - \mathrm{acl}(\emptyset)$ definable over $(c_iA_i : i \leq n)$ for some $n < \omega$. Now $a_1 \mathop{\smile\!\!\!\!\!\!|} (A_i : i \leq n)$ and $\mathrm{tp}(c_i/A_i)$ is an $\emptyset$-conjugate of q. It follows that $\mathrm{tp}(a_1)$ is Q-internal.

Finally, let a_0 be the (finite) set of a-conjugates of a_1. Then a_0 is an imaginary element in $\mathrm{dcl}(a)$. By $\emptyset$-invariance of Q, every a-conjugate of a_1 is Q-internal, as is the tuple, and hence the set, of those conjugates. ⊣

THEOREM 3.5. *Let the group* G *be* $\bigcap$*-definable over* $\emptyset$ *in a small simple theory, and suppose that a generic type* p *of* G *is non-orthogonal to some type* q. *Then there is a normal relatively* $\emptyset$*-definable subgroup* N *of infinite index in* G *such that the quotient* G/N *is* Q*-internal, where* Q *is the collection of* $\emptyset$*-conjugates of* q.

PROOF. By Proposition 3.4, for any realization a of p there is some Q-internal $a_0 \in \mathrm{dcl}(a) - \mathrm{acl}(\emptyset)$, say $a_0 = f(a)$ for some $\emptyset$-definable function f.

Put

$$X := \{g \in G : \exists x\ [\mathrm{lstp}(x) = \mathrm{lstp}(xg) = \mathrm{lstp}(a) \wedge \\ x \mathop{\smile\!\!\!\!\!\!|}\ g \wedge xg \mathop{\smile\!\!\!\!\!\!|}\ g \wedge f(xg) = f(x)]\}.$$

Then X is type-definable over a as in the proof of Lemma 2.10, or in fact over any realization $a' \models \mathrm{lstp}(a)$ (and we choose $a' \perp a$). Suppose $g, g' \in X$ with $g \perp g'$. The Independence Theorem yields the existence of $x \models p$ with $x \perp g, g'$, such that both xg and xg' realize p independently of g, g', with $f(xg) = f(x) = f(xg')$. Since $xg \perp g, g'$ and $xg' \perp g, g'$, both xg and $xg(g^{-1}g')$ realize p independently of $g^{-1}g'$, and $f(xg(g^{-1}g')) = f(xg') = f(xg)$. Hence $g^{-1}g' \in X$. So X^2 is a type-definable subgroup K of G by Lemma 2.8, whose generic types are all contained in X.

Suppose K has bounded index in G. Then X is generic in G, and there is a generic $g \in X$ and independent $x \models p$ with $f(xg) = f(x)$. As $xg \perp x$ by genericity of g, we get $f(xg) \perp xg$, contradicting $f(a) \notin \mathrm{acl}(\emptyset)$. Therefore K has unbounded index in G; since it is type-definable over a', it is contained in a relatively a'-definable supergroup H of infinite index in G by Lemma 3.3 (applied to some definable supergroup of G).

Claim. *$\mathrm{tp}(aH/a')$ is Q-internal.*

Proof of Claim. If $\mathfrak{M}$ is any model containing a' with $\mathfrak{M} \perp a$, then by Fact 2.1.6 for any generic type p' of $G^0_{\mathfrak{M}}$ over $\mathfrak{M}$ there is an infinite Morley sequence $I := (x_i : i < |T|^+)$ in p' independent of a, a', such that $x_i a \models \mathrm{tp}(a/\mathfrak{M})$ for all $i < |T|^+$. We claim that aH is definable over $A := \{a', x_i, f(x_i a) : i < |T|^+\}$. So suppose $\mathrm{tp}(a''/A) = \mathrm{tp}(a/A)$. Then $f(x_i a'') = f(x_i a)$ for all $i < |T|^+$. Since $(x_i a : i < |T|^+)$ and $(x_i a'' : i < |T|^+)$ are also independent sequences by genericity of p', and as any element can fork with at most $|T|$ elements in any independent sequence, there must be some $i < |T|^+$ such that $a^{-1}a''$ is independent of $x_i a$ and of $x_i a''$. Since $f(x_i a(a^{-1}a'')) = f(x_i a'') = f(x_i a)$, we have $a^{-1}a'' \in X$, and $aH = a''H$.

Hence $aH \in \mathrm{dcl}(A) = \mathrm{dcl}(a', I, f(x_i a) : i < |T|^+)$, and $a \perp_{a'} I$; as $\mathrm{tp}(f(x_i a))$ is Q-internal for all $i < |T|^+$, so is $\mathrm{tp}(aH/a')$. ⊣

Claim. *G/H is q-internal.*

Proof of Claim. For any $g \in G$ there is $x \models p$ with $x \perp g, a'$; as $\mathrm{tp}(x^{-1})$ is generic, $gx^{-1} \perp g, a'$, whence $gx^{-1} \perp_{a'} gH$. Now $gH \in \mathrm{dcl}(gx^{-1}, xH)$, as $\mathrm{tp}(xH/a')$ is Q-internal, so is $\mathrm{tp}(gH/a')$ by transitivity of internality. ⊣

Now let $\mathfrak{H}$ be the family of G-conjugates of automorphic images of H. By Fact 2.5 there is a finite intersection $N_0 = \bigcap_{i<n} K^{g_i}$ of subgroups in $\mathfrak{H}$ such that a finite extension N of N_0 is normal in G and invariant under automorphisms stabilizing $\mathfrak{H}$ setwise. But Q-internality of G/H implies Q-internality of G/H'

for all $H' \in \mathfrak{H}$ by $\emptyset$-invariance of Q and definability of the conjugation map; as gN is definable over gN_0, and gN_0 is definable over $(gK^{g_i} : i < n)$ for all $g \in G$, the quotient G/N is Q-internal. Since N must have infinite index in G and be relatively $\emptyset$-definable, we are done. ⊣

REMARK 7. If G is a group $\bigcap$-definable over $\emptyset$ in a small simple theory, and Q is an $\emptyset$-invariant family of types closed under extensions (also forking ones), we can iterate the construction and obtain a sequence $G = G_0 \rhd G_1 \rhd \cdots \rhd G_\alpha$ of groups $\bigcap$-definable over $\emptyset$, such that G_i/G_{i+1} is unbounded and Q-internal for all $i < \alpha$, and $G_\lambda = \bigcap_{i<\lambda} G_i$ for any limit ordinal $\lambda \leq \alpha$, and all generic types of G_α are orthogonal to Q: if a generic type of G_α were non-orthogonal to a type $q \in Q$, we could just iterate one step further; as everything is $\bigcap$-definable over $\emptyset$, we must come to a halt somewhere. If we require G_α to be connected over $\emptyset$, it can be shown that it is unique. We call it the *Q-connected component* G^Q of G.

REMARK 8. It follows from [23, Lemma 3.4.18] that if Q is actually a set of formulas over $\emptyset$ and $G^Q = \{1\}$, then we may choose a finite sequence $G = G_0 \rhd G_1 \rhd \cdots \rhd G_n = \{1\}$ with Q-internal quotients.

QUESTION 1. *Do groups definable in a small simple theory satisfy property* $\mathfrak{R}$ *from* [22]? *Do they have an infinite definable abelian, or finite-by-abelian subgroup?*

§4. Supersimple theories.

From now on, we shall work in a simple theory. In this section and the next one, P will denote an $\emptyset$-invariant family of types closed under non-forking extensions, and On^+ is the class of ordinals together with ∞ (where $\alpha < \infty$ for every ordinal α). We shall quickly review the definition and basic properties of the SU_P-rank defined in [21] or [22].

DEFINITION 6. The (*Lascar*) *rank relative to* P, denoted SU_P, is the smallest function from the collection of all types (over parameters in the monster model) to On^+ satisfying for every ordinal α:

$SU_P(q) \geq \alpha + 1$ if there is an extension $q' \supseteq q$ over some set A, a type $p \in P$ over A, and realizations $a \models q'$ and $b \models p$ with $a \not\perp_A b$ and $SU_P(a/Ab) \geq \alpha$.

Clearly, if q is a nonforking extension of p, then $SU_P(q) = SU_P(p)$, and $SU_P(q) = 0$ if and only if q is hereditarily orthogonal to P. However, q may well be foreign to P and still have nonzero SU_P-rank. In case T is supersimple and P is the family of all types, SU_P-rank is equal to SU-rank. As SU-rank measures forking, SU_P-rank measures forking with respect to types in P. It has similar properties and is useful in particular in simple, not supersimple theories, when we can find an invariant set P of types such that some type q

has ordinal SU_P-rank (for instance, if p is a regular type over $\emptyset$, $P = \{p\}$ and q is a p-simple type in the sense of Hrushovski [7]).

By the usual proofs we obtain:

PROPOSITION 4.1.

1. *Suppose* $SU_P(a/Ab) < \infty$ *and* $SU_P(a/A) \geq SU_P(a/Ab) + \omega^\alpha \cdot n$. *Then* $SU_P(b/A) \geq SU_P(b/Aa) + \omega^\alpha \cdot n$.
2. $SU_P(a/bA) + SU_P(b/A) \leq SU_P(ab/A) \leq SU_P(a/bA) \oplus SU_P(b/A)$.
3. *If* a *and* b *are independent over* A, *then* $SU_P(ab/A) = SU_P(a/A) \oplus SU_P(b/A)$.

In order to avoid working with hyperimaginaries, we shall in fact need a slight variant of 4.1.1.

LEMMA 4.2. *If* $SU_P(a/AB) < \infty$ *and* $SU_P(a/A) \geq SU_P(a/AB) + 1$, *then there is some* $\bar{b} \in B$ *with* $SU_P(\bar{b}/A) \geq SU_P(\bar{b}/Aa) + 1$.

PROOF. By induction on $SU_P(a/AB)$, we may assume that the result holds for smaller SU_P-rank. By definition of SU_P-rank and the assumption of the lemma, there is some $C \supseteq A$ and c realizing some type in P over C such that $a \not\perp_C c$ and $SU_P(a/Cc) \geq SU_P(a/AB)$. We may choose $Cc \perp_{Aa} B$, so $SU_P(\bar{b}/Cca) = SU_P(\bar{b}/Aa)$ for all $\bar{b} \in B$. Now if $B \not\perp_C c$, then there is some $\bar{b} \in B$ with $\bar{b} \not\perp_C c$, and

$$SU_P(\bar{b}/A) \geq SU_P(\bar{b}/C) \geq SU_P(\bar{b}/Cc) + 1 \geq SU_P(\bar{b}/Cca) + 1$$
$$= SU_P(\bar{b}/Aa) + 1.$$

Otherwise $B \perp_C c$. Then since $a \not\perp_{BC} c$, we get

$$SU_P(a/Cc) \geq SU_P(a/AB) \geq SU_P(a/CB) \geq SU_P(a/CcB) + 1,$$

and by inductive hypothesis there is $\bar{b} \in B$ with

$$SU_P(\bar{b}/A) \geq SU_P(\bar{b}/Cc) \geq SU_P(\bar{b}/Cca) + 1$$
$$= SU_P(\bar{b}/Aa) + 1. \qquad \dashv$$

This basically yields finite character of P-forking (where we say that *a P-forks* with B over A if $SU_P(a/A) > SU_P(a/AB)$). Note that if P is the family of all types, then finite character is obvious from the definition of forking.

COROLLARY 4.3. *If* A *is the ascending union of sets* A_i *for* $i < \lambda$ *and* $SU_P(a/A_i) = \alpha < \infty$ *for all* $i < \lambda$, *then* $SU_P(a/A) = \alpha$.

PROOF. Suppose $SU_P(a/A) < \alpha$. Then by Lemma 4.2 there is some $\bar{b} \in A$ with $SU_P(\bar{b}/A_0) \geq SU_P(\bar{b}/A_0 a) + 1$. By Proposition 4.1.1 we get

$$SU_P(a/A_0) \geq SU_P(a/A_0\bar{b}) + 1 \geq SU_P(a/A_i) + 1$$

for any sufficiently big i such that $\bar{b} \in A_i$. But this contradicts our assumption. $\dashv$

We shall call a type *P-minimal* if every forking extension has smaller SU_P-rank.

COROLLARY 4.4. *Every type of ordinal SU_P-rank has a P-minimal extension of the same SU_P-rank.*

PROOF. As any chain of forking extensions has bounded length, this is immediate from Corollary 4.3. ⊣

COROLLARY 4.5. *Every P-minimal type p of ordinal (non-zero) SU_P-rank is non-orthogonal to a regular type.*

PROOF. Let q be a type of minimal SU_P-rank non-orthogonal to p. Let p be realized by a and q by b over some set A. If there is $B \supset A$ with $B \not\perp_A b$ and $B \perp_A a$, we replace A by B; by repeating this process we may assume that b is dominated by a over A.

CLAIM. *q is P-minimal.*

PROOF OF CLAIM. Suppose there is c with $b \not\perp_A c$. Replacing c by a finite bit of a Morley sequence in $\mathrm{tp}(b/Ac)$, we may assume that $\mathrm{tp}(c/A)$ has ordinal SU_P-rank; we may clearly further assume $c \perp_{Ab} a$. By domination $c \not\perp_A a$, whence $SU_P(a/A) \geq SU_P(a/Ac) + 1$ by P-minimality. By Proposition 4.1.1 we get

$$SU_P(c/A) \geq SU_P(c/Aa) + 1 \geq SU_P(c/Aab) + 1 = SU_P(c/Ab) + 1;$$

another application of 4.1.1 yields $SU_P(b/A) \geq SU_P(b/Ac) + 1$. ⊣

Now p is orthogonal to all types of SU_P-rank less than $SU_P(q)$, in particular to all forking extensions of q. As a dominates b over A, this shows that q is orthogonal to all its forking extensions. ⊣

We shall also need a version of Shelah rank:

DEFINITION 7. The *(Shelah) rank relative to P*, denoted D_P, is the smallest function from the class of all formulas to On^+ satisfying

> $D_P(\varphi) \geq \alpha + 1$ if there is some set A of parameters containing those of φ, a realization c of some type in P over A, and a formula $\varphi'(x, c, A)$ forking over A, such that $\varphi' \vdash \varphi$ and $D_P(\varphi') \geq \alpha$.

For a type p we put $D_P(p) = \min\{D_P(\varphi) \colon \varphi \in p\}$. If P is the family of all types, the subscript P is omitted.

One sees inductively that $D_P(\varphi \vee \psi) = \max\{D_P(\varphi), D_P(\psi)\}$, whence $D_P(\varphi) = \max\{D_P(p) \colon \varphi \in p\}$. Clearly, if $\mathrm{tp}(c/A)$ is in P and $a \not\perp_A c$, then $D_P(a/A) > D_P(a/Ac)$. In particular, if q is a forking extension of p, then $D(q) < D(p)$. However, Kim has constructed an example of a non-forking extension q of a type p with $D(q) < D(p)$. In general, $SU_P(p) \leq D_P(p)$ for any type p. (See [14, Section 6] or [23, Section 5.1] for the usual SU- and D-rank in a simple theory.)

§5. Supersimple groups. The assumptions from the last section remain in force: the ambient theory is simple, and P is an $\emptyset$-invariant family of types closed under non-forking extensions.

Since any two generic types in a type-definable group have non-forking extensions such that one type is a translate of the other one, and any type in the group has a non-forking extension which is a translate of an extension of a generic type, and as SU_P-rank is preserved under translation and non-forking extensions, it is clear that to a type-definable group or quotient space we can ascribe an SU_P-rank, namely the rank of any of its generic types (which may be ∞). If H is a type-definable subgroup of bounded index in G, then $SU_P(H) = SU_P(G)$. The converse holds (using Fact 2.1.3) if a generic type of G is P-minimal of ordinal SU_P-rank (and then all generic types are); in this case a type p in G is generic if and only if $SU_P(p) = SU_P(G)$. We shall call such groups *P-connected.* In any case the Lascar inequalities 4.1.2 immediately yield the corresponding inequalities for groups (see [17, Corollaire 6.3] and [22, Proposition 3.6.1] for the stable case): if H is a type-definable subgroup of G, then

$$SU_P(H) + SU_P(G/H) \leq SU_P(G) \leq SU_P(H) \oplus SU_P(G/H).$$

(If H is definable, then G/H consists of ordinary imaginary elements. If H is $\bigcap$-definable, then G/H consists of inverse limits of imaginary elements in the usual way. Lastly, if H is merely type-definable, then G/H consists of hyperimaginary elements.) In fact, these inequalities hold even if G is not a type-definable group containing H, but merely a type-definable union of cosets of H, say of the form KH, where K is some type-definable group (all inside a big type-definable group).

However, it is an open question whether there is a type-definable group G in a simple theory and a generic type p of G with $D_P(p) < D_P(G)$. (There always is some generic type of G with $D_P(G) = D_P(p)$.)

Lemma 5.1. *Let G be a type-definable group in a simple theory, and p a type in G over a model $\mathfrak{M}$, such that for independent realizations a and b of p the product ab^{-1} is independent of a. Then a right translate of (some non-forking extension of) p is a generic type of* $\operatorname{stab}(p)$.

Proof. If a and b are independent realizations of p, then ba^{-1} is independent of b; as $ab^{-1} = (ba^{-1})^{-1}$, we have that ab^{-1} is independent both of a and of b. By definition, $ab^{-1} \in S(p)$, so

$$\begin{aligned} D(S(p), \varphi*', k) &\geq D(ab^{-1}/\mathfrak{M}, \varphi*', k) = D(ab^{-1}/\mathfrak{M}, b, \varphi*', k) \\ &= D(a/\mathfrak{M}, b, \varphi*', k) = D(a/\mathfrak{M}, \varphi*', k) = D(p, \varphi*', k) \end{aligned}$$

for all $\mathcal{L}$-formulas φ and all $k < \omega$. On the other hand, for any $g \in S(p)$ there is by definition some $x \models p$ with $x \mathop{\smile\!\!\!\!\!\!|}_{\mathfrak{M}} g$, $gx \mathop{\smile\!\!\!\!\!\!|}_{\mathfrak{M}} g$ and $gx \models p$. Hence

$$D(g/\mathfrak{M},\varphi*',k)=D(g/\mathfrak{M},x,\varphi*',k)=D(gx/\mathfrak{M},x,\varphi*',k)$$
$$\leq D(gx/\mathfrak{M},\varphi*',k)=D(p,\varphi*',k)$$

for all $\mathcal{L}$-formulas φ and all $k<\omega$; whence $D(S(p),\varphi*',k)\leq D(p,\varphi*',k)$. Thus

$$D(\mathrm{stab}(p),\varphi*',k)=D(S(p),\varphi*',k)=D(ab^{-1}/\mathfrak{M},\varphi*',k)=D(p,\varphi*',k).$$

By Fact 2.1.3 (or rather a right-multiplication version of it) $\mathrm{tp}(ab^{-1}/\mathfrak{M})$ is a generic type of $\mathrm{stab}(p)$, which has a non-forking extension $\mathrm{tp}(ab^{-1}/\mathfrak{M},b)$, a right translate of $\mathrm{tp}(a/\mathfrak{M},b)$, which is a non-forking extension of p. ⊣

LEMMA 5.2. *Let p be a P-minimal type in a type-definable group, all over a model $\mathfrak{M}$. Suppose $SU_P(p)=\omega^{\alpha_1}n_1+\cdots+\omega^{\alpha_k}n_k=:\beta$, where $\alpha_1>\alpha_2>\cdots>\alpha_k$, and $SU_P(ab^{-1}/\mathfrak{M})<\beta+\omega^{\alpha_k}$ for any two independent realizations a and b of p. Then (some non-forking extension of) p is a translate of a generic type of the stabilizer* $\mathrm{stab}(p)$. *In particular the latter is P-minimal as well, and* $SU_P(\mathrm{stab}(p))=\beta$.

PROOF. Choose two independent realizations a and b of p. Since

$$SU_P(ab^{-1}/\mathfrak{M})\geq SU_P(ab^{-1}/\mathfrak{M},b)=SU_P(a/\mathfrak{M},b)=SU_P(a/\mathfrak{M})=\beta,$$

there is $A\supseteq\mathfrak{M}$ with $SU_P(ab^{-1}/A)=\beta$; by Corollary 4.4 we may choose A such that $\mathrm{tp}(ab^{-1}/A)$ is P-minimal. Clearly we may assume $A\mathrel{\raise0.2ex\hbox{$\smile$}\kern-0.7em\lower0.1ex\hbox{$|$}}_{\mathfrak{M},ab^{-1}}a,b$. As $\mathrm{tp}(ab^{-1}/A)$ is also based on a Morley-sequence $(c_i: i<\omega)$ in $\mathrm{lstp}(ab^{-1}/A)$ (as in the proof of Proposition 3.4), we may replace A by such a Morley sequence and assume that every tuple in A has ordinal SU_P-rank over $\mathfrak{M}$. Now $SU_P(\bar a/\mathfrak{M},a,b)=SU_P(\bar a/\mathfrak{M},ab^{-1})$ for all $\bar a\in A$; since

$$SU_P(ab^{-1}/\mathfrak{M})<SU_P(ab^{-1}/\mathfrak{M},\bar a)+\omega^{\alpha_k}$$

by assumption, Proposition 4.1.1 yields first

$$SU_P(\bar a/\mathfrak{M})<SU_P(\bar a/\mathfrak{M},ab^{-1})+\omega^{\alpha_k}=SU_P(\bar a/\mathfrak{M},a,b)+\omega^{\alpha_k},$$

and then

$$SU_P(a,b/\mathfrak{M})<SU_P(a,b/\mathfrak{M},\bar a)+\omega^{\alpha_k}.$$

CLAIM. $a,b\mathrel{\raise0.2ex\hbox{$\smile$}\kern-0.7em\lower0.1ex\hbox{$|$}}_{\mathfrak{M}}A$.

PROOF OF CLAIM. Suppose first $SU_P(a,b/\mathfrak{M},\bar a)<SU_P(a,b/\mathfrak{M})$ for some $\bar a\in A$. Since $a\mathrel{\raise0.2ex\hbox{$\smile$}\kern-0.7em\lower0.1ex\hbox{$|$}}_{\mathfrak{M}}b$, we have $SU_P(a,b/\mathfrak{M})=\beta\oplus\beta$ by Proposition 4.1.3, and the smallest monomial in $SU_P(a,b/\mathfrak{M})$ is ω^{α_k}. So

$$SU_P(a,b/\mathfrak{M},\bar a)\leq\omega^{\alpha_1}(2n_1)+\cdots+\omega^{\alpha_{k-1}}(2n_{k-1})+\omega^{\alpha_k}(2n_k-1)+\cdots,$$

whence

$$SU_P(a,b/\mathfrak{M},\bar a)+\omega^{\alpha_k}\leq\sum_{i=1}^{k}\omega^{\alpha_i}(2n_i)=SU_P(a,b/\mathfrak{M}),$$

contradicting our previous inequality. Therefore

$$SU_P(a,b/\mathfrak{M}) = SU_P(a,b/\mathfrak{M},\bar{a});$$

as p is P-minimal and a,b are two independent realizations of p, also $\mathrm{tp}(a,b/\mathfrak{M})$ is P-minimal, whence $a,b \mathop{\smile\!\!\!\!\!\!|}_{\mathfrak{M}} A$. ⊣

Therefore $ab^{-1} \mathop{\smile\!\!\!\!\!\!|}_{\mathfrak{M}} A$; in particular $\mathrm{tp}(ab^{-1}/\mathfrak{M})$ is P-minimal. Now

$$SU_P(ab^{-1}/\mathfrak{M},b) = SU_P(a/\mathfrak{M},b) = \beta = SU_P(ab^{-1}/A) = SU_P(ab^{-1}/\mathfrak{M});$$

P-minimality of $\mathrm{tp}(ab^{-1}/\mathfrak{M})$ yields $ab^{-1} \mathop{\smile\!\!\!\!\!\!|}_{\mathfrak{M}} b$, and similarly $ab^{-1} \mathop{\smile\!\!\!\!\!\!|}_{\mathfrak{M}} a$.

By Lemma 5.1 a translate of a non-forking extension of p is generic for $\mathrm{stab}(p)$. The rest follows from translation-invariance of SU_P-rank and P-minimality. ⊣

Corollary 5.3. *Let G be a type-definable group in a simple theory, with $SU_P(G) = \omega^{\alpha_1}n_1 + \cdots + \omega^{\alpha_k}n_k$, and put $\beta_i = \omega^{\alpha_1}n_1 + \cdots + \omega^{\alpha_i}n_i$ for $1 \le i \le k$. Then G has a type-definable normal subgroup G_i of rank β_i; it is unique up to commensurativity.*

Proof. If p is a P-minimal type of SU_P-rank β_i over some model $\mathfrak{M}$, it must satisfy the requirements of Lemma 5.2, as for independent realizations a and b of p we get

$$SU_P(ab^{-1}/\mathfrak{M}) \le SU_P(G) < \beta_i + \omega^{\alpha_i}.$$

So $H := \mathrm{stab}(p)$ is a P-connected subgroup of SU_P-rank β_i.

Claim. *H and H^g are commensurate for any $g \in G$.*

Proof of Claim. If $H \cap H^g$ has unbounded index in H, then $SU_P(H \cap H^g) < \beta_i$ by P-minimality; since the smallest monomial in $SU_P(H)$ is ω^{α_i}, we get $SU_P(H) \ge SU_P(H \cap H^g) + \omega^{\alpha_i}$. The Lascar inequality for goups now implies $SU_P(H/(H \cap H^g)) \ge \omega^{\alpha_i}$. Hence

$$\begin{aligned} SU_P(G) &\ge SU_P(HH^g) \ge SU_P(H^g) + SU_P(HH^g/H^g) \\ &\ge SU_P(H^g) + SU_P(H/(H \cap H^g)) \ge \beta_i + \omega^{\alpha_i} > SU_P(G), \end{aligned}$$

a contradiction. Hence $H \cap H^g$ has bounded index in H, and by symmetry in H^g. ⊣

By Lemma 2.11 there is a type-definable normal subgroup G_i of G commensurate with a bounded intersection of G-conjugates of H; as these are all commensurate with H, so is G_i. Hence $SU_P(G_i) = \beta_i$. Finally, if N is another normal subgroup of G with $SU_P(N) = \beta_i$, we can prove commensurativity of G_i and N similarly as we did for H and H^g. ⊣

Remark 9. As G_i is commensurate with all its automorphic images, we can use Lemma 2.10 to find a subgroup N of G of rank β_i which is type-definable over the parameters needed to type-define G. In fact, starting from the normal subgroup G_i, the proof of Lemma 2.10 will also yield a normal subgroup N.

COROLLARY 5.4. *A type-definable simple group, or a division ring, of ordinal SU_P-rank in a simple theory has SU_P-rank $\omega^\alpha n$ for some ordinal α and some $n < \omega$.*

PROOF. Let G be a simple type-definable group in a simple theory, with $SU_P(G) = \omega^\alpha n + \beta$ for some $0 < \beta < \omega^\alpha$. By Corollary 5.3 there is a type-definable normal subgroup H of G with $SU_P(H) = \omega^\alpha n$; contradicting simplicity of G.

If K is a type-definable division ring in a simple theory with $SU_P(K) = \omega^\alpha n + \beta$ where $0 < \beta < \omega^\alpha$, we likewise get a P-minimal type-definable subgroup H of K^+ with $SU(H) = \omega^\alpha n$. If $k \in K^\times$ and H and kH were not commensurate, then $SU_P(H \cap kH) < \omega^\alpha n$ by P-minimality, so $SU_P(H) \geq SU_P(H \cap kH) + \omega^\alpha$. Therefore $SU_P(H/(H \cap kH)) \geq \omega^\alpha$, whence

$$\begin{aligned} SU_P(H + kH) &\geq SU_P(H) + SU_P((H + kH)/kH) \\ &= SU_P(H) + SU_P(H/(H \cap kH)) \geq \omega^\alpha(n+1) > SU_P(K), \end{aligned}$$

a contradiction. Hence H is commensurate with all its K-translates. Applying Lemma 2.10 to $G := K^+ \rtimes K^\times$ with subgroup H, we get a normal subgroup N of G commensurate with some intersection of G-conjugates (i.e., $K^\times$-translates) of H. As those are all commensurate, N is commensurate with H, and so is $N_0 := N \cap K^+$. Hence N_0 is a subgroup of K^+ with $SU_P(N_0) = \omega^\alpha n$ which is invariant under translation by elements in $K^\times$. Thus N_0 is an ideal with $(0) < N_0 < K$, a contradiction. ⊣

We shall now work towards proving that type-definable groups in a supersimple theory are $\bigcap$-definable.

PROPOSITION 5.5. *Let G be a type-definable group in a simple theory, with $SU(G) = \omega^\alpha n$ and $D(G) < \infty$. Then there is a definable supergroup G_0 of G, and definable subgroups G_i of G_0 with $G = \bigcap_i G_i$.*

PROOF. Clearly we may assume that we work over a model $\mathfrak{M}$ of the ambient theory containing all the relevant parameters. By compactness there is a definable superset X_0 of G with $D(X_0) = D(G)$, such that multiplication is defined and associative on X_0 (but may go outside) and containing, for every $x \in X_0$, a unique two-sided inverse x^{-1}. Again by compactness, there is some definable set $X_1 \supseteq G$ with $X_1^2 \subseteq X_0$.

CLAIM 1. $SU(X_1) = \omega^\alpha n$.

PROOF OF CLAIM. Suppose X_1 contains a type p with $SU(p) > \omega^\alpha n$. Let $(x_i : i \in I)$ be a Morley sequence in p and consider the sets $x_i G$, for $i \in I$. Since $X_1^2 \subseteq X_0$, they are all contained in X_0. On the other hand, since $SU(x_i/x_j) > SU(G)$ for $i \neq j$, we get $x_i \notin x_j G$; by compactness and indiscernibility there is a definable superset X' of G contained in X_1 such that $x_i X' \cap x_j X' = \emptyset$ for all $i \neq j$. It follows that $D(X') < D(X_0)$, contradicting $D(G) \leq D(X') < D(X_0) = D(G)$. ⊣

Let X_2 be a definable superset of G with $X_2^2 \subseteq X_1$.

CLAIM 2. *Given a formula $\varphi(x,y)$ with $\varphi(x,y) \vdash x \in X_2$, there is a formula $\vartheta(y)$ over $\mathfrak{M}$ such that $\models \vartheta(m)$ if and only if $\varphi(x,m)$ contains a type of SU-rank $\omega^\alpha n$.*

PROOF OF CLAIM. Let $p \in S_G(\mathfrak{M})$ be a generic type for G. We show that the condition "$\varphi(x,m)$ contains a type of SU-rank $\omega^\alpha n$" is type-definable over $\mathfrak{M}$, and that its negation is type-definable over $\mathfrak{M}$ as well.

If $\varphi(x,m)$ contains a type q with $SU(q) = \omega^\alpha n$, consider a realization $x_0 \models q$ and $g \models p$ with $g \perp_{\mathfrak{M}} x_0, m$. Then $g^{-1}x_0 \models \varphi(gx,m)$, and

$$\begin{aligned}\omega^\alpha n = SU(X_1) &\geq SU(g^{-1}x_0/\mathfrak{M},m) \geq SU(g^{-1}x_0/\mathfrak{M},m,g)\\ &= SU(x_0/\mathfrak{M},m,g) = SU(x_0/\mathfrak{M},m) = SU(p) = \omega^\alpha n,\end{aligned}$$

whence $g^{-1}x_0 \perp_{\mathfrak{M},m} g$ and $\varphi(gx,m)$ does not fork over $\mathfrak{M}\cup\{m\}$. Conversely, suppose $\varphi(gx,m)$ does not fork over $\mathfrak{M}\cup\{m\}$ for some $g \models p$ with $g \perp_{\mathfrak{M}} m$. Then there is $x_0 \models \varphi(gx,m)$ with $x_0 \perp_{\mathfrak{M},m} g$, and

$$\begin{aligned}SU(gx_0/\mathfrak{M},m) &\geq SU(gx_0/\mathfrak{M},m,x_0) = SU(g/\mathfrak{M},m,x_0)\\ &= SU(g/\mathfrak{M},m) = SU(g/\mathfrak{M}) = \omega^\alpha n.\end{aligned}$$

Since $gx_0 \models \varphi(x,m)$, the formula $\varphi(x,m)$ contains a type of SU-rank $\omega^\alpha n$. But the condition "$\exists g \models p\,[g \perp_{\mathfrak{M}} m \wedge \varphi(gx,m)$ does not fork over $\mathfrak{M}\cup\{m\}]$" is type-definable over $\mathfrak{M}$, as we only have to express that there is an independent sequence $(g_i : i < \omega)$ in p with $(g_i : i < \omega) \perp_{\mathfrak{M}} m$ and indiscernible over $\mathfrak{M}\cup\{m\}$, such that $\bigwedge_{i<\omega} \varphi(g_i x, m)$ is consistent, which we can do using local ranks. Hence there is a partial type $\pi(y)$ such that $\models \pi(m)$ if and only if $\varphi(x,m)$ contains a type of SU-rank $\omega^\alpha n$.

Now consider the following condition: "there is an independent sequence $(g_i : i < \omega)$ in p with $(g_i : i < \omega) \perp_{\mathfrak{M}} m$ and indiscernible over $\mathfrak{M}\cup\{m\}$, such that $\bigwedge_{i\leq n} \varphi(g_i x, m)$ is inconsistent". This is again type-definable using local ranks. Suppose it is true of m, as witnessed by some sequence $(g_i : i < \omega)$. Then $\varphi(g_0 x, m)$ forks over $\mathfrak{M}\cup\{m\}$, and the formula $\varphi(x,m)$ cannot contain a type of SU-rank $\omega^\alpha n$.

Conversely, suppose $\varphi(x,m)$ does not contain a type of SU-rank $\omega^\alpha n$, and consider an independent sequence $(g_i : i < \omega)$ in p with $(g_i : i < \omega) \perp_{\mathfrak{M}} m$ and indiscernible over $\mathfrak{M}\cup\{m\}$. Then $\varphi(g_0 x, m)$ forks over $\mathfrak{M}\cup\{m\}$, so $\bigwedge_{i<\omega} \varphi(g_i x, m)$ is inconsistent. If $\bigwedge_{i\leq n} \varphi(g_i x, m)$ is consistent and contains an element x_0, then $x_0 \not\perp_{\mathfrak{M},m} g_i$ for all $i \leq n$. Since $g_i \perp_{\mathfrak{M}} (m, g_j : j < i)$ and hence

$$SU(g_i/\mathfrak{M}, m, g_j : j < i) = SU(g_i/\mathfrak{M}) = \omega^\alpha n,$$

we get

$$SU(g_i/\mathfrak{M}, m, g_j : j < i) \geq SU(g_i/\mathfrak{M}, m, x_0, g_j : j < i) + \omega^\alpha,$$

whence by the Lascar inequality 4.1.1

$$SU(x_0/\mathfrak{M}, m, g_j : j < i) \geq SU(x_0/\mathfrak{M}, m, g_j : j \leq i) + \omega^\alpha,$$

contradicting $SU(x_0/\mathfrak{M}, m) \leq SU(X_2) = \omega^\alpha n$. It follows that $\bigwedge_{i \leq n} \varphi(g_i x, m)$ is inconsistent.

Hence there is a partial type $\pi'(y)$ such that $\models \pi'(m)$ if and only if $\varphi(x, m)$ does not contain a type of SU-rank $\omega^\alpha n$. By compactness, we find the required formula ϑ. ⊣

Now consider a definable superset X of G closed under inverse, with $X^4 \subseteq X_2$. By the previous claim there is a formula $\vartheta(x, y)$ over $\mathfrak{M}$ such that $\models \vartheta(x, y)$ if and only if $x \in X \wedge y \in X^3$ and $xX \cap yX$ contains a type of SU-rank $\omega^\alpha n$.

Claim 3. *ϑ is a stable formula.*

Proof of Claim. Let $(a_i, b_i : i < \omega)$ be an $\mathfrak{M}$-indiscernible sequence such that $\models \vartheta(a_i, b_j)$ if and only if $i \leq j$. Then $a_0 X \cap b_0 X$ contains a type $p(x, a_0, b_0)$ of SU-rank $\omega^\alpha n$, and clearly $p(x, a_0, b_0)$ does not fork over $\mathfrak{M}$. But then $\bigwedge_{i<\omega} p(x, a_i, b_i)$ does not fork over $\mathfrak{M}$ (see [14]) and must have SU-rank $\omega^\alpha n$, so $\models \vartheta(a_i, b_j)$ for all $i, j < \omega$, a contradiction. ⊣

Put $Y = \{y \in X^3 : \bigwedge_i d_{p_i} x\, \vartheta(x, y)\}$, where d_{p_i} runs through the finitely many generic ϑ-definitions for G on X^3 (as ϑ is stable). For all $x, y \in G$ we have $xX \cap yX \supseteq G$, so $\vartheta(x, y)$ holds and $Y \supseteq G$. On the other hand, if $y \in Y, g \in G$, and h is generic for G over $\mathfrak{M} \cup \{y, g\}$, then $g^{-1}h$ is generic for G over $\mathfrak{M} \cup \{y, g\}$; moreover $hX \cap gyX$ contains a type of SU-rank $\omega^\alpha n$ if and only if $g^{-1}hX \cap yX$ does, and $hX \cap gyX \neq \emptyset$ implies $gy \in X^3$. Therefore $\models \vartheta(h, gy)$ for all generic h, and $gy \in Y$. So

$$Z = \{x \in Y : \forall y \in Y\ xy \in Y\}$$

is a definable superset of G closed under multiplication, and the set G_X of invertible elements of Z forms a definable supergroup G_X of G. Finally, as $G_X \subseteq Z \subseteq Y \subseteq X^3$, we have that $\bigcap_{X \supseteq G} G_X = G$. ⊣

Theorem 5.6. *A type-definable group G in a supersimple theory is $\bigcap$-definable.*

Proof. Clearly, we may assume that the ambient model is sufficiently saturated. By compactness, there is a definable superset X of G on which multiplication is defined (but may go outside) and is associative (whenever all partial products are in X), and such that any $x \in X$ has a two-sided inverse in X. We use induction on the number of monomials in $SU(G)$. So suppose $SU(G) = \omega^\alpha n + \beta$ with $\beta < \omega^\alpha$. By Corollary 5.3 there is a type-definable normal subgroup N of G with $SU(N) = \omega^\alpha n$, and N is $\bigcap$-definable by Proposition 5.5, say $N = \bigcap_{i<\alpha} H_i$ for some definable groups

H_i; as $\bigcap_{i<\alpha} H_i = N \subseteq G$, we may assume $GH_i \subseteq X$ for all $i < \alpha$ by compactness. By supersimplicity we may further assume $SU(H_i) = SU(N)$ for all $i < \alpha$, so

$$SU(H_i^g) = SU(N) \leq SU(H_i^g \cap H_i^{g'})$$

for all $g, g' \in G$. Hence for all $i < \alpha$ any two G-conjugates of H_i are commensurate, and thus commensurable by definability; by saturation and compactness this commensurability must be uniform.

By compactness, since $(\bigcap_{i<\alpha} H_i)^G = N^G = N \leq H_0$, there is some $i_0 < \alpha$ such that all G-conjugates of H_{i_0} are contained in H_0. By Fact 2.3 applied to the family of G-conjugates of H_{i_0} inside $\langle H_{i_0}^g : g \in G\rangle \leq H_0$ there is a G-invariant definable group H commensurable with H_{i_0} contained in H_0. Then $N_X(H)/H$ is a definable (imaginary) set containing the type-definable (imaginary) group GH/H. Since H is commensurable with N, we have

$$\begin{aligned} \omega^\alpha n + SU(GH/H) &= SU(H) + SU(GH/H) \leq SU(GH) \\ &= SU(G) = \omega^\alpha n + \beta \leq SU(H) \oplus SU(GH/H) \\ &= \omega^\alpha n \oplus SU(GH/H). \end{aligned}$$

Hence $SU(GH/H) = \beta$, and GH/H is $\bigcap$-definable by the inductive hypothesis. In particular there is a definable group $G_0/H \subseteq N_X(H)/H$ with $GH/H \leq G_0/H$; clearly the pre-image G_0 is a definable group with $G \leq G_0 \subseteq X$. Therefore every definable superset of G contains a definable supergroup of G; it follows that G is $\bigcap$-definable. ⊣

Corollary 5.7. *A type-definable division ring F in a supersimple theory is definable.*

Proof. Clearly we may assume that F is infinite. By Theorem 5.6 the additive and the multiplicative group of F are both $\bigcap$-definable. By supersimplicity and compactness there is a definable multiplicative group M containing $F^\times$ and a definable additive group $A \subseteq M \cup \{0\}$ containing F^+ with $SU(A) = SU(F)$, such that the distributive laws hold on A. By compactness again, there is a definable additive group A' containing F^+ such that $kA' \subseteq A$ for all $k \in F^\times$. Then for every $k \in F^\times$ the group kA' is an isomorphic image of A' inside A; since

$$SU(A) \geq SU(A' + kA') \geq SU(A') = SU(kA') \geq SU(F) = SU(A),$$

this means that $A' + kA'$ is commensurate with A' and kA', and hence commensurable by definability of A'. By compactness A' is uniformly commensurable with its $F^\times$-translates; by Fact 2.3 there is a definable $F^\times$-invariant subgroup A_0 of A commensurable with A' containing $\bigcap_{k\in F^\times} kA'$ and hence F. Let $F_0 := \{m \in A_0 : mA_0 \leq A_0\}$, a definable subgroup of A_0 containing F. Then F_0 is a ring; since $F_0^\times \leq M$, it has no zero divisors. But then for any non-zero $a \in F_0$ the sequence $(a^i F_0 : i < \omega)$ cannot descend infinitely, as otherwise the formula $xA_0 < yA_0$ would define a partial order with infinite chains, contradicting simplicity. Hence $a^iF_0 = a^{i+j}F_0$ for some $i \geq 0, j > 0$, and $a^i(1 - a^jk) = 0$ for some $k \in F_0$. It follows that a is invertible, and F_0 is a division ring.

Starting with a smaller A, we see that F is the intersection of definable division rings F_i; as the additive index $|F_i^+ : F_j^+|$ is infinite for any $F_i > F_j$ and therefore $SU(F_i) > SU(F_j)$, supersimplicity yields $F_i = F_j$, whence $F = F_0$. ⊣

We shall now formulate the appropriate version of the Zil'ber Indecomposability Theorem. (Note that we revert to the case of an arbitrary family P; if the theory is supersimple, we shall get definable groups by Theorem 5.6.)

THEOREM 5.8. *Let G be a type-definable group in a simple theory of rank $SU_P(G) < \omega^{\alpha+1}$, and $\mathfrak{X}$ a family of type-definable sets. Then there is a type-definable subgroup H of G with $H \subseteq X_1^{\pm 1} \dots X_m^{\pm 1}$ for some $X_1, \dots, X_m \in \mathfrak{X}$, such that $SU_P(XH) < SU_P(H) + \omega^\alpha$ for all $X \in \mathfrak{X}$ (meaning that $SU_P(p) < SU_P(H) + \omega^\alpha$ for any type p extending the partial type $x \in XH$).*

REMARK 10. A type-definable subset X of G is called *α_P-indecomposable* if for all type-definable subgroups H of G either X is contained in a single coset of H, or $SU_P(XH) \geq SU_P(H) + \omega^\alpha$ (or equivalently, $SU_P(XH/H) \geq \omega^\alpha$). So if $X \in \mathfrak{X}$ is α_P-indecomposable and contains 1, then X is contained in H; if this holds for all $X \in \mathfrak{X}$, then H is equal to $\langle \mathfrak{X} \rangle$.

PROOF. As $SU_P(G) < \omega^{\alpha+1}$, there must be some maximal $k < \omega$ such that for some $X_1, \dots, X_{m'} \in \mathfrak{X}$ the set $X_1^{\pm 1} \dots X_{m'}^{\pm 1} =: Y$ contains a type of SU_P-rank at least $\omega^\alpha k$. Choose a P-minimal type p containing the formula $x \in Y$ with $SU_P(p) = \omega^\alpha k$, and put $H := \mathrm{stab}(p)$, a type-definable group by Fact 2.1.6. Note that $S(p) \subseteq YY^{-1}$, and $\mathrm{stab}(p) \subseteq (YY^{-1})^2$. By maximality of k, the assumptions of Lemma 5.2 are satisfied, and $SU_P(H) = \omega^\alpha k$. Since k was chosen maximal and $H \subseteq (YY^{-1})^2$, we have $SU_P(XH) < SU_P(H) + \omega^\alpha$ for all $i \in I$; in fact we even have $SU_P(ZH) < SU_P(H) + \omega^\alpha$ for any finite product Z of elements of the family $\mathfrak{X}$. ⊣

REMARK 11. If H' is a conjugate of H under an automorphism stabilizing Y setwise, then by P-connectivity either $SU_P(H \cap H') < \omega^\alpha k$, or H and H' are commensurate. But the first case cannot happen, since then $SU_P(HH') \geq \omega^\alpha(k+1)$, contradicting maximality of k. We can therefore apply Lemma 2.10 to find a subgroup N type-definable over the parameters used for Y and commensurate with H; if all the X are G-invariant subsets, we may obtain a normal subgroup N commensurate with H by Lemma 2.11. In either case the proof of Lemmas 2.10 and 2.11 yields that N is contained in $\langle \mathfrak{X} \rangle$.

In particular, if T is small or supersimple and G and all $X \in \mathfrak{X}$ are type-defined over a finite set, then H can be chosen type-defined over a finite set and $\bigcap$-definable by Lemma 3.3 or Theorem 5.6. If in addition all $X \in \mathfrak{X}$ are definable, then we actually get a big definable subgroup inside $(YY^{-1})^2$ by compactness.

Recall that a group is *simple* if it has no normal subgroups, and *definably simple* if it has no definable normal subgroups. (This terminology has nothing to do with simplicity of a theory.)

COROLLARY 5.9. *Suppose G is a definable, definably simple nonabelian group in a simple theory with $0 < SU_P(G) < \infty$. Then G is simple.*

PROOF. Suppose $SU_P(G) = \omega^\alpha n + \beta$ with $\beta < \omega^\alpha$. We claim first that $SU_P(g^G) \geq \omega^\alpha$ for any $g \neq 1$ in G. So suppose $SU_P(g^G) < \omega^\alpha$. As g^G and $G/C_G(g)$ are in definable bijection, we get $SU_P(G/C_G(g)) < \omega^\alpha$. By Proposition 2.5 applied to the family of G-conjugates of $C_G(g)$ there is a definable normal subgroup N commensurable with some intersection of the form $C_G(g)^{g_0} \cap \cdots \cap C_G(g)^{g_n}$. Now $SU_P(G/C_G(g)^{g_i}) = SU_P(G/C_G(g))$ for all $i \leq n$ and any coset $h \cdot \bigcap_{i \leq n} C_G(g)^{g_i}$ is definable over $hC_G(g)^{g_0} \times \cdots \times hC_G(g)^{g_n}$. Therefore

$$SU_P(G/N) = SU(G/\bigcap_{i \leq n} C_G(g)^{g_i}) \leq \bigoplus_{i \leq n} SU_P(G/C_G(g)^{g_i}) < \omega^\alpha.$$

So N is non-trivial, and must equal G by definable simplicity. Hence $C_G(g)$ has finite index in G, and the intersection of the G-conjugates of $C_G(g)$ is a definable normal subgroup of finite index in G. Again by definable simplicity, $C_G(g) = G$, but then $g \in Z(G)$. As G is non-abelian and definable, $Z(G)$ is a proper non-trivial definable normal subgroup, a contradiction.

Next we claim that G has no type-definable normal subgroup. For if N is type-definable normal minimal (up to commensurativity) and $n \in N$ is nontrivial, then $X := (n^G \cup n^{-G})^k$ contains a type-definable normal subgroup N_1 for big $k < \omega$ by Theorem 5.8; by minimality the index of N_1 in N must be bounded. But this means that finitely many translates of X must cover N, and the subunion of those translates by elements in N must be equal to N. This contradicts definable simplicity. (Note that for supersimple theories, definable and type-definable simplicity are easily seen to be the same by Theorem 5.6 and Fact 2.3; the argument above is needed only for the case of a more general family P.)

Finally, if N is a non-trivial normal subgroup of G and $1 \neq n \in N$, then $(n^G \cup n^{-G})^k$ contains a non-trivial type-definable normal subgroup of G for big $k < \omega$ by Theorem 5.8, which must be proper since it is contained in N. This final contradiction finishes the proof. ⊣

QUESTION 2. *If $SU_P(G) = 1$, does this imply that G has a definable finite-by-abelian subgroup A with $SU_P(A) = 1$?*

REMARK 12. It is shown in [16] that a supersimple division ring is commutative.

§6. One-based groups.

DEFINITION 8. For a tuple a, the *bounded closure* $\mathrm{bdd}(a)$ is the collection of all classes modulo type-definable equivalence relations with only boundedly many conjugates over a.

Note that $\mathrm{bdd}(a) \supseteq \mathrm{acl}(a)$. By compactness, if b is an imaginary element in $\mathrm{bdd}(a)$, then $b \in \mathrm{acl}(a)$.

DEFINITION 9. A structure is *one-based* if any two tuples a and b are independent over $\mathrm{bdd}(a) \cap \mathrm{bdd}(b)$.

A stable one-based group is abelian-by-finite. This is no longer true if the theory is merely simple, or even of SU-rank 1:

EXAMPLE 2. Let V be a vector space over a finite field F from Example 1, with a non-degenerate *skew* bilinear form $\langle .,. \rangle$. It is simple (of SU-rank 1) and one-based [4]. Define a group law on $V \times F$ by $(v, f) * (v', f') = (v + v', f + f' + \langle v, v' \rangle)$. This turns $V \times F$ into a nilpotent group of class 2 with finite centre $\{0\} \times F$, which is not abelian-by-finite; clearly, $V \times F$ is still one-based of SU-rank 1.

PROPOSITION 6.1. *Let G be a one-based $\emptyset$-definable group in a simple theory. Then every definable subgroup H is commensurable with one definable over* $\mathrm{acl}(\emptyset)$.

PROOF. Let H^c be a locally connected component as given by Corollary 2.6, with canonical parameter u. Let h realize a generic type for H^c over u, and g realize a generic type for G over u, h. So $\mathrm{tp}(hg/g, u)$ is generic for $H^c g$, and $\mathrm{tp}(hg/u, h)$ is generic for G.

Now let v be the canonical parameter for the coset $H^c g$. If $(h_i : i < \omega)$ is a Morley sequence in $\mathrm{tp}(hg/g, u)$, then $(h_i : i < \omega)$ is an infinite independent sequence of generic elements of $H^c g$. By local connectivity every distinct coset of a conjugate of H^c intersects $H^c g$ in a coset of infinite index and cannot contain infinitely many independent generic elements. Hence any automorphism of the monster model stabilizing $(h_i : i < \omega)$ must stabilize $H^c g$, and hence v. Therefore $v \in \mathrm{dcl}(h_i : i < \omega)$. As $v \in \mathrm{dcl}(g, u)$, we get $v \perp_{\mathrm{Cb}(hg/g,u)} (h_i : i < \omega)$, whence $v \in \mathrm{acl}(\mathrm{Cb}(hg/g, u))$. On the other hand, one-basedness implies $\mathrm{Cb}(hg/g, u) \subset \mathrm{acl}(hg)$. Thus $v \in \mathrm{acl}(hg)$.

As $H^c = (H^c g)(H^c g)^{-1}$, the canonical parameter u is definable over v, whence $u \in \mathrm{acl}(hg)$. But $hg \perp u$, so $u \in \mathrm{acl}(\emptyset)$, and H^c is definable over $\mathrm{acl}(\emptyset)$. ⊣

COROLLARY 6.2. *If $\mathfrak{H}$ is a family of uniformly definable subgroups of a one-based group definable in a simple theory, then there are only finitely many commensurability classes among members of $\mathfrak{H}$.*

PROOF. Assuming the Corollary does not hold, compactness yields a definable subgroup which is not commensurable with any $\operatorname{acl}(\emptyset)$-definable subgroup, contradicting Proposition 6.1. ⊣

THEOREM 6.3. *Suppose G is a one-based group definable in a simple theory. Then G is finite-by-abelian-by-finite. In fact, G has a subgroup N of finite index such that N' is finite and central in N.*

PROOF. Consider, for $g \in G$, the subgroup $H_g = \{(h, h^g) : h \in G\} < G^2$. Then $\mathfrak{H} := \{H_g : g \in G\}$ is a family of uniformly definable subgroups of G^2 and has only finitely many commensurability classes. But H_g and $H_{g'}$ are commensurable if and only if $C_G(g'g^{-1})$ has finite index in G; note that this index will then be finitely bounded by Lemma 2.4. If we define $Z^* := \{g \in G : |G : C_G(g)| < \omega\}$, then Z^* is a definable subgroup of G and the commensurability classes of $\mathfrak{H}$ correspond to the cosets of Z^* in G. Therefore Z^* has finite index in G; replacing G by Z^* we may assume that every centralizer has finite index in G. But then $[g, G]$ is finite for every $g \in G$, and for independent g, h in G we get

$$[g, h] \in \operatorname{acl}(g) \cap \operatorname{acl}(h) = \operatorname{acl}(\emptyset).$$

But since $C_G(g)$ has finite index in G for every g, for every h there is h' independent from g with $[g, h] = [g, h']$. It follows that there are only finitely many commutators, and G' is finite.

Therefore, if G denotes the original group again, $N = C_{Z^*(G)}(Z^*(G)')$ will do. ⊣

QUESTION 3. *Is there a characterization of one-based simple groups in terms of their definable subsets?*

REFERENCES

[1] GEORGE M. BERGMAN and HENDRIK W. LENSTRA, JR., *Subgroups close to normal subgroups*, ***Journal of Algebra***, vol. 127 (1989), pp. 80–97.

[2] STEVEN BUECHLER, ANAND PILLAY, and FRANK O. WAGNER, *Supersimple theories*, ***Journal of the American Mathematical Society***, vol. 14 (2001), pp. 109–124.

[3] ZOÉ CHATZIDAKIS and EHUD HRUSHOVSKI, *The model theory of difference fields*, ***Transactions of the American Mathematical Society***, vol. 351 (1999), pp. 2997–3071.

[4] GREGORY CHERLIN and EHUD HRUSHOVSKI, *Permutation groups with few orbits on 5-tuples, and their infinite limits*, Preprint, 1995.

[5] PAUL ERDŐS and R. RADO, *A partition calculus in set theory*, ***Bulletin of the American Mathematical Society***, vol. 62 (1956), pp. 427–489.

[6] BRADD HART, BYUNGHAN KIM, and ANAND PILLAY, *Coordinatization and canonical bases in simple theories*, ***The Journal of Symbolic Logic***, vol. 65 (2000), pp. 293–309.

[7] EHUD HRUSHOVSKI, *Contributions to stable model theory*, ***Ph.D. thesis***, University of California at Berkeley, USA, 1986.

[8] ———, *Pseudo-finite fields and related structures*, Preprint, 1991.

[9] EHUD HRUSHOVSKI and ANAND PILLAY, *Groups definable in local fields and pseudo-finite fields*, ***Israel Journal of Mathematics***, vol. 85 (1994), pp. 203–262.

[10] ———, *Definable subgroups of algebraic groups over finite fields*, ***Journal für die Reine und Angewandte Mathematik***, vol. 462 (1995), pp. 69–91.

[11] BYUNGHAN KIM, *Recent results on simple first-order theories*, ***Model theory of groups and automorphism groups*** (David Evans, editor), LMS Lecture Notes 244, Cambridge University Press, Cambridge, United Kingdom, 1997, pp. 202–212.

[12] ———, *Forking in simple unstable theories*, ***Journal of the London Mathematical Society***, vol. 57 (1998), pp. 257–267.

[13] ———, *A note on Lascar strong types in simple theories*, ***The Journal of Symbolic Logic***, vol. 63 (1998), pp. 926–936.

[14] BYUNGHAN KIM and ANAND PILLAY, *Simple theories*, ***Annals of Pure and Applied Logic***, vol. 88 (1997).

[15] ANAND PILLAY, *Definability and definable groups in simple theories*, ***The Journal of Symbolic Logic***, vol. 63 (1998), pp. 788–796.

[16] ANAND PILLAY, THOMAS SCANLON, and FRANK O. WAGNER, *Supersimple division rings*, ***Mathematical Research Letters***, vol. 5 (1998), pp. 473–483.

[17] BRUNO P. POIZAT, ***Groupes stables***, Nur Al-Mantiq Wal-Ma'rifah, Villeurbanne, France, 1987.

[18] G. SCHLICHTING, *Operationen mit periodischen Stabilisatoren*, ***Archiv der Mathematik***, vol. 34 (1980), pp. 97–99.

[19] SAHARON SHELAH, *Simple unstable theories*, ***Annals of Pure and Applied Logic***, vol. 19 (1980), pp. 177–203.

[20] ———, *Toward classifying unstable theories*, ***Annals of Pure and Applied Logic***, vol. 80 (1996), pp. 229–255.

[21] FRANK O. WAGNER, *Hyperstable theories*, ***Logic: from foundations to applications (European Logic Colloquium 1993)*** (Charles Steinhorn, Wilfrid Hodges, Martin Hyland, and John Truss, editors), Oxford University Press, Oxford, United Kingdom, 1996, pp. 483–514.

[22] ———, *Stable groups*, LMS Lecture Notes 240, Cambridge University Press, Cambridge, United Kingdom, 1997.

[23] ———, ***Simple theories***, Kluwer Academic Publishers, Dordrecht, The Netherlands, 2000.

[24] ———, *Hyperdefinable groups in simple theories*, ***Journal of Mathematical Logic***, vol. 1 (2001), no. 1, pp. 152–172.

INSTITUT GIRARD DESARGUES (LYON 1)
UNIVERSITÉ CLAUDE BERNARD
21, AVENUE CLAUDE BERNARD
69622 VILLEURBANNE-CEDEX, FRANCE
E-mail: wagner@desargues.univ-lyon1.fr

PROVABLE RECURSIVENESS AND COMPLEXITY

S. S. WAINER

§0. Introduction. Over fifty years ago, Kreisel (1951, 1952), in the course of developing the no counterexample interpretation, characterized the functions "computable in" (provably recursive in) arithmetic as those definable by recursions over standard primitive recursive well-orderings of order-types less than ε_0. His underlying question, then implicit, later more explicit: "What more do we know if we have proved a theorem by restricted means than if we merely know that it is true ?" became a research programme—reading off the information content of proofs—which has driven much of proof theory and related areas of computer science ever since. In the case of Π^0_2 statements, or "program specifications", their truth asserts the mere existence of an algorithm, whereas a proof contains information from which one can be synthesized, and its complexity estimated, using normalization or cut-elimination or (originally) epsilon-substitution. However, as is well understood, the provably recursive functions of classical (Peano) or intuitionistic (Heyting) arithmetic, though mathematically interesting they may be, are nevertheless in general so fast growing as to be very far from being in any sense feasibly or realistically computable. It took another thirty-five years before Buss' ***Bounded Arithmetic*** (1986) provided the first proof theoretical characterization of the polytime functions. Since then there has been a growing interest in the search for other, natural analogues of traditional theories like PA or HA, without the explicitly-imposed quantifier bounds of BA, but whose provably recursive functions form "more realistic" complexity classes. In particular, Leivant (1995) characterized PTIME and the Grzegorczyk classes E^2, E^3 in theories based on a certain ramified induction scheme formalising his earlier notion of ramified recursion. It is these results which interest us here, and we shall re-work them, but in a quite different setting, and with entirely different methods.

Our aim is to explore analogies with the "traditional" results, by finding ways to make the old, "fast growing" hierarchy of bounding functions become

Partially supported by research grant GR/R 15856/01 of the UK Engineering and Physical Sciences Research Council.

Logic Colloquium '01
Edited by M. Baaz, S. Friedman, and J. Krajíček
Lecture Notes in Logic, 20

"slow growing". This sounds like just a technical exercise, but it provides a rough-and-ready guide for making one's favourite arithmetical theory "more feasible". For if we can restrict arithmetic accordingly (and it turns out that there is a very simple and quite natural modifying principle, proof theoretically analogous to the normal/safe variable separation of Bellantoni and Cook (1992)) then the provably recursive functions will be elementary (E^3) at most, since the slow growing functions below ε_0 are just exponential polynomials; and in the Σ_1 inductive fragment we should obtain only sub-elementary (E^2) functions, since the slow growing functions below ω^ω are polynomials.

We begin with a brief survey of one of the traditional methods for characterizing provable recursiveness—cut elimination with ordinal bounds (see e.g., Weiermann (1996) or Fairtlough and Wainer (1998)). This requires an embedding of the underlying formal system into an infinitary one, which is really nothing other than the truth definition with a cut rule added and ordinal heights assigned. But by carefully controlling the ordinal assignment, using a crucial idea of Buchholz (1987), the usual hierarchies of bounding functions can be read off directly from proofs whose cut rank has been reduced to the computational, i.e., Σ_1, level. Since the infinitary system is complete, and admits cut-reduction, we therefore have a general method for attaching, to any given termination proof, say with proof-theoretic ordinal α, a function B_α which bounds the complexity of the recursive function so defined. It only remains to fit to this framework an appropriate formalism for introducing the recursive functions, in order to produce a system within which recursions, proofs and complexity bounds can all be systematically developed. The case made by Leivant (1994, 1995) for the Herbrand-Gödel-Kleene equation calculus, where computations are just equational derivations, seems incontrovertible, since provable termination of a function $f(\vec{n})$ defined by a system of equations E can be expressed directly (without any coding of the T-predicate) as

$$E,\ \vec{n}\colon N \vdash^\alpha \exists a\ (f(\vec{n}) = a)$$

and the associated bounding function $B_\alpha(\max \vec{n})$ will supply a complexity measure.

§1. Ordinal bounds for recursion. We inductively generate an infinitary system of sequents

$$E,\ n\colon N,\ \Gamma \vdash^\alpha A$$

where (i) E is a (consistent) set of Herbrand-Gödel-Kleene defining equations for partial recursive functions $f, g, h, \ldots$; (ii) n is a bound on the numerical inputs (or more precisely, its representing numeral); (iii) A is a closed formula, and Γ a finite multiset of closed formulas, built up from atomic equations between arbitrary terms t involving the function symbols of E; and (iv) α, β

denote ordinal bounds which we shall be more specific about later (for the time being think of β as being smaller than α "modulo the input n").

The first two rules below are just the input and substitution axioms of the equation calculus, the next two are computation rules for N, and the rest are essentially just formalised versions of the truth definition, with Cut added.

E1: Input axioms:

$$E,\ n\colon N,\ \Gamma \vdash^{\alpha} e(\vec{n})$$

where e is either one of the defining equations of E or an identity $t = t$, and $e(\vec{n})$ denotes the result of substituting, for its variables, numerals for numbers $\leq n$.

E2: Substitution axioms:

$$E,\ n\colon N,\ \Gamma,\ t_1 = t_2,\ e(t_1) \vdash^{\alpha} e(t_2)$$

where $e(t_1)$ is an equation between terms in the language of E, with t_1 occurring as a subterm, and $e(t_2)$ is the result of replacing an occurrence of t_1 by t_2.

N1: $$E,\ n\colon N,\ \Gamma \vdash^{\alpha} m\colon N \quad \text{provided } m \leq n+1$$

N2: $$\frac{E,\ n\colon N,\ \Gamma \vdash^{\beta} n'\colon N \qquad E,\ n'\colon N,\ \Gamma \vdash^{\beta'} A}{E,\ n\colon N,\ \Gamma \vdash^{\alpha} A}$$

Cut: $$\frac{E,\ n\colon N,\ \Gamma \vdash^{\beta} C \qquad E,\ n\colon N,\ \Gamma,\ C \vdash^{\beta'} A}{E,\ n\colon N,\ \Gamma \vdash^{\alpha} A}$$

∃L: $$\frac{E,\ \max(n,i)\colon N,\ \Gamma,\ B(i) \vdash^{\beta_i} A \quad \text{for every } i \in N}{E,\ n\colon N,\ \Gamma,\ \exists b B(b) \vdash^{\alpha} A}$$

∃R: $$\frac{E,\ n\colon N,\ \Gamma \vdash^{\beta} m\colon N \qquad E,\ n\colon N,\ \Gamma \vdash^{\beta'} A(m)}{E,\ n\colon N,\ \Gamma \vdash^{\alpha} \exists a A(a)}$$

∀L: $$\frac{E,\ n\colon N,\ \Gamma \vdash^{\beta} m\colon N \qquad E,\ n\colon N,\ \Gamma,\ \forall b B(b),\ B(m) \vdash^{\beta'} A}{E,\ n\colon N,\ \Gamma,\ \forall b B(b) \vdash^{\alpha} A}$$

∀R: $$\frac{E,\ \max(n,i)\colon N,\ \Gamma \vdash^{\beta_i} A(i) \quad \text{for every } i \in N}{E,\ n\colon N,\ \Gamma \vdash^{\alpha} \forall a A(a)}$$

In addition, there are of course two rules for each propositional symbol, but it is not necessary to list them since they are quite standard. However it should be noted that the rules are formalized in the style of **minimal**, not classical, logic, as on page 65 of Troelstra and Schwichtenberg (1996). This is natural from a computational viewpoint, and does not restrict the classes of functions which can be proved total (though it does incur the cost of much greater sensitivity to the logical structure of induction formulas used in their termination proofs—see e.g., Burr (2000)).

Ordinal Assignment à la Buchholz. The ordinal bounds on sequents above are intensional, "tree ordinals", generated inductively by: 0 is a tree ordinal; if α is a tree ordinal so is $\alpha + 1$; and if $\lambda_0, \lambda_1, \lambda_2, \ldots$ is an ω-sequence of tree ordinals then the function $i \mapsto \lambda_i$ denoted $\lambda = \sup \lambda_i$, is itself also a tree ordinal. Thus tree ordinals carry a specific choice of fundamental sequence to each "limit" encountered in their build-up, and because of this the usual definitions of primitive recursive functions lift easily to tree ordinals. For example addition is defined by:

$$\alpha + 0 = 0, \qquad \alpha + (\beta + 1) = (\alpha + \beta) + 1, \qquad \alpha + \lambda = \sup(\alpha + \lambda_i)$$

and "2 times" is defined by:

$$2 \cdot 0 = 0, \qquad 2 \cdot (\beta + 1) = 2 \cdot \beta + 1 + 1, \qquad 2 \cdot \lambda = \sup 2 \cdot \lambda_i$$

and exponentiation is defined by:

$$2^0 = 1, \qquad 2^{\beta+1} = 2^\beta + 2^\beta, \qquad 2^\lambda = \sup 2^{\lambda_i}.$$

For ω we choose the specific fundamental sequence $\omega = \sup(i + 1)$.

DEFINITION. For each integer i there is a predecessor function given by:

$$P_i(0) = 0, \qquad P_i(\alpha + 1) = \alpha, \qquad P_i(\lambda) = P_i(\lambda_i)$$

and by iterating P_i we obtain, for each non-zero tree ordinal α, the finite set $\alpha[i]$ of all its "i-predecessors" thus:

$$\alpha[i] = \{P_i(\alpha), P_i^2(\alpha), P_i^3(\alpha), \ldots, 0\}.$$

Call a tree ordinal α "structured" if every sub-tree ordinal of the form $\lambda = \sup \lambda_i$ (occurring in the build-up of α) has the property that $\lambda_i \in \lambda[i + 1]$ for all i. Then if α is structured, $\alpha[i] \subset \alpha[i + 1]$ for all i, and each of its sub-tree ordinals β appears in one, and all succeeding, $\alpha[i]$. Thus we can think of a structured α as the directed union of its finite sub-orderings $\alpha[i]$. The basic example is $\omega[i] = \{0, 1, \ldots, i\}$. All tree ordinals used here will be structured ones.

Ordinal Bounds. The condition on ordinal bounds in the above sequents is to be as follows:

- In rules E1, E2, N1, the bound α is arbitrary.
- In rules N2, Cut, $\exists$R, $\forall$L, and propositional rules, $\beta, \beta' \in \alpha[n]$.
- In rules $\exists$L and $\forall$R, $\beta_i \in \alpha[\max(n, i)]$.

LEMMA 1.1 (Weakening). *If $E, n \colon N, \Gamma \vdash^\alpha A$ where $n \leq n'$ and $\alpha[k] \subset \alpha'[k]$ for every $k \geq n$ then $E, n' \colon N, \Gamma \vdash^{\alpha'} A$.*

DEFINITION. The "fast growing" bounding functions B_α are given by the recursion:

$$B_0(n) = n + 1, \qquad B_{\alpha+1}(n) = B_\alpha(B_\alpha(n)), \qquad B_\lambda(n) = B_{\lambda_n}(n).$$

LEMMA 1.2. *$m \leq B_\alpha(n)$ if and only if $n: N \vdash^\alpha m: N$ by the N rules. Hence if α is structured, (i) B_α is strictly increasing and (ii) $B_\beta(n) < B_\alpha(n)$ whenever $\beta \in \alpha[n]$.*

The proof is easy by inductions on α, see Fairtlough and Wainer (1998).

DEFINITION. We shall adopt a somewhat restrictive notion of Σ_1 formula, viz. one of the form $\exists \vec{a} A(\vec{a})$ where A is a conjunction of (atomic) equations between terms. A typical example is $\exists a(t = a)$ which we shall abbreviate to $t\downarrow$.

NOTATION. We signify that a derivation, in the infinitary calculus above, involves only Σ_1 cut formulas C, by attaching a subscript 1 to the proof-gate thus: $E, n: N, \Gamma \vdash_1^\alpha A$. If all cut formulas are atomic equations (or possibly conjunctions of them) we attach a subscript 0 instead.

The careful assignment of ordinal bounds to sequents has the following consequences:

LEMMA 1.3 (Bounding Lemma). *Let Γ consist of (conjunctions of) atomic formulas only, and let k be the greatest numerical parameter occurring in them.*

1. *If $E, n: N, \Gamma \vdash_1^\alpha m: N$ then $m \leq B_\alpha(\max(n, k))$.*
2. *If $E, n: N, \Gamma \vdash_1^\alpha \exists \vec{a} A(\vec{a})$ then there are numbers $\vec{m} \leq B_\alpha(\max(n, k))$ such that $E, \max(n, k): N, \Gamma \vdash_0^{2\cdot\alpha} A(\vec{m})$.*

PROOF. Both parts are dealt with simultaneously by induction on α. Since only Σ_1 cuts are involved, it is only the E, N, Cut and $\exists$ rules which come into play. The E rules require no action at all, and if N1 is applied (in case 1) then we have immediately $m \leq n + 1 \leq B_\alpha(\max(n, k))$.

(N2) Suppose the sequent in 1 or 2 comes about by the N2 rule. Then one of the premises is of exactly the same form, but with n replaced by n' and α replaced by a $\beta' \in \alpha[n]$. Therefore by the induction hypothesis, and re-application of N2 to reduce $n': N$ to $n: N$, the desired result follows since $2 \cdot \beta' \in 2 \cdot \alpha[n]$. But the bound on existential witnesses that we have at this stage is $B_{\beta'}(\max(n', k))$. However the other premise is a sequent of the form 1 with m replaced by n' and α replaced by a $\beta \in \alpha[n]$, so by the induction hypothesis $n' \leq B_\beta(\max(n, k))$. Thus, substituting $B_\beta(\max(n, k))$ for n' we obtain the required bound

$$B_{\beta'}(\max(n', k)) \leq B_{\beta'}(B_\beta(\max(n, k))) \leq B_\alpha(\max(n, k))$$

using the definition of B_α and its majorization properties.

($\exists$) The $\exists$L rule does not apply. If the sequent in 2 comes about by an application of $\exists$R, introducing the outermost quantifier in $\exists \vec{a} A(\vec{a})$, then the induction hypothesis applied to the first premise yields a witness $m \leq B_\beta(\max(n, k))$. The induction hypothesis applied to the second premise yields the desired result, but with bound $B_{\beta'}(\max(n, k))$ on witnesses for the remaining quantifiers. However $B_{\beta'}(\max(n, k))$ and $B_\beta(\max(n, k))$ are both less than the required bound $B_\alpha(\max(n,k))$ since $\beta, \beta' \in \alpha[n]$ and $\alpha[n] \subset \alpha[\max(n,k)]$ by the assumed structuredness of α.

(Cut) Finally, suppose the sequent in 1 or 2 arises by an application of Cut with cut formula $C \equiv \exists\vec{c}D(\vec{c})$. By applying the induction hypothesis to the first premise we obtain witnesses $\vec{\ell}$ no larger than $B_\beta(\max(n,k))$ such that

$$E,\ n\colon N,\ \Gamma \vdash_0^{2\cdot\beta} D(\vec{\ell})$$

and by "inverting" the $\exists\vec{c}$ in the second premise we obtain

$$E,\ \max(n,\vec{\ell})\colon N,\ D(\vec{\ell}),\ \Gamma \vdash_1^{\beta'} F$$

where $F \equiv m\colon N$ or $F \equiv \exists\vec{a}A(\vec{a})$. Now there may be other numerical parameters in C and hence in $D(\vec{\ell})$ which are not displayed, but the system is set up in such a way that these will again be no larger than $B_\beta(\max(n,k))$. Therefore, applying the induction hypothesis to this last sequent, in the case $F \equiv \exists\vec{a}A(\vec{a})$, we obtain numerical witnesses $\vec{m}$ bounded by:

$$B_{\beta'}(\max(n,k,\vec{\ell},B_\beta(\max(n,k)))) \le B_{\beta'}(B_\beta(\max(n,k))) \le B_\alpha(\max(n,k))$$

and

$$E,\ \max(n,\vec{\ell})\colon N,\ D(\vec{\ell}),\ \Gamma \vdash_0^{2\cdot\beta'} A(\vec{m}).$$

Since $\vec{\ell} \le B_\beta(\max(n,k)) \le B_{2\cdot\beta}(\max(n,k))$ we have, by a lemma above,

$$E,\ \max(n,k)\colon N,\ D(\vec{\ell}),\ \Gamma \vdash_0^{2\cdot\beta} \max(n,\vec{\ell})\colon N$$

and so by the N2 rule, with $\gamma = \max(\beta,\beta')$, we obtain

$$E,\ \max(n,k)\colon N,\ D(\vec{\ell}),\ \Gamma \vdash_0^{2\cdot\gamma+1} A(\vec{m}).$$

Then by a cut on $D(\vec{\ell})$, with a weakening, we obtain the required

$$E,\ \max(n,k)\colon N,\ \Gamma \vdash_0^{2\cdot\alpha} A(\vec{m})$$

since $\gamma \in \alpha[\max(n,k)]$ implies $2\cdot\gamma+1 \in 2\cdot\alpha[\max(n,k)]$. The other case $F \equiv m\colon N$ is much simpler. This completes the proof.

LEMMA 1.4. *If E, $\max(n,k)\colon N$, $\Gamma \vdash_0^\gamma A$ as above, where Γ, A consists of (possibly conjunctions of) atomic equations, then the derivation is finite, and the height of its derivation tree is less than $B_{\gamma+1}(\max(n,k))$.*

PROOF. This is easy to see, using the majorization properties of B_γ. The only rule needing any care is N2, where the premises are

$$E,\ \max(n,k)\colon N,\ \Gamma \vdash_0^\beta m\colon N$$

and

$$E,\ m\colon N,\ \Gamma \vdash_0^{\beta'} A$$

with $\beta, \beta' \in \gamma[\max(n,k)]$ and $m \le B_\beta(\max(n,k))$ by the Bounding Lemma above. Assuming inductively that the result holds for each of these, their derivation heights are less than $B_{\beta+1}(\max(n,k))$ and $B_{\beta'+1}(B_\beta(\max(n,k))$ respectively. After applying N2, the whole derivation therefore has height less than the maximum of these plus one, hence less than or equal to $B_{\beta'+1}(B_{\beta+1}(\max(n,k)))$. But since $\beta, \beta' \in \gamma[\max(n,k)]$ this is no larger than $B_\gamma(B_\gamma(\max(n,k)))$ which is just $B_{\gamma+1}(\max(n,k))$.

THEOREM 1.5 (Complexity). *If f is defined by the system of equations E and $E, n \colon N \vdash_1^\alpha f(\vec{n})\downarrow$ where $n = \max \vec{n}$, then by the Bounding Lemma*

$$E,\ n\colon N \vdash_0^{2\cdot\alpha} f(\vec{n}) = m \quad \textit{where} \quad m \leq B_\alpha(n).$$

This is a computation of $f(\vec{n})$ from E, and by the last lemma, the number of computation steps (nodes in the binary branching tree) is less than

$$2^{B_{2\cdot\alpha+1}(n)}.$$

REMARK. For any recursive function f given by a system of equations E, there will be, for each n, a number k_n which bounds both the value of $f(n)$ and the height of its derivation from E. This means that $n\colon N \vdash_0 m\colon N$ and E, $n\colon N \vdash_0 f(n) = m$, both with finite bound k_n. Therefore for each n we have

$$E,\ n\colon N \vdash_0^{k_n+1} f(n)\downarrow$$

and hence $E \vdash_0^\alpha \forall a(f(a)\downarrow)$ with $\alpha = \sup k_i + 1$. Thus every recursive function can be proved total by an infinitary derivation whose tree ordinal bound is of set-theoretic rank just ω. But note that α is merely an encoding of our assumption that f is total, so nothing is gained. This old observation shows that we cannot hope to classify recursive functions by set-theoretic ordinals alone, but must use some intensional notion like "tree ordinal". The task is not to capture just one recursive function at a time, but to classify "naturally occurring" collections of them by "natural" assignments of tree ordinals.

1.1. Provable recursiveness in arithmetic. Suppose f is provably recursive in PA. This means we can prove $\exists a T(x,a)$ where $T(x,a)$ is an elementary-recursive relation expressing the fact that computation of $f(x)$ terminates in a steps. Let E be a system of equations defining the characteristic function of T, and assume that these equations have been added to PA as axioms. If $\exists a T(x,a)$ is provable with classical logic, it is provable using minimal logic only.

LEMMA 1.6 (Embedding). *Suppose $A(\vec{t})$ is provable in arithmetic with minimal logic, where $\vec{t}$ are terms. Then there is a fixed number k such that: substituting arbitrary numerals $\vec{n}$ for the free variables, and letting $\vec{m}$ denote the resulting numerical values of the terms $\vec{t}$, we have*

$$E,\ \max \vec{n}\colon N \vdash^{\omega\cdot k} A(\vec{m})$$

in the infinitary system set out above.

PROOF. The embedding goes fairly directly, and we shall only consider two cases in illustration (suppressing any side-formulas Γ which might occur as assumptions in the antecedent).

If $\exists a A(a, \vec{t})$ comes about by exists introduction from $A(t_0, \vec{t})$ then, by induction on proof-height, we have a k such that for all assignments $\vec{n}$ to the free variables,

$$E,\ \max \vec{n}\colon N \vdash^{\omega \cdot k} A(m_0, \vec{m})$$

where m_0 and $\vec{m}$ are the values of the terms t_0 and $\vec{t}$. Now since m_0 is the value of an elementary-recursive term $t_0(\vec{n})$, and since every elementary-recursive function is bounded by an iterated exponential, and since $B_\omega(n) = n + 2^{n+1}$, it follows easily that for a large enough k, $m_0 \leq B_{\omega \cdot k}(\max \vec{n})$. But this means that $\max \vec{n}\colon N \vdash^{\omega \cdot k} m_0\colon N$ by the N rules. We can therefore apply the $\exists$R rule of the infinitary system, to obtain

$$E,\ \max \vec{n}\colon N \vdash^{\omega \cdot (k+1)} \exists a A(a, \vec{m})$$

as required, because $\omega \cdot k \in \omega \cdot (k+1)[\max \vec{n}]$.

Suppose $\forall a A(a, \vec{t})$ comes about by induction (formalised as an appropriate rule) from $A(0, \vec{t})$ and $A(a, \vec{t}) \vdash A(a+1, \vec{t})$ where variable a does not occur free anywhere else. Then, again by induction, there is a k such that for all numerical assignments i to a, and $\vec{n}$ to the other free variables, we have

$$E,\ \max \vec{n}\colon N \vdash^{\omega \cdot k} A(0, \vec{m})$$

and

$$E,\ \max(i, \vec{n})\colon N,\ A(i, \vec{m}) \vdash^{\omega \cdot k} A(i+1, \vec{m})$$

where $\vec{m}$ are the values of the terms $\vec{t}$. Now by repeated Cuts on $A(i, \vec{m})$ for $i = 0, 1, \ldots, j-1$ we obtain for each j,

$$E,\ \max(j, \vec{n})\colon N \vdash^{\omega \cdot k + j} A(j, \vec{m})$$

and then by the infinitary $\forall$R rule, since $\omega \cdot k + j \in \omega \cdot (k+1)[\max(j, \vec{n})]$,

$$E,\ \max \vec{n}\colon N \vdash^{\omega \cdot (k+1)} \forall a A(a, \vec{m}).$$

This completes the proof.

THEOREM 1.7 (Old results). *A function is provably recursive in PA if and only if it is computable in a number of steps bounded by B_α for some $\alpha \prec \varepsilon_0$. A function is provably recursive in Σ_1-Induction if and only if it is primitive recursive.*

PROOF. If in PA we can prove the termination $\exists a T(x, a)$ of a recursive function $f(x)$, then by the Embedding Lemma there is a k such that for all inputs n,

$$E,\ n\colon N \vdash^{\omega \cdot k} \exists a T(n, a).$$

There will of course be many Cuts in this derivation, possibly of great logical complexity, depending crucially on the complexity of induction formulas used in the original proof in formal arithmetic. But Gentzen-style cut elimination will successively reduce the "cut rank" (maximum "size" of cut formulas) one level at a time, so that they eventually become at worst Σ_1 (counted as rank 1). The price paid for each such reduction is an exponential increase in the ordinal bound. Thus we obtain, for each n, a derivation

$$E,\ n\colon N \vdash_1^{\alpha} \exists a T(n,a)$$

where $\alpha = 2_r(\omega \cdot k) \prec \varepsilon_0$, $r+1$ being the original cut rank, and $2_r(\omega \cdot k) = \omega \cdot k$ if $r = 0$ or $2^{2_{r-1}(\omega \cdot k)}$ if $r > 0$. The Bounding Lemma then gives a bound $B_\alpha(n)$ on the number of steps m needed to satisfy $T(n,m)$, i.e., to compute $f(n)$. If the original proof only uses Σ_1-Induction then we can take $r = 0$ so the bound will be $B_{\omega \cdot k}$ and this is primitive recursive.

The converse follows from Gentzen's result that in PA one can prove transfinite induction to any $\alpha \prec \varepsilon_0$. This is just what's needed to prove that B_α, and consequently any function computable in fewer steps, will be provably total.

§2. Input and output. It is easy to check that for any α, $B_{\alpha+\omega}(n)$ is the 2^{n+1}-times iterate of B_α on n. Thus already at low levels we have $B_{\omega \cdot 2}$ superexponential and B_{ω^2} a version of the Ackermann function. Recall that the aim here is to seek "more feasible" analogues of the results above, by finding simple reformulations of arithmetic whose provably recursive functions are (preferably) polynomially bounded or (at worst) lie below the superexponential level. The reason for developing the older results above in the way that we have, is that they suggest an idea as to how one might go about this, i.e., by exercising stricter control over the ordinal assignment.

The infinitary rules, assigning ordinal bounds to recursions, contain number declarations $n\colon N$ which may occur in the antecedent or succedent of a sequent. These denote "values" computed in the course of a derivation, according to the rules N1, N2. What we shall do is introduce a new, arbitrary but fixed, "induction parameter" $k\colon I$ and insist that in *every* rule, the ordinal bounds β assigned to premises must belong to $\alpha[k]$ where α is the ordinal bound on the conclusion. Intuitively we think of $k\colon I$ as "input" and $n\colon N$ as "output" (though strictly speaking, k will actually measure the size of the input in terms of its binary length). As we shall see, the new bounding functions (below ε_0) then become exponential polynomials. Therefore if we can reformulate arithmetic so that proofs embed into these new input/output sequents, the provably recursive functions will be elementary-recursive at most. These ideas were first worked out by Ostrin (1999).

So, modify each sequent as follows:

- Add a new kind of number declaration $k\colon I$.
- Control the assignment of ordinal bounds by $\beta, \beta', \beta_i \in \alpha[k]$.

Note that throughout such an infinitary derivation, the k remains fixed. Then the computational rules N1, N2 become (with E, Γ suppressed)

$$k\colon I,\ n\colon N \vdash^{\alpha} m\colon N \quad \text{where } m \leq n+1$$

and with $\beta, \beta' \in \alpha[k]$,

$$\frac{k\colon I,\ n\colon N \vdash^{\beta} n'\colon N \qquad k\colon I,\ n'\colon N \vdash^{\beta'} A}{k\colon I,\ n\colon N \vdash^{\alpha} A}$$

DEFINITION. The new "slower growing" bounding functions B_α now have two variables, an input k and an output n, and are given by the recursion:

$$B_0(k;n) = n+1, \quad B_{\alpha+1}(k;n) = B_\alpha(k; B_\alpha(k;n)), \quad B_\lambda(k;n) = B_{\lambda_k}(k;n).$$

LEMMA 2.1. *$m \leq B_\alpha(k;n)$ if and only if, using the new N rules, we can derive $k\colon I, n\colon N \vdash^{\alpha} m\colon N$.*

PROOF. For the "only if" proceed by induction on α. If $\alpha = 0$ the result is immediate by rule N1. If $\alpha > 0$ then it's easy to see that $B_\alpha(k;n) = B_\beta(k; B_\beta(k;n))$ where $\beta = P_k(\alpha)$. So if $m \leq B_\alpha(k;n)$ then $m \leq B_\beta(k;n')$ where $n' = B_\beta(k;n)$. Therefore by the induction hypothesis we have both $k\colon I, n\colon N \vdash^{\beta} n'\colon N$ and $k\colon I, n'\colon N \vdash^{\beta} m\colon N$. Hence $k\colon I, n\colon N \vdash^{\alpha} m\colon N$ by the N2 rule.

The "if" part again follows by induction on α. The result is immediate if the sequent $k\colon I,\ n\colon N \vdash^{\alpha} m\colon N$ comes about by the N1 rule. If it comes about by an application of N2 from premises $k\colon I,\ n\colon N \vdash^{\beta} n'\colon N$ and $k\colon I,\ n'\colon N \vdash^{\beta'} m\colon N$ then by the induction hypothesis $n' \leq B_\beta(k;n)$ and $m \leq B_{\beta'}(k;n')$. Thus $m \leq B_{\beta'}(k; B_\beta(k;n))$ and this is bounded by $B_\alpha(k;n)$ since $\beta, \beta' \in \alpha[k]$.

LEMMA 2.2. *$B_\alpha(k;n) = n + 2^{G(\alpha,k)}$ where $G(\alpha,k)$ is the size of $\alpha[k]$. Furthermore, if $\alpha \prec \varepsilon_0$ then $G(\alpha,k)$ is the exponential polynomial obtained by substituting $k+1$ for ω throughout the Cantor normal form of α.*

PROOF. The first part follows directly from the recursive definition of B_α, by noting also that $G(0,k) = 0$, $G(\alpha+1,k) = G(\alpha,k)+1$ and $G(\lambda,k) = G(\lambda_k,k)$. The second part uses $G(\omega,k) = G(k+1,k) = k+1$ together with the easily verified fact that if $\varphi(\alpha,\beta)$ denotes either $\alpha+\beta$ or $\alpha\cdot\beta$ or α^β then

$$G(\varphi(\alpha,\beta),k) = \varphi(G(\alpha,k), G(\beta,k)).$$

§3. **Elementary arithmetic.** Now we describe a weak system of arithmetic developed in Ostrin and Wainer (2001), analogous to PA or HA, but with two kinds of variables—induction variables and quantifier variables, which play roles corresponding to the normal/safe recursion variables of Bellantoni and Cook (1992). A very appealing feature of Leivant's "intrinsic" theories, which we have adopted here, is that they are based on Kleene's equation calculus, which allows for a natural notion of provable recursiveness, completely free of any coding implicit in the more traditional definition involving the T-predicate. Thus one is allowed to introduce arbitrary partial recursive functions f by means of their equational definitions as axioms, but the logical and inductive power of the theory severely restricts one's ability to prove termination: $f(x)\downarrow$.

The principal logical restriction which must be applied to such theories concerns the $\exists$-introduction and (dually) $\forall$-elimination rules. If arbitrary terms t were allowed as witnesses for $\exists$-introduction, then from the axiom $t = t$ we could immediately deduce $\exists a(t = a)$ and hence, with $t = f(x)$, $f(x)\downarrow$ for every f ! This is clearly not what we want. In order to avoid it we make the restriction that only "basic" terms: variables or 0 or their successors or predecessors, may be used as witnesses. This is not quite so restrictive as it first appears, since the equality rule $t = a, A(t) \vdash A(a)$ yields immediately $t\downarrow$, $A(t) \vdash \exists a A(a)$. Thus a term may be used to witness an existential quantifier only when it has been proven to be defined. In particular, if f is introduced by a defining equation $f(x) = t$ then to prove $f(x)\downarrow$ we first must prove (compute) $t\downarrow$. Thus, provided we formulate the theory carefully enough, proofs in its Σ_1 fragment will correspond to computations in the equation calculus, and bounds on proof-size will yield complexity measures.

3.1. The theory EA(I;O). There will be two kinds of variables: "input" variables denoted $x, y, z, \ldots$, and "output" variables denoted $a, b, c, \ldots$, both intended as ranging over natural numbers. Output variables may be bound by quantifiers, but input variables will always be free. The *basic terms* are: variables of either kind, the constant 0, or the result of repeated application of the successor S or predecessor P. General *terms* are built up in the usual way from 0 and variables of either kind, by application of S, P and arbitrary function symbols $f, g, h, \ldots$ denoting partial recursive functions given by sets E of Herbrand-Gödel-Kleene-style defining equations.

Atomic formulas will be equations $t_1 = t_2$ between arbitrary terms, and formulas $A, B, \ldots$ are built from these by applying propositional connectives and quantifiers $\exists a, \forall a$ over output variables a. The negation of a formula $\neg A$ will be defined as $A \rightarrow \perp$.

As before, we shall work with minimal logic in a sequent style, for example the system G3m as set out on page 65 of Troelstra and Schwichtenberg (1996). As Leivant points out, a classical proof of $f(x)\downarrow \equiv \exists a(f(x) = a)$ can be transformed, by the double-negation interpretation, into a proof in minimal logic.

It is not necessary to list the propositional rules as they are quite standard, and the cut rule (with "cut formula" C) is:

$$\frac{\Gamma \vdash C \qquad \Gamma,\ C \vdash A}{\Gamma \vdash A}$$

where, throughout, Γ is an arbitrary finite multiset of formulas. However, as stressed above, the quantifier rules need restricting. Thus the minimal left-$\exists$ and right-$\exists$ rules are:

$$\frac{\Gamma,\ A(b) \vdash B}{\Gamma,\ \exists a A(a) \vdash B} \qquad \frac{\Gamma \vdash A(t)}{\Gamma \vdash \exists a A(a)}$$

where, in the left-$\exists$ rule the output variable b is not free in Γ, B, and in the right-$\exists$ rule the witnessing term t is basic. The left-$\forall$ and right-$\forall$ rules are:

$$\frac{\Gamma,\ \forall a A(a),\ A(t) \vdash B}{\Gamma,\ \forall a A(a) \vdash B} \qquad \frac{\Gamma \vdash A(b)}{\Gamma \vdash \forall a A(a)}$$

where, in the left-hand rule the term t is basic, and in the right-hand rule the output variable b is not free in Γ.

The logical axioms are, with A atomic,

$$\Gamma,\ A \vdash A$$

and the equality axioms are $\Gamma \vdash t = t$ and, again with $A(.)$ atomic,

$$\Gamma,\ t_1 = t_2,\ A(t_1) \vdash A(t_2)$$

The logic allows these to be generalised straightforwardly to an arbitrary formula A and the quantifier rules then enable us to derive

$$\Gamma,\ t\downarrow,\ A(t) \vdash \exists a A(a)$$

$$\Gamma,\ t\downarrow,\ \forall a A(a) \vdash A(t)$$

for any terms t and formulas A.

Two further principles are needed, describing the data-type N, namely induction and cases (a number is either zero or a successor). We present these as rules rather than their equivalent axioms, since this will afford a closer match between proofs and computations. The induction rule (with "induction formula" $A(.)$) is

$$\frac{\Gamma \vdash A(0) \qquad \Gamma,\ A(a) \vdash A(Sa)}{\Gamma \vdash A(x)}$$

where the output variable a is not free in Γ and where, in the conclusion, x is an input variable, or a basic term on an input variable.

The cases rule is

$$\frac{\Gamma \vdash A(0) \qquad \Gamma \vdash A(Sa)}{\Gamma \vdash A(t)}$$

where t is any basic term. Note that with this rule it is easy to derive $\forall a$ $(a = 0 \lor a = S(Pa))$ from the definition: $P(0) = 0$ and $P(Sa) = a$.

Definition. A k-ary function f is *provably recursive* in EA(I;O) if it can be defined by a system E of equations such that, with input variables $x_1, \ldots, x_k$,

$$\bar{E} \vdash f(x_1, \ldots, x_k)\downarrow$$

where $\bar{E}$ denotes the set of universal closures (over output variables) of the defining equations in E.

3.2. Elementary functions are provably recursive. Let E be a system of defining equations containing the usual primitive recursions for addition and multiplication, and further equations:

$$p_0 = S0, \qquad p_i = p_{i_0} + p_{i_1}, \qquad p_i = p_{i_0} \cdot b$$

defining a sequence $\{p_i : i = 0, 1, 2, \ldots\}$ of polynomials in variables $\vec{b} = b_1, \ldots, b_n$. Henceforth we allow $p(\vec{b})$ to stand for any one of the polynomials so generated (clearly all polynomials can be built up in this way).

DEFINITION. The *progressiveness* of a formula $A(a)$ with distinguished free variable a, is expressed by the formula

$$Prog_a A \equiv A(0) \wedge \forall a\ (A(a) \rightarrow A(Sa))$$

thus the induction principle of EA(I;O) is equivalent to

$$Prog_a A \vdash A(x).$$

The following lemmas derive extensions of this principle, first to any polynomial in $\vec{x}$, then to any finitely iterated exponential. In the next section we shall see that this is the most that EA(I;O) can do.

LEMMA 3.1. *Let $p(\vec{b})$ be any polynomial defined by a system of equations E as above. Then for every formula $A(a)$ we have, with input variables substituted for the variables of p,*

$$\bar{E},\ Prog_a A \vdash A(p(\vec{x})).$$

PROOF. Proceed by induction over the build-up of the polynomial p according to the given equations E. We argue in an informal natural deduction style, deriving the succedent of a sequent from its antecedent.

If p is the constant 1 (that is $S0$) then $A(S0)$ follows immediately from $A(0)$ and $A(0) \rightarrow A(S0)$, the latter arising from substitution of the defined, basic term 0 for the universally quantified variable a in $\forall a(A(a) \rightarrow A(Sa))$.

Suppose p is $p_0 + p_1$ where, by the induction hypothesis, the result is assumed for each of p_0 and p_1 separately. First choose $A(a)$ to be the formula $a\downarrow$ and note that in this case $Prog_a A$ is provable. Then the induction hypothesis applied to p_0 gives $p_0(\vec{x})\downarrow$. Now again with an arbitrary formula A, we can easily derive

$$\bar{E},\ Prog_a A,\ A(a) \vdash Prog_b(a + b\downarrow \wedge A(a + b))$$

because if $a+b$ is assumed to be defined, it can be substituted for the universally quantified a in $\forall a(A(a) \rightarrow A(Sa))$ to yield $A(a+b) \rightarrow A(a+Sb))$. Therefore by the induction hypothesis applied to p_1 we obtain

$$\bar{E},\ Prog_a A,\ A(a) \vdash a + p_1(\vec{x})\downarrow \wedge A(a + p_1(\vec{x}))$$

and hence

$$\bar{E},\ Prog_a A \vdash \forall a(A(a) \to A(a + p_1(\vec{x}))).$$

Finally, substituting the defined term $p_0(\vec{x})$ for a, and using the induction hypothesis on p_0 to give $A(p_0(\vec{x}))$ we get the desired result

$$\bar{E},\ Prog_a A \vdash A(p_0(\vec{x}) + p_1(\vec{x})).$$

Suppose p is $p_1 \cdot b$ where b is a fresh variable not occurring in p_1. By the induction hypothesis applied to p_1 we have as above, $p_1(\vec{x})\downarrow$ and

$$\bar{E},\ Prog_a A \vdash \forall a(A(a) \to A(a + p_1(\vec{x})))$$

for any formula A. Also, from the defining equations E and since $p_1(\vec{x})\downarrow$, we have $p_1(\vec{x}) \cdot 0 = 0$ and $p_1(\vec{x}) \cdot Sb = (p_1(\vec{x}) \cdot b) + p_1(\vec{x})$. Therefore we can prove

$$\bar{E},\ Prog_a A \vdash Prog_b(p_1(\vec{x}) \cdot b\downarrow \wedge A(p_1(\vec{x}) \cdot b))$$

and an application of the EA(I;O)-induction principle on variable b gives, for any input variable x,

$$\bar{E},\ Prog_a A \vdash p_1(\vec{x}) \cdot x\downarrow \wedge A(p_1(\vec{x}) \cdot x)$$

and hence $\bar{E}$, $Prog_a A \vdash A(p(\vec{x}))$ as required.

DEFINITION. Extend the system of equations E above by adding the new recursive definitions:

$$f_1(a,0) = Sa, \qquad f_1(a,Sb) = f_1(f_1(a,b),b)$$

and for each $k = 2, 3, \dots,$

$$f_k(a,b_1,\dots,b_k) = f_1(a, f_{k-1}(b_1,\dots,b_k))$$

so that $f_1(a,b) = a + 2^b$ and $f_k(a,\vec{b}) = a + 2^{f_{k-1}(\vec{b})}$. Finally define

$$2_k(p(\vec{x})) = f_k(0,\dots,0,p(\vec{x}))$$

for each polynomial p given by E.

LEMMA 3.2. *In* EA(I;O) *we can prove, for each k and any formula* $A(a)$,

$$\bar{E},\ Prog_a A \vdash A(2_k(p(\vec{x}))).$$

PROOF. First note that by a similar argument to one used in the previous lemma (and going back all the way to Gentzen) we can prove, for any formula $A(a)$,

$$\bar{E},\ Prog_a A \vdash Prog_b \forall a(A(a) \to f_1(a,b)\downarrow \wedge A(f_1(a,b)))$$

since the $b := 0$ case follows straight from $Prog_a A$, and the induction step from b to Sb follows by appealing to the hypothesis twice: from $A(a)$ we first obtain $A(f_1(a,b))$ with $f_1(a,b)\downarrow$, and then (by substituting the defined

$f_1(a,b)$ for the universally quantified variable a) from $A(f_1(a,b))$ follows $A(f_1(a,Sb))$ with $f_1(a,Sb)\downarrow$, using the defining equations for f_1.

The result is now obtained straightforwardly by induction on k. Assuming $\bar{E}$ and $Prog_a A$ we derive

$$Prog_b \forall a(A(a) \rightarrow f_1(a,b)\downarrow \wedge A(f_1(a,b)))$$

and then by the previous lemma,

$$\forall a(A(a) \rightarrow f_1(a,p(\vec{x}))\downarrow \wedge A(f_1(a,p(\vec{x}))))$$

and then by putting $a := 0$ and using $A(0)$ we have $2_1(p(\vec{x}))\downarrow$ and $A(2_1(p(\vec{x})))$, which is the case $k = 1$. For the step from k to $k+1$ do the same, but instead of the previous lemma use the induction to replace $p(\vec{x})$ by $2_k(p(\vec{x}))$.

THEOREM 3.3. *Every elementary* (E^3) *function is provably recursive in the theory* EA(I;O), *and every sub-elementary* (E^2) *function is provably recursive in the fragment which allows induction only on* Σ_1 *formulas.*

PROOF. Any elementary function $g(\vec{x})$ is computable by a register machine M (working in unary notation with basic instructions "successor", "predecessor", "transfer" and "jump") within a number of steps bounded by $2_k(p(\vec{x}))$ for some fixed k and polynomial p. Let $r_1(c), r_2(c), \ldots, r_n(c)$ be the values held in its registers at step c of the computation, and let $i(c)$ be the number of the machine instruction to be performed next. Each of these functions depends also on the input parameters $\vec{x}$, but we suppress mention of these for brevity. The state of the computation $\langle i, r_1, r_2, \ldots, r_n\rangle$ at step $c+1$ is obtained from the state at step c by performing the atomic act dictated by the instruction $i(c)$. Thus the values of $i, r_1, \ldots, r_n$ at step $c+1$ can be defined from their values at step c by a simultaneous recursive definition involving only the successor S, predecessor P and definitions by cases C. So now, add these defining equations for $i, r_1, \ldots, r_n$ to the system E above, together with the equations for predecessor and cases: $C(0,a,b) = a$, $C(Sd,a,b) = b$, and notice that the cases rule built into EA(I;O) ensures that we can prove $\forall d \forall a \forall b\ C(d,a,b)\downarrow$. Since the passage from one step to the next involves only applications of C or basic terms, all of which are provably defined, it is easy to convince oneself that the Σ_1 formula

$$\exists \vec{a}\ (i(c) = a_0 \wedge r_1(c) = a_1 \wedge \ldots \wedge r_n(c) = a_n)$$

is provably progressive in variable c. Call this formula $A(\vec{x}, c)$. Then by the second lemma above we can prove

$$\bar{E} \vdash A(\vec{x}, 2_k(p(\vec{x})))$$

and hence, with the convention that the final output is the value of r_1 when the computation terminates,

$$\bar{E} \vdash r_1(2_k(p(\vec{x})))\downarrow.$$

Hence the function g given by $g(\vec{x}) = r_1(2_k(p(\vec{x})))$ is provably recursive.

In just the same way, but using only the first lemma above, we see that any sub-elementary function (which, e.g., by Rödding, is register machine computable in a number of steps bounded by just a polynomial of its inputs) is provably recursive in the Σ_1-inductive fragment. This is because the proof of $A(\vec{x}, p(\vec{x}))$ by the first lemma only uses inductions on substitution instances of A, and here, A is Σ_1.

3.3. Provably recursive functions are elementary. Suppose we have a derivation of $\bar{E} \vdash f(\vec{x})\downarrow$ in EA(I;O), and suppose (arbitrary, but fixed) numerals $\vec{n}$ are substituted for the input variables $\vec{x}$ throughout. In the resulting derivation, each application of induction takes the form:

$$\frac{\Gamma \vdash A(0) \qquad \Gamma,\ A(a) \vdash A(Sa)}{\Gamma \vdash A(t(n_i))}$$

where $t(x_i)$ is the basic term appearing in the conclusion of the original (unsubstituted) EA(I;O)-induction. Let m denote the value of $t(n_i)$, so m is not greater than n_i plus the length of term t. Furthermore, let ℓ denote the length of the binary representation of m. Then, given the premises, we can unravel the induction so as to obtain a derivation of

$$\Gamma \vdash A(m)$$

by a sequence of cuts on the formula A, with proof-height $\ell + 1$. To see this we first induct on ℓ to derive

$$\Gamma,\ A(a) \vdash A(S^m a) \quad \text{and} \quad \Gamma,\ A(a) \vdash A(S^{m+1} a)$$

by sequences of A-cuts with proof-height ℓ. This is immediate when $\ell = 1$, and if $\ell > 1$ then either $m = 2m_0$ or $m = 2m_0 + 1$ where m_0 has binary length less than ℓ. So from the result for m_0 we get

$$\Gamma,\ A(a) \vdash A(S^{m_0} a) \quad \text{and} \quad \Gamma,\ A(S^{m_0} a) \vdash A(S^m a)$$

by substitution of $S^{m_0} a$ for the free variable a, and both of these derivations have proof-height $\ell - 1$. Therefore one more cut yields

$$\Gamma,\ A(a) \vdash A(S^m a)$$

as required. The case $A(S^{m+1} a)$ is done in just the same way.

Therefore if we now substitute 0 for variable a, and appeal to the base case of the induction, a final cut on $A(0)$ yields $\Gamma \vdash A(m)$ with height $\ell + 1$ as required.

LEMMA 3.4 (Embedding). *If $\bar{E} \vdash f(\vec{x})\downarrow$ in EA(I;O) there is a fixed number d determined by this derivation, such that: for all inputs $\vec{n}$ of binary length $\leq k$, we can derive*

$$E,\ k{:}\,I,\ 0{:}\,N \vdash^{\omega \cdot d} f(\vec{n})\downarrow$$

in the modified infinitary system. Furthermore the non-atomic cut-formulas in this derivation are the induction-formulas occurring in the EA(I;O) *proof.*

Proof. First, by standard "free cut"-elimination arguments we can eliminate from the given EA(I;O) derivation all non-atomic cut-formulas which are not induction formulas. Then pass through the resulting free-cut-free proof, substituting the numerals for the input variables and translating each step into the new infinitary system in much the same way as before. Each induction is unravelled in the manner described above. Suppose the two premises of the induction have been embedded with ordinal bound $\omega \cdot d$. If the conclusion is $A(t(x_i))$ with t a basic term then, upon substitution of n_i for x_i, we obtain $t(n_i)$ with value $m \leq n_i + c$ where c measures the length of the term t (determined by the given EA(I;O) proof). The binary length of m is then $\ell < k + c$ and we thus obtain a derivation of

$$E,\ k\colon I,\ m\colon N \vdash^{\omega\cdot d+k+c} A(m)$$

whose ordinal bound we can manipulate so that

$$E,\ k\colon I,\ m\colon N \vdash^{\omega\cdot(d+c)+k} A(m).$$

Now $m \leq c + 2^{k+1} = B_\omega(k;c)$ and $c \leq B_{\omega\cdot c}(k;0)$. So $m \leq B_{\omega\cdot(d+c)}(k;0)$ and by the lemma about B,

$$E,\ k\colon I,\ 0\colon N \vdash^{\omega\cdot(d+c)} m\colon N.$$

So by applying the N2 rule,

$$E,\ k\colon I,\ 0\colon N \vdash^{\omega\cdot(d+c)+k+1} A(m)$$

and then since $\omega\cdot(d+c+1)[k] = \omega\cdot(d+c)+k+1[k]$ we have

$$E,\ k\colon I,\ 0\colon N \vdash^{\omega\cdot(d+c+1)} A(m)$$

as required.

Theorem 3.5. *The provably recursive functions of* EA(I;O) *are all elementary* (E^3), *and the provably recursive functions of its* Σ_1 *inductive fragment are all subelementary* (E^2).

Proof. If $\bar{E} \vdash f(x)\downarrow$ in EA(I;O) then for any given input n of binary length k we have, for a fixed d,

$$E,\ k\colon I,\ 0\colon N \vdash^{\omega\cdot d} f(n)\downarrow.$$

By successive cut-reduction, bring the cut rank $r+1$ down to the level of Σ_1. Then

$$E,\ k\colon I,\ 0\colon N \vdash^{\alpha}_{1} f(n)\downarrow$$

where $\alpha = 2_r(\omega\cdot d) \prec \varepsilon_0$. Then by an entirely analogous version of the Complexity Theorem 1.5, modified to the new kind of sequents, we obtain a complexity bound $B_{2\cdot\alpha+1}(k;0)$ which is an elementary function of the binary length of input n. Hence f itself is elementary.

If the original EA(I;O) proof uses only Σ_1 inductions, then it will embed straightaway as a rank 1 derivation:

$$E,\ k\colon I,\ 0\colon N \vdash_1^{\omega\cdot d} f(n)\downarrow$$

and the Bounding Lemma will give a value $m \leq B_{\omega\cdot d}(k;0) = 2^{(k+1)\cdot d}$ such that

$$E,\ k\colon I,\ 0\colon N \vdash_0^{2\cdot\omega\cdot d} f(n) = m.$$

Since all the ordinal bounds assigned in this derivation must belong to $2\cdot\omega\cdot d[k]$, its height is $\leq G(2\cdot\omega\cdot d, k) = (k+1)\cdot 2d$ and the number of nodes (steps in the computation) is therefore $\leq 2^{(k+1)\cdot 2d}$. But k is the binary length of input n, so $2^{k+1} \leq 4\cdot n$ and hence $f(n)$ is computable in a number of steps bounded by a polynomial of n. Thus f is subelementary.

In Ostrin and Wainer (2001) these results are treated in a somewhat different fashion, without need of ordinals, and it is also shown how exponential time is characterized at the level corresponding to Π_2^0 induction. In Cagman, Ostrin and Wainer (2000) a preliminary version of the two–sorted arithmetic is analysed in terms of a functional interpretation.

If EA(I;O) is reformulated as a theory on binary strings, rather than unary numerals, then the Σ_1 inductive fragment captures PTIME. The author is grateful to the referee for bringing to his attention the recent work of Marion (2001) characterizing PTIME in another theory AT, with which our EA(I;O) seems to share some fundamental characteristics. AT is formulated in the "pure" fragment of natural deduction, and has only one kind of variable, but the crucial similarity lies in his restriction of the ∀-elimination rule to so-called "actual" terms, which are the same as our "basic" ones. A further restriction is placed on the "cases" rule, so that no ∀-eliminations are allowed in the derivations of its premises, for otherwise induction variables may become "unguarded" (our two kinds of variables obviate the need for any such restrictions on the structure of proofs). Marion proves that the PTIME functions are just those provably total in the fragment of AT allowing induction (over strings) wherein the induction formulas are conjunctions of atoms.

REFERENCES

[1] S. Bellantoni and S. Cook, *A new recursion theoretic characterization of the polytime functions*, ***Computational Complexity***, vol. 2 (1992), pp. 97–110.

[2] W. Buchholz, *An independence result for $\Pi_1^1 - CA + BI$*, ***Annals of Pure and Applied Logic***, vol. 23 (1987), pp. 131–155.

[3] W. Burr, *Fragments of Heyting arithmetic*, ***The Journal of Symbolic Logic***, vol. 65 (2000), pp. 1223–1240.

[4] N. Cagman, G. E. Ostrin, and S. S. Wainer, *Proof theoretic complexity of low subrecursive classes*, ***Foundations of secure computation*** (F. L. Bauer and R. Steinbrüggen, editors), IOS Press, 2000, pp. 249–285.

[5] M. FAIRTLOUGH and S. S. WAINER, *Hierarchies of provably recursive functions*, ***Handbook of proof theory*** (S. Buss, editor), Elsevier Science BV, 1998, pp. 149–207.

[6] G. KREISEL, *On the interpretation of non-finitist proofs, parts I, II*, ***The Journal of Symbolic Logic***, vol. 16 (1951), pp. 241–267, vol. 17 (1952) pp. 43–58.

[7] D. LEIVANT, *A foundational delineation of poly-time*, ***Information and Computation***, vol. 110 (1994), pp. 391–420.

[8] ———, *Intrinsic theories and computational complexity*, ***Logic and computational complexity*** (D. Leivant, editor), Lecture Notes in Computer Science, vol. 960, Springer-Verlag, 1995, pp. 177–194.

[9] J-Y. MARION, *Actual arithmetic and feasibility*, ***Proceedings of CSL 2001*** (L. Fribourg, editor), Lecture Notes in Computer Science, vol. 2142, Springer-Verlag, 2001, pp. 115–129.

[10] G. E. OSTRIN, *Proof theories of low subrecursive classes*, ***Ph.D. thesis***, Leeds, 1999.

[11] G. E. OSTRIN and S. S. WAINER, *Elementary arithmetic*, Leeds, preprint, 2001, to appear in ***Annals of Pure and Applied Logic***.

[12] D. RÖDDING, *Klassen rekursiver Funktionen*, ***Proceedings of Summer School in Logic, Leeds 1967*** (M. H. Löb, editor), Lecture Notes in Mathematics, vol. 70, Springer-Verlag, 1968, pp. 159–222.

[13] A. S. TROELSTRA and H. SCHWICHTENBERG, ***Basic proof theory***, Cambridge Tracts in Theoretical Computer Science, vol. 43, CUP, 1996.

[14] A. WEIERMANN, *How to characterize provably total functions by local predicativity*, ***The Journal of Symbolic Logic***, vol. 61 (1996), pp. 52–69.

UNIVERSITY OF LEEDS
LS2 9JT, UK
E-mail: s.s.wainer@leeds.ac.uk